LINEAR
INTEGRATED
NETWORKS

BELL LABORATORIES SERIES

Light Transmission Optics, Dietrich Marcuse

Linear Integrated Networks:
Fundamentals, George S. Moschytz

Linear Integrated Networks:
Design, George S. Moschtyz

LINEAR INTEGRATED NETWORKS

Design

GEORGE S. MOSCHYTZ

Bell Telephone Laboratories
presently professor,
Swiss Federal Institute of Technology,
Zurich, Switzerland

VAN NOSTRAND REINHOLD COMPANY

New York/Cincinnati/Toronto/London/Melbourne

Van Nostrand Reinhold Company Regional Offices:
New York   Cincinnati   Chicago   Millbrae   Dallas

Van Nostrand Reinhold Company International Offices:
London   Toronto   Melbourne

Library of Congress Catalog Card Number: 74-13707
ISBN: 0-442-25582-9

Manufactured in the United States of America

Published by Van Nostrand Reinhold Company
450 West 33rd Street, New York, N.Y. 10001

Published simultaneously in Canada by Van Nostrand Reinhold Ltd.

15   14   13   12   11   10   9   8   7   6   5   4   3   2   1

Library of Congress Cataloging in Publication Data

Moschytz, George S
  Linear integrated networks: design.

  (Bell Telephone Laboratories series)
  Includes bibliographical references.
   1. Linear integrated circuits. 2. Hybrid
integrated circuits. 3. Electronic circuit design.
I. Title.
TK7874.M5344   621.381'73   74-13707
ISBN 0-442-25582-9

To My Parents
Who Paved the Way

FOREWORD

It is quite timely for a work to appear on the theory and design of linear integrated networks which emphasizes the growing recognition and importance of hybrid integrated circuits based on silicon semiconductor technology and precision passive film technology. Dr. Moschytz has undertaken a project which provides a sound theoretical basis for understanding linear active networks while stressing the practical active network realizations which were achieved during recent years, particularly at Bell Laboratories where he supervised much of the advanced development and design effort on active linear networks. Many of the examples and illustrations in the work are based on these designs, with careful consideration given to the numerous design and performance trade offs involved in developing high performance active networks using a hybrid technology.

This work will improve the communications and understanding necessary among the specialists in semiconductor device, component, and network technologies, and it effectively complements other recent books by providing practical illustrations of active networks designed to meet special requirements. The author gives special consideration to the unique features of hybrid integrated circuits with regard to small size, low power dissipation, batch fabrication, parasitic effects, network and device sensitivity, functional adjustment, and network performance.

The two volumes, devoted respectively to fundamentals and design, should become a valuable addition to the literature, and will be of considerable benefit to students and to practicing industry engineers and scientists. The volumes are the latest in the Bell Telephone Laboratories series, in which we attempt to bring the engineering and scientific community up to date on subjects of major current interest.

David Feldman
Director
Film and Hybrid Technology Laboratory
Bell Telephone Laboratories

PREFACE

A new technology has emerged with hybrid integrated circuit technology, and it has already drastically affected the design of linear networks such as frequency selective filters, shaping networks, and delay lines. The need to eliminate inductors from these networks has had particularly far-reaching consequences in that much of the classical filter theory and conventional filter know-how that was previously at the designer's disposal can no longer be directly applied. More and more, *active filters* in hybrid integrated form are replacing their passive LCR counterparts in modern communication and control systems, and the need has become evident of documenting the new design rules and guidelines necessary to perform this transition from the old to the new. This transition is, of course, motivated by such very real incentives as more economical (batch) production methods, higher reliability, smaller size, manufacturing processes that are compatible with other (e.g. digital) integrated circuit production methods, and so on. These benefits of the new technology are well known at the present time, and the need to justify more elaborately the trend toward inductorless networks (of which linear integrated networks belong to the most important exponents) should no longer be necessary.

This book attempts to fill the need, expressed above, of documenting the design rules and guidelines necessary for the design, development, and manufacture of linear networks—and in particular of active filters—realized in hybrid integrated form. To do so, it draws heavily on the companion book *Linear Integrated Networks: Fundamentals* in which the prerequisites necessary for the understanding of the material presented here are covered in detail. To facilitate the coordination between the two books, and to provide a self-contained, comprehensive work on the subject of linear integrated networks (LINs), cross references are given wherever necessary.

This book covers the theory of LIN design, while at the same time blending with it the practical aspects of LIN development and manufacture, as encountered by the author in an industrial environment. The close interrelationship between the theory and practice of LIN design is considered a vital aspect of the book; wherever possible practical considerations are given a sound theoretical foundation; likewise, theoretical discussions are firmly tied to practical experience. Development steps that have been successfully proven in practice are enumerated in easy to follow fashion. These steps are based on the detailed theoretical considerations preceding them, as well as on the experience acquired during implementation in practice.

Most of the material covered here appears in book form for the first time; much of it has not been published in any form before. The material has been divided into two parts. Part I, comprising the first three chapters, deals with the synthesis of active networks. This part is essentially independent of the technology utilized for the actual realization and covers the general theory of active network design. After briefly discussing the filter approximation problem, those aspects peculiar to *active* filter design are covered. These encompass the problem of optimum pole-zero pairing, the sensitivity of cascaded second- and third-order networks, the general realization of such networks, as well as their realization in building-block form. A comprehensive classification of active networks using single active devices is given and the most useful multi-amplifier configurations are discussed in detail. Part II, comprising the remaining four chapters, specifically deals with the design of active networks in *hybrid-integrated* form. Here the almost unlimited variety of active networks encompassed in the general treatment of Part I are reconsidered with a view to hybrid-integrated circuit realization. This review takes the parasitic and other nonideal effects inherent in hybrid-integrated circuit technology into account. Using a suitable active network type as an illustrative example, the step-by-step design and development procedure necessary for the realization of hybrid-integrated filter building blocks is covered in detail. The capabilities of such filters are compared with those of other inductorless-filter techniques, and those areas in which active filters seem to hold most promise are mapped out. Finally, modifications necessary to extend the capabilities of hybrid-integrated networks such that they can be fully utilized in, and beyond, those areas are covered in detail.

The material covered in this book is directed at student and engineer alike, and is presented such as to permit self study. For this reason numerous examples are worked out in detail and liberal use is made of figures.

Chapter 1 covers those design aspects common to all active networks, irrespective of the technology used for their realization. These include optimization for maximum inband gain and minimum sensitivity, the sensitivity of cascaded and inductor-simulated active networks and the characterization and sensitivity of second and third-order networks. *Chapter 2* contains a unified theory and classification of all second-order active networks comprising a single active element. This classification permits conclusive generalizations to be made with respect to characteristics, sensitivity, etc., for each network class. In *Chapter 3* the most useful and proven single and multiple-amplifier second-order active filter building blocks are presented in detail. The two topologies (cascade and parallel) in which these building blocks can be used are discussed. In *Chapter 4* those practical aspects are considered that enter the design process when linear active networks are to be realized using hybrid integrated circuit technology. The multitude of active circuits that emerge from the classifications in Chapters 2 and 3 are evaluated from the point of view of hybrid integrated circuit technology, and the interrelationship between network class and technology is demonstrated. The effects of parasitics and other nonideal characteristics are analyzed and tuning methods that enable the ensuing initial errors to be compensated for (or "tuned out") are considered. *Chapter 5* deals with the actual design and development of filter building blocks in hybrid integrated circuit form. This chapter deals with such topics as layout, substrate and film materials, single- and double-substrate assemblies, scribing methods, and the like. Other inductorless filter methods are also reviewed and, based on this review, areas of applications where active filters appear particularly useful are pointed out. In this context the economics of hybrid-integrated circuits are also considered. *Chapter 6* presents the step-by-step optimization and design of a family of hybrid integrated filter building blocks that have proved themselves in production and practice. Partly in cook book fashion, and with extensive tabulated material, this chapter has sufficient detailed design information to permit the reader to follow and understand the intricacies of hybrid integrated building block design. Finally, in *Chapter 7*, some of the latest methods used to extend the frequency and selectivity capabilities of hybrid integrated networks are covered. Methods are discussed of designing hybrid integrated active filters for extreme low, and very high frequency applications (compared to presently available frequency ranges). The various methods of inductor simulation are elaborated on and optimization methods for networks with maximum dynamic range and minimized sensitivity are given. Furthermore some of the latest results are given with respect to negative-feedback-coupled active filter building blocks, and methods are presented of analyzing the improved filter stability thereby obtainable.

Much of the material covered in this book and the companion book (*LINs: Fundamentals*) has been " class-room tested " as a two-semester course with two rather different types of students. First given as an out-of-hours course at Bell Telephone Laboratories, Holmdel, N.J., the students were exclusively engineers who, in some way or another, had use for certain aspects, if not for the outright design, of hybrid-integrated networks in the course of their daily work. The second time the material was taught as a first year graduate course at the Swiss Federal Institute of Technology, Zurich, Switzerland (while on leave of absence from Bell Telephone Laboratories as a visiting professor at the Institute of Telecommunications).

In teaching the material covered in this and the companion book in a two-semester course the following sequence has been found useful:

Chapters 1 to 4 (*Fundamentals*) as well as Chapter 2 (*Design*) are covered in the first semester in order to provide a somewhat rounded off treatment of linear active networks in general. Chapters 5 (*Fundamentals*) and selected topics of chapters 1, 3, 4, 6 and 7 (*Design*) are then dealt with in the second semester, whereby chapters 6, 7 and 8 (*Fundamentals*) are suggested as additional reading material. In this way the second semester covers the integrated circuit aspects of linear network design. Depending on how much of *Fundamentals* it is deemed necessary to teach, this program can, of course, be drastically shortened—if need be to a one semester course.

GEORGE S. MOSCHYTZ

ACKNOWLEDGMENTS

Many of the ideas presented in this book were formed during the exciting and stimulating experience of designing and getting into production hybrid integrated filters for the Bell System in a team effort in which I was involved with numerous colleagues on the staffs of the Bell Telephone Laboratories and the Western Electric Company. To name all the people who contributed to that successful effort is practically impossible, and I must acknowledge my indeptedness to them collectively. However, I wish to acknowledge specifically D. Hirsch and N. E. Snow as well as D. Feldman, C. T. Goddard and D. A. McLean for their encouragement and support during the whole project. Furthermore I wish to acknowledge the main contributors to that project namely W. L. Barton, W. H. D'Zio, J. S. Fisher, M. C. Kolibaba, J. A. Krayesky, E. Lueder, G. Malek, D. G. Marsh, D. R. Means, D. G. Medill, D. O. Melroy, W. H. Orr, N. P. Palumbo, A. L. Pappas, C. J. Steffen, S. E. Sussman, W. Thelen and R. M. Zeigler, of Bell Telephone Laboratories and G. Allerton, T. Breisch, S. Hause, and W. B. Reichard of the Western Electric Company.

I am greatly indebted to T. F. Epley, formerly of the Van Nostrand Reinhold Co., whose enthusiasm and encouragement provided the necessary motivation for this book and the companion book *Linear Integrated Networks: Fundamentals*, and whose guidance through the early stages was invaluable; and to his successor B. R. Nathan, and his colleagues Mrs. A. W. Gordon, J. Stirbis and Mrs. F. Landau, whose much appreciated advice assisted me up to the completion of both books.

I am indebted to many friends, colleagues and students for their valuable comments and suggestions during the preparation of this book. I am grateful for the constructive criticisms of the reviewers at Bell Telephone Laboratories (BTL), in particular to R. K. Even, D. R. Means, E. Lueder, C. J. Steffen and G. Malek. I am especially indebted to R. K. Even and D. R. Means for their very thorough review of the entire manuscript and for their numerous constructive criticisms and helpful advice regarding organization and presentation of the book. I also greatly appreciate the numerous helpful comments from my students at the Swiss Federal Institute of Technology (SFIT) in Zurich, in particular from R. Hammer, P. Horn and M. Kuenzli.

I am most grateful for the kind cooperation of D. Hirsch and E. R. Kretzmer of BTL and of Professor H. Weber of the SFIT for making it possible for the manuscript to progress smoothly while I was in Zurich on a leave of absence from Bell Telephone Laboratories. My deep appreciation also goes to the extensive secretarial assistance of Miss P. Jobes at BTL and Mrs. A. Huebscher and Mrs. L. Hubmann at the SFIT, as well as to the excellent typing skills of Mrs. M. Taolise. Finally, my thanks go to my wife Doris and my daughters Joy, Helen and Miriam for the interest and understanding they showed during this undertaking.

G. S. Moschytz

CONTENTS

PART II
FUNDAMENTALS OF HYBRID INTEGRATED NETWORK DESIGN

4. Network Design Using Hybrid Integrated Circuits

5. The Development of Hybrid Integrated Filter Building Blocks

6. Designing Hybrid Integrated Filter Building Blocks

7. Extending the Capabilities of Active Filters

PART I:

FUNDAMENTALS OF ACTIVE NETWORK SYNTHESIS

1

LINEAR ACTIVE NETWORK DESIGN

INTRODUCTION

In this chapter we first consider the so-called approximation problem briefly, that is, the problem of determining a rational transfer function that will satisfy our system requirements in terms of a frequency response or in the time domain. Having done so, the question must then be answered, how to translate the given transfer function into a suitable active network that not only responds in the required manner but also satisfies various additional constraints such as, for example, providing a specified stability over given ambient conditions, a specified signal-to-noise ratio over a given dynamic range, or a specified current drain over a given supply voltage range. Some answers to these and related questions will be given in this and the following chapters.

1.1 THE THREE STEPS IN LINEAR ACTIVE NETWORK SYNTHESIS

Network synthesis is the process by which a network is obtained that satisfies a set of prescribed electrical characteristics. The required characteristics may be specified in either the frequency domain, the time domain, or possibly even in both. In the case of active network synthesis, a three-step process is used to obtain a desired network (see Fig. 1-1).

Step 1. Approximation. Obtain a rational transfer function in s which approximates with acceptable accuracy the prescribed electrical charac-

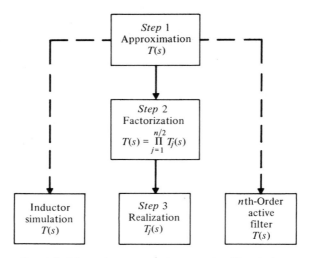

FIG. 1-1. The various approaches to active filter design.

teristics. In the case of a voltage transfer function, for example, this function is of the form

$$T(s) = \frac{\text{output}}{\text{input}} = \frac{b_m s^m + b_{m-1}s^{m-1} + \cdots + b_1 s + b_0}{a_n s^n + a_{n-1}s^{n-1} + \cdots + a_1 s + a_0} \qquad (1\text{-}1)$$

where $s = \sigma + j\omega$ and $n \geq m$.

In most cases such a ratio of polynomials with real coefficients can readily be found. In some cases, however, particularly when designing pulse-shaping networks, the required characteristics are either specified in the time domain or in the form of a transcendental function in s. In such cases the approximation of "nonrational" specifications by a rational function of the type represented by (1-1) may cause some difficulties. If the response of the initially obtained rational function does not lie within the specified limits, optimization techniques[1] must be used, which perturb the individual poles and zeros of the rational function until its response falls within the tolerances specified.

Step 2. Decomposition. Obtain the roots of the numerator polynomial (the zeros) and of the denominator polynomials (the poles). The desired transfer function can then be written as the product of second-order subfunctions as follows:

$$T(s) = \prod_{j=1}^{n/2} T_j(s) = \prod_{j=1}^{n/2} K_j \frac{(s - z_{j1})(s - z_{j2})}{(s - p_j)(s - p_j^*)} \qquad (1\text{-}2)$$

1. G. Szentirmai, Computer aids in filter design: A review, *IEEE Trans. Circuit Theory,* **CT-18,** 35–40 (1971).

Note that odd polynomials have single negative real roots in addition to the conjugate complex pairs.

Step 3. Realization. Select an appropriate synthesis method to realize the individual subfunctions $T_j(s)$.

As simple as these three rules may seem at first sight, putting them into effect may be quite difficult. The approximation problem (step 1) alone warrants extensive treatment far beyond the scope of this book. Detailed information on this subject can be found in the references for this chapter listed at the end of the book. Furthermore, numerous computer programs are now available that match the coefficients of a rational transfer function to the prescribed electrical requirements of a network. Thus, we shall assume in the following that a network function in s as given by (1-1) is at hand; our problem of designing the corresponding active network then takes off from there.

Steps 2 and 3 also require considerable elaboration, which will follow in this and the following chapters. First, however, we shall pursue an alternative path (see Fig. 1-1) that takes off (as do steps 2 and 3) from the existing function $T(s)$ and attempts to realize the corresponding *LCR* network in active, inductorless form, by inductor simulation.

1.2 INDUCTOR SIMULATION

A gyrator is a two-port device with an impedance-inversion property that converts a capacitive load into a simulated inductive reactance at its input terminals.[2]

In Fig. 1-2 a configuration providing the simulation of a grounded inductor is shown; a floating inductor requires two gyrators, as shown in Fig. 1-3.[3] Inductor simulation using a silicon integrated gyrator combined with a chip or thin-film capacitor therefore comes closest to the practical realization of a

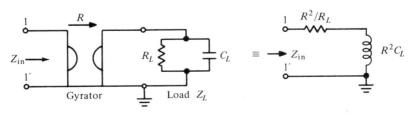

FIG. 1-2. Simulation of a grounded inductor by a gyrator–capacitor combination; $R = $ gyration resistance.

2. See, for example, Chapter 5, Section 5.4.2 of *Linear Integrated Networks: Fundamentals.*
3. If $R_1 \neq R_2$ in the figure, then a transformer must be added to the equivalent circuit (see, for example, Chapter 5, Fig. 5-22 of *Linear Integrated Networks: Fundamentals*).

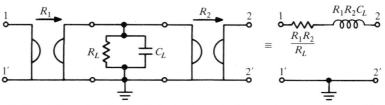

FIG. 1-3. Simulation of a floating inductor by a double gyrator–capacitor combination.

microminiaturized inductor in hybrid integrated form and, when incorporated in an *LC* filter, provides a hybrid integrated version of the equivalent *LC* filter. Herein lies one of the main attractions of this design method: the well established theory as well as the tabulated, and therefore readily available, data on the design of *LC* filters can be directly applied to it. Following this procedure, the optimum *LC* filter meeting the specified electrical requirements is first found using conventional *LC* filter design methods. Subsequently, each grounded or floating inductor is replaced by a single, or double, gyrator–capacitor combination, respectively, in order to obtain the inductor-simulated filter equivalent using gyrators. An example of this conversion from *LC* filter to inductor-simulated filter using gyrators is shown for the case of a notch filter in Fig. 1-4. The gyrators themselves may, of course, in turn be simulated using operational amplifiers, for example. Furthermore other methods of inductor simulation using amplifiers or other devices may be used.

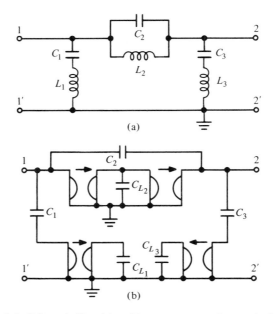

FIG. 1-4. *LC* notch filter (a) and its gyrator–capacitor equivalent (b).

1.2.1 The Argument for Inductor Simulation

A powerful argument for the use of inductor-simulated active filters can be found if the sensitivity of these filters to component variations is examined.[4] Consider, for example, the doubly-terminated reactive LC filter network represented in Fig. 1-5. In general the input and output impedances of the filter will be matched to the source and load resistances R_s and R_L, respectively, at certain frequencies within the passband of the filter. At any frequency the input impedance of the reactive LC network will have the form

$$Z_{in} = R_{in} + jX_{in} \tag{1-3}$$

and the power P_{in} transmitted into the filter network by the voltage source $v_s = \hat{V}_s \sin \omega t$ is

$$P_{in} = \frac{V_s^2}{2} \frac{\text{Re } Z_{in}}{|R_s + Z_{in}|^2} = \frac{V_s^2}{2} \frac{R_{in}}{(R_{in} + R_s)^2 + X_{in}^2} \tag{1-4}$$

Assuming that the LC filter network is composed of lossless (i.e., ideal) components, none of the power P_{in} will be dissipated in the filter; it will all be transmitted to the load resistor R_L. Thus the output power

$$P_L = P_{in} \tag{1-5}$$

One way of establishing the transmission sensitivity of our filter network is to investigate how the output power P_L varies as any component value x_i in the filter is changed from its design value. For any component x_i we have, in general,

$$\frac{dP_L}{dx_i} = \frac{\partial P_L}{\partial R_{in}} \cdot \frac{\partial R_{in}}{\partial x_i} + \frac{\partial P_L}{\partial X_{in}} \cdot \frac{\partial X_{in}}{\partial x_i} \tag{1-6}$$

To find (1-6) we obtain the following expression from (1-4):

$$\frac{\partial P_L}{\partial R_{in}} = \frac{V_s^2}{2} \frac{R_s^2 - R_{in}^2 + X_{in}^2}{[(R_s + R_{in})^2 + X_{in}^2]^2} \tag{1-7}*$$

and

$$\frac{\partial P_L}{\partial X_{in}} = -V_s^2 \frac{R_{in} X_{in}}{[(R_s + R_{in})^2 + X_{in}^2]^2} \tag{1-8}$$

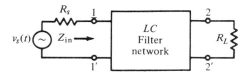

FIG. 1-5. Doubly-terminated LC filter.

4. D. F. Sheahan, Inductorless filters, Tech. Rep. 6560-15, Systems Theory Laboratory, Stanford Univ., Stanford, Calif, September 1967.
* Square brackets are used to emphasize important definitions or results.

In order to attain maximum transfer of power through the filter at frequencies of minimum loss in the passband, the designer must match the input imped-ance Z_{in} to the source resistance R_s as closely as possible. He therefore designs his filter such that $R_{in} = R_s$ and $X_{in} = 0$. The better he achieves this match, the closer he comes to making $\partial P_L/dR_{in} = \partial P_L/\partial X_{in} = 0$. As a conse-quence (see (1-6)), $dP_L/dx_i = 0$, i.e., the sensitivity of the loss to component tolerances is zero at every frequency in the passband to a first approximation.

The preceding discussion is based on the presumption that, at the fre-quencies of minimum loss over the passband, the source delivers its maximum available power to the load. To achieve this the filter output impedance must be matched to the load resistor, as the following simple reasoning makes clear. As we have seen, in order to obtain maximum power from the source, the fil-ter input impedance is matched to the source resistance. Assuming that the filter is dissipationless, the maximum available power is then delivered to the load resistor. Now, any change in the value of the load resistor will not result in a larger output power. Consequently the output impedance of the filter must also be matched to the load resistor. Thus, we can state in general that a doubly terminated LC filter, having input and output impedances matched to the source and load resistances respectively within the passband, will have zero first derivative of the output power, that is, zero sensitivity of the loss with respect to component tolerances at every frequency within the passband.

Consider, for example, the insertion loss vs. frequency of a reactance net-work shown qualitatively in Fig. 1-6. At the frequencies of zero loss the net-work is matched at both ends and the source delivers its maximum available power into the load. At these frequencies any component change, whether increasing or decreasing, can only cause the loss to *increase*. In the vicinity of the correct value of any component, the curve relating loss to the value of that component must therefore be quadratic, and consequently, $d(\text{loss})/d(\text{com-ponent})$ must be zero.

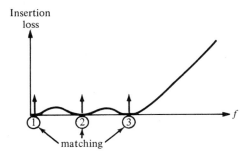

FIG. 1-6. Insertion loss characteristic of typical doubly-terminated LC filter. At minimum-loss points 1, 2, and 3, the source delivers its maximum power into the load due to impedance matching (i.e., input impedance to source resistance, output impedance to load resistance).

This low-sensitivity property of doubly-terminated LC filter networks was explained by Orchard[5] in 1966. He went on to suggest that, in order to retain low sensitivity when designing inductorless filters, the inductor simulation method described above should be used; i.e., a conventional, doubly-loaded LC ladder filter should first be designed to meet specifications, and then each inductor replaced by the appropriate gyrator–capacitor combination. An added bonus to this approach is, of course, that it can utilize the wealth of LC filter theory and design data that has accumulated over the last half century. Whereas these arguments are convincing indeed, and in fact served as a strong incentive for renewed efforts toward the design of improved and economical integrated gyrators or other inductor-simulating devices, the outcome has not been as decisive as one might have expected. Indeed, it was later pointed out* that the actual improvement in sensitivity may not be significant compared to the inherent disadvantages of non-cascaded networks. Be that as it may, inductor simulation did not become the commonly adopted method of designing inductorless filters. Apart from having been used in certain experimental systems, it does not seem to have established itself on a large scale in practice.

Some of the reasons for the lack of widespread usage of inductor-simulated gyrator filters, in particular those based on economical or technical shortcomings of monolithic silicon integrated gyrators, have been discussed elsewhere.[6] It may be that another reason is the very fact that the nature of an inductor-simulated filter using gyrators is so similar to that of a conventional LC filter, not only with respect to low sensitivity but also with respect to other, less desirable characteristics. The conventional LC filter, for example, is an nth-order network with at least n reactive elements inseparably interconnected in a physical unit. The resulting interdependence of the network elements, e.g., the fact that any change in one coil will require a corresponding change to be made on all the other filter coils, makes the production of precision networks expensive and time consuming, not to speak of the accompanying exasperation, involved in tuning high-order filters to tight initial tolerances. This tuning problem is carried over to the inductor-simulated filter. Attempting to solve it by breaking up the filter (i.e., the LC or the inductor-simulated type) into a cascade of lower-order sections is inadvisable, since it can be shown[7] that this deteriorates the favorable sensitivity characteristics. As we shall see shortly, the opposite is true for most amplifier-based methods of active filter design. Another drawback that the passive LC filter designer has to contend with is a given amount of signal attenuation in the passband that must be overcome with gain somewhere else.

* A. G. J. Holt and M. R. Lee, Sensitivity Comparison of Active-Cascade and Inductance-Simulation Schemes, *Proc. IEE*, **119**, No. 3, 277–282 (March 1972).
5. H. J. Orchard, Inductorless filters, *Electron. Lett.*, **2**, 224–225 (1966).
6. Chapter 7, section 7-33, in *Linear Integrated Networks: Fundamentals*.
7. W. H. Holmes, W. Heinlein, and S. Grützmann, Empfindlichkeit von LC-Filtern, aktiven RC-Filtern und aktiven C-Filtern, *Frequenz*, **22**, 2–11 (1968).

The designer of active inductor-simulated filters must accept the same drawback, even though his filter structure contains active elements which could, in different circumstances, provide gain. Here again, standard active-filter techniques, as discussed further on, suffer from no such limitations. Furthermore, the amplifiers are then used to provide not only gain but also isolation between the cascaded low-order filter sections, thereby avoiding the problems related to signal reflection that occur in *LC*, and consequently also in inductor-simulated filter networks.

Since low-cost integrated gyrators have been very slow to appear commercially, there has been an increasing tendency to look to other methods of designing gyrator filters using already available linear monolithic integrated circuits. The most common of these circuits is the operational amplifier, two of which are required to simulate a gyrator. Excellent results have been reported with Riordan's circuit[8] and it has been argued, as a consequence, that with cheap, monolithic operational amplifiers available to simulate them, the need for monolithic gyrators has diminished. On the other hand, since two amplifiers are required for every gyrator, the power consumption for gyrator filters is doubled. The filter shown in Fig. 1-4, for example, would require eight operational amplifiers, which does not seem economical on the basis of device count or of power consumption.

1.2.2 Simulation of Grounded Inductors Using Single Operational Amplifiers

Methods do exist of simulating inductors, albeit only grounded ones, with single operational amplifiers. One general configuration serving this purpose is shown in Fig. 1-7.[9] By straightforward analysis one obtains the input impedance as

$$Z_{in} = \frac{V_1}{I_1} = \frac{Y_2(1 + A) + Y_3}{[Y_2 + Y_3(1 - A)]Y_1} \tag{1-9}$$

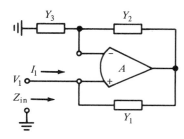

FIG. 1-7. Single-amplifier configuration for the simulation of grounded inductors.

8. H. J. Orchard and D. F. Sheahan, Inductorless bandpass filters, *IEEE J. Solid-State Circuits*, SC-5, 108–118 (1970); see, also Chapter 5, Section 5.4.3 of *Linear Integrated Networks: Fundamentals*.
9. D. Patranabis, Grounded inductor simulation with a differential amplifier, *Int. J. Electron.*, 28, 481–483 (1970).

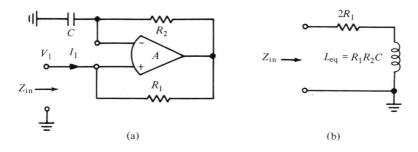

(a)   (b)

FIG. 1-8. Grounded-inductor simulation using the configuration of Fig. 1-7: (a) circuit; (b) equivalent circuit.

and, for unity gain,

$$Z_{in}\big|_{A=1} = \frac{2}{Y_1} + \frac{Y_3}{Y_1 Y_2} \tag{1-10}$$

Letting $Y_1 = 1/R_1$, $Y_2 = 1/R_2$, and $Y_3 = sC$, as shown in Fig. 1-8a we obtain from (1-10):

$$Z_{in} = 2R_1 + sCR_1R_2 \tag{1-11}$$

which corresponds to an inductor $L_{eq} = CR_1R_2$ whose resistive component is $2R_1$ (see Fig. 1-8b). Increasing C and R_2 while leaving R_1 constant, large inductors with high Q can be simulated.

Another, similar configuration[10] is shown in Fig. 1-9. A simple calculation yields the input impedance as

$$Z_{in} = \frac{Y_1 + Y_2 + Y_3}{Y_1[Y_3(1 - \beta) + Y_2]} \tag{1-12}$$

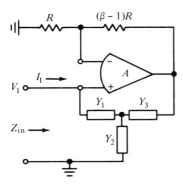

FIG. 1-9. Second single-amplifier configuration for grounded-inductor simulation.

10. A. J. Prestcott, Loss compensated active gyrator using differential-input operational amplifiers, *Electron. Lett.*, 283–284 (1966).

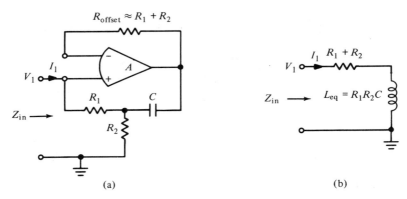

(a)   (b)

FIG. 1-10. Grounded-inductor simulation using the configuration of Fig. 1-9: (a) circuit; (b) equivalent circuit.

With $Y_1 = 1/R_1$, $Y_2 = 1/R_2$, and $Y_3 = sC$ (see Fig. 1-10a) we obtain

$$Z_{in} = \frac{sCR_1R_2 + R_1 + R_2}{sCR_2(1 - \beta) + 1} \tag{1-13}$$

With unity gain (1-12) becomes

$$Z_{in}\big|_{\beta=1} = \frac{1}{Y_1} + \frac{1}{Y_2} + \frac{Y_3}{Y_1 Y_2} \tag{1-14}$$

and (1-13) becomes

$$Z_{in} = R_1 + R_2 + sCR_1R_2 \tag{1-15}$$

This has the same form as (1-11) and provides the same capabilities (see Fig. 1-10b).

The circuit of Fig. 1-9, when used in the unity-gain mode, is based on a general characteristic of unity-gain, positive-feedback circuits of the type shown in Fig. 1-11a. It can be shown quite easily,[11] that the input impedance

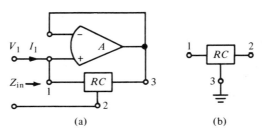

(a)   (b)

FIG. 1-11. Generalized configuration for single-amplifier grounded-inductor simulation: (a) circuit; (b) 3-terminal RC network.

11. See, for example, Chapter 3, Section 3.1.4 of *Linear Integrated Networks: Fundamentals*.

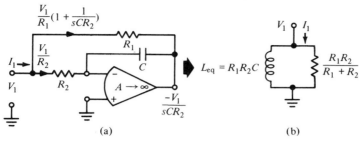

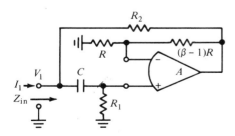

FIG. 1-12. Single-amplifier grounded-inductor simulation: (a) circuit; (b) equivalent circuit.

Z_{in} between terminals 1 and 2 equals the negative of the inverse of the transfer admittance of the RC network by itself with terminals 3 common (Fig. 1-11b). Thus in terms of Fig. 1-11,

$$Z_{in} = -\frac{1}{y_{12}} \tag{1-16}$$

Thus, as we would expect from (1-14), y_{12} of the T-circuit used in Fig. 1-9 is $-(Y_1 Y_2)/(Y_1 + Y_2 + Y_3)$.

Various other methods of grounded inductor simulation using single operational amplifiers have been suggested, two of which are shown in Figs. 1-12 and 1-13. The former[12] provides a grounded inductor with a low Q since R_1 in parallel with R_2 invariably shunts the simulated inductor L_{eq}. Notice also that it uses the operational amplifier as an integrator, i.e., in the open-loop mode with the feedback capacitor C.

The circuit shown in Fig. 1-13 has the following input impedance[13]

$$Z_{in} = \frac{R_2(sCR_1 + 1)}{sC[R_2 + R_1(1 - \beta)] + 1} \tag{1-17}$$

FIG. 1-13. Single-amplifier grounded-inductor simulation.

12. R. L. Ford and F. E. J. Girling, Active filters using simulated inductance, Royal Radar Establishment Techn. Note No. 717, January 1966.
13. D. Berndt and S. C. Dutta Roy, Inductor simulation using a single unity-gain amplifier, *IEEE J. Solid-State Circuits*, **SC-4**, 161–162 (1969).

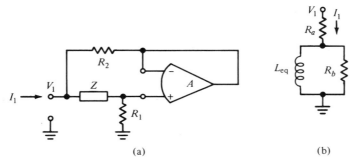

FIG. 1-14. Circuit configuration of Fig. 1-13 using unity-gain amplifier: (a) circuit; (b) equivalent circuit.

For unity gain ($\beta = 1$) and a general reactive impedance Z (Fig. 1-14a) we have

$$Z_{in} = R_2 \frac{Z + R_1}{Z + R_2} \tag{1-18}$$

and with $Z = 1/sC$ this becomes

$$Z_{in} = R_2 \frac{sCR_1 + 1}{sCR_2 + 1} \tag{1-19}$$

If $R_1 > R_2$, (1-19) corresponds to the equivalent RL network shown in Fig. 1-14b, where

$$L_{eq} = CR_2(R_1 - R_2) \tag{1-20}$$

and

$$R_a = R_2 \tag{1-21a}$$

$$R_b = R_1 - R_2 \tag{1-21b}$$

A comparison of this method and that of Fig. 1-10 has shown[14] that the latter is more suitable for high-Q inductors that are reasonably frequency independent. Since, however, the "coil Q" and L_{eq} value cannot be adjusted independently and, in fact, low L_{eq} values are generally associated with low Q, the circuit in Fig. 1-14 is to be preferred where a low L_{eq} with a moderately high Q is required. We shall return to single-amplifier inductor simulation and to a comparison of the circuits given above, in Chapter 7.

Various methods do exist of simulating floating inductors using operational amplifiers, but in general they require a number of amplifiers similar to the four necessary when using two gyrators for the same purpose. Thus the simulated floating inductor is still somewhat problematic and has, in fact, stimulated approaches to filter design in which new alternatives to floating inductors

14. S. C. Dutta Roy and V. Nagarajan, On inductor simulation using a unity-gain amplifier, *IEEE J. Solid-State Circuits*, SC-5, 95–98 (1970).

(e.g., grounded inductors combined with grounded negative capacitors, frequency-dependent negative resistors and conductors, generalized and positive immitance converters[15]) have been found, using operational amplifiers. Nevertheless, it is in the area of active-filter design, in which the operational amplifiers are used as amplifiers rather than to simulate other devices, that the greatest efforts and the greatest progress in terms of actual commercial usage have been made. Rather than simulating known LC filter structures, new methods have been found that are applicable directly to the new, and entirely different hybrid integrated circuit technology. The most common structures to have emerged are single and multiple feedback loops using one or more operational amplifiers in the positive or negative feedback mode. It is with the design of these circuit types that we shall mainly be concerned in this and the following five chapters.

1.3 DECOMPOSING nTH-ORDER NETWORK FUNCTIONS

It is noteworthy that the concept of sensitivity originated in connection with active networks, namely feedback amplifiers, and that it was only with the recent advent of active network synthesis that it appeared in the literature again. This inadvertently conveys the impression that sensitivity must be both higher and more difficult to control in active networks than in passive ones. The closest general proof to this effect was advanced by Orchard[16] (see Section 1.2.1) who demonstrated that, beside being unconditionally stable, resistor-terminated LC ladder networks are also minimum-sensitivity networks. No such general statements can be made for active RC networks whose sensitivity —and stability—can vary over a large range depending on the method of realization used. The unconditional stability of passive LC networks follows directly from the definition of a general passive n-port for which the total energy dissipated between any two terminals must be greater than or equal to zero.[17] Referring to the one-port in Fig. 1-15 this definition of passivity takes on the following form:

$$E(t) = \int_{-\infty}^{t} v(\tau) \cdot i(\tau) \, d\tau = \int_{t_0}^{t} v(\tau) i(\tau) \, d\tau + E(t_0) \geq 0 \qquad [1\text{-}22]$$

FIG. 1-15. One-port used to define activity and passivity in a network.

15. See H. J. Orchard and D. F. Sheahan, loc. cit., and other references at the end of the book.
16. H. J. Orchard, loc. cit.
17. Note that the definition of passivity of an *n-port* is independent of whether active elements are contained in it or not; it is based entirely on the behavior at the terminals. By contrast a *network* is called active only if it contains one or more active elements in it, passive if it contains none.

where

$$0 \leq E(t_0) \leq K < \infty$$

for any t_0 and any $t \geq t_0$. $E(t_0)$ is the energy stored in the one-port at the time t_0. For an n-port, the definition remains, in essence, the same. Since the natural frequencies of a network that dissipates energy must have a finite damping factor, i.e., a finite negative real part, no poles can be located on the $j\omega$ axis to cause oscillation.[18] No such limitation exists for the poles of active networks to which the right-half s-plane is principally as accessible as the left.

Because of the potentially higher sensitivity and the conditional stability of active RC networks, one is compelled to consider certain constraints when designing RC active networks that are not critical with passive ones. One important constraint concerns the complexity, or the order, of an active network. It has been shown for various active-filter approaches that the sensitivity of a network increases rapidly with the order of the network. It therefore becomes impractical, in general, to realize transmission functions of higher than second order.[19] Consequently, to realize it in active RC form, a specified transmission function is broken up into products (i.e., cascade connection) or sums (i.e., parallel connection) of second- and third-order functions.[20] This brings us to the second of the three basic steps in active filter design (i.e., decomposition). The decomposition requirement in no way limits the variety of transmission functions realizable with active RC networks. In fact, as we shall see later, the decomposition into second-order (or at most, with the inclusion of a negative real pole, third-order) networks has many advantages beside sensitivity minimization. These include the capability of filter building-block design compatible with hybrid integrated circuit batch-fabrication methods, ease of tuning if the individual second-order sections are isolated from each other and, finally, because of the isolation of individual stages, filter design simplification by reducing the network order.

1.3.1 Considerations and Criteria for Network Decomposition

The first question to be answered, when decomposing an nth-order network into second-order[21] pole–zero pairs, is by which criterion this selection should be made. Obviously there are many possible combinations and it should not be hard to imagine that the final outcome of this selection process will have a

18. This would not be true of ideal lossless inductors and capacitors, which do not exist physically.
19. The order here refers to the number of complex conjugate poles. Additional negative real poles have no detrimental effect on the sensitivity, as they can be realized by passive RC networks.
20. *Cascade* connections are generally used and are assumed in what follows. In Chapter 3, Section 3-4, we consider parallel connections.
21. In what follows we shall, for brevity, speak only of second-order networks, where it should be understood that third-order networks, having additional negative real poles, may be included as well.

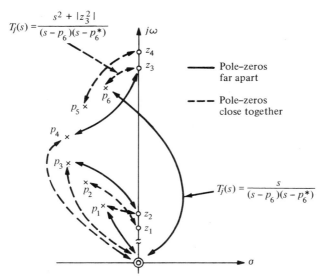

FIG. 1-16. Two possibilities for the pairing of a pole–zero ensemble.

direct bearing on, among other things, the sensitivity, the signal-to-noise capabilities, and the dynamic range of the resulting cascaded nth-order network. The degrees of freedom in this decomposition are threefold:

1. The pole–zero pair selection. If $T(s)$ has n poles and zeros (including the zeros at the origin and at infinity) there are $(n/2)!$ possible combinations of pole–zero pairs. In Fig. 1-16, for example, where the pole–zero distribution in the upper s-plane is shown, two possibilities of pole–zero pairing are shown. The first (solid line) pairs the poles and zeros that are farthest apart; thus, for example, the dominant and therefore most critical pole p_6 is combined with one of the zeros at the origin, resulting in a second-order function:

$$T_6(s) = K_6 \frac{s}{(s - p_6)(s - p_6^*)} \qquad (1\text{-}23)$$

The second combines the poles and zeros that are closest together; thus again starting with the critical pole p_6, it is paired with z_3, so that now

$$T_6(s) = K_6 \frac{(s - z_3)(s - z_3^*)}{(s - p_6)(s - p_6^*)} = K_6 \frac{s^2 + |z_3^2|}{(s - p_6)(s - p_6^*)} \qquad (1\text{-}24)$$

2. The gain distribution. The constants K_j are subject to the constraint that $\prod_j K_j = K$, where K is the gain (i.e., constant multiplicand) of the nth order function $T(s)$. Apart from this constraint the distribution of the individual K_j is arbitrary and, as we shall see presently, may therefore be optimized to satisfy a given criterion.

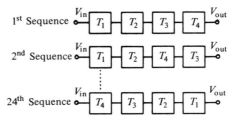

FIG. 1-17. Some of the 24 possible cascading sequences of a four-section network.

3. The cascading sequence. Having established the individual second-order functions $T_j(s)$ by steps 1 and 2 we still have a degree of freedom left in the sequence in which the functions are to be cascaded. For example an eighth-order function $T(s) = \prod_{j=1}^{4} T_j(s)$ can be obtained by arranging the 4 sections in any of 24 (i.e., 4!) possible sequences, some of which are shown in Fig. 1-17.

Having established the degrees of freedom at our disposal, the question arises according to which criterion, or criteria, a network is to be optimized by means of these degrees of freedom. Here again there is more than one possibility, depending to some extent on the application and the characteristics of the available active devices. To name the most important criteria, we have the following:

1. Maximizing the dynamic range. This is important if the active filter is required to process signals of large amplitude while the signal distortion is to remain acceptably small. Clearly, the input signals can be well controlled against overload of the first stage, but overload of individual sections *within* the active filter may not readily be recognized. The output signal level of any section $T_j(s)$ may overload the following stage without any noticeable drastic effects at the output of the overall filter, while the overall frequency response, if measured carefully, might be found to be far from the specified and designed characteristic.

2. Minimizing inband losses. This criterion is closely akin to criterion 1. It may be that, in order to overcome in-band losses or to actually provide gain, the gain per filter stage is increased to the point of overloading following stages, thereby again causing signal distortion. It is to be expected that there is an optimum decomposition of $T(s)$ into pole–zero pairs, an optimum distribution of the constants K_j, and an optimum sequence of the cascaded sections T_j such as to minimize inband losses, or even to provide gain, without introducing signal distortion.

3. Maximizing the signal-to-noise ratio. This will be of importance when the incoming signal levels are very low and degradation due to noise generated within the active element of the network may therefore ensue.[22]

22. One of the disadvantages of active, as compared to passive, networks is of course that the active elements within the former generate noise. Thus the signal-to-noise ratio is an important parameter here, where the noise referred to is (here and in the remainder of this chapter) the noise generated *within* the active network.

4. Minimizing overall transmission sensitivity. It will be shown that in a certain network type the pole–zero pairing has a direct effect on the transmission sensitivity.

5. Minimizing DC offset. Active filters, in contrast to passive ones, may introduce an unacceptable DC offset voltage at the output. For example, this may be the case with low-pass filters which are used to filter out unwanted harmonics of a signal whose zero crossings are critical.

6. Simplifying the tuning procedure. Some second-order active filters are simpler to tune than others, depending on their transfer function and method of realization. A bandpass filter may be simpler to tune (or occur more often in a given system) then a low-pass or high-pass filter. Thus, confronted, for example, with the problem of decomposing the transfer function

$$T(s) = K \frac{s^2}{\left(s^2 + \dfrac{\omega_{p_1}}{q_{p_1}} s + \omega_{p_1}^2\right)\left(s^2 + \dfrac{\omega_{p_2}}{q_{p_2}} s + \omega_{p_2}^2\right)} \tag{1-25}$$

whose pole–zero plot is shown in Fig. 1-18, the decomposition

$$T_1(s) = K_1 \frac{s^2}{s^2 + \dfrac{\omega_{p_1}}{q_{p_1}} s + \omega_{p_1}^2} \tag{1-26a}$$

$$T_2(s) = \frac{K_2}{s^2 + \dfrac{\omega_{p_2}}{q_{p_2}} s + \omega_{p_2}^2} \tag{1-26b}$$

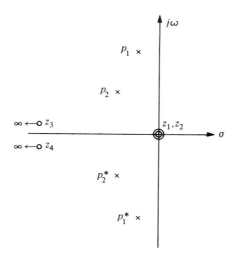

FIG. 1-18. Pole–zero plot of (1-25).

may be less desirable than

$$T_1'(s) = K_1' \frac{s}{s^2 + \dfrac{\omega_{p_1}}{q_{p_1}} s + \omega_{p_1}^2} \tag{1-27a}$$

and

$$T_2'(s) = K_2' \frac{s}{s^2 + \dfrac{\omega_{p_2}}{q_{p_2}} s + \omega_{p_2}^2} \tag{1-27b}$$

The difference in frequency response between the corresponding high-pass–low-pass decomposition and the bandpass–bandpass decomposition is qualitatively shown in Fig. 1-19.

Other criteria for optimization may be required in any given system, but the ones listed above occur most often and therefore may be considered the most important. The point to be emphasized is that, in general, any criterion can result in a different "optimum" decomposition, i.e., in different $T_j(s)$ functions and K_j values and a different cascading sequence. Thus, confronted with the knowledge that we cannot optimize a network simultaneously with respect to all criteria of the kind listed above, we must ask ourselves which one, or ones, to settle for.

In trying to resolve this question, some qualifying considerations should be taken into account. A number of the criteria listed above depend not only on

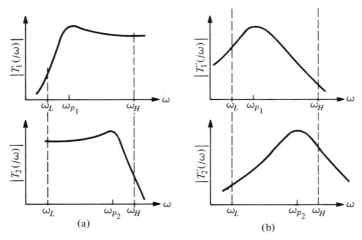

FIG. 1-19. Frequency responses corresponding to different pole–zero pairing of (1-25). (a) Pairing for (1-26); (b) pairing for (1-27).

an optimum decomposition but also on the characteristics of the available active device (e.g., operational amplifier). Thus for a given application, a judicious choice must be made between input voltage-swing capabilities, DC offset voltage, noise performance, bandwidth, and so on. In this way it may be possible to satisfy some of the criteria in advance, and to use an optimization method to satisfy the others. Besides, we should recall that at present only very few methods of optimized decomposition are known. The most important one determines, in turn, the pole–zero pairs, the K_j constants, and the cascading sequence in order to maximize the dynamic range while minimizing in-band losses and distortion. Thus it serves to satisfy our criteria 1, 2 and, in a sense, also 3. This method, which originates from Lueder[23] will be discussed next. Some considerations pertinent to the fulfillment of criterion 4 will follow in Section 1.3.3.

1.3.2 Pole–Zero Pairing for Maximum Dynamic Range and Inband Gain

Transfer Function with Complex Zeros and Complex Poles   In order to find a method of decomposing an nth-order transfer function consisting exclusively of complex zeros and complex poles into a product of second-order functions $T_j(s)$, such that the permissible voltage swing at the input of each corresponding second-order network is as high as possible without creating distortion, while the in-band losses are as low as possible, we proceed as follows. Consider the following general second-order function with complex zeros and poles:

$$T_j(s) = K_j \frac{s^2 + \dfrac{\omega_z}{q_z} s + \omega_z^2}{s^2 + \dfrac{\omega_p}{q_p} s + \omega_p^2} \tag{1-28}$$

Normalizing with respect to K_j, we obtain the magnitude of $T_j(j\omega)$ as a function of frequency:

$$\left| \frac{T_j(j\omega)}{K_j} \right|^2 = F_j(\omega) = \frac{(\omega_z^2 - \omega^2)^2 + (\omega_z/q_z)^2 \omega^2}{(\omega_p^2 - \omega^2)^2 + (\omega_p/q_p)^2 \omega^2} \tag{1-29}$$

Beside a possible extremum at the origin ($\omega = 0$), $F_j(\omega)$ can have at most, two additional extrema at finite frequencies: a maximum at a frequency ω_m, say,

23. E. Lueder, A decomposition of a transfer function minimizing distortion and in-band losses, *Bell Syst. Tech. J.* **49**, 455–470 (1970); also, Cascading of *RC*-active two-ports in order to minimize in-band losses and to avoid distortion, *Proc. Int. Conf. Commun.*, Boulder, Colorado, June 9–11, 1969.

and a minimum at ω_0. Taking the derivative of (1-29) we obtain the frequencies ω_m and ω_0 from the equation

$$\left(\frac{\omega}{\omega_p}\right)^2 =$$

$$\frac{c^4 - 1 \pm \left[(c^4 - 1)^2 - c^2\left\{2(c^2 - 1) + \frac{1}{q_z^2} - \frac{c^2}{q_p^2}\right\}\left\{2(c^2 - 1) + \frac{1}{q_p^2} - \frac{c^2}{q_z^2}\right\}\right]^{1/2}}{2(c^2 - 1) + \frac{1}{q_p^2} - \frac{c^2}{q_z^2}}$$

(1-30)

where $c = \omega_z/\omega_p$. Furthermore there is an extremum at $\omega = \infty$ where

$$F_j(\infty) = 1 \tag{1-31a}$$

The value at the origin is given by

$$F_j(0) = c^4 \tag{1-31b}$$

If ω_m and ω_0, obtained from (1-30), are real and nonnegative, the corresponding extema $F_j(\omega_m)$ and $F_j(\omega_0)$ are obtained from (1-29) and (1-30) as follows:

$$F_j(\omega_{m,0}) = \frac{\left\{c^2 - \left(\frac{\omega_{m,0}}{\omega_p}\right)^2\right\}^2 + \frac{c^2}{q_z^2}\left(\frac{\omega_{m,0}}{\omega_p}\right)^2}{\left\{1 - \left(\frac{\omega_{m,0}}{\omega_p}\right)^2\right\}^2 + \left(\frac{\omega_{m,0}}{\omega_p}\right)^2 \cdot \frac{1}{q_p^2}} \tag{1-32}$$

In general, $F_j(\omega)$ will have either the shape of Fig. 1-20a or b, i.e., $F_j(\omega_m)$ and $F_j(\omega_0)$ may be either larger, smaller, or equal to $F_j(\infty) = 1$.

In order to optimize the dynamic range of a cascade of second-order networks with gain functions of the form (1-29), Lueder reasoned as follows. The dynamic range and in-band losses are primarily of interest within a prescribed frequency band ranging from ω_1 to ω_2 (Fig. 1-20). In general, $|T_j(\omega)/K_j|$ is required to be close to some average in-band response, e.g., unity, for $\omega_1 < \omega < \omega_2$. Outside the band, i.e. for $\omega < \omega_1$ and for $\omega > \omega_2$, $|T_j(\omega)/K_j|$ will be required to be attenuated with respect to that average. We assume now that the maximum of $F_j(\omega)$, $F_{j\,max}(= F_j(\omega_m))$, and its minimum $F_{j\,min}(= F_j(\omega_0))$, both lie within the bounds ω_1 and ω_2. Then, optimizing for the dynamic range means first, that $F_{j\,max}$ must be as small as possible, i.e., exceed the average in-band response as little as possible. This is because at ω_m the maximum in-band signal is most likely to occur (assuming an input signal to the jth stage having a constant in-band spectrum) and overloading of the input to the $(j + 1)$th stage is therefore risked most.

At ω_0 the signal will be attenuated most, namely by $F_{j\,min}$. In order to overcome this attenuation by subsequent amplification, we risk, once again,

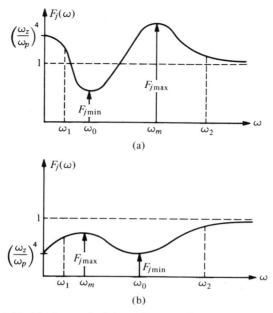

FIG. 1-20. Maxima and minima of a general second-order function.

overloading the succeeding stage and, possibly, deteriorating the signal-to-noise ratio. Not to do so would result in high in-band losses. Thus, our second stipulation says that $F_{j\,min}$ must be as large as possible, i.e., it also should be as close to the average in-band response as possible. Combining our two conditions we require that the ratio $F_{j\,max}/F_{j\,min}$ be as close to unity or, expressed logarithmically, that

$$d_j = \log{(F_{j\,max}/F_{j\,min})} \qquad (1\text{-}33)$$

be as close to zero, as possible. Clearly, either $F_{j\,max}$ or $F_{j\,min}$ can coincide with either of the band-edge responses (i.e., $F_j(\omega_1)$ or $F_j(\omega_2)$), without in any way invalidating our discussion.

What if either $F_{j\,max}$ or $F_{j\,min}$ lie outside the ω_1–ω_2 frequency band? If $F_{j\,min}$ lies outside the band, then the lowest response within the band will be at either one of the band edges, i.e., $F_{j\,min}$ will be replaced by either $F_j(\omega_1)$ or $F_j(\omega_2)$ in (1-33). If $F_{j\,max}$ lies outside the band (e.g., at $F_j(0)$) the situation is somewhat different. We could, of course, argue that the maximum in-band response will also be at either band edge and therefore replace $F_{j\,max}$ in (1-33) by either $F_j(\omega_1)$ or $F_j(\omega_2)$. This, however, ignores the fact that certain signals, occurring in or out of band, may have frequency components in the vicinity of the out-of-band frequency ω_m. These may be large enough, when multiplied by $F_{j\,max}$, to overdrive the succeeding filter stage, shifting its bias point and causing distortion of signals *in* the passband. To avoid this possibility we stipulate that the value of $F_{j\,max}$ in (1-33) must be the maximum response of $F_j(\omega)$, whether ω_m is inside or outside the $\omega_1-\omega_2$ frequency band.

Having defined the term d_j in (1-33), we can now find this number for every possible pole–zero pair assignment,[24] i.e., for every second-order function $T_j(s)$ contained in a given set of n poles and n zeros (including the zeros at infinity). Thus, there will be a different set of $n/2$ values of d_j, where $j = 1, 2, \ldots, n/2$, for every possible pole–zero assignment of $T(s)$. It is then our task to find that particular pole–zero assignment whose maximum d_j is a minimum, i.e., for which

$$\max \{d_j\}_{j=1, 2, \ldots, n/2} \to \min \qquad\qquad [1\text{-}34]$$

To demonstrate the procedure outlined above, consider, for example, the pole–zero plot shown in Fig. 1-21. Pairing the poles (p_2, p_2^*) with the zeros (z_1, z_1^*) we find the corresponding F_{max}/F_{min} for this particular pair, designating its value by $d_{2,1}$. We continue to pair the remaining pole–zero pairs; $d_{\nu,\mu}$ will designate the F_{max}/F_{min} for the pair p_ν, z_μ. Notice that we are only considering complex conjugate poles and zeros for the time being, leaving the case of negative real roots for later. Hence we need consider only the pairing of the roots in the upper s-plane; the complex conjugate roots below must, of necessity, follow suit. All possible assignments (p_ν, z_μ), with their associated $d_{\nu,\mu}$ are then listed in a table, as shown, in Fig. 1-22. The number of possible entries is $(n/2)^2$.

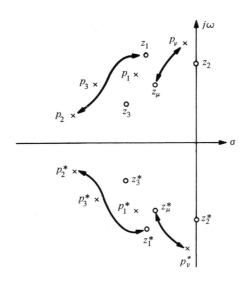

FIG. 1-21. Pairing a pole–zero ensemble.

24. To avoid confusion we shall designate the pairing of a single pole pair with a single zero pair as a "pole–zero *pair* assignment," whereas the assignment of all the pole pairs and zero pairs of a given function $T(s)$ into pole–zero pairs simply as its "pole–zero assignment." The latter is identical with the factorization of $T(s)$ into $\prod_{j=1}^{n/2} T_j(s)$, whereas the former corresponds to any one of the single functions $T_J(s)$.

	z_1	z_2	- - -	- - -	z_μ	- - -	$z_{n/2}$
p_1	$d_{1,1}$	$d_{1,2}$	- - -	- - -	$d_{1,\mu}$		
p_2	$d_{2,1}$	$d_{2,2}$	- - -	- - -	$d_{2,\mu}$		
⋮	⋮	⋮	⋮		$d_{3,\mu}$		
					⋮		
p_ν	$d_{\nu,1}$	$d_{\nu,2}$	$d_{\nu,3}$		$d_{\nu,\mu}$		$d_{\nu,n/2}$
⋮	⋮	⋮	⋮		⋮		
$p_{n/2}$	$d_{n/2,1}$	$d_{n/2,2}$	$d_{n/2,3}$		$d_{n/2,\mu}$		$d_{n/2,n/2}$

FIG. 1-22. The general $d_{\nu,\mu}$ table.

Having obtained the $d_{\nu,\mu}$ *table* (Fig. 1-22) we proceed to plot the values $d_{\nu,\mu}$ graphically as shown in Fig. 1-23. The zeros are distributed (for convenience in equal intervals) on the abscissa, while the ordinate represents $d_{\nu,\mu}$. The $d_{\nu,\mu}$ values in each column of our $d_{\nu,\mu}$ table are now plotted along a vertical line going through the z_μ corresponding to that column. To each point of a vertical line we assign the number $1, 2, \ldots, \nu, \ldots, n/2$ of the corresponding pole (i.e., the *pole number*). Thus, each column in the $d_{\nu,\mu}$ table is represented by a corresponding column in the $d_{\nu,\mu}$ graph. The $d_{\nu,\mu}$ set whose maximum is less than the maximum in any other set (i.e., whose maximum is a minimum) is then found by drawing the lowest possible line (the *assignment line*) parallel to the abscissa underneath which each pole number (i.e., $1, 2, \ldots, \nu, \ldots, n/2$) appears in a *different* column at least once. Since each of these numbers belongs to a pole, the solution consists of the assignments of these

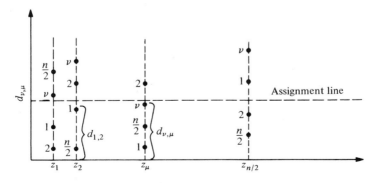

FIG. 1-23. The general $d_{\nu,\mu}$ graph.

poles to the zeros at the bases of the columns. The "minimized maximum" of the resulting $d_{v,\mu}$ set is the highest $d_{v,\mu}$ value ($d_{1\,max}$) below the assignment line.

If some of the pole numbers appear below the assignment line in more than one column, then for those columns a corresponding choice of pole–zero assignments still exists. To establish the optimum assignment among them, a second assignment line, which *excludes* the pole–zero pair corresponding to the highest $d_{v,\mu}$ value, can be drawn, and so on, until, by steps, a unique pole–zero assignment is obtained. Clearly, if two or more $d_{v,\mu}$ values on different columns are identical, then the corresponding pole–zero pairs can be interchanged without any effect on the dynamic range, as long as their interchanged collective $d_{v,\mu}$ values are also the same.

EXAMPLE: Let us illustrate the foregoing discussion by an example. Consider an eighth-order transfer function with four complex conjugate pairs of poles and zeros, respectively. We can derive the $d_{v,\mu}$ table as shown, e.g., in Fig. 1-24, and obtain the corresponding $d_{v,\mu}$ graph, which we shall assume to be as shown in Fig. 1-25a. The assignment line is the lowest horizontal line beneath which each pole number appears in a different column at least once; thus, in our example the first assignment line is determined by pole p_3 in the z_3 column. This prescribes the combination of pole pair (p_3, p_3^*) with zero pair (z_3, z_3^*), and the corresponding $d_{3,3}$ (the maximum $d_{v,\mu}$ of this set) may be designated $d_{1\,max}$. The remaining pole–zero assignments are not yet uniquely established, however, since the remaining pole numbers, i.e., 1, 2, and 4, occur more than once in the other three columns. Excluding our p_3–z_3 pole–zero pair from the set, we plot the remaining poles and zeros in the $d_{v,\mu}$ graph shown in Fig. 1-25b and immediately obtain our next assignment, p_1–z_2, since $d_{1,2}$ is clearly the maximum $d_{v,\mu}$ value ($d_{2\,max}$) in this subset. In the third assignment (Fig. 1-25c) there is no freedom of selection left and an assignment line is actually unnecessary. Notice that in spite of the fact that $d_{2,1}$ and $d_{4,4}$ are identical, pole p_2 cannot be combined with z_4 nor p_4 with z_1, since the resulting $d_{v,\mu}$ values are very large (Fig. 1-25a). This interchangeability is only permitted if the collective $d_{v,\mu}$ values remain the same,

	z_1	z_2	z_3	z_4
p_1	$d_{1,1}$	$d_{1,2}$	$d_{1,3}$	$d_{1,4}$
p_2	$d_{2,1}$	$d_{2,2}$	$d_{2,3}$	$d_{2,4}$
p_3	$d_{3,1}$	$d_{3,2}$	$d_{3,3}$	$d_{3,4}$
p_4	$d_{4,1}$	$d_{4,2}$	$d_{4,3}$	$d_{4,4}$

FIG. 1-24. The $d_{v,\mu}$ table of an eighth-order function.

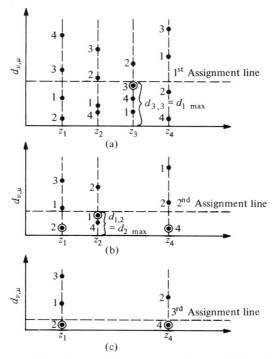

FIG. 1-25. Example of the three $d_{v,\mu}$ graphs obtained for an eighth-order function.

e.g., as in the situation shown in Fig. 1-26. Thus, whereas in our example (Fig. 1-25), the pole–zero pairing must be (p_3, z_3), (p_1, z_2), (p_2, z_1), and (p_4, z_4), in the situation shown in Fig. 1-26, either (p_4, z_1), (p_2, z_4), or (p_4, z_4) and (p_2, z_1) would be permissible.

In some cases the assignment of one particular pole to one particular zero may be prescribed from the outset for one reason or another (e.g., to acco-modate the realization procedure used, or to conform to existing filter sections). If so, this assignment is simply established in advance by leaving it out of the

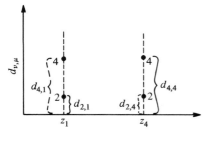

FIG. 1-26. Pole–zero interchangeability in a $d_{v,\mu}$ graph.

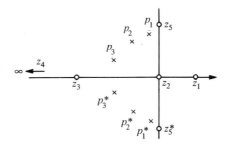

FIG. 1-27. Pole–zero ensemble containing real zeros.

assignment process. If, on the other hand, a particular pole–zero pair assignment is undesirable, then a high $d_{\nu,\,\mu}$ value is assigned to it, thereby preventing this pair from being combined.

Transfer Function with Real and Complex Zeros and Complex Poles

So far we have considered only complex conjugate pole and zero pairs. Let us now examine how to proceed when some of the zeros of $T(s)$ are on the real axis. As before, all n poles are assumed to be complex (Fig. 1-27); thus n is even. Including the zeros at infinity, $T(s)$ then also has n zeros. If the number of zeros on the real axis (including the origin and infinity) is r_z, then r_z is also even, since both n and the number of zeros not on the real axis are even. The r_z zeros on the real axis can now be paired in a number of ways. For example, the four real zeros z_1, z_2, z_3, z_4 in Fig. 1-27 can be paired in the following three ways: $(z_1 z_2)\,(z_3 z_4)$; $(z_1 z_3)\,(z_2 z_4)$; or $(z_1 z_4)\,(z_2 z_3)$. In general, with r_z different zeros on the real axis, where r_z is even, there are α_0 possible ways of pairing the zeros, where[25]

$$\alpha_0 = (r_z - 1)(r_z - 3)(r_z - 5)\cdots 5\cdot 3\cdot 1 \tag{1-35}$$

In the case of multiple zeros on the real axis there are fewer than α_0 possibilities. For example, if in Fig. 1-27 z_1 and z_2 coincide, then there are only two possibilities: $(z_1 z_2)\,(z_3 z_4)$ or $(z_1 z_3)\,(z_2 z_4)$. The same two possibilities exist if in addition, z_3 and z_4 coincide as well.

Having established all possible pairs of real-axis zeros we must incorporate them into the $d_{\nu,\,\mu}$ graph. Selecting one of the possible sets of zero pairs (e.g., the set $(z_1 z_2)(z_3 z_4)$ in our example above) each pair is added to the pairs of complex conjugate zeros in the graph and treated exactly as they are. Thus, the $d_{\nu,\,\mu}$ graph of the pole–zero plot in Fig. 1-27 may have the form shown in Fig. 1-28a. This does not exhaust all possible assignments, however, since there are the two other ways of combining the four real zeros of Fig. 1-27,

25. To prove this, we start out with r_z zeros, where r_z is even. Selecting one zero, we can form $(r_z - 1)$ pairs with the other zeros. There are $(r_z - 2)$ zeros left, from which we can again pick one and form $(r_z - 3)$ possible pairs, and so on. Applying the multiplication rule of combinatorial analysis (see, e.g., J. Riordan, *An Introduction to Combinatorial Analysis* (New York: John Wiley & Sons, 1958)) we obtain α_0 possible pairwise assignments as given by (1-35).

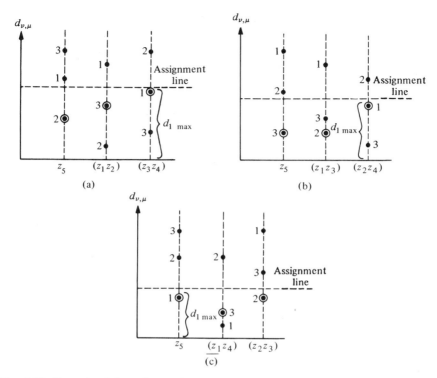

FIG. 1-28. Example of three $d_{v,\mu}$ graphs obtained for the pole–zero ensemble in Fig. 1-27.

corresponding to the two additional $d_{v,\mu}$ graphs shown, for example, in Fig. 1-28b and c. The graph whose $d_{1\,max}$ is smallest provides the solution to the assignment problem. In our example, the assignment given in Fig. 1-28c is therefore the one to select. Clearly, if the r_z real-axis zeros are simple, there are α_0 different $d_{v,\mu}$ graphs to select the assignment solution from, corresponding to the α_0 sets of zero pairs given by (1-35). If some of the real-axis zeros occur in multiples, there will be accordingly less, as discussed above.

Transfer Function with Real and Complex Zeros and Real and Complex Poles   Finally we must examine the most general case in which some of the poles also lie on the negative real axis. Consider first the most simple and, in practice, very common case in which we have one negative real pole in addition to the r_z real-axis zeros. The real-axis pole can be assigned to one of the real-axis zeros; there are r_z possible ways of doing this. According to (1-35) the $(r_z - 1)$ zeros left can be combined pairwise in $\alpha_1 = (r_z - 2)(r_z - 4) \cdots$ $5 \cdot 3 \cdot 1$ ways, where, again, each pair of the α_1 sets is treated as a complex conjugate pair of zeros. Hence we have $r_z \alpha_1$ sets of zeros to be assigned to the poles, which now means $(r_z \alpha_1)$ $d_{v,\mu}$ graphs from which to select the assignment with the smallest $d_{1\,max}$. If some of the r_z zeros occur in multiples we have somewhat less than $(r_z \alpha_1)$ possible graphs, as discussed above.

If there are r_p negative real poles, where $r_p > 1$, the procedure becomes more complicated but is not any different in nature from the preceding steps described. If, for example, r_p is odd, the first step is to assign one of the zeros to one of the poles and the second step is to pair the remaining real-axis zeros and poles. There will be several possibilities for both of these steps. For each set of real-axis pole–zero pairs a $d_{v,\mu}$ table can be derived and the corresponding $d_{v,\mu}$ graph, pertaining to the complex pole–zero pairs, extended accordingly. The $d_{v,\mu}$ graph with the smallest $d_{1\,max}$ again provides the assignment solution. If r_p is even, the first step mentioned above is omitted.

Third-Order Pole Zero Assignments   We have made reference, repeatedly, to the fact that third-order networks are as acceptable as second-order networks since none of the factors requiring the decomposition into lower-order networks are thereby affected. This is because the third, necessarily negative real, pole is realizable by a simple resistor–capacitor combination which may be attached to the active second-order network so as to produce the desired third-order network function (see Fig. 1-29) while leaving such critical factors as network sensitivity, building-block design, and even ease of tuning (all of which necessitate the decomposition in the first place) virtually unchanged. As a consequence, in the interest of economy with respect to the number of amplifiers required, there is no justification for combining any pair of negative real poles into a second-order network (consisting of a passive RC network and a buffer amplifier). Each negative real pole occurring in a given nth order transfer function should be combined with a complex conjugate pole pair, so that the final cascade consists exclusively of third- and second-order networks, each of which contains a complex conjugate pole–pair. This presumes that there are more complex conjugage pole pairs than there are negative real poles, which, in practice, is generally the case.

As we did for second-order networks, we must now find a strategy for the optimum combination of third- and second-order networks, where the existence of the former depends on the existence of negative real poles. Whereas sensitivity remains virtually unaffected here, the dynamic range and in-band

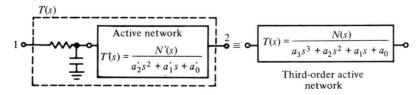

FIG. 1-29. Realization of a third-order network.

gain still depend on an optimum assignment. Using the same criterion of goodness as he did for the second-order case, Lueder has also solved this assignment problem.[26] The large number of assignment possibilities that exists in the third-order case requires a mathematical strategy that cannot possibly be coped with manually. In essence, though, the computer routine used for its solution is based on the solution to the same "bottleneck assignment problem" that was solved for the second-order case.[27] The algorithm used was adapted by S. Halfin[28] to the pole–zero assignment problem both for second- and third-order networks[29] and a method thereby obtained of listing by computer the optimal decomposition of a given pole–zero distribution (of which in some cases there may be more than one).

Lacking a computer program to solve the third-order assignment problem, the following guidelines may be used to solve it in practice:

1. In most practical situations there will be much fewer negative real than complex poles.

2. The second-order pole–zero assignment can be carried out for all the complex poles, combined with the maximum number of zeros n'_z, such that $n'_z \le n'_p$, where n'_p is the total number of complex poles. If the total number of zeros (including those at the origin and infinity) is n_z, where $n_z > n'_z$, care should be taken to include any existing negative real and complex zeros in the n'_z zeros.

3. Having obtained $n'_p/2$ second-order functions using the second-order assignment strategy, the remaining $(n_p - n'_p)$ negative real poles must be combined with the corresponding number of second-order functions. By increasing the order of some (or all) of the $n'_p/2$ second-order sections by one, it can in general be said that the dynamic range of the overall filter will, if anything, be improved. The reason for this is that the additional pole is formed by an RC low-pass section (Fig. 1-29) which tends to decrease the likelihood of amplifier overdrive. Thus, if the dynamic range is satisfactory for the $n'_p/2$ second-order sections, it will be still more so, after the order of $(n_p - n'_p)$ of them has been increased to three.

4. Since the overall dynamic range will not be grossly affected by the conversion of some of the networks from second to third order, convenience, in terms of ease of final trimming, may be the most useful criterion to follow here. With this in mind, second-order *low-pass* sections should be the first to

26. E. Lueder, Der maximale Dynamikbereich bei *RC*-aktiven Filtern mit Teilvierpolen dritten Grades, *Arch. Elek. Übertragung*, **26**, 487–492 (1972).
27. O. Gross, The bottleneck assignment problem: An algorithm, *Rand Paper*, March 6, 1959, p. 1630.
28. S. Halfin, Optimization method for cascading filters, *Bell Syst. Tech. J.* **49**, 185–190 (1970).
29. Clearly, the third-order assignment must also encompass the second-order case. All possible third-order assignments are first determined by assigning every available negative real pole to a complex pole pair and these, in turn, to zeros. The remaining complex second-order poles are then assigned to zeros in the manner described above.

receive the additional, third pole, followed by the other all-pole types (high-pass and bandpass). These may be followed by second-order sections containing complex conjugate zeros, so that the order of second-order sections containing negative real zeros is least likely to be increased. Clearly, some criterion other than ease of final network adjustment may dictate the distribution of the $(n_p - n_p')$ negative real poles among the $n_p'/2$ second-order functions with complex conjugate poles (e.g., commonality with other existing networks). Since the addition of these poles invariably corresponds to the addition of a simple RC network to a second-order section, this distribution can be carried out with relative ease and without the fear of detrimental consequences on the performance of the completed filter.

A NUMERICAL EXAMPLE: The pole–zero assignment procedure that has been outlined above and that is epitomized by (1-34) can be summarized in words as follows:

In order to maximize the dynamic range and the in-band gain of an nth-order filter network, the poles and zeros of the individual second- or third-order sections should be chosen such that the response of each section is guaranteed to be *as flat as possible* within the frequency band of interest.

We shall illustrate this by a numerical example. Consider the following fifth-order transfer function:

$$T(s) = K \cdot \frac{(s^2 + \omega_{z_1}^2)(s^2 + \omega_{z_2}^2)}{(s - p_0)(s - p_1)(s - p_1^*)(s - p_2)(s - p_2^*)} \tag{1-36}$$

where

$$\omega_{z1} = 29.2 \text{ krad/sec}$$
$$\omega_{z2} = 43.2 \text{ krad/sec}$$
$$p_0 = 16.8 \text{ krad/sec}$$
$$p_1, p_1^* = -9.7 \pm j17.5 \text{ krad/sec}$$
$$p_2, p_2^* = -2.36 \pm j22.4 \text{ krad/sec}$$

The pole–zero plot is shown in Fig. 1-30 and the frequency response $|T(j\omega)|$ in Fig. 1-31. With the following pole–zero assignment (solid lines in Fig. 1-30):

$$T_0(s) = \frac{16.8}{s - p_0} \tag{1-37a}$$

$$T_1(s) = 0.47 \, \frac{s^2 + \omega_{z_1}^2}{(s - p_1)(s - p_1^*)} \tag{1-37b}$$

$$T_2(s) = 0.27 \, \frac{s^2 + \omega_{z_2}^2}{(s - p_2)(s - p_2^*)} \tag{1-37c}$$

we obtain the frequency responses shown in Fig. 1-32. Notice that the coefficients K_0, K_1, and K_2 were selected such that $|T_j(0)| = 1$. This makes a

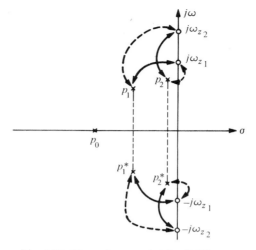

Fig. 1-30. The pole–zero plot for (1-36).

comparison of the frequency responses clearer. Applying the Lueder–Halfin algorithm, we obtain the pole–zero assignment indicated by broken lines in Fig. 1-30. This results in the transfer functions

$$T_0(s) = \frac{16.8}{s - p_0} \tag{1-38a}$$

$$T_1'(s) = 0.21 \frac{(s^2 + \omega_{z_2}^2)}{(s - p_1)(s - p_1^*)} \tag{1-38b}$$

$$T_2'(s) = 0.6 \frac{s^2 + \omega_{z_1}^2}{(s - p_2)(s - p_2^*)} \tag{1-38c}$$

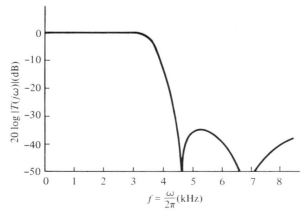

FIG. 1-31. Frequency response corresponding to (1-36).

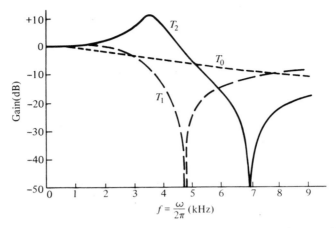

FIG. 1-32. Frequency responses corresponding to pole–zero assignment (1-37).

and the frequency responses plotted in Fig. 1-33. A comparison of the responses in Figs. 1-32 and 1-33 shows that by combining the high-Q poles p_2, p_2^* with the zeros lying *closest* to them (i.e., ω_{z1}, ω_{z1}^*), $|T_2'(j\omega)|$ becomes flatter in the passband (between 0 Hz and 3 kHz). $T_0(s)$ remains the same in each case and can, of course, be combined with either one of the complex pole pairs to form a third-order function. Its effect on the frequency response will not be very noticeable in any case.

The example just given was simple enough to have permitted the optimum pole–zero assignment to be obtained without going through the steps outlined

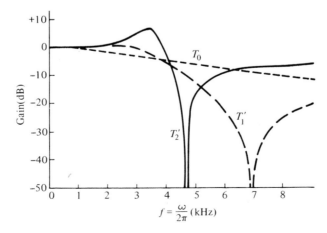

FIG. 1-33. Frequency responses corresponding to pole–zero assignment (1-38).

above. By contrast we shall subsequently (after discussing the various other aspects pertaining to the decomposition of $T(s)$ into subfunctions $T_j(s)$) go through the pole–zero assignment of a higher-order function by this method. There the indispensibility of this pole–zero assignment procedure will immediately become apparent.

1.3.3 Pole–Zero Pairing for Minimum Sensitivity

In pairing the zeros of $T(s)$ with its poles, with a view to minimizing loss and maximizing dynamic range, we are giving up a valuable degree of freedom in active filter design that might have come in useful for sensitivity minimization. This is not to say that the former objective is not an important one. Clearly, one of the less attractive characteristics of an active filter is the presence of an active device which may place critical constraints on the maximum attainable dynamic range. This disadvantage is often countered by the argument that active devices are capable of providing gain. However, if the pole–zero pairs are combined improperly, this is no longer true. The resulting loss may be quite high when many second-order sections are in cascade, with a consequent deterioration in signal-to-noise ratio. Thus the question is not so much *whether* poles and zeros should be paired for minimum distortion, but rather *how the resulting assignments affect the overall transmission sensitivity*. This question will be discussed next.

Broadly speaking, selecting poles and zeros for "maximum flatness" in the passband in order to minimize distortion results in *combining the high-Q poles of $T(s)$ with zeros that are as close to them as possible*. This was shown to be the case for the example of Fig. 1-30. Also, for the example in Fig. 1-16, it is reasonable to assume that the Lueder–Halfin algorithm would combine the poles and zeros interconnected by the dashed, rather than by the solid lines. This is more rule-of-thumb than rigorous, though, since the assignment depends on the initial pole–zero distribution, which may conceivably prescribe a deviation from this general rule in some cases. It does, however, provide a basis from which to ask what effect pole–zero pairing (or more specifically, the pole–zero distance of the individual second-order networks of a filter cascade) has on the sensitivity of the overall filter function $T(s)$. In general the individual second-order sections are isolated from each other, either by additional or by internal buffering, so that any interaction between sections becomes negligible. The question therefore becomes very much simpler, since we can assume that *the overall sensitivity of an nth-order network is minimized by minimizing the sensitivity of the individual second-order sections*. To go about this, we must first examine the effect of the pole and zero location on the sensitivity of a general second-order network.

Transmission Sensitivity of Second-Order Networks[30]   Consider the general second-order network with complex poles

$$T(s) = \frac{N(s)}{D(s)} = K \cdot \frac{(s - z_1)(s - z_2)}{(s - p)(s - p^*)} = K \cdot \frac{s^2 + 2\sigma_z s + \omega_z^2}{s^2 + 2\sigma_p s + \omega_p^2} \qquad (1\text{-}39)$$

Rewriting (1-39) in the equivalent bilinear form we obtain

$$T(s) = \frac{N(s)}{D(s)} = \frac{a_0 A(s) + x b_0 B(s)}{u_0 U(s) + x v_0 V(s)} \qquad (1\text{-}40)$$

where

$$A(s) = s^2 + a_1 s + a_2 \qquad (1\text{-}41a)$$

$$B(s) = s^2 + b_1 s + b_2 \qquad (1\text{-}41b)$$

$$U(s) = s^2 + u_1 s + u_2 \qquad (1\text{-}41c)$$

$$V(s) = s^2 + v_1 s + v_2 \qquad (1\text{-}41d)$$

a_0, b_0, u_0, and v_0 are real constants associated with the scaling factor K in (1-39), and x is the network parameter with respect to which we are examining the transmission sensitivity. The roots of $A(s)$, $B(s)$, $U(s)$, and $V(s)$ can be negative real or complex conjugate, depending on the decomposition used to obtain the bilinear form. Huelsman has shown[31] that there are 26 possible root-locus formations for the roots of $N(s)$ and $D(s)$ with respect to the variable x, depending on the relative signs of a_0, b_0, u_0, v_0, and x and on the root locations of $A(s)$, $B(s)$, $U(s)$, and $V(s)$. The roots of the binomials given by (1-41) are thus interpreted in terms of the root loci of $N(s)$ and $D(s)$ with respect to variable x values. Thus, according to this interpretation, the roots of $U(s)$ and of $V(s)$ define the roots of $D(s)$ corresponding to $x = 0$ (i.e., "open-loop poles") and $x = \infty$ (i.e., "open-loop zeros"), respectively. The corresponding root locus may, for example, have the form shown in Fig. 1-34. Referring to this figure we have

$$U(s) = (s - P_1)(s - P_2) \qquad (1\text{-}42)$$

and

$$V(s) = (s - \bar{P}_1)(s - \bar{P}_2) \qquad (1\text{-}43)$$

where P_i denotes an open-loop pole, $\bar{P}_i$ an open-loop zero.[32] Based on an analogous root locus for the roots of $N(s)$ (see Fig. 1-34) we obtain

$$A(s) = (s - Z_1)(s - Z_2) \qquad (1\text{-}44)$$

and

$$B(s) = (s - \bar{Z}_1)(s - \bar{Z}_2) \qquad (1\text{-}45)$$

30. The following discussion is based on G. S. Moschytz, Second-order pole–zero pair selection for *n*th-order, minimum sensitivity networks, *IEEE Trans. Circuit Theory* **CT-17**, 527–534 (1970).
31. L. P. Huelsman, Stability criteria for active *RC* synthesis techniques, *Proc. Nat. Electronics Conf.*, 1964, pp. 731–736.
32. The terms "open-loop" and "closed-loop" are not necessarily meant literally here. Rather they are derived from the origin of the respective roots in the bilinear form. Naturally if x is the active element in an active network then the terms apply accurately.

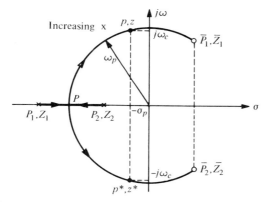

FIG. 1-34. Root-locus corresponding to decomposition of a second-order polynominal.

where, in the figure, Z_1, Z_2 and $\bar{Z}_1$, $\bar{Z}_2$ corresponds to P_1, P_2 and $\bar{P}_1$, $\bar{P}_2$, respectively.

In Table 1-1, 13 of the 26 possible root locus formations pertaining to a second-order polynomial are shown. The remaining 13 formations are the duals of the ones shown, in that the zeros of $U(s) = 0$ and those of $V(s) = 0$ are interchanged. Thus the geometrical configurations are the same but the direction of increasing x is reversed. In Chapter 2 it will be shown that for the purposes of active RC filter design these 26 root-locus formations can be reduced to four basic types which represent the foundation for a classification of second-order active networks using one active element.

Assuming simple poles and zeros of $T(s)$, the transmission sensitivity of (1-39) is given by:[33]

$$S_x^{T(s)} = S_x^K + \frac{\mathscr{S}_x^p}{s - p} + \frac{\mathscr{S}_x^{p*}}{s - p^*} - \frac{\mathscr{S}_x^{z_1}}{s - z_1} - \frac{\mathscr{S}_x^{z_2}}{s - z_2} \qquad (1\text{-}46)$$

and the corresponding zero and pole sensitivities by

$$\mathscr{S}_x^{z_1} = a_0 \cdot \frac{(z_1 - Z_1)(z_1 - Z_2)}{z_1 - z_2} \qquad (1\text{-}47)$$

$$\mathscr{S}_x^{z_2} = a_0 \cdot \frac{(z_2 - Z_1)(z_2 - Z_2)}{z_2 - z_1} \qquad (1\text{-}48)$$

$$\mathscr{S}_x^{p} = u_0 \cdot \frac{(p - P_1)(p - P_2)}{2j\omega_c} \qquad (1\text{-}49)$$

$$\mathscr{S}_x^{p*} = u_0 \cdot \frac{(p^* - P_1)(p^* - P_2)}{2j\omega_c} \qquad (1\text{-}50)$$

33. See, for example, Chapter 4, equations (4-6), (4-92), and (4-93) of *Linear Integrated Networks: Fundamentals*.

TABLE 1-1, DECOMPOSITIONS AND ROOT LOCI OF A SECOND-ORDER POLYNOMIAL

Case	Sign of u_0, xv_0	Roots of $U(s)=0$	Roots of $V(s)=0$	Root Locus
1	same	complex	complex	
2	same	real	complex	
3	same	$P_1 = \infty$ P_2 real	complex	
4	same	$P_1, P_2 = \infty$	complex	
5	same	real	real	
6	same	$P_1 = \infty$ P_2 real	real	
7	same	real	$\bar{P}_1, \bar{P}_2 = \infty$	
8	opposite	complex	complex	

38

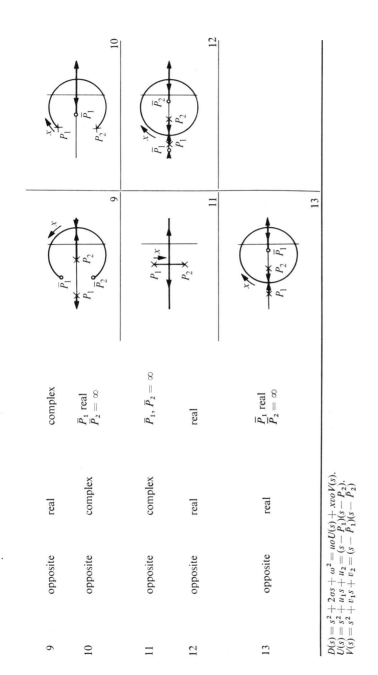

9	opposite	real	complex
10	opposite	complex	$\bar{P}_1$ real $\bar{P}_2 = \infty$
11	opposite	complex	$\bar{P}_1, \bar{P}_2 = \infty$
12	opposite	real	real
13	opposite	real	$\bar{P}_1$ real $\bar{P}_2 = \infty$

$D(s) = s^2 + 2\sigma s + \omega^2 = uoU(s) + xvoV(s)$.
$U(s) = s^2 + u_1 s + u_2 = (s - P_1)(s - P_2)$.
$V(s) = s^2 + v_1 s + v_2 = (s - \bar{P}_1)(s - \bar{P}_2)$

39

where $p = -\sigma_p + j\omega_c$ and $\omega_c = \sqrt{\omega_p^2 - \sigma_p^2}$. Note that the zero sensitivities depend only on z_1, z_2, Z_1 and Z_2, the pole sensitivities only on p, p^*, P_1 and P_2. Thus the zero sensitivities are independent of the poles of $T(s)$ and vice versa.

Second-Order Allpole and Quasi-Allpole Networks[34]   Since the changes of $T(s)$ on the imaginary axis in the vicinity of the pole p are in general of interest, we examine the transmission sensitivity for $s = j\omega_c$. In this case all the terms in (1-46) except the one involving p may be neglected. This also includes the term S_x^K, which may be disregarded anyway, since changes in K involve only the gain level, which is of little consequence in a high-selectivity filter. Thus with (1-46) and (1-49) we obtain

$$S_x^{T(s)} \approx \frac{\mathscr{S}_x^p}{s - p} = u_0 \frac{(p - P_1)(p - P_2)}{2j\omega_c(s + \sigma_p - j\omega_c)} \qquad [1\text{-}51]$$

For $s = j\omega_c$,

$$S_x^{T(j\omega_c)} \approx -j\frac{Q}{\omega_c^2} \cdot u_0(p - P_1)(p - P_2) \qquad [1\text{-}52]$$

where we have assumed a high-Q pole and therefore

$$Q = \frac{\omega_p}{2\sigma_p} \approx \frac{\omega_c}{2\sigma_p} \qquad [1\text{-}53]$$

The equation (1-52) gives the transmission sensitivity of a high-selectivity second-order network, i.e., a network with high-Q poles. The sensitivity is proportional to Q, a not unexpected result. Perhaps less expected is the fact that it is altogether independent of the zeros of $T(s)$, i.e., *no one selection of pole–zero combination will decrease it more than any other.*

As pointed out earlier, P_1 and P_2 are the open-loop poles of $T(s)$ corresponding to the roots of $D(s) = 0$ when $x = 0$. Referring to Fig. 1-35, the product of the distances between p, P_1, and P_2 must therefore be minimized in order to minimize S_x^T. It can be shown[35] that this product becomes smaller the closer the open-loop poles P_1 and P_2 are to the closed-loop poles p and p^* along the corresponding root locus. If P_1 and P_2 are the poles of a passive RC network and, consequently, restricted to the negative real axis, minimum sensitivity is obtained when P_1 and P_2 coincide as a double pole, e.g., at the coalescence point P in Fig. 1-34. In the event that the parameter x represents the active element of a second-order network $T(s)$, $U(s)$ will be associated with the driving point or transfer function of an RC network. Thus, minimum

34. By "quasi-allpole" networks we understand networks whose zeros, although finite, are distant from the poles and from the $j\omega$ axis.
35. G. S. Moschytz, loc. cit. pp. 533–534.

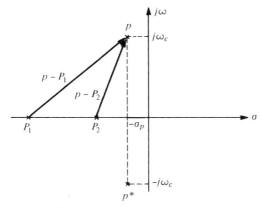

FIG. 1-35. Diagram for the minimization of S_x^T.

sensitivity with respect to variations in x can then be obtained only in the limit when $U(s)$ has a double root. This corresponds to the decomposition for minimum sensitivity obtained by Horowitz.[36]

EXAMPLE: To demonstrate the above by an example, consider the following denominator polynomial of a second-order filter:

$$D(s) = s^2 + 2s + 5 \tag{1-54}$$

whose roots are $p, p^* = -1 \pm j2$. $D(s)$ can now be decomposed into a bi-linear form in any number of ways, of which we shall consider the following three:

$$D(s) = s^2 + 2s + 5 = U(s) + xV(s)$$

$U(s) = (s - P_1)(s - P_2)$	$xV(s)$	
$(s + 5)(s + 2)$	$-x(5s + 5)$	(1-55a)
$(s + 2.24)^2$	$-x(2.48s)$	(1-55b)
$(s + 1)^2$	$+4x$	(1-55c)

In each of these decompositions, the nominal value of x is unity. Since the roots of $U(s)$ are the open-loop poles and the roots of $V(s)$ the open-loop zeros, it is simple to draw the root loci corresponding to the three decompositions given by (1-55); they are shown in Fig. 1-36. In our discussion above it was pointed out that minimum sensitivity occurs when the open-loop poles

36. I. M. Horowitz, Optimization of negative-impedance conversion methods of active *RC* synthesis, *IRE Trans. Circuit Theory*, CT-6, 296–303 (1959).

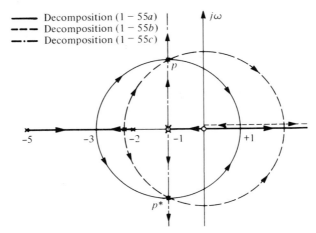

FIG. 1-36. Root loci corresponding to (1-55).

P_1 and P_2 are closest to the closed-loop pole pair p, p^*. Thus by inspection of the root loci in Fig. 1-36, we expect the transmission sensitivity to improve progressively from decomposition (1-55a) to (1-55c). Whether this is, in fact, true can be easily determined from (1-51). It is sufficient to calculate the pole sensitivity $\mathscr{S}_x^p = (p - P_1)(p - P_2)/2j\omega_c$, the remaining terms being the same for all three decompositions and u_0 being unity. For the case of (1-55a) we obtain

$$\mathscr{S}_x^p = (-1 + j2 + 5)(-1 + j2 + 2)/4j = 2.5$$

and the other two decompositions, in precisely the same way, we obtain for

$$\mathscr{S}_x^p = 1.38 \ \underline{/26°30'} \qquad \text{and} \qquad \mathscr{S}_x^p = 1 \ \underline{/90°}$$

respectively. As expected from inspection of the root loci, the pole sensitivity, and with it the transmission sensitivity, is smallest for the third decomposition, (1-55c), and increases from (1-55b) to (1-55a). If we recall that the argument of the pole sensitivity as given for these three cases is the same as the angle of the tangent to the root locus at the pole p, then we see as well that the root locus at p progresses horizontally for the first case, at an angle of 26°30' toward the $j\omega$ axis in the second, and in a vertical direction in the third. A glance at Fig. 1-36 of course bears this out. We see, as well, that the pole sensitivity is informative, not only with respect to sensitivity, but, (with its argument) with respect to stability as well.

Second-Order Networks with High-Q Poles and Zeros   The sensitivity expression given by (1-52) is based on an approximation in which the sensitivity contributions of the zeros and of the complex conjugate pole have been neglected. These terms must be considered, however, if the specified second-order function consists of a pole and zero pair that are close to one another and close to the $j\omega$ axis; in such a case the zero sensitivity contributions are no longer negligibly small.

In the vicinity of interest, i.e., for $s = j\omega_c$, the sensitivity term of the complex conjugate pole p^* becomes

$$\left.\frac{\mathscr{S}_x^{p*}}{s - p^*}\right|_{s = j\omega_c} = u_0 \frac{(p^* - P_1)(p^* - P_2)}{4\omega_c^2 - 2j\omega_c\sigma} \tag{1-56a}$$

For $\sigma \ll \omega_c$

$$\left.\frac{\mathscr{S}_x^{p*}}{s - p^*}\right|_{s = j\omega_c} \approx u_0 \frac{(p^* - P_1)(p^* - P_2)}{4\omega_c^2} \tag{1-56b}$$

This expression is still negligible compared to the main sensitivity contribution given by (1-52), even for medium Q values.

The zero sensitivity terms in (1-46) become

$$\left.\frac{\mathscr{S}_x^{z_1}}{s - z_1}\right|_{s = j\omega_c} = a_0 \frac{(z_1 - Z_1)(z_1 - Z_2)}{(z_1 - z_2)(j\omega_c - z_1)} \tag{1-56c}$$

and

$$\left.\frac{\mathscr{S}_x^{z_2}}{s - z_2}\right|_{s = j\omega_c} = a_0 \frac{(z_2 - Z_1)(z_2 - Z_2)}{(z_2 - z_1)(j\omega_c - z_2)} \tag{1-56d}$$

If z_1 and z_2 are complex conjugate, the corresponding zero sensitivity values $\mathscr{S}_x^{z_1}$ and $\mathscr{S}_x^{z_2}$ will be so as well. Furthermore, by reasoning in exactly the same way as for the pole sensitivity term (1-52), the zero terms (1-56c) and (1-56d) will become smaller the closer Z_1 and Z_2 are located to the closed-loop zeros z_1 and z_2 along the corresponding root locus.

It is clear from inspection of (1-56c) and (1-56d) that the contributions of the zero sensitivities $\mathscr{S}_x^{z_1}$ and $\mathscr{S}_x^{z_2}$ to $S_x^{T(j\omega_c)}$ will be smaller the farther away the two zeros z_1 and z_2 are located from the pole frequency ω_c. However, decreasing the individual zero sensitivity terms by selecting maximally distant pole–zero pairs represents only one way of minimizing the overall transmission sensitivity, and may not necessarily be the most effective one at that. Furthermore, as was pointed out earlier, the resulting pole–zero combinations will very likely conflict with the pole–zero combinations for minimum distortion and maximum dynamic range where pole–zero pairs are generally required to be as close together as possible.

If the pole p has a high Q, we have seen that, when compared with (1-52), (1-56b) can be neglected in (1-46). Similarly, unless either z_1 or z_2 is close to p, (1-56c) and (1-56d) are also negligible. On the other hand, if z_1 is close to p, then z_2 will equal z_1^* and be close to p^*; (1-56d) can then be neglected but (1-56c) cannot be. Thus, when a zero pair is located close to a high-Q pole pair, we obtain in the vicinity of $s = j\omega_c$, from (1-52) and (1-56c)

$$S_x^{T(j\omega_c)} \approx -\left[a_0 \frac{(z_1 - Z_1)(z_1 - Z_2)}{(z_1 - z_2)(j\omega_c - z_1)} + ju_0 \frac{Q}{\omega_c^2}(p - P_1)(p - P_2)\right] \tag{1-57}$$

By inspection of (1-46), and (1-57) the transmission sensitivity has the general form:

$$S_x^{T(s)} = S_x^K + \underbrace{(f_1 + f_2)}_{\text{poles}} - \underbrace{(f_3 + f_4)}_{\text{zeros}} \qquad (1\text{-}58a)$$

or, for high-Q poles and zeros,

$$S_x^{T(s)} = S_x^K + f_1 - f_3 \qquad (1\text{-}58b)$$

where the f_i represent the individual pole and zero terms in the expressions for the transmission sensitivity. Consequently, depending on the polarities of the corresponding residues, the resulting transmission sensitivity may actually be smaller, due to cancellation of individual pole and zero terms, than it would be by minimizing the zero terms alone. Which of the two methods of sensitivity minimization is preferable (or even possible) must be decided individually for worst-case conditions and for each network in question.

Pole-Zero Pair Selection for Minimum-Sensitivity nth-Order Networks

For an nth-order network with m zeros we have

$$S_x^{T(s)} = S_x^K + \sum_{i=1}^{n} \frac{\mathcal{S}_x^{p_i}}{s - p_i} - \sum_{i=1}^{m} \frac{\mathcal{S}_x^{z_i}}{s - z_i} = S_x^K + F_1 - F_2 \qquad [1\text{-}59]$$

Let us now assume that the pole pair $p, p^* = -\sigma_p \pm j\omega_c$ is the dominant pole pair in the nth-order transfer function $T(s)$. The stability of the overall network, consisting of a cascade of $n/2$ second-order networks, will then be primarily determined by the stability of this pole pair, which in turn will determine the frequency behavior of the network in the vicinity of ω_c. Consequently, referring to (1-56c) and (1-56d), one way of minimizing the overall sensitivity function will be to select the pole-zero pairs for the high-Q second-order functions as far apart as possible. In so doing, one will achieve a minimum contribution from the zero terms (i.e., F_2) in the vicinity of the high-Q pole frequencies.

Another method of minimizing the overall sensitivity function is to ensure partial or even total cancellation of the pole–zero pair terms in (1-59), i.e., to require that $(F_1 - F_2) \rightarrow 0$. This can be achieved by an appropriate method of polynomial decomposition of individual second-order functions such that, in (1-58), $f_1 + f_2 - f_3 - f_4 \rightarrow 0$ for each second-order section. This in turn will lead to a method of synthesizing the second-order networks required in cascade to realize $T(s)$. Some guidelines for the polynomial decomposition of second-order functions providing this cancellation will now be discussed.

Starting out with $T(s)$ given by the product of bilinear, second-order functions, we have

$$T(s) = \prod_{j=1}^{n/2} T_j(s) = \prod_{j=1}^{n/2} \frac{N_j(s)}{D_j(s)} = \prod_{j=1}^{n/2} \frac{A_j(s) + x_j B_j(s)}{U_j(s) + x_j V_j(s)} \qquad (1\text{-}60)$$

where for simplicity the individual coefficients a_{0j}, b_{0j}, u_{0j}, and v_{0j} are assumed equal to unity. The transmission sensitivity of any individual network function $T_j(s)$ to its most variable parameter x_j then follows directly.[37] Referring to (1-39) and expressing the polynomials in (1-60) in terms of their roots, we find that, in order to minimize the sensitivity of a given second-order transfer function $N_j(s)/D_j(s)$, we should attempt to decompose the two second-order polynomials $N_j(s)$ and $D_j(s)$ such that either one of the following conditions is fulfilled:[38]

$$\frac{A_j(s)}{U_j(s)} = \frac{(s - Z_{1j})(s - Z_{2j})}{(s - P_{1j})(s - P_{2j})} = t_{14j}(s)$$

$$= \frac{(s - z_j)(s - z_j^*)}{(s - p_j)(s - p_j^*)} = \frac{N_j(s)}{D_j(s)} \quad \text{[1-61]}$$

or

$$\frac{B_j(s)}{V_j(s)} = \frac{(s - \bar{Z}_{1j})(s - \bar{Z}_{2j})}{(s - \bar{P}_{1j})(s - \bar{P}_{2j})}$$

$$= \frac{(s - z_j)(s - z_j^*)}{(s - p_j)(s - p_j^*)} = \frac{N_j(s)}{D_j(s)} \quad \text{[1-62]}$$

Conversely, for a given decomposition, i.e., when a prescribed network realization is given, the pole–zero pairs that form the second-order polynomials $N_j(s)$ and $D_j(s)$ should be selected such as to satisfy either (1-61) or (1-62). Thus, for example, if to realize $T_j(s)$ a circuit configuration is used for which the decomposition with respect to the most variable element x_j results in $B_j(s)$ being equal to $V_j(s)$, then, in order to satisfy (1-62), pole–zero pairs should be selected as close together as possible. If this is inconvenient, sensitivity can be minimized by selecting pole–zero pairs to satisfy (1-61).

Before we attempt to do this, however, let us take a closer look at our two conditions (1-61) and (1-62). The condition (1-61) says, in effect, that the leakage path $t_{14}(s)$ should have the same transfer function as our desired transfer function $T_j(s)$, or, referring to Fig. 1-34, that the open-loop poles of the numerator (Z_{1j} and Z_{2j}) and of the denominator (P_{1j} and P_{2j}) should be equal to the closed-loop zeros z_j, z_j^* and poles p_j, p_j^*, respectively, of $T_j(s)$. This is a trivial statement; it says that, in order to minimize the transmission sensitivity with respect to an element x_j, $T_j(s)$ should simply be realized independently of x_j. On the face of it (1-62) is not much more helpful. It states that for minimum transmission sensitivity, the open-loop zeros of the numerator ($\bar{Z}_{1j}$ and $\bar{Z}_{2j}$) and of the denominator ($\bar{P}_{1j}$ and $\bar{P}_{2j}$) should coincide with the closed-loop zeros and poles, respectively, of our desired transfer function

37. The transmission sensitivity is given by $S_{T(s)}^x = [U(s)/D(s)] - [A(s)/N(s)]$ or $S_x^{T(s)} = -x$ $\{[V(s)/D(s)] - [B(s)/N(s)]\}$. See, for example, Chapter 4, equation (4-19a) and (4-19b), respectively, of *Linear Integrated Networks: Fundamentals*.
38. The term t_{14j} is the leakage path in the essential flow graph of $T_J(s)$ with respect to x_j.

$T_j(s)$. This is only possible for infinite values of x_j; thus in the process of satisfying (1-62), the parameter x_j is being increased beyond all measure and the transmission sensitivity[39] will certainly not be small. We must remember, however, that our conditions need not be satisfied for all s, and that therefore the identities just mentioned between open-loop and closed-loop roots need not hold accurately. Rather, we must treat conditions (1-61) and (1-62) as guidelines for pole–zero selection such that, over a given frequency range, the following errors will be minimized:

$$\varepsilon_1 = \left| \frac{B(j\omega)}{V(j\omega)} \right| - \left| \frac{N(j\omega)}{D(j\omega)} \right| \to \min \qquad [1\text{-}63]$$

or

$$\varepsilon_2 = \left| \frac{A(j\omega)}{U(j\omega)} \right| - \left| \frac{N(j\omega)}{D(j\omega)} \right| \to \min \qquad [1\text{-}64]$$

In this way, the (closed-loop) pole–zero assignment may be carried out to minimize conditions (1-63) or (1-64) for a given type of network realization (i.e., for a given open-loop pole–zero distribution). Conversely a network realization may be selected whose open-loop poles and zeros (i.e., of each subnetwork $T_j(s)$) are matched to the specified set of closed-loop poles and zeros such as to satisfy (1-63) or (1-64).

For the special case where we require (1-61) or (1-62) to be satisfied mainly at low frequencies, we can apply (1-61) or (1-62) to the origin of the $j\omega$ axis and obtain

$$\frac{Z_{1j} Z_{2j}}{P_{1j} P_{2j}} = \frac{z_j z_j^*}{p_j p_j^*} = \frac{\omega_{zj}^2}{\omega_{pj}^2} \qquad [1\text{-}65]$$

and

$$\frac{\bar{Z}_{1j} \bar{Z}_{2j}}{\bar{P}_{1j} \bar{P}_{2j}} = \frac{z_j z_j^*}{p_j p_j^*} = \frac{\omega_{zj}^2}{\omega_{pj}^2} \qquad [1\text{-}66]$$

For example, (1-65) specifies that the DC gain of the leakage path $t_{14}(s)$ must be equal to the DC gain of $T_j(s)$. (The constants a_0, b_0, u_0, v_0, and K_j must of course then also be taken into account.) Equation (1-66) makes the same statement for the passive network obtained when $x \to \infty$. A similar approximation can be derived from (1-61) and (1-62) for high frequencies when we let $s \to \infty$. We find, for example, from (1-61) that $t_{14}(s \to \infty)$ must equal unity, or to be more exact, must equal K_j. Equations (1-63) and (1-64) provide the conditions necessary to minimize the transmission sensitivity for frequency ranges within these two frequency extremes.

It will be clear from the preceding discussion that it is impossible to state a general pole–zero assignment rule for the sensitivity minimization of cascaded

---

39. The transmission sensitivity as given by $S_x^{T(s)} = -x\{[V(s)/D(s)] - [B(s)/N(s)]\}$.

nth-order networks, without considering how the individual second-order sections are to be realized. Although some of the various existing guidelines for the minimization of transmission sensitivity[40] have been given above, which of these possibilities yields best results must be investigated on an individual basis.

Pole-Zero Pair Selection for Type I and Type II Networks   In contrast to the nonunique nature of the general pole–zero pairing problem for sensitivity minimization, there does exist a type of network referred to as type I networks[41] (in contrast to all other networks, which have been designated type II) for which a pole–zero pairing rule may be stated explicitly and, as it turns out, simply. This network class will be discussed in more detail in what follows.

Referring to the bilinear network function given by (1-40), we make the following definition:

A type I network is a network for which $a_0 A(s)$ or $b_0 B(s)$ equals zero in the bilinear form of the transmission function $T(s)$ with respect to the most variable element x.

It then follows that the transmission sensitivity $S_x^{T(s)}$ of a type I network depends only on the denominator $D(s)$ of $T(s)$, i.e., on its poles; in fact, the poles of $S_x^{T(s)}$ coincide with the poles of the corresponding transmittance $T(s)$. We can therefore make the following statement:

The transmission sensitivity of type I networks is independent of the transmission zeros. Therefore pole–zero pairing has no influence on the overall sensitivity and need not be considered for sensitivity minimization.

For a cascade of $n/2$ second-order type I networks, we have

$$S_x^{T(s)} = S_x^K + \sum_{i=1}^{n} \frac{\mathscr{S}_x^{p_i}}{s - p_i} \tag{1-67}$$

and, for a second-order network in the vicinity of the pole frequency ω_c,

$$S_x^{T(j\omega_c)} = \frac{u_0 Q}{\omega_c^2} \left[\frac{(p^* - P_1)(p^* - P_2)}{4Q - j} - j(p - P_1)(p - P_2) \right] \qquad [1\text{-}68]$$

For a high Q (i.e., $\omega_c \approx \omega_p$), this expression equals that given by (1-52).

It is useful at this point to relate the conditions defining a type I network to the branches of the basic flow graph of a general network. This is shown in Fig. 1-37 in terms of the bilinear form given by (1-40). Inspection of Fig. 1-37 shows that type I networks either possess no leakage path ($A(s) = 0$) or *only*

40. See also E. Lueder, A decomposition of pp. 426-427 a transfer function minimizing sensitivity, *IEEE Trans. Circuit Theory*, **CT-17**, pp. 426–427 (1970), in which the overall sensitivity is minimized by the branch-and-bound method (see Section 1.3.5 below).
41. See also Chapter 4, Sections 4.1.1 and 4.3.2 of *Linear Integrated Networks: Fundamentals*.

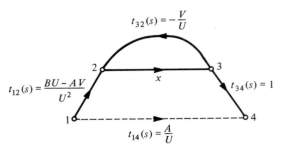

$$t_{32}(s) = -\frac{V}{U}$$

$$t_{12}(s) = \frac{BU - AV}{U^2}$$

$$t_{34}(s) = 1$$

$$t_{14}(s) = \frac{A}{U}$$

FIG. 1-37. Signal-flow graph of a general network.

a leakage path $(B(s) = 0)$. Either way, as shown in Fig. 1-38a and b, there is only *one forward path* between source and sink.

In general, the stability of the characteristics of a high-selectivity active network ultimately depends on the transmission sensitivity with respect to one or more of its *active* elements.[42] It is therefore important to consider the two classes of networks that are characterized by the nonexistence (type I) and by the existence (type II) of more than one forward path in the basic flow graph with respect to that particular *active* element x whose changes cause the largest variations in the network characteristics of interest.

Type I networks can be represented by a *single* active feedback loop with respect to x. The transmission zeros of these networks are independent of x and the transmission sensitivity is the reciprocal of the return difference with

$$a_0 A(s) = 0: \quad t_{14} = 0$$

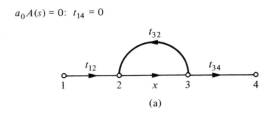

(a)

$$b_0 B(s) = 0: \quad t_{12} t_{34} = t_{14} t_{32}$$

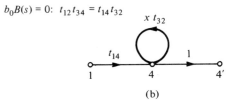

(b)

FIG. 1-38. Signal-flow graphs of type I networks: (a) with $A(s) = 0$; (b) with $B(s) = 0$.

42. As we shall see in Chapter 3 some synthesis methods for the realization of high-selectivity second-order networks use more than one active element. The sensitivity with respect to each of the active elements, or with respect to the most critical active element which determines the overall stability of the network, must then be investigated.

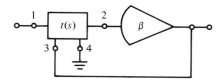

FIG. 1-39. Type I network.

respect to x. Assuming that primarily the sensitivity with respect to the active element x determines the stability of the network characteristics, the sensitivity is not influenced by the location of the transmission zeros.

As an example consider the general network configuration shown in Fig. 1-39. The sensitivity with respect to the voltage gain β is of interest. With the voltage transfer parameters given by

$$t_{12} = \left.\frac{V_2}{V_1}\right|_{V_3=0} = \frac{n_{12}}{d_{12}} \tag{1-69}$$

and

$$t_{32} = \left.\frac{V_2}{V_3}\right|_{V_1=0} = \frac{n_{32}}{d_{32}} \tag{1-70}$$

we obtain:

$$T(s) = \frac{V_3}{V_1} = \frac{\beta t_{12}}{1 - \beta t_{32}} = \frac{d_{32}}{d_{12}} \cdot \frac{\beta n_{12}}{d_{32} - \beta n_{32}} \tag{1-71}$$

If $d_{12} = d_{32} = d$, we find, using (1-40),

$$a_0 A(s) = 0 \tag{1-72a}$$

$$b_0 B(s) = n_{12} \tag{1-72b}$$

$$u_0 U(s) = d \tag{1-72c}$$

$$v_0 V(s) = -n_{32} \tag{1-72d}$$

Clearly, then, the configuration shown in Fig. 1-39 is a type I network with respect to the gain element β, and $S_{\beta}^{T(s)}$ is independent of pole–zero pairing.

Type II networks may be characterized by the general signal-flow graph shown in Fig. 1-37. Such networks have two forward paths, namely one through the element x and a leakage path that bypasses x. Since these are perfectly general networks, their transmission sensitivity depends on both the poles and zeros of the specified transmission $T(s)$. Thus sensitivity minimization must follow one of the procedures outlined in this section.

As an example of a type II network, we can convert the single-ended amplifier configuration of Fig. 1-39 into the differential configuration shown in Fig. 1-40. Using the voltage transfer parameters given by (1-69) and (1-70) we obtain the voltage transfer function

$$T(s) = \frac{V_3}{V_1} = \frac{1 - \beta(1 - t_{12})}{1 - \beta t_{32}} = \frac{d_{32}}{d_{12}} \frac{d_{12} - \beta(d_{12} - n_{12})}{d_{32} - \beta n_{32}} \tag{1-73}$$

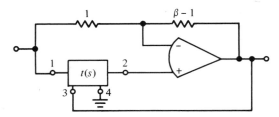

FIG. 1-40. Type II network.

If $d_{12} = d_{32} = d$, the elements of the bilinear form result as follows:

$$a_0 A(s) = d \qquad (1\text{-}74a)$$

$$b_0 B(s) = -d(1 - t_{12}) \qquad (1\text{-}74b)$$

$$u_0 U(s) = d \qquad (1\text{-}74c)$$

$$v_0 V(s) = -n_{32} \qquad (1\text{-}74d)$$

The differential amplifier configuration of Fig. 1-40 therefore clearly represents a type II network. Note that $a_0 A(s) = u_0 U(s)$. Thus, from (1-61) one way of minimizing $S_\beta^{T(s)}$ is to pair the closest poles and zeros together.

Summary   The methods of sensitivity minimization applicable to the two network classes discussed above are summarized in Table 1-2. Note that when using type I networks, we can always combine the poles and zeros for optimum dynamic range, since sensitivity is not thereby effected. With type II networks we must be more careful; in general we must settle for a pole–zero assignment that is optimum either with respect to sensitivity or dynamic range. If we wish to minimize the effect of the zeros on sensitivity by placing them far apart from the poles, the resulting assignment will generally not coincide with the assignment optimizing dynamic range.

In many practical cases, only a few high-Q poles will occur in a given distribution of poles and zeros. We may then select those zeros minimizing the sensitivity of the high-Q poles, and pair the rest for optimum dynamic range. In order to find those zeros minimizing sensitivity, one may start out by selecting zeros far away from the high-Q poles. The transmission sensitivity is then virtually independent of the zero sensitivity (see (1-51)). If it is still too high, one can attempt to minimize it by combining a zero pair with the pole pair in question such that the effect of the resulting zero sensitivity cancels that of the pole sensitivity. If a zero very close to the pole is selected, we have the terms given in (1-57) which, if we are fortunate, may cancel sufficiently. Otherwise, we may take the more general approach of selecting a pole–zero combination such that (1-63) or (1-64) are minimized over the frequency range of interest.

TABLE 1-2, POLE–ZERO PAIR SELECTION FOR MINIMIZATION OF
TRANSMISSION SENSITIVITY

1. *Minimizing pole terms* (dominant-pole or type I network).
   Select open-loop poles P_1, P_2 as close to closed-loop poles p, p^* as possible:

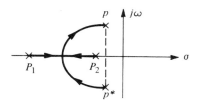

2. *Minimizing zero terms* (type II network).
   Select zeros as far away from pole as possible:

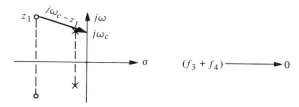

3. *Minimizing transmission sensitivity* (type II network).
   Select pole–zero pairs so that pole and zero terms cancel:

$$S_x^{T(s)} \approx \underbrace{(f_1 + f_2)}_{\text{Poles}} - \underbrace{(f_3 + f_4)}_{\text{Zeros}} = 0$$

1.3.4 Optimizing the Gain Distribution of Cascaded Networks

In decomposing our nth order function $T(s)$ into the product of second- or third-order functions $T_j(s)$ as given by (1-2), we now assume that the poles of each function $T_j(s)$ have been combined with zeros so as to optimize the dynamic range, the transmission sensitivity, or both. Our next task is to assign a constant factor K_j to each function $T_j(s)$, such that

$$\prod_{j=1}^{n/2} K_j = K \qquad [1\text{-}75]$$

where K is the gain factor associated with the overall function $T(s)$. The constraint (1-75) is, of course, by no means conclusive; there are an infinite number of ways of distributing K_j factors such that (1-75) is satisfied. Thus, we have again the choice of optimizing this gain distribution, and shall show

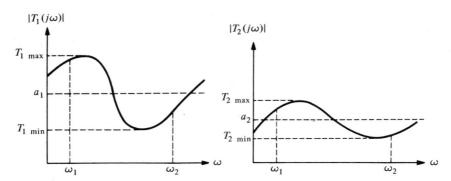

FIG. 1-41. Two frequency responses.

how to do so for maximum dynamic range or for minimum sensitivity. Considering the former first, we proceed according to the method proposed by Lueder.[43]

To understand the optimization with respect to the dynamic range, let us consider the two frequency responses in Fig. 1-41. If $T(s) = T_1(s) \cdot T_2(s)$, then lowering the level of $T_1(s)$ will impell us to raise the level of $T_2(s)$ accordingly. Any shift in these levels amounts to proportional changes in the gain factors K_1 and K_2. Our task is to find a judicious criterion for shifting these levels, i.e., for assigning appropriate K_j's, according to the optimization of our choice. If overdrive in either section is our main concern (and the overdrive characteristics of both sections are the same, which may in general be assumed), then the levels should be shifted such that $T_{1\,max} = T_{2\,max}$, where

$$T_{j\,max} = \max |T_j(j\omega)| \tag{1-76a}$$

and

$$T_{j\,min} = \min |T_j(j\omega)| \tag{1-76b}$$

for frequencies ω within the band of interest, i.e., $\omega_1 \leq \omega \leq \omega_2$. If, on the other hand, we wish to maximize the signal-to-noise ratio because of persistent high noise levels generated within a network, then we must make $T_{1\,min} = T_{2\,min}$ and allow for this minimum value to be as large as possible within the constraint of (1-75).

To take both overloading *and* noise into account, we specify the *average* level of each stage and prescribe that all these levels should be equal. Thus, setting

$$T_j(j\omega) = K_j \cdot t_j(j\omega) \tag{1-77}$$

the average level is given by

$$a_j = \frac{T_{j\,max} + T_{j\,min}}{2} = K_j \frac{t_{j\,max} + t_{j\,min}}{2} \tag{1-78}$$

43. E. Lueder, Eine Kettenschaltung *RC*-aktiver Vierpole mit maximalem Dynamikbereich, *Arch. Elek. Übertragung*, **24**, 467–470 (1970).

where $t_{j\,max}$ and $t_{j\,min}$ are defined in the same way as $T_{j\,max}$ and $T_{j\,min}$, respectively. Requiring equal average levels in the passband for our two functions in Fig. 1-41 therefore means $a_1 = a_2$, or, for the $n/2$ sections making up an nth order network, we require specifically

$$K_1 \frac{t_{1\,max} + t_{1\,min}}{2} = K_j \frac{t_{j\,max} + t_{j\,min}}{2} \tag{1-79}$$

for $j = 1, 2, \ldots, n/2$. Taking the product of all $n/2$ averages a_j as given by (1-78) we obtain, using (1-79) and (1-75),

$$\left[K_1 \frac{t_{1\,max} + t_{1\,min}}{2} \right]^{n/2} = \prod_{j=1}^{n/2} K_j \cdot \prod_{j=1}^{n/2} \left(\frac{t_{j\,max} + t_{j\,min}}{2} \right)$$

$$= K \prod_{j=1}^{n/2} \left(\frac{t_{j\,max} + t_{j\,min}}{2} \right) \tag{1-80}$$

Solving for K_1 we obtain

$$K_1 = \frac{\left[K \prod_{j=1}^{n/2} (t_{j\,max} + t_{j\,min}) \right]^{2/n}}{t_{1\,max} + t_{1\,min}} \tag{1-81}$$

and with (1-79):

$$K_j = \frac{\left[K \prod_{j=1}^{n/2} (t_{j\,max} + t_{j\,min}) \right]^{2/n}}{t_{j\,max} + t_{j\,min}} \tag{1-82}$$

This expression provides all the gain constants K_j, for $j = 1, 2, \ldots, n/2$, optimizing the dynamic range according to the criterion (1-79).

We consider briefly, now, what effect the gain distribution may have on the network *sensitivity*. In general, K_j will be at least partially determined by the ratio of two resistors, say R_{j1} and R_{j2}. Then

$$K_j = \frac{R_{j1}}{R_{j2}} K_j' \tag{1-83}$$

and, using (1-2),

$$T_j(s) = \frac{R_{j1}}{R_{j2}} K_j' \frac{(s - z_{j1})(s - z_{j2})}{(s - p_j)(s - p_j^*)} = \frac{R_{j1}}{R_{j2}} T_j'(s) \tag{1-84}$$

$T_j'(s)$ has the same shape as $T_j(s)$ but a different level. If $T_j'(s)$ is independent of R_{j1} and R_{j2} then the transmission sensitivity of $T(s)$ to any other component x will be

$$S_x^{T_j(s)} = S_x^{T'_j(s)} \tag{1-85}$$

i.e., the sensitivity to some critical (active) component x will remain unchanged when we vary R_{j1} or R_{j2} to obtian some K_j value. Assuming that it exists, this independence of $T_j'(s)$ with respect to R_{j1} or R_{j2} clearly represents a big advantage. When it does not exist, we are in a situation similar to the one encountered with Type II networks with respect to the pole–zero assignment: we must make a choice between optimizing for dynamic range or minimizing sensitivity.

1.3.5 Optimizing the Sequence of Cascaded Networks

The last remaining step in our decomposition of $T(s)$ into second- and third-order networks is to establish the order in which the individual sections should be cascaded. Since we can, in general, assume that the individual stages are isolated from each other, the cascading sequence will not have any effect on the overall sensitivity, and we can proceed to optimize the sequence with respect to the optimum dynamic range without any misgivings. This optimization was also developed by Lueder,[44] whose procedure we shall now discuss.

Consider the $n/2$ stages in Fig. 1-42. The input voltage spectrum to the $(i + 1)$th stage is $V_i(j\omega)$; the maximum voltage occurring in each spectrum is $V_{i\,\text{max}}$. $V_{\text{in}}(j\omega)$ is the input spectrum to the cascade and therefore not subject to any constraints. Since any amplitude at any frequency may occur at the input, it is reasonable to assume a constant input spectrum, i.e., $V_{\text{in}}(j\omega) =$ constant, at least for $\omega_1 < \omega < \omega_2$. As in the previous optimizations for maximum dynamic range, we shall also assume equal input voltage-swing capabilities for the amplifiers in the cascade. In order to prevent overdriving any one of them, we are interested in the sequence of $n/2$ stages in which the $V_{i\,\text{max}}$ to occur at any input is minimal, or

$$V_{i\,\text{max}} \equiv \max | V_i(j\omega)| \tag{1-86}$$

for $0 < \omega < \infty$ and $i = 1, 2, \ldots, [(n/2) - 1]$, should be as small as possible.

Incidentally, in any given network realization, it may not actually be the input voltage that first overdrives an amplifier but a voltage at some internal node of a section. Nevertheless, the internal voltage must be linearly related

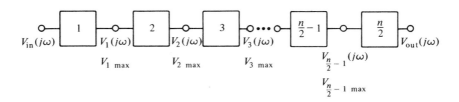

FIG. 1-42. Cascade of $n/2$ second-order networks.

44. E. Lueder, op. cit.

to the input voltage, and therefore minimizing the input voltage will in turn minimize the internal voltage threatening to overdrive the amplifier. One should, however, keep in mind the fact that the input capabilities of an amplifier may not be the same as the input capabilities of the filter stage in which it is located. Due to some internal amplification, the voltage appearing at the input terminals of the amplifier may be larger than the voltage at the input terminals of the filter stage. To prevent amplifier overdrive, the signal level at the input terminals of the filter stage must then be proportionately reduced. If all the filter stages of a cascade have the same topology (and use the same amplifiers), then the reduction in permissible input signal level applies to all stages. In some cases, however, the topology of various stages may be the same, but the magnitude of the internal voltage causing overdrive in any given stage may depend on, say, the Q of the pole pair realized by that stage. In such cases the sections most "sensitive" to overdrive should either be safeguarded with a resistive input pad, or located separately within the cascade sequence, in positions where the signal levels are known to be low.

1. Optimizing Dynamic Range *and* Signal-to-Noise Ratio

Beside being concerned about not overdriving individual filter sections in order not to cause signal distortion, one must, in general, be sure to maximize the signal-to-noise ratio of the overall filter. This means that the smallest signal voltages in the passband should be as large as possible, i.e.,

$$V_{i\,\min} \equiv \min | V_i(j\omega)| \qquad\qquad (1\text{-}87)$$

for $\omega_1 \leq \omega \leq \omega_2$ and $i = 1, 2, \ldots, [(n/2) - 1]$, should be as large as possible. Notice that signals outside the passband bounded by ω_1 and ω_2, are not taken into consideration here; since the signal-to-noise ratio is measured at the *output* of a section it may be assumed that out-of-band signals are greatly attenuated and do not affect the signal-to-noise ratio. In contrast the overdrive condition (1-86) includes signals outside the passband, since signals from neighboring channels can overdrive a filter section at its *input*, thereby changing the biasing point of its amplifier and causing distortion of signals *within* the passband.

Our requirement that $V_{i\,\max}$ be as small as possible and $V_{i\,\min}$ as large as possible can be satisfied by stipulating that the ratio

$$d_i = \log \frac{V_{i\,\max}}{V_{i\,\min}} \rightarrow \min(\text{sequence}) \qquad\qquad [1\text{-}88]$$

In other words we are looking for that particular cascading sequence of our $n/2$ second-order sections in which the maximum ratio d_i to occur is a minimum. Thus

$$d_{\max} \equiv \max \{d_i\} \rightarrow \min(\text{sequence}) \qquad\qquad [1\text{-}89]$$

$$i = 1, 2, \ldots, (n/2) - 1$$

Note the similarity between (1-89) and the criterion (1-34), according to which the pole–zero assignment for maximum dynamic range was accomplished. This similarity is not coincidental. In the assignment problem we endeavor to maintain as flat a frequency response of the individual sections in the passband as possible; here we endeavor to cascade the sections such that, at the interface between any two sections, the signal spectrum over the passband remains as flat as possible. The selection of the constant K'_j for the gain distribution (as discussed in Section 1.3.4) then attempts to shift the average levels of the resulting maximally flat signal spectra so as to permit the largest possible signal swing (without distortion) at the highest average level, thereby maximizing the signal-to-noise ratio. The reader will perceive that these three steps, undertaken to maximize the dynamic range, are not necessarily independent of each other, even though, by treating them separately, we have assumed that they are. Some comments on this point will follow the present discussion.

Let us now return to our problem of finding the optimum sequence for maximum dynamic range, while assuming that the signal level at the input of a filter stage (rather than at the input terminals of its amplifier) is crucial. We begin by selecting a starting sequence that seems, at least intuitively, to solve our problem. We then rearrange this sequence, with the help of an optimization strategy known in linear programming as the "branch and bound" method,[45] to improve on it. The latter obviates the necessity of examining all $(n/2)!$ possible sequences of the $n/2$ sections that actually exist.

To obtain a good starting sequence we proceed as follows: We select the first stage as the one with the flattest frequency response $T_i(j\omega)$ for $\omega_1 \le \omega \le \omega_2$, i.e., the stage with the smallest value of $(T_{i\,max} - T_{i\,min})$ as defined by (1-76a) and (1-76b). The second stage has the next flattest response and so on. Having obtained this starting sequence we can calculate the corresponding d_i values according to (1-88) and mark the largest d_i value occurring in that sequence as $d_{1\,max}$. Any other sequence in which the bound $d_{1\,max}$ is exceeded is inferior to our starting sequence and need not be considered further. To examine whether any sequence exists with a bound *lower* than $d_{1\,max}$, we develop the tree shown in Fig. 1-43. The input signal $V_{in}(j\omega)$ (whose frequency response we assume to be flat) represented by the first node, is successively fed into each of the $n/2$ sections, producing the outputs $V_{11}(j\omega)$, $V_{12}(j\omega)$ ..., $V_{1,n/2}(j\omega)$, and the corresponding d values obtained from (1-88), namely, d_{11}, d_{12}, ..., $d_{1,n/2}$. In these designations, the first subscript denotes the position of a section in the cascade, the second the number of the particular section used. Those nodes with $d_{1v} > d_{1\,max}$ cannot lead to a better solution and are discarded. We therefore continue only at those nodes where $d_{1v} \le d_{1\,max}$. Applying the signal $V_{1v}(j\omega)$ occurring at such a node to all the stages not yet in the cascade, the development of the

45. E. Lawler and D. Wood, Branch and bound methods: A survey, *Oper. Res.*, **1966**, 699–719.

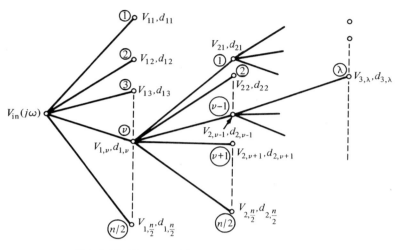

Fig. 1-43. Tree pertaining to sequence optimization.

tree in Fig. 1-43 is continued. Instead of continuing at all the nodes with values $d_{1\nu} \le d_{1\ max}$, it is simpler to pick out only one, building up a tree from it. If at any point the $d_{1\ max}$ bound is exceeded, that particular trial is abandoned and another one begun, starting at one of the previous nodes for which it was found that $d_{1\nu} \le d_{1\ max}$. If all the trials are unsuccessful, then the starting sequence is the best one. If an improved solution is found, then we use its maximum value $d_{2\ max}$ as our new bound and attempt to improve on it in the manner described above. We continue until after, say, μ successful trials we obtain a sequence whose maximum d_i value $d_{\mu\ max}$ is smaller than any of the preceding bounds, i.e.,

$$d_{\mu\ max} < d_{\mu-1\ max} < \cdots d_{2\ max} < d_{1\ max} \qquad (1\text{-}90)$$

If there are no further untried nodes from which to develop a segment of the tree, then $d_{\mu\ max}$ cannot be improved on and the corresponding sequence is the optimum one. The number of trials required to find it naturally depends on how close to the optimum the starting sequence was. In practice, the starting sequence, obtained in the manner described, is often the optimum sequence.[46]

In some cases, considerations other than dynamic-range optimization may have to be taken into account for the filter sequence. It may, for example, be important to have a low-pass or a bandpass section for the first stage, in order to keep high frequencies from the amplifiers, and thereby avoid slew-rate problems.[47] Similarly, it may be desirable to have a high-pass or a bandpass characteristic associated with the last stage in order to remove power-

46. As a useful rule of thumb this sequence is the one with increasing pole Q's (i.e. $q_{p1} < q_{p2} < \ldots < q_{pn/2}$).
47. See, for example, Chapter 7, Section 7.2.7 of *Linear Integrated Networks: Fundamentals*.

supply ripple, or noise created by the amplifiers themselves, from the output signal. If such constraints are imposed on the first and last stages, for instance, the remaining sequence can still be determined as described above.

2.   Optimizing Dynamic Range *or* Signal-to-Noise Ratio

Often the dynamic range may be of more concern than the signal-to-noise ratio. If so, we can replace (1-88) by

$$d_i = V_{i\,max} \tag{1-91}$$

where

$$V_{i\,max} \equiv \max | V_i(j\omega)| \tag{1-92}$$

for $0 < \omega < \infty$ and $i = 1, 2, \ldots, [(n/2) - 1]$, and search for the sequence of sections for which the maximum voltage $V_{i\,max}$ is a minimum. Thus, instead of (1-89), we now require that

$$d_{max} \equiv \max \{d_i\} = \max V_{i\,max} \to \min(\text{sequence}) \tag{1-93}$$

for $i = 1, 2, \ldots, (n/2) - 1$. With the newly defined d_i values, the procedure, using the branch and bound method, is exactly the same as it is to satisfy (1-89). If, on the other hand, the signal-to-noise ratio is of primary concern, then it follows from the preceding discussion that we set

$$d_i = V_{i\,min} \tag{1-94}$$

and proceed as above, with the difference that we now wish to *maximize* $V_{i\,min}$, i.e., we search for a sequence such that

$$d_{min} \equiv \min \{d_i\} = \min V_{i\,min} \to \max(\text{sequence}) \tag{1-95}$$

where

$$V_{i\,min} \equiv \min | V_i(j\omega)| \tag{1-96}$$

for $\omega_1 \leq \omega \leq \omega_2$ and $i = 1, 2, \ldots, [(n/2) - 1]$. Frequently the optimum sequence for maximum dynamic range of a filter turns out to be the one in which the pole Q of the cascaded second-order networks increases from input to output.

3.   The Dynamic Range:

To obtain the actual dynamic range resulting from the optimization procedures outlined above, let us assume that the maximum voltage V_{max} in a filter cascade occurs at the frequency ω_m, where, using (1-92),

$$V_{max} = \max \{V_{i\,max}\}, \qquad i = 1, 2, \ldots, (n/2) - 1 \tag{1-97}$$

It is this maximum voltage that determines the dynamic range of the overall filter. The gain g of the filter, from the signal at the input $| V_{in}(j\omega_m)|$, to the terminal in the sequence at which V_{max} occurs, is

$$g = \frac{V_{max}}{| V_{in}(j\omega_m)|} \tag{1-98}$$

Since the $V_{i\,max}$ values must be known anyway, either for (1-89) or (1-93), V_{max} is available in either case and can be obtained by inspection according to (1-97).

Assume, now, that each filter section has the same input dynamic range V_{amp}, where the subscript implies that this range is determined by the input characteristics of the (identical) amplifiers used. Then the highest permissible voltage at the input to the overall filter, or its dynamic range V_{dyn}, is given by

$$V_{dyn} = \frac{V_{amp}}{g} = \frac{V_{amp}}{V_{max}} \mid V_{in}(j\omega_m) \mid \qquad (1\text{-}99)$$

If we recall that the pole–zero assignment and the gain distribution preceding the sequence optimization were designed expressly to minimize the gain g by flattening the individual frequency responses in the passband, then (1-99) helps to point out how all three steps, when optimized for that purpose, contribute to the maximization of the dynamic range.

1.3.6. Concluding Remarks on the Dynamic-Range Optimization of Cascaded Networks.

The last point made above leads to some additional comments on the nature of the three-part optimization for maximum dynamic range. We have assumed all along that the pole–zero assignment, gain distribution, and sequence optimization are independent of each other, and we therefore optimized each separately. In actual fact, this independence is a simplification, if not a very gross one, and all three optimization steps are interrelated. Thus, for example, the optimum distribution of the constants K_j may well depend on both the pole–zero assignment and the cascade sequence, and so on. The advantage of treating these steps separately is, of course, that they can be coped with more easily in this way. Nevertheless in order to obtain a still "better" optimum, the three steps can be combined into a single procedure (i.e., computer program) in which they are dealt with simultaneously.[48] The amount of improvement thereby obtained, vs. the previous three-step approach, has not been established. However, since the "three-step" and the "simultaneous" approaches are very similar (the individual steps simply being carried out iteratively in the "simultaneous" approach), the improvement cannot be expected to be drastic. The penalty for any improvement obtained is the dependence on an available computer program, since any attempt at coping with the three optimization steps simultaneously by hand, even for relatively low-order filters, would seem ill advised.

Clearly, with the availability of active devices, (e.g., operational amplifiers) with ever improving input signal-handling capabilities, the urgency for

48. S. Halfin, unpublished Bell Telephone Laboratories memorandum; R. M. Zeigler, unpublished Bell Telephone Laboratories memorandum; W. Schoeller and E. Lueder, unpublished report, Stuttgart University.

optimization with respect to dynamic range as compared, say, with sensitivity (where possible), may lessen somewhat. On the other hand, *large signal-handling capability requires proportionately high supply-voltages and, consequently, results in increased power dissipation and heat generation.* Where these effects are to be minimized, supply voltages must be low (e.g., ± 6 V or even ± 3 V instead of ± 12 V) and the dynamic range must be optimized in order to avoid overdrive, and consequently distortion, while maintaining an acceptable signal-to-noise ratio.

1.4 SOME CHARACTERISTICS OF SECOND–ORDER NETWORK FUNCTIONS

For reasons given in the previous section, active networks generally consist of second- and third-order networks combined in some prescribed way (generally in cascade) to give the desired overall transmission. Because they generate complex conjugate pole pairs, the characteristics of second-order networks, in particular their transmission and root sensitivities, take on particular importance. Thus, to give some idea of the transmission characteristics obtainable with second-order networks, the frequency characteristics of the most common ones are summarized in Table 1-3. The examples given are special cases of the general second-order function

$$T(s) = K \frac{s^2 + \dfrac{\omega_z}{q_z} s + \omega_z^2}{s^2 + \dfrac{\omega_p}{q_p} s + \omega_p^2} \tag{1-100}$$

Note that there are essentially six basic second-order functions that can be derived from (1-100). The corresponding second-order filter networks are:

1. Low-pass network (LPN)
2. High-pass network (HPN)
3. Bandpass network (BPN)
4. Frequency-emphasizing network (FEN)
5. Frequency-rejection network (FRN)
6. All-pass network (APN)

The remaining two functions in Table 1-3 are special cases of those listed above, that is, the infinite-null network is an FRN with infinite q_z, the resonator a combination of a low-pass and a bandpass network. The latter is sometimes referred to as a *low-pass resonator*. A final case, the combination of a high-pass network and a bandpass network (thus the numerator is $s^2 + \omega_z s$) follows in precisely the same way, it is sometimes referred to as a *high-pass resonator*.

TABLE 1-3, THE SECOND-ORDER NETWORK FUNCTIONS

Network type	Transmission function	Pole-zero diagram	Frequency response
1. Low-pass filter $\omega_z \to \infty$ $K \to 0$ $K\omega_z^2 = K_1 = const$ $K\omega_z = 0$	$$T(s) = \frac{K_1}{s^2 + \dfrac{\omega_p}{q_p}s + \omega_p^2}$$ $$\lvert T(j\omega)\rvert = \frac{K_1}{\dfrac{\omega_p}{q_p}\omega_c} \cdot \sin\phi(\omega)$$ $$\omega_p = \sqrt{\omega_c^2 + \sigma^2}$$		 $$\lvert T(j\omega_p)\rvert = T_{\omega_p} = K_1\frac{q_p}{\omega_p^2} = q_p T_0$$ $$\omega_m = \sqrt{\omega_c^2 - \sigma^2}$$ $$x_\varrho = 1 + \frac{4q_p}{4q_p^2 - 1} \approx 1 + \frac{1}{q_p}$$
2. High-pass filter $(\omega_z = 0)$	$$T(s) = K\frac{s^2}{s^2 + \dfrac{\omega_p}{q_p}s + \omega_p^2}$$		 $$T_{\omega_p} = K \cdot q_p = q_p \cdot T_\infty$$

(Continued)

TABLE 1-3 (Continued)

Network type	Transmission function	Pole-zero diagram	Frequency response

3. Bandpass filter

$$T(s) = \frac{\omega_k s}{s^2 + \dfrac{\omega_p}{q_p}s + \omega_p^2}$$

$$\left(\begin{array}{c} q_z = 0 \\ \omega_z = 0 \\ \omega_z/q_z = \omega_k = \text{const} \end{array}\right)$$

$\omega_m = \omega_p$

4. Frequency emphasizing network

$$T(s) = K\frac{s^2 + \dfrac{\omega_z}{q_z}s + \omega_z^2}{s^2 + \dfrac{\omega_p}{q_p}s + \omega_p^2}$$

$$\omega_z^2 = \sqrt{\omega_{cz}^2 + \sigma_z^2}$$
$$\omega_p^2 = \sqrt{\omega_{cp}^2 + \sigma_p^2}$$

$$\left(\begin{array}{c} \omega_z \lessgtr \omega_p \\ q_p > q_z \end{array}\right)$$

$\omega_m = \omega_z = \omega_p$

5. Frequency rejection network

$$T(s) = K\frac{s^2 + \dfrac{\omega_z}{q_z}s + \omega_z^2}{s^2 + \dfrac{\omega_p}{q_p}s + \omega_p^2}$$

$$\left(\begin{array}{c} \omega_z \lessgtr \omega_p \\ q_p < q_z \end{array}\right)$$

$\omega_0 = \omega_z = \omega_p$

TABLE 1-3 (Continued)

Network type	Transmission function	Pole-zero diagram	Frequency response
6. All-pass filter $\left(\begin{array}{c}\omega_z = \omega_p \\ q_p = -q_z > 0\end{array}\right)$	$T(s) = K\dfrac{s^2 - \frac{\omega_p}{q_p}s + \omega_p^2}{s^2 + \frac{\omega_p}{q_p}s + \omega_p^2}$		
7. Frequency rejection network with infinite null $\omega_z \gtreqless \omega_p$ $q_z \to \infty$	$T(s) = K\dfrac{s^2 + \omega_z^2}{s^2 + \frac{\omega_p}{q_p}s + \omega_p^2}$		
8. Low-pass resonator $\omega_{z_2} \to \infty$	$T(s) = K\dfrac{s + \omega_{z_1}}{s^2 + \frac{\omega_p}{q_p}s + \omega_p^2}$		

TABLE 1-4, $F(\omega_m)$ AND $F(\omega_0)$ FOR SECOND-ORDER FUNCTIONS WHERE $F(\omega) = \left| \dfrac{T(j\omega)}{K} \right|^2$

$T(s)$	ω_m^2	$F(\omega_m)$	ω_0	$F(\omega_0)$	ω_m real if	$F(0)$	$F(\infty)$
$\dfrac{K}{s^2 + \frac{\omega_p}{q_p}s + \omega_p^2}$	$\omega_p^2\left(1 - \dfrac{1}{2q_p^2}\right)$	$\dfrac{q_p}{\omega_p^4\left(1 - \dfrac{1}{4q_p^2}\right)}$	∞	0	$q_p^2 \geq \frac{1}{2}$	$\dfrac{1}{\omega_p^2}$	0
$K\dfrac{s^2}{s^2 + \frac{\omega_p}{q_p}s + \omega_p^2}$	$\dfrac{\omega_p^2}{1 - \dfrac{1}{2q_p^2}}$	$\dfrac{q_p^2}{1 - \dfrac{1}{4q_p^2}}$	0	0	$q_p^2 \geq \frac{1}{2}$	0	1
$K\dfrac{s}{s^2 + \frac{\omega_p}{q_p}s + \omega_p^2}$	ω_p^2	q_p^2/ω_p^2	$0, \infty$	0	always real	0	0
$K\dfrac{s^2 + \frac{\omega_p}{q_z}s + \omega_p^2}{s^2 + \frac{\omega_p}{q_p}s + \omega_p^2}$	ω_p^2 if $q_p > q_z$	$\left(\dfrac{q_p}{q_z}\right)^2$	ω_p^2 if $q_p < q_z$	$\left(\dfrac{q_p}{q_z}\right)^2$	always real	1	1

$T(s)$	ω_m^2	$F(\omega_m)$	ω_0	$F(\omega_0)$	ω_m real if	$F(0)$	$F(\infty)$
$K\dfrac{s^2+\omega_z^2}{s^2+\dfrac{\omega_p}{q_p}s+\omega_p^2}$	$\dfrac{\left(\dfrac{\omega_z}{\omega_p}\right)^2\left(1-\dfrac{1}{2q_p^2}\right)-1}{\left(\dfrac{\omega_z}{\omega_p}\right)^2\dfrac{1}{2q_p^2}+1}\,\omega_p^2$	$\dfrac{(\omega_z^2-\omega_m^2)^2}{(\omega_p^2-\omega_m^2)^2+\left(\dfrac{\omega_p}{q_p}\right)^2\omega_m^2}$	ω_z^2	0	$\dfrac{\omega_z}{\omega_p}>1:\ q_p^2\geq\dfrac{\left(\dfrac{\omega_z}{\omega_p}\right)^2}{2\left[\left(\dfrac{\omega_z}{\omega_p}\right)^2-1\right]}$ $\dfrac{\omega_z}{\omega_p}<1:\ q_p^2\geq\dfrac{1}{2\left[1-\left(\dfrac{\omega_z}{\omega_p}\right)^2\right]}$	$\left(\dfrac{\omega_z}{\omega_p}\right)^2$	1
$K\dfrac{s(s+\omega_z)}{s^2+\dfrac{\omega_p}{q_p}s+\omega_p^2}$	$\left\{-1\pm\sqrt{\left[\left(\dfrac{\omega_z}{\omega_p}\right)^2+1\right]^2-\left(\dfrac{\omega_z}{\omega_p}\right)^2\dfrac{1}{q_p^2}}\right\}\omega_p^2$	$\dfrac{\omega_z^2(\omega_m^2+\omega_z^2)}{(\omega_p^2-\omega_m^2)^2+\left(\dfrac{\omega_p}{q_p}\right)^2\omega_m^2}$	0	0	$q_p^2\geq\dfrac{1}{\left(\dfrac{\omega_z}{\omega_p}\right)^2+2}$	0	1
$K\dfrac{s+\omega_z}{s^2+\dfrac{\omega_p}{q_p}s+\omega_p^2}$	$\left\{-\left(\dfrac{\omega_z}{\omega_p}\right)^2\pm\sqrt{\left[\left(\dfrac{\omega_z}{\omega_p}\right)^2+1\right]^2-\left(\dfrac{\omega_z}{\omega_p}\right)^2\dfrac{1}{q_p^2}}\right\}\omega_p^2$	$\dfrac{\omega_m^2+\omega_z^2}{(\omega_p^2-\omega_m^2)^2+\left(\dfrac{\omega_p}{q_p}\right)^2\omega_m^2}$	∞	0	$q_p^2\geq\dfrac{\left(\dfrac{\omega_z}{\omega_p}\right)^2}{2\left[\left(\dfrac{\omega_z}{\omega_p}\right)^2+1\right]}$	$\dfrac{\omega_z^2}{\omega_p^2}$	0

To carry out the pole–zero assignment discussed in Section 1.3.2, we require the magnitude function $F(\omega)$ as defined by (1-29) for every possible second-order function. More specifically, we need to know the peak value of $F(\omega)$, which occurs at frequency ω_m and the valley, occurring at ω_0. These values, as well as the conditions for the existence of a peak (no comparable constraint exists for the existence of a valley) are listed in Table 1-4 for the most important second-order functions.

With the expressions in Table 1-4 we are in a position to carry out the pole–zero assignment for any given nth order function in order to obtain the corresponding product of second-order functions. We shall illustrate the procedure, as described in Section 1.3.2, by the following detailed example.[49]

1.4.1 Example: Decomposition of a Tenth–Order Network Function

Consider the tenth-order bandpass filter (used, for example, in a single-sideband modem for data transmission in a telephone group band) whose frequency response is shown in Fig. 1-44. The poles and zeros realizing this response are

$$z_1 = \pm j0.32233523 \cdot 10^{-6}$$
$$z_2 = \pm j0.36742346 \cdot 10^{6}$$
$$z_3 = -0.31480 \cdot 10^{5} \pm j0.3132295 \cdot 10^{6}$$
$$z_4 = 0 \text{ (twice)}$$
$$z_5 = \infty \text{ (twice)}$$
$$p_1 = -0.276100 \cdot 10^{5} \pm j0.2961048 \cdot 10^{6}$$
$$p_2 = -0.31480 \cdot 10^{5} \pm j0.3132295 \cdot 10^{6}$$
$$p_3 = -0.8706 \cdot 10^{4} \pm j0.314697 \cdot 10^{6}$$
$$p_4 = -0.93340 \cdot 10^{5} \pm j0.18670202 \cdot 10^{6}$$
$$p_5 = -0.25280 \cdot 10^{5} \pm j0.62888167 \cdot 10^{5}$$

The corresponding pole–zero plot is shown in Fig. 1-45. Notice that the zero z_3 and the pole p_2 coincide; such a pole–zero pair (sometimes referred to as a *phantom pair*) is required when the active filter realization is based on the FEN-decomposition to be described in Chapter 3 (Section 3.1.1).

Following the procedure outlined in Section 1.3.2, we observe that the four zeros on the real axis, z_4, z_4', z_5, and z_5' can be combined pairwise in two ways: (z_4, z_4'), (z_5, z_5') or (z_4, z_5), (z_4', z_5'). For convenience, we designate the first two combinations (z_4, z_4') and (z_5, z_5') by $z_{4,1}$ and $z_{5,1}$, respectively, and the second two, namely, (z_4, z_5) and (z_4', z_5') by z_{42} and z_{52}, respectively. Using the expressions in Table 1-4 we can now calculate the d_j values accord-

49. E. Lueder, loc. cit.

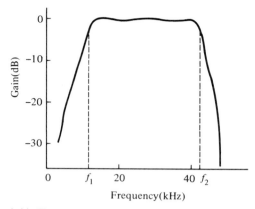

FIG. 1-44. Frequency response of tenth-order bandpass filter.

ing to (1-33b) and obtain the $d_{v,\mu}$ table given in Fig. 1-46. With the zeros z_1, z_2, z_3, z_{41}, and z_{51} we obtain the $d_{v,\mu}$ graph shown in Fig. 1-47a and with the zeros z_1, z_2, z_3, z_{42}, and z_{52} the $d_{v,\mu}$ graph in Fig. 1-47b. The first assignment line is shown for each case. For Fig. 1-47a the corresponding $d_{1\,max} = 0.56$; the $d_{j\,max}$ values resulting from the successive four assignment lines are $d_{2\,max} = 0.4$, $d_{3\,max} = 0.3$, $d_{4\,max} = 0.25$, and $d_{5\,max} = 0.18$. The corresponding pole–zero pairs are (z_1, p_3), (z_2, p_2), (z_3, p_1) (z_{41}, p_5) and (z_{51}, p_4). Since the first assignment line for this case is lower than that in Fig. 1-47b (where $d_{1\,max} = 1.55$), this is the desired pole–zero assignment. Note, incidentally, that because of the symmetry of $d_{5,42}$ and $d_{5,52}$ as well as of $d_{4,42}$ and $d_{4,52}$ in Fig. 1-47b, we have the situation illustrated previously in Fig. 1-26, i.e., the combinations (p_5, z_{42}), (p_4, z_{52}) or (p_4, z_{42}), (p_5, z_{52}) are equivalent; the actual selection may therefore be based on some other than dynamic-range considerations.

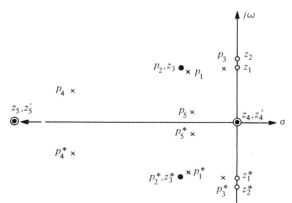

FIG. 1-45. Pole–zero plot.

	z_1	z_2	z_3	z_{41}	z_{51}	z_{42}	z_{52}
p_1	0.15	0.57	0.25	3.7	1.36	2.47	2.47
p_2	0.18	0.3	0	3.8	1.49	2.69	2.69
p_3	0.18	1.25	1.1	4.9	2.5	3.77	3.77
p_4	1.87	1.4	1.6	1.9	0.56	0.87	0.87
p_5	4.1	3.3	3.7	0.4	2.64	1.5	1.5

FIG. 1-46. $d_{v,\mu}$ table pertaining to pole–zero plot of Fig. 1-45.

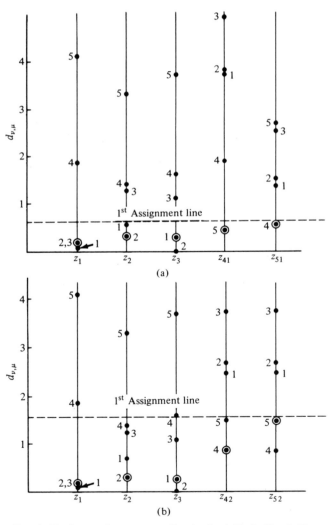

(a)

(b)

FIG. 1-47. $d_{v,\mu}$ graphs corresponding to $d_{v,\mu}$ table in Fig. 1-46.

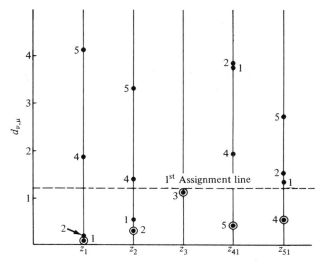

FIG. 1-48. Alternative to first assignment in Fig. 1-47a.

Frequently some additional constraint may predetermine one or more pole–zero pairs of an assignment. Using the FEN synthesis method, for example, pole p_3 must be assigned to zero z_3, whereas the remaining assignments are free. It is clear from inspection, that the assignment in Fig. 1-47a will still be smaller than that in Fig. 1-47b with p_3 assigned to z_3. Therefore, redrawing Fig. 1-47a in Fig. 1-48 we obtain the assignments (z_1, p_1), (z_2, p_2), (z_3, p_3), (z_{41}, p_5), and (z_{51}, p_4) with $d_{1\,max} = 1.1$, $d_{2\,max} = 0.56$, $d_{3\,max} = 0.4$, $d_{4\,max} = 0.3$, and $d_{5\,max} = 0.15$. The corresponding second-order transfer functions are

$$T_1(s) = 0.85 \frac{s^2 + 0.1039 \cdot 10^{12}}{s^2 + 0.5522 \cdot 10^5 s + 0.8844 \cdot 10^{11}} \qquad \text{(FRN)}$$

$$T_2(s) = 0.75 \frac{s^2 + 0.135 \cdot 10^{12}}{s^2 + 0.06296 \cdot 10^6 s + 9.911 \cdot 10^{10}} \qquad \text{(FRN)}$$

$$T_3(s) = \frac{s^2 + 0.06296 \cdot 10^6 s + 9.911 \cdot 10^{10}}{s^2 + 1.7412 \cdot 10^4 s + 9.911 \cdot 10^{10}} \qquad \text{(FEN)}$$

$$T_4(s) = 3.277 \cdot 10^{10} \frac{1}{s^2 + 1.8668 \cdot 10^5 s + 4.357 \cdot 10^{10}} \qquad \text{(LPN)}$$

$$T_5(s) = 0.834 \frac{s^2}{s^2 + 5.056 \cdot 10^4 s + 4.594 \cdot 10^9} \qquad \text{(HPN)}$$

The corresponding frequency responses are shown in Fig. 1-49. Notice that the factors K_j, $j = 1, 2, \ldots, 5$ have been calculated from (1-82) such that the overall response is 0 dB in the passband, as shown in Fig. 1-44. The optimum

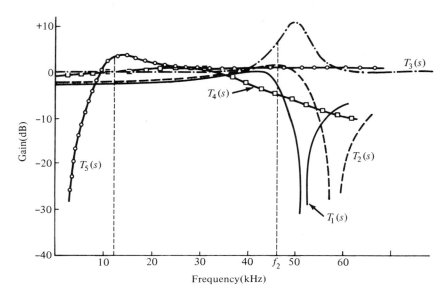

FIG. 1-49. Frequency responses corresponding to final pole–zero assignments in Figs. 1-47 and 1-48.

sequence for the five sections can be obtained according to the procedure outlined in Section 1.3.5. Often, on the other hand, the sequence can be determined directly, by considerations such as the following. If high-frequency harmonics, or noise, are to be suppressed and kept from the other sections, the low-pass section $T_4(s)$ should be the first in the sequence. The FEN $T_3(s)$ has a peak in the vicinity of the valleys of the FRNs $T_1(s)$ and $T_2(s)$; thus, to prevent the peak of $T_3(s)$ from becoming too large, it may be preceded by $T_1(s)$ and $T_2(s)$ to attenuate the signal first. Finally, it is common for a certain amount of 60-Hz noise from the power supplies feeding the amplifiers to accumulate through a number of sections. It is therefore useful to conclude the sequence with a high-pass network (e.g., $T_5(s)$) if one is available. A useful sequence is therefore

$$T_4(s)T_1(s)T_2(s)T_3(s)T_5(s)$$

With this sequence, as well as the gain distribution and pole–zero assignment derived above, the dynamic range of the amplifiers can be fully utilized and the passband gain maintained with ease. By contrast, using five sections that do not result from this optimization, the in-band attenuation may, typically, be 20 dB or more and the dynamic range of the overall filter reduced significantly even though the same amplifiers are used.

1.5 THE SENSITIVITY OF SECOND-ORDER NETWORKS

It was pointed out earlier that for active RC networks to compare favorably, in terms of their sensitivity to component variations, with their LC counterparts, they must be decomposed into a cascade of second-order networks.[50] We have mentioned also (and we shall pursue this point in more detail in later chapters), that this decomposition has numerous other advantages beside that of lowering sensitivity (e.g., building-block design, ease of tuning etc.). One particular advantage with respect to overall sensitivity manifests itself if the individual sections are isolated from one another. This, incidentally, can almost always be assumed in practice, particularly if operational amplifier configurations are used. In such a cascade we have

$$T(s) = \prod_i T_i(s) = \prod_i \frac{N_i(s)}{D_i(s)} \tag{1-101}$$

where

$$\text{degree } N_i(s) \leq \text{degree } D_i(s) \leq 3$$

Then the transmission sensitivity $S_x^{T(s)}$ of the overall filter to variations of a component x in the kth section is

$$S_x^{T(s)} = \sum_i S_x^{T_i(s)} = \sum_i (S_x^{N_i(s)} - S_x^{D_i(s)}) \tag{1-102}$$

However, since x is only in the kth section, and this section is isolated from all the others, it follows that

$$S_x^{T_i(s)} = 0 \qquad \text{for} \qquad i \neq k \tag{1-103}$$

and therefore (1-102) simplifies to

$$S_x^{T(s)} = S_x^{T_k(s)} = S_x^{N_k(s)} - S_x^{D_k(s)} \tag{1-104}$$

Thus, in examining the overall filter sensitivity characteristics of the commonly used cascade of isolated second-order[51] sections, it suffices for us to examine those of each second-order section in turn. Second-order sections therefore have a natural priority, also with respect to sensitivity considerations.

In what follows we shall apply well known sensitivity concepts to second-order networks in order to derive certain expressions found to be very useful for this special class of networks. In particular we shall examine the gain, phase, and delay sensitivities of these networks, these being in general the quantities of most interest in a practical application.

---

50. As we shall see in Chapter 7, a certain degree of coupling between the second-order stages by means of negative feedback may still further decrease the sensitivity. This does not alter the fact, however, that we start out with a cascade of second-order stages that are isolated from each other.
51. We shall restrict ourselves, in what follows, to second-order networks, having pointed out before that the additional negative real pole of a third-order network has negligible effect on sensitivity.

1.5.1 Gain, Phase, and Delay Sensitivity[52]

The transmission sensitivity of a network whose transfer function is $T(j\omega)$ to variations of a component x can be written as[53]

$$S_x^{T(j\omega)} = \mathcal{S}_x^{\alpha(\omega)} + j\mathcal{S}_x^{\phi(\omega)} \tag{1-105}$$

where

$$T(j\omega) = |T(j\omega)|\, e^{j\phi(\omega)} = e^{\alpha(\omega)+j\phi(\omega)} \tag{1-106a}$$

$$S_x^{T(j\omega)} = \frac{d[\ln T(j\omega)]}{d[\ln x]} \tag{1-106b}$$

$$\mathcal{S}_x^{\alpha(\omega)} = \frac{d\alpha(\omega)}{dx/x} = \frac{d[\ln |T(j\omega)|]}{d[\ln x]} \tag{1-106c}$$

$$\mathcal{S}_x^{\phi(\omega)} = \frac{d\phi(\omega)}{dx/x} = \frac{d[\arg T(j\omega)]}{dx/x} \tag{1-106d}$$

$\alpha(\omega)$ and $\phi(\omega)$ are the gain and phase functions of $T(j\omega)$, and $\mathcal{S}_x^{\alpha(\omega)}$ and $S_x^{\phi(\omega)}$ are the respective gain- and phase-sensitivity functions.

Gain Sensitivity   The function $\mathcal{S}_x^{\alpha(\omega)}$ gives the change in gain, $d\alpha(\omega)$, caused by a relative change in a component x. Since $\alpha(\omega)$ is expressed as the natural logarithm of $|T(j\omega)|$, $\mathcal{S}_x^{\alpha(\omega)}$ expresses the gain variation in nepers due to a percentage change in x:

$$\Delta\alpha(\omega) = \mathcal{S}_x^{\alpha(\omega)} \frac{\Delta x}{x} [\text{nepers}] \tag{1-107}$$

Since

$$20 \log_{10}|T(j\omega)| = 20 \frac{\ln |T(j\omega)|}{\ln 10} = 8.68 \ln |T(j\omega)| \tag{1-108}$$

$\Delta\alpha(\omega)$ can be expressed in decibels:

$$\Delta\alpha(\omega) = 8.68\,\mathcal{S}_x^{\alpha(\omega)} \frac{\Delta x}{x} [\text{dB}] \tag{1-109}$$

If, for example the gain sensitivity is unity at a frequency ω_0, a 1% change in x will cause a gain change of 0.0868 dB, or of approximately 0.1 dB. Thus, in general,

$$\Delta\alpha(\omega)\Big|_{\frac{\Delta x}{x}=1\%} \approx 0.1 \cdot \mathcal{S}_x^{\alpha(\omega)} [\text{dB}] \tag{1-110}$$

---

52. This section is largely based on D. Hilberman, An approach to the sensitivity and statistical variability of biquadratic filters, *IEEE Trans. Circuit Theory*, **CT-20**, 382–390 (1973).
53. See Chapter 4, Section 4.1.6 of *Linear Integrated Networks: Fundamentals*.

Calculating $\alpha(\omega) = \ln |T(j\omega)|$ for a general second-order function given in the form

$$T(s) = \frac{b_2 s^2 + b_1 s + b_0}{a_2 s^2 + a_1 s + a_0} \qquad (1\text{-}111)$$

we obtain

$$\alpha(\omega) = \ln | - b_2 \omega^2 + jb_1\omega + b_0| - \ln | - a_2 \omega^2 + ja_1\omega + a_0| \qquad (1\text{-}112)$$

and

$$\mathscr{S}_x^{\alpha(\omega)} = x \frac{d\alpha(\omega)}{dx} = x \frac{d}{dx} [\ln | - b_2 \omega^2 + jb_1\omega + b_0|]$$

$$- x \frac{d}{dx} [\ln | - a_2 \omega^2 + ja_1\omega + a_0|] \qquad (1\text{-}113)$$

The Normalized g-Functions It can be shown[54] that a general expression can be derived from (1-113) that expresses the gain sensitivity in terms of the gain-to-coefficient sensitivities and the coefficient-to-component sensitivities as follows:

$$\mathscr{S}_x^{\alpha(\omega)} = \sum_{j=0}^{2} \mathscr{S}_{b_j}^{\alpha(\omega)} S_x^{b_j} - \sum_{j=0}^{2} \mathscr{S}_{a_j}^{\alpha(\omega)} S_x^{a_j}$$

$$= g_{b_2} S_x^{b_2} + g_{b_1} S_x^{b_1} + g_{b_0} S_x^{b_0} - g_{a_2} S_x^{a_2} - g_{a_1} S_x^{a_1} - g_{a_0} S_x^{a_0} \qquad [1\text{-}114]$$

where the coefficient-to-component sensitivities (or, briefly, the coefficient sensitivities) $S_x^{a_j}$ are defined as[55]

$$S_x^{a_j} = \frac{d[\ln a_j]}{d[\ln x]} = \frac{da_j/a_j}{dx/x} \qquad (1\text{-}115)$$

and the gain-to-coefficient sensitivity functions g_{a_j} (or, briefly, the g functions) are

$$g_{a_2} = \mathscr{S}_{a_2}^{\alpha(\omega)} = \frac{-a_2 \omega^2(a_0 - a_2 \omega^2)}{(a_0 - a_2 \omega^2)^2 + (a_1\omega)^2} \qquad [1\text{-}116a]$$

$$g_{a_1} = \mathscr{S}_{a_1}^{\alpha(\omega)} = \frac{(a_1\omega)^2}{(a_0 - a_2 \omega^2)^2 + (a_1\omega)^2} \qquad [1\text{-}116b]$$

$$g_{a_0} = \mathscr{S}_{a_0}^{\alpha(\omega)} = \frac{a_0(a_0 - a_2 \omega^2)}{(a_0 - a_2 \omega^2)^2 + (a_1\omega)^2} \qquad [1\text{-}116c]$$

The g_{b_j} functions have the same form, where the coefficients b_j, $j = 0, 1, 2$ are used. Notice that

$$\sum_{j=0}^{2} g_{a_j} = \sum_{j=0}^{2} g_{b_j} = 1 \qquad [1\text{-}117]$$

---

54. The following derivations are also given by D. Åkerberg, Technical Report No. 31, Royal Inst. of Technology, Stockholm, Sweden, September 1970.
55. Note that $S_x^{a_j}$ is a real number, while $\mathscr{S}_{a_j}^{\alpha(\omega)}$ is a function of frequency.

Since the form of the g functions of the numerator and denominator of $T(s)$ are identical, we can proceed by considering only a general second-order polynomial of the form

$$h_2 s^2 + h_1 s + h_0 \tag{1-118}$$

Setting $s = j\omega$ and making the substitutions

$$\Omega = \omega\sqrt{h_2/h_0} \tag{1-119a}$$

and

$$q = \sqrt{h_2 h_0/h_1} \tag{1-119b}$$

(1-118) is normalized to

$$1 - \Omega^2 + j\Omega/q \tag{1-120}$$

The normalized g functions then result from (1-116) as

$$g_0 = \frac{1 - \Omega^2}{(1 - \Omega^2)^2 + (\Omega/q)^2} \tag{1-121a}$$

$$g_1 = \frac{(\Omega/q)^2}{(1 - \Omega^2)^2 + (\Omega/q)^2} \tag{1-121b}$$

and

$$g_2 = -\frac{\Omega^2(1 - \Omega^2)}{(1 - \Omega^2)^2 + (\Omega/q)^2} \tag{1-121c}$$

These normalized functions are plotted in Fig. 1-50a, b, and c for representative values of q.

We have shown above that the gain sensitivity of a general second-order network function can be expressed as a linear combination of the functions g_0, g_1, and g_2. These functions are multiplied by the sensitivities of the coefficients a_j and $b_j, j = 0, 1, 2$ to component variations. Since these coefficients depend on the particular method of realization used, this representation of $\mathcal{S}_x^{\alpha(\omega)}$ permits a comparison to be made, in terms of gain sensitivity, of different active networks realizing the same transfer function. Such a comparison entails merely a comparison of the corresponding coefficient sensitivities, weighted by g_0, g_1, and g_2.

The most important feature of the $\mathcal{S}_x^{\alpha(\omega)}$ representation in terms of the g functions is that the latter functions are quite general, i.e., they are independent of the network realization used. In addition, the following observations should be made.

1. For high Q it can readily be shown that the maximum values are

$$[g_0]_{max} \approx \pm \frac{Q}{2} \quad \text{at} \quad \Omega = 1 \mp \frac{1}{2Q} \quad (1\text{-}122a)$$

$$[g_1]_{max} = 1 \quad \text{at} \quad \Omega = 1 \quad (1\text{-}122b)$$

$$[g_2]_{max} \approx \pm \frac{Q}{2} \quad \text{at} \quad \Omega = 1 \pm \frac{1}{2Q} \quad (1\text{-}122c)$$

This implies that, for equal effects on $\Delta\alpha$, the components contained in the coefficient h_1 of (1-118) may vary about $Q/2$ times more than those affecting h_0 or h_2. In other words, a coefficient-sensitivity ratio $S_x^{h_1} : S_x^{h_0}$ of $Q/2$ may be tolerated. Actually the ratio g_0/g_1 is frequency dependent as plotted in Fig. 1-50d. Thus, at $\Omega = 1 \pm 1/2Q$ the ratio of permissible coefficient sensitivities for equal effects on the gain is more like $Q : 1$. As the distance from the resonant frequency is increased, this ratio becomes even larger.

2. Each g function is zero when its corresponding coefficient is zero. This follows directly from (1-116). Thus, in a simple bandpass circuit in which b_1 is the only nonzero coefficient in the numerator we have

$$b_0 = b_2 = g_{b_0} = g_{b_2} = 0 \quad (1\text{-}123a)$$

and therefore, from (1-117),

$$g_{b_1} = 1 \quad (1\text{-}123b)$$

Similar relations may be derived for any of the second-order functions listed in Table 1-3. Thus, for example, considering case 7 of that table, in which $q_z = \infty$, i.e., $b_1 = \omega_z/q_z = 0$, $b_2 = 1$, and $b_0 = \omega_z^2$, we obtain from (1-116)

$$g_{b_0} = \frac{b_0}{b_0 - b_2\omega^2} = \frac{\omega_z^2}{\omega_z^2 - \omega^2} \quad (1\text{-}124a)$$

$$g_{b_1} = 0 \quad (1\text{-}124b)$$

$$g_{b_2} = -\frac{b_2\omega^2}{b_0 - b_2\omega^2} = -\frac{\omega^2}{\omega_z^2 - \omega^2} \quad (1\text{-}124c)$$

At the notch frequency where $\omega = \omega_z$, g_{b_0}, and g_{b_2} take on infinite values.

The g_ω and g_q Functions In referring to Table 1-3, the reader will recognize the convenience of deriving $\mathscr{S}_x^{\alpha(\omega)}$ for the form of $T(s)$ as given by (1-100), since this is more commonly used than (1-111). We then obtain an expression equivalent to (1-114), namely,

$$\mathscr{S}_x^{\alpha(\omega)} = S_x^K + g_{\omega_z} S_x^{\omega_z} + g_{q_z} S_x^{q_z} - g_{\omega_p} S_x^{\omega_p} - g_{q_p} S_x^{q_p} \quad [1\text{-}125]$$

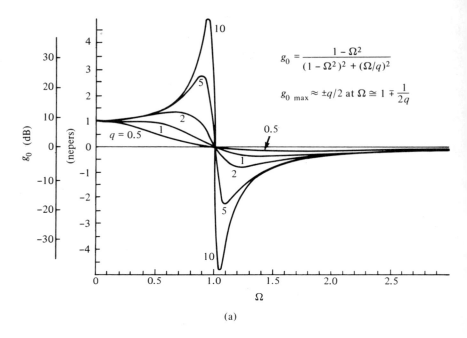

$$g_0 = \frac{1 - \Omega^2}{(1 - \Omega^2)^2 + (\Omega/q)^2}$$

$$g_{0\ max} \approx \pm q/2 \text{ at } \Omega \cong 1 \mp \frac{1}{2q}$$

(a)

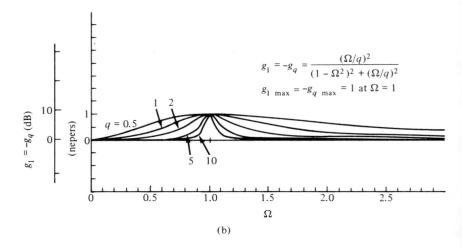

$$g_1 = -g_q = \frac{(\Omega/q)^2}{(1 - \Omega^2)^2 + (\Omega/q)^2}$$

$$g_{1\ max} = -g_{q\ max} = 1 \text{ at } \Omega = 1$$

(b)

FIG. 1-50. Plots of normalized gain-to-coefficient sensitivity functions: (a) $g_0(\Omega, q)$; (b) $g_1(\Omega, q)$; (c) $g_2(\Omega, q)$; (d) $g_0/g_1(\Omega, q)$.

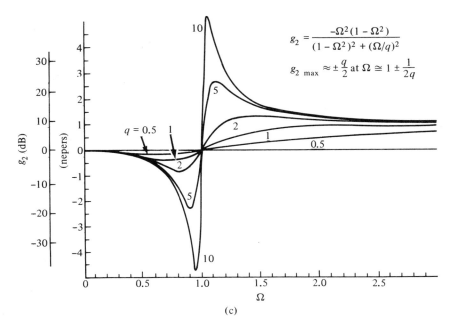

$$g_2 = \frac{-\Omega^2(1 - \Omega^2)}{(1 - \Omega^2)^2 + (\Omega/q)^2}$$

$$g_{2\ max} \approx \pm \frac{q}{2} \text{ at } \Omega \cong 1 \pm \frac{1}{2q}$$

(c)

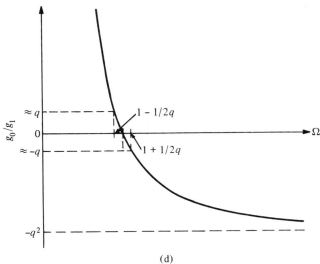

(d)

FIG. 1-50 *Continued*

The new g functions are simple combinations of the old ones:

$$g_\omega = \mathscr{S}_\omega^{\alpha(\omega)} = g_1 + 2g_0 \frac{(\Omega/q)^2 + 2(1 - \Omega^2)}{(\Omega/q)^2 + (1 - \Omega^2)^2} \qquad [1\text{-}126]$$

and

$$g_q = \mathscr{S}_q^{\alpha(\omega)} = -g_1 \frac{-(\Omega/q)^2}{(\Omega/q)^2 + (1 - \Omega^2)^2} \qquad [1\text{-}127]$$

where, for a polynomial of the form

$$P(s) = \left(\frac{s}{\omega_n}\right)^2 + \frac{1}{q}\frac{s}{\omega_n} + 1 \qquad (1\text{-}128)$$

the normalized frequency is given by

$$\Omega = \omega/\omega_n \qquad (1\text{-}129)$$

The new function g_ω is plotted in Fig. 1-51a, and the ratio g_ω/g_q in Fig. 1-51b. From our discussion in observation 1 above and from Fig. 1-51b it follows immediately that the gain is more sensitive to variations in frequency (i.e. ω_z or ω_p) than it is to changes in Q (i.e., q_z or q_p). Thus, for example, for equal effects on $\Delta\alpha$, we can tolerate approximately a Q sensitivity (i.e., $S_x^{q_z}$ or $S_x^{q_p}$) Q-times-higher than the frequency sensitivity (i.e., $S_x^{\omega_z}$ or $S_x^{\omega_p}$) at the frequencies $\Omega = 1 \pm 1/2q$. We shall come back to this important conclusion later.

From (1-126) and (1-127) it is evident that for high Q values the function g_0 is primarily associated with the frequency sensitivity of $\alpha(\omega)$, and g_1 with the Q sensitivity. In considering the effects of pole–zero pairing on sensitivity these functions therefore provide some additional valuable insight. The choice of which pole to combine with which zero for minimum gain sensitivity can now be viewed as the decision on how close g_{0z} may be to g_{0p}, where g_{0z} and g_{0p} are the g_0 functions associated with the zeros and poles, respectively, of the second-order function $T(s)$. If we ignore, for a moment, the effect of the weighting by the coefficient sensitivities, we can assume a frequency ω_z that is so close to ω_p that g_{0z} will contribute to $\mathscr{S}_x^{\alpha(\omega)}$ near ω_p. Calling this frequency ω_z', we can then conclude that as long as ω_z is further away from ω_p than ω_z' the sensitivity will not depend on where ω_z is located.

Other methods of minimizing the gain sensitivity also become apparent from expressions (1-114) and (1-125). Clearly, the coefficient sensitivities or the ω and Q sensitivities should be minimized for minimum $\mathscr{S}_x^{\alpha(\omega)}$. These values depend on the method of network realization used. In any event, the minimization of the *frequency* sensitivity, or, in terms of (1-111), of the a_0 and b_0 sensitivity, is the most crucial, as a comparison of the g_ω and g_q functions clearly shows.

The relationship between the coefficient sensitivity and the ω and Q sensitivities of a second-order network can readily be obtained. Comparing

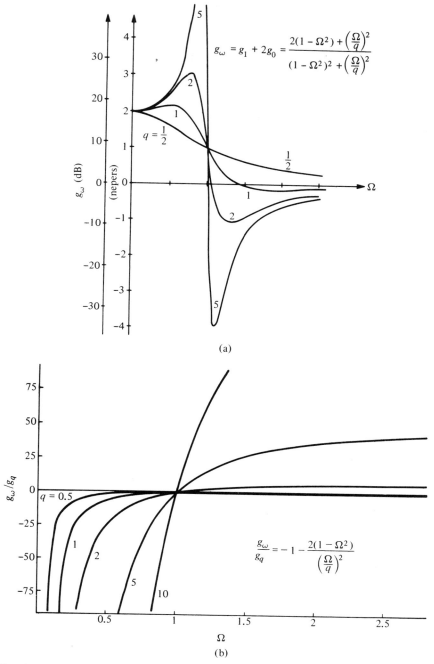

$$g_\omega = g_1 + 2g_0 = \frac{2(1 - \Omega^2) + \left(\frac{\Omega}{q}\right)^2}{(1 - \Omega^2)^2 + \left(\frac{\Omega}{q}\right)^2}$$

(a)

$$\frac{g_\omega}{g_q} = -1 - \frac{2(1 - \Omega^2)}{\left(\frac{\Omega}{q}\right)^2}$$

(b)

FIG. 1-51. Plots of normalized gain-sensitivity functions: (a) $g_\omega(\Omega, q)$; (b) $g_\omega/g_q(\Omega, q)$.

(1-128) with the denominator in (1-111) and assuming that $a_2 = 1$ (a_2 can be included in the multiplicative constant of $T(s)$) we obtain

$$S_x^q = \tfrac{1}{2}S_x^{a_0} - S_x^{a_1} \tag{1-130}$$

and

$$S_x^{\omega_n} = \tfrac{1}{2}S_x^{a_0} \tag{1-131}$$

If $T(s)$ is given in the form of (1-39), in which $2\sigma = \omega_n/q$, then it follows immediately that

$$\mathscr{S}_x^{\alpha(\omega)} = S_x^K + 2g_{0z}S_x^{\omega_z} + g_{1z}S_x^{\sigma_z} - 2g_{0p}S_x^{\omega_p} - g_{1p}S_x^{\sigma_p} \tag{1-132}$$

where g_{0z}, g_{0p}, and g_{1z}, g_{1p} are the g_0 and g_1 functions plotted in Fig. 1-50 applied to the zeros and poles, respectively, of $T(s)$.

EXAMPLE: To give a simple example of the discussion above, consider the *LCR* network shown in Fig. 1-52. The voltage transfer function is

$$T(s) = \frac{V_2}{V_1} = \frac{1/LC}{s^2 + (1/RC)s + (1/LC)} \tag{1-133}$$

The corresponding K, ω, and q sensitivities are obtained as follows:

Parameter x	S_x^K	$S_x^{\omega_p}$	$S_x^{q_p}$
R	0	0	1
L	-1	$-\tfrac{1}{2}$	$-\tfrac{1}{2}$
C	-1	$-\tfrac{1}{2}$	$\tfrac{1}{2}$

With (1-125) we then obtain

$$\mathscr{S}_R^{\alpha(\omega)} = -g_{q_p} = g_{1p} \tag{1-134}$$

$$\mathscr{S}_L^{\alpha(\omega)} = \tfrac{1}{2}(g_{\omega_p} + g_{q_p}) - 1$$
$$= g_{0p} - 1 \tag{1-135}$$

$$\mathscr{S}_C^{\alpha(\omega)} = \tfrac{1}{2}(g_{\omega_p} - g_{q_p}) - 1$$
$$= g_{1p} + g_{0p} - 1 \tag{1-136}$$

The fact that the gain sensitivities are independent of the component values here is due to the simplicity of the circuit rather than to any general rule.

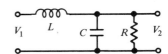

FIG. 1-52. *LCR* low-pass network.

Phase and Delay Sensitivity

The p Functions   The phase sensitivity (being the imaginary part of the transmission sensitivity) can also be derived as a sum of general, frequency-dependent functions, designated *p* functions here, that are weighted by the coefficient sensitivities. Referring to (1-111) one obtains

$$\mathcal{S}_x^{\phi(\omega)} = p_{b_2} S_x^{b_2} + p_{b_1} S_x^{b_1} + p_{b_0} S_x^{b_0} - p_{a_2} S_x^{a_2} - p_{a_1} S_x^{a_1} - p_{a_0} S_x^{a_0} \qquad [1\text{-}137]$$

where

$$p_{a_2} = \mathcal{S}_{a_2}^{\phi(\omega)} = \frac{a_1 a_2 \omega^3}{(a_0 - a_2 \omega^2)^2 + (a_1 \omega)^2} \qquad [1\text{-}138a]$$

$$p_{a_1} = \mathcal{S}_{a_1}^{\phi(\omega)} = \frac{a_1 \omega (a_0 - a_2 \omega^2)}{(a_0 - a_2 \omega^2)^2 + (a_1 \omega)^2} \qquad [1\text{-}138b]$$

$$p_{a_0} = \mathcal{S}_{a_0}^{\phi(\omega)} = \frac{- a_1 a_0 \omega}{(a_0 - a_2 \omega^2)^2 + (a_1 \omega)^2} \qquad [1\text{-}138c]$$

Note that

$$p_{a_2} + p_{a_1} + p_{a_0} = 0 \qquad [1\text{-}139]$$

The p_{b_j} functions have the same form except that the coefficients $b_j, j = 0, 1, 2$ replace the a_j above. The normalized functions p_1 and p_0 are plotted in Fig. 1-53a and b for some typical q values, where Ω and q are defined as in (1-119). Naturally the functions p_ω and p_q, corresponding to the gain functions g_ω and g_q discussed earlier, can be obtained as linear combinations of the functions p_1 and p_0; $p_\omega = p_1 + 2p_0$ and $p_q = -p_1$. The former is plotted in Fig. 1-53c and the ratio p_ω/p_q in Fig. 1-53d. We see that, independently of q, the phase is more sensitive to variations in frequency than it is to variations in q.

The t Functions   Where the phase sensitivity is the change in phase, in radians, resulting from a percentage change in some variable x, the delay sensitivity is the change in delay, given in seconds, for a percentage change in x:

$$\mathcal{S}_x^{\tau(\omega)} = \frac{d\tau(\omega)}{dx/x} \qquad (1\text{-}140)$$

where

$$\tau(\omega) = - \frac{d\phi(\omega)}{d\omega} \qquad (1\text{-}141)$$

Thus

$$\mathcal{S}_x^{\tau(\omega)} = - \frac{d}{d\omega} \left[x \frac{d\phi(\omega)}{dx} \right]$$

$$= t_{b_2} S_x^{b_2} + t_{b_1} S_x^{b_1} + t_{b_0} S_x^{b_0} - t_{a_2} S_x^{a_2} - t_{a_1} S_x^{a_1} - t_{a_0} S_x^{a_0} \qquad [1\text{-}142]$$

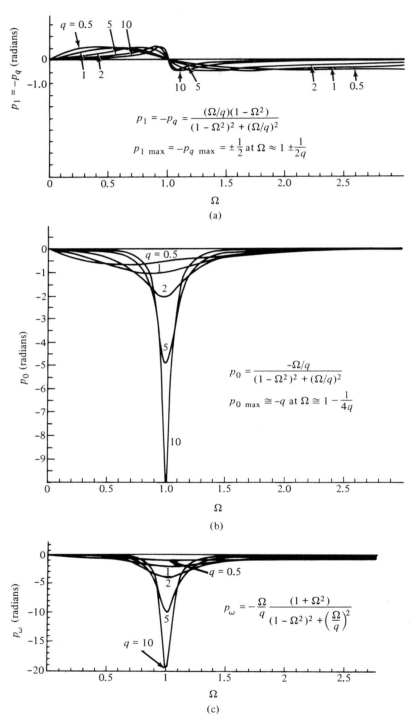

(a)

$$p_1 = -p_q = \frac{(\Omega/q)(1 - \Omega^2)}{(1 - \Omega^2)^2 + (\Omega/q)^2}$$

$$p_{1\ max} = -p_{q\ max} = \pm\frac{1}{2} \text{ at } \Omega \approx 1 \pm \frac{1}{2q}$$

(b)

$$p_0 = \frac{-\Omega/q}{(1 - \Omega^2)^2 + (\Omega/q)^2}$$

$$p_{0\ max} \cong -q \text{ at } \Omega \cong 1 - \frac{1}{4q}$$

(c)

$$p_\omega = -\frac{\Omega}{q} \frac{(1 + \Omega^2)}{(1 - \Omega^2)^2 + \left(\frac{\Omega}{q}\right)^2}$$

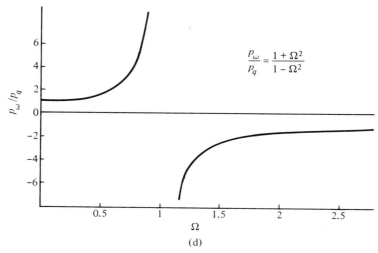

$$\frac{p_\omega}{p_q} = \frac{1 + \Omega^2}{1 - \Omega^2}$$

FIG. 1-53. Plots of normalized phase-sensitivity functions: (a) $p_1(\Omega, q)$; (b) $p_0(\Omega, q)$; (c) $p_\omega(\Omega, q)$; (d) $p_\omega/p_q(\Omega, q)$.

where

$$t_{a_2} = \mathscr{S}_{a_2}^{\tau(\omega)} = \frac{d}{d\omega}(-p_{a_2})$$

$$= -\frac{a_1 a_2 \omega^2 [(a_0 - a_2 \omega^2)(3a_0 + a_2 \omega^2) + (a_1 \omega)^2]}{[(a_0 - a_2 \omega^2)^2 + (a_1 \omega)^2]^2} \qquad [1\text{-}143\text{a}]$$

$$t_{a_1} = \mathscr{S}_{a_1}^{\tau(\omega)} = \frac{d}{d\omega}(-p_{a_1})$$

$$= -\frac{a_1(a_0 + a_2 \omega^2)[(a_0 - a_2 \omega^2)^2 - (a_1 \omega)^2]}{[(a_0 - a_2 \omega^2)^2 + (a_1 \omega)^2]^2} \qquad [1\text{-}143\text{b}]$$

$$t_{a_0} = \mathscr{S}_{a_0}^{\tau(\omega)} = \frac{d}{d\omega}(-p_{a_0})$$

$$= \frac{a_0 a_1[(a_0 - a_2 \omega^2)(a_0 + 3a_2 \omega^2) - (a_1 \omega)^2]}{[(a_0 - a_2 \omega^2)^2 + (a_1 \omega)^2]^2} \qquad [1\text{-}143\text{c}]$$

Here again,

$$t_{a_2} + t_{a_1} + t_{a_0} = 0 \qquad [1\text{-}144]$$

and, replacing the a_j by b_j, the t_b functions are obtained. The normalized t_1 and t_0 functions are plotted in Fig. 1-54a and b. The curves show that an all-pass section has a delay, in seconds, whose maximum change is the same for equal changes in the coefficients h_1 and h_0 (see 1-118). The functions t_ω and t_q corresponding to g_ω and g_q follow as $t_\omega = t_1 + 2t_0$ and $t_q = -t_1$. They have been plotted in Fig. 1-54c and d.

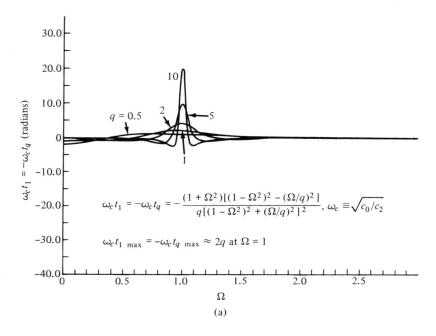

$$\omega_c t_1 = -\omega_c t_q = -\frac{(1 + \Omega^2)[(1 - \Omega^2)^2 - (\Omega/q)^2]}{q[(1 - \Omega^2)^2 + (\Omega/q)^2]^2}, \quad \omega_c \equiv \sqrt{c_0/c_2}$$

$$\omega_c t_{1 \text{ max}} = -\omega_c t_{q \text{ max}} \approx 2q \text{ at } \Omega = 1$$

(a)

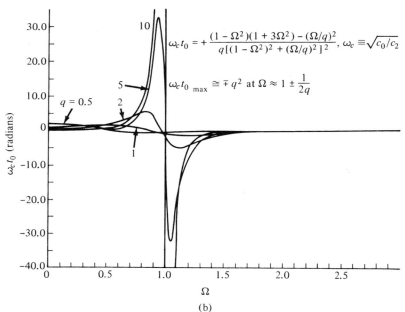

$$\omega_c t_0 = +\frac{(1 - \Omega^2)(1 + 3\Omega^2) - (\Omega/q)^2}{q[(1 - \Omega^2)^2 + (\Omega/q)^2]^2}, \quad \omega_c \equiv \sqrt{c_0/c_2}$$

$$\omega_c t_{0 \text{ max}} \cong \mp q^2 \text{ at } \Omega \approx 1 \pm \frac{1}{2q}$$

(b)

Fig. 1-54. Plots of normalized delay-sensitivity functions: (a) $t_1(\Omega, q)$; (b) $t_0(\Omega, q)$; (c) $t_\omega(\Omega, q)$; (d) $t_q(\Omega, q)$.

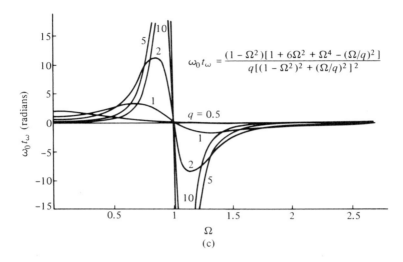

$$\omega_0 t_\omega = \frac{(1 - \Omega^2)[1 + 6\Omega^2 + \Omega^4 - (\Omega/q)^2]}{q[(1 - \Omega^2)^2 + (\Omega/q)^2]^2}$$

(c)

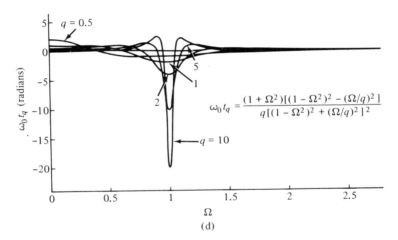

$$\omega_0 t_q = \frac{(1 + \Omega^2)[(1 - \Omega^2)^2 - (\Omega/q)^2]}{q[(1 - \Omega^2)^2 + (\Omega/q)^2]^2}$$

(d)

FIG. 1-54 *Continued*

Summary   In conclusion, we have seen that the sensitivity of the response (gain, phase, or delay) of a network can be expanded into a sum of factors each of which is, in turn, the product of two sensitivities. The first sensitivity, the response-to-coefficient sensitivity, is dependent only on the given transfer function and is a function of frequency. It provides a weighting of the second sensitivity, the coefficient-to-component sensitivity, which is a real number. The latter depends on the circuit realization and can therefore be minimized by a judicious circuit selection. More will be said on the interrelationship between the choice of circuit (e.g., positive or negative feedback type) and the available technology (e.g., discrete or integrated components) in Chapter 4.

The coefficient-to-component sensitivity functions can be converted to frequency-to-component and Q-to-component sensitivity functions with the help of (1-130) and (1-131). Using a coefficient or an ω–q sensitivity table, (as given for (1-133)), it is then a straightforward matter to obtain the gain, phase, or delay sensitivities as functions of frequency for any given second-order network function.

1.5.2 The Sensitivity of High-Q Networks

From the well known expression for an nth order transfer function with simple poles and zeros

$$T(s) = K \frac{\prod\limits_{i=1}^{m}(s - z_i)}{\prod\limits_{i=1}^{n}(s - p_i)} \tag{1-145}$$

we readily obtain the total differential

$$\frac{dT(s)}{T(s)} = S_K^{T(s)} \frac{dK}{K} + \sum_{i=1}^{m} S_{z_i}^{T(s)} \frac{dz_i}{z_i} + \sum_{i=1}^{n} S_{p_i}^{T(s)} \frac{dp_i}{p_i} \tag{1-146}$$

To obtain the sensitivities of $T(s)$ to its zeros and poles we have

$$\frac{dT(s)}{dz_i} = -K \frac{\prod\limits_{\substack{v=1 \\ v \neq i}}^{m}(s - z_v)}{\prod\limits_{i=1}^{m}(s - p_i)} = -\frac{T(s)}{s - z_i} \tag{1-147}$$

Consequently

$$S_{z_i}^{T(s)} = -\frac{z_i}{s - z_i} \tag{1-148}$$

Similarly

$$\frac{dT(s)}{dp_i} = K \frac{\prod\limits_{i=1}^{m}(s - z_i)}{\prod\limits_{\substack{v=1 \\ v \neq i}}^{n}(s - p_v)(s - p_i)^2} = \frac{T(s)}{s - p_i} \tag{1-149}$$

Thus

$$S_{p_i}^{T(s)} = \frac{p_i}{s - p_i} \tag{1-150}$$

Furthermore

$$S_K^{T(s)} = 1 \tag{1-151}$$

therefore (1-146) becomes

$$\frac{dT(s)}{T(s)} = \frac{dK}{K} - \sum_{i=1}^{m} \frac{dz_i}{s - z_i} + \sum_{i=1}^{n} \frac{dp_i}{s - p_i} \qquad [1\text{-}152]$$

Dividing both sides by dx/x we obtain the fundamental transmission-sensitivity expression given by (1-59). However, since we are interested in the behavior of $dT(j\omega)/T(j\omega)$ in the vicinity of a high-Q pole, we shall, in what follows examine (1-152) more closely in its present form.

Amplitude and Phase Sensitivity in the Vicinity of a High-Q Pole

Whether we are considering an ensemble of n poles and examining the behavior of $T(j\omega)$ only in the vicinity of its dominant pole p, or we are considering only a second-order network with a high-Q pole pair ($p, p^* = -\sigma_p \pm j\omega_c$), in either case we can write from (1-152)

$$d[\ln T(j\omega)] = d\alpha(\omega) + jd\phi(\omega) \approx \frac{dp}{s - p}\bigg|_{s \approx j\omega_c} \qquad [1\text{-}153]$$

With

$$dp = -d\sigma_p + jd\omega_c \qquad (1\text{-}154a)$$

where

$$\omega_c^2 = \omega_p^2 - \sigma_p^2 \qquad (1\text{-}154b)$$

we obtain from (1-153)

$$d\alpha(\omega) \approx \frac{-\sigma_p \, d\sigma_p + (\omega - \omega_c)d\omega_c}{\sigma_p^2 + (\omega - \omega_c)^2} \qquad (1\text{-}155a)$$

and

$$d\phi(\omega) \approx \frac{\sigma_p \, d\omega_c + (\omega - \omega_c)d\sigma_p}{\sigma_p^2 + (\omega - \omega_c)^2} \qquad (1\text{-}155b)$$

Since

$$\omega_c = \omega_p \sqrt{1 - (1/4Q^2)} \qquad (1\text{-}156)$$

where Q is defined by (1-53), it follows that

$$\omega_c|_{Q \gg 1} \approx \omega_p \qquad (1\text{-}157)$$

At the pole frequency ω_p we then have

$$d\alpha(\omega_p) \approx -\frac{d\sigma_p}{\sigma_p} = \frac{dQ}{Q} - \frac{d\omega_p}{\omega_p} \qquad [1\text{-}158a]$$

and

$$d\phi(\omega_p) \approx 2Q \frac{d\omega_p}{\omega_p} \qquad [1\text{-}158b]$$

These results also follow from the g and p curves discussed in the previous section. They show that $\alpha(\omega)$ is equally and directly dependent on frequency and Q variations at the pole frequency ω_p. On the other hand the phase $\phi(\omega)$

depends very strongly, magnified $2Q$ times, on frequency variations at ω_p. Clearly, since the phase is so sensitive to frequency variations at ω_p, the phase is an excellent indicator, compared with the amplitude $\alpha(\omega)$, for the tuning of the pole frequency ω_p. We shall come back to this point when discussing the tuning of second-order active networks in Chapter 4.

EXAMPLE: A SECOND-ORDER BANDPASS NETWORK   The frequency characteristic of a function $T(s)$ in the vicinity of a high-Q pole along the $j\omega$ axis can be closely approximated by the frequency response of a second-order bandpass function. It is therefore of interest to examine a second-order, high-Q bandpass function, since we can derive from it some of the properties of a general function $T(s)$ in the vicinity of a high-Q pole.

Consider the bandpass function

$$T(s) = K \frac{s}{s^2 + (\omega_p/Q)s + \omega_p^2} \tag{1-159}$$

whose amplitude and phase responses are shown in Fig. 1-55. Since $T(s)$ is dimensionless (e.g., a voltage transfer function) K must have the dimension of a frequency and we can write

$$K = K'\omega_p \tag{1-160}$$

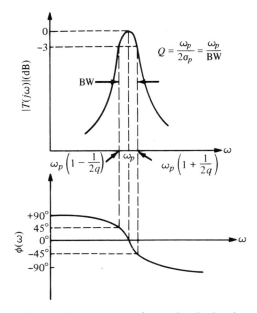

FIG. 1-55. Gain and phase response of second-order bandpass function.

Assuming a high-Q pole we then have from (1-152)

$$\frac{dT(s)}{T(s)} \approx \frac{dK}{K} + \frac{dp}{s-p} \tag{1-161}$$

Making the same assumptions as we did to obtain (1-158), we then have for the gain function

$$d\alpha(\omega_p) \approx \frac{dQ}{Q} \tag{1-162}$$

Thus, in the vicinity of ω_p, $\alpha(\omega)$ *is independent of frequency variations* $d\omega_p/\omega_p$ *and directly proportional to variations in Q.* The behavior of the phase response is as given by (1-158b).

Referring to Fig. 1-55 we can now consider the frequencies at which the amplitude is 3 dB lower than at ω_p, namely, $\omega_p \pm \sigma_p$ or, in terms of the pole Q, $\omega_p \pm (1/2Q)$. Here we have, from (1-155) and (1-161),

$$d\alpha(\omega_p \pm \sigma_p) \approx \frac{-d\sigma_p \pm d\omega_p}{2\sigma_p} = \frac{1}{2}\left(\frac{dQ}{Q} - \frac{d\omega_p}{\omega_p}\right) \pm Q\frac{d\omega_p}{\omega_p}$$

$$\approx \frac{1}{2}\frac{dQ}{Q} \pm Q\frac{d\omega_p}{\omega_p} \tag{1-163a}$$

and

$$d\phi(\omega_p \pm \sigma_p) \approx \frac{\pm d\sigma_p + d\omega_p}{2\sigma_p} = \pm\frac{1}{2}\left(\frac{d\omega_p}{\omega_p} - \frac{dQ}{Q}\right) + Q\frac{d\omega_p}{\omega_p}$$

$$\approx \mp\frac{1}{2}\frac{dQ}{Q} + Q\frac{d\omega_p}{\omega_p} \tag{1-163b}$$

This is in agreement with our discussion in the previous section: at the -3 dB frequencies both the amplitude and phase are $2Q$ *times* more sensitive to variations in frequency, than to variations in Q. *Thus, for high-Q circuits, frequency stability is far more (namely, at least $2Q$ times more) crucial, both for the amplitude and phase response of a second-order network, than Q stability.*[56] This point must be emphasized, since it contradicts the generally accepted practice of stressing the importance of Q stability while paying all too little attention to frequency stability. The reason for this tendency is historical and well understandable: in the early days of active networks, the major problem confronting the active-network designer was the prevention of self-oscillation, i.e., poles drifting toward the RHP or their Q's becoming

---

56. The plot of g_ω/g_q in Fig. 1-51b actually demonstrated that beyond the 3 dB frequencies the sensitivity to frequency variations is even larger than $2Q$ times the sensitivity with respect to Q. However, in general the characteristic of a network is particularly important in the vicinity of each pole frequency (i.e., over the corresponding 3 dB band) so that our factor of $2Q$ serves as a useful rule of thumb.

infinitely large. Today, with the vastly improved (hybrid integrated) components at one's disposal, and the great deal of additional knowledge pertaining to active circuits, the drastic effects leading to self-oscillation are well under control and therefore preventable. In this new situation (1-163) shows us that the stabilization of frequency must be given priority over the stabilization of Q, at least as long as

$$2Q \frac{d\omega_p}{\omega_p} > \frac{dQ}{Q} \tag{1-164a}$$

or

$$2Q S_x^{\omega_p} > S_x^Q \tag{1-164b}$$

(1-164b) suggests that the product $Q S_x^{\omega_p}$ rather than $S_x^{\omega_p}$ be compared with S_x^Q, in order to establish which of the two sensitivities is more critical. As we shall see in Chapter 4, the minimization of $S_x^{\omega_p}$ is mainly a technological problem, since ω_p generally depends only on resistors and capacitors. Thus, although in the past most effort has been expended on the improvement of S_x^Q by considering ever more elaborate circuit configurations, *once a certain degree of Q stability has been attained, all these configurations are equally well, or poorly, suited to the realization of high-Q networks, depending on the characteristics of the passive components with which they are built and the frequency stability that can be guaranteed with them.*

A simple example will serve to illustrate these remarks. Consider again the high-Q bandpass network in Fig. 1-55. Clearly, a change in the center frequency ω_p, which may seem only an insignificantly small percentage of ω_p, may be a very large percentage of the bandwidth BW. Thus, it is really the change $\Delta\omega_p$ in the center frequency as a fraction of the bandwidth BW, that is of interest. This ratio can be expressed simply as[57]

$$\frac{\Delta\omega_p}{BW} = Q \frac{\Delta\omega_p}{\omega_p} \tag{1-165}$$

where $Q = \omega_p/BW$. This representation contains the same multiplication of the frequency variation by Q as (1-163a). If, for example, the center frequency changes by 0.5% and the Q is 100, then, according to (1-165), the center frequency will change by 50% of the bandwidth, i.e., the frequency response will have shifted according to the dashed curve in Fig. 1-56. Note the resulting attenuation at the lower -3 dB frequency, ω_l. Even if a less realistic (and hardly achievable) worst-case shift in the location of the pole frequency of 0.1% is assumed, the change in center frequency will still cover 10% of the bandwidth. *In most cases then, the maximum Q's feasible with active net-*

---

57. D. R. Means, private communication.

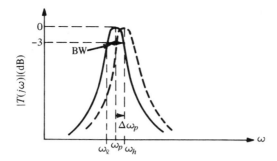

FIG. 1-56. Effect of frequency shift on high-Q bandpass function.

works will be limited by the attainable stability of the pole frequency, long before they are limited by the attainable stability of the pole Q. Notable exceptions to this rule are inductorless active networks in which crystals, rather than RC networks, determine the pole frequency and its stability.[58] Unfortunately, the scope of this book does not allow us to examine this interesting topic.

Transmission and Pole Variation in the Vicinity of a High-Q Pole

We can now use the results obtained in (1-158) and (1-162) to proceed one step further. Using (1-153) we can write[59]

$$\frac{dT(j\omega_p)}{T(j\omega_p)} = d\alpha(\omega_p) + jd\phi(\omega_p) \approx \frac{dQ}{Q} + j2Q\frac{d\omega_p}{\omega_p} \qquad [1\text{-}166]$$

We recall now that[60]

$$\left.\frac{dp}{p}\right|_{Q\gg 1} \approx \frac{d\omega_p}{\omega_p} - j\frac{1}{2Q}\frac{dQ}{Q} \qquad (1\text{-}167)$$

Multiplying both sides of (1-167) by $j2Q$ we therefore find[61]

$$\frac{dT(j\omega_p)}{T(j\omega_p)} \approx j2Q\frac{dp}{p} \qquad [1\text{-}168]$$

58. G. S. Moschytz, Inductorless filters: A survey, Parts I and II, *IEEE Spectrum*, August and September 1970, pp. 30–36 and pp. 63–75, respectively.
59. It has been shown by D. R. Means (Inductorless Active Filter Design Using Discrete or Monolithic Coupled Piezoelectric Resonators, Tech. Rept. 400-223 New York Univ., 1972) that for a finite frequency shift $\Delta\,\omega_p/\omega_p$, $\Delta T(j\omega_p)/T(j\omega_p)$ is given by the Taylor series expansion $\sum_1^\infty (j2Q\,\Delta\omega_p/\omega_p)^n$. Consequently (1-66) is valid only when $\Delta\,\omega_p/\omega_p \ll \frac{1}{2}Q$ and $(2Q\Delta\omega_p/\omega_p)^2 \ll \Delta\,Q/Q$. The second, more severe condition is unlikely to hold as Q becomes very large. For that case we obtain $\Delta T(j\omega_p)/T(j\omega_p) = \Delta\,Q/Q + [(j2Q\Delta\omega_p/\omega_p) - 4Q^2(\Delta\omega_p/\omega_p)^2 - j8Q^3(\Delta\omega_p/\omega_p)^3 + \cdots]$. This does not alter the conclusion, though, that $T(j\omega_p)$ is at least $2Q$ times more sensitive to a variation in ω_p than it is to an equal variation in Q.
60. See Chapter 4, Section 4.2.3, equation (4-129) of *Linear Integrated Networks: Fundamentals*.
61. G. S. Moschytz, A note on pole, frequency, and Q sensitivity, *IEEE J. Solid-State Circuits*, SC-6, 267–269 (1971) and SC-6, 423 (1971).

Proceeding in like manner with [1-163] we obtain

$$\frac{dT(j\omega)}{T(j\omega)}\bigg|_{\omega_p \pm \sigma_p} = \frac{1}{2}\left(\frac{dQ}{Q} - \frac{d\omega_p}{\omega_p}\right)(1 \mp j) + Q\frac{d\omega_p}{\omega_p}(j \pm 1)$$

$$\approx \frac{1}{2}\frac{dQ}{Q}(1 \mp j) + Q\frac{d\omega_p}{\omega_p}(j \pm 1) \qquad [1\text{-}169]$$

We must remember that these expressions, as well as the other derivations in this section, are valid only when $Q \gg 1$.

Transmission, Gain, and Phase Sensitivity and the Open-Loop Poles
We now consider the effect of the open-loop poles P_1 and P_2 on the gain and phase sensitivity of high-Q second-order networks (or of nth-order networks in the vicinity of a dominant, high-Q pole). Starting out with (1-52)[62] and letting $\omega_c \approx \omega_p$, we have

$$S_x^{T(j\omega_p)} \approx -ju_0 \frac{Q}{\omega_p^2}(p - P_1)(p - P_2) \qquad (1\text{-}170)$$

With the open-loop poles given by

$$P_1 = -\mu_1 + jv_1 \qquad (1\text{-}171a)$$

and

$$P_2 = -\mu_2 + jv_2 \qquad (1\text{-}171b)$$

(1-170) becomes

$$S_x^{T(j\omega_p)} \approx -ju_0 \frac{Q}{\omega_p^2}[(\mu_1 - \sigma_p) + j(\omega_p - v_1)][(\mu_2 - \sigma_p) + j(\omega_p + v_2)] \qquad (1\text{-}172)$$

P_1 and P_2 are either complex conjugate or negative real. When they are complex conjugate we have

$$\mu_1 = \mu_2 = \mu \qquad (1\text{-}173a)$$

and

$$v_1 = -v_2 = v \qquad (1\text{-}173b)$$

The gain sensitivity at ω_p then results from (1-172) as

$$\mathscr{S}_x^{\alpha(\omega_p)} = u_0\left(\frac{Q}{Q_\mu} - 1\right) \qquad [1\text{-}174a]$$

where

$$Q_\mu = \frac{\omega_r}{2\mu} \qquad [1\text{-}174b]$$

---

62. This expression, it will be remembered, was derived for high-Q type I networks, or type II networks whose high-Q pole is located far from any zeros.

Similarly, the phase sensitivity is given by

$$\mathscr{S}_x^{\phi(\omega_p)} = -u_0 Q\left[\frac{1}{4Q_{\mu p}^2} + \left(\frac{v}{\omega_p}\right)^2 - 1\right]$$ [1-175a]

where

$$Q_{\mu p} = \frac{\omega_p}{2(\mu - \sigma_p)}$$ [1-175b]

When P_1 and P_2 are negative real, i.e.,

$$P_1 = -\mu_1; \qquad P_2 = -\mu_2; \qquad v_1 = v_2 = 0$$ (1-176)

then the gain sensitivity becomes

$$\mathscr{S}_x^{\alpha(\omega_p)}\bigg|_{v_1=v_2=0} = u_0\left(\frac{Q}{Q_{\mu 12}} - 1\right)$$ [1-177a]

where

$$Q_{\mu 12} = \frac{\omega_p}{\mu_1 + \mu_2}$$ [1-177b]

and

$$\mathscr{S}_x^{\phi(\omega_p)}\bigg|_{v_1=v_2=0} \approx -u_0 Q\left(\frac{1}{4Q_{\mu 1 p}Q_{\mu 2 p}} - 1\right)$$ [1-178a]

where

$$Q_{\mu 1 p} = \frac{\omega_p}{2(\mu_1 - \sigma_p)}; \qquad Q_{\mu 2 p} = \frac{\omega_p}{2(\mu_2 - \sigma_p)}$$ [1-178b]

With (1-41c) μ_i and v_i, $i = 1, 2$, are related to $U(s)$ by

$$-(\mu_1 + \mu_2) = -2\mu = u_1$$ (1-179a)

and

$$\mu^2 + v^2 = \mu_1\mu_2 = u_2.$$ (1-179b)

The main point resulting from expressions (1-174) to (1-178) is that the gain and phase sensitivity decreases as the Q of the passive networks (e.g., Q_μ, $Q_{\mu 12}$ etc.) is increased. As pointed out before, this can be achieved in the decomposition outlined by equations (1-40) to (1-45) by selecting the open-loop poles P_1 and P_2 as close together as possible when on the negative real axis, and as close to the closed-loop poles as possible when complex conjugate.

1.5.3 The Derivation of Root Sensitivity from Coefficient Sensitivity

When calculating the transfer function of a second-order network it is common to obtain its denominator in the form

$$D(s) = s^2 + a_1 s + a_0 \tag{1-180}$$

where a_1 and a_0 are functions of the components of the network. In general a coefficient a_2 will precede the s^2-term but it is a simple matter to normalize the coefficients a_1 and a_0 such that a_2 is unity. Having the coefficients a_1 and a_0, their sensitivities to the network components can readily be obtained—generally by inspection. We have seen in the preceding sections, however, that the pole (or zero) sensitivity may often be more desirable, since it leads directly to the physically meaningful, because measurable, frequency and Q sensitivities. It is the object of this brief section to provide the relationships between the readily calculatable coefficient sensitivities of a second-order polynomial and its root sensitivities.

With the pole $p = -\sigma + j\omega_c$, the corresponding natural undamped frequency $\omega_p = \sqrt{\sigma^2 + \omega_c^2}$ and the pole Q, $q_p = \omega_p/2\sigma$, we have

$$\mathscr{S}_x^p = \frac{dp}{dx/x} = \mathscr{S}_x^{-\sigma} + j\mathscr{S}_x^{\omega_c} \tag{1-181}$$

With (1-181) we immediately have the pole sensitivity S_x^p by dividing both sides by p and can then readily derive the ω_p and Q sensitivities.

Equating the two forms of $D(s)$,

$$s^2 + 2\sigma s + (\sigma^2 + \omega_c^2) = s^2 + a_1 s + a_0 \tag{1-182}$$

we obtain

$$S_x^\sigma = S_x^{a_1} \tag{1-183a}$$

and

$$\mathscr{S}_x^\sigma = \frac{a_1}{2} S_x^{a_1} \tag{1-183b}$$

Similarly,

$$S_x^{\omega_c} = \frac{2a_0 S_x^{a_0} - a_1^2 S_x^{a_1}}{4a_0 - a_1^2} \tag{1-184a}$$

and

$$\mathscr{S}_x^{\omega_c} = \frac{1}{2} \frac{2a_0 S_x^{a_0} - a_1^2 S_x^{a_1}}{\sqrt{4a_0 - a_1^2}} \tag{1-184b}$$

The relationship between the (ω_p, q_p) and the (ω_c, σ) sensitivities is also readily obtainable:

$$S_x^\sigma = S_x^{\omega_p} - S_x^{q_p} \qquad [1\text{-}185a]$$

$$\mathscr{S}_x^\sigma = \frac{\omega_p}{2q_p}(S_x^{\omega_p} - S_x^{q_p}) \qquad [1\text{-}185b]$$

$$S_x^{\omega_c} = S_x^{\omega_p} + \frac{S_x^{q_p}}{4q_p^2 - 1} \qquad [1\text{-}186a]$$

and

$$\mathscr{S}_x^{\omega_c} = \frac{\omega_p}{2q_p}\left[\sqrt{4q_p^2 - 1}\, S_x^{\omega_p} + \frac{S_x^{q_p}}{\sqrt{4q_p^2 - 1}}\right] \qquad [1\text{-}186b]$$

Finally, for the sake of completeness, we repeat the relationship between the (ω_p, q_p) and the coefficient sensitivities given earlier, namely,

$$S_x^{\omega_p} = \tfrac{1}{2}S_x^{a_0} \qquad [1\text{-}187a]$$

and

$$S_x^{q_p} = \tfrac{1}{2}S_x^{a_0} - S_x^{a_1} \qquad [1\text{-}187b]$$

Since the $\mathscr{S}_x^{\omega_p}$, $S_x^{q_b}$ forms are rarely if ever used, they have been omitted here.

With the set of equations (1-183) through (1-187) it is a simple matter, in theory, to derive the root sensitivities of a second-order network from its coefficient sensitivities. In practice the actual calculations may still become quite lengthy, in which case the aid of a desk calculator or even of a simple computer routine should prove very welcome.

1.5.4 The Effects of the Open–Loop Poles and Zeros on the Sensitivity of a Network

Closed-Loop Pole Sensitivity as a Function of Open-Loop Pole and Zero Locations   We return here to the representation of the network function $T(s)$ as a bilinear function of x as given by (1-40). Here the roots are interpreted in terms of the root loci of $N(s)$ and $D(s)$, for a given network function $T(s) = N(s)/D(s)$. Considering only the denominator $D(s)$, the root locus of the poles of $T(s)$ is defined by the roots of $D(x, s)$, i.e., by

$$D(x, s) = 0 \qquad (1\text{-}188)$$

Deriving the characteristic equation from (1-40), p is a pole of $T(s)$ if it satisfies the condition that

$$x\left.\frac{v_0\, V(s)}{u_0\, U(s)}\right|_{s=p} = -1 \qquad (1\text{-}189)$$

where $s = p$ represents a point on the root locus with respect to the variable x. The resulting root locus of a typical second-order function was shown in Fig. 1-34. Equation (1-189) defines the loop gain $L(s)$ of $T(s)$ with respect to x. Thus, generalizing the representation of (1-42) and (1-43) we have

$$L(s) = x \frac{V(s)}{U(s)} = x \frac{\prod\limits_{i=1}^{m} (s - \bar{P}_i)}{\prod\limits_{i=1}^{n} (s - P_i)} \qquad (1\text{-}190)$$

where the multiplicative constant v_0/u_0 is combined with the variable x. For a pole $s = p$, we have

$$x = -\frac{\prod\limits_{i=1}^{n} (p - P_i)}{\prod\limits_{i=1}^{m} (p - \bar{P}_i)} \qquad (1\text{-}191)$$

To find $\mathscr{S}_x^p$, we take the logarithm of (1-191) and then calculate the derivative of p with respect to $\ln x$. We obtain

$$\mathscr{S}_x^p = \frac{dp}{d \ln x} = \left[\sum_{i=1}^{n} \frac{1}{p - P_i} - \sum_{i=1}^{m} \frac{1}{p - \bar{P}_i} \right]^{-1} \qquad [1\text{-}192]$$

For a second-order network this simplifies to

$$\mathscr{S}_x^p = \frac{1}{\dfrac{1}{p - P_1} + \dfrac{1}{p - P_2} - \dfrac{1}{p - \bar{P}_1} - \dfrac{1}{p - \bar{P}_2}} \qquad [1\text{-}193]$$

Naturally, equivalent expressions are valid for a zero z of $T(s)$. Equation (1-193) suggests that the pole sensitivity of a second-order network can be made to approach zero if one of the open-loop poles P_1 or P_2, or one of the open loop zeros $\bar{P}_1$ or $\bar{P}_2$, can be made to coincide, or at least to lie as close as possible to the closed-loop pole p. The same applies to Z_1, Z_2, $\bar{Z}_1$, and $\bar{Z}_2$ and the closed-loop zero z.[63] This, of course, is in agreement with the conclusions reached in Section 1.3.3.

In the second-order case (1-191) becomes

$$x = -\frac{(p - P_1)(p - P_2)}{(p - \bar{P}_1)(p - \bar{P}_2)} \qquad (1\text{-}194)$$

Thus the value of x necessary to generate p will be *smaller* the closer P_1 and P_2 are to p (and its conjugate p^*) along the corresponding root locus; conversely, x will be *larger* the closer the open-loop zeros $\bar{P}_1$ and $\bar{P}_2$ are to p.

63. Phung Le, private communication.

Another representation of $\mathscr{S}_x^p$, in terms of the open-loop poles and zeros can be obtained as follows. Taking the logarithm of (1-191) and then calculating the derivative with respect to $d(\ln R_j)$, where R_j is the jth resistor in the network, we obtain

$$\frac{d(\ln x)}{d(\ln R_j)} = \sum_{i=1}^{n} \frac{dP_i/d(\ln R_j)}{p - P_i} - \sum_{i=1}^{m} \frac{d\bar{P}_i/d(\ln R_j)}{p - \bar{P}_i} \qquad (1\text{-}195)$$

From (1-192) we can replace $d(\ln x)$ by

$$d(\ln x) = \sum_{i=1}^{n} \frac{dp}{p - P_i} - \sum_{i=1}^{m} \frac{dp}{p - \bar{P}_i} \qquad (1\text{-}196)$$

Substituting (1-196) into (1-195) we obtain

$$pS_{R_j}^p \left[\sum_{i=1}^{n} \frac{1}{p - P_i} - \sum_{i=1}^{m} \frac{1}{p - \bar{P}_i} \right] = \sum_{i=1}^{n} \frac{P_i S_{R_j}^{P_i}}{p - P_i} - \sum \frac{\bar{P}_i S_{R_j}^{\bar{P}_i}}{p - \bar{P}_i} \qquad (1\text{-}197)$$

where we have used the identity $\mathscr{S}_x^F \equiv FS_x^F$ for p, P_i, and $\bar{P}_i$. Assuming that the network contains r resistors, we obtain an equation of the form of (1-197) for each of the r resistors. Adding the resulting r equations and remembering that

$$\sum_{j=1}^{r} S_{R_j}^p = \sum_{j=1}^{r} S_{R_j}^{P_i} = \sum_{j=1}^{r} S_{R_j}^{\bar{P}_i} = -1 \qquad (1\text{-}198)$$

we obtain

$$p \left[\sum_{i=1}^{n} \frac{1}{p - P_i} - \sum_{i=1}^{m} \frac{1}{p - \bar{P}_i} \right] = \sum_{i=1}^{n} \frac{P_i}{p - P_i} - \sum_{i=1}^{m} \frac{\bar{P}_i}{p - \bar{P}_i} \qquad [1\text{-}199]$$

Comparing this with (1-192) we see immediately that

$$\mathscr{S}_x^p = p \left[\sum_{i=1}^{n} \frac{P_i}{p - P_i} - \sum_{i=1}^{m} \frac{\bar{P}_i}{p - \bar{P}_i} \right]^{-1} \qquad [1\text{-}200]$$

For a second-order network this becomes

$$\mathscr{S}_x^p = \frac{p}{\dfrac{P_1}{p - P_1} + \dfrac{P_2}{p - P_2} - \dfrac{\bar{P}_1}{p - \bar{P}_1} + \dfrac{\bar{P}_2}{p - \bar{P}_2}} \qquad [1\text{-}201]$$

In contrast to (1-49), (1-192) and (1-200) express the closed-loop pole sensitivity as a function of *both* open-loop poles *and* open-loop zeros.

Closed-Loop Pole Displacement as a Function of Open-Loop Pole and Zero Displacements

Equation (1-192) provides the pole sensitivity with respect to variations in x as a function of the open-loop poles and zeros.

Let us now examine the sensitivity of p, not only to variations in x, but to variations in the open-loop poles and zeros as well. With this in mind we take the total derivative of (1-190), obtaining

$$dL = \frac{\partial L}{\partial s} ds + \frac{\partial L}{\partial x} dx + \sum_{i=1}^{m} \frac{\partial L}{\partial \bar{P}_i} d\bar{P}_i + \sum_{i=1}^{n} \frac{\partial L}{\partial P_i} dP_i \qquad (1\text{-}202)$$

On the root locus the loop gain is constant, namely -1, so that the total differential dL is zero for $s = p$. Thus, solving for dp, we have

$$dp = -\frac{1}{(\partial L/\partial s)_{s=p}} \left[\left(\frac{\partial L}{\partial x}\right) dx + \sum_{i=1}^{m} \left(\frac{\partial L}{\partial \bar{P}_i}\right) d\bar{P}_i + \sum_{i=1}^{n} \left(\frac{\partial L}{\partial P_i}\right) dP_i \right]_{s=p} \qquad (1\text{-}203)$$

Since p is also a function of x, $\bar{P}_i$, and P_i, we can take the total differential of of p, obtaining

$$dp = \mathcal{S}_x^p \frac{dx}{x} + \sum_{i=1}^{m} \frac{\partial p}{\partial \bar{P}_i} d\bar{P}_i + \sum_{i=1}^{n} \frac{\partial p}{\partial P_i} dP_i \qquad (1\text{-}204)$$

Comparing (1-203) with (1-204) we obtain

$$\mathcal{S}_x^p = -x \left(\frac{\partial L/\partial x}{\partial L/\partial s}\right)_{s=p} \qquad (1\text{-}205)$$

$$\frac{\partial p}{\partial \bar{P}_i} = -\left(\frac{\partial L/\partial \bar{P}_i}{\partial L/\partial s}\right)_{s=p} \qquad (1\text{-}206a)$$

and

$$\frac{\partial p}{\partial P_i} = -\left(\frac{\partial L/\partial P_i}{\partial L/\partial s}\right)_{s=p} \qquad (1\text{-}206b)$$

Now, from (1-190) we have

$$\left(\frac{\partial L}{\partial x}\right)_{s=p} = \frac{\prod_{i=1}^{m}(s - \bar{P}_i)}{\prod_{i=1}^{n}(s - P_i)}\Bigg|_{s=p} = \left(\frac{L}{x}\right)_{s=p} \qquad (1\text{-}207)$$

and because $L(p) = -1$, (1-207) yields

$$\left(\frac{\partial L}{\partial x}\right)_{s=p} = -\frac{1}{x} \qquad (1\text{-}208)$$

Equation (1-205) therefore becomes

$$\mathcal{S}_x^p = \frac{1}{(\partial L/\partial s)_{s=p}} \qquad (1\text{-}209)$$

Similarly, from (1-190),

$$\left(\frac{\partial L}{\partial \bar{P}_i}\right)_{s=p} = -x \left.\frac{\prod\limits_{i=1}^{m-1}(s - \bar{P}_i)}{\prod\limits_{i=1}^{n}(s - P_i)}\right|_{s=p}$$

$$= -\left.\frac{L}{s - \bar{P}_i}\right|_{s=p} = \frac{1}{p - \bar{P}_i} \qquad (1\text{-}210)$$

Thus, using (1-206) and (1-209) we obtain the differential of a closed-loop pole p with respect to an open-loop zero $\bar{P}_i$ as

$$\frac{\partial p}{\partial \bar{P}_i} = -\frac{1/(p - \bar{P}_i)}{(\partial L/\partial s)_{s=p}} = -\frac{\mathscr{S}^p_x}{p - \bar{P}_i} \qquad [1\text{-}211a]$$

Similarly, the differential of p with respect to an open-loop pole P_i is

$$\frac{\partial p}{\partial P_i} = \frac{1/(p - P_i)}{(\partial L/\partial s)_{s=p}} = \frac{\mathscr{S}^p_x}{p - P_i} \qquad [1\text{-}211b]$$

where x is an element in the network and $\mathscr{S}^p_x$ the (closed-loop) pole sensitivity with respect to x. It follows that whatever properties are found for the closed-loop pole sensitivity apply also to these pole differentials.

Substituting (1-211) into (1-204), we obtain[64]

$$dp = \mathscr{S}^p_x \left[\frac{dx}{x} - \sum_{i=1}^{m} \frac{d\bar{P}_i}{p - \bar{P}_i} + \sum_{i=1}^{n} \frac{dP_i}{p - P_i}\right] \qquad [1\text{-}212]$$

This expression gives the closed-loop pole displacement as a function of the pole sensitivity with respect to the element x and as a function of the open-loop roots and their displacements. Notice that in order to minimize pole displacements dp, caused by shifts of the open-loop poles P_i or zeros $\bar{P}_i$, the distance from the open-loop roots to the closed-loop pole should be as large as possible. If we recall now that $\mathscr{S}^p_x$ as given by (1-192) (that is the sensitivity of p to some network element x), requires the opposite condition, namely the open-loop roots as close to p as possible, then it is clear that there is some optimum decomposition for each network realization, depending on the circuit and the characteristics of the components used. If the variation of some element x, say the gain, dominates, then (1-192) should be minimized by selecting the P_i and/or $\bar{P}_i$ as close to p as possible; if the variations of the open-loop roots dominate, then the decomposition should locate them far away from p.

A useful relationship can be derived for the differentials of (1-211). Consider the construction of the root locus as defined by (1-189). If all open-loop

64. H. Ur, Root locus properties and sensitivity relations in control systems, *IRE Trans. Autom. Control*, **AC-5**, 57–65 (1960).

zeros and poles are displaced by the same amount δ, then all closed-loop roots are displaced by the same amount δ. Thus if $d\bar{P}_i = dP_i = \delta$ for all i, then $dp = \delta$ for all i. Therefore if in (1-204) we set $dp = d\bar{P}_i = dP_i = \delta$ and $dx = 0$, we obtain

$$\sum_{i=1}^{m} \frac{dp}{d\bar{P}_i} + \sum_{i=1}^{n} \frac{dp}{dP_i} = 1 \qquad [\text{1-213}]$$

Let us now assume that the open-loop poles and zeros of $T(s)$ are generated by a passive RC network and that x corresponds to the active network element G. Furthermore we assume that both the resistors and the capacitors of the network track closely with ambient variations (e.g., temperature, humidity, aging). Then we can write[65]

$$dP_i = -P_i(1 + \delta)\frac{dR}{R} \qquad [\text{1-214a}]$$

and

$$d\bar{P}_i = -\bar{P}_i(1 + \delta)\frac{dR}{R} \qquad [\text{1-214b}]$$

where

$$\frac{dC}{C} = \delta\frac{dR}{R} \qquad [\text{1-214c}]$$

Similarly, the displacement of the closed-loop pole as a result of variations of the passive elements and the gain element G is given by

$$dp = \mathscr{S}_G^p \frac{dG}{G} - p\frac{dR}{R}(1 + \delta) \qquad [\text{1-215}]$$

Substituting (1-214) and (1-215) into (1-212) we obtain (1-200), with x replaced by G. Note, however, that (1-200) was derived for general networks irrespective of whether the passive components track or not.

In order to eliminate the variations of x from (1-212) and to obtain dp as a function of only the open-loop pole and zero displacements dP_i and $d\bar{P}_i$, we take the total derivative of x as given by (1-191):

$$dx = \sum_{i=1}^{n} \frac{\partial x}{\partial P_i} dP_i + \sum_{i=1}^{m} \frac{\partial x}{\partial \bar{P}_i} d\bar{P}_i \qquad (1\text{-}216)$$

where

$$\frac{\partial x}{\partial P_i} = \frac{\prod_{\substack{j=1 \\ j \neq i}}^{n-1} (p - P_j)}{\prod_{i=1}^{m} (p - \bar{P}_i)} = \frac{x}{p - P_i} \qquad (1\text{-}217a)$$

65. See Chapter 4, equation (4-120) of *Linear Integrated Networks: Fundamentals*.

and

$$\frac{\partial x}{\partial \overline{P}_i} = -\frac{\prod\limits_{i=1}^{n}(p-P_i)}{\prod\limits_{\substack{j=1 \\ j \neq i}}^{m-1}(p-\overline{P}_j)(p-\overline{P}_i^2)} = -\frac{x}{p-\overline{P}_i} \tag{1-217b}$$

Thus,

$$\frac{dx}{x} = \sum_{i=1}^{n}\frac{dP_i}{p-P_i} - \sum_{i=1}^{m}\frac{d\overline{P}_i}{p-\overline{P}_i} \tag{1-218}$$

Substituting (1-200) and (1-216) through (1-218) into

$$\frac{dp}{p} = \frac{1}{p}\mathscr{S}_x^p\frac{dx}{x} \tag{1-219}$$

we obtain

$$\frac{dp}{p} = \frac{\sum\limits_{i=1}^{n}\dfrac{dP_i}{p-P_i} - \sum\limits_{i=1}^{m}\dfrac{d\overline{P}_i}{p-\overline{P}_i}}{\sum\limits_{i=1}^{n}\dfrac{P_i}{p-P_i} - \sum\limits_{i=1}^{m}\dfrac{\overline{P}_i}{p-\overline{P}_i}} \tag{1-220}$$

This, then, is the variation of the closed-loop pole as a function of the variations of all open-loop roots.

Expressions for Second-Order Networks   Some of the expressions derived above can be simplified significantly for second-order networks. From (1-49) and (1-211) we obtain

$$\frac{dp}{d\overline{P}_1} = -\frac{1}{2j\omega_c}\frac{(p-P_1)(p-P_2)}{p-\overline{P}_1} \tag{1-221a}$$

$$\frac{dp}{d\overline{P}_2} = -\frac{1}{2j\omega_c}\frac{(p-P_1)(p-P_2)}{p-\overline{P}_2} \tag{1-221b}$$

$$\frac{dp}{dP_1} = \frac{p-P_2}{2j\omega_c} \tag{1-221c}$$

$$\frac{dp}{dP_2} = \frac{p-P_1}{2j\omega_c} \tag{1-221d}$$

To minimize the closed-loop pole differential with respect to an open-loop zero, the pole should therefore be as far away from the zero as possible. Furthermore, the closed-loop pole should be as close to the open-loop poles as possible. The latter requirement is true for all the pole differentials listed above. Thus, in minimizing pole sensitivity with respect to an (active) element x (see (1-49)), by minimizing the distances between closed- and open-loop poles, the pole differentials with respect to open-loop roots are also minimized.

Applying (1-213) to the sum of the expressions given by (1-221) yields the following relationship between the open-loop poles and zeros of a second-order network:

$$\left[\frac{1}{p - P_1} - \frac{1}{p - \bar{P}_1} \right] + \left[\frac{1}{p - P_2} - \frac{1}{p - \bar{P}_2} \right] = 2j\omega_c \qquad (1\text{-}222)$$

where the prescribed closed-loop pole p equals $-\sigma_p + j\omega_c$. Rewriting this expression, we obtain

$$\frac{P_1 - \bar{P}_1}{(p - P_1)(p - \bar{P}_1)} + \frac{P_2 - \bar{P}_2}{(p - P_2)(p - \bar{P}_2)} = 2j\omega_c \qquad [1\text{-}223]$$

This provides two equations for the open-loop roots (the real part of the left side equals zero, the imaginary part equals $2\omega_c$).

Finally, assuming the common situation that the open-loop poles P_1 and P_2 are negative real (i.e., generated by an RC network) and the open-loop zeros complex conjugate and therefore much closer to the closed-loop pole p than P_1 and P_2, we have

$$\frac{d\bar{P}_i}{p - \bar{P}_i} \gg \frac{dP_i}{p - P_i} \qquad (1\text{-}224\text{a})$$

$$\frac{\bar{P}_i}{p - \bar{P}_i} \gg \frac{P_i}{p - P_i} \qquad (1\text{-}224\text{b})$$

and with $m = n = 2$ from (1-220),

$$\frac{dp}{p} \approx \frac{d\bar{P}_1(p - \bar{P}_2) + d\bar{P}_2(p - \bar{P}_1)}{\bar{P}_1(p - \bar{P}_2) + \bar{P}_2(p - \bar{P}_2)} \qquad [1\text{-}225]$$

Since $\bar{P}_2 = (\bar{P}_1)^*$, $\bar{P}_1$ will be much closer to p than $\bar{P}_2$. Consequently,

$$\frac{dp}{p} \approx \frac{d\bar{P}_1}{\bar{P}_1} \qquad (1\text{-}226)$$

Thus, the closer $\bar{P}_1$ is to p, the more p will vary as $\bar{P}_1$ does. $\bar{P}_1, p$ becomes a dipole that shifts together. Clearly, the element x (e.g., the gain G) must be very large in this situation; if it is large enough and stable (the latter situation achieved by distributing the total gain among more than one active element, as will be discussed in Chapter 3) then the stability of the closed-loop pole p will depend primarily on the stability of the RC network generating $\bar{P}_1$ and $\bar{P}_2$. This topic was discussed elsewhere.[66]

66. See Chapter 8, Section 8.4, of *Linear Integrated Networks: Fundamentals*.

1.6 THE SENSITIVITY OF THIRD–ORDER NETWORKS

Whenever negative real poles occur in the pole-zero ensemble of an nth order function $T(s)$, it is generally desirable, from an economic standpoint, to combine them with complex conjugate pole pairs, thereby forming third-order networks. One or two negative real poles almost routinely occur in filters of reasonably high order (say higher than four). We shall therefore briefly present here the expressions for the pole sensitivity of third-order networks given by Soderstrand and Mitra.[67]

Consider the denominator of a third-order function,

$$D(s) = a_3 s^3 + a_2 s^2 + a_1 s + a_0 \qquad (1\text{-}227)$$

Assuming, for simplicity, that $a_3 = 1$ (a_3 can be included in the multiplicative constant of $T(s)$) we can write $D(s)$ in the form

$$D(s) = (s + \gamma)\left(s^2 + \frac{\omega_p}{q_p} s + \omega_p^2\right) \qquad (1\text{-}228)$$

$$= (s + \gamma)(s - p)(s - p^*)$$

Then

$$a_2 = \frac{\omega_p}{q_p} + \gamma \qquad (1\text{-}229\text{a})$$

$$a_1 = \omega_p^2 + \frac{\omega_p \gamma}{q_p} \qquad (1\text{-}229\text{b})$$

$$a_0 = \omega_p^2 \gamma \qquad (1\text{-}229\text{c})$$

The coefficient sensitivities can then be related to the sensitivities of ω_p, q_p, and γ as follows:

$$
\begin{bmatrix} S_x^{a_2} \\[2ex] S_x^{a_1} \\[2ex] S_x^{a_0} \end{bmatrix}
=
\begin{bmatrix}
\dfrac{1}{a_2}\dfrac{\omega_p}{q_p} & -\dfrac{1}{a_2}\dfrac{\omega_p}{q_p} & \dfrac{\gamma}{a_2} \\[2ex]
\dfrac{1}{a_1}\left[2\omega_p^2 + \dfrac{\omega_p \gamma}{q_p}\right] & -\dfrac{\gamma}{a_1}\dfrac{\omega_p}{q_p} & \dfrac{\gamma}{a_1}\dfrac{\omega_p}{q_p} \\[2ex]
\dfrac{2\gamma}{a_0}\omega_p^2 & 0 & \dfrac{\gamma}{a_0}\omega_p^2
\end{bmatrix}
\cdot
\begin{bmatrix} S_x^{\omega_p} \\[2ex] S_x^{q_p} \\[2ex] S_x^{\gamma} \end{bmatrix}
\qquad [1\text{-}230]
$$

67. M. A. Soderstrand and S. K. Mitra, Sensitivity analysis of third-order filters, *Int. J. Electron.* **30**, 265–272 (1971).

Solving for $S_x^{\omega_p}$, $S_x^{q_p}$ and S_x^{γ} as functions of $S_x^{a_2}$, $S_x^{a_1}$, and $S_x^{a_0}$, we obtain,

$$S_x^{\omega_p} = \frac{(q_p\gamma - \omega_p)a_0 S_x^{a_0} + q_p\omega_p^2 a_1 S_x^{a_1} - q_p\omega_p^2 a_2 S_x^{a_2}\gamma}{2\omega_p^2[q_p(\omega_p^2 + \gamma^2) - \omega_p\gamma]} \qquad [1\text{-}231a]$$

$$S_x^{q_p} = \frac{a_0(2\omega_p q_p^2 + \gamma q_p - \omega_p)S_x^{a_0}}{2\omega_p^2[(\omega_p^2 + \gamma^2)q_p - \omega_p\gamma]}$$
$$+ \frac{a_1(\omega_p^2 q_p - 2\omega_p\gamma q_p^2)S_x^{a_1} + a_2\omega_p^2(q_p\gamma - 2\omega_p q_p^2)S_x^{a_2}}{2\omega_p^2[(\omega_p^2 + \gamma^2)q_p - \omega_p\gamma]} \qquad [1\text{-}231b]$$

$$S_x^{\gamma} = \frac{q_p\left[\dfrac{a_0}{\gamma} S_x^{a_0} - a_1 S_x^{a_1} + a_2\gamma S_x^{a_2}\right]}{(\omega_p^2 + \gamma^2)q_p - \omega_p\gamma} \qquad [1\text{-}231c]$$

Consider, for example, the third-order network shown in Fig. 1-57. The transfer function is given as

$$T(s) = \frac{\beta G_1}{C_1 C_3(1 - \beta)} \frac{s(sC_3 + G_2 + G_3)}{s^3 + a_2 s^2 + a_1 s + a_0} \qquad (1\text{-}232)$$

where

$$a_2 = \frac{C_1 C_2(G_2 + G_3)(1 - \beta) + (C_1 + C_2)G_2 C_3 + C_2 G_1 C_3}{C_1 C_2 C_3(1 - \beta)} \qquad (1\text{-}233a)$$

$$a_1 = \frac{(C_1 + C_2)G_2 G_3(1 - \beta) + C_2 G_1(G_2 + G_3) + C_3 G_1 G_2}{C_1 C_2 C_3(1 - \beta)} \qquad (1\text{-}233b)$$

$$a_0 = \frac{G_1 G_2 G_3}{C_1 C_2 C_3} \qquad (1\text{-}233c)$$

With the normalized element values

$$G_2 = G_3 = C_1 = C_2 = 1$$
$$G_1 = C_3 = 2 \qquad (1\text{-}234)$$
$$\beta = -(4Q - 1)$$

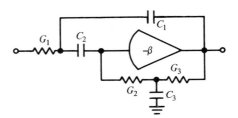

FIG. 1-57. Third-order network.

the transfer function simplifies to

$$T(s) = -\left(\frac{4Q-1}{2Q}\right) \frac{s(s+1)}{s^3 + \left(\frac{1}{Q}+1\right)s^2 + \left(\frac{1}{Q}+1\right)s + 1}$$

$$= -\left(\frac{4Q-1}{2Q}\right) \frac{s}{s^2 + \frac{1}{Q}s + 1} \tag{1-235}$$

Comparing with (1-228) we have $\omega_p = \gamma = 1$, $a_2 = a_1 = 1 + 1/Q$, and $a_0 = 1$. Substituting into (1-231) we therefore have

$$S_x^{\omega_p} = \frac{(Q-1)S_x^{a_0} + (Q+1)(S_x^{a_1} - S_x^{a_2})}{2(2Q-1)} \tag{1-236a}$$

$$S_x^{Q} = \frac{(Q+1)[S_x^{a_0} - S_x^{a_1} - S_x^{a_2}]}{2} \tag{1-236b}$$

$$S_x^{\gamma} = \frac{QS_x^{a_0} + (Q+1)(S_x^{a_2} - S_x^{a_1})}{(2Q-1)} \tag{1-236c}$$

In order to find these sensitivities with respect to the gain β of the amplifier we find from (1-233)

$$S_\beta^{a_0} = 0 \tag{1-237a}$$

$$S_\beta^{a_2} = S_\beta^{a_1} = -\frac{(4Q-1)}{4Q(Q+1)} \tag{1-237b}$$

and consequently with (1-236)

$$S_\beta^{\omega_p} = S_\beta^{\gamma} = 0 \tag{1-238a}$$

$$S_\beta^{Q} = 1 - \frac{1}{4Q} \tag{1-238b}$$

Notice that S_β^{Q} increases only from 0.5 to 1 as Q increases from 0.5 to infinity. However, the gain β increases in direct proportion to Q, as seen in (1-234). Consequently the *gain-sensitivity product* Γ is also proportional to Q, namely:

$$\Gamma = \beta S_\beta^{Q} = -\frac{(4Q-1)^2}{4Q} \approx -4Q \tag{1-239}$$

The significance of the gain-sensitivity product will be discussed in detail under Section 4.1.2 in chapter 4.

CHAPTER

2

THE SYNTHESIS OF SECOND-ORDER ACTIVE NETWORKS WITH SINGLE ACTIVE ELEMENTS

INTRODUCTION

There has been a vast amount of literature on active filter techniques in the last few years and a general classification which permits a distinction between the major concepts in active filter synthesis as compared to the many minor variations of these concepts can help significantly to avoid the rather confusing picture that otherwise results. In what follows we shall be more concerned with those general concepts than with the details of individual methods, in the belief that a thorough understanding of the fundamental underlying principles of active network design will provide a sound basis from which any specific applications, and variations thereof, can be easily derived and understood.

2.1 A FEEDBACK-ORIENTED NETWORK CLASSIFICATION

It has been shown by Horowitz and Branner[1] that the most common methods of active filter synthesis can be systematically classified, based on the type of polynomial decomposition from which they are derived. To demonstrate this let us consider the general signal-flow graph shown in Fig. 2-1. In this representation we have the feedback path t_{32} with respect to a forward branch t_{23} which alone contains a particular network parameter on which attention is to be focused. In the present case we assume the parameter of

1. I. M. Horowitz and G. R. Branner, A unified survey of active RC synthesis techniques, *Proc. Nat. Electron. Conf.*, **23**, 257–261 (1967).

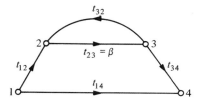

Fig. 2-1. General signal-flow graph.

interest, which shall be designated by β, to be the gain, transconductance, or
any other characteristic of the active device used in the network. If there is
more than one active device in the network, the devices can often be lumped
into a single element β, which again results in the representation of Fig. 2-1.
If the active elements do not act as a single collective unit, then the feedback
network $t_{32}(s)$ may be an *active* network whose transfer function has been
realized separately using a suitable active synthesis technique. In this case
a basic flow graph of the form shown in Fig. 2-1 applies to each active subnetwork whose active element cannot be included in the main collective
unit β. Either way, we can assume that the remaining branches t_{ij} represent
passive *RC* networks whose poles are negative real.

In order to realize a general network function $T(s) = N(s)/D(s)$, we must
now determine the branch gains $t_{ij}(s)$ of the general signal-flow graph such
that

$$T(s) = \frac{N(s)}{D(s)} = t_{14}(s) + \beta \frac{t_{12}(s)t_{34}(s)}{1 - \beta t_{32}(s)} \tag{2-1}$$

The functions

$$t_{ij}(s) = \frac{n_{ij}(s)}{d_{ij}(s)} \tag{2-2}$$

are the transfer functions of passive *RC* networks. As such, the roots of the
denominators $d_{ij}(s)$ must be on the negative real axis; the roots of the numerators $n_{ij}(s)$ may, theoretically, be located anywhere in the s-plane. In practice
the *RC* networks realizing the $t_{ij}(s)$ functions will generally have a common
ground (i.e., three-terminal networks), so that the location of the zeros of
$n_{ij}(s)$ will have to satisfy the Fialkow–Gerst conditions.[2] If the *RC* networks
are minimum-phase, the zeros of $n_{ij}(s)$ will be restricted to the left half of
the s-plane; however, this is not necessarily a requirement. The gain parameter β may be considered to be a positive or negative constant, depending
on whether positive or negative feedback is used. If β is not constant, i.e.,
if it is a function of s (due to parasitics or the like) then that part of it which
is s-dependent may be removed and absorbed in $t_{32}(s)$.

In deriving a network classification based on the general flow graph of
Fig. 2-1, the trend in the literature has been to consider a general classification

---

2. See Chapter 1, Section 1.5, of *Linear Integrated Networks: Fundamentals*.

first and to follow this with the simpler classification applicable to second-order networks. Since in practice the procedure of decomposing higher-order networks into second- or at most third-order networks has been almost universally adopted, the initial general classification does not seem to serve any useful purpose, and, indeed, tends to confuse what otherwise becomes a clear and concise system of basic active network types.

Addressing ourselves immediately to the problem of realizing general second-order networks, we find from (2-1) and (2-2) that the denominator $D(s)$ must be realizable in the form

$$D(s) = 1 \pm \beta t_{32}(s) = [d_{32}(s) \pm \beta n_{32}(s)] \frac{1}{d_{32}(s)} \tag{2-3}$$

The positive sign in (2-3) applies to negative feedback (i.e., negative β), the negative sign to positive feedback. The term $t_{32}(s)$ is assumed to be the transmission of a second-order passive RC network (since β is the parameter characterizing the only active element assumed in the network). Thus the denominator of $t_{32}(s)$ has the form

$$d_{32}(s) = (s - P_1)(s - P_2) = s^2 + \frac{\omega_0}{\hat{Q}} s + \omega_0^2 \tag{2-4}$$

where $\hat{Q} < 0.5$, i.e., the roots of $d_{32}(s)$ lie on the negative real axis.* The general form of the desired second-order function $T(s)$ has the form

$$T(s) = \frac{N(s)}{D(s)} = \frac{N(s)}{s^2 + \frac{\omega_p}{Q} s + \omega_p^2} \tag{2-5}$$

where $Q > 0.5$, i.e., the poles of $T(s)$ are assumed to be complex conjugate. Thus, *in purely analytical terms*, the fundamental problem of active network synthesis is *to decompose a second-order polynomial with complex roots into the sum or difference of two polynomials, one of which has negative real roots*, namely, using (2-3),

$$s^2 + \frac{\omega_p}{Q} s + \omega_p^2 = s^2 + \frac{\omega_0}{\hat{Q}} s + \omega_0^2 \pm \beta n_{32}(s) \tag{2-6}$$

The object of the decomposition is *to find the unknowns ω_0, $\hat{Q}$, β and $n_{32}(s)$ as a function of the given parameters ω_p and Q, with the constraint that $\hat{Q} < 0.5$*. Since this problem is under-defined, additional conditions for optimum performance, such as minimum sensitivity, can be taken into consideration.

---

* In this and subsequent chapters the overscript "$\wedge$" implies that the term in question characterizes a *passive* RC network.

In feedback terms, the active synthesis problem is one of selecting a feed-back transmission of the form

$$t_{32}(s) = \frac{n_{32}(s)}{d_{32}(s)} = \frac{(s - \bar{P})(s - \bar{P}')}{(s - P_1)(s - P_2)} = \frac{(s - \bar{P})(s - \bar{P}')}{s^2 + \dfrac{\omega_0}{\hat{Q}} s + \omega_0^2} \tag{2-7}$$

such that the roots of the return difference $1 \pm \beta t_{32}(s)$ *provide the desired poles of* $T(s)$ *for some value of* β. The poles of $t_{32}(s)$ must be on the negative real axis, while the zeros of $t_{32}(s)$ may be real or complex conjugate. In these terms a classification of synthesis methods depends on two factors: (i) the sign of β and (ii) the location of the zeros of $t_{32}(s)$.

The sign of β divides all single-element synthesis techniques into two basic classes: (1) $\beta < 0$ which corresponds to negative feedback, and (2) $\beta > 0$, which corresponds to positive feedback. The location of the zeros of $t_{32}(s)$ then identifies several subclasses under these two major divisions. As shown in what follows, these subclasses can readily be obtained if we relate the zeros and poles of $t_{32}(s)$ to the root loci of the corresponding feedback configurations.

It will be recalled from root-locus theory[3] that the poles and zeros of $t_{32}(s)$ are the open-loop poles and zeros of the root locus defined by the character-istic equation $\beta t_{32}(s) = \mp 1$ where -1 applies to negative feedback, $+1$ to positive feedback. With $t_{32}(s)$ given by the general form of (2-7) the open-loop poles are restricted to the negative real axis (see Fig. 2-2a). Thus, the remaining problem is to establish where the roots of $n_{32}(s)$ (of which there can be at most two) may be located for the positive and negative feedback cases, bearing in mind that the objective is to obtain root loci providing *complex* poles.

2.1.1 Network Classes Based on Negative Feedback

Considering negative feedback first ($\beta < 0$) we recall that the root locus may occupy a section of the real axis only when the total number of critical frequencies to the *right* of the section is *odd*.

Since we are interested in generating complex conjugate closed-loop poles, starting out the root locus from the negative real pole pair P_1 and P_2 (see Fig. 2-2a), the root locus must exist *between* P_1 and P_2 on the negative real axis. In this way the root locus will coalesce between P_1 and P_2 and then take off along a pair of complex conjugate branches, as implied in Fig. 2-2a. With a little thought it should become clear to the reader that the root locus would be tied to the negative real axis if it existed anywhere other than between P_1 and P_2 on the negative real axis. Thus, it follows that, when real, the two zeros $\bar{P}_i$, $\bar{P}_i'$ of our second-order function $t_{32}(s)$, which represent the open-

---

3. See Chapter 2, Section 2.1 of *Linear Integrated Networks: Fundamentals*.

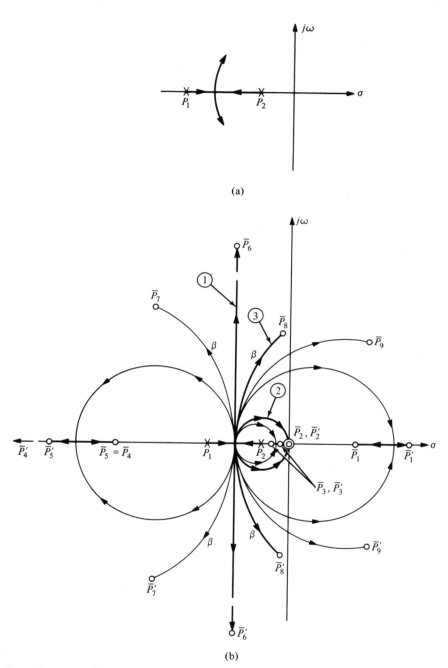

(a)

(b)

Fig. 2-2. Root loci for active *RC* networks based on negative feedback: (a) on the real axis, (b) in the *s*-plane.

loop zeros of our root locus, are not free to lie *anywhere* on the real axis, but are constrained as follows (see Fig. 2-2b):

1. They must lie outside the stretch P_1P_2.
2. When to the right of P_1P_2 they must occur as a finite *pair* (e.g., $\bar{P}_1, \bar{P}_1'$; $\bar{P}_2, \bar{P}_2'$; $\bar{P}_3, \bar{P}_3'$).
3. They may lie anywhere to the left of P_1P_2 including infinity (see $\bar{P}_4$, $\bar{P}_4'$; $\bar{P}_5, \bar{P}_5'$; $\bar{P}_6, \bar{P}_6'$).
4. When the open-loop zeros are complex conjugate they may be located anywhere in the s-plane (e.g., $\bar{P}_7, \bar{P}_7'$; $\bar{P}_8, \bar{P}_8'$; $\bar{P}_9, \bar{P}_9'$).

A superficial glance at Fig. 2-2b may lead one to the conclusion that there is a large variety of functions $t_{32}(s)$, (corresponding to the many possible locations of the open-loop zeros i.e., of the roots of $n_{32}(s)$), compatible with the negative-feedback realization of complex poles. However, on closer examination it becomes evident that the number of useful functions $t_{32}(s)$ is only very small. To begin with, any open-loop zeros to the left of P_1, i.e., those whose real coordinates are more negative than P_1 (e.g., $\bar{P}_4, \bar{P}_4'$; $\bar{P}_5$, $\bar{P}_5'$; $\bar{P}_7, \bar{P}_7'$ in Fig. 2-2b) are of no interest to us since, after reaching a modest maximum, the achievable pole Q will actually decrease with increasing β (e.g., gain). Except for very special applications, this is unlikely to be of any practical value and will not be considered further here. The first case for which the pole Q is guaranteed to increase monotonically with β occurs when $\bar{P}_i, \bar{P}_i'$ have moved infinitely far away from P_1 and P_2, and the circular root locus has degenerated into a vertical line as indicated by the locus 1 corresponding to $\bar{P}_6, \bar{P}_6'$ in Fig. 2-2b. This case has special significance since it represents the boundary between loci whose pole-Q does increase (to the right of 1) or does not increase (to the left of 1) monotonically with β; furthermore it corresponds to a function $t_{32}(s)$ that can be realized easily in practice. We shall therefore emphasize this significance by referring to any network whose root locus follows curve 1 as a *class 1 network*.

The Q's of the poles traversing the root loci to the right of 1 in Fig. 2-2b increase with β; strictly speaking the corresponding loci need not, therefore, be subdivided in any way. However, the network types realizing $t_{32}(s)$ functions with real zero pairs are different from those with complex conjugate zero pairs; hence it is practically, if not theoretically, meaningful to distinguish between the former (*class 2 networks*) and the latter (*class 3*). If, now, in addition to stipulating that the feedback function $t_{32}(s)$ be realized by a passive RC network, we add that the network should be unbalanced, then because of the Fialkow–Gerst conditions we must eliminate those forms of $t_{32}(s)$ having zeros on the positive real axis (e.g. $\bar{P}_1, \bar{P}_1'$ in Fig. 2-2b). Thus the $t_{32}(s)$ function for class 2 networks must have a zero pair at, or to the left of, the origin. These three classes, whose root loci all really belong to the same ensemble of curves, will now be discussed separately. We would do well to remember, however, that this division is somewhat artificial and that, in essence, if we consider the closed-loop poles generated along a root locus, the three negative-feedback classes really constitute one basic class.

Class 1 Networks   For this network class we have $\beta < 0$ and the two open-loop zeros of the feedback network located at infinity. This corresponds to

$$n_{32}(s) = \text{const} = k\omega_0^2 \qquad \text{[2-8a]}$$

and

$$t_{32}(s) = k \frac{\omega_0^2}{s^2 + \dfrac{\omega_0}{\hat{Q}} s + \omega_0^2} \qquad \text{[2-8b]}$$

where k is a scaling factor depending on the passive RC realization of $t_{32}(s)$. The root locus with respect to β corresponding to the open-loop poles and zeros determined by (2-8b) is shown in Fig. 2-3. To obtain the Q of the complex pole pair corresponding to this negative feedback scheme, we can write the polynomial decomposition corresponding to (2-6), namely,

$$s^2 + \frac{\omega_p}{Q} s + \omega_p^2 = s^2 + \frac{\omega_0}{\hat{Q}} s + \omega_0^2 + \beta k \omega_0^2 \qquad (2-9)$$

Solving for ω_p and Q we have

$$\omega_p = \omega_0 \sqrt{1 + k\beta} \qquad \text{[2-10a]}$$

and

$$Q = \hat{Q} \sqrt{1 + k\beta} \qquad \text{[2-10b]}$$

Thus complex poles are obtained by a multiplication of $\hat{Q}$ (which is less than 0.5) by a factor $\sqrt{1 + k\beta} > 2$. The gain β required for a specified Q is proportional to $(Q/\hat{Q})^2$.

Class 2 Networks   For this network class we have $\beta < 0$ and the two open-loop zeros of the feedback network are real and to the right of P_1 and P_2. Thus we have

$$n_{32}(s) = k(s - \bar{P}_1)(s - \bar{P}_2) \qquad \text{[2-11a]}$$

and

$$t_{32}(s) = k \frac{(s - \bar{P}_1)(s - \bar{P}_2)}{s^2 + \dfrac{\omega_0}{\hat{Q}} s + \omega_0^2} \qquad \text{[2-11b]}$$

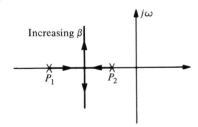

FIG. 2-3. Root locus for class 1 network.

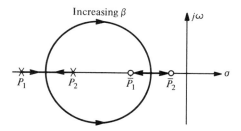

FIG. 2-4. Root locus for class 2 network.

where, referring to Fig. 2-2a, $\bar{P}_1$ and $\bar{P}_2$ lie to the right of P_1 and P_2. A typical root locus corresponding to this case is shown in Fig. 2-4. Proceeding in the same way as in the preceding case by considering the polynomial decomposition corresponding to (2-6), and solving for the equivalent frequency ω_p and Q we obtain

$$\omega_p = \omega_0 \sqrt{\frac{1 + (k\beta\bar{P}_1\bar{P}_2/\omega_0^2)}{1 + k\beta}} \qquad [2\text{-}12\text{a}]$$

and

$$Q = \hat{Q}\frac{\sqrt{[1 + (k\beta\bar{P}_1\bar{P}_2/\omega_0^2)](1 + k\beta)}}{1 - [k\beta\hat{Q}(\bar{P}_1 + \bar{P}_2)/\omega_0]} \qquad [2\text{-}12\text{b}]$$

$\bar{P}_1$ and $\bar{P}_2$ are assumed positive here, hence the negative sign in the denominator of (2-12b). Recall, though, that positive real zeros are unachievable with unbalanced networks (Fialkow–Gerst condition); thus $\bar{P}_1$ and $\bar{P}_2$ will generally be zero or negative and the denominator of (2-12b) will therefore be positive.

For the common case that the two zeros are at the origin, (2-12a) and (2-12b) simplify to

$$\omega_p\big|_{P_1 = P_2 = 0} = \frac{\omega_0}{\sqrt{1 + k\beta}} \qquad [2\text{-}13\text{a}]$$

and

$$Q\big|_{P_1 = P_2 = 0} = \hat{Q}\sqrt{1 + k\beta} \qquad [2\text{-}13\text{b}]$$

These equations show that complex poles are obtained here in essentially the same way as in the previous case, i.e., by a multiplication of Q by $\sqrt{1 + k\beta}$. Furthermore a comparison of (2-12b) and (2-13b) shows that nothing is to be gained by shifting the zeros $\bar{P}_1$ and $\bar{P}_2$ away from the origin since this results in more gain β required to obtain a given Q and with it a larger sensitivity of Q to β; (this becomes immediately obvious if only one zero is shifted from the origin).

Class 3 Networks   Here we have $\beta < 0$ and the open-loop zeros of the feedback network are complex. Thus we obtain:

$$n_{32}(s) = k\left(s^2 + \frac{\omega_z}{q_z}s + \omega_z^2\right)$$ [2-14a]

and

$$t_{32}(s) = k \cdot \frac{s^2 + \dfrac{\omega_z}{q_z}s + \omega_z^2}{s^2 + \dfrac{\omega_0}{\hat{Q}}s + \omega_0^2}$$ [2-14b]

where

$$q_z > 0.5$$ [2-15]

and, of course, $q_z > \hat{Q}$. The corresponding root locus is shown in Fig. 2-5.

The expressions for ω_p and Q derived as above are

$$\omega_p = \omega_0\sqrt{\frac{1 + k\beta(\omega_z/\omega_0)^2}{1 + k\beta}}$$ [2-16a]

and

$$Q = \hat{Q}\frac{\sqrt{[1 + k\beta(\omega_z/\omega_0)^2](1 + k\beta)}}{1 + k\beta(\hat{Q}/q_z)(\omega_z/\omega_0)}$$ [2-16b]

Note that for very large β values, ω_p approaches ω_0, which depends only on passive components. At the same time, from (2-16b) it follows that Q then equals q_z. Thus with increasing gain the poles of the active network are realizable by the complex zeros of the passive feedback network $t_{32}(s)$. This of course also follows from the root-locus diagram in Fig. 2-5.

It is useful to examine the last expression for Q more closely. If we let

$$v = \frac{\hat{Q}}{q_z}$$ (2-17)

then

$$Q = \hat{Q}\frac{\sqrt{[1 + k\beta(\omega_z/\omega_0)^2](1 + k\beta)}}{1 + k\beta v(\omega_z/\omega_0)}$$ [2-18]

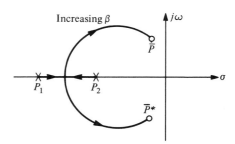

FIG. 2-5. Root locus for class 3 network.

We can now distinguish between three special cases in this class:

    a. $v > 0$: complex zeros in the LHP
    b. $v = 0$: complex zeros on the $j\omega$ axis
    c. $v < 0$: complex zeros in the RHP

Case a essentially corresponds to the decomposition occurring in class 2 networks. The value of β required for a given Q increases with v, i.e., with the distance the complex zeros are moved to the left of the $j\omega$ axis. Conversely, as the zeros are shifted onto the $j\omega$ axis ($v = 0$) and if, in addition we let $\omega_z = \omega_0$, we obtain

$$\omega_p\Big|_{\substack{\omega_z = \omega_0 \\ v = 0}} = \omega_0 \qquad (2\text{-}19a)$$

and

$$Q\Big|_{\substack{\omega_z = \omega_0 \\ v = 0}} = \hat{Q}(1 + k\beta) \qquad (2\text{-}19b)$$

Thus, by shifting the zeros onto the $j\omega$ axis, a linear instead of a quadratic relationship between β and Q is obtained; furthermore ω_p becomes independent of β. Shifting the zeros into the right half s-plane ($v < 0$), β becomes still smaller for a given Q. Assuming $\omega_z = \omega_0$ for convenience, we obtain for this case c

$$Q\Big|_{\substack{\omega_z = \omega_0 \\ v < 0}} = \hat{Q}\,\frac{(1 + k\beta)}{1 - k\beta v} \qquad (2\text{-}20)$$

In general we have

$$\beta\Big|_{\omega_z = \omega_0} = \frac{1}{k} \cdot \frac{(Q/\hat{Q}) - 1}{1 \mp v(Q/\hat{Q})} \qquad \begin{array}{l} (-)\colon \text{LHP}, v > 0 \\ (+)\colon \text{RHP}, v < 0 \end{array} \qquad (2\text{-}21)$$

This clearly shows how, for a given Q, the required value of β can be influenced by the location of the complex zeros in the s-plane. As we shall see later, there may be disadvantages in placing the zeros in the RHP in order to obtain smaller β values, due to an increase in pole sensitivity. This is evident from the fact that RHP zeros correspond to a degree of *positive* feedback in the network (depending on how large the negative v is chosen).

    As we shall see in Section 2-3, when discussing the realization of $t_{32}(s)$ by an RC network, in one aspect class 1 and 2 networks differ fundamentally from those of class 3. For the former, $t_{32}(s)$ must have zeros on the negative real axis including the origin and infinity; it can therefore be realized by a ladder network. For the latter $t_{32}(s)$ must have complex zeros; this can only be accomplished by a bridged or a parallel ladder network. On the other hand the three network classes have one thing in common, namely that Q values larger than 0.5 (i.e., complex poles) are obtained by Q multiplication, i.e., multiplication of the Q pertaining to the basic RC feedback network $t_{32}(s)$ (designated $\hat{Q}$), by the factor β or its square root. β is obtained from the active element in the network. We shall see now, that with *positive* feedback the complex poles are obtained in a fundamentally different way.

2.1.2 Network Classes Based on Positive Feedback

The root locus of a positive-feedback system ($\beta > 0$) may occupy portions of the real axis lying to the *left* of any *even* number of critical frequencies and occupies the entire portion of the real axis to the right of the rightmost critical frequency. Let us start out again with the two negative real poles shown in Fig. 2-2a while bearing in mind that we are again interested only in generating complex poles. It then follows, as in the negative-feedback case, that the root locus must lie *between* P_1 and P_2 on the negative real axis; only then can a coalescence point, and consequently a pair of symmetrical root-locus branches in the s-plane, be obtained. These branches then provide the desired complex conjugate pole pair. In order for the root-locus to exist between P_1 and P_2, the location of the zeros of $n_{32}(s)$ is restricted as follows:

1. One real zero must lie to the right of P_1 and P_2.
2. The second zero must lie to the left of P_1 and P_2, including at infinity.

Clearly these restrictions prohibit the existence of a complex zero pair and permit only one basic pole–zero configuration which, together with a typical corresponding root locus, is shown in Fig. 2-6a. In general the zero $\bar{P}_1$ is at infinity, which does not change the nature of the root locus in any way (see Fig. 2-6b). Thus, we have only *one* basic class of networks in the positive-feedback category; the special cases with $\bar{P}_1$ at infinity and/or $\bar{P}_2$ at the origin represent variations that neither add any significant feature to the resulting root loci nor require basically different network types to realize the corresponding $t_{32}(s)$ functions. The reason for this is that one or both of the zeros of the $t_{32}(s)$ functions *must* lie on the negative real axis, and we recall that RC ladder networks are well qualified for this purpose.

Class 4 Networks   For this network class we have $\beta > 0$ while one open-loop zero of the feedback network lies to the right, the other to the left of the open-loop poles P_1 and P_2. Thus we can write

$$n_{32}(s) = k(s - \bar{P}_1)(s - \bar{P}_2) \qquad [2\text{-}22a]$$

and

$$t_{32}(s) = k \frac{(s - \bar{P}_1)(s - \bar{P}_2)}{s^2 + \dfrac{\omega_0}{\bar{Q}} s + \omega_0^2} \qquad [2\text{-}22b]$$

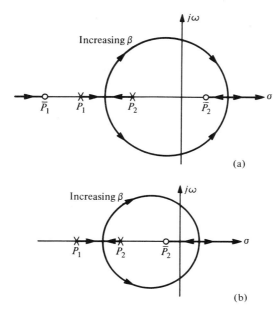

FIG. 2-6. Root loci for class 4 networks; (a) $\bar{P}_1$ finite; (b) $\bar{P}_1$ at infinity.

where, referring to Fig. 2-6a, $\bar{P}_1$ and $\bar{P}_2$ lie on either side of the stretch P_1–P_2 on the real axis. As a consequence, $\bar{P}_2$ may be either positive or negative, but $\bar{P}_1$, on the other hand, may only be negative, since it must lie left of P_1 and P_2. However, since in general $t_{32}(s)$ is an unbalanced network, $\bar{P}_2$ may not be positive; it will then be restricted to the negative real axis, namely to the right of the stretch P_1–P_2 up to, and including, the origin.

To obtain ω_p and Q the polynomial decomposition corresponding to (2-6) becomes

$$s^2 + \frac{\omega_p}{Q}s + \omega_p^2 = s^2 + \frac{\omega_0}{Q}s + \omega_0^2 - k\beta(s - \bar{P}_1)(s - \bar{P}_2) \qquad (2\text{-}23)$$

Solving for ω_p and Q we obtain

$$\omega_p = \omega_0 \sqrt{\frac{1 - k\beta(\bar{P}_1\bar{P}_2/\omega_0^2)}{1 - k\beta}} \qquad (2\text{-}24a)$$

and

$$Q = \hat{Q} \frac{\sqrt{[1 - k\beta(\bar{P}_1\bar{P}_2/\omega_0^2)](1 - k\beta)}}{1 + k\beta\hat{Q}\dfrac{\bar{P}_1 + \bar{P}_2}{\omega_0}} \qquad (2\text{-}24b)$$

Notice that the expressions are the same as those obtained for general class 2 networks (2-12) except that the polarity of $k\beta$ is here reversed. As in that case, $\bar{P}_1$ and $\bar{P}_2$ have been assumed positive, hence the positive sign in the denominator of (2-24b). However, since $\bar{P}_1$ will always be negative and $\bar{P}_2$ either negative or zero (assuming an unbalanced network for $t_{32}(s)$), the denominator of (2-24b) will be less than unity. The signs due to the term $\bar{P}_1\bar{P}_2$ in the numerator of (2-24a) and (2-24b) will of course remain unchanged if no positive real $\bar{P}_2$ values are permitted. With this assumption (i.e., unbalanced networks for $t_{32}(s)$, therefore $\bar{P}_1$ and $\bar{P}_2$ negative real) we can rewrite (2-22b) in the form

$$t_{32}(s) = k \, \frac{s^2 + \dfrac{\omega_z}{q_z} s + \omega_z^2}{s^2 + \dfrac{\omega_0}{\hat{Q}} s + \omega_0^2} \tag{2-25a}$$

where, because of the location of $\bar{P}_1$ and $\bar{P}_2$ with respect to P_1 and P_2, we have

$$q_z < \hat{Q} \tag{2-25b}$$

Thus, in contrast to class 3 networks, where $t_{32}(s)$ must be a frequency-rejection function (see (2-14b)), here $t_{32}(s)$ must be realized by a frequency-emphasizing network.

Let us now consider the three special cases that follow directly from (2-22b).

a. $\bar{P}_1$ At Infinity Since k must here have the dimension of radians per second we shall use the designation ω_k instead of k in (2-22b), and obtain

$$t_{32}(s) = \omega_k \, \frac{s - \bar{P}_2}{s^2 + \dfrac{\omega_0}{\hat{Q}} s + \omega_0^2} \tag{2-26}$$

From (2-23) it then follows that

$$\omega_p = \omega_0 \sqrt{1 + \beta \frac{\omega_k \bar{P}_2}{\omega_0^2}} \tag{2-27a}$$

and

$$Q = \hat{Q} \, \frac{\sqrt{1 + \beta(\omega_k \bar{P}_2/\omega_0^2)}}{1 - \beta(\omega_k/\omega_0)\hat{Q}} \tag{2-27b}$$

The signs in the numerators of (2-27a) and (2-27b) are positive when $\bar{P}_2$ is in the RHP, negative when it is in the LHP. Since the former situation is unlikely (unbalanced network) it follows that Q will be decreased by the expression under the square root, and that, in order for Q to remain real,

$$\beta < \frac{\omega_0^2}{\omega_k \bar{P}_2} \tag{2-27c}$$

Furthermore, for stability we require the $Q > 0$, so that

$$0 < \beta < \frac{\omega_0/\omega_k}{\hat{Q}} \tag{2-27d}$$

where the more stringent of the two conditions is decisive. Because of the reduction in Q, however, this case is of little practical significance.

b. $\bar{P}_2$ At Origin   For this case we have

$$t_{32}(s) = k \frac{s(s - \bar{P}_1)}{s^2 + \dfrac{\omega_0}{\hat{Q}} s + \omega_0^2}$$ (2-28)

From (2-24) we then obtain

$$\omega_p = \frac{\omega_0}{\sqrt{1 - k\beta}}$$ [2-29a]

and

$$Q = \hat{Q} \frac{\sqrt{1 - k\beta}}{1 + k\beta\hat{Q}(\bar{P}_1/\omega_0)}$$ [2-29b]

Since $\bar{P}_1$ must be negative, it follows from (2-29b) that for $Q > 0$ we require that

$$\beta < \frac{\omega_0}{\bar{P}_1} \cdot \frac{1}{k\hat{Q}}$$ (2-30a)

and, to guarantee Q remaining real, we require that

$$\beta < \frac{1}{k}$$ (2-30b)

Here again, whichever of the two conditions is most stringent will be the decisive one. Either way, because of the difference in the numerator of (2-29b), this case is also of little practical value.

c. $\bar{P}_1$ At Infinity; $\bar{P}_2$ At Origin   For this case we have, from (2-26),

$$t_{32}(s) = \omega_k \cdot \frac{s}{s^2 + \dfrac{\omega_0}{\hat{Q}} s + \omega_0^2}$$ (2-31)

Thus the circular root locus of Fig. 2-6 is centered at the origin. From (2-27) it then follows that

$$\omega_p = \omega_0$$ [2-32a]

and

$$Q = \frac{\hat{Q}}{1 - \beta \dfrac{\omega_k}{\omega_0} \hat{Q}}$$ [2-32b]

This case provides the simplest expressions and, in fact, is the most important of the three special cases considered here. The reasons for this are numerous. For one thing $t_{32}(s)$ is realizable by a simple bandpass network; for another the frequency ω_p is independent of β here, which is generally a desirable feature. (Note that of the negative-feedback networks only class 3 networks can also provide this; see (2-19a).) Furthermore, the Q sensitivity to β is smallest here since, in this case alone, β does not appear in the numerator of the expression for Q and there is no additional contribution from the numerator to S_β^Q. Notice also that only in case c is Q proportional to $\hat{Q}$ directly, without still being reduced by a term smaller than unity (see square roots in the numerators of (2-27b) and (2-29b) and remember that $\bar{P}_2$ is negative). It is therefore not surprising that cases a and b are rarely, if ever, used. For this reason we shall, in what follows, refer to networks belonging to class 4c simply as class 4 networks, unless otherwise stated.

Equations (2-27b) and (2-32b) show the fundamental method by which complex poles (i.e., $Q > 0.5$) are obtained with positive feedback, i.e., by dividing $\hat{Q}$ of the passive RC network by the difference between two relatively large (compared to the resulting difference) quantities. In order to guarantee stability (LHP poles) the gain β is limited in both cases by

$$0 < \beta < \frac{\omega_0/\omega_k}{\hat{Q}} \tag{2-33}$$

The upper limit of β is generally in the order of one to three.

2.1.3 Summary

In summary, it has been shown above that there exist only four distinct classes to be considered when examining the realization of a pair of complex poles using a single active element (or a single effective collection of active elements) in combination with a passive RC network. The four classes are summarized in Table 2-1. Only the cases in each class with open-loop zeros at the origin or infinity are given in the table (with the exception of class 3), because, as was mentioned above, they correspond to the minimum-sensitivity configurations and are therefore most commonly used.

Although some methods of single-active-element synthesis *can* produce more than one pair of complex poles, experience and theoretical considerations have shown the practical limit per active device to be a single pole pair with at most an additional real pole. All higher-order networks are thus realized by a cascade of second- or third-order active networks using synthesis procedures based on any one of the four classes discussed above or on multi-amplifier building blocks, as discussed in Chapter 3. In the

TABLE 2-1. POLYNOMIAL DECOMPOSITION FOR COMPLEX POLES

Class	Decomposition	Q; ω_p	Feedback function $t_{32}(s)$	Root locus
		Negative feedback ($\beta < 0$)		
1	$\left(s^2 + \dfrac{\omega_0}{Q}s + \omega_0^2\right) + \beta k\omega_0^2$	$Q\sqrt{1+k\beta}$; $\omega_0\sqrt{1+k\beta}$	$k\,\dfrac{\omega_0^2}{s^2 + \dfrac{\omega_0}{Q}s + \omega_0^2}$ [Low-pass function]	
2	$\left(s^2 + \dfrac{\omega_0}{Q}s + \omega_0^2\right) + \beta k s^2$	$Q\sqrt{1+k\beta}$; $\omega_0/\sqrt{1+k\beta}$	$k\,\dfrac{s^2}{s^2 + \dfrac{\omega_0}{Q}s + \omega_0^2}$ [High-pass function]	
3	$\left(s^2 + \dfrac{\omega_0}{Q}s + \omega^2\right) +$ $\beta k\left(s^2 + \dfrac{\omega_z}{q_z}s + \omega_z^2\right)$	$Q\dfrac{\{[1 + k\beta(\omega_z/\omega_0)^2](1+k\beta)\}^{1/2}}{1 + k\beta\dfrac{Q}{q_z}\cdot\dfrac{\omega_z}{\omega_0}}$; $\omega_0\left[\dfrac{1 + k\beta\left(\dfrac{\omega_z}{\omega_0}\right)^2}{1 + k\beta}\right]^{1/2}$	$k\,\dfrac{s^2 + \dfrac{\omega_z}{q_z}s + \omega_z^2}{s^2 + \dfrac{\omega_0}{Q}s + \omega_0^2}$ [Band-rejection function]	
		Positive feedback ($\beta > 0$)		
4	$\left(s^2 + \dfrac{\omega_0}{Q}s + \omega_0^2\right) - \beta\omega_k s$	$\dfrac{Q}{1 - Q\dfrac{\omega_k}{\omega_0}\beta}$; ω_0	$\omega_k\,\dfrac{s}{s^2 + \dfrac{\omega_0}{Q}s + \omega_0^2}$ [Bandpass function]	

next section it will be demonstrated how various synthesis techniques which are representative of the most common network-design methods using single active elements belong to one or the other of the four network classes described above.

2.2 SOME REPRESENTATIVE SYNTHESIS METHODS

Although the first active RC filter in the form of a frequency-selective amplifier goes back to the late 1930s[4], the present widespread interest was triggered almost twenty years later by Linvill, Sallen, and Key.[5] The many schemes that have been suggested for active filter design since then are based on the use of four basically different active circuits or devices:

1. The negative impedance converter (NIC).
2. The gyrator.
3. The operational amplifier (OpAmp).
4. The phase-locked loop (PLL).

It would hardly be possible, nor would it serve any useful purpose, to discuss the large number of synthesis methods using these circuits. Instead, it will be demonstrated in what follows that typical synthesis methods using the first three circuit types fall into one of the network classes discussed in the previous section. Only the PLL is based on other than linear feedback concepts;[6] thus it is excluded from our network classification. For this reason, and because of its rather restricted applicability as a frequency-selective network the PLL will not be dealt with here.

2.2.1 NIC Synthesis

The Cascaded NIC Structure   Consider two passive RC two-ports N_A and N_B connected in cascade as shown in Fig. 2-7. The open circuit transfer impedance $\hat{Z}_{21}$ of the two networks in cascade is given by

$$\hat{Z}_{21} = \frac{V_2}{I_1}\bigg|_{I_2=0} = \frac{z_{21A}z_{21B}}{z_{22A}+z_{11B}} \qquad [2\text{-}34]$$

where z_{21A}, z_{21B} are the transfer impedances, z_{22A} and z_{11B} the driving point impedances of networks N_A and N_B, respectively. The poles of $\hat{Z}_{21}$ are here

4. H. H. Scott, A new type of selective circuit and some applications, *Proc. IRE*, **26**, 226–235 (1938).
5. J. G. Linvill, *RC* active filters, *Proc. IRE*, **42**, 555–564 (1954); R. P. Sallen and E. L. Key, A practical method of designing *RC* active filters, *IRE Trans. Circuit Theory* **CT-2**, 78–85 (1955).
6. G. S. Moschytz, Miniaturized *RC* filters using the phase-locked loop, *Bell Systems Tech. J.* **44**, 823–870 (1965).

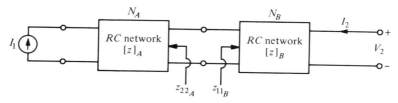

FIG. 2-7. Cascade connection of two RC networks.

determined by the zeros of the sum of the two RC driving point impedances z_{22A} and z_{11B}. These two impedances can be expressed as the ratio of two rational polynominals:

$$z_{22A} = \frac{n_{22A}(s)}{d_{22A}(s)} \tag{2-35a}$$

and

$$z_{11B} = \frac{n_{11B}(s)}{d_{11B}(s)} \tag{2-35b}$$

To satisfy the conditions required for RC driving point impedances the polynomials $n_{22A}(s)$, $d_{22A}(s)$, $n_{11B}(s)$, and $d_{11B}(s)$ must have simple negative real roots (corresponding to interlacing poles and zeros) and the degree of the denominators $d_{22A}(s)$ and $d_{11B}(s)$ must be greater than or equal to the degree of the corresponding numerators (see Chapter 1, Section 1.3.1 of *Linear Integrated Networks: Fundamentals.*)

The poles of the network consisting of the two RC networks N_A and N_B in cascade will, of course, be negative real, just as are those of the two individual RC networks. However, it can be shown that *complex* poles can be obtained by taking the *difference* of z_{22A} and z_{11B} in the denominator of $\hat{Z}_{21}$ instead of the sum. This fact is based on properties of the so-called (RC): $(-RC)$ *decomposition* which states that under certain conditions the ratio of two polynomials $P(s)$ and $Q(s)$ can be realized by the difference of two functions $\hat{Z}_A(s)$ and $\hat{Z}_B(s)$ possessing RC driving point impedance characteristics[7]. More explicitly, if the roots of $Q(s)$ have only distinct negative real roots and the degree of $Q(s)$ is greater than or equal to the degree of $P(s)$, then we can write:

$$\frac{P(s)}{Q(s)} = \hat{Z}_A(s) - \hat{Z}_B(s) \tag{2-36}$$

$P(s)$ may be an arbitrary polynomial (with complex roots) and $\hat{Z}_A(s)$ and $\hat{Z}_B(s)$ are realizable as RC driving point impedances.[8]

7. S. K. Mitra, *Analysis and Synthesis of Linear Active Networks*, (New York: John Wiley and Sons, 1969), pp. 98-102.
8. An equivalent statement applies with respect to two RC driving point admittances $\hat{Y}_A(s)$ and $\hat{Y}_B(s)$: $P(s)/Q(s) = \hat{Y}_A(s) - \hat{Y}_B(s)$. In this case the degree of $P(s)$ may be less than or equal to one plus the degree of $Q(s)$.

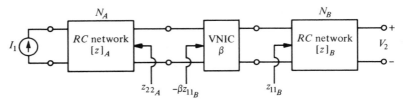

FIG. 2-8. Two *RC* networks and a VNIC in cascade.

Interconnecting a voltage-inversion-type negative impedance converter (VNIC) with a conversion factor β between the two networks of Fig. 2-7, as shown in Fig. 2-8, a difference of driving point impedances in the form of (2-36) is obtained in the denominator of the corresponding transfer impedance:

$$Z_{21} = \frac{V_2}{I_1}\bigg|_{I_2=0} = \frac{z_{21A} z_{21B}}{z_{22A} - \beta z_{11B}} \qquad [2\text{-}37]$$

Thus, in order to obtain a specified transfer impedance $Z_{21}(s)$ with complex poles, the *RC* networks N_A and N_B must be realized, using well known *RC* synthesis techniques, to provide the required transfer and driving point impedances required by (2-37). The two networks are then interconnected as shown in Fig. 2-8 with a VNIC providing the required conversion factor β. For simplicity β is very often chosen equal to unity.

Rewriting (2-37) as follows:

$$Z_{21} = \frac{Z(s)}{1 - \beta(z_{11B}/z_{22A})} \qquad [2\text{-}38]$$

where $Z(s) = z_{21A} z_{21B}/z_{22A}$, it becomes clear that network synthesis based on the NIC configuration shown in Fig. 2-8 belongs to the class 4, i.e., positive-feedback, discussed in the previous section. Indeed, since the poles and zeros of z_{11B}/z_{22A} are simple and negative, the equivalent feedback network $t_{32}(s)$ is realizable by an *RC ladder* network (whose pole–zero locations, it will be recalled, are restricted in the same way). Thus an equivalent *feedback* representation of the NIC configuration in Fig. 2-8 is that shown in Fig. 2-9. It consists of a positive amplifier in the forward path and

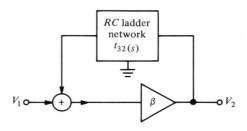

FIG. 2-9. Feedback representation of NIC circuit of Fig. 2-8.

a grounded RC ladder in the feedback network. The corresponding voltage transfer function is

$$T(s) = \frac{V_2}{V_1} = \frac{N(s)}{D(s)} = \frac{\beta}{1 - \beta t_{32}(s)} \tag{2-39}$$

Naturally, added circuitry must be included to obtain the desired numerator $Z(s)$. This can be done in a variety of ways, as will be dealt with later in the discussion on network synthesis using feedback amplifiers.

Linvill first described the NIC synthesis method discussed above in 1954 and, as pointed out before, it was his paper, together with that of Sallen and Key a year later, that started the ensuing widespread interest in linear active networks. Interestingly enough the paper by Sallen and Key covered a positive-feedback scheme with an RC-ladder feedback network just as illustrated in Fig. 2-9, except that there it was used to characterize the NIC method suggested by Linvill. *Thus these two classical papers described active networks belonging to class 4.*

The Parallel NIC Structure   Numerous other methods of NIC network synthesis have been suggested since Linvill's first paper. One that was considered to be particularly useful is due to Yanagisawa.[9] In essence, this method is closely related to Linvill's. It is based on the *parallel* connection of two passive RC networks N_A and N_B, as shown in Fig. 2-10. The corresponding voltage-transfer ratio is given by

$$\hat{T}(s) = \frac{V_2}{V_1}\bigg|_{I_2 = 0} = -\frac{y_{21AB}}{y_{22AB}} = -\frac{y_{21A} + y_{21B}}{y_{22A} + y_{22B}} \tag{2-40}$$

The Yanagisawa NIC configuration is a result of the same reasoning as was used in Linvill's case. Equation (2-40) characterizes a passive RC network and therefore $\hat{T}(s)$ possesses only negative real poles. The $(RC):(-RC)$

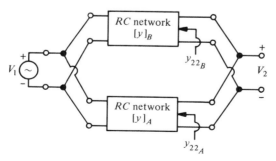

FIG. 2-10. Parallel connection of two RC networks.

9. T. Yanagisawa, RC active networks using current inversion type negative impedance converters, *IRE Trans. Circuit Theory*, **CT-4**, 140–144 (1957).

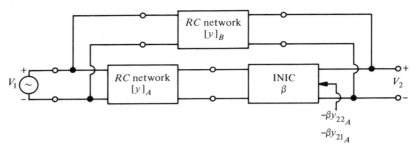

FIG. 2-11. Parallel connection of two *RC* networks and an INIC.

decomposition (this time the *RC* driving point *admittance* parameter version) implies that the *difference* of driving point admittances in the denominator of (2-40) instead of the sum, can provide complex poles. This difference of admittance parameters can be obtained by inserting a *current*-inversion-type negative-impedance converter (INIC) with a conversion ratio β into the configuration of Fig. 2-10 in the manner shown in Fig. 2-11. The corresponding voltage transfer function $T(s)$ is given by

$$T(s) = \frac{V_2}{V_1}\bigg|_{I_2 = 0} = -\frac{y_{21B} - \beta y_{21A}}{y_{22B} - \beta y_{22A}} \qquad [2\text{-}41]$$

By realizing networks N_A and N_B by known *RC* synthesis techniques to obtain the required transfer and driving point admittances, any real rational voltage transfer function can be obtained. Note that, contrary to Linvill's technique, the conversion ratio β appears in the numerator as well as in the denominator of the transfer function. Thus both the poles *and* the zeros of $T(s)$ depend on the active element of the network. In terms of sensitivity then, Yanagisawa's is a type II network, whereas Linvill's is type I. It will be recalled that the latter provides greater freedom with regard to pole–zero pairing of *n*th-order networks than the former. Another difference is that Linvill's method uses balanced passive *RC* structures such as bridged-*T*'s or twin-*T*'s to provide complex zeros; Yanagisawa's method produces complex zeros in the same way that both methods use to produce complex poles, i.e., by taking the difference between admittances, one of which is multiplied by the active conversion factor β. In terms of hybrid integrated circuits, zeros produced by highly stable (e.g., thin-film) passive *RC* structures are generally preferable to those depending on active-device parameters, since the former can be made more stable with respect to aging, temperature, and humidity effects than the latter. This is not true when conventional discrete *RC* components are used, in which case Yanagisawa's method is preferable.

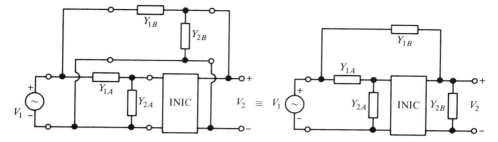

FIG. 2-12. NIC network with driving point admittances.

Inspection of (2-41) shows that Yanagisawa's method of synthesis also belongs to the positive-feedback type of network (class 4), since $T(s)$ can be written in the form

$$T(s) = \frac{V_2}{V_1}\bigg|_{I_2=0} = \frac{F(s)}{1 - \beta(y_{22A}/y_{22B})} \tag{2-42}$$

Consequently the equivalent positive-feedback amplifier scheme shown in Fig. 2-9 also applies here.

One very attractive aspect of Yanagisawa's NIC synthesis method is that by using inverted L sections for networks N_A and N_B, only driving point admittances need be synthesized. Referring to the resulting configuration shown in Fig. 2-12 the voltage-transfer function simplifies to

$$T(s) = \frac{V_2}{V_1}\bigg|_{I_2=0} = \frac{Y_{1B} - \beta Y_{1A}}{(Y_{1B} - \beta Y_{1A}) + (Y_{2B} - \beta Y_{2A})} \tag{2-43}$$

This expression only involves driving point admittance parameters. Furthermore the configuration in Fig. 2-12 is independent of loading effects on the two networks N_A and N_B, since these are simply considered a part of Y_{2A} and Y_{2B}.

It was mainly on synthesis techniques using the NIC that the early literature on active networks concentrated, because this method lent itself particularly well to elegant theoretical solutions to the active-filter synthesis problem. It was not until practical implementations of NIC filter designs were actually attempted that their many inherent problems emerged. Not the least of these was the realization of a stable NIC with a well controlled conversion factor β. As a result, filter design using the NIC gradually receded in the mid-1960s and has not been actively pursued since. It remains of significant academic interest, though, because it demonstrates some of the basic ideas in active-filter synthesis very clearly. Also, since it was for some time generally considered to be the most useful method, various aspects of NIC filter design, including optimization with respect to maximum stability and minimum sensitivity, were studied and reported on. Many of the ideas and concepts that evolved in conjunction with these studies remain useful and may be extended to other schemes of active-filter design.

Because of its virtual obsolesence today, nothing more will be said about NIC filter synthesis and design in this text. This is mainly because, after having introduced the basic concepts and tools of active filter design, we are attempting to concentrate on those methods that either have already proven themselves, both practically and economically, or that show promise of doing so in the near future. The main purpose of the preceding discussion on NIC filter design was to demonstrate that the NIC filter belongs to the positive-feedback class of networks and can be replaced by a positive-feedback amplifier with an *RC* ladder network in the feedback loop. Interestingly enough, the latter network type has an important advantage over the NIC filter when in hybrid integrated form, i.e., that high-quality, low-cost silicon monolithic integrated operational amplifiers have been successfully mass produced and are readily available on the market place. Perhaps if the device industry had put as much effort into the design of NICs as it did into the design of operational amplifiers, NIC filter design would today still be actively pursued.

2.2.2 Gyrator Synthesis

The Cascaded Gyrator Structure   It was shown in the previous section that NIC synthesis of active networks is based on properties of the (RC): $(-RC)$ polynomial decomposition. In this section we shall show that gyrator synthesis is based on properties of the (RC): (RL) decomposition.[10] This decomposition states that under certain conditions the ratio of two polynomials $P(s)$ and $Q(s)$ can be realized by the *sum* of two functions $\hat{Z}_A(s)$ and $\mathscr{Z}_B(s)$ which satisfy RC and RL driving point impedance conditions, respectively. Thus

$$\frac{P(s)}{Q(s)} = \hat{Z}_A(s) + \mathscr{Z}_B(s) \qquad [2\text{-}44]$$

Like the poles of $\hat{Z}_A(s)$ and $\mathscr{Z}_B(s)$, $Q(s)$ may have only distinct, negative real roots and the degree of $Q(s)$ must be equal to or less by one than the degree of $P(s)$. $P(s)$ may be an arbitrary polynomial with complex roots and may therefore represent the denominator of a general network function. Consequently a general network with complex poles can be synthesized by cas-

10. D. A. Calahan, Sensitivity minimization in active *RC* synthesis, *IEEE Trans. Circuit Theory*, **CT-9**, 38–42 (1962).

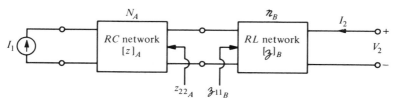

FIG. 2-13. Cascade connection of RC and RL networks.

cading an RC network N_A with an RL network $\mathcal{N}_B$ as shown in Fig. 2-13. The corresponding open circuit transfer impedance Z_{21} is given by

$$Z_{21} = \frac{V_2}{I_1}\bigg|_{I_2=0} = \frac{z_{21A}\,z_{21B}}{z_{22A} + z_{11B}} \qquad [2\text{-}45]$$

In contrast to (2-34), which characterizes a passive RC network, the transfer impedance (2-45) characterizes an RLC network. Its denominator, decomposed in the manner given by (2-44) into the sum of an RC and an RL driving point impedance, possesses complex poles.

Whereas the realization of a general transfer function by an RC and an RL network in cascade presents no difficulties, it clearly does not satisfy the objective we have set ourselves, namely of synthesizing general linear networks *without* using inductors. We shall now see, however, that once the denominator of the desired network function is given in the $(RC):(RL)$ decomposed form, only a simple step remains in order to meet our stated objective.

It can be shown,[11] that the driving point *impedance* function of an RL network has identical properties to the driving point *admittance* of its dual, which is an RC network. Consider for example the RL impedance in Fig. 2-14a and its dual in Fig. 2-14b. The impedance of the RL network is

$$\mathscr{Z}_{RL} = R + sL \qquad (2\text{-}46a)$$

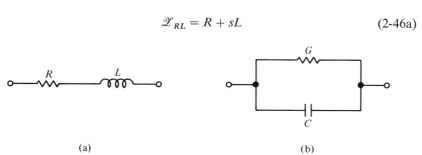

(a)

(b)

FIG. 2-14. RL one-port (a) and its dual (b).

11. See, for example, Chapter 1, Section 1.3.3, of *Linear Integrated Networks: Fundamentals*.

and the admittance of its dual is

$$Y_{RC} = \frac{1}{Z_{RC}} = G + sC \tag{2-46b}$$

Clearly the properties of (2-46a) are identical to those of (2-46b) and we can find a scaling impedance r such that

$$\mathcal{Z}_{RL} = r^2 Y_{RC} \tag{2-47}$$

Consequently in order to convert the RLC network of Fig. 2-13 into an RC-gyrator network we must take the dual of the RL network $\mathcal{N}_B$; this is an RC network and will be designated N_B. Then we must provide for an inversion of the input driving point impedance z_{11B} of N_B to obtain an RC driving point admittance y_B such that (for no load, i.e., $I_2 = 0$) $r^2 y_B = r^2/z_{11B}$ $= x_{11B}$. This inversion can be achieved with a gyrator whose gyration impedance is r. The resulting RC–gyrator equivalent of the RLC network in Fig. 2-13 is shown in Fig. 2-15. The overall open-circuit transfer impedance of this network can be obtained directly as follows. Referring to Fig. 2-15 we have

$$V_2 \Big|_{I_2 = 0} = -z_{21B} I_B \tag{2-48a}$$

and, with the open-circuit impedance parameters of an ideal gyrator (i.e., $z_{11} = z_{22} = 0$, $z_{12} = -r$, $z_{21} = r$),

$$I_B = -\frac{V_A}{r} \tag{2-48b}$$

Since the load impedance of network N_A is $r^2/z_{11B} = r^2 y_B$ for the open-circuit case considered here (i.e., $I_2 = 0$), we have

$$\frac{V_A}{I_1} \Big|_{I_2 = 0} = \frac{z_{21A} r^2 y_B}{z_{22A} + r^2 y_B} \tag{2-48c}$$

Combining (2-48a), (2-48b), and (2-48c) the overall open-circuit transfer impedance of the network in Fig. 2-15 becomes

$$Z_{21} = \frac{V_2}{I_1} \Big|_{I_2 = 0} = r y_B \cdot \frac{z_{21A} z_{21B}}{z_{22A} + r^2 y_B} \tag{2-49}$$

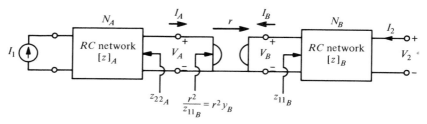

FIG. 2-15. Gyrator in cascade with two RC networks

Since $y_B = 1/z_{11B}$ is the driving point admittance of an RC network, the denominator of (2-49) is of the same basic form as that of the RLC network characterized by (2-45).

The synthesis method described above, in which two RC networks separated by a gyrator are in cascade, was described by Shenoi.[12] A modification thereof, using a reactive gyrator (i.e., $z_{11} = z_{22} = 0$, $z_{12} = -r(s)$, $z_{21} = r(s)$) was published shortly thereafter.[13] Topologically this method is closely akin to Linvill's NIC synthesis method (see Fig. 2-8). The actual synthesis procedure is also similar, in that networks N_A and N_B must be synthesized to provide appropriate driving point and transfer parameters so as to realize a desired transfer function. Here as there, the synthesis of networks N_A and N_B involves known RC synthesis procedures that will not be dwelt on here. The basic difference between the two methods, apart from the former using an NIC and the latter a gyrator, is in the class of feedback networks to which the two methods belong.

To find the equivalent feedback network of the gyrator configuration in Fig. 2-15, we can rewrite (2-49) in the form

$$Z_{21} = \frac{Z(s)}{1 + r^2(y_B/z_{22A})} \qquad [2\text{-}50]$$

where $Z(s) = r y_B z_{21A} z_{21B}/z_{22A}$. Since r^2 characterizes the active element of the network, namely the gyrator, and y_B/z_{22A} the two passive RC networks N_A and N_B in Fig. 2-15, it is evident that this method of gyrator synthesis belongs to one of the *negative*-feedback classes listed in Table 2-1. Being the driving point admittance and impedance, respectively, of an RC network, y_B and z_{22A} must have simple and negative real poles and zeros. Consequently this method may belong to either class 1 or class 2, but not to class 3. An equivalent feedback configuration may have the form shown in Fig. 2-16. It consists of a negative amplifier in the forward path and a grounded ladder in the feedback path. The overall voltage-transfer function is

$$T(s) = \frac{V_2}{V_1} = \frac{-\beta}{1 + \beta t_{32}(s)} \qquad (2\text{-}51)$$

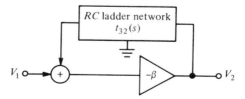

FIG. 2-16. Feedback representation of gyrator circuit of Fig. 2-15.

12. B. A. Shenoi, Practical realization of a gyrator circuit and RC–gyrator filters, *IEEE Trans. Circuit Theory*, **CT-12**, 374–380. (1965).
13. S. K. Mitra and W. G. Howard, Jr., The reactive gyrator—A new concept and its application to active network synthesis, *IEEE Intern. Conv. Rec.*, **14** (Part 7) 319–326 (1966).

where $\beta t_{32}(s) = r^2 y_B/z_{22A}$. Clearly the negative-feedback configuration only accounts for the mechanism by which complex poles are obtained. Additional circuitry is required to obtain specified zeros. This may involve any one of a number of passive RC synthesis techniques, some of which will be described later.

Parallel Gyrator Structure   In the same way that a topological analogy to Linvill's cascaded NIC structure exists in gyrator synthesis, there is also a gyrator analogy to the parallel NIC structure of Yanagisawa (see Fig. 2-11). As might be expected, this has the form shown in Fig. 2-17. As with the parallel NIC configuration of Fig. 2-11 it is most convenient to characterize networks N_A and N_B by their short-circuit admittance parameters and the gyrator by its gyration conductance g. With the admittance parameters of the ideal gyrator (i.e., $y_{11} = y_{22} = 0$, $y_{12} = g$, $y_{21} = -g$) and of N_A and N_B, and following a similar derivation to that used in the cascade gyrator structure, we obtain the voltage transfer function

$$T(s) = \left.\frac{V_2}{V_1}\right|_{I_2 = 0} = -\frac{y_{21B} + gy_{21A}/y_{22A}}{y_{22B} + g^2/y_{22A}} \qquad [2\text{-}52]$$

Since N_A and N_B are RC networks, g^2/y_{22A} is an RL driving point admittance function so that here again we have an $(RC):(RL)$ decomposition (the admittance version of (2-44)) in the denominator of $T(s)$, which, as we know, is capable of realizing complex poles. As in the corresponding NIC configuration, the zeros as well as the poles of $T(s)$ here depend on the active network parameter g, so that this is a type II network with respect to sensitivity. As with the parallel NIC case, this configuration has the advantage of being realizable with only driving point admittances when N_A and N_B have the form of inverted L sections, as shown in Fig. 2-18. The corresponding voltage transfer function results as follows:

$$T(s) = \left.\frac{V_2}{V_1}\right|_{I_2 = 0} = \frac{Y_{1B} + [gY_{1A}/(Y_{1A} + Y_{2A})]}{Y_{1B} + Y_{2B} + [g^2/(Y_{1A} + Y_{2A})]} \qquad (2\text{-}53)$$

Here $T(s)$ is determined only by driving point admittances; this simplifies the synthesis of N_A and N_B significantly.

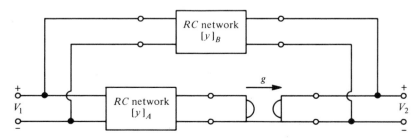

FIG. 2-17. Parallel connection of two RC networks and a gyrator.

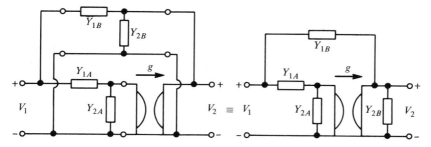

FIG. 2-18. Gyrator network with driving point admittances.

The parallel gyrator configuration of Fig. 2-17 was proposed by Holmes[14] and a modified version thereof using an ideal *negative* gyrator (i.e., $y_{11} = y_{22} = 0$, $y_{12} = y_{21} = g$) some time later.[15] Inspection of the corresponding transfer functions shows that the parallel gyrator structure also belongs to the negative-feedback network group (i.e. either class 1 or class 2) and may therefore also be represented by a negative-feedback configuration of the kind shown in Fig. 2-16. Indeed it is this similarity between the gyrator synthesis methods discussed here and those methods using simple negative-feedback amplifier structures that may well account for the present lack of interest in the former; the negative-feedback amplifier required for the basic equivalent structure of the kind shown in Fig. 2-16 is readily available as an inexpensive operational amplifier in the form of a silicon integrated chip. (As pointed out earlier, similar reasoning may also be partially responsible for the decline of interest in NIC synthesis methods.) Since the resulting circuits may be expected to perform as well as their gyrator counterparts, there has been little incentive, from this point of view at least, to develop viable silicon integrated gyrators. The fact that the gyrator has not completely disappeared from the active-filter scene in the same way that the NIC has, is due not to its use in the synthesis methods discussed above. It is due primarily to the usefulness of the gyrator in simulating inductors in conventional passive *RLC* filter designs, as was discussed in Chapter 1, and also to the ease with which it can be realized by operational amplifier pairs.[16]

2.3 OPERATIONAL AMPLIFIER SYNTHESIS

In the foregoing discussion on NIC and gyrator synthesis methods, it was possible to show that each method could be reduced to a simple, equivalent feedback configuration. However, NIC and gyrator synthesis methods were not conceived as feedback networks, and their originators did not usually

14. W. H. Holmes, A new method of gyrator–RC filter synthesis, *Proc. IEEE* (Corresp.) **54**, 1459–1460 (1966).
15. V. Ramachandran and M. N. S. Swamy, Realization of an arbitrary voltage transfer function using an ideal negative gyrator, *Proc. 11th Mid-West Symp. Circuit Theory*, May, 1968.
16. See, for example, Chapter 5, Section 5.4.3, of *Linear Integrated Networks: Fundamentals*.

point out this analogy, since it has little significance other than for the purpose of network classification and, as we shall see later, network sensitivity.

The situation is very different with synthesis methods using operational amplifiers. These devices are used almost exclusively as negative or positive (i.e. inverting or noninverting) feedback amplifiers and it is to be expected that their use in active network design can be derived directly from our preceding feedback-oriented classification. Indeed, we shall see in what follows that each feedback class listed in Table 2-1 leads directly to a basic operational-amplifier synthesis approach that may not originally have been, but certainly could have been, derived in this way. Thus, rather than describing various synthesis methods and then tracing them back to their generic feedback class (as we did for NIC and gyrator methods) we shall here proceed in the reverse order, starting out with the network class and deriving one of its typical exponents.[17] It is hoped that the reader will then be capable on his own of so classifying any other single-amplifier network he may encounter, and that he may thereby better understand the basic principle upon which it is founded.

In essence, any feedback-amplifier realization of the decompositions listed in Table 2-1 will have the basic form shown in Fig. 2-19. The amplifier gain β is positive or negative depending on whether a positive or negative feedback class is being considered. Defining the following voltage-transfer functions:

$$t_{12} = \frac{V_2}{V_1}\bigg|_{V_3 = 0} \qquad [2\text{-}54]$$

and

$$t_{32} = \frac{V_2}{V_3}\bigg|_{V_1 = 0} \qquad [2\text{-}55]$$

we obtain for the overall voltage transfer function of the network of Fig. 2-19:

$$T(s) = \frac{V_3}{V_1} = \pm\,\beta\,\frac{t_{12}}{1 \mp \beta t_{32}} \qquad [2\text{-}56]$$

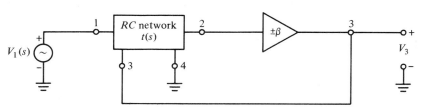

FIG. 2-19. Feedback representation of general active RC network using one operational amplifier.

17. A similar approach was taken by S. K. Mitra, Filter design using integrated operational amplifiers, pp. 4-1, WESCON, San Francisco, California, August 1969.

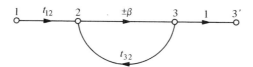

FIG. 2-20. Signal-flow graph representation of Fig. 2-19.

Notice that t_{12} corresponds to the forward transfer function of $T(s)$ and t_{32} to its feedback function. The network of Fig. 2-19 can therefore be represented by the familiar signal-flow graph shown in Fig. 2-20. As will be explained shortly, t_{12} determines the zeros of $T(s)$, β and t_{32} its poles. The type of network used for t_{32} will depend on the class of network decided upon,[18] whereas t_{12} will vary according to the specified requirements. In many cases, the desired zeros defined by t_{12} can be obtained by entering the RC network $t(s)$ with a driving source as shown in Fig. 2-19 (i.e., $V_1(s)$). In so doing, care must be taken not to alter the nature of the established t_{32} function, as seen within the feedback loop. Either a voltage or current source can be used as the driving, or input source. However, more than one second-order network will frequently be connected in cascade, in which case the driving source should have the same characteristics as the output characteristics of an operational amplifier. Since these are practically equivalent to those of an ideal voltage source, only this kind of driving source will be considered.

2.3.1 Class 1 Networks: Low–Pass Network in Negative–Feedback Loop (LPNF)

From Table 2-1 we find the required feedback function $t_{32}(s)$ for a class 1 network to be that of a second-order low-pass RC network (see, for example, Table 1-3). Therefore the corresponding negative feedback configuration must have the form shown in Fig. 2-21a. The conductance G_3 is included in order to account for a finite input impedance of the amplifier. Entry into the network from a driving voltage source can be made from various points without altering the basic low-pass nature of the feedback network $t_{32}(s)$. If, for example, a bandpass network is required (one zero at the origin) entry can be made in series with C_1, as shown in Fig. 2-21b. If a low-pass network is required entry can be made via an additional conductance $\bar{G}_1$ in parallel with C_1, as shown in Fig. 2-21c. The latter configuration has been described in the literature.[19] Notice that in this case the functions t_{12} and t_{32}

18. Not infrequently, publications appear describing "new" active network topologies that contain nothing more than realizations of the more general $t_{32}(s)$ functions described in Section 2.1. Thus where we are, in general, limiting ourselves to the $t_{32}(s)$ functions listed in Table 2-1, a newly published circuit might realize the function $t_{32}(s) = (s - \bar{P}_1)(s - \bar{P}_2)/d_{32}(s)$ instead of the class 2 function $t_{32}(s) = s^2/d_{32}(s)$ to which we shall limit ourselves in that class. The differences between the two do not seem significant enough to warrant separate treatment and have been discussed sufficiently in general terms in Section 2.1.
19. P. L. Taylor, Flexible design method for active RC two-ports, *Proc. IEE* (London), **110**, 1607–1616 (1963).

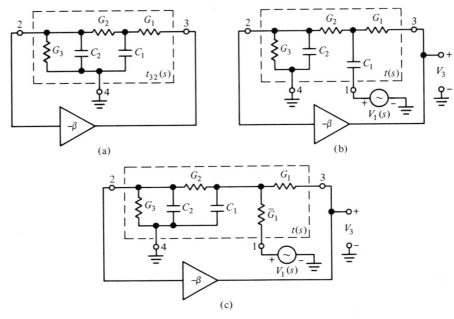

FIG. 2-21. Class 1 networks; (a) basic form; (b) and (c) with driving voltage source.

as defined by (2-54) and (2-55) are both low-pass functions, whereas in Fig. 2-21b, t_{12} has bandpass character. Calculating t_{12} and t_{32} for Fig. 2-21c and substituting the resulting functions into (2-56) we obtain the desired low-pass function

$$T(s) = \frac{K}{s^2 + \dfrac{\omega_p}{Q} s + \omega_p^2} \tag{2-57}$$

where

$$\omega_p = \left[\frac{G_2(\bar{G}_1 + G_1) + G_3(\bar{G}_1 + G_1 + G_2)}{C_1 C_2} \right]^{1/2}$$
$$\cdot \left[1 + \beta \cdot \frac{G_1 G_2}{G_2(\bar{G}_1 + G_1) + G_3(\bar{G}_1 + G_1 + G_2)} \right]^{1/2} \tag{2-58a}$$

$$Q = \frac{[C_1 C_2 \{G_2(\bar{G}_1 + G_1) + G_3(\bar{G}_1 + G_1 + G_2)\}]^{1/2}}{C_2(\bar{G}_1 + G_1 + G_2) + C_1(G_2 + G_3)}$$
$$\cdot \left[1 + \beta \frac{G_1 G_2}{G_2(\bar{G}_1 + G_1) + G_3(\bar{G}_1 + G_1 + G_2)} \right]^{1/2} \tag{2-58b}$$

and

$$K = -\beta \frac{\bar{G}_1 G_2}{C_1 C_2} \tag{2-58c}$$

As is to be expected the expressions for ω_p and Q in (2-58a) and (2-58b) are identical in form to those of (2-10a) and (2-10b), respectively. Furthermore, by comparing coefficients we can immediately obtain $t_{32}(s)$ in the form of (2-8b).

It is clear from the expressions in (2-58) that the number of unknown circuit elements exceeds the number of given coefficients, of which there are three, namely K, ω_p, and Q. In the absence of any additional constraints,[20] we can therefore select the additional elements for convenience or circuit simplicity and proceed with the design. A design example is given in Table 2-2, where, for a normalized frequency $\omega_p = 1$, the components are:[21]

$$\bar{G}_1 = G_1 = G_2 = G_3 = 1$$
$$C_1 = C_2 = 5Q \qquad (2\text{-}59)$$
$$\beta = 25Q^2 - 5$$

The relevance of the Q sensitivity and gain–sensitivity product, which are also given in this table will be explained in Chapter 4. Note that, as already pointed out in Section 2.1, the required gain is proportional to Q^2. As we shall see, this limits class 1 networks to low-pole-Q applications.

2.3.2 Class 2 Networks: High–Pass Network in Negative Feedback Loop (HPNF)

By inspection of Table 2-1 we see that the basic difference between this and the previous network class is that $t_{32}(s)$ here requires two zeros at the origin instead of at infinity. Thus $t_{32}(s)$ is a second-order high-pass network in the negative-feedback loop, as shown in Fig. 2-22a. Class 2 networks are therefore dual, in the $(RC):(CR)$ sense, to class 1 networks.

We proceed here in precisely the same way as in the previous case, by inserting the driving source $V_1(s)$ into $t(s)$ without changing the high-pass nature of $t_{32}(s)$ in the feedback loop. Thus, a high-pass network can be obtained by adding capacitor $\bar{C}_1$ as shown in Fig. 2-22b. This results in both $t_{32}(s)$ and $t_{12}(s)$ being high-pass functions by analogy to the case in Fig. 2-21c, where both were low-pass. Obtaining a bandpass characteristic for $t_{12}(s)$ (and therefore for $T(s)$) without interfering with the high-pass nature of $t_{32}(s)$ is particularly simple, as shown in Fig. 2-22c, where the driving source is connected in series with G_1. This circuit has the transfer function[22]

$$T(s) = K \cdot \frac{s}{s^2 + \dfrac{\omega_p}{Q}s + \omega_p^2} \qquad (2\text{-}60)$$

20. Such constraints, optimizing the circuit in some way (e.g., for a particular technology, for use in a particular application, for minimum component range, for maximum production yields, for minimum sensitivity, etc.), will be discussed in later chapters.
21. The first three examples in Table 2-2 are taken from S. K. Mitra, op. cit.
22. R. P. Sallen and E. L. Key, op. cit.

TABLE 2-2. FOUR REPRESENTATIVE SINGLE-AMPLIFIER, SECOND-ORDER FILTER NETWORKS

Class		Example	β	S_β^Q	Gain-sensitivity product Γ
1. Low-pass network in negative-feedback loop	Low-pass filter		$25Q^2 - 5$	$\dfrac{\beta}{2(5+\beta)} < \dfrac{1}{2}$	$\dfrac{(25Q^2 - 5)^2}{50Q^2} \approx 12Q^2 - 3$
2. High-pass network in negative-feedback loop	Bandpass filter		$9Q^2 - 1$	$\dfrac{\beta}{2(1+\beta)} < \dfrac{1}{2}$	$\dfrac{(9Q^2 - 1)^2}{18Q^2} \approx 4Q^2 - 1$
3. Band-rejection network in negative-feedback loop	Low-pass filter		A	$\dfrac{3Q^2}{A}$	$3Q^2$
4. Bandpass network in positive feedback loop	Bandpass filter		$4.02 - \dfrac{2.01}{Q}$	$2.01Q - 1$	$8.1Q + 2/Q - 8$

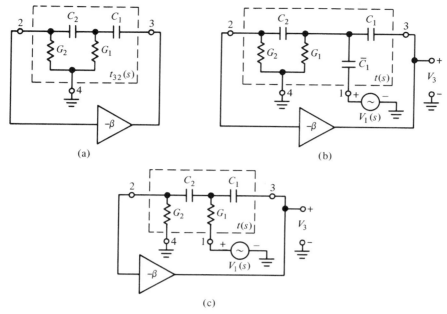

FIG. 2-22. Class 2 networks; (a) basic form; (b) and (c) with driving voltage source.

where

$$\omega_p = \frac{\sqrt{G_1 G_2 / C_1 C_2}}{\sqrt{1 + \beta}} \tag{2-61a}$$

$$Q = \frac{\sqrt{G_1 G_2 C_1 C_2}}{G_1 C_2 + G_2 C_1 + G_2 C_2} \sqrt{1 + \beta} \tag{2-61b}$$

and

$$K = -\frac{\beta}{1 + \beta} \cdot \frac{G_1}{C_1} \tag{2-61c}$$

A comparison of (2-61a) and (2-61b) with their equivalents in (2-13a) and (2-13b) provides the coefficients of $t_{32}(s)$ in terms of (2-11b) for the case that $\bar{P}_1 = \bar{P}_2 = 0$. As a design example of this circuit, consider the case for which

$$G_1 = G_2 = 1$$

$$C_1 = C_2 = \frac{1}{3Q} \tag{2-62}$$

$$\beta = 9Q^2 - 1$$

The corresponding circuit[23] along with its Q sensitivity and gain–sensitivity product is shown in Table 2-2. Here again the required gain β is proportional to Q^2, which also limits this circuit to low-Q applications.

23. P. R. Geffe, *RC*-amplifier resonators for active filters, *IEEE Trans. Circuit Theory*, **CT-15**, 415–419 (1968).

2.3.3 Class 3 Networks : Frequency–Rejection Network in Negative–Feedback Loop (FRNF)

From Table 2-1 we see that the t_{32}-function in this class of networks requires complex conjugate zeros whose Q is higher than the Q of the poles (i.e., $q_z > \hat{Q}$). We recall from Table 1-3 that such a function defines a frequency rejection network and furthermore* that complex zeros cannot be obtained with RC ladder networks but (restricting ourselves to three-terminal networks) only with cascaded bridged-T structures (Dasher's method) or with parallel ladder structures (Guillemin's method). Since we require only a single-zero pair, the former method results in a single bridge-T, the latter in a twin-T network.

A Bridged-T Feedback Network   Let us first consider the more frequently used bridged-T network for the realization of $t_{32}(s)$. We then obtain the two dual[24] feedback configurations shown in Fig. 2-23. There are numerous possibilities of entering either t_{32} or t'_{32} without changing the nature of either feedback loop. By entering in series with G_2 in Fig. 2-23a or with C'_1 in Fig. 2-23b the two dual bandpass networks shown in Fig. 2-24a and b are obtained. The conductances $\bar{G}_2$ and $\bar{G}'_1$ are sometimes added to modify the peak gain of the frequency response. By entering t_{32} with a capacitor $\bar{C}_2$ in parallel to G_2, or t'_{32} with a conductance $\bar{G}'_2$ in parallel with C'_2, a high-pass or a low-pass network is obtained, respectively, as shown in Figs. 2-24c and d.

Straightforward analysis of any one of the networks in Fig. 2-24 verifies that the frequency characteristics become increasingly independent of the

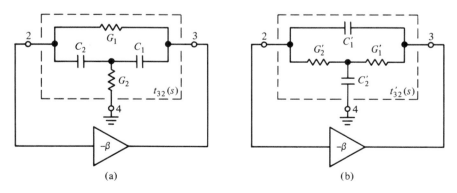

FIG. 2-23. Basic forms of class 3 networks using bridged-T.

* See, for example, Section 1.5.2, Chapter 1 of *Linear Integrated Networks: Fundamentals*.
24. Duality is here to be understood in the sense of the $(RC):(CR)$ transformation; see, for example, Chapter 3, Section 3.4, of *Linear Integrated Networks: Fundamentals*.

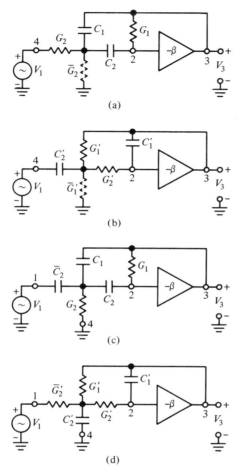

FIG. 2-24. Class 3 networks with bridged-T in feedback loop driven by voltage source from various entry points; (a) and (b) bandpass filters; (c) high-pass filter; (d) low-pass filter.

amplifier gain β (thereby remaining dependent only on the passive circuit components) the larger β gets. This indicates that the operational amplifier used for β should be operated in the open-loop mode. If the open-loop gain is sufficiently large, the open-loop zeros (which are the zeros of the bridged-T network) become the closed-loop poles of $T(s)$.

Let us, for example, examine the low-pass network in Fig. 2-24d.[25] Its

25. This network has been described in various forms by more than one author, e.g., M. H. Nichols and L. Rausch, *Radio Telemetry* (New York: John Wiley & Sons, Inc., 1956) p. 396 ("Rausch filter"); R. Brennan and A. Bridgman, Simulation of transfer functions using only one operational amplifier, *IRE WESCON Record*, Vol. 1, pp. 273–277, 1957; E. J. Foster Active low-pass filter design, *IEEE Trans. Audio.*, **AU-13**, 104–111 (1965).

transfer function must be the same as (2-57). For very large β values the coefficients of (2-57) are found to be:

$$\omega_p = \sqrt{G_1' G_2' / C_1' C_2'} \tag{2-63a}$$

$$Q = \frac{\sqrt{G_1' G_2' \, C_2' / C_1'}}{G_1' + G_2' + \bar{G}_2'} \tag{2-63b}$$

and

$$K = -\frac{G_2' \, \bar{G}_2'}{C_1' \, C_2'} \tag{2-63c}$$

Notice that all three coefficients are independent of β. Thus, given sufficiently stable passive components, the frequency characteristics of these networks are theoretically very stable. In practice, however, β is limited, giving rise to error terms in the three coefficients of (2-63a), (2-63b), and (2-63c), which in turn depend on β, and limit Q to medium values. Let us now assume the following design values:

$$G_1' = G_2' = \bar{G}_2' = 1$$

$$C_1' = \frac{1}{3Q}, \qquad C_2' = 3Q \tag{2-64}$$

$$\beta = A$$

where A is the open-loop gain of the operational amplifier. This gain will be considered large enough to permit the approximations leading to (2-63a), (2-63b), and (2-63c). The resulting circuit, Q sensitivity, and gain–sensitivity product are shown in Table 2-2.

The basic bridged-T networks in Fig. 2-23 provide circuits whose Q is given by (2-20) except that $v > 0$ and therefore the sign in the denominator is positive. In order to obtain $v \le 0$, a twin-T network must be used in the feedback loop and the same methods of inserting the input voltage source $V_1(s)$, as those shown in Fig. 2-24, may then be carried out. Thus, one of the advantages of a twin-T is that the required gain for a given Q may be reduced by letting $v \le 0$ (see (2-21)). As discussed later (Section 2.4.1), by using a differential amplifier and adding positive feedback to the existing negative feedback, the same effect is obtained with a bridged-T, as if a twin-T with RHP zeros (i.e., $v < 0$) were used.

A Twin-T Feedback Network   Let us now examine the class 3 network with a twin-T in the feedback loop in more detail. We recall[26] that the general twin-T can be designed to have a complex zero pair on the $j\omega$ axis or on either

26. See Chapter 8, Section 8.12 of *Linear Integrated Networks: Fundamentals.*

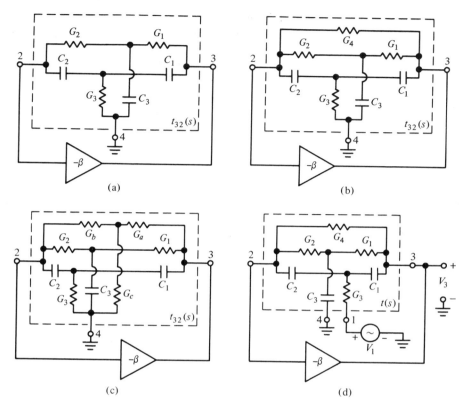

FIG. 2-25. Class 3 networks with twin-T in feedback loop: (a)-(c) various basic forms; (d) with driving source.

side of it, while maintaining cancellation of the third (negative real) zero and pole. Thus we can easily generate the three cases $v \gtrless 0$ discussed in Section 2.1.1 with a general twin-T. Whether RHP zeros are tolerable or not depends on the circuit. Clearly, if the amplifier in Fig. 2-25a is operated with open-loop gain in order to simulate infinite gain, then the closed-loop poles lie very close to (and under ideal conditions coincide with) the open-loop zeros, in which case RHP or even $j\omega$-axis twin-T zeros must be avoided. If on the other hand, the amplifier gain has a well controlled and stable medium value, then RHP zeros may be advantageous, since they provide a degree of positive feedback (see (2-21)) which, if judiciously chosen, may actually improve the overall sensitivity of the network. The latter situation will be discussed in more detail in conjunction with FEN design in Chapter 6.

Even with high-gain amplifiers there is no objection, in principle, to using a twin-T for $t_{32}(s)$ to provide RHP open-loop zeros, as long as the gain can be maintained stable enough to guarantee LHP closed-loop poles. If this is not the case, then open-loop zeros in the RHP or on the $j\omega$ axis can be avoided by bridging the twin-T with a finite conductance G_4, as shown in Fig. 2-25b. As a rule, G_4 will be very small, i.e., the corresponding resistor R_4 very large. If the size of R_4 becomes too large for thin- or thick-film implementation, then a resistive T network using sufficiently low resistor values can be used to replace R_4, as shown in Fig. 2-25c. In terms of the resistive T,

$$R_4 = \frac{1}{G_4} = R_a + R_b + \frac{R_a R_b}{R_c} \qquad (2\text{-}65)$$

This can be varied arbitrarily[27] by varying the ratio $R_a R_b/R_c$. Signal source insertion is carried out here in the same way as before. Thus the nature of the twin-T feedback network is not changed at all in the configuration shown in Fig. 2-25d, in which the signal source is connected in series with G_3 in order to obtain a bandpass network.

Bridged-T vs. Twin-T in Class 3 Networks   It may be appropriate, at this point, to ask ourselves what difference it makes whether we use a bridge-T as in Fig. 2-24 or a twin-T as in Fig. 2-25 to realize the complex conjugate open-loop zero pair required for class 3 operation. Clearly, the bridged-T is a simpler circuit while the twin-T affords more versatility in zero placement (i.e., $j\omega$ axis and RHP zeros). The difference goes deeper than that, however, as we shall see if we take another look at the bridged-T with a view to class 3 operation. Consider the bridged-T used as the feedback network $t_{32}(s)$ in Fig. 2-23a. Letting $R_1 = 1/G_1$ and $R_2 = 1/G_2$, we find

$$t_{32}(s) = \frac{s^2 + \dfrac{\omega_N}{q_z} s + \omega_N^2}{s^2 + \dfrac{\omega_N}{\hat{Q}} s + \omega_N^2} \qquad (2\text{-}66)$$

where

$$\omega_N = \frac{1}{\sqrt{R_1 R_2 C_1 C_2}} \qquad (2\text{-}67a)$$

$$q_z = \frac{\sqrt{C_1/C_2}}{1 + C_1/C_2} \sqrt{R_1/R_2} \qquad (2\text{-}67b)$$

$$\hat{Q} = \frac{\sqrt{R_1/R_2} \sqrt{C_1/C_2}}{(R_1/R_2) + (C_1/C_2) + 1} \qquad (2\text{-}67c)$$

27. Actually, as discussed in Chapter 3, Section 3.1.1, the selection of the resistors R_a, R_b, and R_c is not arbitrary; most important, the loading of the network (G_{34} of the equivalent P_t network) on node 3 must be considered. Furthermore, if β is finite the transfer function is affected, and if $\beta = A(s)$ the loop gain change may have to be considered.

Comparing (2-66) with (2-14b) we find that with the bridged-T, $\omega_z = \omega_0 = \omega_N$ and $k = 1$. Thus, from (2-16b) we obtain the pole Q of the active feedback network as

$$Q = \hat{Q} \frac{1 + \beta}{1 + \dfrac{\beta \hat{Q}}{q_z}} = \frac{1 + \beta}{\dfrac{1}{\hat{Q}} + \dfrac{\beta}{q_z}} \qquad [2\text{-}68]$$

To obtain a high pole Q, it follows from the second expression in (2-68) that both $\hat{Q}$ and q_z should be as large as possible. In the limit, for $q_z = \infty$ (i.e., $j\omega$-axis zeros) and $\hat{Q} = 0.5$ (i.e., double poles on the negative real axis) we obtain $Q_{max} = (1 + \beta)/2$. Let us see how close to this maximum we can actually come with the bridged-T.

Considering q_z first, it is easy to see from (2-67b) that for a large q_z we must let $R_1 \gg R_2$. Letting

$$r = \sqrt{R_1/R_2} \qquad (2\text{-}69a)$$

and

$$c = \sqrt{C_1/C_2} \qquad (2\text{-}69b)$$

we have

$$q_z = \frac{rc}{1 + c^2} \qquad (2\text{-}70)$$

This expression is proportional to c for small c values, proportional to $1/c$ for large c values. Taking the derivative with respect to c we find that $q_{z\,max}$ occurs for $C_1 = C_2$ thus:

$$q_{z\,max} = q_z\big|_{C_1 = C_2} = \tfrac{1}{2}\sqrt{R_1/R_2} \qquad (2\text{-}71)$$

It is immediately clear from (2-71), where one of the problems will be when using the bridged-T. When used in the "infinite-gain" mode of class 3 networks, in which the closed-loop poles coincide with the open-loop zeros, a high-Q closed-loop pole pair requires an open-loop zero pair with the same high q_z (since, ideally $q_z = Q$). Since $R_1/R_2 = 4q_z^2$, we thereby impose a significant strain on any hybrid integrated circuit implementation since, even for a pole Q of 20, we already require a spread between R_1 and R_2 of 1600. Of course, it is possible to use a scheme similar to that shown in Fig. 2-25c, in which R_1 is replaced by a resistive T such that an expression of the form (2-65) results, allowing $R_1 \gg R_2$ with a smaller spread in resistor values. However, as pointed out earlier, this also entails certain problems, beside increasing the complexity of the circuit.

Let us now examine the attainable values of $\hat{Q}$. From (2-67c) and (2-69) it follows that

$$\hat{Q} = \frac{rc}{r^2 + c^2 + 1} \qquad (2\text{-}72)$$

Taking the derivative with respect to r (or c, since the two have an equivalent effect in the expression for $\hat{Q}$) we obtain

$$\hat{Q}_{max} = \hat{Q}\bigg|_{r^2 = c^2 + 1} = \frac{c}{2\sqrt{c^2 + 1}} \qquad (2\text{-}73)$$

For the case that $c = 1$ we have $\hat{Q}_{max} = 1/2\sqrt{2} \approx 0.35$, which is relatively high; however, at the same time, $q_z = 1/\sqrt{2} \approx 0.7$. We see that $\hat{Q}$ and q_z are not independent; let us find the dependence between the two. It is easy to find from (2-70) and (2-72) that

$$\hat{Q} = \frac{q_z}{1 + q_z^2[1 + (C_2/C_1)]} \qquad [2\text{-}74a]$$

or, for $q_z \gg 1$,

$$q_z\hat{Q} \approx \frac{1}{1 + (C_2/C_1)} \qquad [2\text{-}74b]$$

From (2-74a) it can be seen that for small values of q_z, $\hat{Q}$ is proportional to q_z, while for large values of q_z, $\hat{Q}$ decreases as $\{q_z[1 + (C_2/C_1)]\}^{-1}$. This is shown qualitatively in Fig. 2-26a. It follows readily that the maxima of $\hat{Q}$ occur for

$$q_z\big|_{\hat{Q}max} = \frac{c}{\sqrt{1 + c^2}} \leq 1 \qquad (2\text{-}75a)$$

Then

$$\hat{Q}_{max} = \frac{1}{2}q_z = \frac{1}{2}\frac{c}{\sqrt{1 + c^2}} \qquad (2\text{-}75b)$$

From these expressions we see that q_z and $\hat{Q}$ cannot approach their respective maximum values at the same time; the larger q_z is selected, the smaller the corresponding Q becomes or, seen in the s-plane (Fig. 2-26b), the closer to the $j\omega$ axis the bridged-T zeros are moved, the further apart the poles move away from each other. The ensuing difficulty in obtaining a high Q in (2-68) will be obvious. We can overcome the incompatibility between high $\hat{Q}$ and q_z to some extent, as Fig. 2-26a shows, by selecting $c \gg 1$. This results in Q decreasing more slowly with increasing q_z, yet it does so at the expense of a correspondingly large spread in the capacitors C_1 and C_2.

The preceding digression from our network classification was intended to show the basic difference between the bridged-T and twin-T for use in class 3 networks. The bridged-T is limited by the constraint that $\hat{Q} \cdot q_z = \text{const}$; such a constraint does not exist for the twin-T. For the latter, $\hat{Q}$ and q_z are

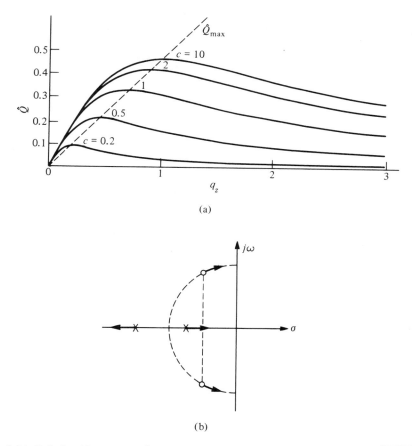

FIG. 2-26. Relationship between $\hat{Q}$ and q_z of bridged-T for various values of $c = \sqrt{C_1/C_2}$.

virtually independent of each other, at least over the ranges of $\hat{Q}$ and q_z that are of interest here. Furthermore, in the case of the bridged-T, any, even reasonably large, values of q_z or $\hat{Q}$ can only be obtained at the cost of component spread or added complexity; the same is true, but to a far lesser degree, in the case of the twin-T. For example, a high q_z value, either positive or negative, can easily be obtained with the twin-T, while maintaining $\hat{Q}$ values between say 0.25 and 0.4, without any drastic increase in component spread. As we have seen above, the same is not possible with the bridged-T, where, for example, the resistor spread increases as the square of the desired q_z.

An RC Voltage-Divider Feedback Network   A third RC network providing frequency rejection, if only to a limited degree, is the RC voltage divider in the class 3 network shown in Fig. 2-27a.

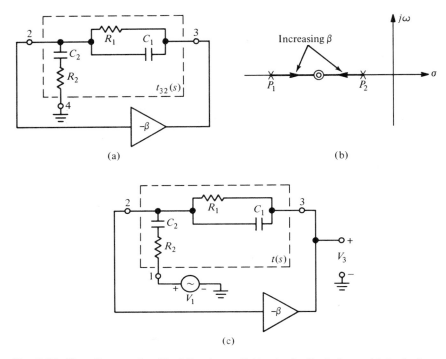

FIG. 2-27. Class 3 network with RC voltage divider in feedback loop; (a) basic form; (b) root locus; (c) with driving source.

Referring to Fig. 2-27a we obtain a transfer function $t_{32}(s)$ of the form given by (2-14b) where

$$\omega_z = \omega_0 = (R_1 R_2 C_1 C_2)^{-1/2} \tag{2-76a}$$

$$q_z = \frac{\sqrt{R_1 R_2 C_1 C_2}}{R_1 C_1 + R_2 C_2} \tag{2-76b}$$

$$\hat{Q} = \frac{\sqrt{R_1 R_2 C_1 C_2}}{R_1 C_1 + R_2 C_2 + R_1 C_2} \tag{2-76c}$$

and

$$k = 1 \tag{2-76d}$$

Notice that $q_z > \hat{Q}$ so that we have a frequency-rejecting function as required. In many respects, though, this is a rudimentary class 3 network in that the zeros obtainable with $t_{32}(s)$, although of higher Q than the poles, are still restricted to the negative real axis (notice that $t_{32}(s)$ is realized by a ladder network). For equal time constants (i.e., $R_1 C_1 = R_2 C_2$) a double zero is obtained. Thus the root locus for this network with respect to β may typically have the form shown in Fig. 2-27b. Using an operational amplifier with very high gain, double poles on the negative real axis may therefore be approximated with the circuit shown in Fig. 2-27c. Clearly, in this form, the configuration has very little, if anything, to offer; it is mentioned, however, because with added positive feedback its Q limitations can be readily overcome, as will be discussed in Section 2.4.1.

2.3.4 Class 4 Networks: Bandpass Network in Positive–Feedback Loop (BPPF)

Referring to the decomposition for this network class, which is the only positive-feedback class of the four in Table 2-1, we find that the t_{32} function must have either frequency-emphasizing, bandpass or resonator character.

Bandpass Feedback Networks   Considering the most frequent (the bandpass) case first (i.e., class 4c) the feedback loop can have either of the two basic forms shown in Fig. 2-28a and b, as well as their $(RC):(CR)$ duals e.g., replacing each conductance in Fig. 2-28a by a capacitor and vice versa, results in the configuration of Fig. 2-28c.

As pointed out earlier, parallel elements can be added to the feedback network, as long as the basic bandpass character of the resulting t_{32} function, as specified for a class 4c network, is not changed. Since two dual T networks in parallel can produce bandpass characteristics, the feedback configuration shown in Fig. 2-28d is also permitted. The same flexibility exists with respect to the insertion of the driving source V_1 (as long as the constraint with respect to the nature of t_{32} is observed). Thus, for example, V_1 can be inserted in series with C_1 of the configuration in Fig. 2-28a to provide a high-pass network (see Fig. 2-28e) and in series with an added conductance $\bar{G}_1$ to provide a bandpass section (see Fig. 2-28f). In order to obtain a bandpass network with the configuration in Fig. 2-28c, a capacitor $\bar{C}_1$ must be connected in parallel with G_1 (see Fig. 2-28g). By inserting V_1 in series with G_1 in Fig. 2-28c, a low-pass network (i.e., the dual of the high-pass obtained in Fig. 2-28e) is obtained (see Fig. 2-28h). As a final example, inserting the driving source at the ground terminal 4 of Fig. 2-28d results in a twin-T network in the forward loop (without affecting the bandpass in the feedback), as shown in Fig. 2-28i.

Clearly, numerous additional circuit configurations can be obtained by manipulating the circuit elements along the same lines as above while taking care to maintain the basic bandpass character of $t_{32}(s)$. In doing so here, as well as with the feedback networks in the other three classes, a comprehensive catalog of active filter networks, similar to that published by R. P. Sallen and E. L. Key in 1955, could be derived. As we shall see, however, in Chapters 3 and 4, when we consider the sensitivity and related features of the four network classes discussed above, not all of them are equally useful, particularly if we restrict ourselves to hybrid integrated circuit implementation.

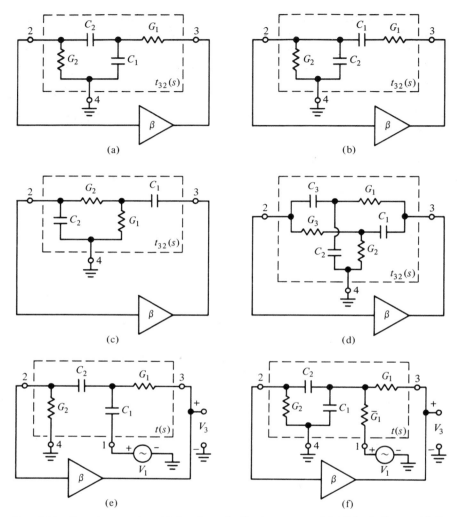

FIG. 2-28. Class 4 networks with bandpass feedback response; (a)-(d) basic forms; (e)-(i) with driving voltage source.

Let us now consider a numerical design example for the network type discussed here. Taking the bandpass network in Fig. 2-28f, we obtain the transfer function given by (2-60), where

$$\omega_p = \left[\frac{(G_1 + \bar{G}_1)G_2}{C_1 C_2} \right]^{1/2} \tag{2-77a}$$

$$Q = \frac{\hat{Q}}{1 - \beta \dfrac{\omega_k}{\omega_p} \hat{Q}} \tag{2-77b}$$

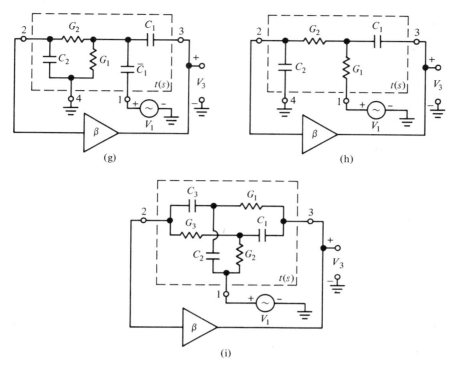

FIG. 2-28 (*Continued*)

$$\hat{Q} = \frac{[(G_1 + \bar{G}_1)G_2 C_1 C_2]^{1/2}}{(G_1 + \bar{G}_1)C_2 + G_2(C_1 + C_2)} \qquad (2\text{-}77c)$$

$$\omega_k = \frac{G_1}{C_1} \qquad (2\text{-}77d)$$

and

$$K = \beta \cdot \frac{\bar{G}_1}{C_1} \qquad (2\text{-}77e)$$

Here the coefficients of the bandpass transfer function have been derived in terms of the general feedback function $t_{32}(s)$ as given by (2-31) and (2-32). Since the zero of $t_{32}(s)$ is at the origin, ω_p is independent of β and is equal to the undamped natural frequency ω_0 of $t_{32}(s)$. Letting

$$G_1 = \bar{G}_1 = 5$$
$$G_2 = 0.1$$
$$C_1 = 10 \qquad (2\text{-}78)$$
$$C_2 = 0.1$$

we obtain

$$\hat{Q} = \frac{1}{2.01} \tag{2-79a}$$

and

$$\beta = 4.02 - \frac{2}{Q} \tag{2-79b}$$

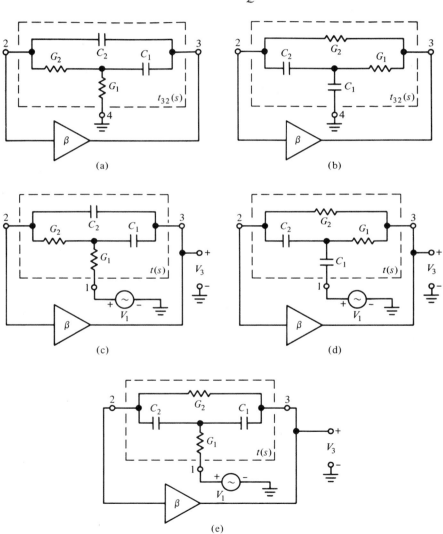

FIG. 2-29. Class 4 networks with resonator feedback response; (a) and (b) basic forms; (c)–(e) with driving voltage source.

The resulting network with the corresponding Q sensitivity and gain–sensitivity product is shown in Table 2-2. Notice that the required gain β for a given Q is low, while the Q sensitivity on the other hand is proportional to Q. This is typical of positive-feedback (i.e., class 4) networks.

Resonator Feedback Networks   Although they are rarely used (for the reasons mentioned in Section 2.1.2), let us, in the interest of completeness, at least demonstrate the feasibility of a network belonging to class 4a or class 4b. Consider first the network shown in Fig. 2-29a. Calculating the transfer function of the feedback network $t_{32}(s)$ we obtain

$$t_{32}(s) = \frac{s(s + \alpha)}{s^2 + \alpha s + \omega_p^2} \tag{2-80}$$

where

$$\alpha = \frac{G_2}{C_2} + \frac{G_1 + G_2}{C_1} \tag{2-81a}$$

and

$$\omega_p = \left(\frac{G_1 G_2}{C_1 C_2}\right)^{1/2} \tag{2-81b}$$

A network providing the transfer function (2-80) is sometimes referred to as a high-pass resonator. Comparing (2-80) with (2-28) we see that this circuit belongs to class 4b and that

$$\bar{P}_1 = -\alpha \tag{2-82a}$$

and

$$P_1 + P_2 = -\alpha = \bar{P}_1 \tag{2-82b}$$

Thus, the open-loop roots are located as shown qualitatively in Fig. 2-30a. Inserting a voltage source in series with G_1 (Fig. 2-29c) we obtain a low-pass network[28] whose pole frequency and Q are given by (2-29a) and (2-29b), where $k = 1$. If we now let

$$\frac{\alpha}{\omega_0} = \frac{1}{\hat{Q}} \tag{2-83}$$

then we have

$$\omega_p = \frac{\omega_0}{\sqrt{1 - \beta}} \tag{2-84a}$$

and

$$Q = \frac{\hat{Q}}{\sqrt{1 - \beta}} \tag{2-84b}$$

28. R. M. Inigo, Active filter realization using finite-gain voltage amplifiers, *IEEE Trans. Circuit Theory*, **CT-17**, 445–447 (1970).

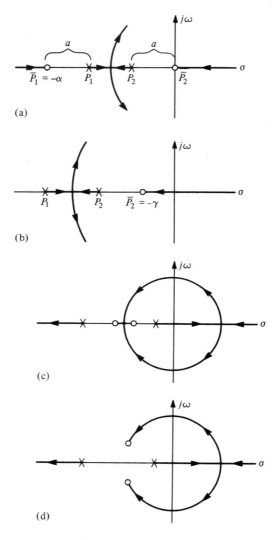

FIG. 2-30. Root loci for class 4 networks; (a) and (b) with resonator feedback response; (c) and (d) with frequency-rejection feedback response.

Thus, by assuming (2-83), we have eliminated the reduction of Q by the term $\sqrt{1 - \beta}$ in the numerator (see (2-29b)). However, the poor sensitivity of Q to variations in β remains. It is readily found that

$$S_\beta^Q = \frac{1}{2} \left[\left(\frac{Q}{\bar{Q}} \right)^2 - 1 \right] \tag{2-85}$$

which, as we shall see later, is much higher than that of class 4c circuits; furthermore, ω_p remains dependent on β.

Let us now consider the $RC:CR$ dual of Fig. 2-29a, as shown in Fig. 2-29b. Calculating the transfer function of the feedback network, we obtain

$$t_{32}(s) = \frac{\omega_p^2}{\gamma} \cdot \frac{s + \gamma}{s^2 + \dfrac{\omega_p^2}{\gamma} s + \omega_p^2} \qquad (2\text{-}86)$$

where

$$\gamma = \frac{G_1 G_2}{G_1 C_2 + G_2(C_1 + C_2)} \qquad (2\text{-}87)$$

and ω_p is given by (2-81b). A network providing this transfer function is sometimes referred to as a low-pass resonator. Comparing (2-86) with (2-26) we recognize that this circuit belongs to class 4a, where

$$\bar{P}_2 = -\gamma \qquad (2\text{-}88a)$$

$$\omega_k = \frac{\omega_p^2}{\gamma} \qquad (2\text{-}88b)$$

and

$$\frac{1}{P_1} + \frac{1}{P_2} = -\frac{1}{\gamma} = \frac{1}{\bar{P}_2} \qquad (2\text{-}88c)$$

The open-loop roots are therefore located as shown qualitatively in Fig. 2-30b. Inserting a voltage source V_1 in series with C_1 (Fig. 2-29b) we obtain a high-pass network[29] whose pole frequency and Q are given by (2-27a) and (2-27b). Letting

$$\omega_0 = \omega_p \qquad (2\text{-}89a)$$

and

$$\frac{\gamma}{\omega_p} = \hat{Q} \qquad (2\text{-}89b)$$

we obtain

$$\omega_p = \omega_0 \sqrt{1 - \beta} \qquad (2\text{-}90)$$

and Q is again given by (2-84b). Thus the same comments apply here as those following (2-84).

Finally, it is interesting to note what happens if a symmetrical bridged-T is used in the feedback path in order to obtain, say, a bandpass network (Fig. 2-29e). This configuration violates our stipulation for positive-feedback networks; the zeros of a bridged-T cannot have a lower Q than its poles, the circuit is therefore always a frequency-rejection network and cannot be designed for frequency emphasis. Thus, the corresponding root locus has the

29. Ibid.

form shown, for example, in Fig. 2-30c or d. It would seem that complex conjugate poles can be obtained, however, only for gains far larger than the
realizability or stability conditions allow; thus, in the region of realizable β
values, the pole Q *decreases* with gain until the rightmost pole crosses the
$j\omega$-axis along the real-axis; at this point the circuit becomes unstable.

2.4 DIFFERENTIAL AMPLIFIER SYNTHESIS

In the preceding section on operational amplifier synthesis the amplifiers
were used as single-ended input, single-ended output devices. Although the
commercially available operational amplifier is itself a differential input
device, it can readily be used as a single-ended device by grounding the
unused terminal. This is evident from Table 2-2, particularly for the examples
related to the first three network classes. The positive-feedback or class-4
network, which requires the operational amplifier to be in the noninverting
mode, can in a limited sense be classified as a differential rather than a single-
ended network, inasmuch as part of the *output* signal is fed back to both

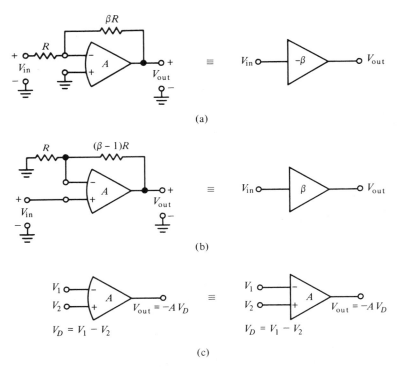

FIG. 2-31. The operational amplifier in various modes; (a) inverting mode; (b) noninverting
mode; (c) differential mode.

input terminals of the operational amplifier.[30] In practice, the resistive, gain-determining feedback network in the noninverting mode has come to be accepted as part of the amplifier itself, although it does not have to be, as will be shown in an example shortly. In general, though, we shall continue, as we have heretofore, to assume the inverting and noninverting operational amplifier modes to be single-ended at the input, with a negative or positive closed-loop gain of β, as illustrated in Fig. 2-31a and b. In the differential mode (Fig. 2-31c) we shall assume that the operational amplifier is used without any feedback (other than that required for frequency stabilization), or, in other words, that it is operating in the open-loop mode.

2.4.1 Dual-Feedback Single-Ended Input

Using the operational amplifier in the differential mode, we can apply both positive *and* negative feedback simultaneously (referred to here as dual feedback) by combining any one of the three forms of negative feedback in our classification (see Table 2-1) with a positive feedback network. Consider the network shown in Fig. 2-32.

The *RC* networks $t(s)$ and $t'(s)$ are connected to the inverting and non-inverting terminals, respectively, of the differential-input amplifier. If we consider the case for which the input voltage source V_1 is feeding into $t(s)$ at terminal 1 and terminal 1' of $t'(s)$ is grounded, we obtain the transfer function

$$T(s) = \frac{V_3}{V_1} = -A \cdot \frac{t_{12}}{1 + A(t_{32} - t'_{32})} \approx \frac{-t_{12}}{t_{32} - t'_{32}} \qquad [2\text{-}91]$$

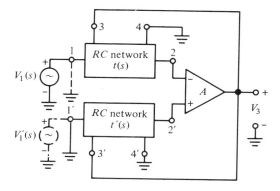

FIG. 2-32. Opamp network: the general dual-feedback single-ended input configuration.

30. In general, the *input* signal, or parts thereof, are assumed to be connected to *both* input terminals for the operational amplifier to be considered in the differential mode. We shall refer to this later, as the "double-ended input mode."

Similarly, connecting the voltage source V_1' to terminal $1'$ with terminal 1 grounded, we obtain

$$T'(s) = \frac{V_3}{V_1'} = A \cdot \frac{t_{12}'}{1 + A(t_{32} - t_{32}')} \approx \frac{t_{12}'}{t_{32} - t_{32}'} \qquad [2\text{-}92]$$

In both cases the transfer functions t_{ij} are defined in the same way as (2-54) and (2-55),[31] and it is assumed that $A \gg 1$. Notice that the denominators in (2-91) and (2-92) are identical. Only the numerator (the zeros) of the overall voltage transfer function will depend on which of the two networks $t(s)$ or $t'(s)$ contains the input signal source. Thus, for either case, the decomposition corresponding to (2-6) follows as

$$D(s) = s^2 + \frac{\omega_p}{Q} s + \omega_p^2 = d(s) + A[n_{32}(s) - n_{32}'(s)] \qquad (2\text{-}93)$$

where we have assumed that

$$d_{32}(s) = d_{32}'(s) = d(s) = s^2 + \frac{\omega_0}{Q} s + \omega_0^2 \qquad (2\text{-}94)$$

If this assumption were not made, $t_{32}'(s)$ would be required to be at least a fourth-order network for the decomposition to be possible at all. This would complicate the overall circuit greatly and require many more than the minimum number of resistors and capacitors necessary for a canonical second-order network. Even as it is, a separate second-order RC network is required for $t_{32}(s)$ and $t_{32}'(s)$, which may complicate the network unjustifiably.

One simple method of utilizing the dual-feedback scheme of Fig. 2-32 is to let the network to which the input source is connected (e.g., $t(s)$ in Fig. 2-32) determine the basic frequency function, and to use the other network to provide a *constant* amount of feedback of opposite polarity. This is discussed in the following.

Combining Constant Positive Feedback with Frequency-Dependent Negative Feedback   Consider the network configuration shown in Fig. 2-33. Here

$$t_{32}' = \alpha \qquad (2\text{-}95)$$

where $0 \le \alpha \le 1$. With the corresponding transfer function (2-91) we obtain

$$T(s) = -A \cdot \frac{t_{12}}{1 + A(t_{32} - \alpha)} \qquad [2\text{-}96]$$

For very large (open-loop) gain this simplifies to

$$T(s)\bigg|_{A \gg 1} = -\frac{t_{12}}{t_{32} - \alpha} \qquad [2\text{-}97]$$

31. With the obvious exception that with V_1 at terminal 1 of $t(s)$, the transfer parameter $t_{32}'(s)$ of $t'(s)$ is defined with terminals $4'$ *and* $1'$ grounded, and vice versa for $t_{32}(s)$.

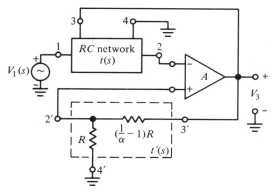

FIG. 2-33. Opamp network combining constant positive feedback with frequency-dependent negative feedback.

We must now determine what kind of function $t_{32}(s)$ must be in order to obtain a complex conjugate pole pair in (2-97). Assuming that $t_{32} = n_{32}/d_{32}$ is realized by a passive RC network in the negative feedback path of the configuration shown in Fig. 2-33, it would be reasonable to assume that a network of any one of our three basic negative-feedback classes can be combined accordingly with constant positive feedback. By examining the root locus of the poles of (2-97) with respect to α we can readily see if this is indeed the case.

The root locus of the poles of (2-97) is defined by the characteristic equation

$$\alpha \frac{d_{32}(s)}{n_{32}(s)} = \alpha \frac{(s - P_1)(s - P_2)}{(s - \bar{P}_1)(s - \bar{P}_2)} = 1 \tag{2-98}$$

Since $t_{32}(s)$ is to be realized by an RC network, the roots P_1 and P_2 of $d_{32}(s)$, which represent the open-loop zeros in (2-98), must be negative real. The open-loop poles of (2-98) are the roots $\bar{P}_1$ and $\bar{P}_2$ of $n_{32}(s)$, and we must find where they must be located in order to obtain a complex conjugate closed-loop pole pair. Assuming an unbalanced RC network for $t_{32}(s)$ and bearing in mind the rules of 180° root-locus construction, it follows that the zeros of $t_{32}(s)$ must either be complex conjugate or lie in the interval (P_1, P_2) on the negative real axis. Thus we obtain the two root loci—the only two types possible—with respect to α shown in Fig. 2-34. It therefore follows that $t_{32}(s)$ *must* have the form

$$t_{32}(s) = k \frac{s^2 + \dfrac{\omega_z}{q_z} s + \omega_z^2}{s^2 + \dfrac{\omega_0}{\hat{Q}} s + \omega_0^2} \tag{2-99}$$

where

$$q_z > \hat{Q}$$

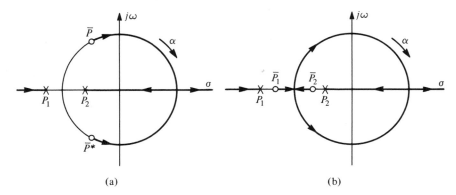

(a)   (b)

FIG. 2-34. Root locus of $\alpha/t_{32}(s) = \alpha d_{32}(s)/n_{32}(s) = 1$ when $d_{32}(s)$ has negative real roots and $n_{32}(s)$ has (a) complex roots or (b) negative real roots between those of $d_{32}(s) = 0$.

This of course defines a frequency-rejection network. Furthermore it implies that constant positive feedback may be combined *only* with class 3 networks in order to obtain complex poles with the configuration of Fig. 2-33. This will be illustrated by the following examples.

Class 3 Networks with Constant Positive Feedback.   Let us start out with one of the basic class 3 networks that was shown in Fig. 2-24, e.g., the band-pass network shown in Fig. 2-24a. Combining this with a positive resistive feedback network in the manner shown in Fig. 2-33 we obtain the network shown in Fig. 2-35.[32] Substituting (2-14b) into the denominator of (2-96), where the differential open-loop gain A is used instead of β, and expanding the decomposition corresponding to (2-93), we obtain, for the two desired pole parameters ω_p and Q,

$$\omega_p = \omega_0 \left[\frac{1 - kA\alpha + kA(\omega_z/\omega_0)^2}{1 - kA\alpha + kA} \right]^{1/2} \qquad (2\text{-}100a)$$

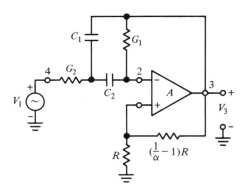

FIG. 2-35. Class 3 network with positive feedback.

32. T. Deliyannis, High-Q factor circuit with reduced sensitivity, *Electronics Lett.*, **4**, 577 (1968).

and

$$Q = \hat{Q} \frac{\{[1 - kA\alpha + kA(\omega_z/\omega_0)^2][1 - kA\alpha + kA]\}^{1/2}}{1 - kA\alpha + kA(\omega_z/\omega_0)(\hat{Q}/q_z)} \qquad (2\text{-}100b)$$

For the case that $\omega_z = \omega_0$ in t_{32} (see 2-14b), the pole frequency ω_p equals the zero frequency of the passive null network, i.e.,

$$\omega_p \bigg|_{\omega_z = \omega_0} = \omega_0 \qquad (2\text{-}101a)$$

and

$$Q \bigg|_{\omega_z = \omega_0} = \frac{1 - kA\alpha + kA}{1 - kA\alpha + (kA\hat{Q}/q_z)} \cdot \hat{Q} \qquad (2\text{-}101b)$$

If in addition $A \gg 1$, then Q simplifies to

$$Q \bigg|_{A \gg 1} = q_z \frac{1 - \alpha}{1 - \alpha(q_z/\hat{Q})} \qquad (2\text{-}102)$$

In order for Q to remain finite, the denominator of (2-102) must remain larger than zero, so that

$$0 < \alpha < \frac{\hat{Q}}{q_z} \qquad (2\text{-}103)$$

More specifically, for given Q, $\hat{Q}$, and q_z, the required amount of positive feedback α results as

$$\alpha = \frac{(Q/q_z) - 1}{(Q/\hat{Q}) - 1} \approx \hat{Q}\left(\frac{1}{q_z} - \frac{1}{Q}\right) \qquad (2\text{-}104)$$

By definition,

$$\frac{\hat{Q}}{q_z} < 1 \qquad (2\text{-}105)$$

Therefore, for (2-104) to remain positive, the following inequality must hold:

$$\hat{Q} < q_z \leq Q \qquad (2\text{-}106)$$

When q_z is approximately equal to $\hat{Q}$ (i.e., ≤ 0.5) α may be close to unity; when q_z is large, α may be very much smaller than unity and (2-102) simplifies to

$$Q \bigg|_{\substack{A \gg 1 \\ \alpha \ll 1}} \approx \frac{q_z}{1 - \alpha(q_z/\hat{Q})} \qquad (2\text{-}107)$$

This expression shows clearly that the positive feedback characterized by α manifests itself in the denominator of Q in exactly the same way as in all other positive-feedback schemes, i.e., by generating the difference between two quantities. The more positive feedback we have, i.e., the larger α becomes, the closer this difference comes to being zero. The interplay between negative and positive feedback is further illustrated qualitatively by the root-locus

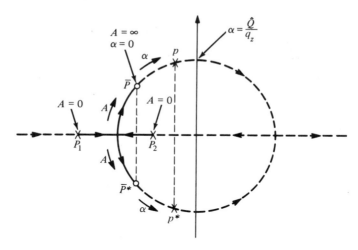

FIG. 2-36. Root locus of a class 3 network with positive feedback.

diagram in Fig. 2-36. If we separate the effects of the positive and negative feedback path, then we can think of the closed-loop poles p and p^* being generated as follows. The poles and zeros of the bridged-T in the negative feedback path are P_1, P_2 and $\bar{P}$, $\bar{P}^*$, respectively. Consequently, for infinite open-loop gain, i.e., $A = \infty$, the closed-loop poles would coincide with $\bar{P}$ and $\bar{P}^*$ were it not for the effect of the positive feedback, which shifts the closed-loop poles still closer to the $j\omega$ axis, i.e., to the desired location p, p^*. Clearly this scheme is useful in overcoming the basic limitation of the bridged-T, as discussed in Section 2.3.3. We pointed out there that the objectives of obtaining a high q_z and a $\hat{Q}$ close to 0.5 were incompatible with the bridged-T, and that, even when moderately approached, resulted in large component spreads that are difficult to realize with anything but discrete components. From Fig. 2-36 we see that the bridged-T can now be dimensioned to provide a relatively high $\hat{Q}$ (P_1 and P_2 relatively close together on the negative real axis), while q_z must of necessity be small. The resulting closed-loop pole Q would also be small, except that by adding the positive feedback this limitation is overcome, and a higher closed-loop pole Q is readily obtained.

In designing this type of dual-feedback network, the question must be resolved as to how much Q enhancement should be obtained from positive, and how much from negative, feedback. In other words, how much of a given Q value should be obtained from α and how much from q_z? The value of q_z, which is the zero Q of the passive feedback network $t(s)$, can be made arbitrarily large, but the penalty is a large spread of passive component values if a bridged-T is used. On the other hand, as we shall see in Chapter 4, α, which determines the amount of positive feedback, increases the pole sensitivity.[33] As so often in engineering problems, the optimum solution requires a compromise. The criteria for this compromise will depend to some extent on the application, but even more so on the technology used for the network realization. Thus, a general answer to this question cannot be given. However, the optimization of this and the other networks discussed so far will be dealt with, from the point of view of hybrid integrated, and discrete-component implementation, in Chapter 3.

It is interesting to note, that whereas in the single-loop feedback case (i.e., single-ended amplifier input) the positive feedback loop must have bandpass character, this is no longer true for the differential amplifier case, where, as we have just seen, a frequency-independent, positive-feedback network may be used in conjunction with an appropriate negative-feedback network $t_{32}(s)$. The same scheme can be used with any one of the circuits shown in Fig. 2-24[34] as well as any others that can be derived in network class 3, such as the Wien bridge network.

The Wien bridge Network.   One class-3 network that particularly benefits from the dual-feedback scheme of Fig. 2-33 is the simple voltage-divider configuration shown earlier in Fig. 2-27a. It will be recalled that because this network was not capable of providing pole Q's higher than 0.5, even with an infinite-gain amplifier, it was called a rudimentary class 3 network. Adding frequency-independent positive feedback, as shown in Fig. 2-37a, complex poles can readily be obtained, by the same mechanism as was illustrated by the root locus in Fig. 2-36. In this case the poles would be confined to the negative real axis (Fig. 2-37b), were it not for the positive feedback α that shifts them off and into the s-plane as complex conjugate pairs in the manner illustrated by the root locus of Fig. 2-34b. The configuration in Fig. 2-37a will be recognized as a Wien bridge network. Whether an oscillator or a filter results depends on the amount of positive feedback, which is characterized by α.

Consider, for example, the Wien bridge bandpass filter[35] shown in Fig. 2-37c. Calculating $t_{12}(s)$ and $t_{32}(s)$ we obtain

$$t_{12}(s) = \frac{\omega_{12}}{s^2 + \dfrac{\omega_0}{\hat{Q}} s + \omega_0^2} \qquad (2\text{-}108)$$

33. If a twin-T is used, q_z can be selected arbitrarily large, with no increase in component spread; in this case the advantages of adding positive feedback to boost Q, as in (2-102) are doubtful, since Q can be increased by RHP zeros (see (2-20)).
34. J. J. Friend, A single op-amp biquadratic filter section, 1970 *IEEE Int. Symp. Circuit Theory, Digest of Tech. Papers*, pp. 179–180.
35. J. W. Craig, Practical designs for *RC* active filters using operational amplifiers, *Tech. Note* 1968-26, Lincoln Laboratories, August 21, 1968.

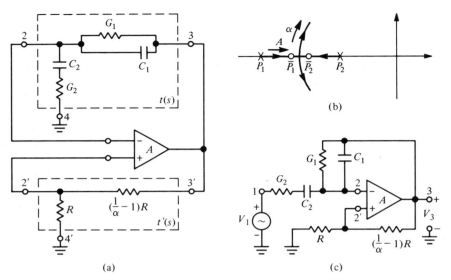

FIG. 2-37. Wien bridge network; (a) basic circuit; (b) root locus; (c) with driving source .

and

$$t_{32}(s) = \frac{s^2 + \dfrac{\omega_z}{q_z} s + \omega_z^2}{s^2 + \dfrac{\omega_0}{\hat{Q}} s + \omega_0^2} \qquad (2\text{-}109)$$

where

$$\omega_{12} = \frac{G_2}{C_1}; \qquad \omega_0 = \omega_z = \left(\frac{G_1 G_2}{C_1 C_2}\right)^{1/2} \qquad (2\text{-}110)$$

$$q_z = \frac{\sqrt{G_1 G_2 C_1 C_2}}{G_1 C_2 + G_2 C_1}; \qquad \hat{Q} = \frac{\sqrt{G_1 G_2 C_1 C_2}}{G_1 C_2 + G_2 C_1 + G_2 C_2}$$

As can be seen by inspection, $q_z > \hat{Q}$ and therefore a frequency-rejection network is guaranteed. Furthermore since $1/q_z$ can be written in the form

$$\frac{1}{q_z} = x + \frac{1}{x} \qquad (2\text{-}111)$$

where $x = [G_1 C_2 / G_2 C_1]^{1/2}$, it follows directly that

$$q_{z\,\text{max}} = 0.5 \qquad (2\text{-}112)$$

This value is obtained when $x = 1$, i.e., when $G_1 C_2 = G_2 C_1$.

Assuming a high-gain amplifier for the configuration in Fig. 2-37c we can substitute $t_{12}(s)$ and $t_{32}(s)$ into (2-97) and obtain

$$\omega_p = \omega_z = \omega_0 \tag{2-113a}$$

and

$$Q = q_z \frac{1 - \alpha}{1 - \alpha(q_z/\hat{Q})} \tag{2-113b}$$

These expressions are identical with (2-101a) and (2-102), which apply to the circuit in Fig. 2-35. Before we jump to any conclusions about the performance of the two circuits, however, we must remember that in the circuit of Fig. 2-37c q_z is limited, i.e., $0 < q_z \leq 0.5$, whereas for the circuit of Fig. 2-35, $0 < q_z < \infty$. Since it is primarily the choice of the constants α, q_z, and $\hat{Q}$ that determines the stability of the final network with respect to a specified Q, and since the choice of q_z is much more limited for the configuration of Fig. 2-37 than for that of Fig. 2-35, the former is only rarely used. In particular, the positive feedback α will generally be required to be larger (i.e., closer to unity) for the Wien bridge network than for the network of Fig. 2-35; this may result in a less stable network. We shall come back to these questions in the following chapters. We can note in summary here, though, that in terms of q_z flexibility, the L network of Fig. 2-37a is much more limited than the bridged-T, and this in turn is inferior in terms of q_z flexibility to the twin-T. It is because of the flexibility of the twin-T that additional positive feedback is unnecessary; the twin-T is capable of providing positive feedback on its own when its zeros are located in the RHP (i.e., $v < 0$ as in (2-20)).

The Duality between Class 4 Networks and Class 3 Networks with Constant Positive Feedback   We pointed out earlier that the operational-amplifier realizations of class 4 networks are, strictly speaking, differential-amplifier configurations. This becomes clear if we redraw a class 4 network using an operational amplifier in terms of Fig. 2-32. This is shown in Fig. 2-38. Inspection of this figure shows that we have here a case of frequency-dependent positive feedback combined with constant negative feedback. This is the opposite of the special case shown in Fig. 2-33. In complete analogy with the former case, we have here

$$t_{32} = \alpha \tag{2-114}$$

with $0 \leq \alpha \leq 1$. Substituting into the corresponding transfer function (2-92) we obtain

$$T'(s) = A \cdot \frac{t'_{12}}{1 + A(\alpha - t'_{32})} \tag{2-115}$$

For high open-loop gain values this becomes

$$T'(s)\Big|_{A \gg 1} = \frac{t'_{12}}{\alpha - t'_{32}} \tag{2-116}$$

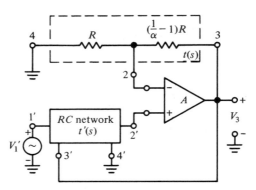

FIG. 2-38. General class 4 network.

Comparing this with (2-97), we observe that the two expressions are virtually the same, except that here α gives a measure of the amount of *negative* gain being fed back through the inverting input terminal of the amplifier. It is customary to use the negative gain β, rather than the negative feedback α, in (2-116); β may either be used to designate the term $1/\alpha$ or $(1/\alpha - 1)$. If we use the former sense of β, i.e.,

$$\beta = \frac{1}{\alpha} \tag{2-117}$$

(2-116) takes on the familiar form

$$T'(s)\Big|_{A \gg 1} = \frac{\beta t'_{12}}{1 - \beta t'_{32}} \tag{2-118}$$

This will be recognized as the transfer function characterizing a positive-feedback network using an operational amplifier in the noninverting mode.

The analogy between the positive- and negative-feedback cases can be carried still further. In conjunction with the latter, we showed that only a class 3 negative-feedback network could be combined with a *constant* positive-feedback network $t_{32} = \alpha$. A class 3 network, it will be remembered, requires a frequency-rejection network in its feedback path. Thus we have there the feedback configuration shown qualitatively in Fig. 2-39a. Similarly, we showed that there is basically only one possible type of positive-feedback network, the class 4 network, and that it requires a bandpass, or frequency-emphasizing, network in the positive feedback loop. This is shown qualitatively in Fig. 2-39b. Thus the positive feedback network, which is actually a positive dual-feedback network, is the exact inverse of the negative dual-feedback network shown in Fig. 2-39a.

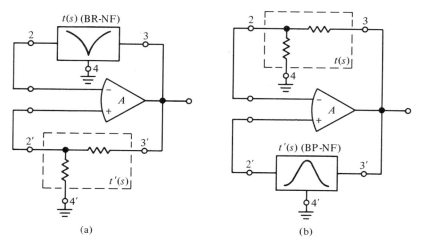

FIG. 2-39. Demonstration of the duality between (a) class 3 networks with positive feedback and (b) class 4 networks.

2.4.2 The Noninverting Input Mode with Frequency–Dependent Negative Feedback

So far the examples given for the general network configuration of Fig. 2-32 have all dealt with input driving sources connected to the inverting side of the differential amplifier (see $V_1(s)$ in Fig. 2-32). However, one of the useful features of the noninverting terminal of the amplifier is the extremely high impedance it provides so that any RC network connected to it is virtually operating in an unloaded, or open-circuit, state. This is utilized in the configuration shown in Fig. 2-40a, which results directly from the general configuration in Fig. 2-32. Since the noninverting feedback loop to the RC network is absent, the corresponding voltage transfer function $t'(s)$ here coincides with $t'_{12}(s)$, whereas $t'_{32}(s) \equiv 0$. Thus, from (2-92),

$$T(s) = \frac{V_3}{V'_1} = A \cdot \frac{t'_{12}}{1 + At_{32}} \qquad [2\text{-}119]$$

and for large A

$$T(s)\big|_{A \gg 1} = \frac{t'_{12}}{t_{32}} \qquad [2\text{-}120]$$

Letting

$$t'_{12} = \frac{n'_{12}}{d'_{12}} \qquad (2\text{-}121a)$$

and

$$t_{32} = \frac{n_{32}}{d_{32}} \qquad (2\text{-}121b)$$

(2-120) becomes

$$T(s)\big|_{A \gg 1} = \frac{n'_{12}}{n_{32}} \cdot \frac{d_{32}}{d'_{12}} \tag{2-122}$$

The networks t'_{12} and t_{32} are passive RC, and since they are separate they are independent of one another. Nevertheless, it is simple enough to realize them in such a way that

$$d'_{12} = d_{32} = d \tag{2-123}$$

Then (2-122) simplifies to

$$T(s)\big|_{A \gg 1} = \frac{n'_{12}}{n_{32}} \tag{2-124}$$

This expression has the interesting property that the zeros are realized by the zeros of the voltage transfer function t'_{12} of $t'(s)$, whereas the poles are realized by the zeros of the voltage transfer function $t_{32}(s)$ of the entirely separate RC network realizing $t(s)$.

Cascading RC Networks with Class 3 Networks for Pole–Zero Cancellation   The realization of poles and zeros by the voltage transfer functions of separate RC networks as obtainable from the configuration in Fig. 2-40a provides a very useful feature. Restricting ourselves as heretofore to the realization of second-order networks having complex poles, it follows from (2-124) that $t(s)$ in Fig. 2-40a must be realized either by a bridged-T[36] or a twin-T[37] network. This is shown in Fig. 2-40b and c, respectively. The voltage transfer function of either network is of the form

$$t_{32}(s) = \frac{s^2 + \dfrac{\omega_{32}}{q_{32}} s + \omega_{32}^2}{s^2 + \dfrac{\omega_0}{Q} s + \omega_0^2} \tag{2-125}$$

The zeros of (2-125) become the poles of $T(s)$ in (2-124). The twin-T in Fig. 2-40c is capable of providing poles with higher Q's than the bridged-T network in Fig. 2-40b; it can be designed to provide low-Q poles by bridging it with the resistance R_4. Since R_4 will be required to be quite large as Q values increase (assuming that the twin-T is tuned for $j\omega$-axis zeros) it can be replaced by a resistor T network, as was shown in Fig. 2-25c.

We shall see in Chapter 3 that it is not advisable to use single-amplifier second-order networks for high pole Q's if reasonable Q stability is to be maintained. Restricting ourselves to low pole Q's (say, less than 25) the

36. D. Wolff, Two structures to minimize sensitivity in active RC networks, *Electronics Lett.* **2**, 152–153 (1966). Also: H. Burkhardt and N. Fliege, Synthesis of RC active filters with prescribed complex pole sensitivity, *Nachr. Tech. Z.* **9**, 468–472 (1971).
37. T. A. Hamilton and A. S. Sedra, A new low sensitivity, high-Q active filter configuration using a single IC op. amp., 1970 IEEE Intern. Symp. Circuit Theory, Atlanta, Georgia.

bridged-T network of Fig. 2-40b may be perfectly adequate, and even preferable, to the twin-T, because of its lower degree of complexity.[38] Comparing Fig. 2-40b with the basic class 3 configurations of Fig. 2-23 shows that they are in fact identical, insofar as pole realization is concerned. (The dual configuration to that of Fig. 2-40b, which corresponds to Fig. 2-23b can, of course, also be used). It will be recalled that the class 3 networks derived from the configurations shown in Fig. 2-23 require as high a gain as possible, in other words open-loop-mode amplifier operation, in order for the closed-loop poles to depend as closely as possible on the zeros of the bridged-T of the feedback network. This same requirement has been tacitly assumed for the configurations of Fig. 2-40, which, like most differential amplifier networks, operate with maximum available gain.

It is in the zero realization of the overall network function $T(s)$ that the configuration in Fig. 2-40a and the class 3 networks of Fig. 2-23 differ fundamentally. Where the zeros in the networks shown in Fig. 2-24 are obtained by appropriate signal-source insertions into the bridged-T feedback network, they are independently obtained here by the voltage-transfer-function zeros of a separate function, $t'_{12}(s)$. Thus, for example, a low-pass network can be obtained with the configuration of Fig. 2-40b by using an RC network with a low-pass transfer function for $t'_{12}(s)$. The only constraint on $t'_{12}(s)$ is that its denominator (i.e., poles) be identical to that of $t_{32}(s)$, as given in (2-125). Thus

$$t'_{12}(s) = \left.\frac{V'_2}{V'_1}\right|_{V_4'=0} = \frac{K'}{s^2 + \dfrac{\omega'_0}{\hat{Q}'}s + \omega_0'^2} \tag{2-126}$$

where, for pole–zero cancellation in (2-122), it is required that

$$\omega_0 = \omega'_0 \tag{2-127a}$$

and

$$\hat{Q} = \hat{Q}' \tag{2-127b}$$

The corresponding low-pass network is shown in Fig. 2-41. For the coefficients in (2-125) we obtain

$$\omega_{32} = \omega_0 = \sqrt{\frac{G_1 G_2}{C_1 C_2}} \tag{2-128a}$$

$$q_{32} = \frac{\sqrt{G_1 G_2 C_1 C_2}}{G_1(C_1 + C_2)} \tag{2-128b}$$

---

38. On the other hand, the spread of component values necessary to obtain $q_z = 25$ with a reasonable value of $\hat{Q}$ (say 0.25, hence $v = 0.01$) may not be acceptable for hybrid integrated circuit technology.

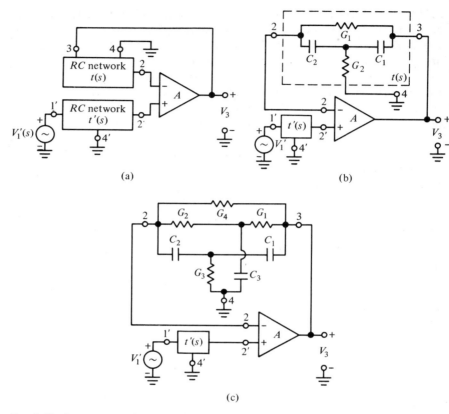

FIG. 2-40. Opamp networks combining noninverting input mode with frequency-dependent, negative feedback; (a) basic configuration; (b) using class 3 network with bridged-T; (c) using class 3 network with twin-T.

and

$$\hat{Q} = \frac{\sqrt{G_1 G_2 C_1 C_2}}{G_1(C_1 + C_2) + G_2 C_2} \tag{2-128c}$$

For the coefficients in (2-126) we obtain

$$\omega_0' = \sqrt{K'} = \sqrt{G_1' G_2' / C_1' C_2'} \tag{2-129a}$$

and

$$\hat{Q}' = \frac{\sqrt{G_1' G_2' C_1' C_2'}}{G_2'(C_1' + C_2') + G_1' C_2'} \tag{2-129b}$$

To obtain a low-pass function

$$T(s) = \frac{V_3}{V_1} = \frac{K}{s^2 + \dfrac{\omega_p}{Q} s + \omega_p^2} \tag{2-130}$$

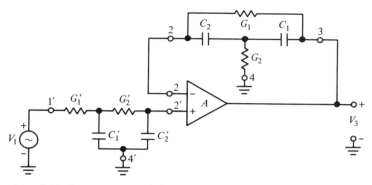

FIG. 2-41. Low-pass network based on the configuration of Fig. 2-40b.

with the configuration of Fig. 2-41, expressions (2-128a) and (2-129a), as well as (2-128c) and (2-129b) must be set equal. In this way the conditions in (2-127) are satisfied and pole–zero cancellation is obtained in (2-122). Substituting the numerators of (2-125) and (2-126) into (2-124) we obtain

$$T(s) = \frac{K'}{s^2 + \dfrac{\omega_{32}}{q_{32}} s + \omega_{32}^2} \tag{2-131}$$

Comparing coefficients with the desired function (2-130) we find

$$K' = K = \frac{G_1 G_2}{C_1 C_2} \tag{2-132a}$$

$$\omega_{32} = \omega_p = \sqrt{G_1 G_2 / C_1 C_2} \tag{2-132b}$$

and

$$q_{32} = Q = \frac{\sqrt{G_2/G_1}}{\sqrt{C_1/C_2} + \sqrt{C_2/C_1}} \tag{2-132c}$$

Following precisely the same procedure, any other second-order, minimum-phase network can be obtained with the configuration of Fig. 2-40a. For example, a bandpass, high-pass, and frequency-rejection network is shown in Fig. 2-42. Non-minimum-phase networks (i.e., RHP zeros) can be obtained within limits ($|q_z| > 1$) with a twin-T network, and consequently with the twin-T circuit shown in Fig. 2-42. Normally, however, the twin-T is used to generate $j\omega$-axis zeros as part of a frequency-rejection network.

Clearly any realizable second-order passive RC network can be used for $t'(s)$ in the manner shown in Fig. 2-42 with the resulting variety of second-order transfer functions $T(s)$. The poles of any network type are always generated in the same way, namely by a class 3, i.e., a bridged-T or twin-T, negative-feedback arrangement; the zeros which determine the basic character

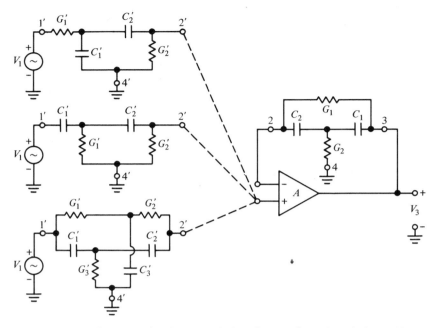

FIG. 2-42. Various second-order networks based on configuration of Fig. 2-40b.

of the given function $T(s)$ are generated by the passive RC network $t'(s)$. It should be noted however, that the components of the bridged-T or twin-T must be modified according to (2-132c) for each newly specified pole Q of $T(s)$; as a result $\hat{Q}$, given by (2-128c), will vary at the same time. In order to satisfy (2-127b) this also means a change in $\hat{Q}'$, or in the values of the passive input network $t'(s)$. Thus, the component values of *both* networks depend on the specified pole Q.

2.4.3 Dual–Feedback Double–Ended Input

One of the disadvantages of the single-ended, dual-feedback networks typified by the basic configuration shown in Fig. 2-32 is that they require two separate RC networks $t(s)$ and $t'(s)$. It is perhaps for this reason that the most common types of dual-feedback networks in that category are the ones comprising a second-order RC network in one feedback loop and a resistive voltage divider in the other. The latter provides constant feedback of the opposite polarity to that used in the RC-network loop. Another simple, and perhaps obvious way of eliminating one of the two RC networks in Fig. 2-32 is to combine both of them into one by joining them together at the input. The two networks thereby have a common node and can be considered to be

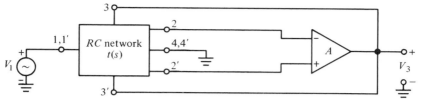

FIG. 2-43. Opamp network: the general dual-feedback, double-ended input configuration.

parts of a single, second-order RC network, as shown in Fig. 2-43. Using voltage-transfer parameters similar to those used in the single-ended input case, namely

$$t_{ij} = \frac{V_j}{V_i}\bigg|_{\substack{V_\mu=V_{\mu'}=0 \\ \mu \neq i \\ \mu \neq j}} \tag{2-133a}$$

and

$$t'_{ij} = \frac{V'_j}{V'_i}\bigg|_{\substack{V_\mu=V_{\mu'}=0 \\ \mu' \neq i' \\ \mu' \neq j'}} \tag{2-133b}$$

we obtain the overall voltage transfer function

$$T(s) = \frac{V_3}{V_1} = -A\frac{(t_{12} - t'_{12})}{1 + A(t_{32} - t'_{32})} \tag{2-134}$$

For $A \gg 1$ this simplifies to

$$T(s)\bigg|_{A \gg 1} \approx -\frac{t_{12} - t'_{12}}{t_{32} - t'_{32}} \tag{2-135}$$

The first important point to note here is that the form of the denominator of (2-134) is identical to that of the single-ended case [see (2-91) and (2-92)]. Thus, as one would expect, the denominator, or the poles, of $T(s)$ can be realized by the same types of dual-feedback as those discussed in the preceding section. The basic difference between (2-135) and either (2-91) or (2-92) appears in the numerator. In the latter two cases, the zeros of $T(s)$ are the zeros of t_{12}; in (2-135) they are the zeros of $(t_{12} - t'_{12})$.

This difference is a significant one. If t_{12} is the voltage-transfer function of an unbalanced passive RC network (which is assumed throughout this chapter), then, as we know, complex zeros can only be obtained with t_{12} if the corresponding network comprises parallel ladder networks or cascaded bridged-Ts. The location in the s-plane is further constrained by the Fialkow–Gerst conditions. Thus, the further in the right half-plane (or, more specifically, the closer to the positive real axis) the zeros are to be, the higher must be the order of the network realizing t_{12}, i.e., the more resistors and capacitors will be required. An increase in the number of capacitors is generally undesirable. Certainly, for complex zeros, the *RC* network realizing t_{12} may not be a ladder network, no matter what its complexity. As we know the zeros of the latter are confined to the negative real axis. Thus with the network classes—and the configurations derived from them—that we have considered so far (e.g., Fig. 2-32) we have no convenient way of generating arbitrary complex conjugate zero pairs, these being determined by the transmission zeros of *RC* networks. Of course the twin-*T* is very useful when zeros on or close to the *jω* axis are required and the bridged-*T* can be used to provide relatively low-*Q* LHP zeros. Yet when well defined complex zeros are to be located accurately at a distance from the *jω* axis, as is the case with all-pass networks, then even the twin-*T* is not very practical, since zeros are limited in the RHP to q_z values larger than unity. Furthermore, the precision tuning of these zeros becomes all the more difficult, the further from the *jω* axis they are required to be.

This situation is completely changed with the configuration of Fig. 2-43. Complex zeros are now no longer determined by the transmission zeros of *RC* networks but by the *difference* of two *RC* voltage-transfer functions t_{12} and t'_{12}. In fact t_{12} and t'_{12} need not have complex zeros of their own; they are arbitrary *RC* transfer functions and may therefore be realized by, among other things, simple *RC* ladder networks. This follows directly from the $(RC):(-RC)$ decomposition discussed in Section 2.2.1 if one considers the similarity between the characteristics of the transfer functions of *RC* ladder networks and those of the driving point functions of general *RC* networks. Thus, in what follows, we shall examine the realization of second-order transfer functions with complex zeros based on networks of the type characterized by Fig. 2-43 and by the expression (2-135).

Combining Constant Positive Feedback with Frequency-Dependent Negative Feedback   Consider the network shown in Fig. 2-44. One half of the network consists of a *T*-network with the conductances Y'_1, Y'_2, and Y'_3. By inspection we find that

$$t'_{12} = \frac{Y'_1}{Y'_1 + Y'_2 + Y'_3} \tag{2-136a}$$

and

$$t'_{32} = \frac{Y'_2}{Y'_1 + Y'_2 + Y'_3} \tag{2-136b}$$

Assuming that $A \gg 1$ and substituting into (2-135) we obtain

$$T(s)\bigg|_{A \gg 1} = -\frac{t_{12}(Y'_1 + Y'_2 + Y'_3) - Y'_1}{t_{32}(Y'_1 + Y'_2 + Y'_3) - Y'_2} \tag{2-137}$$

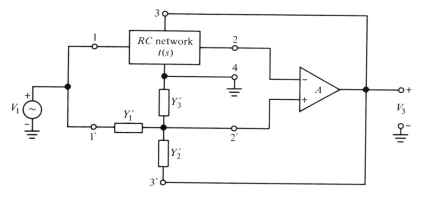

FIG. 2-44. Network configuration based on Fig. 2-43.

Considering now the special case of constant, i.e., frequency-independent, positive feedback, namely (see Fig. 2-45)

$$Y_1' = \frac{\alpha'}{1 - \alpha'} G', \; Y_2' = 0, \; Y_3' = G' \tag{2-138a}$$

we obtain

$$t_{12}' = \alpha', \qquad t_{32}' = 0 \tag{2-138b}$$

where $0 \leq \alpha' \leq 1$. Then (2-137) simplifies to

$$T(s)\Big|_{A \gg 1} = -\frac{t_{12} - \alpha'}{t_{32}} \tag{2-139}$$

Here the poles of $T(s)$ are realized by the zeros of $t_{32}(s)$; the zeros of $T(s)$ are obtained by modifying the zeros of $t_{12}(s)$ by the constant α'. Notice the similarity between this case and that represented by the configuration in

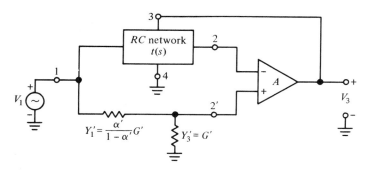

FIG. 2-45. Simplified version of Fig. 2-44.

Fig. 2-33. In fact, the general configuration in Fig. 2-44 encompasses both these forms. Fig. 2-33 is obtained by setting

$$Y_1' = 0, \qquad Y_2' = \alpha(1 - \alpha')^{-1}G', \qquad Y_3' = G' \qquad (2\text{-}140a)$$

From (2-136)

$$t_{31} = 0, \qquad t_{32}' = \alpha' \qquad (2\text{-}140b)$$

Substituting into (2-137) results in the expression obtained in (2-96). Thus with *single-ended input* ($Y_1' = 0$, $t_{12}' = 0$) and *dual feedback* of the kind shown in Fig. 2-33, only the *poles* are modified by the feedback voltage divider α; with *double-ended input* and *no positive feedback* ($Y_2' = 0$, $t_{32}' = 0$) as shown in Fig. 2-45, only the *zeros* are effected by the feedforward voltage divider α'.

EXAMPLE: A SECOND-ORDER NETWORK WITH COMPLEX ZEROS: To demonstrate the difference between the single-ended-input, dual-feedback configuration of Fig. 2-33 and the double-ended-input, single-feedback configuration of Fig. 2-45, we shall consider a practical circuit example. For the sake of a comparison between the two configurations, we shall select the same RC network $t(s)$ as was used in Fig. 2-35, and simply connect the voltage divider characterized by α according to Fig. 2-45. The resulting circuit is shown in Fig. 2-46. By inspection of Fig. 2-46, t_{12} has bandpass character and can therefore be represented in the form

$$t_{12}(s) = \frac{n_{12}(s)}{\hat{d}(s)} = k_{12} \cdot \frac{\omega_0 s}{s^2 + \dfrac{\omega_0 s}{\hat{Q}} + \omega_0^2} \qquad (2\text{-}141a)$$

As we have already seen from (2-139), the poles of $T(s)$ are realized by the zeros of $t_{32}(s)$. Thus, if $T(s)$ is to have complex poles, $t_{32}(s)$ must have the form

$$t_{32}(s) = \frac{n_{32}(s)}{\hat{d}(s)} = k_{32} \frac{s^2 + \dfrac{\omega_z}{q_z} s + \omega_z^2}{s^2 + \dfrac{\omega_0}{\hat{Q}} s + \omega_0^2} \qquad (2\text{-}141b)$$

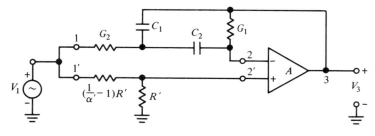

FIG. 2-46. Class 3 network in the double-ended-input, single-feedback mode.

that is, $t_{32}(s)$ is realized by a frequency rejection network and the pole generation corresponds to that of a class 3 network. Straightforward analysis gives

$$k_{12} = \left[\frac{G_2 C_2}{G_1 C_1}\right]^{1/2}; \qquad k_{32} = 1; \qquad \omega_0 = \omega_z = \left[\frac{G_1 G_2}{C_1 C_2}\right]^{1/2}$$

$$q_z = \left[\frac{G_2}{G_1} \cdot \frac{C_1 C_2}{(C_1 + C_2)^2}\right]^{1/2}; \qquad \hat{Q} = \frac{[G_1 G_2 C_1 C_2]^{1/2}}{G_1(C_1 + C_2) + G_2 C_2} \qquad (2\text{-}142)$$

Since t_{12} and t_{32} both apply to $t(s)$ they will have the same denominator or poles. Substituting (2-141a) and (2-141b) into (2-139) we obtain

$$T(s) = \alpha' \frac{s^2 + \omega_0\left(\frac{1}{\hat{Q}} - \frac{k_{12}}{\alpha'}\right)s + \omega_0^2}{s^2 + \frac{\omega_0}{q_z}s + \omega_0^2} \qquad (2\text{-}143)$$

Depending on k_{12}/α' being larger or smaller than $1/\hat{Q}$, RHP or LHP zeros can be obtained.

The dependence of the zeros of $T(s)$ on α' becomes clear if we consider the root locus of the zeros with respect to α'. With the expressions in (2-141) we have

$$T(s) = \frac{n_{12} - \alpha'\hat{d}}{n_{32}} \qquad (2\text{-}144)$$

The zeros of $T(s)$ are the roots of the characteristic equation

$$n_{12} - \alpha'\hat{d} = 0 \qquad (2\text{-}145a)$$

or

$$\alpha' \frac{\hat{d}}{n_{12}} = \alpha' \frac{(s - P_1)(s - P_2)}{\omega_0 s} = 1 \qquad (2\text{-}145b)$$

which corresponds to the zero-degree or positive-feedback locus shown in Fig. 2-47. Clearly by varying only α' (i.e., a single resistor in the circuit of Fig. 2-46) zeros anywhere in the s-plane can be obtained. In particular, for $\alpha' = 0$ we obtain a bandpass network, for $\alpha' = \alpha_1$ an all-pass network,[39] and for $\alpha' = \alpha_2$ a frequency-rejection network. From (2-143) we find

$$\alpha_1 = k_{12} \frac{q_z \hat{Q}}{q_z + \hat{Q}} \qquad (2\text{-}146a)$$

and

$$\alpha_2 = k_{12} \hat{Q} \qquad (2\text{-}146b)$$

39. T. Deliyannis, RC active all-pass sections, Electronics Lett. 5, 59 (1969).

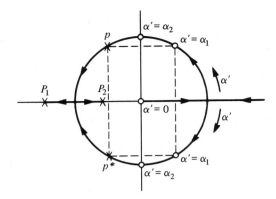

FIG. 2-47. Root locus corresponding to (2-145b).

As we shall see in Chapter 3 this network can be modified by including positive feedback (i.e., $Y'_2 \neq 0$ in Fig. 2-44) to provide an additional degree of freedom in locating the poles and zeros of (2-143).

A general comparison of the circuits shown in Figs. 2-33 and 2-45 is given in Table 2-3. Starting out with the general network in which dual feedback and a double-ended input is used, the dual-feedback mode is obtained by setting $\alpha'_1 = 0$, the double-ended input mode, by setting $\alpha'_2 = 0$. If, in the dual-feedback mode, t_{12} is a ladder network (see, e.g., Fig. 2-34), then $T(s)$ can have only negative real zeros (the origin included). On the other hand, the poles which are primarily determined by the zeros of $t_{32}(s)$ and may therefore be complex, may be further modified by the positive-feedback factor α'_2. In the double-ended-input case ($\alpha'_2 = 0$), the situation is reversed. The negative real zeros of t_{12} can be placed anywhere in the s-plane because of the subtraction of α' in the numerator. The poles, on the other hand, are determined by the zeros of $t_{32}(s)$ only. Thus, the combination of both, as shown for the general circuit in Table 2-3, in which $\alpha'_1 \neq 0$ and $\alpha'_2 \neq 0$, provides the widest flexibility as far as pole and zero realizations in the s-plane are concerned.

Combining Constant Negative Feedback with Frequency-Dependent Positive Feedback   We shall examine here briefly what might be considered the dual or inverse version of Fig. 2-44. This is shown in Fig. 2-48. We need not dwell long on this case since the resulting expressions resemble the previous ones closely. Unprimed, but otherwise identical terms to those in (2-136) are obtained for t_{12} and t_{32} and the overall transfer function corresponding to (2-137) results as

$$T(s)\bigg|_{A \gg 1} = - \frac{Y_1 - t'_{12}(Y_1 + Y_2 + Y_3)}{Y_2 - t'_{32}(Y_1 + Y_2 + Y_3)} \qquad [2\text{-}147]$$

TABLE 2-3. ZERO AND POLE MODIFICATION WITH DUAL FEEDBACK AND DOUBLE-ENDED INPUT

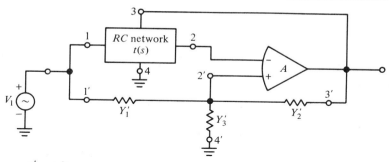

$$Y_1' = \frac{\alpha_1'}{1-\alpha_1'}\frac{1}{R}$$

$$Y_2' = \frac{\alpha_2'}{1-\alpha_2'}\frac{1}{R}$$

$$Y_3' = \frac{1}{R}$$

$$T(s)\big|_{A\gg1} = -\frac{t_{12}(Y_1'+Y_2'+Y_3')-Y_1'}{t_{32}(Y_1'+Y_2'+Y_3')-Y_2'} = -\frac{t_{12}(1-\alpha_1'\alpha_2')-\alpha_1'(1-\alpha_2')}{t_{32}(1-\alpha_1'\alpha_2')-\alpha_2'(1-\alpha_1')}$$

Dual feedback: $\alpha_1' = 0$ Double-ended input: $\alpha_2' = 0$

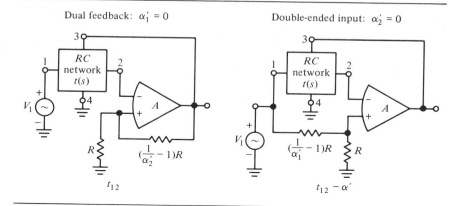

t_{12} $t_{12}-\alpha'$

Consider now the case of constant or frequency independent negative feedback as shown in Fig. 2-49. By inspection we have

$$Y_1 = G, \qquad Y_2 = \frac{\alpha}{1-\alpha}G, \qquad Y_3 = 0 \qquad (2\text{-}148a)$$

and

$$t_{12} = (1-\alpha), \qquad t_{32} = \alpha \qquad (2\text{-}148b)$$

Substituting into (2-135) (or (2-147)), we obtain

$$T(s)\bigg|_{A\gg1} = \frac{\alpha-1+t_{12}'}{\alpha-t_{32}'} \qquad [2\text{-}149]$$

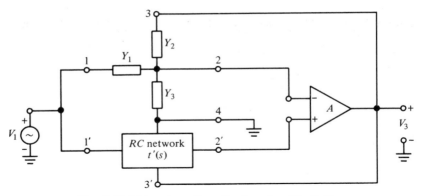

FIG. 2-48. Network configuration based on Fig. 2-43.

Letting $\beta = 1/\alpha$ (see (2-117))

$$T(s)\bigg|_{A \gg 1} = \frac{1 + \beta(t'_{12} - 1)}{1 - \beta t'_{32}} \qquad [2\text{-}150]$$

As expected, the conductive divider characterized by α here affects both the numerator *and* denominator of $T(s)$, since the configuration in Fig. 2-49 comprises both dual feedback and a double-ended input. A comparison of Figs. 2-49 and 2-38, showing the similarity between the two configurations, is given in Table 2-4. Both are contained in the general form shown in Fig. 2-48. Fig. 2-38 can be obtained by setting

$$Y_1 = 0, \qquad Y_2 = \frac{\alpha}{1 - \alpha} G, \qquad Y_3 = G \qquad (2\text{-}151\text{a})$$

so that

$$t_{12} = 0, \qquad t_{32} = \alpha \qquad (2\text{-}151\text{b})$$

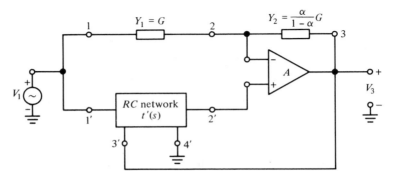

FIG. 2-49. Simplified version of Fig. 2-48.

TABLE 2-4. ZERO AND POLE MODIFICATION WITH DUAL FEEDBACK AND DOUBLE-ENDED INPUT

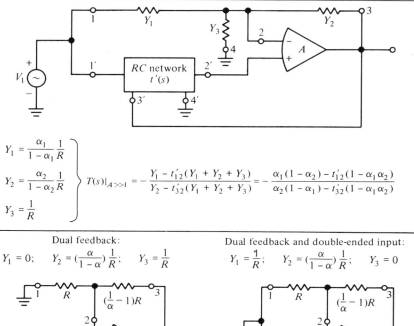

$$Y_1 = \frac{\alpha_1}{1-\alpha_1}\frac{1}{R}$$

$$Y_2 = \frac{\alpha_2}{1-\alpha_2}\frac{1}{R}$$

$$Y_3 = \frac{1}{R}$$

$$T(s)|_{A\gg1} = -\frac{Y_1 - t'_{12}(Y_1+Y_2+Y_3)}{Y_2 - t'_{32}(Y_1+Y_2+Y_3)} = -\frac{\alpha_1(1-\alpha_2) - t'_{12}(1-\alpha_1\alpha_2)}{\alpha_2(1-\alpha_1) - t'_{32}(1-\alpha_1\alpha_2)}$$

Dual feedback:	Dual feedback and double-ended input:
$Y_1 = 0; \quad Y_2 = (\frac{\alpha}{1-\alpha})\frac{1}{R}; \quad Y_3 = \frac{1}{R}$	$Y_1 = \frac{1}{R}; \quad Y_2 = (\frac{\alpha}{1-\alpha})\frac{1}{R}; \quad Y_3 = 0$

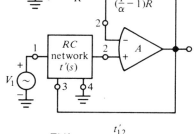 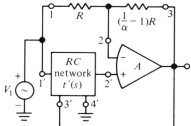

$$T(s)|_{A\gg1} = \frac{t'_{12}}{\alpha - t'_{32}}$$

$$= \frac{\beta t'_{12}}{1 - \beta t'_{32}}$$

$$T(s)|_{A\gg1} = \frac{\alpha - 1 + t'_{12}}{\alpha - t'_{32}}$$

$$= \frac{1 + \beta(t'_{12} - 1)}{1 - \beta t'_{32}}$$

Substituting into (2-135), (2-116) results. The denominators in (2-116) and (2-149) are identical since, as far as the feedback is concerned, both circuits behave in the same way. It is the double-ended input—or the lack of it— that affects the numerator, as a comparison of the two expressions clearly shows.

EXAMPLE: A SECOND-ORDER NETWORK WITH COMPLEX ZEROS: Since the zeros can be influenced by α in the network topology of Fig. 2-49 much as they are in the (dual) topology of Fig. 2-45, it is to be expected that, while heeding this duality, complex zeros may be obtained in a very similar fashion.

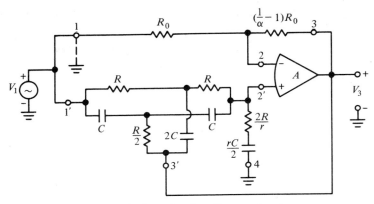

FIG. 2-50. Second-order all-pass network.

The duality of the two topologies corresponds to that already pointed out in Fig. 2-39. There we compared a class 3 network with constant positive feedback with a class 4 network, with constant negative feedback. Here the dual to the class 3 network with constant positive feedforward (Fig. 2-46) must be a class 4 network with constant negative feedforward.[40] Thus, where t_{12} and t_{32} in the configuration of Fig. 2-45 are, respectively, bandpass and frequency-rejection functions (see (2-141)) in order to obtain complex zeros, t'_{12} and t'_{32} for the configuration of Fig. 2-48 must be, respectively, frequency-rejection and bandpass functions. Thus, we have

$$t'_{12} = k_{12} \cdot \frac{s^2 + \omega_z^2}{s^2 + \dfrac{\omega_0}{\hat{Q}} s + \omega_0^2} \qquad (2\text{-}152a)$$

and

$$t'_{32} = k_{32} \frac{\omega_0 s}{s^2 + \dfrac{\omega_0}{\hat{Q}} s + \omega_0^2} \qquad (2\text{-}152b)$$

A class 4 network with negative feedforward (and feedback) that provides the required functions t'_{12} and t'_{32} as given by (2-152) is shown in Fig. 2-50.[41] Here t'_{12} of the form of (2-152a) is evidently realized by a symmetrical twin-T loaded by a series RC network and t'_{32} has bandpass character as specified by (2-152b). Thus, with the designations of Fig. 2-50 we have

$$k_{12} = 1; \qquad \omega_z = \omega_0 = \frac{1}{RC}; \qquad 0 < \alpha < 1$$

$$k_{32} = 4; \qquad \hat{Q} = \frac{1}{4 + r} \qquad (2\text{-}153)$$

40. The fact that here we also have constant negative feedback is of no significance, since this does not affect the zeros of the network.
41. G. S. Moschytz, A general all-pass network based on the Sallen–Key circuit, *IEEE Trans. Circuit Theory*, **CT-19**, pp. 392–394 (1972).

Substituting these expressions into (2-149) we obtain

$$
T(s)\bigg|_{A \gg 1} = \frac{s^2 + \dfrac{\alpha - 1}{\alpha}\dfrac{\omega_0}{\hat{Q}}s + \omega_0^2}{s^2 + \left(1 - 4\dfrac{\hat{Q}}{\alpha}\right)\dfrac{\omega_0}{\hat{Q}}s + \omega_0^2}
\tag{2-154}
$$

Here again we can select α to provide a given pole–zero configuration. To obtain an all-pass network, for instance, we require that

$$
\alpha\big|_{\text{all-pass}} = \tfrac{1}{2}(4\hat{Q} + 1)
\tag{2-155}
$$

and, with (2-153)

$$
\alpha\bigg|_{\text{all-pass}} = \frac{8 + r}{2(4 + r)}
\tag{2-156}
$$

More will be said about this and related networks in Chapter 3. The important thing to recognize again here is that by using the differential- or double-ended-input mode, complex conjugate zeros can readily be obtained. By simple reasoning and application of root-locus theory it becomes clear that where t_{12} for the topology of Fig. 2-46 has to have bandpass character in order to obtain complex zeros in the s-plane, for the topology of Fig. 2-49, t'_{12} must be realized by a band-rejection network.[42] Naturally a twin-T or a bridged-T network can be used for the latter. Either way, the zeros of t'_{12} are shifted into the left- or right-half s-plane according to the value selected for α. The corresponding root locus of the zeros of the numerator of (2-150) with respect to β (where $\beta = 1/\alpha$) coincides with that in Fig. 2-47, except that with increasing β the locus is traversed in the opposite direction. At the same time, because the α network also provides negative *feedback*, its value also affects the poles of the network.

Let us now examine the circuit of Fig. 2-50 when terminal 1 is grounded (as shown by the dashed line) instead of being connected to the input voltage source V_1. We then have the circuit configuration of Fig. 2-38 and the transfer function (2-116). With $t'_{12}(s)$ and $t'_{32}(s)$ as given by (2-152) and (2-153) we then obtain

$$
T(s)\bigg|_{A \gg 1} = \frac{1}{\alpha}\frac{s^2 + \omega_0^2}{s^2 + \left(1 - 4\dfrac{\hat{Q}}{\alpha}\right)\dfrac{\omega_0}{\hat{Q}}s + \omega_0^2}
\tag{2-157}
$$

Note that the zeros of $T(s)$ are the zeros of $t'_{12}(s)$ and, being realized by a balanced twin-T, they lie on the $j\omega$ axis; the poles are the same as those of (2-154). The resulting network defines a frequency-rejection network (see Chapter 1, Table 1-3). The all-pass and frequency-rejection networks con-

---

42. This can also be shown by considering the circuit configurations that result when the input and ground terminals of a single-amplifier network are interchanged; see D. Hilberman, Input and ground as complements in active filters, *IEEE Trans. Circuit Theory*, **CT-20** pp. 540–547 (1973); also see Chapter 3, Section 3.1.4 of *Linear Integrated Networks: Fundamentals*.

tained in Fig. 2-50 therefore correspond to the two circuit types compared in the lower half of Table 2-4. With the dual-feedback, single-ended input network, complex zeros are determined exclusively by the transmission zeros of the passive RC network; with the dual-feedback double-ended input network the complex zeros are the result of a differential-amplifier subtraction of the signal $t'_{12} V_1$ from the signal $t_{12} V_1$ (see (2-135)). In either case a frequency-rejection network can be obtained. However, where the single-ended input method will be preferable when highly accurate and stable passive components are available, the double-ended input method may be used when a high-quality differential amplifier with a large common-mode range and a high common-mode rejection ratio is at hand.

With zeros on, or close to, the $j\omega$ axis and complex conjugate poles, the frequency-rejection network is very useful in filter design and therefore very commonly used. It provides transmission loss at the frequency ω_0 and is required in the design of so-called elliptic filters (also sometimes called filters with finite zeros). The skirts of all-pole filters (e.g., Butterworth, Chebyshev) roll off monotonically; their zeros are at infinity.[43]

Second-Order Networks Based on the Synthesis of Driving Point Admittances   An obvious step after the general configurations given in Figs. 2-44 and 2-48 is to retain only the conductances Y_i and Y'_i as shown in Fig. 2-51. Substituting (2-136a) and (2-136b) and identical unprimed expressions into (2-135) we obtain:[44]

$$T(s)\bigg|_{A \gg 1} = \frac{Y'_1(Y_1 + Y_2 + Y_3) - Y_1(Y'_1 + Y'_2 + Y'_3)}{Y_2(Y'_1 + Y'_2 + Y'_3) - Y'_2(Y_1 + Y_2 + Y_3)} \qquad [2\text{-}158]$$

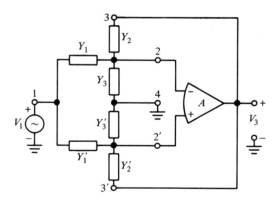

FIG. 2-51. Network based on synthesis of driving point admittances.

43. For the theory of these filter types see, e.g., L. Weinberg, *Network Analysis and Synthesis*, (New York: McGraw-Hill, 1962); Chapter 11.
44. S. K. Mitra, Active *RC* filters employing a single operational amplifier as the active element, *Proc. Hawaii Intern. Conf. System Sci.*, Honolulu, Hawaii, Jan. 1968. J. S. Brugler, *RC* Synthesis with differential-input operational amplifiers, in *Papers on Integrated Circuit Synthesis*, Tech. Rep. 6560-4, SEL 66-058, Stanford Electronics Lab., Stanford, California, June 1966.

Constraining the network elements such that

$$Y_1 + Y_2 + Y_3 = Y_1' + Y_2' + Y_3' \tag{2-159}$$

(2-158) simplifies to

$$T(s) = \frac{Y_1' - Y_1}{Y_2 - Y_2'} \tag{2-160}$$

Note that condition (2-159) implies that the same admittance is seen looking into the RC network from terminal 2 to ground as from 2′ to ground. With that condition satisfied, zeros and poles are obtained by synthesizing Y_1', Y_1, Y_2', and Y_2 in such a way that the differences in (2-160) are zero at the desired locations in the s-plane. This method has the advantage that only driving point admittances need be synthesized to obtain a given transfer function.

Let us illustrate the procedure required to synthesize the admittances of (2-160) such as to realize a given transfer function $T(s) = N(s)/D(s)$. As might be expected by the form of (2-160), this synthesis is based directly on the admittance version of the $(RC):(-RC)$ decomposition. We first select a polynominal $Q(s)$ with distinct negative real roots whose degree-plus-one is equal to or larger than the degree of $N(s)$ or $D(s)$, whichever of the two is larger. Because of the properties of the $(RC):(-RC)$ decomposition we can then express $T(s)$ in the form

$$T(s) = \frac{N(s)/Q(s)}{D(s)/Q(s)} = \frac{Y_A - Y_B}{Y_C - Y_D} \tag{2-161}$$

where the Y's are passive RC driving point admittances. Comparing (2-160) and (2-161) we have

$$Y_1' = Y_A; \qquad Y_1 = Y_B; \qquad Y_2 = Y_C; \qquad Y_2' = Y_D \tag{2-162}$$

The remaining two admittances follow from (2-159):

$$\begin{aligned} Y_3 &= Y_1' + Y_2' \\ Y_3' &= Y_1 + Y_2 \end{aligned} \tag{2-163}$$

Notice that (2-159) remains satisfied if any common terms on the right-hand side of (2-163) are subtracted. As we shall see below, this can be used to reduce the complexity of Y_3 and Y_3'.

EXAMPLE: We are now ready to go through a numerical example. Let us realize the all-pass function with double poles and zeros given by

$$T(s) = \frac{(s-2)^2}{(s+2)^2}$$

Selecting $Q(s) = s + 2$ and decomposing in the form of (2-161) we obtain

$$\frac{N(s)}{Q(s)} = \frac{s^2 - 4s + 4}{s + 2} = s + 2 - \left(\frac{8s}{s + 2}\right)$$

$$\frac{D(s)}{Q(s)} = \frac{(s + 2)^2}{s + 2} = s + 2$$

With (2-162) it therefore follows that:

$$Y_1' = s + 2; \qquad Y_1 = \frac{8s}{s + 2}$$

$$Y_2 = s + 2; \qquad Y_2' = 0$$

and, from (2-163),

$$Y_3 = s + 2; \qquad Y_3' = s + 2 + \frac{8s}{s + 2}$$

Since the term $s + 2$ is common to Y_3' and Y_3 it can be subtracted from both and we obtain

$$Y_3 = 0; \qquad Y_3' = \frac{8s}{s + 2}$$

The resulting network is shown in Fig. 2-52.

An example involving a specified amplitude function may also be of interest. Let us realize a frequency-rejection network whose transfer function is given by

$$T(s) = \frac{N(s)}{D(s)} = \frac{s^2 + 2}{s^2 + 0.1s + 1}$$

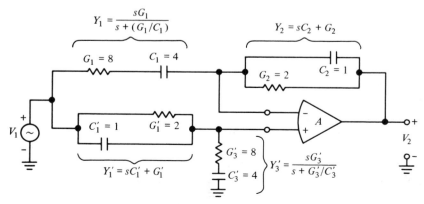

FIG. 2-52. All-pass network based on configuration of Fig. 2-50.

Selecting a polynominal $Q(s) = s + 1$ and going through the same steps as above, we obtain

$$\frac{N(s)}{Q(s)} = \frac{s^2 + 2}{s + 1} = s + 2 - \left(\frac{3s}{s + 1}\right)$$

$$\frac{D(s)}{Q(s)} = \frac{s^2 + 0.1s + 1}{s + 1} = s + 1 - \frac{1.9s}{(s + 1)}$$

Thus, with (2-162),

$$Y_1' = s + 2; \qquad Y_1 = \frac{3s}{s + 1}; \qquad Y_2 = s + 1; \qquad Y_2' = \frac{1.9s}{s + 1}$$

and with (2-163)

$$Y_3 = s + 2 + \frac{1.9s}{s + 1}; \qquad Y_3' = s + 1 + \frac{3s}{s + 1}$$

Subtracting $s + 1 + 1.9s/(s + 1)$ from Y_3 and Y_3', we obtain

$$Y_3 = 1.0; \qquad Y_3' = \frac{1.1s}{s + 1}$$

The final circuit is given in Fig. 2-53. Notice that the networks connected to the operational-amplifier terminals in Fig. 2-52 and 2-53 are RC ladder networks. As such their transmission zeros lie on the negative real axis and can, at best, coincide to form double zeros. However, by subtracting the output signals of the two respective ladder networks in the manner indicated by the $(RC):(-RC)$ decomposition, complex zeros are readily obtained anywhere in the s-plane.

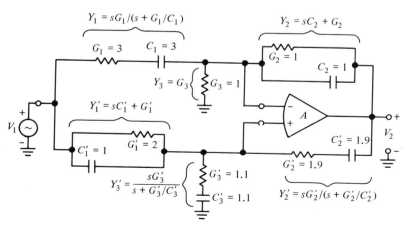

FIG. 2-53. Frequency-rejection network based on configuration of Fig. 2-50.

CHAPTER

3

ACTIVE FILTER BUILDING BLOCKS

INTRODUCTION

In Chapter 1 we discussed the fact that high-order filters are broken down into second-order subsections, when realized in active *RC* form. In a way, then, the second-order network is in itself an active-filter building-block with which higher-order networks can be obtained. Nevertheless, the large variety of second-order networks that emerged in the preceding chapter will make it clear to the reader that an additional selection is required in order to find a useful building block—or family of building blocks—with which to assemble any conceivable *n*th-order filter.

With integrated filter implementation in mind, the desirability of such a building-block should require little explanation. After all, the success that integrated *digital* systems have met with is due, in large measure, to the fact that they are broken down into simple modules, or building blocks: high-quantity, and consequently low-cost, products are thereby guaranteed. The same reasoning applies here. The large variety of filter types must be drastically reduced by breaking down complex filter networks into small families of basic second-order building blocks. Assuming that the building blocks are sufficiently versatile to be used in a large number of low-run applications, the demand for them will increase sufficiently to justify the high costs of integrated-circuit production; the initially supposed economical unfeasibility of linear integrated active filters can thereby be overcome.

The close tie between the economic feasibility of integrated active filters and their practical implementation was recognized in the mid-sixties[1] and a

---

1. G. S. Moschytz, unpublished Bell Telephone Laboratories Memorandum, 1965.

method of active-filter design using general-purpose building blocks was proposed and subsequently developed.[2] Various other methods were suggested soon thereafter. The most important will be discussed in this chapter. Those using single amplifiers are, of course, direct derivatives of the circuits discussed in the preceding chapter. Most of the *multiple*-amplifier building blocks have different origins.

No attempt will be made in this chapter to qualify the various building blocks described. We must remember that the choice of a particular circuit is closely related both to the technology used for its implementation and to the application for which it is intended. Since these aspects will not be discussed here, any qualification would have little meaning. On the other hand, when we focus our attention on hybrid integrated implementation in the following chapters, we shall find that certain building-block types are particularly well suited to that technology.

3.1 SINGLE-AMPLIFIER BUILDING BLOCKS

A filter building block must be capable of realizing the general biquadratic transfer function

$$T(s) = \frac{N(s)}{D(s)} = \frac{b_2 s^2 + b_1 s + b_0}{a_2 s^2 + a_1 s + a_0} = K \frac{s^2 + \dfrac{\omega_z}{q_z} s + \omega_z^2}{s^2 + \dfrac{\omega_p}{q_p} s + \omega_p^2} \qquad [3\text{-}1]$$

in all its possible variations (see Table 1-3). In its most general form, it is characterized by five parameters, namely K, ω_z, q_z, ω_p, and q_p. Assuming an amplifier gain of β, each of these parameters may depend on all r resistors R_i, all c capacitors C_i, and on β. If ω_z and q_z are independent of β, the network is type I, in terms of its transmission sensitivity with respect to β. Since the amplitude response is considerably more sensitive to variations in ω_p than to variations in q_p (see Chapter 1, Section 1.5.2) it is important that ω_p be as stable as possible. An obvious step in this direction is to use only circuits in which ω_p is independent of β.[3] This immediately disqualifies class 1 or class 2 networks as building-block candidates. Whether it is for this reason or because they are inherently less versatile,[4] these two network

2. G. S. Moschytz, Miniaturized filter building blocks using frequency emphasizing networks *Proc. Nat. Conf.* **23**, 364–369 (1967).
3. Any remaining measure to stabilize ω_p is in the technological sphere, i.e., it will depend on the quality of the passive components used. Thus when considering means of stabilizing the characteristics of building blocks in this and the following chapters we shall, in general, be concerned with the stabilization of q_p.
4. Class 2 networks are essentially the $(RC):(CR)$ duals of class 1 networks. By contrast class 3 and class 4 networks contain their own dual networks within their respective classes, thereby significantly increasing the number of possible networks in each class.

classes have not, in fact, ever been suggested for use as building blocks. This leaves us with class 3 and class 4 networks, whose versatility and similarity has already been pointed out in Chapter 2.

When we speak of "versatility" we are introducing a characteristic not generally associated with networks, but vital to the success of a building block in integrated form. In this context, a *versatile* second-order network is one whose parameters K, ω_z, q_z, ω_p, and q_p can be modified easily or, more specifically a network that provides every possible second-order transfer function (over a given frequency range) *while requiring a minimum number of circuit modifications to do so.*

We recall from Chapter 1 that "every possible" second-order network function in essence comprises the following six basic networks:

1. Low-pass network (LPN)
2. High-pass network (HPN)
3. Bandpass network (BPN)
4. Frequency-emphasizing network (FEN)
5. Frequency rejection network (FRN)
6. All-pass network (APN)

as listed in Chapter 1, Table 1-3. Various additional functions, such as the low-pass resonator $(N(s) = s + \omega_z)$ or the high-pass resonator $(N(s) = s(s + \omega_z))$ may also be required; given the capability of realizing the six just mentioned, the realization of these additional functions follows readily. Third-order realization may also be required and, as we shall see, can be readily obtained; the third pole, being negative real, may be realized by an additional first-order RC low-pass section. Thus, without any loss in generality, we can restrict ourselves, for the time being, to finding building-blocks capable of realizing the six given functions. The final selection of a building-block may then depend, among other things, on the ease with which it can be modified to third-order operation or to the generation of negative real zeros, as required by a resonator.

A useful second-order building block may have the form shown in Fig. 3-1a, where each second-order function is available at a different terminal, and a separate resistor[5] is available for the fine adjustment of each parameter. Very often it is impractical to have a different terminal for each function; the fixtures used for processing and tuning the building blocks should remain unaltered, irrespective of the function used, as should the sockets or other containers for the finished product. The concept illustrated in Fig. 3-1b satisfies these constraints more readily. Here, the locations of the input and output terminals (as well as the supply, ground, and any other external terminals) are independent of the function used; the function is determined

---

5. We recall that only integrated *resistors* can readily be fine-adjusted (see Chapter 6 of *Linear Integrated Networks: Fundamentals*).

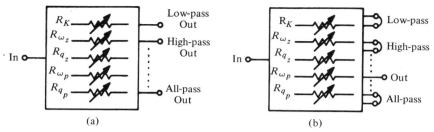

FIG. 3-1. General building block concepts: (a) one output terminal per function; (b) one output terminal, one internal connection per function.

by some internal or external physical interconnections (e.g., conductors on the substrate or on the printed-wiring board beneath, wire straps, socket interconnections). These considerations may appear to be "hardware trivia" to the network designer, until he discovers that the choice of a network in any given application may well be determined by nothing more significant than that.

3.1.1 Class 3 Building Blocks

Bridged-T Feedback Networks (The Single-Amplifier Biquad)

The Circuit   Consider the circuit shown in Fig. 3-2. We will, no doubt, recognize it as a dual-feedback type, of the kind shown in Chapter 2, Fig. 2-33, in which the bridged-T is in the negative-feedback path, corresponding to class 3 operation, and a resistive feedback network (G_b, G_d) provides positive feedback.[6] This is, in fact, precisely the type of network proposed by Deliyannis (see Fig. 2-35) and discussed under 2.4.1. By adding the conductors G_4, G_6, G_7 and G_c (shown by the dashed lines) to the circuit, and connecting these to the input voltage source V_1 (with the exception of G_7, which loads the bridged-T) the circuit is converted into a dual-feedback, double-ended network of the type discussed in Section 2.4.3 (see Fig. 2-44) of which Deliyannis' all-pass network (Fig. 2-46) is simply a special case. This circuit, redrawn in Fig. 3-3, was proposed by Friend as a general-purpose second-order filter building block.[7] He refers to it as a single-amplifier biquad (SAB) because it realizes an arbitrary biquadratic transfer function.[8] Assuming

---

6. For a class 3 building block combining negative feedback with the double-ended input mode see P. O. Brackett, A new single op amp active *RC* network configuration, *Proc. 16th Midwest Symp. Circuit Theory*, Waterloo, Ontario, Canada, 1973.
7. J. J. Friend, A single operational-amplifier biquadratic filter section, *1970 IEEE Int. Symp. Circuit Theory, Digest of Tech. Papers*, pp. 179–180, December 1970.
8. From this point of view, any general-purpose building block capable of realizing any second-order transfer function can be referred to as a biquad (this being the characteristic of a general-purpose building-block). For convenience however, and because the name was coined for this, and the circuit described in Section 3.3.1 (originally for the latter) we shall refer here to the "single-amplifier biquad" and to the circuit of 3.3.1 as the "multi-amplifier biquad"; the other biquads in this chapter will be referred to, wherever possible, by the designations introduced by their originators.

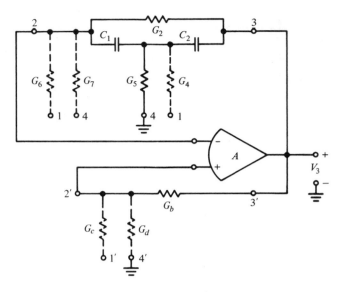

FIG. 3-2. Class 3 network with frequency-independent positive feedback.

infinite amplifier gain, we obtain the transfer function $T(s) = V_2/V_1$ in the form of (3-1), where

$$b_2 = \frac{G_c}{G_a} \tag{3-2a}$$

$$b_1 = \frac{1}{G_a} \left\{ \left(\frac{1}{C_1} + \frac{1}{C_2} \right) [G_c(G_2 + G_7) - G_6(G_b + G_d)] \right.$$

$$\left. + \frac{1}{C_2} [G_c G_5 - G_4(G_b + G_d)] \right\} \tag{3-2b}$$

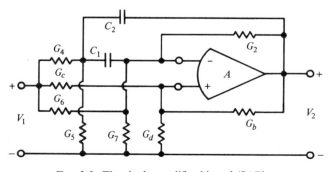

FIG. 3-3. The single-amplifier biquad (SAB).

$$b_0 = \frac{G_4 + G_5}{C_1 C_2 G_a} [G_c(G_2 + G_7) - G_6(G_b + G_d)] \tag{3-2c}$$

$$a_2 = 1 \tag{3-2d}$$

$$a_1 = \frac{1}{G_a} \left[(G_2 G_a - G_b G_3) \left(\frac{1}{C_1} + \frac{1}{C_2} \right) - \frac{G_b G_1}{C_2} \right] \tag{3-2e}$$

$$a_0 = \frac{G_1}{C_1 C_2 G_a} (G_2 G_a - G_b G_3) \tag{3-2f}$$

and

$$G_a \equiv G_c + G_d \tag{3-3a}$$

$$G_1 \equiv G_4 + G_5 \tag{3-3b}$$

$$G_3 \equiv G_6 + G_7 \tag{3-3c}$$

Equating the coefficients b_i, $i = 0, 1, 2$ to the appropriate values, the design equations and constraints for the desired functions are obtained. Thus, for example, for a bandpass network we have

$$\text{BPN:} \qquad G_6 = G_7 = G_c = 0 \tag{3-4}$$

Similarly, for frequency-rejection and all-pass networks,

$$\text{FRN } (\omega_z > \omega_p): \qquad G_6 = 0 \tag{3-5}$$

$$\text{FRN } (\omega_z < \omega_p): \qquad G_7 = 0 \tag{3-6}$$

$$\text{APN:} \qquad G_b = G_5 = G_6 = G_7 = 0 \tag{3-7}$$

A high-pass network is obtained by setting $b_1 = b_0 = 0$ and solving the resulting expressions in such a way that none of the components turn out to be negative. Naturally, this latter constraint must be heeded for all the filter functions. In the case of the high-pass filter, though, it is probably easier to add an additional capacitor to the circuit (see Fig. 2-24c), producing the configuration shown in Fig. 3-4. The low-pass section actually requires a

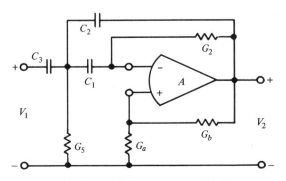

FIG. 3-4. A SAB high-pass network.

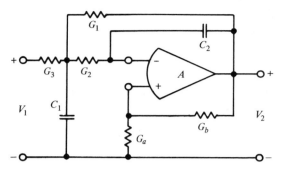

FIG. 3-5. A SAB low-pass network.

separate configuration, (corresponding to Fig. 2-24d, in which the RC-dual bridged-T is used) since it cannot be obtained from the general circuit of Fig. 3-3. The resulting circuit is shown in Fig. 3-5.

Circuit Transforms   Each of the functions realizable by the general-purpose circuit of Fig. 3-3 are restricted to some extent by realizability conditions and by constraints on the available pass-band gain. Realizability conditions generally become problematic when they require large component spreads or component values. The solution to this problem is to select reasonable capacitor values and to transform the configuration of those resistors whose values are required to be large, in order to obtain acceptable values. This is done by using the well known delta–Y transform. Referring to Fig. 3-6, the Y impedances in terms of the delta impedances are

$$Z_A = \frac{Z_{AB} Z_{AC}}{Z_T}; \qquad Z_B = \frac{Z_{AB} Z_{BC}}{Z_T}; \qquad Z_C = \frac{Z_{AC} Z_{BC}}{Z_T} \qquad [3\text{-}8]$$

where

$$Z_T = Z_{AB} + Z_{AC} + Z_{BC}$$

Referring to Fig. 3-3, if R_2 or R_7 is too large (e.g., in an FRN) a phantom resistor R_x may be added to the output of the amplifier (Fig. 3-7a), thereby

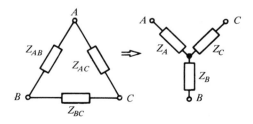

FIG. 3-6. The delta–Y transform.

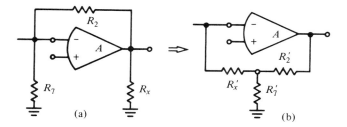

FIG. 3-7. Delta–Y transform applied to decrease R_2 or R_7.

forming the resistive delta R_7, R_2, R_x. Assuming that R_x does not load down the amplifier, the characteristics of the basic circuit (Fig. 3-3) will not be changed. Carrying out the delta–Y transform we obtain Fig. 3-7b.[9] To minimize the largest resistor in the Y, R_x should be equal to the smaller of R_2 and R_7.

Similarly, if R_2 or R_6 becomes too large (e.g., in the high-pass or FRN) a resistor R_x may be added from the input to the output of the network (Fig. 3-8a) without affecting its characteristics (i.e., no overloading of the amplifier). The transformed circuit is shown in Fig. 3-8b. Again, for optimal results R_x should equal the smaller of R_2 and R_6. Exactly the same situation applies if R_1 and/or R_3 is too large (e.g., in the low-pass section). R_3 then takes the place of R_6 and R_1 that of R_2.

In the circuit configurations in which either R_6 or R_7 (see Fig. 3-3) is left out (e.g., bandpass or all-pass circuits) R_2 may become very large. It can be reduced by connecting R_x across the amplifier input terminals (Fig. 3-9a) and, with R_b, forming a delta configuration that can then be transformed into a Y (Fig. 3-9b). R_b can usually be chosen at will, since the positive-feedback side of the circuit is totally resistive, and its impedance level may be changed without introducing increases in the capacitor values.

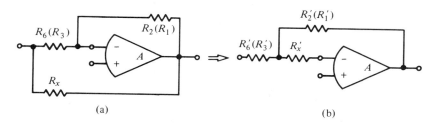

FIG. 3-8. Delta–Y transform applied to decrease R_2 or R_6.

9. Primes are added to the resistors in the transformed network.

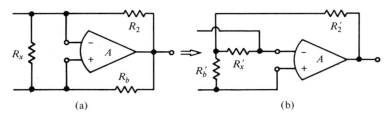

(a)   (b)

FIG. 3-9. Delta–Y transform applied to decrease R_6 or R_7.

Finally, a large R_4 and/or R_5 may be reduced by adding a phantom resistor from the input to ground (Fig. 3-10a) and then transforming to a Y (Fig. 3-10b). This case corresponds to the transformation discussed in Chapter 2, Section 2.3.3.

In all of these cases it must be determined whether the phantom resistor R_x appreciably loads the circuit. This is most likely to happen when R_x bridges the opamp input and output (and consequently also connects the output of the preceding stage, through R_x, to the output of the stage under consideration). If, for example, the (negative) gain of the stage under consideration is 100, and R_x is 10 kΩ, then the load on the previous stage will be less than 100 Ω.

The other problem mentioned above is that the user of the general circuit of Fig. 3-3, may be restricted, by the design and realizability equations, to pass-band gains well below those actually desired. Here again, a type of transform can come to the designers aid. Consider the partial circuit shown in Fig. 3-11a. It shows the output voltage determined by the voltage source V_0, and the voltages V_i and V_j at the ends of two arbitrary admittances y_{0i} and y_{0j}. If, now, we wish to have a μ-times-larger voltage at the output (i.e., μV_0) without affecting the voltages V_i and V_j at the nodes i and j, then the admittances leaving the voltage source (μV_0) must be divided by μ. For this purpose, additional admittances to the common reference are necessary such that the admittances y_{ii} and y_{jj} remain unchanged (Fig. 3-11b).[10] Obviously, in order to guarantee positive real admittances, we require that $\mu \geq 1$. Applying this transform to the RC combination shown in Fig. 3-12a, for example, we

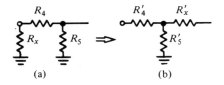

(a)   (b)

FIG. 3-10. Delta–Y transform applied to decrease R_4 or R_5.

10. W. Thelen, unpublished memorandum, Bell Telephone Laboratories, 1967.

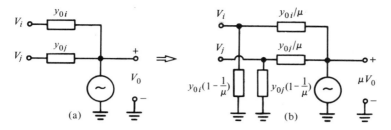

FIG. 3-11. General output voltage-level transform such that input levels remain unchanged.

obtain the configuration in Fig. 3-12b. Notice that, although the number of capacitors is increased by one, the *total* capacitance value has remained unchanged. In contrast, the two resistors are larger than R; their sum is $\mu^2 R/(\mu - 1)$. The minimum value for the total resistance occurs at $\mu = 2$; it is $4R$. If this increase in resistance is undesirable a modification of the transform shown in Fig. 3-11 can be carried out as follows. The resistor R (Fig. 3-12a) is broken up into two parts R_0 and $R - R_0$ (Fig. 3-13a) and the transform only carried out on one of the parts, e.g., on R_0 (see Fig. 3-13b). By a judicious choice of R_0, the resulting resistive T can be made to have (i) a minimum value for the largest resistor, (ii) a minimum value for the total resistance, or (iii) both of the above. Of course, these considerations are only of interest if the implementation is sensitive to large resistor values (as for example, with thin-film technology).

Instead of using the transforms shown in Figs. 3-11 to 3-13, an alternative method of obtaining an output voltage μ times larger than the voltage fed back within a circuit can be used. With a relatively low-impedance voltage divider at the amplifier output, $1/\mu$ times the output voltage can be tapped off and fed back into the circuit (Fig. 3-14). Since the amplifier's output impedance is roughly equal to the Thévenin-equivalent output impedance divided by the open-loop gain, the driving capability of the amplifier will not be significantly affected as long as the loop gain is sufficiently large. However, the open-loop gain is now decreased by a factor of $1/\mu$, so that the error

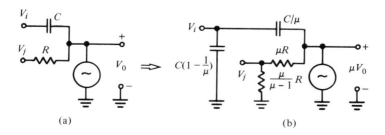

FIG. 3-12. Voltage-level transform applied to RC network.

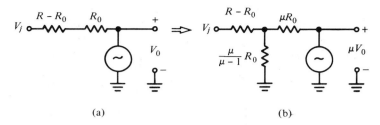

FIG. 3-13. Modified voltage-level transform.

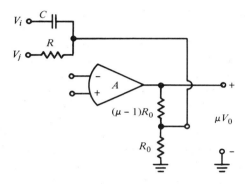

FIG. 3-14. Alternate output voltage-level transform.

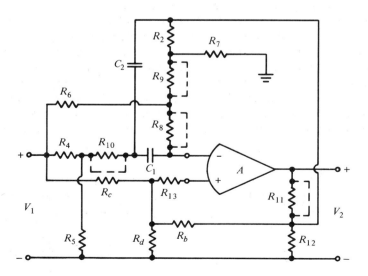

FIG. 3-15. The single-amplifier biquad after delta–Y and voltage-level transforms.

introduced by finite amplifier gain is increased. Consequently, this scheme should only be used for relatively small μ values corresponding to a gain enhancement of, say, less than 10 dB.

The transforms discussed above, both those for overcoming unacceptably high resistor values as well as those for obtaining pass-band gains not available with the original circuits, can, of course, be applied to any active RC networks incorporating operational amplifiers. They were discussed in conjunction with the circuit of Fig. 3-3 because without them this particular circuit would be severely limited. Applying them to Fig. 3-3 we obtain the circuit shown in Fig. 3-15. The reader will recognize the resistors R_4, R_5, and R_{10} as constituting the T transform of R_4 and R_5 of Fig. 3-3; likewise R_6 and R_2 are T transformed into R_6–R_8–R_2 (R_9 is shorted), and R_7, R_2 into R_9–R_7–R_2. R_{11} and R_{12} provide the output gain enhancement demonstrated in Fig. 3-14. When it is not required, R_{11} is short-circuited and R_{12} opened. R_{13} is inserted when necessary for DC offset compensation. Since all resistors are easily opened, only the ones normally shorted are indicated by (dashed) shorting conductors. Notice that the low-pass configuration requires a relocation of various components including capacitors C_1 and C_2, according to the configuration shown in Fig. 3-5.

Sensitivity Considerations For simplicity, we shall demonstrate the sensitivity characteristics of the single-amplifier biquad using the bandpass case as an example. This implies no lack of generality, as the poles of all the network types of a class are generated in the same way and it is primarily the pole sensitivity that determines the overall sensitivity characteristics of a network. For a bandpass function the circuit of Fig. 3-3 simplifies to the configuration shown in Fig. 3-16. (This corresponds exactly to the circuit discussed in Chapter 2, Section 2.4.1 (Fig. 2-35), except that the component

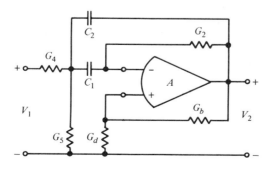

FIG. 3-16. A SAB bandpass network.

designations here conform to those of Fig. 3-3). Assuming infinite gain, a straightforward calculation gives us the following voltage transfer function

$$T(s) = \frac{V_2}{V_1} = -\frac{\dfrac{G_4}{C_2}\left(1 + \dfrac{G_b}{G_d}\right)s}{s^2 + \left[G_2\left(\dfrac{1}{C_1} + \dfrac{1}{C_2}\right) - \dfrac{G_b(G_4 + G_5)}{G_d C_2}\right]s + \dfrac{G_2(G_4 + G_5)}{C_1 C_2}}$$

$$= K\frac{\dfrac{\omega_p}{q_p}s}{s^2 + \dfrac{\omega_p}{q_p}s + \omega_p^2} \tag{3-9}$$

Notice that there are seven components to be chosen but only three equatious that need to be satisfied. Referring to (3-1) and (3-9) these are

$$b_1 = K\frac{\omega_p}{q_p} = \frac{G_4}{C_2}\left(1 + \frac{G_b}{G_d}\right) \tag{3-10a}$$

$$a_1 = \frac{\omega_p}{q_p} = G_2\left(\frac{1}{C_1} + \frac{1}{C_2}\right) - \frac{G_b(G_4 + G_5)}{G_d C_2} \tag{3-10b}$$

$$a_0 = \omega_p^2 = \frac{G_2(G_4 + G_5)}{C_1 C_2} \tag{3-10c}$$

Additional conditions for the remaining components can be obtained from sensitivity considerations. As discussed in Chapter 1, one useful measure of circuit sensitivity is provided by the coefficient sensitivities. These can be readily derived from (3-9) and are listed in Table 3-1.[11] One method of minimizing circuit sensitivity is to assume various component values and to find a set that minimizes the coefficient sensitivities. This is normally carried out by computer. Here, however, we shall demonstrate the effect on coefficient sensitivities (and with them, on the transmission sensitivity) of two special component sets.

First let us consider the component set

$$C_1 = C_2 = C \tag{3-11a}$$

and

$$G_2 = G_4 + G_5 = G \tag{3-11b}$$

11. Tables 3-1 to 3-3 were derived by J. Friend in unpublished notes at Bell Telephone Laboratories.

TABLE 3-1. COEFFICIENT SENSITIVITIES OF SINGLE-
AMPLIFIER BIQUAD USED AS A BPN

Component x	Coefficient sensitivity		
	$S_x^{a_1}$	$S_x^{a_0}$	$S_x^{b_1}$
C_1	$-\dfrac{G_2}{a_1 C_1}$	-1	0
C_2	$-1 - S_{C_1}^{a_1}$	-1	-1
R_2	$-\dfrac{G_2}{a_1}\left(\dfrac{1}{C_1} + \dfrac{1}{C_2}\right)$	-1	0
R_4	$\dfrac{G_b G_4}{G_d C_2 a_1}$	$-\dfrac{G_4}{G_4 + G_5}$	-1
R_5	$\dfrac{G_5}{G_4} S_{R_4}^{a_1}$	$-\dfrac{G_5}{G_4 + G_5}$	0
R_b	$S_{R_4}^{a_1} + S_{R_5}^{a_1}$	0	$-\dfrac{G_b}{G_b + G_d}$
R_d	$-S_{R_4}^{a_1} - S_{R_5}^{a_1}$	0	$\dfrac{G_b}{G_b + G_d}$

From (3-10) we then have

$$G = \omega_p C \tag{3-12a}$$

$$\frac{G_b}{G_d} = 2 - \frac{1}{q_p} \tag{3-12b}$$

$$G_4 = G\left(\frac{K}{3q_p - 1}\right) \tag{3-12c}$$

With this component set we obtain the coefficient sensitivities listed in Table 3-2.

By contrast, let us now consider a set with which there is no positive feedback, thus

$$G_b = 0 \tag{3-13a}$$

and then select

$$C_1 = C_2 = C \tag{3-13b}$$

We then obtain from (3-10)

$$G_4 = a_1 K C \tag{3-14a}$$

$$G_2 = a_1 C / 2 \tag{3-14b}$$

$$G_5 = [(2a_0/a_1) - a_1 K]C \tag{3-14c}$$

TABLE 3-2. COEFFICIENT SENSITIVITIES OF THE SINGLE AMPLIFIER BIQUAD FOR TWO SPECIAL BPN'S

Component	$G_2 = G_4 + G_5 = G$, $C_1 = C_2 = C$			$G_b = 0$, $C_1 = C_2 = C$		
x	S_x^{a1}	S_x^{ao}	S_x^{b1}	S_x^{a1}	S_x^{ao}	S_x^{b1}
C_1	$-q_p$	-1	0	$-\frac{1}{2}$	-1	0
C_2	$-1+q_p$	-1	-1	$-\frac{1}{2}$	-1	-1
R_2	$-2q_p$	-1	0	-1	-1	0
R_4	$K\left(\dfrac{2q_p-1}{3q_p-1}\right)$	$-\dfrac{K}{3q_p-1}$	-1	0	$\dfrac{K}{2q_p^2}$	-1
R_5	$(3q_p-1-K)\left(\dfrac{2q_p-1}{3q_p-1}\right)$	$-\dfrac{3q_p-1-K}{3q_p-1}$	0	0	$-1-\left(\dfrac{K}{2q_p^2}\right)$	0
R_b	$2q_p-1$	0	$-\dfrac{2q_p-1}{3q_p-1}$	0	0	0
R_d	$-(2q_p-1)$	0	$+\dfrac{2q_p-1}{3q_p-1}$	0	0	0

The corresponding coefficient sensitivities are also listed in Table 3-2. A comparison of the two cases shows that the circuit with positive feedback results in coefficient sensitivities on the order of $2q_p$ whereas the circuit without it produces sensitivities on the order of unity. Neither of these two circuits are optimized, yet the resulting difference in sensitivities is indicative of the effect that added positive feedback has on the general circuit. Extensive sensitivity studies have shown that the wider the tolerances and the poorer the tracking of the components, the less positive feedback should be used for minimum variations of pole Q. On the other hand, the less positive feedback used, the larger the component spread may be required to be. For example, for case 2 (i.e., without positive feedback) the ratio of R_2/R_5 is $4q_p^2$; thus already for a q_p of 10, R_2 will be 200 kΩ if R_5 is 500 Ω. While this will cause no difficulties for a circuit using discrete components, it will hardly be tolerable for a circuit realized in thin film. In the latter case, the various circuit transformations previously mentioned must be used to bring the resistor values down. Care must be taken in doing so, however, since, as pointed out before, certain transformations may either decrease the equivalent open-loop gain (e.g., voltage divider at the amplifier output) or load down the amplifier, thereby deteriorating its signal-handling capabilities.

So far we have assumed that the amplifier used in the building block has infinite gain. With finite gain A, the coefficients of the transfer function (3-9) become

$$b_1' = \frac{G_4/C_2}{1 - \alpha + \dfrac{1}{A}} \tag{3-15a}$$

$$a_1' = G_2\left(\frac{1}{C_1} + \frac{1}{C_2}\right) - \frac{G_b(G_4 + G_5)}{G_d C_2}\left(\frac{1 - \alpha - \dfrac{G_d/G_b}{A}}{1 - \alpha + \dfrac{1}{A}}\right) \tag{3-15b}$$

$$a_0' = \frac{G_2(G_4 + G_5)}{C_1 C_2} \tag{3-15c}$$

where $(1 - \alpha) = G_d/(G_b + G_d)$. Note that $a_0 = \omega_p^2$ is unaffected by the amplifier gain in the bandpass configuration. However this is not true in the general circuit nor in the BPN when the gain is a function of frequency. The pole Q is decreased by the finite gain. To find the optimum component set minimizing the total variation in q_p the new coefficient sensitivities can be calculated from (3-15) (see Table 3-3) and a computer search carried out. It is then found that the case for $G_b = 0$ is no longer optimum; a certain amount of positive feedback is now required.

TABLE 3-3. COEFFICIENT SENSITIVITIES OF THE SINGLE-AMPLIFIER BIQUAD WITH FINITE OPAMP GAIN USED AS A BPN

Component	Coefficient sensitivity*		
x	$S_x^{a'_1}$	$S_x^{a'_0}$	$S_x^{b'_1}$
C_1	$-\dfrac{G_2}{C_1 a'_1}$	-1	0
C_2	$-1 - S_{C_1}^{a'_1}$	-1	-1
R_2	$\dfrac{G_2}{a'_1}\left(\dfrac{1}{C_1} + \dfrac{1}{C_2}\right)$	-1	0
R_4	$\dfrac{G_4\left[\dfrac{G_b}{G_d} - \dfrac{\rho}{A}\right]}{C_2 a'_1(1 + \rho/A)}$	$-\dfrac{G_4}{G_4 + G_5}$	-1
R_5	$\dfrac{G_5}{G_4} S_{R_4}^{a'_1}$	$-\dfrac{G_5}{G_4 + G_5}$	0
R_b	$\dfrac{G_b(G_4 + G_5)}{G_d C_2 a'_1(1 + \rho/A)^2}$	0	$-\dfrac{K_b}{1 + \rho/A}$
R_d	$-S_{R_b}^{a'_1}$	0	$\dfrac{K_b}{1 + \rho/A}$
A	$\dfrac{(G_4 + G_5)\rho^2}{C_2 a'_1 A(1 + \rho/A)^2}$	0	$-\dfrac{\rho/A}{1 + \rho/A}$

* $K_b \equiv G_b/(G_b + G_d)$, $\quad \rho \equiv 1 + G_b/G_d$

An alternative to this optimization approach is to examine, in more detail, the general expressions for ω_p and q_p that were given for this network type in Chapter 2 (see (2-100a) and (2-100b)). In general, the bridged-T feedback network t_{32} will be unloaded; therefore we will have $\omega_p = \omega_z = \omega_0$ (see (2-101a)). In this case the q_p expression given by (2-101b) is valid, and we can write

$$q_p = q_N \cdot q_P \tag{3-16}$$

where q_N is the Q factor due to negative feedback, q_P that due to positive feedback. The factor q_N is given by

$$q_N = \hat{q} \, \frac{1 + kA}{1 + (kA\hat{q}/q_z)} = q_z \, \frac{1 + (1/kA)}{1 + (q_z/\hat{q}kA)} \tag{3-17}$$

Notice that q_N has exactly the form of a general class 3 network (see (2-15b) when $\omega_z = \omega_0$). The second term, q_P, is given by

$$q_P = \frac{1 - \dfrac{kA\alpha}{1 + kA}}{1 - \dfrac{kA\alpha}{1 + kA\hat{q}/q_z}} \tag{3-18}$$

This expression becomes unity when there is no positive feedback (i.e., $\alpha = 0$). Since $kA \gg 1$, (3-18) can be simplified as follows:

$$q_P \approx \frac{1 - \alpha}{1 - \dfrac{kA\alpha}{1 + kA\hat{q}/q_z}} \tag{3-19}$$

For a specified q_p, we now have two parameters to select corresponding to q_P and q_N, respectively. The parameter responsible for q_N is q_z (or, more accurately, $\hat{q}/q_z$, since $\hat{q}$ depends directly on q_z, see (2-74)), and the parameter responsible for q_P is α, i.e., the amount of positive feedback. One of the two conditions required for the determination of q_z and α is, of course, (3-16), where q_p is specified; the other can be selected with a view to minimizing the Q sensitivity. One method of accomplishing the latter is to specify that the total variation of q_P (i.e., with respect to the drift in all circuit components) be equal to the total variation of q_N. However, as we shall see in Chapter 4 (and as a numerical evaluation of the coefficient sensitivities listed in Tables 3-1 to 3-3 shows), the Q sensitivity of negative-feedback networks to variations in the passive components is small; on the other hand, because near-open-loop operation of the amplifier is required, the variation of q_P with respect to variations in gain may be quite large.

A useful second condition for our circuit may therefore be to specify that the variation of q_P with respect to drift in the gain A should equal the corresponding variation of q_N, i.e.,

$$V_A^{q_P} = V_A^{q_N} \tag{3-20a}$$

Since we can cancel $\Delta A/A$ in this expression we obtain the requirement that

$$S_A^{q_P} = S_A^{q_N} \tag{3-20b}$$

From (3-17) we have

$$S_A^{q_N} = \frac{kA}{1 + kA}\left[\frac{1 - (\hat{q}/q_z)}{1 + (kA\hat{q}/q_z)}\right] \tag{3-21a}$$

and from (3-19)

$$S_A^{q_P} = \frac{kA\alpha}{\left[1 + kA\left(\dfrac{\hat{q}}{q_z} - \alpha\right)\right]\left(1 + kA\dfrac{\hat{q}}{q_z}\right)} \tag{3-21b}$$

Equating (3-21a) with (3-21b) and solving for α we obtain

$$\alpha = \left(\frac{q_z}{\hat{q}} - 1\right)\left[\frac{1 + (kA\hat{q}/q_z)}{1 + kA}\right] \tag{3-22}$$

With (3-17) this becomes

$$\alpha = \frac{1}{q_N}(q_z - \hat{q}) = \frac{q_P}{q_p}(q_z - \hat{q}) \tag{[3-23]}$$

Substituting (3-23) into (3-18) and eliminating q_P with (3-16), we can solve for q_N and subsequently for q_P, such that $V_A^{qP} = V_A^{qN}$ while $q_N q_P = q_p$. The resulting values for q_N and q_P are functions of $q_z/\hat{q}$, which, in turn, is given by the bridged-T in the feedback loop. Remember that (for a bridged-T) q_z and $\hat{q}$ are interdependent according to (2-74). Alternatively we can assume that $q_N \approx q_z$, and then solve for the corresponding α, q_z, and q_P values.

The FEN Decomposition: A Pole–Zero Cancellation Technique
Before going on to the next type of class 3 single-amplifier building block we must briefly digress in order to discuss a functional decomposition that was derived especially for building-block design.[12]
    We start out with the general second-order function (3-1) and decompose it as follows:

$$T(s) = \frac{N(s)}{D(s)} = K\frac{s^2 + \dfrac{\omega_z}{q_z}s + \omega_z^2}{s^2 + \dfrac{\omega_p}{q_p}s + \omega_p^2}$$

$$= T_L(s) \cdot T_H(s) \tag{[3-24]}$$

where

$$T_L(s) = \frac{N(s)}{D_L(s)} = K_L \cdot \frac{s^2 + \dfrac{\omega_z}{q_z}s + \omega_z^2}{s^2 + \dfrac{\omega_p}{q_L}s + \omega_p^2} \tag{[3-25a]}$$

and

$$T_H(s) = \frac{N_H(s)}{D(s)} K_H = \cdot \frac{s^2 + \dfrac{\omega_p}{q_L}s + \omega_p^2}{s^2 + \dfrac{\omega_p}{q_p}s + \omega_p^2} \tag{[3-25b]}$$

Notice that

$$D_L(s) \equiv N_H(s) \tag{[3-25c]}$$

12. G. S. Moschytz, Miniaturized filter building blocks using frequency emphasizing networks, *Proc. Nat. Electronics Conf.*, **23**, 364–369 (1967).

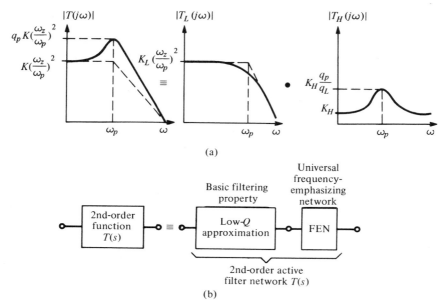

FIG. 3-17. The FEN decomposition: (a) a low-pass response; (b) the general case.

and, to satisfy (3-24),

$$K_L \cdot K_H = K \qquad [3\text{-}25\text{d}]$$

The general second order function (3-1) is therefore decomposed into the product of two functions $T_L(s)$ and $T_H(s)$. $T_L(s)$ is an exact replica of the desired function $T(s)$ except that its pole Q, q_L, is selected lower than the desired q_p. Thus $T_L(s)$ provides the asymptotic characteristic of $T(s)$, but, in the vicinity of the pole frequency ω_p its response is overdamped, i.e., the frequency response $|T_L(j\omega)|$ exhibits less peaking than $|T(j\omega)|$. The second function $T_H(s)$ is a correcting function that has to make up for the lack of peaking provided by $T_L(s)$; it *emphasizes* the frequencies in the vicinity of ω_p (while flattening out toward high and low frequencies) and is realized by a *frequency emphasizing network* or FEN. The *FEN decomposition* is illustrated for the case of a low pass filter in Fig. 3-17a and summarized for the general case in Fig. 3-17b. Its important feature is that, irrespective of the nature of $T(s)$, $T_H(s)$ must always provide the same frequency response, namely that of an FEN. It is defined by a frequency emphasizing characteristic providing the required resonance properties in the vicinity of the undamped natural frequency ω_p.

Another way of interpreting the FEN decomposition is to say that a pair of low Q "phantom" poles are introduced by $T_L(s)$ which are subsequently cancelled by the "phantom" zeros of the correcting function $T_H(s)$. It is

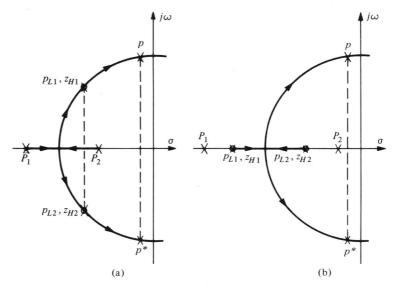

FIG. 3-18. The FEN decomposition as a pole–zero cancellation technique with phantom pole–zero pair: (a) complex; (b) negative real.

therefore also referred to as a *pole–zero cancellation technique*. Seen in this light the poles p, p^*, specified by $T(s)$ may be realized, for example, by a class 3 feedback network whose root locus is shown in Fig. 3-18a. Instead of the feedback network shifting the closed loop poles all the way from P_1, P_2 to p, p^*, two networks are used, the first of which moves them up to p_{L1}, p_{L2}, and the second from there to the specified location p, p^*. In this respect the pole–zero cancellation technique bears a similarity to the dual-feedback technique previously discussed, in which negative feedback is used to shift the closed loop poles part of the way along the root locus and positive feedback is then used to shift them the rest of the way (see, e.g., Fig. 2-34a). Thus (3-24) can be written in the form

$$T(s) = K \frac{N(s),}{(s - p)(s - p^*)}$$

$$= K_L \frac{N(s)}{(s - p_{L1})(s - p_{L2})} \cdot K_H \frac{(s - z_{H1})(s - z_{H2})}{(s - p)(s - p^*)} \qquad [3\text{-}26]$$

where $p_{L1} = z_{H1}$ and $p_{L2} = z_{H2}$. If p_{L1} and p_{L2} are on the negative real axis then $T_L(s)$ is a passive RC network and $q_L < 0.5$. In that case the locus of the poles is as shown in Fig. 3-18b. An example of this case is shown in Fig. 3-19. On the left is an LCR low-pass network whose pole Q is assumed to be higher than 0.5. On the right the realization is shown, consisting of a passive RC approximation in cascade with the active correcting function possessing FEN characteristics.

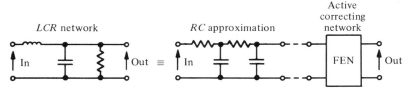

FIG. 3-19. The FEN decomposition applied to a low-pass network.

What have we achieved by the FEN decomposition? By splitting up any second-order function into a low-Q section realizing the actual filtering property (e.g., low-pass, bandpass, bandreject, etc.) and a universal correcting function, or FEN, by which the actual (high) pole Q is to be realized, we are able to treat the problem of *filter realization* and *filter stability* separately. The realization of the basic filter is initially a "low-Q" problem in which stability problems, at least with respect to pole Q, do not occur. The problem of stabilizing the pole *frequency* remains with us, of course, since it is independent of Q and must be solved by an appropriate choice of resistors, capacitors, their TCR's, TCCs, aging properties, and so on. This is a technological problem that is independent not only of pole Q but also of the filter realization used (assuming, as we do, that only realizations in which ω_p is independent of gain are used). The stability problem, which relates pole Q to gain and component drift, is then solved separately by designing a network having FEN characteristics whose pole Q is sufficiently stable over a given range of Q and frequency. There are numerous circuits capable of providing stable FEN characteristics; however, they must also be conveniently cascadable with the $T_L(s)$ network, and also permit simple frequency and Q adjustments. In what follows we shall discuss a single-amplifier FEN and later, in conjunction with the multi-amplifier building blocks, a high-Q FEN consisting of two amplifiers.

Twin-T Feedback Networks (The Single-Amplifier FEN)[13]

The Circuit   Consider the circuit shown in Fig. 3-20. Assuming that $q_z > q_L$ this is a class 3 feedback network, specifically a special case of the circuits shown in Fig. 2-40, in which $t'(s) \equiv 1$. Thus we have (from (2-119))

$$T_H(s) = \frac{V_3}{V_1} = \frac{A}{1 + At_{32}}$$

$$= \frac{A}{1 + A} \cdot \frac{s^2 + \dfrac{\omega_N}{q_L} s + \omega_N^2}{s^2 + s\dfrac{\omega_N}{q_z}\left[\dfrac{A}{1 + A} + \dfrac{q_z}{q_L(1 + A)}\right] + \omega_N^2} \tag{3-27}$$

13. T. A. Hamilton and A. S. Sedra, Some new configurations for active filters, *IEEE Trans. Circuit Theory*, CT-19, 25–33 (1972).

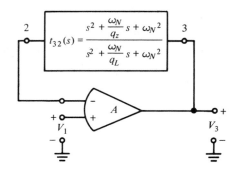

FIG. 3-20. A single-amplifier FEN.

Notice that we use the subscript H for $T(s)$ because (3-27) is an FEN-type function. If $t_{32}(s)$ is a passive RC network then, according to our notation for RC networks, q_L becomes $\hat{q}$ and since $A \gg 1$, (3-27) becomes

$$T_H(s) = \frac{1}{1 + \dfrac{1}{A}} \cdot \left. \frac{s^2 + \dfrac{\omega_N}{\hat{q}} s + \omega_N^2}{s^2 + \dfrac{\omega_N}{q_p} s + \omega_N^2} \right|_{A \gg 1} \approx \frac{1}{t_{32}(s)} \qquad (3\text{-}28)$$

where

$$q_p = q_z \left. \frac{\dfrac{1 + A}{A}}{1 + \dfrac{q_z}{A\hat{q}}} \right|_{A \gg q_z/\hat{q}} \approx q_z \qquad (3\text{-}29)$$

Since the input impedance at the noninverting terminal of the amplifier in Fig. 3-20 is very high, we can precede the FEN by an RC network $T_R(s)$ whose poles cancel the zeros of (3-28) and whose zeros are the zeros of the specified function $T(s)$ (see Fig. 3-21). The null frequency ω_N of the feedback

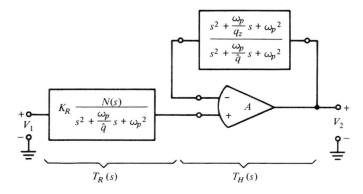

FIG. 3-21. Second-order filter realization using the single-amplifier FEN.

network $t_{32}(s)$ must here equal the required pole frequency ω_p. Various networks suitable for the realization of $T_R(s)$ for commonly used second-order functions are listed in Table 3-4.

The circuit type characterized by Fig. 3-21 has already been discussed in general terms in Chapter 2 (see Section 2.4.1). Here we are interested in it as a potential candidate for building-block design. For this purpose, we must examine its sensitivity characteristics.

Sensitivity Considerations   From (3-29) we obtain the sensitivity of q_p to variations of A:

$$S_A^{q_p} = \left.\frac{\dfrac{q_z}{A\hat{q}}}{1 + \dfrac{q_z}{A\hat{q}}}\right|_{q_z/A\hat{q} \ll 1} \approx \frac{q_z}{A\hat{q}} \tag{3-30}$$

Thus, to minimize the deviation in q_p, (3-30) must be minimized. For a given q_z (which, according to (3-29), is approximately equal to q_p) and a given gain A, the network realizing $t_{32}(s)$ in Fig. 3-20 must have a maximum $\hat{q}$, i.e., it should come as close to 0.5 as possible.

Since $T_H(s)$ is a class 3 network, $t_{32}(s)$ will typically be a bridged-T or a twin-T network. The zeros of $t_{32}(s)$ determine the poles of $T(s)$. Assuming high enough gain (so that $q_p = q_z$), a symmetrical bridged-T can be dimensioned using the equations given in Chapter 2 (Section 2.3.3). For high q_p values we select $C_1 = C_2 = C$. With (2-71) we then have $R_1 = 4q_z^2 R_2 \approx 4q_p^2 R_2$, resulting in the network shown in Fig. 3-22a. We then find from (2-67) that

$$\omega_p = \frac{1}{2q_p RC} \tag{3-31a}$$

$$\hat{q} = \frac{q_p}{2q_p^2 + 1} \approx \frac{1}{2q_p} \quad (q_p \gg 1) \tag{3-31b}$$

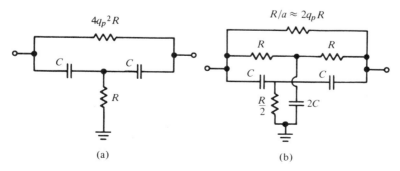

(a)

(b)

Fig. 3-22. Feedback networks used for the single-amplifier FEN: (a) bridged-T; (b) bridged twin-T.

TABLE 3-4. SOME $T_R(s)$–NETWORKS SUITABLE FOR SINGLE–AMPLIFIER FEN DESIGN

Voltage-transfer function $T_R(s)$	RC network	Parameter–component relations
LPN $$K\dfrac{\omega_p^2}{s^2+\dfrac{\omega_p}{\hat{q}}s+\omega_p^2}$$		$$K=\dfrac{r_2}{r_2+r_4+r_2 r_4}$$ $$\omega_p=\sqrt{\dfrac{r_2+r_4+r_2 r_4}{c_4}}\,\dfrac{1}{RC}$$ $$\hat{q}=\dfrac{\sqrt{c_4(r_2+r_4+r_2 r_4)}}{c_4(1+r_2)+r_2+r_4}$$
HPN $$K\dfrac{s^2}{s^2+\dfrac{\omega_p}{\hat{q}}s+\omega_p^2}$$		$$K=\dfrac{c_2}{c_2+c_4+c_2 c_4}$$ $$\omega_p=\sqrt{\dfrac{r_4}{c_2+c_4+c_2 c_4}}\,\dfrac{1}{RC}$$ $$\hat{q}=\dfrac{\sqrt{r_4(c_2+c_4+c_2 c_4)}}{r_4(1+c_2)+c_2+c_4}$$
BPN $$K\dfrac{\omega_p s}{s^2+\dfrac{\omega_p}{\hat{q}}s+\omega_p^2}$$		$$K=\sqrt{\dfrac{c_2}{r_4(1+r_2)}}$$ $$\omega_p=\sqrt{\dfrac{r_4(1+r_3)}{c_2}}\,\dfrac{1}{RC}$$ $$\hat{q}=\dfrac{\sqrt{r_4 c_2(1+r_3)}}{c_2(1+r_3+r_4)+r_4}$$
FRN $$K\dfrac{s^2+\dfrac{\omega_z}{q_z}s+\omega_z^2}{s^2+\dfrac{\omega_p}{\hat{q}}s+\omega_p^2}$$		$$K=\dfrac{1}{1+c_4}$$ $$\omega_z=\sqrt{\dfrac{r_3}{c_2}}\,\dfrac{1}{RC}\;;\quad q_z=\dfrac{\sqrt{c_2 r_3}}{1+c_2}$$ $$\omega_p=\sqrt{\dfrac{1+r_4}{1+c_4}}\,\omega_z$$ $$\hat{q}=\sqrt{(1+r_4)(1+c_4)r_3 c_2}$$ $$\times\dfrac{(1+c_2)}{(1+c_2)^2(1+r_4)+c_2 r_3(1+c_2+c_4)}$$

and, with (3-30),

$$S_A^{q_p} = \frac{2q_p^2}{A} \tag{3-32}$$

Thus the sensitivity is proportional to q_p^2. This limits the usefulness of the circuit to low-Q applications.

As already discussed in Section 2.3.3, the twin-T gives us more flexibility in that we can vary q_z independently of $\hat{q}$. If, for example, we consider the bridged twin-T in Fig. 3-22b we obtain the transfer function

$$t_{32}(s) = \frac{s^2 + \dfrac{\omega_N}{q_z} s + \omega_N^2}{s^2 + \dfrac{\omega_N}{\hat{q}} s + \omega_N^2} \tag{3-33}$$

where

$$\omega_N = \frac{\sqrt{1 + 2a}}{RC} \tag{3-34a}$$

$$q_z = \frac{\sqrt{1 + 2a}}{2a} \tag{3-34b}$$

$$\hat{q} = \frac{1}{2} \frac{\sqrt{1 + 2a}}{2 + a} \tag{3-34c}$$

To minimize (3-30), $\hat{q}$ should be as close to its maximum value of 0.5 as possible. Now, the maximum value of $\hat{q}$ as a function of a, as given in (3-34c), occurs when $a = 1$. Then

$$\hat{q}_{max}\big|_{a=1} = \frac{1}{2} \frac{\sqrt{3}}{3} = 0.29 \tag{3-35a}$$

and

$$q_z\big|_{a=1} = \frac{\sqrt{3}}{2} = 0.866 \tag{3-35b}$$

Clearly q_z will generally be required to be much larger than 0.866, in which case $a \ll 1$ and

$$\omega_N\big|_{a \ll 1} \approx \frac{1}{RC} \tag{3-36a}$$

$$q_z\big|_{a \ll 1} \approx \frac{1}{2a} \tag{3-36b}$$

$$\hat{q}\big|_{a \ll 1} \approx 0.25 \tag{3-36c}$$

Note that q_z can take on very large values independently of $\hat{q}$, which approaches a reasonable value of 0.25. From (3-30) we then have

$$S_A^{q_p} = \frac{4q_p}{A} \tag{3-37}$$

where we have assumed that A is large, so that $q_p = q_z$. Equation (3-37) is proportional to q_p rather than to q_p^2; consequently, the bridged twin-T is to be preferred over the bridged-T as the feedback network of the single-amplifier FEN.

One of the disadvantages of the bridged twin-T is that the bridging resistor R/a becomes very large for high q_p (or q_z) values, namely $2q_p R$. This is not as large as the $4q_p^2$ of the bridged-T, but it may still become prohibitive in hybrid integrated implementation. As discussed previously (see Sections 2.3.3 and the discussion of circuit transforms earlier in this section), a resistive T can be used to lower the total resistance, at the cost, however, of complicating the circuit. Since the zeros of the bridged twin-T become the poles of $T_H(s)$ and therefore of $T(s)$, they must be placed accurately. This can be done by tuning the unbridged twin-T for zeros as close to the $j\omega$ axis as possible and subsequently, with the bridging resistor, bringing the zeros back into the left half-plane until the desired null depth $v = \hat{q}/q_z$ is obtained. Unfortunately ω_N does not remain constant during this process, but will vary from $1/RC$ to $\sqrt{1 + 2a}/RC$ as the bridging resistor is connected. This must be accounted for by initially tuning the unbridged twin-T for a null at the frequency $\omega_N/\sqrt{1 + 2a}$. Some of the problems involving the accurate placement of the twin-T zeros in the left half-plane would be avoided if the twin-T could be dimensioned to provide independent adjustment for ω_N and q_z. This can indeed be done—in fact, a method has been developed[14] of synthesizing a general twin-T providing an arbitrary pair of zeros in the complex s-plane (outside the 60° sector on either side of the positive real axis).

The Single-Amplifier FEN and Biquad: A Comparison   The reason for comparing these two building blocks is that they are so closely related that we do well to recognize where the essential differences lie. Both are, of course derivatives of class 3 networks, and therefore both require an RC network providing complex zeros in the feedback loop. The operational amplifiers operate in the (frequency-compensated) open-loop mode in order for the gain A to be large enough for those complex zeros to become the closed-loop poles of the complete network. In other words, the ω_N and q_z in the

14. E. Lueder: The general second-order twin-T and its application to frequency emphasizing networks, *Bell Syst. Tech. J.* **51**, 301–316 (1972); Chapter 8, Section 8.1.2. of *Linear Integrated Networks: Fundamentals*.

numerator of the feedback function (e.g., (3-33)) become the ω_p and q_p of $T(s)$. If q_z (which becomes q_p) is relatively high, then the pole Q of $t_{32}(s)$, namely $\hat{q}$, is small if a bridged-T is used as the feedback network. This, in turn, increases the sensitivity of q_p to variations in the gain A, as was shown in the case of the single-amplifier FEN (3-30); it follows directly for the biquad as well. If, for example, we consider only the Q term resulting from negative feedback (3-17) we obtain from (3-21a), for $A \gg 1$,

$$S_A^{q_N} \approx \frac{q_z}{kA\hat{q}} \left(1 - \frac{\hat{q}}{q_p}\right) \approx \frac{q_z}{kA\hat{q}} \qquad (3\text{-}38)$$

This is the same as the FEN expression (3-30). The second approximation in (3-38) is valid if we recognize that with a bridged-T, $\hat{q} \approx (2q_z)^{-1} = (2q_N)^{-1}$ (see (3-31b)). For the same reason (3-38) is also proportional to q_z^2 or q_N^2 (see (3-32)). The biquad remedies this situation by the selection of a smaller q_z than the required q_p, i.e., $q_z = q_N < q_p$, and making up the difference with the positive-feedback multiplicand q_P. In this way $S_A^{q_N}$ is decreased, but the new term $S_A^{q_P}$ is added to it, making up the total $S_A^{q_P} = S_A^{q_N} + S_A^{q_P}$. The single-amplifier FEN solves the problem differently in that it utilizes the twin-T instead of the bridged-T to attain the desired independence between q_z and $\hat{q}$. A Q sensitivity is thereby obtained that is proportional to q_p rather than to q_p^2—albeit at the expense of added components in a circuit that, to begin with, requires more components (in particular capacitors) than the biquad. The increased component count brings with it individual control of the zeros (with the network T_R) and of the poles (with T_H). We must not forget, though, that the two networks T_R and T_H are mutually constrained by design equations that guarantee the pole–zero cancellation in theory; this requires tuning steps that permit the cancellation in practice.

Instead of adding positive feedback for added flexibility and lower Q sensitivity, the biquad can, of course, also use a twin-T circuit in the feedback loop. As was done for the bridged-T (Fig. 3-3), appropriate methods of entry into the twin-T from a driving voltage source must then be found that will provide the desired transfer function from input to output.

A building block circuit based on this scheme is shown as a class 3 network in Fig. 3-23a.[15] In Fig. 3-23b it is redrawn in a more conventional form. A comparison with the bridged-T biquad of Fig. 3-3 shows that, beside not including positive feedback, the twin-T biquad uses the operational amplifier in the single-ended input mode. Thus, in contrast to the biquad, this circuit belongs to the type I category as far as the transmission sensitivity with

15. T. A. Hamilton and A. S. Sedra, A single-amplifier biquad active filter, *IEEE Trans. Circuit Theory* **CT-19**, 398–403 (1972).

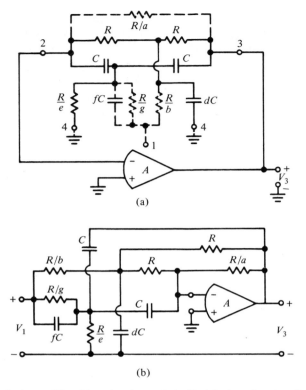

(a)

(b)

FIG. 3-23. A single-amplifier biquad using a twin-T: (a) class 3 network; (b) general building block.

respect to gain is concerned. Analyzing the circuit in Fig. 3-23 we obtain the general second-order transfer function (3-1), where

$$\omega_z = \sqrt{\frac{b}{f}} \frac{1}{RC} \tag{3-39a}$$

$$q_z = \frac{\sqrt{bf}}{g} \tag{3-39b}$$

$$\omega_p = \frac{\sqrt{1 + a(2 + b)}}{RC} \tag{3-39c}$$

$$q_p = \frac{\sqrt{1 + a(2 + b)}}{ad} \tag{3-39d}$$

$$K = f \tag{3-39e}$$

These design equations hold provided that

$$b + 2 = g + e \qquad (3\text{-}39f)$$

and

$$f + 2 = d \qquad (3\text{-}39g)$$

With these expressions it is possible to calculate the coefficient sensitivities as was done for the bridged-T biquad, and to select a set of parameters that will minimize them.

The main minimum-phase second-order functions can be obtained from the twin-T biquad as follows:

$$\text{LPN:} \quad f = g = 0; \quad d = 2; \qquad b + 2 = e \qquad (3\text{-}40)$$

$$\text{HPN:} \quad g = b = 0; \quad e = 2; \qquad 2 + f = d \qquad (3\text{-}41)$$

$$\text{BPN:} \quad f = b = 0; \quad d = 2; \qquad g + e = 2 \qquad (3\text{-}42)$$

$$\text{FRN:} \quad g = 0; \quad e = b + 2; \qquad d = f + 2 \qquad (3\text{-}43)$$

The corresponding coefficients K, ω_z, q_z, ω_p, and q_p can be obtained by inspection from (3-39). Non-minimum-phase functions cannot be obtained with the circuit of Fig. 3-23 unless additional circuitry, providing differential operation at the input of the opamp, is used.

Taking the finite gain of the amplifier into account we find that the zeros of $T(s)$ remain unaffected. This is because of the single-ended-input or type I operation of the circuit; the gain only enters the numerator of the transfer function as a multiplicand. Thus, expressing the numerator in the bilinear form $N(s) = A(s) + xB(s)$, where x is the gain, $A(s) \equiv 0$. The poles, of course, are affected by finite and frequency-dependent gain A, and it can be shown that, for high q_p, the Q sensitivity to A takes on values comparable to those of the single-amplifier FEN (i.e., (3-37)).

Just as the biquad can use a twin-T instead of a bridged-T, the FEN circuit can combine a bridged-T in the negative-feedback path with positive feedback as shown in Fig. 3-24a. Note that the network T_R no longer sees a virtual open circuit but is loaded by the feedback resistors R_a and R_b in parallel. This can be accounted for, as for example in the high-pass filter shown in Fig. 3-24b. Whether this configuration is preferable to the single-amplifier FEN using a twin-T may be questionable however, since the Q-sensitivity is now increased (as for the corresponding biquad), by a term $S_A^{q_p}$. On the other hand, the requirement that (3-39f) and (3-39g) hold may result in more stringent requirements on the passive components in the twin-T biquad, and in a more complicated tuning procedure.

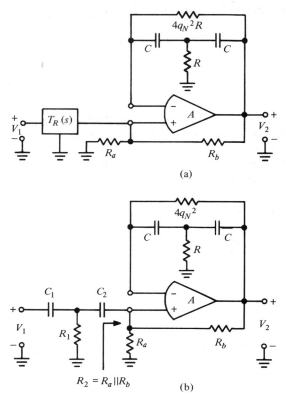

FIG. 3-24. A single-amplifier FEN using a bridged-T and positive feedback: (a) general configuration; (b) high-pass network.

3.1.2 Class 4 Building Blocks

Bandpass Feedback Networks (The Twin-T Single-Amplifier Building Block (TT-SABB))[16]

The General Circuit   Class 4 circuits, i.e., single-amplifier circuits with positive feedback, principally exist only in one form, namely with the amplifier used as a controlled source and with the feedback network possessing the characteristics of either a bandpass network or of an FEN.

The general configuration is shown in Fig. 3-25. The four-terminal network $t(s)$ is passive RC. An operational amplifier in the noninverting mode is

16. G. S. Moschytz, A universal single-amplifier building block based on the Sallen–Key circuit topology, *IEEE Trans. Circuit Theory*, **CT-20**, 37–47 (1973).

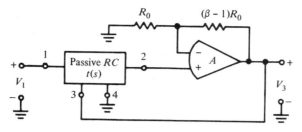

FIG. 3-25. General configuration of a class 4 building block.

used as a voltage amplifier with closed-loop gain β; $t(s)$ is characterized by the following voltage-transfer parameters:

$$t_{12} = \frac{V_2}{V_1}\bigg|_{V_3=0} = \frac{n_{12}}{d_{12}} \tag{3-44a}$$

and

$$t_{32} = \frac{V_2}{V_3}\bigg|_{V_1=0} = \frac{n_{32}}{d_{32}} \tag{3-44b}$$

Since $d_{12} = d_{32} = d$, the overall transfer function is given by

$$T(s) = \frac{V_3}{V_1} = \frac{\beta t_{12}}{1 - \beta t_{32}} = \frac{n_{12} \cdot \beta}{d - \beta n_{32}} \tag{3-45}$$

Adding a path between A and A' and one between B and B' to the loaded twin-T shown in Fig. 3-26 and using this as the four-terminal RC network $t(s)$, one obtains the circuit shown in Fig. 3-27. This configuration provides every possible second-order amplitude response merely by disconnecting appropriate components or interconnection paths.[17] As mentioned earlier, this

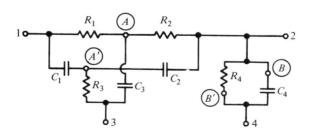

FIG. 3-26. The loaded twin-T used in a class 4 building block.

17. G. S. Moschytz and W. Thelen, Design of hybrid-integrated filter building blocks, *IEEE J. Solid-State Circuits*, **SC-5**, 99–107 (1970).

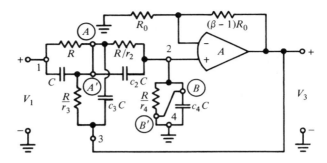

FIG. 3-27. The twin-T single-amplifier building block (TT-SABB).

feature is necessary for hybrid integrated circuit design, where a common topology permits batch processing and handling, irrespective of the actual function required. A single, general-purpose filter building block containing all the components and conducting paths required can then be fabricated in large quantities. The component and conductor disconnections necessary to generate a particular network function are subsequently carried out by some process that is compatible with HIC technology (e.g., scribing, etching, cutting, etc.).

A summary of all possible amplitude functions derived from the general circuit shown in Figure 3-27 (by deleting the appropriate components) is shown in Table 3-5. As in Figure 3-27, all resistors R_i and capacitors C_i are referred to a given resistor R and capacitor C by corresponding factors r_i and c_i, respectively. The functions listed are special cases of the general second-order function. Note that the undamped natural frequencies ω_z and ω_p as well as the zero Q, q_z, are functions of the R_i and C_i only. Only the pole Q, q_p, depends on β, by a function of the form

$$q_p = \frac{\kappa_1}{\kappa_2 - \kappa_3 \beta} \qquad [3\text{-}46]$$

where the κ_i are functions of the R_i and C_i. Design equations for the coefficients corresponding to each particular network function as well as the Q sensitivities to variations in the voltage gain β are also given in Table 3-5.

It is of interest to note that the pole Q of the passive RC network $t(s)$, namely $\hat{q}$, results from (3-46) by setting β equal to zero, thus

$$\hat{q} = q_p|_{\beta=0} = \frac{\kappa_1}{\kappa_2} < 0.5 \qquad [3\text{-}47]$$

The inequality in (3-47) sets a constraint on the κ_1 and κ_2 values given in Table 3-5.

TABLE 3-5. DESIGN EQUATIONS FOR TWIN-T SINGLE-AMPLIFIER BUILDING BLOCK (TT-SABB)*

	Low-pass	High-pass	Bandpass Type A
Circuit			
Func.	$K\,\dfrac{\omega_p^2}{s^2 + \dfrac{\omega_p}{q_p}s + \omega_p^2}$	$K\,\dfrac{s^2}{s^2 + \dfrac{\omega_p}{q_p}s + \omega_p^2}$	$K\,\dfrac{\omega_p s}{s^2 + \dfrac{\omega_p}{q_p}s + \omega_p^2}$
Design Equations	$K = \dfrac{r_2}{r_2 + r_4 + r_2 r_4}\,\beta$ $\omega_p = \sqrt{\dfrac{r_2 + r_4 + r_2 r_4}{c_4}}\cdot\dfrac{1}{RC}$ $\kappa_1 = \sqrt{c_4(r_2 + r_4 + r_2 r_4)}$ $\kappa_2 = c_4(1+r_2) + r_2 + r_4$ $\kappa_3 = r_2$	$K = \dfrac{c_2}{c_2 + c_4 + c_2 c_4}\,\beta$ $\omega_p = \sqrt{\dfrac{r_4}{c_2 + c_4 + c_2 c_4}}\cdot\dfrac{1}{RC}$ $\kappa_1 = \sqrt{r_4(c_2 + c_4 + c_2 c_4)}$ $\kappa_2 = r_4(1+c_2) + c_2 + c_4$ $\kappa_3 = c_2$	$K = \dfrac{\beta}{\sqrt{r_4 c_4(1+r_3)}}$ $\omega_p = \sqrt{\dfrac{r_4(1+r_3)}{c_4}}\cdot\dfrac{1}{RC}$ $\kappa_1 = \sqrt{c_4 r_4(1+r_3)}$ $\kappa_2 = (1+r_3)(1+c_4) + r_4$ $\kappa_3 = r_3$
	Special case: $r_2 = 1,\ r_4 = \dfrac{r}{2},$ $c_4 = \dfrac{c}{2}$ $K = \dfrac{\beta}{1+r}$ $\omega_p = \sqrt{\dfrac{1+r}{c}}\cdot\dfrac{1}{RC}$ $\kappa_1 = \sqrt{c(1+r)}$ $\kappa_2 = 2 + r + c$ $\kappa_3 = 2$	**Special case:** $r_4 = \dfrac{r}{4},\ c_2 = 1,$ $c_4 = \dfrac{c}{2}$ $K = \dfrac{\beta}{1+c}$ $\omega_p = \sqrt{\dfrac{r}{1+c}}\cdot\dfrac{1}{RC}$ $\kappa_1 = \sqrt{r(1+c)}$ $\kappa_2 = 2 + r + c$ $\kappa_3 = 2$	**Special case:** $r_3 = 2,\ r_4 = \dfrac{r}{2},$ $c_4 = \dfrac{c}{2}$ $K = \dfrac{\beta}{2\sqrt{3rc}}$ $\omega_p = \sqrt{\dfrac{3r}{c}}\,\dfrac{1}{RC}$ $\kappa_1 = \sqrt{3rc}$ $\kappa_2 = 6 + r + 3c$ $\kappa_3 = 4$
$S_\beta^{\omega_p}$†	$\kappa = \dfrac{2+r+c}{\sqrt{c(1+r)}};\ \kappa_{min}\big|_{c=2+r} = 2\sqrt{\dfrac{2+r}{1+r}}$	$\kappa = \dfrac{2+r+c}{\sqrt{r(1+c)}};\ \kappa_{min}\big|_{r=2+c} = 2\sqrt{\dfrac{2+c}{1+c}}$	$\kappa = \dfrac{6+r+3c}{\sqrt{3rc}};\ \kappa_{min}\big|_{r=6+3c} = 2\sqrt{\dfrac{6+3c}{3c}}$

(continued)

221

TABLE 3-5. DESIGN EQUATIONS FOR TWIN-T SINGLE-AMPLIFIER BUILDING BLOCK (TT-SABB)* (continued)

	Bandpass Type B	Bandpass Type $\bar{B}$	Low-pass resonator	Complex zeros
Circuit				

Func.

Bandpass Type B:
$$K\,\frac{\omega_p s}{s^2 + \frac{\omega_p}{q_p}s + \omega_p^2}$$

Bandpass Type $\bar{B}$:
$$K\,\frac{\omega_p s}{s^2 + \frac{\omega_p}{q_p}s + \omega_p^2}$$

Low-pass resonator:
$$K\,\frac{s + \omega_z}{s^2 + \frac{\omega_p}{q_p}s + \omega_p^2}$$

Complex zeros:
$$K\,\frac{s^2 + \frac{\omega_z}{q_z}s + \omega_z^2}{s^2 + \frac{\omega_p}{q_p}s + \omega_p^2}$$

Design Equations

Bandpass Type B:
$$K = \sqrt{\frac{c_2}{r_4(1+r_3)}}\cdot\beta$$
$$\omega_p = \sqrt{\frac{r_4(1+r_3)}{c_2}}\cdot\frac{1}{RC}$$
$$\kappa_1 = \sqrt{r_4 c_2(1+r_3)}$$
$$\kappa_2 = c_2(1+r_3+r_4) + r_4$$
$$\kappa_3 = r_3 c_2$$

Special case: $r_3 = 1,\ r_4 = \frac{r}{2}$,
$$c_2 = c$$
$$K = \beta\sqrt{c/r}$$
$$\omega_p = \sqrt{\frac{r}{c}}\cdot\frac{1}{RC}$$
$$\kappa_1 = \sqrt{rc}$$
$$\kappa_2 = c\left(2+\frac{r}{2}\right)+\frac{r}{2}$$
$$\kappa_3 = c$$

Bandpass Type $\bar{B}$:
$$K = \sqrt{\frac{r_2}{c_4(1+c_3)}}\cdot\beta$$
$$\omega_p = \sqrt{\frac{c_4(1+c_3)}{r_2}}\cdot\frac{1}{RC}$$
$$\kappa_1 = \sqrt{r_2 c_4(1+c_3)}$$
$$\kappa_2 = r_2(1+c_3+c_4) + c_4$$
$$\kappa_3 = r_2 c_3$$

Special case: $r_2 = r,\ c_3 = 1$,
$$c_4 = \frac{c}{2}$$
$$K = \beta\cdot\sqrt{\frac{c}{r}}$$
$$\omega_p = \sqrt{\frac{r}{c}}\cdot\frac{1}{RC}$$
$$\kappa_1 = \sqrt{rc}$$
$$\kappa_2 = r\left(2+\frac{c}{2}\right)+\frac{c}{2}$$
$$\kappa_3 = r$$

Low-pass resonator:
$$K = \frac{\beta}{c_4}\cdot\frac{1}{RC}$$
$$\omega_z = \frac{r_3}{RC}$$
$$\omega_p = \sqrt{\frac{r_3}{c_4}}\cdot\frac{1}{RC}$$
$$\kappa_1 = \sqrt{r_3 c_4}$$
$$\kappa_2 = 1 + r_3 + r_3 c_4$$
$$\kappa_3 = r_3$$

Special case: $r_3 = 2$,
$$c_4 = \frac{c}{2}$$
$$K = \frac{2\beta}{cRC}$$
$$\omega_z = \frac{2}{RC}$$
$$\omega_p = \frac{\omega_z}{\sqrt{c_4}}$$
$$\kappa_1 = \sqrt{c}$$
$$\kappa_2 = 3 + c$$
$$\kappa_3 = 2$$

Complex zeros:
$$K = \frac{\beta}{1+c_4}$$
$$\omega_z = \sqrt{\frac{r_3}{c_2}}\cdot\frac{1}{RC}$$
$$q_z = \frac{\sqrt{c_2 r_3}}{1+c_2}$$
$$\omega_p = \sqrt{\frac{1+r_4}{1+c_2}\cdot\frac{r_3}{c_4}\cdot\frac{c_2}{2}}\cdot\frac{1}{RC}$$
$$\kappa_1 = \sqrt{(1+r_4)(1+c_4)r_3 c_2(1+c_2)}$$
$$\kappa_2 = (1+c_2)^2(1+r_4)$$
$$\kappa_3 = (1+c_2)c_2 r_3$$

Special case: $c_2 = 1,\ c_4 = 1$
$$K = \frac{\beta}{2}$$
$$\omega_z = \sqrt{r_3}\cdot\frac{1}{RC}$$
$$q_z = \frac{\sqrt{r_3}}{2}$$
$$\omega_p = \sqrt{\frac{(1+r_4)r_3}{2}}\cdot\frac{1}{RC}$$

$S_\beta^{q_p}$†

Bandpass Type B:
$$\kappa = \frac{c(4+r)+r}{2\sqrt{rc}}\,;\ \ \kappa_{\min}\Big|_{r=\frac{4c}{1+c}} = 2\sqrt{1+c}$$

Bandpass Type $\bar{B}$:
$$\kappa = \frac{r(4+c)+c}{2\sqrt{rc}}\,;\ \ \kappa_{\min}\Big|_{c=\frac{4r}{1+r}} = 2\sqrt{1+r}$$
$$= 2\sqrt{1+r}$$

Low-pass resonator:
$$\kappa = \frac{3+c}{\sqrt{c}}\,;\ \ \kappa_{\min}\big|_{c=3} = 2\sqrt{3}$$

TABLE 3-5. DESIGN EQUATIONS FOR TWIN-T SINGLE-AMPLIFIER BUILDING BLOCK (TT-SABB)* (continued)

Frequency-rejection networks (i.e., $j\omega$-axis zeros)

	Standard FRN, Parallel Twin-T Load	Standard FRN, Series Twin-T Load	Split-feedback FRN, parallel twin-T load
Circuit			

Func.

$$K\;\frac{s^2 + \omega_z^2}{s^2 + \dfrac{\omega_p}{q_p}s + \omega_p^2}$$

Design Equations

Standard FRN, Parallel Twin-T Load	Standard FRN, Series Twin-T Load	Split-feedback FRN, parallel twin-T load
$K = \dfrac{\beta}{1+c}$	$K = \beta,\; r = c$	$K = \dfrac{\beta}{1+c}$
$\omega_z = \dfrac{1}{RC}$	$\omega_z = \dfrac{1}{RC}$	$\omega_z = \dfrac{1}{RC}$
$\dfrac{\omega_p}{\omega_z} = \sqrt{\dfrac{1+r}{1+c}}$	$\omega_p = \omega_z$	$\dfrac{\omega_p}{\omega_z} = \sqrt{\dfrac{1+r}{1+c}}$
$\kappa_1 = \sqrt{(1+r)(1+c)}$	$\kappa_1 = 1$	$\kappa_1 = \sqrt{(1+r)(1+c)}$
$\kappa_2 = 4 + r + c$	$\kappa_2 = 4 + r$	$\kappa_2 = 4 + r + c$
$\kappa_3 = 4$	$\kappa_3 = 4$	$\kappa_3 = 2$

Special case (Parallel Twin-T Load):
$$r = c$$
$$K = \frac{\beta}{1+r}$$
$$\omega_p = \omega_z = \frac{1}{RC}$$
$$\kappa_1 = 1+r$$
$$\kappa_2 = 4 + 2r$$
$$\kappa_3 = 4$$

Special case (Split-feedback):
$$r = c$$
$$K = \frac{\beta}{1+r}$$
$$\omega_p = \omega_z = \frac{1}{RC}$$
$$\kappa_1 = 1+r$$
$$\kappa_2 = 4 + 2r$$
$$\kappa_3 = 2$$

$S_\beta^{q_p}$†		
$\kappa = \dfrac{4+2r}{1+r}$	$\kappa = 4 + r$	$\kappa = \dfrac{4+2r}{1+r}$

* $q_p = \dfrac{\kappa_1}{\kappa_2 - \kappa_3\beta}$; $\quad\beta = \dfrac{\kappa_2}{\kappa_3} - \dfrac{\kappa_1}{\kappa_3}\cdot\dfrac{1}{q_p}$; $\quad S_\beta^{q_p} = \kappa q_p - 1$; $\quad\kappa = \dfrac{\kappa_2}{\kappa_1} \geq 2$; $\quad\beta\kappa = \dfrac{\kappa_2}{\kappa_3}\left(\dfrac{\kappa_2}{\kappa_1} - \dfrac{1}{q_p}\right)$.

† $S_\beta^{q_p}$ is calculated for the "special case" given for each circuit type.

223

Most of the network types listed in Table 3-5 follow directly from the class 4 circuits discussed in Chapter 2 (see Fig. 2-28). They belong to the category of networks first suggested by Sallen and Key.[18] The bandpass network types A and B result from the fact that the second-order bandpass network t_{32} can be realized in different ways. Thus, type A corresponds to Fig. 2-28b, type B to Fig. 2-28f. The RC dual to the type B circuit (see Fig. 2-28g) results in the type $\tilde{B}$ bandpass filter.

Because they are so frequently used, and because they are somewhat more complicated, we shall in what follows discuss in more detail the realization of two of the network functions using the TT-SABB; the circuits in question are the frequency-rejection network (FRN) and the all-pass network (the latter is not included in Table 3-5).

The FRN[19]   Consider the general circuit model of the TT-SABB in Fig. 3-28 when used as a frequency-rejection network. It consists of an active (potentially symmetrical) twin-T, a loading network $Y_L = G_L + sC_L$, and an output voltage source. Straightforward analysis gives us the following transfer function:

$$T(s) = \frac{V_2}{V_1} = -\beta \frac{y_{21}}{y_{22} + Y_L} \tag{3-48}$$

where the y parameters of the active twin-T are

$$y_{11} = \frac{sC_1(sC_2 + G_3)}{s(C_1 + C_2) + G_3} + \frac{G_1(sC_3 + G_2)}{sC_3 + G_1 + G_2} \tag{3-49a}$$

$$y_{12} = -\left[\frac{sC_1(sC_2 + \beta_2 G_3)}{s(C_1 + C_2) + G_3} + \frac{G_1(s\beta_1 C_3 + G_2)}{sC_3 + G_1 + G_2}\right] \tag{3-49b}$$

$$y_{21} = -\left[\frac{s^2 C_1 C_2}{s(C_1 + C_2) + G_3} + \frac{G_1 G_2}{sC_3 + G_1 + G_2}\right] \tag{3-49c}$$

$$y_{22} = \frac{s^2 C_1 C_2 + sC_2 G_3(1 - \beta_2)}{s(C_1 + C_2) + G_3} + \frac{sG_2 C_3(1 - \beta_1) + G_1 G_2}{sC_3 + G_1 + G_2} \tag{3-49d}$$

Assuming, now, that the twin-T is tuned to provide $j\omega$-axis zeros, we obtain the desired FRN-type transfer function:

$$T(s) = \frac{V_2}{V_1} = -\frac{y_{21}}{y_{22}} = K \frac{s^2 + \omega_z^2}{s^2 + \frac{\omega_p}{q_p} s + \omega_p^2} \tag{3-50}$$

18. R. P. Sallen, E. L. Key, op. cit.
19. G. S. Moschytz, Sallen and Key filter networks with amplifier gain larger than or equal to unity, *IEEE J. of Solid-State Circuits*, **SC-2**, 114–116 (1967).

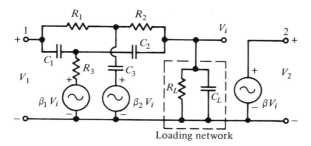

FIG. 3-28. General circuit model of the TT-SABB.

where[20]

$$K = \frac{\beta}{1 + c} \tag{3-51a}$$

$$\omega_z^2 = \frac{1}{R_1 R_2 C_s C_3} = \frac{1}{R_s R_3 C_1 C_2} \tag{3-51b}$$

$$C_s = \frac{C_1 C_2}{C_1 + C_2}; \qquad R_s = R_1 + R_2 \tag{3-51c}$$

$$\omega_p = \omega_z \sqrt{\frac{1 + r}{1 + c}} \tag{3-51d}$$

$$r = \frac{R_s}{R_L}; \qquad c = \frac{C_L}{C_s} \tag{3-51e}$$

$$q_p = \frac{\alpha \sqrt{(1 + r)(1 + c)(1 - \lambda)(1 - \eta)}}{(1 - \beta_1)(1 - \eta) + \alpha^2(1 - \beta_2)(1 - \lambda) + (1 - \lambda)(1 - \eta)(r + c\alpha^2)}$$

$$= \hat{q}_T \frac{\sqrt{(1 + r)(1 + c)}}{1 + \hat{q}_T \left(\dfrac{r}{\alpha} + c\alpha\right) - \dfrac{\beta_1(1 - \eta) + \beta_2 \alpha^2(1 - \lambda)}{\alpha^2(1 - \lambda) + (1 - \eta)}} \tag{3-51f}$$

$$\lambda = \frac{R_1}{R_1 + R_2}; \qquad \eta = \frac{C_2}{C_1 + C_2}; \qquad \omega_1 = \frac{1}{R_3 C_p} = \frac{1}{R_p C_3}$$

$$R_p = \frac{R_1 R_2}{R_1 + R_2}; \qquad C_p = C_1 + C_2; \qquad \alpha = \frac{\omega_1}{\omega_p} \tag{3-51g}$$

and

$$\hat{q}_T = q_{\text{twin}-T} = \frac{\alpha(1 - \lambda)(1 - \eta)}{\alpha^2(1 - \lambda) + (1 - \eta)} \tag{3-51h}$$

20. See Chapter 8, Section 8.1.2 of *Linear Integrated Networks: Fundamentals*.

From (3-51f) we can readily obtain the pole Q of the passive twin-T *including the loading network*, by setting $\beta_1 = \beta_2 = 0$. Hence

$$\hat{q} = q_p|_{\beta_1 = \beta_2 = 0} = \hat{q}_T \frac{\sqrt{(1+r)(1+c)}}{1 + \hat{q}_T\left(\dfrac{r}{\alpha} + c\alpha\right)}$$

$$= \frac{\alpha(1-\lambda)(1-\eta)\sqrt{(1+r)(1+c)}}{\alpha^2(1-\lambda) + (1-\eta) + (1-\lambda)(1-\eta)(r+c\alpha^2)} \tag{3-52}$$

Note that, of all the parameters in (3-50) beside K, only q_p depends on the active quantities β_1 and β_2. Depending on the voltage sources $\beta_1 V_i$ and $\beta_2 V_i$ we can now distinguish between two modes of operation for the FRN, namely the standard FRN[21] and the split-feedback FRN.[22]

THE STANDARD FRN

This is characterized by

$$\beta_1 = \beta_2 = \beta \tag{3-53}$$

and corresponds to the circuit shown in Fig. 3-29a; feedback is applied to both legs of the twin-T. Because of its numerous advantages—which will become apparent presently—a potentially symmetrical twin-T is used here. The transfer function is given by (3-50), and the parameters are now given by

$$K = \frac{\beta}{1+c} \tag{3-54a}$$

$$\omega_z = \frac{1}{RC} \tag{3-54b}$$

$$\omega_p = \omega_z \sqrt{\frac{1+r}{1+c}} \tag{3-54c}$$

$$q_p = \hat{q}_T \frac{\sqrt{(1+r)(1+c)}}{1 + (r+c)\hat{q}_T - \beta} \tag{3-54d}$$

$$\hat{q}_T = \frac{1}{2}\frac{\rho}{1+\rho} \tag{3-54e}$$

and

$$\hat{q} = \hat{q}_T \frac{\sqrt{(1+r)(1+c)}}{1 + (r+c)\hat{q}_T} \tag{3-54f}$$

21. G. S. Moschytz, op. cit. (see footnote 19 of this chapter); also R. N. G. Piercy, Synthesis of active *RC* filter networks, *ATE J.*, **21**, 61–75 (1965).
22. G. S. Moschytz and W. Thelen, The design of hybrid integrated filter building blocks, *IEEE J. Solid-State Circuits*, SC-5, 99–107 (1970): also W. J. Kerwin and L. P. Huelsman, The design of high performance active *RC* bandpass filters, *IEEE Int. Conv. Rec.* 1966.

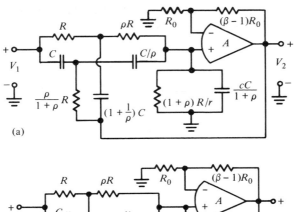

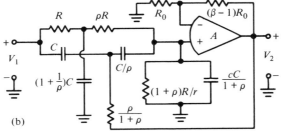

FIG. 3-29. The TT-SABB as a frequency-rejection network (FRN): (a) the standard FRN; (b) the split-feedback FRN.

Furthermore,

$$r = (1 + \rho) \frac{R}{R_L}; \qquad c = (1 + \rho) \frac{C_L}{C} \qquad (3\text{-}54\text{g})$$

THE SPLIT FEEDBACK FRN

This is characterized by

$$\beta_1 \neq 0; \qquad \beta_2 = 0 \qquad (3\text{-}55)$$

and corresponds to the circuit shown in Fig. 3-29b; feedback is applied only to the resistive leg of the twin-T. Of course, the third case ($\beta_1 = 0$ and $\beta_2 \neq 0$), in which feedback is only applied to the capacitive leg of the twin-T, is also possible, and can be derived directly from the general expression (3-51f), since only q_p is affected by the type of feedback used. Again, a potentially symmetrical twin-T is assumed in Fig. 3-29b and all the expressions in (3-54) apply except that q_p is now given by

$$q_p = \hat{q}_T \frac{\sqrt{(1 + r)(1 + c)}}{1 + \hat{q}_T(r + c) - \beta_1/2} \qquad (3\text{-}56)$$

Note that for identical RC networks and for a given q_p, $\beta_1 = 2\beta$. The expressions for q_p are compared for the two circuits, both for the general case and for the cases $\omega_p \gtrless \omega_z$, in Table 3-6. The conditions for stability with

TABLE 3-6. COMPARISON OF STANDARD AND SPLIT-FEEDBACK FRN

Parameter	Standard FRN $\beta_1 = \beta_2 = \beta$	Split-feedback FRN $\beta_1 \neq 0; \quad \beta_2 = 0$
q_p	$\hat{q}_T \dfrac{\sqrt{(1+r)(1+c)}}{1 + \hat{q}_T\left(\dfrac{r}{\alpha} + c\alpha\right) - \beta}$ $= \dfrac{\hat{q}}{1 - \dfrac{\beta}{1 + \hat{q}_T\left(\dfrac{r}{\alpha} + c\alpha\right)}}$	$\hat{q}_T \dfrac{\sqrt{(1+r)(1+c)}}{1 + \hat{q}_T\left(\dfrac{r}{\alpha} + c\alpha\right) - \beta_1 \dfrac{(1-\eta)}{\alpha^2(1-\lambda) + (1-\eta)}}$ $= \dfrac{\hat{q}}{1 - \beta_1 \dfrac{\hat{q}_T}{\alpha(1-\lambda)} \cdot \dfrac{1}{1 + \hat{q}_T\left(\dfrac{r}{\alpha} + c\alpha\right)}}$
$q_p\big|_{\omega_p = \omega_z}$ $r = c$	$\dfrac{1}{2}\dfrac{1+r}{2+r-2\beta}$ $\left(\beta \leq \dfrac{2+r}{2}\right)$	$\dfrac{1}{2}\dfrac{1+r}{2+r-\beta_1}$ $(\beta_1 \leq 2+r)$
$q_p\big|_{\omega_p > \omega_z}$ $r \neq 0; \; c = 0$	$\dfrac{\sqrt{1+r}}{4+r-4\beta}$ $\left(\beta \leq \dfrac{4+r}{4}\right)$	$\dfrac{\sqrt{1+r}}{4+r-2\beta_1}$ $\left(\beta_1 \leq \dfrac{4+r}{2}\right)$
$q_p\big|_{\omega_p < \omega_z}$ $r = 0; \; c \neq 0$	$\dfrac{\sqrt{1+c}}{4+c-4\beta}$ $\left(\beta \leq \dfrac{4+c}{4}\right)$	$\dfrac{\sqrt{1+c}}{4+c-2\beta_1}$ $\left(\beta_1 \leq \dfrac{4+c}{2}\right)$

respect to the gains β and β_1 are also given. Note that, with the constraint that operational amplifiers in the noninverting modes must have gains greater than or equal to unity, the standard FRN requires a loading network ($r \neq 0$, $c \neq 0$) for stability, while the split-feedback circuit requires none. We can calculate the sensitivity of q_p to variations in gain and find, from (3-54d) and (3-56),

$$S_\beta^{q_p} = S_{\beta_1}^{q_p} = \frac{q_p}{\hat{q}} - 1 \qquad (3\text{-}57)$$

where $\hat{q}$ is given by (3-52). Thus the q_p sensitivity to gain is the same for both the standard and the split-feedback FRNs. Clearly, in order to minimize this sensitivity $\hat{q}$ should be as large as possible.

We noted above that the standard FRN requires a loading network, so that for it $\hat{q} \neq \hat{q}_T$, i.e., the pole Q of the total passive RC network (twin-T plus loading network) is not equal to the pole Q of the twin-T alone. The split-feedback FRN, on the other hand, is stable without a loading network, in which case $\hat{q} = \hat{q}_T$. The question is whether there is anything to be gained by including the loading network all the same. There would be if $\hat{q} > \hat{q}_T$, in which case (3-57) would be reduced. On the face of it, $\hat{q}$ is indeed larger than $\hat{q}_T$, but this answer is misleading; as $\hat{q}$ increases (with increasing r for example) the required β_1 for a given q_p increases even more rapidly and the q_p stability is worsened. This is because (as will be shown in Chapter 4), it is the gain–sensitivity product $(\beta \cdot S_\beta^{q_p})$ that must be minimized, particularly in the case of hybrid integrated networks. Calculating this product for the split-feedback case, we find

$$\beta_1 = \left(1 - \frac{\hat{q}}{q_p}\right)\left(\frac{\alpha(1-\lambda)}{\hat{q}_T}\right)\left[1 + \hat{q}_T\left(\frac{r}{\alpha} + c\alpha\right)\right] \tag{3-58}$$

and

$$\beta_1 S_{\beta_1}^{q_p} = \frac{\alpha(1-\lambda)}{\hat{q}_T}\left[1 + \hat{q}_T\left(\frac{r}{\alpha} + c\alpha\right)\right]\left(\frac{q_p}{\hat{q}} + \frac{\hat{q}}{q_p} - 2\right) \tag{3-59}$$

This expression decreases as r and c approach zero, i.e., as the loading network is removed. Consequently there is no advantage in adding the loading network in the split-feedback FRN unless it is otherwise required (e.g., to swamp out the effects of a frequency-dependent and not well controllable input impedance of the amplifier).

The Standard FRN with Unity Gain

The gain of an operational amplifier in the noninverting mode is most stable with respect to variations in the open-loop gain, and its bandwidth is the widest, when the closed-loop gain is unity. Obviously, it is also advantageous with respect to the gain–sensitivity product mentioned above if the gain can be minimized (i.e., unity) for a given sensitivity $S_\beta^{q_p}$. The split-feedback FRN cannot operate with unity gain; even with an unloaded twin-T its gain is approximately two. It is therefore of interest, to what extent the standard FRN can operate with unity gain, i.e., what limitations are thereby imposed on the achievable values of q_p and ω_p/ω_z.

Letting

$$\Omega = \frac{\omega_p}{\omega_z} \tag{3-60}$$

we have, from (3-54d),

$$q_p = \hat{q}_T \frac{(1+c)\Omega}{1 + (r+c)\hat{q}_T - \beta} \tag{3-61}$$

Eliminating r and solving for β, we obtain

$$\beta = 1 + \hat{q}_T\left[\Omega^2(1+c) + c - \frac{\Omega}{q_p}(1+c) - 1\right] \tag{3-62}$$

For $\beta \geq 1$ it follows that

$$\Omega^2(1 + c) + c \geq \frac{\Omega}{q_p}(1 + c) + 1 \tag{3-63}$$

or, solving for c,

$$c \geq \frac{q_p(1 - \Omega^2) + \Omega}{q_p(\Omega^2 + 1) - \Omega} \tag{3-64a}$$

and for r,

$$r \geq \frac{q_p(\Omega^2 - 1) + \Omega}{q_p(\Omega^2 + 1) - \Omega} \tag{3-64b}$$

Within the limits established by (3-64), either r or c can be chosen arbitrarily, and the other then determined by (3-54c). The required β follows from (3-62) and is guaranteed to be greater than or equal to unity. For $\beta = 1$ we have

$$q_p\big|_{\beta=1} = \frac{\sqrt{(1 + r)(1 + c)}}{r + c} \tag{3-65}$$

and the lower bounds in (3-64) apply, i.e.,

$$c = \frac{q_p(1 - \Omega^2) + \Omega}{q_p(\Omega^2 + 1) - \Omega} \tag{3-66a}$$

and

$$r = \frac{q_p(\Omega^2 - 1) + \Omega}{q_p(\Omega^2 + 1) - \Omega} \tag{3-66b}$$

The obtainable frequency ratio Ω and pole Q, q_p, are now limited to that range of values for which r and c remain greater than zero, i.e., are realizable by passive resistors and capacitors. This range of permissible values is plotted in Fig. 3-30. Since any values corresponding to $q_p < 0.5$ can be obtained with passive networks, the region of interest is bounded by the curves $r = 0$, $c = 0$, and $q_p = 0.5$. Note that when eliminating one degree of freedom and setting $\beta = 1$, the range of realizable frequency ratios Ω and of q_p is limited. The boundaries $r = 0$ and $c = 0$ correspond to the network with only a capacitor C_L or a resistor R_L, respectively. In these two cases there is no independence at all between Ω and q_p; they are related according to the boundary expressions shown in Fig. 3-30. For values of $\Omega < 1$ we have

$$\Omega\big|_{\substack{\beta=1 \\ r=0}} = \frac{\sqrt{4q_p^2 + 1} - 1}{2q_p} \tag{3-67a}$$

and, for $\Omega > 1$,

$$\Omega\big|_{\substack{\beta=1 \\ c=0}} = \frac{\sqrt{4q_p^2 + 1} + 1}{2q_p} \tag{3-67b}$$

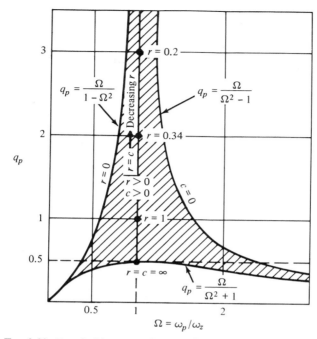

FIG. 3-30. Permissible range of q_p and Ω values for unity-gain FRN.

Thus, when β is unity, the realizable range of Ω on either side of unity decreases rapidly as q_p is increased. On the other hand, the special case of $\Omega = 1$, i.e., $\omega_p = \omega_z$,[23] can be realized with unity gain without any limitations on the achievable q_p. In this case $r = c$ and

$$q_p\Big|_{\substack{\beta=1 \\ \Omega=1}} = \frac{1+r}{2r} \tag{3-68}$$

For the symmetrical or potentially symmetrical twin-T the transfer function can then be written as

$$T(s)\Big|_{\substack{\beta=1 \\ \Omega=1}} = \frac{1}{1 + r(s + \omega_p)^2/(s^2 + \omega_p^2)} \tag{3-69}$$

where $\omega_p = 1/RC$, and the root locus with respect to r can be obtained by inspection (Fig. 3-31). Notice that the pole Q, q_p, increases with decreasing r and, for $r = 0$, becomes infinite. This is, of course, the reason why the standard FRN requires a loading network in the first place.

---

23. As we shall see in Section 3.3.2, this mode of operation is required in the design of high-selectivity FENs.

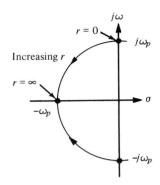

FIG. 3-31. Root locus of poles of unity-gain FRN with respect to r.

Low q_p values become impractical with the standard FRN since r, and therefore the resistor and capacitor spread, may become intolerably large. To prevent this, R_L can be realized as a voltage divider[24] (Fig. 3-32) in such a way that the equivalent gain $\beta' = \beta R''_L/(R'_L + R''_L)$ and $K' = \beta'/(1 + r)$. Notice that β can now be selected larger than unity, whereas β' may be less than or equal to unity; at the same time, r can be decreased.

THE STANDARD FRN WITH A FINITE NULL

So far we have assumed that the twin-T used in the FRN is tuned for an infinite null, i.e., for $j\omega$-axis zeros. In practice the null will, of course, be finite and, in fact, may purposely be placed slightly to the right or left of the $j\omega$ axis. To analyze this case, we start out with the general configuration shown in Fig. 3-33 where the twin-T is given by its y parameters. Then we obtain the transfer function as follows:

$$T(s) = \cfrac{-\beta \dfrac{y_{21}}{y_{22}}}{1 - \beta\left(\dfrac{y_{21}}{y_{22}} + 1\right) + \dfrac{Y_L}{y_{22}}} \tag{3-70}$$

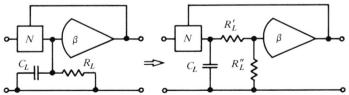

FIG. 3-32. Modification of FRN for low q_p values.

24. J. W. Watterson, Comment on "Sallen and Key filter networks with amplifier gain larger than or equal to unity." *IEEE J. Solid-State Circuits*, **SC-4**, 52–53 (1969); also *IEEE J. Solid-State Circuits*, **SC-4**, 424–425 (1969).

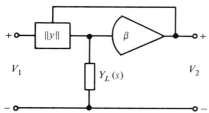

FIG. 3-33. The general FRN configuration.

Assuming, now, that we have a twin-T with a finite-null, then its voltage-transfer function can be written as

$$-\frac{y_{21}}{y_{22}} = \frac{(s + \omega_1)\left(s^2 + \dfrac{\omega_z}{q_z} s + \omega_z^2\right)}{(s + \omega_1')\left(s^2 + \dfrac{\omega_z}{\hat{q}_T} s + \omega_z^2\right)} \tag{3-71}$$

It can be shown[25] that for perturbations near the perfect null, ω_1 equals ω_1', as long as $R_1 C_1 = R_2 C_2$. Thus, assuming this pole–zero cancellation, (3-71) can be considered a second-order function and the twin-T null depth can be defined as

$$|T(j\omega_z)| = v = \frac{\hat{q}_T}{q_z} \tag{3-72}$$

To differentiate between RHP and LHP zeros we can, furthermore, denote

$$\text{RHP Null:} \qquad v_r = \frac{\hat{q}_T}{q_{zr}} \tag{3-73a}$$

$$\text{LPH Null:} \qquad v_l = \frac{\hat{q}_T}{q_{zl}} \tag{3-73b}$$

$$\text{Deepest Null:} \qquad v_0 = \frac{\hat{q}_T}{q_{z0}} \tag{3-73c}$$

By the deepest null, we mean the deepest practically attainable null depth, independent of its location with respect to the $j\omega$ axis. Typically, v is a very small number: for a null depth of -60 dB, $v = 0.001$; for a null depth of -80 dB, $v = 0.0001$.

Assuming that the finite-null twin-T remains second-order, we can write the following equation:

$$y_{22} = C_s \frac{s^2 + \dfrac{\omega_z}{\hat{q}_T} s + \omega_z^2}{s + \omega_1} \tag{3-74}$$

25. G. S. Moschytz, A general approach to twin-T design and its application to hybrid-integrated linear active networks, *Bell Syst. Tech. J.* **49**, 1105–1149; (1970) see also Chapters 8, Section 8.1.2 of *Linear Integrated Networks: Fundamentals*.

where C_s and ω_1 are given in (3-51c) and (3-51g), respectively. Substituting (3-71) and (3-74) into (3-70) we obtain a function of the form

$$T'(s) = K \frac{s^2 \pm \dfrac{\omega_z}{q_z} s + \omega_z^2}{s^2 + \dfrac{\omega_p}{q_p'} s + \omega_p^2} \qquad \begin{matrix} + : \text{LHP} \\ - : \text{RHP} \end{matrix} \qquad (3\text{-}75)$$

where the prime denotes the use of a finite-null twin-T; K, ω_z, and ω_p remain unchanged and are given by (3-54a), (3-54b), and (3-54c) respectively; q_z depends exclusively on the null depth of the passive twin-T; and

$$q_p' = \frac{\alpha\sqrt{(1 + r)(1 + c)(1 - \lambda)(1 - \eta)}}{[\alpha^2(1 - \lambda) + (1 - \eta)][1 - \beta(1 - v)] + (r + c\alpha^2)(1 - \lambda)(1 - \eta)} \qquad (3\text{-}76a)$$

$$= \frac{q_p}{1 \pm Kv \dfrac{q_p}{\hat{q}_T} \left(\dfrac{\omega_z}{\omega_p} \right)} \qquad \begin{matrix} + : \text{LHP} \\ - : \text{RHP} \end{matrix} \qquad (3\text{-}76b)$$

where $\hat{q}_T$ is given by (3-51h). Notice that we can obtain Q enhancement with RHP zeros, i.e., $q_p'(v = v_r) > q_p$. More will be said about the FRN in Section 3-3.2 and in Chapter 6.

The All-pass Network[26]   To obtain an all-pass network we utilize the basic configuration shown in Fig. 3-34. This configuration differs from that of Fig. 3-25 only with respect to the β-feedback network; R_0 is no longer connected to ground but to the input terminal. Thus the single-ended amplifier mode, used to provide minimum-phase network functions, has been changed to a differential-input mode, in order to provide nonminimum phase functions.[27]

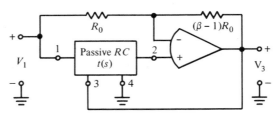

FIG. 3-34. The general all-pass configuration.

26. G. S. Moschytz, A general all-pass network based on the Sallen–Key circuit, *IEEE Trans. Circuit Theory*, **CT-19**, pp. 392–394 (1972).
27. At first glance this may appear to be a disadvantage, because the all-pass network is a type II network whose pole–zero assignment cannot be selected for minimum distortion and maximum dynamic range. We must remember, however, that the all-pass function is characterized by a mirror-image pole–zero configuration with respect to the $j\omega$ axis, which is best maintained as such. Thus any question of optimal pole–zero pairing generally excludes all-pass networks.

Using the voltage-transfer parameters defined by (3-44), the voltage-transfer function for Fig. 3-34 becomes

$$T(s) = \frac{V_3}{V_1} = \frac{1 - \beta(1 - t_{12})}{1 - \beta t_{32}} = \frac{d(1 - \beta) + \beta n_{12}}{d - \beta n_{32}} \qquad (3\text{-}77)$$

A comparison with (3-45) shows that the poles are realized in exactly the same way as those of the amplitude sections.

In order to obtain an all-pass function, it was shown in Chapter 2, Section 2.4.3 that t_{12} must have the characteristics of a null network (loaded or otherwise), so that

$$t_{12} = k_{12} \frac{s^2 + \dfrac{\omega_z}{q_z} s + \omega_z^2}{s^2 + \dfrac{\omega_p}{\hat{q}} s + \omega_p^2} \qquad (3\text{-}78)$$

As a class 4 network, t_{32} must have bandpass character, so that

$$t_{32} = k_{32} \frac{\omega_p s}{s^2 + \dfrac{\omega_p}{\hat{q}} s + \omega_p^2} \qquad (3\text{-}79)$$

Note that t_{12} and t_{32} are presumed to have the same poles, since they are generated by the same RC network. Substituting (3-78) and (3-79) into (3-77) we obtain the function

$$T(s) = K_0 \frac{s^2 - \dfrac{\omega_{0z}}{q_{0z}} s + \omega_{0z}^2}{s^2 + \dfrac{\omega_{0p}}{q_{0p}} s + \omega_{0p}^2} \qquad (3\text{-}80)$$

where

$$K_0 = 1 - \beta + \beta k_{12} \qquad (3\text{-}81a)$$

$$\frac{\omega_{0z}}{q_{0z}} = \frac{[\omega_p/\hat{q}(1 - \beta)] + \beta k_{12}(\omega_z/q_z)}{1 - \beta + \beta k_{12}} \qquad (3\text{-}81b)$$

$$\omega_{0z}^2 = \frac{\omega_p^2(1 - \beta) + \beta k_{12} \omega_z^2}{1 - \beta + \beta k_{12}} \qquad (3\text{-}81c)$$

$$\frac{\omega_{0p}}{q_{0p}} = \omega_p\left(\frac{1}{\hat{q}} - \beta k_{32}\right) \qquad (3\text{-}81d)$$

and

$$\omega_{0p} = \omega_p \qquad (3\text{-}81e)$$

In order for (3-80) to represent an all-pass function the following two conditions must be met:

$$\omega_{0z} = \omega_{0p} = \omega_0 \tag{3-82a}$$

and

$$q_{0z} = -q_{0p} = -q \tag{3-82b}$$

It therefore follows from (3-81c) that we must let

$$\omega_p = \omega_z = \omega_0 \tag{3-83}$$

Then we have, from (3-81b),

$$q_{0z} = \hat{q}\,\frac{1 - \beta + \beta k_{12}}{1 - \beta + \beta k_{12}(\hat{q}/q_z)} \tag{3-84a}$$

and, from (3-81d),

$$q_{0p} = \frac{\hat{q}}{1 - \beta k_{32}\hat{q}} \tag{3-84b}$$

For a given network $t(s)$ we have the corresponding functions $t_{12}(s)$ and $t_{32}(s)$, and combining (3-84) with (3-82) we obtain two implicit conditions that must be fulfilled in order to realize an all-pass function. Since the network realizing $t_{12}(s)$ must be a symmetrical null network (i.e., $\omega_p = \omega_z$) it follows that two basic realizations are possible: one using a twin-T, the other using a bridged-T. These will be discussed next.

THE TWIN-T ALL-PASS NETWORK

Let us first consider a nulled symmetrical twin-T with a parallel RC loading network for $t(s)$. The corresponding circuit is shown in Fig. 3-35a. We then obtain the following two voltage-transfer functions:

$$t_{12} = \frac{n_{12}}{d} = \frac{1}{1 + c}\,\frac{s^2 + \omega_z^2}{s^2 + \dfrac{\omega_p}{\hat{q}}s + \omega_p^2} \tag{3-85a}$$

and

$$t_{32} = \frac{n_{32}}{d} = \frac{1}{1 + c}\,\frac{4s\omega_z}{s^2 + \dfrac{\omega_p}{\hat{q}}s + \omega_p^2} \tag{3-85b}$$

where ω_z and ω_p are given by (3-54b) and (3-54c), respectively. Furthermore $q_z = \infty$,

$$k_{12} = \frac{1}{1 + c} \tag{3-86a}$$

$$k_{32} = \frac{4}{1 + c} \tag{3-86b}$$

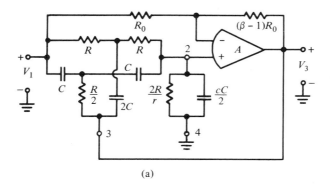

(a)

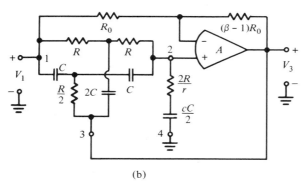

(b)

FIG. 3-35. The twin-T all-pass network: (a) parallel loaded; (b) series loaded.

and, since $\hat{q}_T = 0.25$, we have, from (3-54f),

$$\hat{q} = \frac{\sqrt{(1+r)(1+c)}}{4+r+c} \tag{3-86c}$$

where $\hat{q}$ is the pole Q of the passive network $t(s)$ shown in Fig. 3-35a, while $\hat{q}_T$ is the pole Q of the twin-T alone. To satisfy (3-82a), it follows from (3-54c) that

$$r = c \tag{3-87}$$

Then, from (3-84a), we have

$$q_{0z} = \frac{\hat{q}(1-\beta+\beta')}{1-\beta} = \frac{1+r(1-\beta)}{2(2+r)(1-\beta)} \tag{3-88a}$$

and

$$q_{0p} = \frac{\hat{q}}{1-4\beta'\hat{q}} = \frac{1+r}{4(1-\beta)+2r} \tag{3-88b}$$

where

$$\beta' = \frac{\beta}{1 + c}\bigg|_{r = c} \tag{3-89}$$

The factor K_0 is given by

$$K_0 = 1 - \beta + \beta' \tag{3-90}$$

Equations (3-82b), (3-88a), and (3-88b) define the β and r values required for a given q. Solving for β and r, respectively, two implicit quadratic functions are obtained:

$$\beta^2(8q^2 + 4q) - \beta(8q^2 + 10q - 1) + 4q = 0 \tag{3-91}$$

and

$$r^2(4q^2 - 1) + r(8q^2 - 6q - 1) - 8q = 0 \tag{3-92}$$

Solving (3-91) and (3-92) and retaining only the positive solutions, the $\beta(q)$ and $r(q)$ functions plotted in Fig. 3-36a and b are obtained. Substituting the solutions into (3-90) the corresponding factor K_0, which represents the gain of the all-pass section, is obtained as a function of q (Fig. 3-36c). It shows that the all-pass section with a parallel-loaded twin-T invariably has some loss, which increases for low q values. Often, this loss may be compensated for by gain in an amplitude-shaping filter section. In spite of the disadvantage of loss at low q values, the configuration of Fig. 3-35a has an important advantage; the RC combination in parallel with the high-impedance input terminal of the noninverting operational amplifier can compensate for any parasitic input capacitance present at this terminal. In fact, the internal, parasitic phase of the amplifier can also be compensated for by selecting a proper imbalance of r and c. The correct imbalance is difficult to calculate accurately, but can be adjusted for, functionally, with the circuit in the active mode. A value of r that is, say, 15 % higher than the nominal value is progressively decreased until the amplitude characteristic of the network is flat.

By rearranging the loading network of the twin-T from a parallel to a series configuration, as shown in Fig. 3-35b, an all-pass network with zero loss is obtained. The resulting voltage-transfer parameters are of the same forms as (3-85a) and (3-85b) except that

$$\hat{q} = (4 + r)^{-1} \tag{3-93}$$

and the factor $(1 + c)^{-1}$ is now replaced by unity. In order to fulfill (3-82a), (3-87) must hold. We then obtain the new root Q's

$$q_{0z} = [(4 + r)(1 - \beta)]^{-1} \tag{3-94}$$

and

$$q_{0p} = (4 + r - 4\beta)^{-1} \tag{3-95}$$

The factor K_0 equals unity.

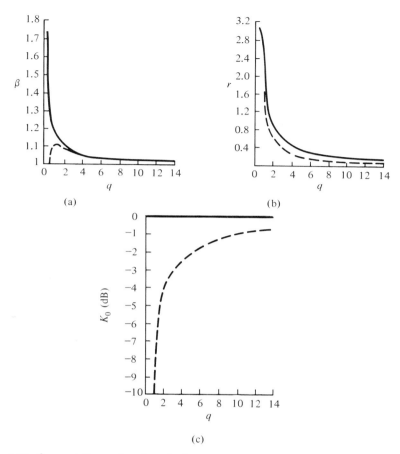

FIG. 3-36. β, r, and K_0 as a function of q for the all-pass networks of Fig. 3-35. Solid lines: series load; broken lines: parallel load.

Equations (3-82b), (3-94), and (3-95) define the implicit functions $\beta(q)$ and $r(q)$, which are now given by

$$4q\beta^2 - \beta(4q - 1) - 2 = 0 \tag{3-96}$$

and

$$qr^2 + r(4q - 1) - 8 = 0 \tag{3-97}$$

Solving these equations for β and r, the solutions for series loading, plotted in Fig. 3-36a and b, are obtained. Notice that for $q > 1$, the required β values for a given q are almost identical to those for parallel loading. The r values for parallel loading are almost half those required for series loading; thus, the parallel-load resistor is twice as large as the series-load resistor. Most significant is the difference in K_0, which is constant and equal to unity for

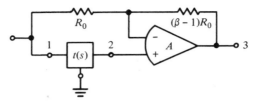

FIG. 3-37. General all-pass configuration providing pole–zero pairs on the real axis.

series loading (Fig. 3-36c). This no-loss advantage is offset by the fact that it is more difficult to compensate the series-loading network for parasitic phase introduced by the β amplifier. If $r \neq c$, the parameters t_{12} and t_{32} are third-order, i.e., there is a parasitic, noncanceling pole–zero pair on the negative real axis. In the parallel-loading case, this pole–zero pair is automatically cancelled with a balanced twin-T, irrespective of r and c, so that the loading components can be adjusted to compensate for phase parasitics of the amplifier. Consequently the parallel-load configuration can be used at higher frequencies than the series-load configuration.

It is evident from Fig. 3-36a and b that neither of the configurations shown in Fig. 3-35 is very practical for q values less than 0.5, i.e., for poles and zeros on the real axis.[28] However, a simple modification to the basic topology of Fig. 3-34 provides poles and zeros on the real axis very easily (Fig. 3-37). The RC network is now a grounded three-terminal network, with the voltage transfer function $t(s)$. The overall transfer function is then given by

$$T(s) = \frac{V_3}{V_1} = \beta t(s) - (\beta - 1) \tag{3-98}$$

Let $t(s)$ be an RC null network (e.g., twin-T or bridged-T) with the transfer function

$$t(s) = \frac{s^2 + \dfrac{\omega_z}{q_z} s + \omega_z^2}{s^2 + \dfrac{\omega_p}{\hat{q}} s + \omega_p^2} \tag{3-99}$$

Substituting (3-99) into (3-98), we obtain

$$T(s) = \frac{s^2 + s\left[\beta\left(\dfrac{\omega_z}{q_z} - \dfrac{\omega_p}{\hat{q}}\right) + \dfrac{\omega_p}{\hat{q}}\right] + \omega_p^2 + \beta(\omega_z^2 - \omega_p^2)}{s^2 + \dfrac{\omega_p}{\hat{q}} s + \omega_p^2} \tag{3-100}$$

28. As it is, the parallel-loading case cannot be used at all, since β becomes less than unity, a situation which is not realizable with an operational amplifier in the noninverting mode. This can be overcome, however, by using the modification shown in Fig. 3-32.

To satisfy the conditions (3-82a) and (3-82b) we let

$$\omega_z = \omega_p \qquad (3\text{-}101a)$$

Then, to satisfy (3-82b), we let

$$q_z/\hat{q} = \beta/(\beta - 2) \qquad (3\text{-}101b)$$

If $t(s)$ is an unloaded null network then $\hat{q} = \hat{q}_T$ and, with (3-72), we obtain

$$\beta = 2/(1 - v) \qquad (3\text{-}102)$$

For a balanced twin-T, v equals zero and β equals 2. The pole Q of the all-pass network is identical to the pole Q of $t(s)$, namely $\hat{q}$. The zero Q equals $-\hat{q}$. Generating right-half-plane zeros with a twin-T, v becomes negative and β in (3-102) becomes less than 2. In general, the choice of the zeros of $t(s)$ has some bearing on the overall sensitivity of the network; here, however, where the pole Q's are very low, it can be ignored. For the circuit configurations of Figs. 3-25 and 3-34 this choice will be discussed further in Chapter 6.

THE BRIDGED-T ALL-PASS NETWORK

Where the loaded twin-T all-pass configuration (Fig. 3-35) has the advantage of similarity with the FRN (Fig. 3-29), an all-pass circuit using a bridged-T has the advantage of circuit simplicity and ease of tuning—besides, the bridged-T all-pass can also readily be derived from the general building-block circuit of Fig. 3-27.

Consider the circuit and its equivalent diagram shown in Fig. 3-38a and 3-38b respectively. The corresponding $t_{12}(s)$ and $t_{32}(s)$ functions have the form of (3-78) and (3-79), respectively, where

$$k_{12} = 1 \qquad (3\text{-}103a)$$

$$k_{32} = \alpha \cdot \frac{r_b}{c_b} \qquad (3\text{-}103b)$$

$$\omega_z = \omega_p = \frac{1}{\sqrt{R_1 R_2 C_1 C_2}} \qquad (3\text{-}103c)$$

$$q_z = \frac{r_b c_b}{1 + c_b^2} \qquad (3\text{-}103d)$$

$$\hat{q} = \frac{r_b c_b}{1 + r_b^2 + c_b^2} \qquad (3\text{-}103e)$$

and

$$r_b = \sqrt{R_1/R_2}; \qquad c_b = \sqrt{C_1/C_2} \qquad (3\text{-}103f)$$

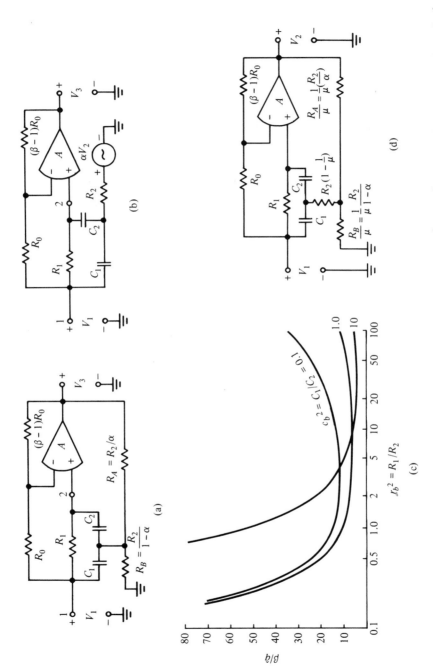

FIG. 3-38. The bridged-T all-pass network: (a) circuit;   (b) equivalent diagram;   (c) plot of $\beta/\dot{q}$ vs. $r_b{}^2$ (see 3-108b) for different values of $c_b{}^2$ for the circuit of Fig. 3-38a;   (d) circuit of Fig. 3-38a modified for low-frequency application.

Thus with (3-84a) we obtain

$$q_{0z} = \frac{r_b c_b}{1 + c_b^2 + (1 - \beta)r_b^2} = \frac{\hat{q}}{1 - \beta \dfrac{r_b}{c_b} \hat{q}} \tag{3-104a}$$

and with (3-84b) we obtain

$$q_{0p} = \frac{r_b c_b}{1 + c_b^2 + (1 - \alpha\beta)r_b^2} = \frac{\hat{q}}{1 - \alpha\beta \dfrac{r_b}{c_b} \hat{q}} \tag{3-104b}$$

We now have the two conditions contained in (3-82b) that must be satisfied by (3-104). The first, namely $q_{0z} = -q_{0p}$, results in the requirement that

$$\alpha = \frac{2}{\beta \dfrac{r_b}{c_b} \cdot \hat{q}} - 1 \tag{3-105}$$

Since $0 < \alpha < 1$, it follows that

$$1 + \frac{1 + c_b^2}{r_b^2} < \beta < 2\left(1 + \frac{1 + c_b^2}{r_b^2}\right) \tag{3-106}$$

It is advisable, from a gain-stability point of view, to let β be as close to unity as possible; in the limit it should not exceed two. Furthermore, since here again the Q sensitivity is given by (3-57), $\hat{q}$, given by (3-103e) should be as large, i.e., as close to 0.5, as possible. Combining both considerations, we require that the ratio $\beta/\hat{q}$ be minimized.[29] Now the second condition contained in (3-82b) states that $q_{0z} = -q$, where q is the required pole and zero Q of the all-pass network. Thus from (3-104a) we have

$$\beta = \frac{c_b}{r_b}\left(\frac{1}{q} + \frac{1}{\hat{q}}\right) = \frac{r_b c_b + q(1 + r_b^2 + c_b^2)}{q r_b^2} \tag{3-107}$$

and with (3-103e), we obtain

$$\frac{\beta}{\hat{q}} = (1 + r_b^2 + c_b^2)\frac{r_b c_b + q(1 + r_b^2 + c_b^2)}{q r_b^3 c_b} \tag{3-108a}$$

which, for high q values, simplifies to

$$\left.\frac{\beta}{\hat{q}}\right|_{q \gg 1} \approx \frac{1 + r_b^2 + c_b^2}{r_b^3 c_b} \tag{3-108b}$$

Computing the expression (3-108a) for various r_b, c_b, and q values, we find that for $q > 2$, $(\beta/\hat{q})$ is virtually independent of q, i.e., (3-108b) is then valid.

---

29. This, in effect, minimizes the gain–sensitivity product $[\beta \cdot S_\beta^{q_{0p}}]$.

Plotting (3-108b) as a function of r_b^2 with parameter c_b^2, we obtain the curves shown in Fig. 3-38c. Note that the minima of $(\beta/\hat{q})$ are relatively flat, i.e., the optimum values for r_b for a given capacitor ratio c_b^2 is not overly critical. Nevertheless, to obtain $r_{b\ opt}$ accurately, we can take the derivative of (3-108b) with respect to r_b. For the useful case that $C_1 = C_2$, i.e., $c_b = 1$, we then obtain

$$r_{b\ opt}|_{c_b=1} = \sqrt{2 + \sqrt{10}} \approx 2.3 \tag{3-109a}$$

From (3-107) we then have

$$\beta|_{c_b=1} = \frac{r_b + q(2 + r_b^2)}{qr_b^2} \tag{3-109b}$$

and

$$\beta\Big|_{\substack{r_b=2.3 \\ c_b=1}} = 1.4 + \frac{0.43}{q} \tag{3-109c}$$

This is within the limits given by (3-106), these being 1.385 and 2.77, respectively.

To obtain $\alpha(r_b, c_b, q)$ we substitute (3-103e) and (3-107) into (3-105) and obtain

$$\alpha = \frac{1 - (\hat{q}/q)}{1 + (\hat{q}/q)} = \frac{q(1 + r_b^2 + c_b^2) - r_b c_b}{q(1 + r_b^2 + c_b^2) + r_b c_b} \tag{3-110a}$$

and, with $c_b = 1$,

$$\alpha|_{c_b=1} = \frac{q(2 + r_b^2) - r_b}{q(2 + r_b^2) + r_b} \tag{3-110b}$$

For the case that $c_b = 1$ and $r_b = 2.3$ we obtain $\hat{q} = 0.32$ and

$$\alpha\Big|_{\substack{r_b=2.3 \\ c_b=1}} \approx \frac{q - 0.315}{q + 0.315} \tag{3-111}$$

For increasing q, α approaches unity.

For very low frequencies, R_2 will become large and the two resistors R_2/α and $R_2/(1 - \alpha)$ in Fig. 3-38a still larger, resulting in difficulties with the practical (e.g., hybrid integrated) implementation. This can be prevented by using the configuration of Fig. 3-38d in which a third resistor is added to the leg of the bridged-T. In scaling down R_A and R_B of Fig. 3-38a by the factor μ the third resistor has the value $R_2 [1 - (1/\mu)]$. Where the total resistance in the leg of the bridged-T in Fig. 3-38a is $R_t = R_2 [1/\alpha(1 - \alpha)]$, it can be reduced to a value ρR_t in Fig. 3-38d with an appropriate value of μ given by

$$\mu = \frac{R_t - R_2}{\rho R_t - R_2} = \frac{1 - \alpha(1 - \alpha)}{\rho - \alpha(1 - \alpha)} \tag{3-112a}$$

The amount ρ by which R_t can be reduced is limited, for given α, by

$$\alpha(1 - \alpha) < \rho < 1 \tag{3-112b}$$

The equivalent diagram of the circuit in Fig. 3-38d is again given by Fig. 3-38b, and the design equations are given by (3-103) to (3-111).

Some final comments on the difference between the configuration of Fig. 3-35a and Fig. 3-38a. Apart from the null network used, the former is loaded by an RC network which is required to be balanced (i.e., $r = c$) so that the all-pass condition $\omega_{0z} = \omega_{0p}$ can be fulfilled. This may introduce tuning problems, since only the condition $R_L = 2R/r$ can be adjusted for, while $C_L = cC/2$ is generally fixed. We require both to be adjustable, though, since there are otherwise not enough adjustable parameters to go round. This becomes clear if we recall that $\omega_p = \sqrt{(1 + r)/(1 + c)} \cdot \omega_z$. Now, in order to satisfy (3-82a), we require one element, say r, to make $\omega_p = \omega_z$. That only leaves β to satisfy (3-82b), which is sufficient with ideal components, but will not be in practice if tight specifications are to be met. The only remaining available component is C_L (assuming that the twin-T has been tuned for a null at ω_z, after which it is not to be readjusted), but, as we know, in hybrid integrated form it is hardly possible to adjust capacitors. Another disadvantage of the twin-T configuration in Fig. 3-35a is that K_0, being equal to $1 - \beta + \beta k_{12}$, is less than unity (see Fig. 3-36c). In delay lines, in which numerous all-pass sections may be cascaded, this can add up to considerable attenuation. The series-loaded circuit (Fig. 3-35b) does not have this disadvantage, but it does also present tuning problems.

By and large, the bridged-T circuit of Fig. 3-38a and d overcomes both the tuning and the attenuation problems. By attenuating the feedback signal by the amount α in the feedback path[30] (using the method presented in Fig. 3-11) instead of in the forward path, no RC balance needs to be carried out, nor does any attenuation in the forward path occur. Moreover, the circuit is less complex than the twin-T configuration, while its topology is compatible with that of the general circuit in Fig. 3-27. More will be said on the subject of practical realization and tuning in Chapter 6.

Sensitivity Considerations   Using the same rationale as we did for the previously discussed building blocks, we shall derive some of the general sensitivity characteristics of the TT-SABB by examining one particular function, the low-pass configuration.

Consider, then, the low-pass circuit shown in Fig. 3-39. We obtain a function of the form

$$T(s) = K \frac{\omega_p^2}{s^2 + \dfrac{\omega_p}{q_p} s + \omega_p^2} \tag{3-113}$$

---

30. Suggested by C. J. Steffen, Bell Telephone Laboratories, Holmdel, N.J.

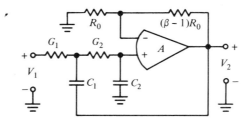

FIG. 3-39. A TT-SABB low-pass network.

where

$$K = \beta \tag{3-114a}$$

$$\omega_p = \sqrt{G_1 G_2 / C_1 C_2} \tag{3-114b}$$

$$q_p = \frac{\sqrt{G_1 G_2 C_1 C_2}}{G_1 C_2 + G_2(C_1 + C_2) - \beta G_2 C_1} \tag{3-114c}$$

By inspection, it follows that

$$S_{G_i}^{\omega_p} = -S_{C_i}^{\omega_p} = \frac{1}{2} \qquad i = 1, 2 \tag{3-115a}$$

and

$$S_\beta^K = 1 \tag{3-115b}$$

It should be noted that the form of (3-115a) is valid for all the circuit types of this class. For the q_p sensitivities we obtain

$$S_{G_1}^{q_p} = -S_{R_1}^{q_p} = \frac{1}{2} - \sqrt{\frac{G_1 C_2}{G_2 C_1}}\, q_p \tag{3-115c}$$

$$S_{G_2}^{q_p} = -S_{R_2}^{q_p} = \frac{1}{2} - \sqrt{\frac{G_2}{G_1 C_1 C_2}}\,(C_1 + C_2 - \beta C_1)\, q_p \tag{3-115d}$$

$$S_{C_1}^{q_p} = \frac{1}{2} - \sqrt{\frac{G_2 C_1}{G_1 C_2}}\,(1 - \beta) q_p \tag{3-115e}$$

$$S_{C_2}^{q_p} = \frac{1}{2} - \sqrt{\frac{C_2}{G_1 G_2 C_1}}\,(G_1 + G_2) q_p \tag{3-115f}$$

$$S_\beta^{q_p} = \beta \sqrt{\frac{G_2 C_1}{G_1 C_2}}\, q_p \tag{3-115g}$$

Notice that all the q_p sensitivities are proportional to q_p. This is true in general and follows from the fact that q_p can be represented by the general form of

(3-46) (see (3-114c) and Table 3-5). Thus, letting $G_1 = 1/R$, $G_2 = r_2/R$, $G_4 = 0$, $C_1 = C$, and $C_2 = c_2 C$, we obtain the values of κ_1, κ_2, and κ_3 given in Table 3-5 for the low-pass circuit. For the general case we obtain

$$S_x^{q_p} = S_x^{\kappa_1} - \frac{q_p}{\hat{q}} S_x^{\kappa_2} + \left(\frac{q_p}{\hat{q}} - 1\right) S_x^{\kappa_3 \beta} \qquad [3\text{-}116]$$

where $\hat{q}$ is given by (3-47) and x may be any component in the expression for q_p. We shall examine the sensitivity of these networks in more depth in Chapters 4 and 6.

Third-Order Networks Throughout our discussion in this and the last chapter we have ignored the fact that in any odd-order function $T(s)$, at least one third-order network will be required. It was pointed out previously that it is not, in general, overly critical to which of the available second-order circuits the odd, negative real pole is assigned; nevertheless the capability should exist of assigning it to most if not necessarily to all of the second-order function types that may occur in a given filter.

Class 4 building blocks lend themselves very well to third-order realization by expanding the basic circuit in the manner shown in Fig. 3-40. Naturally, the added negative real pole will not exactly equal $-1/RC$; due to the lack of isolation it will depend on R, C, and the network used to realize $t(s)$. If we restrict ourselves to third-order *all-pole* networks, $t(s)$ will be a ladder network and the third-order circuit will have the form shown in Fig. 3-41a. This corresponds to an active ladder network, as shown in Fig. 3-41b. Analyzing the circuit[31] we obtain a transfer function $T(s) = V_2/V_1 = N(s)/D(s)$, where

$$N(s) = \beta \qquad (3\text{-}117a)$$

and

$$\begin{aligned}
D(s) = &\ Z_1 Y_1 Z_2 Y_2 Z_3 Y_3 + (1 - \beta)[Z_1 Y_1 Z_2 Y_2 + Y_2(Z_1 + Z_2)] \\
&+ Z_1 Y_3(Y_1 Z_3 + Y_1 Z_2 + Y_2 Z_3) + Z_2 Y_2 Z_3 Y_3 \\
&+ Z_1(Y_1 + Y_3) + Y_3(Z_2 + Z_3) + 1
\end{aligned} \qquad (3\text{-}117b)$$

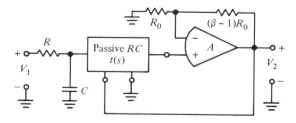

FIG. 3-40. Expanding the TT-SABB for third-order operation.

31. See Chapter 3, Section 3.3 of *Linear Integrated Networks: Fundamentals*.

TABLE 3-7. THIRD-ORDER ALL-POLE NETWORKS

	Low-pass	High-pass	Bandpass (i)
Circuit	(circuit diagram)	(circuit diagram)	(circuit diagram)
Function* $\dfrac{V_2}{V_1}$	$\dfrac{\beta}{a_3 s^3 + a_2 s^2 + a_1 s + 1}$	$\dfrac{R_1 R_2 R_3 C_1 C_2 C_3 \beta s^3}{a_3 s^3 + a_2 s^2 + a_1 s + 1}$	$\dfrac{R_3 C_3 \beta s}{a_3 s^3 + a_2 s^2 + a_1 s + 1}$

Design Equations

Low-pass:

$a_3 = R_1 R_2 R_3 C_1 C_2 C_3$

$a_2 = R_1 R_3[C_2 C_3 + C_1 C_3(1-\beta)] + R_2 R_3[C_2 C_3 + C_1 C_2] + R_1 R_2 C_1 C_2$

$a_1 = R_1[C_2 + C_1(1-\beta)] + R_2 C_2 + R_3[C_2 + C_3] + C_1(1-\beta)$

High-pass:

$a_3 = R_1 R_2 R_3 C_1 C_2 C_3$

$a_2 = R_1 R_3 C_2 C_3 + R_2 R_3(1-\beta) \times (C_2 C_3 + C_1 C_2) + R_1 R_3 C_1(C_2 + C_3) + R_1 R_2 C_1 C_2$

$a_1 = C_2[R_1 + R_2(1-\beta)] + C_1(R_1 + R_3) + R_3 C_3$

Bandpass (i):

$a_3 = R_1 R_2 R_3 C_1 C_2 C_3$

$a_2 = R_1 R_2 \cdot [C_1 C_2 + C_2 C_3 + C_1 C_3(1-\beta)] + R_2 R_3 C_2(C_1 + C_3)$

$a_1 = R_1[C_2 + C_1(1-\beta)] + R_2 C_2 + R_3[C_2 + C_3 + C_1(1-\beta)]$

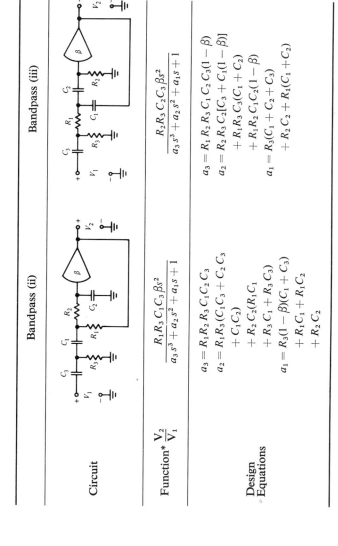

	Bandpass (ii)	Bandpass (iii)
Circuit		
Function* $\dfrac{V_2}{V_1}$	$\dfrac{R_1 R_3 C_1 C_3 \beta s^2}{a_3 s^3 + a_2 s^2 + a_1 s + 1}$	$\dfrac{R_2 R_3 C_2 C_3 \beta s^2}{a_3 s^3 + a_2 s^2 + a_1 s + 1}$
Design Equations	$a_3 = R_1 R_2 R_3 C_1 C_2 C_3$ $a_2 = R_1 R_3 (C_1 C_3 + C_2 C_3$ $\qquad + C_1 C_2)$ $\qquad + R_2 C_2 (R_1 C_1$ $\qquad + R_3 C_1 + R_3 C_3)$ $a_1 = R_3 (1 - \beta)(C_1 + C_3)$ $\qquad + R_1 C_1 + R_1 C_2$ $\qquad + R_2 C_2$	$a_3 = R_1 R_2 R_3 C_1 C_2 C_3 (1 - \beta)$ $a_2 = R_2 R_3 C_2 [C_3 + C_1 (1 - \beta)]$ $\qquad + R_1 R_3 C_3 (C_1 + C_2)$ $\qquad + R_1 R_2 C_1 C_2 (1 - \beta)$ $a_1 = R_3 (C_1 + C_2 + C_3)$ $\qquad + R_2 C_2 + R_1 (C_1 + C_2)$

* $p_{1,2} = -\sigma \pm j\omega_c; \quad p_3 = -\alpha; \quad \sigma^2 + \omega_c^2 = \omega_p^2; \quad a_3 = 1/\alpha \omega_p^2; \quad a_2 = (1 + 2\sigma/\alpha)/\omega_p^2; \quad a_1 = 1/\alpha + 2\sigma/\omega_p^2$

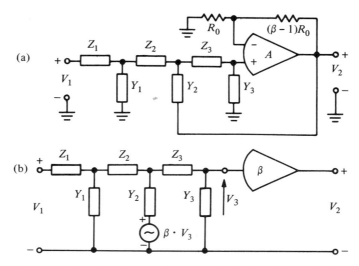

FIG. 3-41. The third-order all-pole network as an active ladder: (a) circuit; (b) equivalent diagram.

If an all-pole high-pass type of circuit is required it is more convenient to use the admittance of the series and the impedance of the shunt elements. This corresponds to multiplying (3-117a) and b) by $Y_1 Z_1 Y_2 Z_2 Y_3 Z_3$. We then obtain the transfer function $T'(s) = N'(s)/D'(s)$, where

$$N'(s) = \beta Y_1 Z_1 Y_2 Z_2 Y_3 Z_3 \qquad (3\text{-}118a)$$

and

$$\begin{aligned} D'(s) &= Y_1 Z_1 Y_2 Z_2 Y_3 Z_3 + (1 - \beta)[Z_1 Y_3 Z_3 (Y_1 + Y_2) + Y_3 Z_3] \\ &\quad + Y_1 Z_1 Z_2 (Y_2 + Y_3) + Y_2 Z_2 Y_3 (Z_1 + Z_3) + Y_1 Z_1 + Z_1 Y_2 + Z_2 Y_3 \\ &\quad + Y_2 Z_2 + 1 \end{aligned} \qquad (3\text{-}118b)$$

Some of the basic all-pole second-order networks, expanded into third-order networks, are listed in Table 3-7[32]. The design equations, in addition to being nonlinear, contain more unknowns than can be solved for explicitly. One way of overcoming this problem is to use the computer[33] to generate families of solutions to the simultaneous nonlinear equations of a network; the family of solutions corresponds to various sets of acceptable values for the number of components exceeding the number of circuit equations. Another is to assume suitable values for the excess components and to solve the available network equations for the rest. In Table 3-8 this has been done for the third-order low-pass network.

32. Derived by S. E. Sussman, Bell Telephone Laboratories, Holmdel, N.J.
33. For example, SECANT, Numerical Analysis Routines, General Electric Desk-Side Time-Sharing Service Program—*Library Users' Guide*, May 1968, pp. 70–72.

TABLE 3-8. DESIGN EQUATIONS FOR THE GENERAL THIRD-ORDER LOW-PASS FILTER

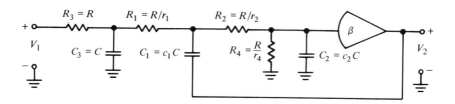

$$T(s) = \frac{\beta}{s^3 + a_2 s^2 + a_1 s + a_0}$$

$a_3 = R_1 R_2 R_3 C_1 C_2 C_3$

$a_2 = R_1 R_3 C_3 [C_2 + C_1(1 - \beta)] + R_2 R_3 C_2(C_1 + C_3) + R_1 R_2 C_1 C_2 + \dfrac{R_1 R_2 R_3}{R_4} C_1 C_3$

$a_1 = R_1[C_2 + C_1(1 - \beta)] + R_2 C_2 + R_3[C_2 + C_3 + C_1(1 - \beta)]$
$\qquad + \dfrac{1}{R_4}[R_3 C_3(R_1 + R_2) + R_2 C_1(R_1 + R_3)]$

$a_0 = 1 + \dfrac{1}{R_4}(R_1 + R_2 + R_3)$

$a_2' = a_2 RCc_1c_2 = (1 - \beta)r_2 c_1 + (r_1 + r_2)c_2 + (1 + r_1)c_1c_2 + r_4 c_1$

$a_1' = a_1(RC)^2 c_1 c_2 = (1 - \beta)(1 + r_1)r_2 c_1 + (r_1 + r_2 + r_1 r_2)c_2 + r_1 r_2 + r_4[r_1 + r_2 + c_1$
$\qquad (1 + r_1)]$

$a_0' = a_0(RC)^3 c_1 c_2 = r_1 r_2 + r_4(r_1 + r_2 + r_1 r_2)$

Given: R, C, c_1, c_2, r_4, a_i $(i = 0, 1, 2)$

Solve r_1 from the cubic equation

$$r_1^3\{c_2[1 + r_4 + c_1(1 + r_4)]\}$$
$$- r_1^2[a_2'(1 + r_4) - r_4 c_2 - c_1 c_2(2 + 3r_4) + r_4^2]$$
$$- r_1[a_2'(1 + 2r_4) - c_1 c_2(1 + 3r_4) + a_0' - a_1'(1 + r_4)]$$
$$- r_4(a_2' + a_0' - a_1' - c_1 c_2) = 0$$

(Note: When $r_4 = 0$, the equation becomes quadratic)

Solve r_2 from

$$r_2 = \frac{a_0' - r_1 r_4}{r_1 + r_4 + r_1 r_4}$$

Solve β from

$$\beta = \frac{r_2 + r_4 + (1 + r_1)c_2}{r_2} + \frac{(r_1 + r_2)c_2 - a_2'}{r_2 c_1}$$

Higher-Order Networks In Chapter 1 we made the point that active networks of order higher than three are invariably realized by cascading the appropriate second- and third-order subnetworks. One reason for this is to guarantee adequate Q stability, the latter becoming ever more critical with increasing Q. However, in many applications, filters of order higher than three are required whose pole Q's are quite low. In such networks the added cost and power requirements of a cascade of second-order active networks (each of which requires an active element) is unnecessary.

Class 4 second-order *all-pole* networks are particularly suitable for expansion to higher-order networks. Beside having the advantage of requiring only one amplifier, the networks are canonic (i.e., an nth-order network requires only n resistors and n capacitors). A low-Q nth-order network very commonly required is the low-pass. The general class 4 nth-order configuration is shown in Fig. 3-42a. Clearly this requires only a simple modification of the corresponding second-order building block. The voltage-transfer function of this configuration has the form $T(s) = \beta/D_n(s)$, where $D_n(s)$ is an nth-order polynomial. Recursive formulas have been developed[34] that determine the coefficients of $D_n(s)$ from those of the $(n-1)$th- and $(n-2)$th-order polynomials $D_{n-1}(s)$ and $D_{n-2}(s)$. Thus, from

$$D_{n-2}(s) = a_{n-2} s^{n-2} + a_{n-3} s^{n-3} + \cdots + a_1 s + 1 \qquad (3\text{-}119a)$$

and

$$D_{n-1}(s) = b_{n-1} s^{n-1} + b_{n-2} s^{n-2} + \cdots + b_1 s + 1 \qquad (3\text{-}119b)$$

we obtain

$$D_n(s) = d_n s^n + d_{n-1} s^{n-1} + d_{n-2} s^{n-2} + \cdots + d_1 s + 1 \qquad (3\text{-}119c)$$

where

$$d_i = (b_i - a_i)\frac{R_n}{R_{n-1}} + b_i + b_{i-1} R_n C_n - \delta_{1_i}\left[\frac{(-1)^n + 1}{2}\right] R_n C_n \beta \qquad (3\text{-}119d)$$

and

$$\delta_{1_i} = \begin{cases} 0 & \text{for } i \neq 1 \\ 1 & \text{for } i = 1 \end{cases}$$

Assuming all equal resistors or capacitors in the configuration of Fig. 3-42a, it is now possible to determine the remaining network components by solving the corresponding set of nonlinear equations. The coefficients can, for example, be equated to those of the appropriate Butterworth, Thomson, or Chebyshev polynomials[35] resulting in the component values for filters up to sixth order listed in Fig. 3-42b.

34. R. S. Aikens and W. J. Kerwin, Single amplifier, minimal *RC*, Butterworth, Thompson and Chebyshev filters to 6th order, *Intern. Filter Symp.*, Santa Monica, California, April 1972.
35. L. Weinberg, *Network Analysis and Synthesis*, (New York; McGraw-Hill Book Co., 1962).

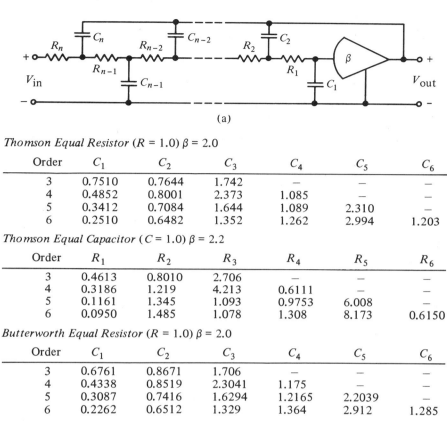

(a)

Thomson Equal Resistor (R = 1.0) β = 2.0

Order	C_1	C_2	C_3	C_4	C_5	C_6
3	0.7510	0.7644	1.742	—	—	—
4	0.4852	0.8001	2.373	1.085	—	—
5	0.3412	0.7084	1.644	1.089	2.310	—
6	0.2510	0.6482	1.352	1.262	2.994	1.203

Thomson Equal Capacitor (C = 1.0) β = 2.2

Order	R_1	R_2	R_3	R_4	R_5	R_6
3	0.4613	0.8010	2.706	—	—	—
4	0.3186	1.219	4.213	0.6111	—	—
5	0.1161	1.345	1.093	0.9753	6.008	—
6	0.0950	1.485	1.078	1.308	8.173	0.6150

Butterworth Equal Resistor (R = 1.0) β = 2.0

Order	C_1	C_2	C_3	C_4	C_5	C_6
3	0.6761	0.8671	1.706	—	—	—
4	0.4338	0.8519	2.3041	1.175	—	—
5	0.3087	0.7416	1.6294	1.2165	2.2039	—
6	0.2262	0.6512	1.329	1.364	2.912	1.285

Butterworth Equal Capacitor (C = 1.0) β = 2.2

Order	R_1	R_2	R_3	R_4	R_5	R_6
3	0.4538	1.0086	2.185	—	—	—
4	0.2696	1.491	3.6435	0.6826	—	—
5	0.1072	1.5131	1.1421	1.0955	4.927	—
6	0.08391	1.599	1.081	1.476	7.040	0.6635

Chebyshev $\frac{1}{2}$ dB Equal Resistor β = 2.0

Order	C_1	C_2	C_3	C_4	C_5	C_6
3	0.5483	0.8891	2.0512	—	—	—
4	0.3176	0.8063	2.698	1.316	—	—
5	0.3087	0.7416	1.629	1.217	2.204	—
6	0.1495	0.5203	1.154	1.555	3.461	2.070

(b)

FIG. 3-42. Expanding the TT-SABB low-pass network for higher-order operation: (a) network; (b) component values for filters to 6th order. Values shown are for $d_n = 1$ in $D_n(s)$.

3.2 DUAL–AMPLIFIER EXPANSION

The reason for introducing more than one amplifier into a second-order network is generally to improve the transmission sensitivity with respect to the active element(s). The following simple discussion may help to demonstrate how this is achieved.

Consider the root locus with respect to the gain β shown in Fig. 3-43. In a typical second-order network[36] the location of the pole pair is determined by the value of β; the larger β, the closer the pole pair will be to the $j\omega$ axis and the higher will be the pole Q. It has been shown[37] that the pole variation in terms of pole frequency ω_p and Q is given by

$$\frac{dp}{p} = \frac{d\omega_p}{\omega_p} - j\frac{dQ/Q}{\sqrt{4Q^2 - 1}} \tag{3-120}$$

For the building blocks under consideration the pole frequency ω_p is a function only of passive elements; frequency stability is therefore independent of gain. Q, on the other hand, depends on the gain β, and its stability (as we shall see in Chapter 4) is proportional to the gain–sensitivity product:

$$\Gamma = \beta \cdot S_\beta^Q \tag{3-121}$$

The smaller Γ, the more stable the pole Q. If, now, a gain β_0 is required to place the closed-loop pole at a specified location in the s-plane (see Fig. 3-43), then

$$\Gamma_0 = \beta_0 S_{\beta_0}^Q \tag{3-122}$$

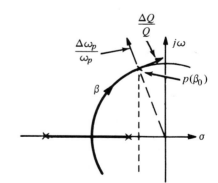

FIG. 3-43. Root locus of pole p with respect to gain β.

36. This argument is not restricted to negative-feedback circuits (i.e., classes 1 to 3) when operational amplifiers are used as the gain elements; the positive finite gain used in class 4 networks is also a result of negative feedback around the open-loop gain.
37. G. S. Moschytz, A note on pole, frequency and Q sensitivity, *IEEE J. Solid-State Circuits*, **SC-6**, 267–269, (1971); see also Chapter 4, Section 4.2.3 of *Linear Integrated Networks: Fundamentals*.

If, instead of one amplifier to obtain the required gain β_0, we distribute β_0 between m amplifiers in cascade, each of which has some gain β_i, then

$$\beta_0 = \prod_{i=1}^{m} \beta_i \qquad (3\text{-}123a)$$

Assuming that the open-loop gain (and its drift due to ambient influences) of all m amplifiers is the same, then our new gain–sensitivity product Γ_1 is given by

$$\Gamma_1 = \sum_{i=1}^{m} \beta_i S_{\beta_i}^Q \qquad (3\text{-}123b)$$

If now the m amplifiers also have equal closed-loop gain, that is $\beta_i \equiv \beta_1$ for $i = 1, \ldots m$, then

$$\Gamma_1 = m\beta_1 S_{\beta_1}^Q = m\beta_0^{1/m} S_{\beta_0^{1/m}}^Q = m^2 \beta_0^{1/m} S_{\beta_0}^Q \qquad [3\text{-}124a]$$

and

$$\frac{\Gamma_1}{\Gamma_0} = \frac{m^2 \beta_0^{1/m}}{\beta_0} = m^2 \beta_0^{(1/m)-1} \qquad [3\text{-}124b]$$

This relationship has been plotted for up to five amplifiers and various total gains β_0 in Fig. 3-44. Notice that there is no advantage in using more than one amplifier if the total required gain β_0 is less than 16; for $\beta_0 < 16$ the corresponding Γ actually increases when two or more amplifiers are used. For $\beta_0 > 16$ there is an optimum number of amplifiers m_{opt} for which the resulting Γ_1/Γ_0 ratio has a minimum. Setting the derivative with respect to m of (3-124b) equal to zero we obtain

$$m_{opt} = \tfrac{1}{2} \ln \beta_0 \qquad [3\text{-}125a]$$

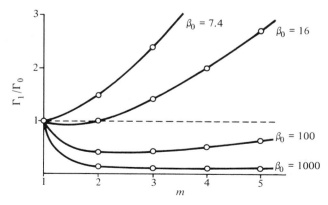

FIG. 3-44. Plot of Γ_1/Γ_0 vs. m for various values of β_0.

and, for $m = m_{\text{opt}}$,

$$\left(\frac{\Gamma_1}{\Gamma_0}\right)_{\min} = \frac{e^2}{4} \frac{(\ln \beta_0)^2}{\beta_0} \qquad \text{[3-125b]}$$

Obviously the nearest integer must be used for the m_{opt} resulting from (3-125a). For example, for $\beta_0 = 100$, (3-125a) results in the value $m_{\text{opt}} = 2.3$ which must be rounded down to 2. This means that two amplifiers with a gain of 10 each will provide a minimum gain–sensitivity product Γ_1; according to (3-124b) it will be 60 % less than if only one amplifier with a closed-loop gain $\beta_0 = 100$ were used.

A glance at Fig. 3-44 shows that by far the most significant improvement in Q stability (i.e., the largest decrease in Γ) occurs when we go from one to two amplifiers. It is perhaps for this reason, among others, that numerous suggestions for adding a second amplifier to a single-amplifier building block have been made. Some of these will be discussed in what follows.

3.2.1 Expanding Single–Amplifier to Dual–Amplifier Networks[38]

A Systematic Method of Dual-Amplifier Expansion   A general scheme for adding a second amplifier to a single-amplifier configuration of the kind shown in Fig. 3-45a is given in Fig. 3-45b. The amplifier is assumed to have infinite input, and zero output impedance and finite gain β. The passive RC network $t(s)$ is an n-pole network of which the first few terminals are shown; the remaining $(n - 4)$ poles are enclosed within the passive section.

Using Nathan's method of analysis for networks with finite-gain operational amplifiers[39] we obtain the admittance matrix for the circuit of Fig. 3-45b as follows:

$$[y] = \begin{matrix} & 1 & 2 & 3 \\ & \begin{bmatrix} \psi_{11} & y_{12} + \beta_1 y_{13} - \beta_1\beta_2 y_{14} & \psi_{15} & \cdots \\ \psi_{21} & y_{22} + \beta_1 y_{23} - \beta_1\beta_2 y_{24} & \psi_{25} & \cdots \\ \psi_{51} & y_{52} + \beta_1 y_{53} - \beta_1\beta_2 y_{54} & \psi_{55} & \cdots \\ \psi_{61} & y_{62} + \beta_1 y_{63} - \beta_1\beta_2 y_{64} & \psi_{65} & \cdots \\ \cdots & \cdots\cdots\cdots\cdots\cdots\cdots\cdots\cdots\cdots\cdots & \cdots & \end{bmatrix} \end{matrix} \qquad \text{(3-126a)}$$

The admittances ψ_{i2} of column 2 have the general form

$$\psi_{i2} = y_{i2} + \beta_1 y_{i3} - \beta_1\beta_2 y_{i4} \qquad \text{(3-126b)}$$

---

38. Much of this discussion is based on A. G. J. Holt, and M. R. Lee, General rule for the compensation of the sensitivity of a network to its active devices, *Electronics Lett.*, **5**, 324–325 (1969.)
39. See Chapter 3, Section 3.2.1 of *Linear Integrated Networks: Fundamentals.*

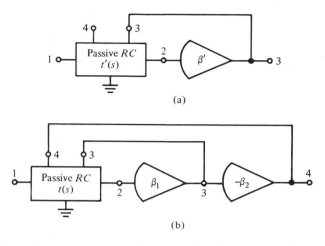

Fig. 3-45. Expanding (a) a single-amplifier network to (b) a dual-amplifier configuration.

Since β_1 and β_2 appear only in this column, all other admittances will have zero sensitivity to the amplifier gains. Letting

$$\beta_1 = \beta \tag{3-127a}$$

and

$$\beta_2 = a\beta \tag{3-127b}$$

we obtain the sensitivity of ψ_{i2} to variations in β as follows:

$$S_\beta^{\psi_{i2}} = \frac{d\psi_{i2}}{\psi_{i2}}\frac{\beta}{d\beta} = \frac{\beta(y_{i3} - 2a\beta y_{i4})}{y_{i2} + \beta y_{i3} - a\beta^2 y_{i4}} \tag{3-128}$$

This is zero when

$$\frac{y_{i3}}{y_{i4}} = 2a\beta \tag{3-129}$$

If this relationship is satisfied for all the admittances in the network, all admittances in the matrix (3-126a) will have zero sensitivity to simultaneous and similar changes in the amplifier gains. As a consequence, all network functions (including their poles and zeros) will have zero sensitivity to such changes. Note that the above analysis is independent of the sign of β.

Denoting the admittances of the original network with a prime and using Nathan's analysis we can now find the relationship between the admittances of the circuits given in Fig. 3-45a and Fig. 3-45b, such that the input and output parameters of the two networks are identical. However, we need only consider the admittances in the principal diagonal and those in column 2; y'_{i3}, for example, which is the admittance between nodes i and 3 of the original network, appears only in the principal diagonal and column 2. The same is true when a cascade of two amplifiers is substituted, as in Fig. 3-45b, with admittances y_{i3} and y_{i4}. Thus, in order for the matrices of the two networks to have equal elements in the principal diagonal, it is only necessary that

$$y_{i3} + y_{i4} = y'_{i3} \qquad (3\text{-}130)$$

since every admittance ψ_{ii} of the new circuit (Fig. 3-45b) now includes the *two* admittances $y_{i3} + y_{i4}$ instead of, as in the original circuit, only y'_{i3}. Except for the elements of column 2, all the other matrix elements for the two circuits remain unchanged. In order for the elements of column 2 for the two matrices to be the same it is necessary that

$$\psi'_{i2} = \psi_{i2} \qquad (3\text{-}131\text{a})$$

or, with Nathan's rule for circuits with finite-gain voltage sources,

$$y_{i2} + \beta' y'_{i3} = y_{i2} + \beta y_{i3} - a\beta^2 y_{i4} \qquad (3\text{-}131\text{b})$$

The last expression results from (3-126a) to (3-127b) and the fact that $y'_{i2} = y_{i2}$. Solving (3-129), (3-130), and (3-131b) for the quantities β, y_{i3}, and y_{i4} of the new circuit we obtain

$$y_{i3} = \frac{2a\beta}{1 + 2a\beta} y'_{i3} \qquad (3\text{-}132\text{a})$$

$$y_{i4} = \frac{1}{1 + 2a\beta} y'_{i3} \qquad (3\text{-}132\text{b})$$

and

$$\beta = \beta' + \sqrt{\beta'\left(\beta' + \frac{1}{a}\right)} \qquad (3\text{-}132\text{c})$$

In terms of resistors and capacitors for the admittances we obtain

$$C_{i3} = \frac{2a\beta}{1 + 2a\beta} C'_{i3}; \qquad C_{i4} = \frac{C'_{i3}}{1 + 2a\beta} \qquad (3\text{-}133\text{a})$$

or

$$R_{i3} = \frac{1 + 2a\beta}{2a\beta} R'_{i3}; \qquad R_{i4} = (1 + 2a\beta) R'_{i3} \qquad (3\text{-}133\text{b})$$

Example: To demonstrate the procedure outlined above we shall go through the steps required to expand the class 3 band-pass network of Fig. 3-46a into the dual-amplifier network of Fig. 3-46b. The transfer function of the original network (Fig. 3-46a) is

$$T'(s) = \frac{V'_3}{V'_1} = \frac{\beta' R'_2 C'_2 s}{s^2 C'_1 C'_2 R'_1 R'_2 (1 + \beta') + s(R'_1 C'_1 + R'_1 C'_2 + R'_2 C'_2) + 1} \qquad (3\text{-}134\text{a})$$

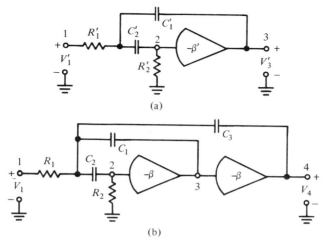

FIG. 3-46. Expanding (a) a class 3 bandpass network into (b) a dual-amplifier bandpass.

With $R_1' = R_2' = R$, $C_1' = C_2' = C$ this simplifies to

$$T'(s) = \frac{\beta' CRs}{s^2 R^2 C^2 (1 + \beta') + 3RCs + 1} \tag{3-134b}$$

The transmission sensitivity to β' follows as

$$S_{\beta'}^{T'(s)} = 1 - \frac{s^2 R^2 C^2 \beta'}{s^2 R^2 C^2 (1 + \beta') + 3RCs + 1} \tag{3-135}$$

The three steps involved in the expansion to the dual-amplifier circuit corresponding to Fig. 3-46b now follow.

1. Replace each voltage amplifier in the original network by a cascade of two amplifiers, in which the first amplifier has the same phase difference as the original amplifier and the second is *always* inverting. The gain of the first amplifier is calculated from (3-132c); the gain of the second (i.e., the factor a) is arbitrary.

Letting $a = 1$, β follows from (3-132c):

$$\beta = \beta'[1 + \sqrt{1 + (1/\beta')}] \tag{3-136}$$

and both amplifiers in the new circuit are in the inverting mode.

2. Where the output of an amplifier in the original network is connected to node i through an admittance y_{i3}', in the expanded network this node is connected to the outputs of the first and second cascaded amplifiers through admittances y_{i3} and y_{i4}, respectively (see (3-132a) to (3-133b)).

From (3-133a) we have, for our example,

$$C_1\bigg|_{\substack{C_1'=C\\a=1}} = \frac{2bC}{1+2b}; \qquad C_3\bigg|_{\substack{C_1'=C\\a=1}} = \frac{C}{1+2b} \qquad (3\text{-}137)$$

where $b = \beta$. Note that in these expressions b is a coefficient with the numerical value of the gain β; however, it is not subject to the same variations.

3. All admittances not connected to the output of an amplifier remain unchanged.

It follows that

$$C_2 = C_2' = C; \qquad R_1 = R_2 = R \qquad (3\text{-}138)$$

The transfer function of the expanded circuit is $T(s) = V_4/V_1 = N(s)/D(s)$, where

$$N(s) = sR_2 C_2 \beta^2 \qquad (3\text{-}139a)$$

and

$$\begin{aligned} D(s) = s^2[C_1 C_2(1 + \beta) &+ C_2 C_3(1 - \beta^2)]R_1 R_2 \\ &+ s[R_1(C_1 + C_2 + C_3) + R_2 C_2] + 1 \end{aligned} \qquad (3\text{-}139b)$$

With (3-137) and (3-138) we obtain

$$N(s) = sRC\beta^2 \qquad (3\text{-}140a)$$

and

$$D(s) = s^2 R^2 C^2 \frac{(1 + 2b + 2b\beta - \beta^2)}{1 + 2b} + 3sRC + 1 \qquad (3\text{-}140b)$$

The transmission sensitivity with respect to β follows as

$$S_\beta^{T(s)} = 2 - \left(\frac{2b\beta - 2\beta^2}{1 + 2b}\right)\frac{s^2 R^2 C^2}{D(s)} = 2 \qquad (3\text{-}141)$$

This is not zero because the β^2 term in (3-140a) does not appear in the admittance matrix (3-126a). The latter accounts for the transfer function when the output is taken from terminal 2 and not from the output of the cascaded amplifiers. The 2 in (3-141) is the sensitivity of the transfer function to the amplifier cascade. This constant value is small, however, compared with what the sensitivity of the denominator of $T(s)$ would normally be. Using the procedure outlined here, this latter sensitivity has been reduced to zero.

Note that if β' is substituted for β in (3-140a) and (3-140b) the transfer function of the original network (Fig. 3-46a) is obtained, multiplied by a factor due to the network output being taken from the output of the second amplifier. Thus the rules given here result in circuits with equal impedance levels but different gain requirements. An alternative procedure would result in equal gain requirements but different impedance levels.

Other Methods of Dual-Amplifier Expansion   Any single-amplifier network in which the amplifier is used as a voltage-controlled voltage source can readily be expanded into a dual-amplifier circuit, using the rules given above in order to minimize the transmission sensitivity with respect to simultaneous (and similar) variations in gain. Similar rules can be derived for other types of controlled source but, as we have already seen, those occur rarely in circuits incorporating operational amplifiers. Thus, other methods of expanding single-amplifier circuits to dual-amplifier circuits are also limited to circuits in which the amplifiers are used as controlled voltage sources. The difference between these other methods and the one described above is that they are less methodical. They generally concern themselves with one or more specific single-amplifier network functions (e.g., low-pass, bandpass, etc.), pointing out how to insert a second amplifier in such a way that the particular function in question becomes less vulnerable to variations in gain.

The most common single-amplifier circuits to have been selected for dual-amplifier expansion are class 1 and class 2 circuits. These circuits have very low sensitivities to the active and passive components; however, the gains required are proportional to Q^2 (see, e.g., Table 2-2). Thus, in spite of the low sensitivity, the gain–sensitivity product, and therefore the Q stability, is poor for anything but low Q. To improve this, a *noninverting* amplifier is introduced in series with the original amplifier; thus the feedback polarity remains unchanged but the gain is distributed between two amplifiers in order to obtain the benefits discussed in conjunction with (3-124a) and (3-124b).

Consider for example, the class 2 bandpass network shown in Fig. 3-47a.[40] For the element values $R_1 = R_2 = 1$, $C_1 = C_2 = 1/3Q$ and $\beta = 9Q^2 - 1$ we obtain a normalized bandpass network with $q_p = Q$. The Q sensitivities are

$$S_{R_1}^Q = -S_{R_2}^Q = -S_{C_1}^Q = S_{C_2}^Q = -\frac{1}{6} \tag{3-142}$$

and

$$S_\beta^Q = \frac{1}{2}\left(1 - \frac{1}{9Q^2}\right) \tag{3-143}$$

Nevertheless, for the gain–sensitivity product (see Table 2-2), we have

$$\Gamma \approx 4Q^2 - 1 \tag{3-144}$$

40. P. R. Geffe, A Q invariant active resonator, *Proc. IEEE*, **57**, 1442 (1969); also M. A. Soderstrand and S. K. Mitra, Extremely low-sensitivity active RC filter, *Proc. IEEE.*, **57**, 2175–2176 (1969).

The proportionality to Q^2 limits this circuit to low Q. By adding the non-inverting amplifier, as shown in Fig. 3-47a, we obtain the transfer function

$$T(s) = \frac{\beta_1 \beta_2 R_2 C_2 s}{(1 - \beta_1\beta_2)R_1 R_2 C_1 C_2 s^2 + (R_1 C_1 + R_2 C_2)s + 1} \qquad (3\text{-}145a)$$

so that

$$Q = \frac{\sqrt{(1 - \beta_1\beta_2)R_1 R_2 C_1 C_2}}{R_1 C_1 + R_2 C_2} \qquad (3\text{-}145b)$$

Letting

$$R_1 = R_2 = 1$$
$$C_1 = C_2 = 1/2Q \qquad (3\text{-}146)$$
$$\beta_1 = -\beta_2 = \sqrt{4Q^2 - 1}\,|_{Q \gg 1} \approx 2Q$$

we obtain a normalized bandpass function (i.e., $\omega_p = 1$, $q_p = Q$). Notice that the gains β_1 and β_2 are now proportional only to Q instead of to Q^2, which results in a gain–sensitivity product also proportional only to Q.

An example showing the expansion of a class 1 network to a dual-amplifier circuit is shown in Fig. 3-47b[41] (see also Table 2-2). There again a noninverting

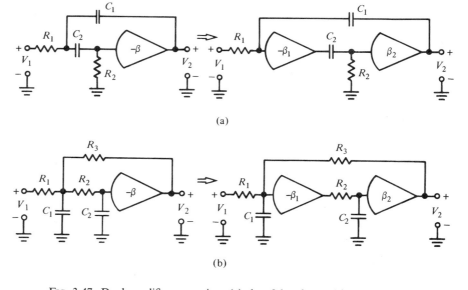

(a)

(b)

Fig. 3-47. Dual amplifier expansion: (a) class 2 bandpass; (b) class 1 low-pass.

41. T. Deliyannis, A low-pass filter with extremely low sensitivity, *Proc. IEEE*, **58**, 1366–1367 (1970).

amplifier is in cascade with the original inverting amplifier. The other possible class 1 and class 2 networks are expanded in exactly the same way (e.g., replacing R_1 in Fig. 3-47b by a capacitor, a dual-amplifier bandpass network is obtained).

Numerous other suggestions for the dual-amplifier expansion have been made; most of them involve the replacement of one high-gain amplifier by a cascade of two amplifiers with moderate gain. Others[42] add positive to negative feedback in order to circumvent some of the problems inherent in dual-amplifier expansion. Some of these problems will now be discussed.

3.2.2 A Closer Look at Dual–Amplifier Expansion

The intention of dual-amplifier expansion is clear enough; the high gain required by various low-sensitivity networks is distributed between two amplifiers in order to obtain the benefits implied by (3-124a) and (3-124b). However, the implementation is rather problematic.

The main difficulty encountered in dual-amplifier expansion is caused by the limited gain–bandwidth product of operational amplifiers. Let us consider once again the basic configuration shown in Fig. 3-45a; naturally we must assume that the operational amplifier has been frequency stabilized in such a way that it rolls off smoothly at -6 dB per octave, or, at least, that the rate of closure, i.e., the intersection of the closed-loop gain β with the frequency-compensated open-loop gain is 6 dB/oct. In cascading two amplifiers with equal gain, say $\beta_1 = \beta_2 = \beta$, so that their product is β', the rolloff of the composite, or dual amplifier, will be -12 dB/oct and will require additional frequency-compensating networks to prevent instability, even though both $\beta_1(\omega)$ and $\beta_2(\omega)$ roll off smoothly at -6 dB/oct (see Fig. 3-48a). Beside adding to the complexity of the circuit, the additional frequency-compensating components will depend on the particular loop gain required, which in turn will depend on the specified q_p. To avoid this problem, which conflicts with our stated objective of designing general-purpose building blocks, we can specify that the closed-loop gain of one of the amplifiers, say β_1, be unity, so that the gain of $\beta_2 = \beta'$. In this way, having been frequency-compensated individually for unity-gain operation, neither amplifier requires any additional frequency compensation. However, one further precaution must be taken even then to ensure absolute stability of the dual-amplifier circuit. Owing to the wide tolerances of the open-loop bandwidth of semiconductor integrated operational amplifiers, the unity-gain cutoff frequency of β_2 may be located relative to that of β_1 in any one of the three ways shown in Fig. 3-48b. The closed-loop gain of β_2 is shown relative to three β_1 amplifiers, each of which

42. For example, P. R. Geffe, *RC*-amplifier resonators for active filters, *IEEE Trans. Circuit Theory*, **CT-15**, 415–419 (1968).

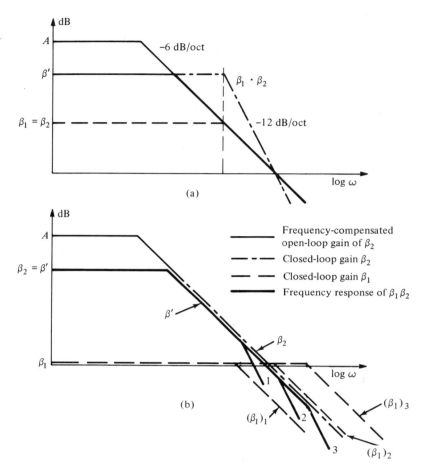

FIG. 3-48. (a) The cause of instability in a dual-amplifier loop. (b) The unconditional stabilization of a dual-amplifier loop.

has a different unity-gain bandwidth: the bandwidth of $(\beta_1)_1$ is less than, that of $(\beta_1)_2$ equal to, and that of $(\beta_1)_3$ greater than the unity-gain bandwidth of amplifier β_2. Only in case 3, in which the unity-gain bandwidth of β_1(i.e., $(\beta_1)_3$) exceeds that of β_2, is the absolute stability of the dual-amplifier circuit guaranteed. Thus, in summary, the two following conditions must be met:

1. The dual-amplifier circuit must be realized by two amplifiers, one of which has unity (or close-to-unity) gain.
2. The bandwidth of the unity-gain amplifier must exceed the unity-gain bandwidth of the high-gain amplifier.

As mentioned in condition 1, the gain of β_1 may be somewhat larger than unity (e.g., two). However, any increase in β_1 potentially deteriorates the stability of the circuit; we must therefore examine the bandwidth characteristics of the amplifiers to ascertain the limits on β_1. One way of enabling β_1 to be somewhat greater than unity is to select a wide-bandwidth, medium-gain amplifier for it, and a high-gain amplifier for β_2 whose unity-gain bandwidth is guaranteed to be lower than that of β_1. If two identical amplifiers are used, the β_2 amplifier must be more tightly frequency compensated than β_1, in order for stability to be guaranteed. In all cases the useful frequency range of the dual-amplifier circuit is determined by the unity-gain bandwidth of β_2.

One way of permitting the full frequency range of β_2 to be available is to use an emitter-follower (or Darlington pair) for β_1. With a high-frequency transistor the f_T will easily exceed the widest cut-off frequency available with an operational amplifier. The voltage offset and other problems ensuing with such an approach are dealt with in conjunction with the discussion of high-frequency FEN building blocks in Chapter 7.

At this stage the reader will understandably ask himself what is the point of the dual-amplifier expansion in the sense, say, of Fig. 3-47, if, in the end, one of the amplifiers must provide all the gain and the other serves only as a unity-gain buffer stage. The answer is that, indeed, because of the frequency limitations mentioned, the methods in question are not nearly as effective in stabilizing Q as might have been anticipated. To be sure, at very low frequencies, where it may be possible to distribute the gains between the two amplifiers equally, nothing in the efficiency of the method has been lost. However, even at higher frequencies, where β_1 must equal unity or thereabout, something can be gained by adding a buffer stage to the circuit. If we consider once more the general configuration of a single-amplifier network, we recall from Chapter 2 that the feedback function $t_{32}(s)$ will have the form

$$t_{32}(s) = k_{32}\ \frac{n_{32}(s)}{s^2 + \dfrac{\omega_p}{\hat{q}}\, s + \omega_p^2} \tag{3-147}$$

where the feedback network t_{32} is assumed passive RC and the maximum value of $\hat{q}$ therefore limited to 0.5. We have briefly pointed out already (and we will come back to this in the following chapters) that it is advantageous to design t_{32} in such a way that $\hat{q}$ is as close to 0.5 as possible. For one thing, this means that the open-loop poles of the feedback system are correspondingly close together, resulting in less gain required for a given pole Q (see Chapter 1, Section 1.3.3); for another, the sensitivity of Q to β can generally be shown to decrease with increasing $\hat{q}$ (see, e.g., (3-57) or (3-116)). The question, then, is how to design second-order RC networks with given ω_p and maximum $\hat{q}$. This problem is discussed in Appendix B of this book. It is shown there that in order to maximize $\hat{q}$, the two sections of an RC ladder-type network providing a pole pair should be isolated from each other as much as possible. One way of obtaining this isolation is to impedance-mismatch the two RC sections making up the feedback network, i.e., to start out with two identical pole sections and to impedance-scale the second section upward with respect to the first. The object of the resulting impedance mismatch is to minimize the loading effect of the second stage on the first. Consider for example the general ladder network in the feedback path of the circuit shown in Fig. 3-49a. If we assume that network n_a defines a pole p_1, then the two poles generated by n_a and n_b in cascade will approach a double pole at p_1 as $\rho \gg 1$. At the same time $\hat{q}$ will approach 0.5. A second method of achieving isolation between the two stages is, of course, by inserting a unity-gain buffer stage between n_a and n_b (see Fig. 3-49b). In this way the component spread inherent in impedance mismatching can be avoided. Because of its suitability as a buffer stage, the noninverting amplifier is generally used as the unity-gain stage β_1, leaving the inverting stage β_2 to provide the gain β'.

Whether the increase of $\hat{q}$ from some value less than 0.5 to 0.5 is worth the expense of an additional amplifier may be questionable, particularly since, even with the limited spread of component values attainable with hybrid integrated circuits, $\hat{q}$ values of 0.45 and even higher can be obtained by impedance mismatching. Furthermore, even if the frequencies are low enough to permit equal or near-equal gains to be distributed between the two amplifiers β_1 and β_2, it can be shown[43] that the sensitivities to changes in some of the passive components in the feedback loop (i.e., those associated with the second amplifier) are thereby increased. Thus, in looking for ways of overcoming the stability problems of high-Q networks the methods discussed above will often prove to be unsatisfactory. It will therefore come as no sur-

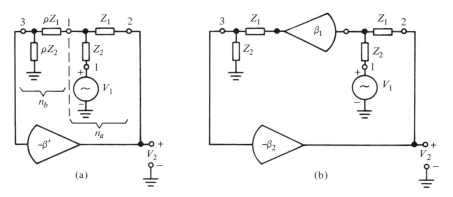

FIG. 3-49. Increasing $\hat{q}$ of a feedback ladder network: (a) by impedance mismatching; (b) with a buffer amplifier.

43. P. R. Geffe, Passive sensitivities of gain compensated networks, *IEEE Trans. Circuit Theory*, **CT-18**, 302–304 (1971). M. A. Soderstrand, Comments on "Passive sensitivities of gain compensated networks, *IEEE Trans. Circuit Theory*, **CT-19**, 107–108 (1972).

prise that the methods which do indeed solve those problems and which have resulted in the multiple-amplifier building blocks to be described next, have entirely different origins.

3.3 MULTIPLE-AMPLIFIER BUILDING BLOCKS

3.3.1 The Analog-Computer Approach

Consider the RLC low-pass filter shown in Fig. 3-50a, for which we obtain the equations

$$V_2 = \frac{1}{C_2} \frac{1}{s + \dfrac{1}{RC_2}} I_2 \tag{3-148a}$$

$$I_2 = \frac{V_1 - V_2}{sL_2} \tag{3-148b}$$

$$V_1 = \frac{I_1 - I_2}{sC_1} \tag{3-148c}$$

$$I_1 = \frac{V_0 - V_1}{sL_1} \tag{3-148d}$$

and the corresponding signal-flow graph shown in Fig. 3-50b. Note that the presence of the reactive components L and C means that the process of integration is taking place in the filter. In fact, according to the equations (3-148), each reactance in the filter may be replaced by an integrator $1/s$ or a

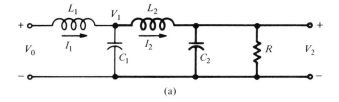

(a)

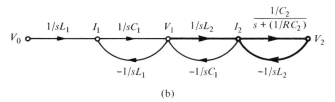

(b)

FIG. 3-50. An RLC low-pass filter: (a) network; (b) signal-flow graph.

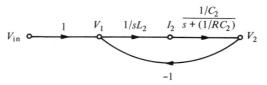

FIG. 3-51. Signal-flow graph of a second-order *RLC* low-pass network.

lossy integrator $1/(s + a)$ (see (3-148a)). Thus, in an analog-computer simulation of the network in Fig. 3-50a we require one integrator per reactive
element.

Restricting ourselves to a second-order network we have only those components and that part of the signal-flow graph in Fig. 3-50 which are drawn
with thick lines; the circuit comprises L_2, C_2, and R. By inspection of the
signal-flow graph in Fig. 3-50b we can shift the endpoint of the feedback
branch $-1/sL_2$ to the node V_1 and obtain the signal-flow graph shown in
Fig. 3-51. Note that this second-order graph consists of an integrator, a lossy
integrator, and an inverter in the feedback loop. We shall see in what follows
that these are the essential elements of a second-order system.

Instead of simulating an *LCR* network directly by the elements of an
analog computer, we can also simulate its transfer function. As with the network, we expect to carry out this simulation with integrators and inverters.
Consider, for example, the following second-order low-pass function, which
is normalized with respect to the pole frequency ω_p:

$$T(s) = \frac{K}{s^2 + \dfrac{\omega_p}{q_p}s + \omega_p^2} = \frac{K/\omega_p^2}{p^2 + \dfrac{p}{q_p} + 1} \qquad (3\text{-}149)$$

where

$$p = s/\omega_p \qquad (3\text{-}150)$$

To realize this function by a system with 100% negative feedback, as specified by the signal-flow graph of Fig. 3-51, it can be rewritten in the form

$$T(p) = \frac{g(p)}{1 + g(p)} = \frac{1}{1 + \dfrac{1}{g(p)}} \qquad (3\text{-}151)$$

where, from (3-149),

$$g(p) = \frac{1}{p\left(p + \dfrac{1}{q_p}\right)} \qquad (3\text{-}152)$$

Our low-pass function can therefore be realized by the block diagram shown
in Fig. 3-52. As is to be expected, this again represents an ideal integrator, a

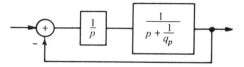

lossy integrator, and an inverter in a loop. The signal-flow graph of Fig. 3-51 corresponds exactly to the block diagram of Fig. 3-52 if

$$\omega_p^2 = \frac{1}{L_2 C_2} \tag{3-153a}$$

and

$$q_p = R\sqrt{C_2/L_2} \tag{3-153b}$$

Notice that the inverse of the pole Q, $1/q_p$, is equal to the loss of the nonideal integrator, which is also equal to the open-loop, negative real pole of the feedback loop.

Using operational amplifiers as inverters and integrators[44] the analog-computer realization of Fig. 3-52, which represents the simulation of a second-order low-pass function, is given by Fig. 3-53[45]. The circuit requires an inverter (A_3), since the two integrators A_1 and A_2 each perform an inversion by themselves. The ratio of the feedback resistor to the input resistor of the lossy integrator determines the damping of the natural modes of the second-order system or, in our terminology, the pole Q, q_p. The pole frequency ω_p is given by $1/RC$.

The derivation of the circuit shown in Fig. 3-53 should leave no doubt as to why this circuit is often referred to as the "analog-computer circuit." It results directly from a simulation, as if by analog-computer, i.e., using integrators to simulate the reactive elements of a second-order *LCR* network. (As we shall see shortly, only the output terminals differ if functions other

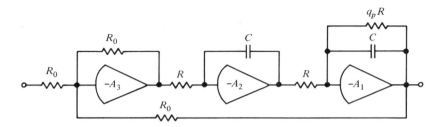

FIG. 3-53. The analog-computer realization of Fig. 3-52.

44. See Chapter 5, Table 5-7, of *Linear Integrated Networks: Fundamentals*.
45. F. E. J. Girling and E. F. Good, Active filters, unpublished Royal Radar Establishment memo, June 1963.

than the second-order low-pass are simulated.) As such, the circuit is almost as old as the analog computer itself and has been reported on frequently[46] being used both as an active filter and as an oscillator.[47] More recently, using state-variable analysis techniques, the circuit has been thoroughly analyzed from a sensitivity standpoint, and general design equations have been derived.[48] This work was expanded on shortly thereafter by L. C. Thomas[49] who, referring to the circuit as "the biquad," described its versatility as a second-order general-purpose building block and as a multipurpose, second-order filtering system (permitting various functions to be obtained at different outputs). In what follows we shall describe some of the important features of this circuit, when it is used as a "biquad" building block. We should point out, however, that this is only one of the many variations of the original analog-computer circuit (Fig. 3-53) that has been suggested since publication of the paper by Kerwin, Huelsman, and Newcomb in 1967.

The Multi-amplifier Biquad Building Block   *The Circuit*   The circuit drawing of the (multi-amplifier) biquad as used to generate all-pole functions, is shown in Fig. 3-54. Note that, but for a rearrangement of the three amplifiers (resulting, among other things, in improved dynamic-range capabilities) the circuit is identical with the original analog-computer circuit of Fig. 3-53. Taking the output at terminal b we obtain the bandpass function

$$T_b(s) = -K_b \frac{\omega_p}{q_p} \frac{s}{s^2 + \frac{\omega_p}{q_p} s + \omega_p^2} \tag{3-154a}$$

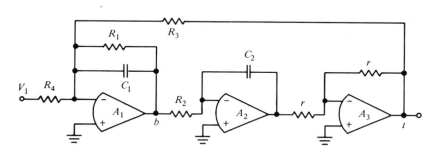

FIG. 3-54. The biquad used to generate all-pole functions.

46. For example, M. J. Somerville and G. H. Tomlinson, Filter synthesis using active *RC* networks, *J. Electron. Control,* **12**, 401–420 (1962); H. Sutcliffe, Tunable filter for low frequencies using operational amplifiers, *Electronic Engng,* **36**, 399–403 (1964); P. R. Geffe, Make a filter out of an oscillator, *Electronic Design,* **15**, (10) 56–58 (1967).
47. E. F. Good, A two-phase low-frequency oscillator, *Electronic Engng.,* **29**, 164–169 (1957) and **29**, 210–213 (1957).
48. W. J. Kerwin, L. P. Huelsman, and R. W. Newcomb: State-variable synthesis for insensitive integrated circuit transfer functions, *IEEE J. Solid-State Circuits,* **SC-2**, 87–92 (1967).
49. L. C. Thomas, The Biquad: Part I, Some practical design considerations, *IEEE Trans. Circuit Theory,* **CT-18**, 350–357 (1971) and Part II, A multipurpose active filtering system, **CT-18**, 358–361 (1971).

and at terminal l the low-pass function

$$T_l(s) = -K_l \frac{\omega_p^2}{s^2 + \frac{\omega_p}{q_p} s + \omega_p^2} \qquad (3\text{-}154b)$$

where

$$K_b = R_1/R_4 \qquad (3\text{-}155a)$$

$$K_l = R_3/R_4 \qquad (3\text{-}155b)$$

$$\omega_p = \frac{1}{\sqrt{R_2 R_3 C_1 C_2}} \qquad (3\text{-}155c)$$

and

$$q_p = R_1 \sqrt{C_1/R_2 R_3 C_2} \qquad (3\text{-}155d)$$

Note that K_b is the peak gain of the bandpass network; it occurs at ω_p and can be regulated by R_4. Similarly, the gain of the low-pass function at ω_p is $q_p \cdot K_l$, and can also be controlled by R_4.

It can be shown[50] that the sensitivity of q_p to variations in the finite operational-amplifier gain can be minimized if $R_2 C_2 = R_3 C_1$. Then we obtain from (3-155c) and (3-155d)

$$\omega_p = \frac{1}{R_3 C_1} = \frac{1}{R_2 C_2} \qquad (3\text{-}156a)$$

and

$$q_p = \frac{R_1}{R_3} \qquad (3\text{-}156b)$$

respectively, while K_b and K_l remain as given by (3-155a) and (3-155b). Note the simplicity of these relations. For each function the voltage gain at the pole frequency ω_p can now be independently determined by R_4; R_3 (and/or R_2) can be used to adjust ω_p independently and, with $R_2 C_2 = R_3 C_1$, q_p can be adjusted independently by R_1. Assuming that the proper sequence for the adjustments is used (i.e., $\omega_p \to q_p \to K_b$ or K_l) the three parameters can be adjusted independently of each other. This is one of the most significant features of the analog-computer circuit.

Expanding the basic loop of Fig. 3-54 by means of a fourth (summing) amplifier to which the weighted sum of V_1, V_b, and V_l are applied as shown

50. A. J. L. Muir and A. E. Robinson, Design of active *RC* filters using operational amplifiers, *Syst. Technol.*, Apr. 1968, pp. 18–30.

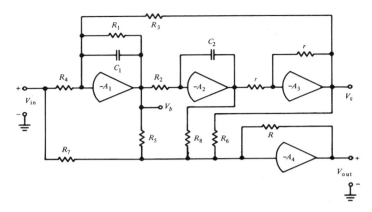

FIG. 3-55. The biquad building block.

in Fig. 3-55, we obtain a general second-order transfer function with an arbitrary pole and zero pair. By inspection we obtain

$$\frac{V_{out}}{V_{in}} = -\frac{R}{R_7} - \frac{R}{R_5} T_b(s) + \left(\frac{R}{R_8} - \frac{R}{R_6}\right) T_l(s) \qquad (3\text{-}157)$$

With (3-154) this becomes

$$\frac{V_{out}}{V_{in}} = -\frac{R}{R_7} \frac{s^2 + s\dfrac{\omega_p}{q_p}\left(1 - \dfrac{R_7}{R_5}K_b\right) + \omega_p^2\left[1 + \left(\dfrac{R_7}{R_8} - \dfrac{R_7}{R_6}\right)K_l\right]}{s^2 + \dfrac{\omega_p}{q_p}s + \omega_p^2} \qquad (3\text{-}158)$$

The poles are realized in precisely the same way as with the all-pole functions; therefore ω_p and q_p are given by (3-155c) and (3-155d) or by (3-156a) and (3-156b). The zeros depend also on the ratios of the weighting resistors R_5, R_6, R_7, and R_8. The dependence on resistor ratios is, of course, advantageous in hybrid integrated circuit implementation, because of the close resistor tracking that can be expected. On the other hand, the ratios may become large (see, e.g., (3-156b), where the required resistor spread equals q_p), in which case the resistive-T must be used, as discussed in Section 2.3.3.

By appropriate selection of the weighting resistors in (3-158), finite zeros of any type (e.g., for an FRN or an all-pass network) can be obtained. A glance at (3-158) shows that the zeros also depend on the pole quantities ω_p and q_p, which in turn depend on the gains in the main loop. Thus, the zero generation follows the type II network category with respect to the transmission sensitivity to gain variations. As discussed in Chapter 1, consideration must therefore be given to the optimum pole–zero pairing, from both dynamic-range and transmission-sensitivity points of view.

The biquad circuit of Fig. 3-55 contains many more components than there are parameters, even when realizing the most general second-order function. Referring to (3-1) the circuit can therefore be dimensioned as follows:[51]

$$R_1 = \frac{q_p/\omega_p}{C_1}; \qquad R_2 = \frac{k_1/\omega_p}{C_2}; \qquad R_3 = \frac{1}{k_1 \omega_p C_1}$$

$$R_4 = \left[k_2 K \left(\frac{\omega_p}{q_p} - \frac{\omega_z}{q_z} \right) C_1 \right]^{-1}; \qquad R_5 = k_2 R \qquad (3\text{-}159)$$

$$R_6 = \frac{k_2}{k_1} \frac{(\omega_p/q_p) - (\omega_z/q_z)}{\omega_p^2 - \omega_z^2} \omega_p R; \qquad R_7 = \frac{R}{K}$$

where C_1, C_2, and R are arbitrary, R_8 has been omitted, and k_1 and k_2 are arbitrary and positive. If k_1 or k_2 may be chosen equal to zero, the corresponding resistors are left out. If (3-156a) is to be satisfied, it follows that $k_1 = 1$. Naturally the other arbitrary component values will be selected for convenience, e.g., with respect to the technology used for circuit implementation. Complete design and realizability conditions have been published in the literature.[52]

Designing for Maximum Dynamic Range   In the preceding paragraphs we found that the design equations for the biquad are not fully determined by the specification of a second-order network function; various network components remain to be selected arbitrarily. This arbitrariness can be somewhat limited if we add the important requirement of maximum dynamic range to the other specifications of the network. Such a requirement will optimize the signal-to-noise capabilities of the biquad, since its maximum signal-handling capability is determined by the maximum voltage swing of the amplifiers, and, for all practical purposes, the output noise consists essentially of the noise generated by the amplifiers themselves.

In specifying a maximum dynamic range, we require the biquad to be dimensioned in such a way that the maximum voltage swings of its amplifiers are equal. The resulting circuit will provide the optimum signal-to-noise conditions. Consider, for example, the all-pole loop of Fig. 3-54. We require that the maximum voltage swings of the three amplifiers A_1, A_2, and A_3 be equal. From (3-154a) we have the maximum output voltage $|V_{1\,\text{max}}|$ of A_1, which occurs at the frequency ω_p, as

$$|V_{1\,\text{max}}| = K_b \cdot V_{\text{in}} \qquad (3\text{-}160)$$

51. J. Tow, Active *RC* filters—A state-space realization, *Proc. IEEE* (Corresp.), **56**, 1137–1139 (1968).
52. J. Tow, Design formulas for active *RC* filters using operational amplifier biquad, *Electronics Lett.*, **5**, 339–341 (1969); also, W. J. Kerwin, L. P. Huelsman, and R. W. Newcomb, op. cit.

The output of amplifiers A_2 and A_3 are equal in magnitude; thus it is sufficient for us to compute the maximum output voltage $|V_{2\,max}|$ of A_2. We have

$$V_2 = V_{in} \frac{K_l \omega_p^2}{s^2 + \dfrac{\omega_p}{q_p} s + \omega_p^2} \qquad (3\text{-}161)$$

This is a low-pass function; the maximum amplitude and corresponding frequency ω_m for this and other second-order functions were listed in Chapter 1, Table 1-4. In terms of (3-161) we obtain

$$\omega_m = \omega_p \sqrt{1 - \frac{1}{2q_p}} \qquad (3\text{-}162)$$

and the maximum amplitude as

$$|V_{2\,max}| = \frac{q_p \cdot K_l}{\sqrt{1 - (1/4q_p^2)}} \cdot V_{in} \qquad (3\text{-}163)$$

Equating (3-160) and (3-163) we obtain the condition

$$\frac{K_b}{K_l} = \frac{q_p}{\sqrt{1 - (1/4q_p^2)}} \qquad (3\text{-}164)$$

With the design equations given in (3-155) we can solve (3-164), obtaining

$$\frac{1}{R_2 C_2} = \omega_p \sqrt{1 - (1/4q_p^2)} \qquad (3\text{-}165a)$$

and, with (3-155c),

$$\frac{1}{R_3 C_1} = \frac{\omega_p}{\sqrt{1 - (1/4q_p^2)}} \qquad (3\text{-}165b)$$

By satisfying (3-165a) and (3-165b) we ensure a maximum dynamic range in the all-pass loop; the maximum voltage swing in the three amplifiers of the loop is equal. Unfortunately we thereby conflict with the condition $R_2 C_2 = R_3 C_1$ which was suggested for minimum sensitivity of q_p to amplifier gain variations. Nevertheless, as q_p increases and the requirement for reduced q_p sensitivity becomes more important, the two conditions converge and, in the limit, are identical.

Assuming, now, that equal capacitors are used, the two conditions (3-165) combined with the equations (3-155) provide a sufficient number of design equations to determine the components of the all-pole loop. If the complete biquad of Fig. 3-55 is to be used, then the computations for maximum dynamic range outlined above must include the fourth (summing) amplifier. The computations thereby remain the same in principle, although they become somewhat more complicated in fact. Rather than solve the corresponding design equations in closed form, it may then be preferable to solve them interactively by computer.

Sensitivity Considerations From (3-155) we can readily calculate the sensitivities of the pole quantities ω_p and q_p to the passive components. We obtain

$$
S_{R_1}^{\omega_p} = 0; \qquad S_{R_2}^{\omega_p} = S_{R_3}^{\omega_p} = -\tfrac{1}{2}
$$
$$
S_{R_1}^{q_p} = 1; \qquad S_{R_2}^{q_p} = S_{R_3}^{q_p} = -\tfrac{1}{2}
$$
(3-166)

These and the preceding calculations have assumed ideal operational amplifiers. For finite gains in the two integrators A_1 and A_2 (finite gain in the unity-gain inverter A_3 has negligible effect on the circuit) we obtain

$$
\omega'_p = \omega_p \left[\frac{1 + \dfrac{1}{q_p A_2}}{1 + \dfrac{1}{A_1} + \dfrac{1}{A_2}} \right]^{1/2}
$$
(3-167a)

and

$$
q'_p = q_p \frac{\sqrt{\left(1 + \dfrac{1}{q_p A_2}\right)\left(1 + \dfrac{1}{A_1} + \dfrac{1}{A_2}\right)}}{1 + \left(\dfrac{1}{A_1} + \dfrac{1}{A_2}\right)(1 + q_p)}
$$
(3-167b)

where the primes indicate actual, as opposed to ideal, quantities. Naturally, the larger the open-loop amplifier gains A_1 and A_2 are, the closer ω'_p and q'_p will be to their ideal values. The corresponding sensitivities for ω'_p result as

$$
S_{A_1}^{\omega_p'} = \frac{1}{2} \frac{1}{1 + \dfrac{A_1}{A_2} + A_1}
$$
(3-168a)

and

$$
S_{A_2}^{\omega_p'} = \frac{1}{2} \frac{1 - \left(1 + \dfrac{1}{A_1}\right)\dfrac{1}{q_p}}{\left(1 + \dfrac{A_2}{A_1} + A_2\right)\left(1 + \dfrac{1}{A_2 q_p}\right)}
$$
(3-168b)

To obtain the sensitivity of q'_p to amplifier gain we assume that $A_1 = A_2 = A$ and, for the time being, that both amplifiers have infinite bandwidth. Taking all terms of the form $1/A^2$ and $1/A$ as negligible in comparison to q_p/A we then have the simplified expression

$$q'_p = \frac{q_p}{1 + \dfrac{2q_p}{A}} \tag{3-169}$$

and consequently

$$S_A^{q_p'} = \frac{2q'_p}{A} \tag{3-170}$$

This is half as large as the corresponding sensitivity of class 3 networks using a twin-T in the feedback loop. Assuming, for example, a variation of 50% for A we obtain

$$V_A^{q_p'} = \frac{\Delta q'_p}{q'_p} = 100 \frac{q'_p}{A} (\%) \tag{3-171}$$

For an acceptable q'_p variation of 5% we require that $A \geq 20q'_p$. This is a reasonable demand, at least at frequencies well below the unity-gain frequency of an operational amplifier.

If a high forward gain is required using the integrator loop of Fig. 3-54, then an additional term enters into (3-170).[53] For the case of a bandpass filter where the peak gain is $K_b = R_1/R_4$ we obtain, instead of (3-170),

$$S_A^{q_p'} = \frac{1}{A}(2q_p + K_b) \tag{3-172}$$

In cases where the input signal-to-noise ratio is poor, it can be improved by increasing the gain of the first integrator of the loop, i.e., by increasing K_b. The penalty is then the increased sensitivity according to (3-172). If, for example, a gain of 40 dB is required for a bandpass section with a q_p of 100, then its sensitivity will be equal to that of a low-gain section with a q_p of 150.

The requirement that the open-loop gain of the operational amplifiers be high imposes the same high-frequency limitations on the biquad as exist for all other active-filter schemes using operational amplifiers. On the other hand, the inherently low sensitivity of this circuit to changes in the values of its components, both active (at low frequencies) and passive, is one of its significant advantages. This low sensitivity can be explained by the fact that the circuit is derived from the analog-computer simulation of a comparable passive LCR network, whose sensitivities are similarly low.

If we add the effect of finite bandwidth to the finite gain of the two amplifiers used as integrators in Fig. 3-54, we have

$$A(s) = A \frac{\Omega}{s + \Omega} \tag{3-173}$$

53. L. C. Thomas, op. cit.

and, setting $A_1 = A_2 = A(s)$, we obtain from (3-167b)

$$q_p' = \frac{q_p}{1 + \dfrac{2q_p}{A\Omega}(\Omega - 2\omega_p)} \tag{3-174}$$

Thus, as ω_p approaches the limit frequency

$$\Omega_L = \frac{\Omega}{2}\left(\frac{A}{2q_p} + 1\right) \tag{3-175}$$

the denominator of (3-174) approaches zero and the circuit becomes unstable. Notice that Ω_L is all the lower, the higher q_p is; it is also proportional to Ω, the 3 dB cutoff frequency of the operational amplifier. The value of Ω depends on how tightly the operational amplifiers must be frequency compensated for stability of the loop. This is an interesting question, since the capacitor of the integrator itself serves to compensate the operational amplifier.[54] It provides leading phase compensation of the feedback network resulting in a -6 dB/oct closed-loop slope. Taking the negative real pole of the lossy integrator into account, any additional frequency compensation can be relatively light, enabling the circuits to be used at higher frequencies than might, at first, seem feasible.

Beside the frequency-dependent gain, two other factors limiting the frequency range of the biquad are the limited slew rate and the nonlinear operation with increasing amplitude of the amplifiers. As a result, the circuit becomes conditionally stable as the operating frequency is increased. This instability manifests itself in the following way. At some high frequency the circuit is stable as long as the input signal level remains low. When the signal level is increased the pole Q, q_p, increases (" Q enhancement"), and at some level the circuit breaks into oscillation. When the signal level is decreased, the circuit generally returns to its stable state. It can be shown that the primary cause of this conditional instability is the nonlinear behavior of the operational amplifier, which introduces an equivalent lagging phase into the loop.[55] This increases with signal level until oscillation occurs. One way of solving this problem is by introducing a controlled and predictable form of nonlinearity into the network, e.g., a diode-limiter network across the capacitor of the lossy integrator. However, this nonlinear compensation has the disadvantage of limiting the dynamic range of the biquad. A more satisfactory alternative may be to compensate the operational amplifier appropriately.

The second cause of conditional instability, slew-rate limiting, can be eliminated by reducing the capacitive loading of the operational amplifier by the integrating feedback capacitor (i.e., C_1 and C_2 in Fig. 3-54). One way

54. See Chapter 7, Table 7-6, type 7, of *Linear Integrated Networks: Fundamentals*.
55. M. Baumwolspiner, Stability considerations in nonlinear feedback structures as applied to active networks, *Bell Syst. Tech. J.* **51**, 2029–2063 (1972).

of achieving this is to introduce a resistor (typically 50 Ω) between the amplifier output terminal and the feedback capacitor. Another is to use amplifiers with sufficiently high slew-rate capabilities, i.e., current driving capabilities at high frequencies. The latter method is to be preferred, since the former further deteriorates the characteristics of the integrators.

Other Analog-Computer Building Blocks *The Switched and Multifunction Biquads* The lack of interaction existing between center-frequency, bandwidth, and forward-gain adjustments of the biquad renders this circuit particularly suitable for applications requiring a number of optional filtering characteristics that differ in shape and frequency, but requiring only a subset of these filtering characteristics at any given time. If, for example, it is required to change the center frequency of a filter function, say of a second-order bandpass network, then only the resistor R_3 need be changed; similarly, to change only the bandwidth (given by $1/R_1 C_1$) or the peak gain (given by R_1/R_4), only R_1 or R_4, respectively, need be changed. This lack of interdependence between frequency, bandwidth, and gain has led to the concept of a switched filter, as shown in Fig. 3-56. Resistors R_1, R_3,[56] and R_4 are replaced by resistive-T networks with electronic, mechanical, or other types of switches S_1, S_3, and S_4. The next step to a voltage-variable filter (e.g., using

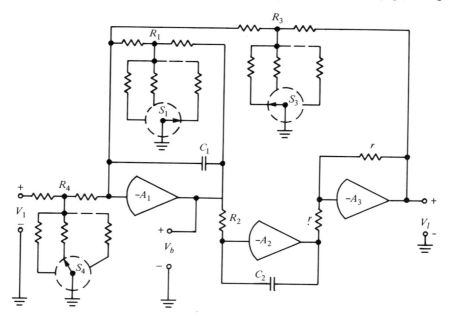

FIG. 3-56. The biquad as a switched-filter network.

56. If the pole frequency is to be varied over large ranges it will be necessary to vary R_2 together with R_3; this will also permit the condition $R_3 C_1 = R_2 C_2$ to be maintained.

voltage-controlled switches) is obvious, and need not be expanded on here. The same applies to continuously variable filters, in which case voltage- or current-variable resistors can be used instead of switched resistors. The concept can be expanded to include variable zeros, in which case one or more of the weighting resistors R_5, R_6, and R_7 must also be varied in discrete steps or continuously.

The biquad can also be used as a second-order multifunction filtering section. Referring to Fig. 3-55, V_b/V_{in}, V_l/V_{in}, and V_{out}/V_{in} simultaneously provide a bandpass, low-pass, and frequency-rejection or -emphasizing network, all with identical poles. Thus, for example, a narrow band of frequencies can be separated from a voice signal by designing a bandpass network which passes the narrow band of frequencies at the bandpass output (V_b), and rejects the narrow frequency band, passing only the voice signal, at the FRN output terminal (V_{out}). Similarly, the output summing amplifier A_4 can also be used to augment the multifunctional character of the filter section. Since this amplifier is outside the natural-mode loop generating the poles, it can be used to provide such nonlinear functions as signal switching, rectification or threshold detection by connecting an appropriate nonlinear device (e.g., diode) across the feedback resistor R. Furthermore, connecting a capacitor across R, a negative real pole may be obtained in addition to the complex pole pair generated in the loop.

Using the Noninverting Integrator   The price to be paid by the biquad, as well as by most other analog-computer circuits, for ease of tuning (at least for all-pole circuits) and low sensitivity, is the number of operational amplifiers required per second-order section. However, one of the amplifiers, the inverter, is required only to provide the phase inversion necessary for negative feedback; it can be dispensed with by using a noninverting integrator. One such integrator is shown in Fig. 3-57a. Its transfer function is given by

$$\frac{V_2}{V_1} = \frac{1}{sR_2 C_2} \cdot \frac{s + (1/R_2' C_2')}{s + (1/R_2 C_2)} \tag{3-176}$$

Thus, to yield a perfect integrator, this circuit depends on a pole–zero cancellation, obtained by balancing $R_2 C_2 = R_2' C_2'$. The effect on q_p of an imbalance in these time constants is given by

$$q_p' = \frac{q_p}{1 + \dfrac{q_p}{2}(\delta - 1)} \tag{3-177}$$

where the imbalance $R_2' C_2'/R_2 C_2$ is designated by δ. Nominally, δ is equal to unity. We can now compute the sensitivity of q_p' to δ and obtain

$$S_\delta^{q_p'} \approx -\frac{1}{2} q_p \tag{3-178}$$

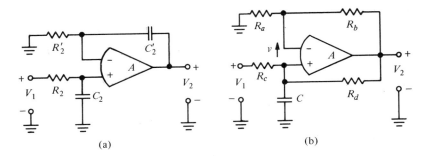

FIG. 3-57. Noninverting integrators: (a) two-capacitor circuit; (b) one-capacitor circuit.

For high q_p, the sensitivity to time-constant imbalance is therefore large. Thus, in a given circuit the adjustment for accurate balance is critical. It can be effected by disconnecting R_2' from ground and driving R_2 and R_2' with a common source at ω_p. Such an arrangement requires that DC be injected to balance the offset voltages that invariably occur. The transfer function corresponding to this tuning arrangement, evaluated at $s = j\omega_p$, is

$$|T(j\omega_p)| \approx \frac{1}{\sqrt{2}} \left(\frac{R_2' C_2'}{R_2 C_2} - 1 \right) \tag{3-179}$$

Thus for $q_p' = q_p$ we desire a null in (3-179) at ω_p. The required depth of the null for a known error in q_p can be derived by comparing (3-177) with (3-179). From (3-177), we define the error in q_p as

$$\varepsilon_{q_p} = \frac{1}{2} q_p \left(\frac{R_2' C_2'}{R_2 C_2} - 1 \right) \tag{3-180}$$

Then

$$|T(j\omega_p)| = \frac{\sqrt{2\varepsilon_{q_p}}}{q_p} \tag{3-181}$$

For example, for $q_p = 100$ and $\varepsilon_{q_p} = 0.01$, we require a 77 dB null in $T(j\omega_p)$. It can be obtained by adjusting R_2.

A noninverting integrator using only one (grounded) capacitor, is shown in Fig. 3-57b.[57] For ideal operation the resistor bridge R_a, R_b, R_c, R_d must be balanced, i.e., $R_a R_d = R_b R_c$. We can again define a measure of imbalance, namely $\rho = R_a R_d / R_c R_d$, which nominally must equal unity. For a specified q_p the actual value is then

$$q_p' = \frac{q_p}{1 + q_p(1 - \rho)} \tag{3-182}$$

57. See Chapter 5, Section 5.6 of *Linear Integrated Networks: Fundamentals*.

Notice the similarity between this expression and (3-177). The sensitivity to ρ is, of course, also similar, i.e., for $\rho \approx 1$,

$$S_\rho^{q_p{'}} \approx - q_p \tag{3-183}$$

and the error in q_p, $\bar\varepsilon_{q_p}$, is now given by

$$\bar\varepsilon_{q_p} = q_p(1 - \rho) \tag{3-184}$$

Since the circuit consists essentially of a resistive bridge, tuning can be carried out on a DC basis. The bridge is obtained by omitting the operational amplifier and grounding R_c. Referring to Fig. (3-57b) the bridge balance is then given by

$$v = \tfrac{1}{2}V_2(1 - \rho) \tag{3-185}$$

resulting in a desired bridge null for a given $\bar\varepsilon_{q_p}$ of

$$\frac{v}{V_2} = \frac{\bar\varepsilon_{q_p}}{4q_p} \tag{3-186}$$

Thus, for the same example as above, i.e., $q_p = 100$ and $\bar\varepsilon_{q_p} = 0.01$, the DC null required is 92 dB.

It will be clear from the above discussion that the price for the elimination of an amplifier in analog-computer circuits is not trivial. An extra, critical step is added to the tuning procedure and the sensitivity to the passive components is considerably increased.

The Feed-Forward Biquad   As we have seen so far, once the basic concept of the analog-computer version of a second-order network had evolved, numerous practical modifications followed. One promising modification of the biquad requires three instead of four amplifiers to generate complex transmission zeros.[58] It does so by using an input feed-forward technique instead of the summation technique used by the four-amplifier biquad. The configuration and design equations of this feed-forward or three-amplifier biquad are summarized in Table 3-9. Note that all but the FEN and resonator network types can be realized directly by this configuration. Thus the configuration shown covers the following functions:

low pass: $m = c = 0$
bandpass: $m = d = 0$
high pass: $c = d = 0$
FRN with perfect null: $m > 0, \quad c = 0, \quad d > 0$
FRN with finite zeros: $m > 0, \quad c \neq 0, \quad d > 0, \quad$ and $\quad (ma - c) \geq 0$
all pass: $c = - ma, \quad d = mb$

58. P. E. Fleischer and J. Tow, Design formulas for biquad active filters using three operational amplifiers, *Proc. IEEE* (Corresp.), **61**, 662–663 (1973).

TABLE 3-9. DESIGN EQUATIONS FOR THE FEED-FORWARD BIQUAD

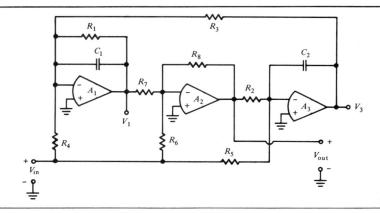

Transfer function

$$\frac{V_{out}}{V_{in}}(s) = -\frac{ms^2 + cs + d}{s^2 + as + b}$$

$$\frac{V_1}{V_{in}}(s) = -k_2 \frac{(ma-c)s + (mb-d)}{s^2 + as + b}$$

$$\frac{V_3}{V_{in}}(s) = -\frac{1}{k_1} \frac{\dfrac{d-mb}{\sqrt{b}}s + \dfrac{ad-bc}{\sqrt{b}}}{s^2 + as + b}$$

$$\left\{\begin{array}{l} m = \dfrac{R_8}{R_6} \\[2mm] c = \dfrac{1}{R_1 C_1}\left[\dfrac{R_8}{R_6} - \dfrac{R_1 R_8}{R_4 R_7}\right] \\[2mm] d = \dfrac{R_8}{R_7} \cdot \dfrac{1}{R_3 R_5 C_1 C_2} \\[2mm] a = 1/R_1 C_1 \\[2mm] b = \dfrac{R_8}{R_7} \cdot \dfrac{1}{R_2 R_3 C_1 C_2} \end{array}\right.$$

Design equations

$$R_1 = \frac{1}{aC_1}; \qquad R_2 = \frac{k_1}{\sqrt{b}\,C_2}; \qquad R_3 = \frac{1}{k_1 k_2} \cdot \frac{1}{\sqrt{b}\,C_1}$$

$$R_4 = \frac{1}{k_2(ma-c)C_1}; \qquad R_5 = \frac{k_1\sqrt{b}}{dC_2}; \qquad R_6 = \frac{1}{m}R_8; \qquad R_7 = k_2 R_8$$

Free parameters:

The free parameters are: C_1, C_2, R_8, k_1, and k_2, where

C_1, C_2, and R_8 control impedance levels; they are chosen to yield convenient element values

k_1 and k_2 may be chosen to establish the maximum voltage levels at the three amplifiers. Alternatively, for minimum sensitivity design (i.e., $R_3 C_1 = R_2 C_2$), set $k_1^2 k_2 = 1$.

Note: depending on the numerator coefficients, some of the "feed-in" resistors R_4, R_5, or R_6 may become infinite.

To obtain an FEN whose zeros are either complex or on the negative-real axis (i.e., $(ma - c) < 0$, $d \geq 0$), a positive-feedback resistor must be added between the output of A_2 and the input of A_1; if one zero is on the negative real axis, the other on the positive real axis (i.e., m and d are of opposite signs), then R_8 and C_2 must be interchanged, a positive-feedback resistor added from A_3 to A_2, and the output taken from A_3. Finally, to obtain a low-pass resonator, i.e., a numerator containing only one negative real zero (i.e., $m = 0$, c and d of opposite signs) the circuit shown in the table is used as it is, but the output is taken from A_3.

The Analog-Computer Approach Based on Positive Feedback: The All-Pass Loop

One of the problems that may arise in the realization of the analog-computer building block as based on the block diagram of Fig. 3-52 is related to the perfect integrator that is required. A perfect integrator requires a feedback capacitor (e.g., C_2 in Fig. 3-54) whose leakage current is less than the bias current of the amplifier. This requirement may strain the capabilities of thin-film capacitors. Various other parasitic effects may deteriorate the performance of an integrator[59] so that its characteristics may be far from perfect, particularly at frequencies above a few kiloHertz. Clearly it would be advantageous to design the circuit for lossy integrators and, still more important, for low-gain amplifiers so that, for a given gain–bandwidth product, a wider frequency range can readily be covered. The "allpass loop" is a circuit providing these features; it will be discussed next.

An alternative to the use of lossy integrators in the loop of Fig. 3-52 results when we reconsider the form of the transfer function (expressed in s) in (3-151) and the corresponding block diagram in Fig. 3-58. As pointed out earlier, according to this representation, analog-computer circuits may be based on networks $g(s)$, inserted in loops possessing 100 % negative feedback; the corresponding poles occur when $g(s) = -1$. However, it is also possible to derive analog-computer-type circuits from networks $g(s)$ inserted in loops possessing 100 % *positive* feedback. In this case, the poles occur when $g(s) = +1$. The higher the pole Q, the closer the roots of either equation (that is, the poles of the corresponding transfer function) will be to the $j\omega$ axis.

If we consider the positive-feedback alternative of the analog-computer approach we must arrange to have $0°$ phase shift and unity gain in the feedback loop at the specified pole frequency. Instead of cascading two

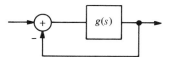

FIG. 3-58. Block diagram of the analog-computer network.

59. R. Stata, Minimizing integrator errors, *Electro-Technol.*, Oct. 1968, pp. 46–50.

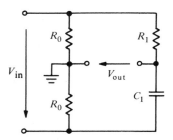

FIG. 3-59. A 90°-phase-shifting network.

integrators and an inverter in a loop (negative feedback) we can, for example, cascade two 90° phase shifters, one of which also inverts the signal (i.e., 270° phase shift), in a positive-feedback loop. A useful 90°-phase-shifting circuit with constant-amplitude response is shown in Fig. 3-59. We obtain the transfer function

$$g_1(s) = \frac{V_{out}}{V_{in}} = -k_1 \frac{s - \alpha_1}{s + \alpha_1} \qquad (3\text{-}187\text{a})$$

where $\alpha_1 = 1/R_1 C_1$ and $k_1 = \frac{1}{2}$: This corresponds to a *first-order all-pass* circuit. At the pole frequency α_1, the output signal is shifted by 270° with respect to the input signal. Interchanging R_1 and C_1 in Fig. 3-59, and designating the interchanged resistor and capacitor by R_2 and C_2, respectively, we obtain the desired 90° phase shift at the pole frequency, since

$$g_2(s) = k_2 \frac{s - \alpha_2}{s + \alpha_2} \qquad (3\text{-}187\text{b})$$

where $\alpha_2 = 1/R_2 C_2$ and $k_2 = \frac{1}{2}$. Letting $g(s) = k_3 g_1(s) g_2(s)$ we obtain, from Fig. 3-58,

$$T(s) = \frac{g(s)}{1 - g(s)}$$

$$= -\frac{k_1 k_2 k_3 (s - \alpha_1)(s - \alpha_2)}{(s + \alpha_1)(s + \alpha_2) + k_1 k_2 k_3 (s - \alpha_1)(s - \alpha_2)} \qquad (3\text{-}188)$$

Using an active version of the first-order all-pass circuit in Fig. 3-59 it is possible to obtain a variety of k_i and α_j values (where $i = 1, 2, 3$, and $j = 1, 2$) such that the poles of (3-188) take on any desired complex conjugate values. Indeed, we shall now show that using either one of two different active first-order all-pass networks, the all-pass loop (much as the integrator loop of Fig. 3-54) can be used as the pole-generating circuit of a general-purpose second-order filter building block. The active all-pass circuits to be considered (one uses amplifiers with differential, the other with single-ended output) satisfy our stated objective: they both require unity or close-to-unity gain.

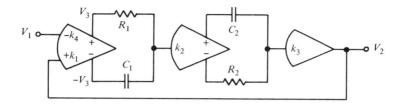

FIG. 3-60. The all-pass loop using differential amplifiers.

Using Amplifiers with Differential Output Using differential-input, differential-output amplifiers to realize the phase shifters of the type shown in Fig. 3-59, in series with a noninverting buffer amplifier, we obtain the circuit shown in Fig. 3-60.[60] The corresponding transfer function is given by (3-188), except that k_1 is now replaced by k_4 in the numerator. We therefore obtain a function of the form of (3-1), where

$$K = \frac{k_2 k_3 k_4}{1 + k_1 k_2 k_3} \tag{3-189a}$$

$$\omega_z = \omega_p = \sqrt{\alpha_1 \alpha_2} \tag{3-189b}$$

$$q_z = -\frac{\sqrt{\alpha_1 \alpha_2}}{\alpha_1 + \alpha_2} \tag{3-189c}$$

and

$$q_p = \frac{\sqrt{\alpha_1 \alpha_2}}{\alpha_1 + \alpha_2} \cdot \frac{1 + k_1 k_2 k_3}{1 - k_1 k_2 k_3} \tag{3-189d}$$

For maximum q_p one selects $\alpha_1 = \alpha_2 = \alpha$, so that

$$k_1 k_2 k_3 = \frac{1 - \dfrac{1}{2q_p}}{1 + \dfrac{1}{2q_p}} \tag{3-190}$$

For high q_p (i.e., > 100) this can be approximated by

$$k_1 k_2 k_3 \approx 1 - \frac{1}{q_p} \tag{3-191}$$

Thus, high q_p values require that the loop-gain be slightly less than unity. Since the gain–bandwidth product of operational amplifiers, as of any other

60. R. Tarmy and M. S. Ghausi, Very high-*Q* insensitive active *RC* networks, *IEEE Trans. Circuit Theory*, **CT-17**, 358–366 (1970).

amplifiers, is limited, this circuit can therefore be used at much higher frequencies than can the negative-feedback-type analog-computer circuit, which
requires high gain. Notice the analogy between positive and negative feedback networks of the analog-computer types with networks using single
amplifiers. In both, the positive-feedback versions require close-to-unity gain,
whereas their negative-feedback counterparts require, ideally, infinite gain.

It is convenient to select

$$k_2 = k_3 = 1 \tag{3-192a}$$

Then, from (3-190),

$$k_1 = \frac{1 - \dfrac{1}{2q_p}}{1 + \dfrac{1}{2q_p}} \approx 1 - \frac{1}{q_p} \tag{3-192b}$$

The value of k_4 depends on the desired forward gain.

To obtain the two different gain values k_1 and k_4 from the same differential-input, differential-output amplifier the configuration shown in Fig. 3-61
can be used. Assuming an infinite input impedance and a gain A, we can readily derive its transfer function:

$$V_3 = \frac{-V_1 \dfrac{R_d}{R_c}\left(1 + \dfrac{R_a}{R_b}\right) + V_2 \dfrac{R_a}{R_b}\left(1 + \dfrac{R_d}{R_c}\right)}{2 + \dfrac{R_a}{R_b} + \dfrac{R_d}{R_c} + \dfrac{1}{A}\left(1 + \dfrac{R_d}{R_c}\right)\left(1 + \dfrac{R_a}{R_b}\right)} \tag{3-193}$$

Letting

$$\mu_1 = \frac{R_d}{R_c}$$
$$\mu_2 = \frac{R_b}{R_a} \tag{3-194}$$

and assuming high open-loop gain we obtain

$$\frac{V_3}{V_1} = -k_4 = -\frac{\mu_1(1 + \mu_2)}{1 + 2\mu_2 + \mu_1\mu_2} \tag{3-195a}$$

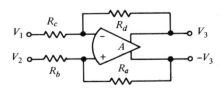

FIG. 3-61. A differential-input, differential-output configuration.

and

$$\frac{V_3}{V_2} = k_1 = \frac{1 + \mu_1}{1 + 2\mu_2 + \mu_1 \mu_2} \tag{3-195b}$$

Thus from (3-192b) we obtain

$$q_p = \frac{1}{1 - k_1} = \frac{1 + 2\mu_2 + \mu_1 \mu_2}{2\mu_2 - \mu_1 + \mu_1 \mu_2} \tag{3-196}$$

The values of the resistor ratios μ_1 and μ_2 are not unique and can, for example, be selected for minimum sensitivity of q_p to variations of these resistor ratios. The sensitivities result directly from (3-196); they are small when

$$\mu_1 \ll 1$$
$$\mu_2 \ll 1 \tag{3-197}$$

i.e.,

$$S_{\mu_1}^{q_p} = \frac{\mu_1(1 + \mu_2)}{(1 + 2\mu_2 + \mu_1\mu_2)^2} q_p \bigg|_{\substack{\mu_1 \ll 1 \\ \mu_2 \ll 1}} \approx \mu_1 q_p \tag{3-198a}$$

and

$$S_{\mu_2}^{q_p} = -\frac{(2\mu_2 + \mu_1\mu_2)(1 + \mu_1)}{(1 + 2\mu_2 + \mu_1\mu_2)^2} q_p \bigg|_{\substack{\mu_1 \ll 1 \\ \mu_2 \ll 1}} \approx -2\mu_2 q_p \tag{3-198b}$$

With (3-197) we then have, from (3-196),

$$q_p \bigg|_{\substack{\mu_1 \ll 1 \\ \mu_2 \ll 1}} \approx \frac{1}{2\mu_2 - \mu_1} \tag{3-199}$$

and, letting $\mu_1 = \mu_2 = \mu$,

$$q_p = \frac{1}{1 - k_1} \bigg|_{\mu_1 = \mu_2 = \mu} \approx \frac{1}{\mu} \tag{3-200}$$

Finally, normalizing the value of the amplitude response $|T(j\omega)|$ to unity at the resonance frequency $\omega_p = \omega_z = \omega_0$ we obtain, with $\alpha_1 = \alpha_2 = \alpha = 1/RC$,

$$|T(j\omega_0)| \bigg|_{\substack{\alpha_1 = \alpha_2 = \alpha \\ \omega_p = \omega_z = \omega_0}} = K \frac{q_p}{q_z} = \frac{k_2 k_3 k_4}{1 - k_1 k_2 k_3} = 1 \tag{3-201}$$

From (3-192) it then follows that

$$k_4 = \frac{1}{q_p} \tag{3-202}$$

The circuit of Fig. 3-60 is capable of providing a high-Q pole pair but the zeros, being of the form given by (3-188), are far from general; they are

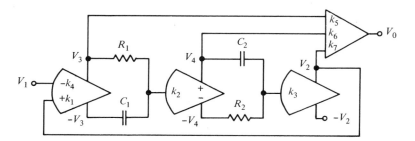

FIG. 3-62. The all-pass-loop building block using differential amplifiers.

negative real. In order to satisfy our building-block criterion, i.e., the capability of providing an arbitrary zero (as well as pole pair), a summing amplifier must be added, much as was done in the case of the biquad. The resulting circuit is shown in Fig. 3-62. From above we have the following design equations:

$$R_1 C_1 = R_2 C_2 = \alpha$$

$$k_1 = 1 - \frac{1}{q_p}$$

$$k_4 = \frac{1}{q_p}$$ (3-203)

$$k_2 = k_3 = 1$$

Then the transfer function of Fig. 3-62 has the numerator

$$N(s) = \frac{1}{q_p}\left\{(-\mu_3 + \mu_4 + \mu_2)s^2 - 2\left[\left(1 + \frac{1}{q_p}\right)\mu_3 + \mu_2\right]s - \mu_3 - \mu_4 - \mu_2\right\}$$ (3-204a)

and the normalized denominator

$$D(s) = 2\left(s^2 + \frac{1}{q_p}s + 1\right)$$ (3-204b)

For high q_p values the factor of s in the linear term in (3-204a) can be simplified to $-2(\mu_3 + \mu_2)$. The proper choice of μ_2, μ_3, and μ_4 produces the desired gain and zeros. Note that negative values of μ_2, μ_3, and μ_4 can be achieved by making connections to the appropriate negative output terminals of the amplifiers in the all-pass loop.

Because of their similarity, the parameter sensitivities to component variations of the analog computer networks based on positive feedback are comparable to those of the analog-computer circuits based on negative feedback. Also, the tuning of the two circuits is similar. The pole frequency of the former can be adjusted by varying R_1 or R_2 with negligible effect on q_p, whereas q_p

can be adjusted by tuning any one of the gains k_1, k_2, or k_3 (or all of them) with negligible effect on the pole frequency. An advantage of the positive-feedback circuit is the fact that its loop gain $|k_1 k_2 k_3| < 1$, which guarantees a high degree of stability, regardless of the excess phase in the amplifiers. This feature makes the circuit particularly attractive for high-frequency applications, and permits its being combined with high-frequency components such as crystals, which would hardly be feasible with most other circuits. Using crystal resonators in the place of C_1 and C_2 (Fig. 3-60) for example, highly stable active filters have been built successfully with center frequencies as high as 20 MHz.[61]

The fact that the analog-computer method based on positive feedback requires amplifier gains of unity or thereabout has the important advantage that the maximum bandwidth of the amplifiers can be utilized and that the optimum gain stability can be acheived. On the other hand, one of the disadvantages of the circuit realization given in Fig. 3-60 is the fact that it requires operational amplifiers with differential output terminals. Quite apart from the fact that such amplifiers are in less demand, and therefore more expensive, than their single-ended counterparts, it has been shown[62] that because of the high sensitivity to unsymmetrical loading of the differential-output amplifiers, this circuit is itself similarly sensitive. Thus, because of its various advantages, foremost among which is its high-frequency capability, a single-ended equivalent to the circuit is desirable, and, as we shall now demonstrate, can readily be derived.

Using Amplifiers with Single-Ended Output   The circuit in Fig. 3-60 was interpreted as a cascade of two first-order all-pass networks of opposite sign in a 100% positive-feedback loop. The corresponding block diagram is shown in Fig. 3-63a. If both all-pass networks have the same polarity, then

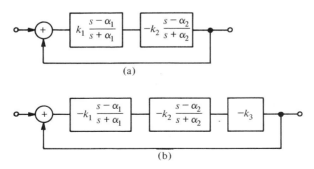

(a)

(b)

FIG. 3-63. Block diagram of the all-pass-loop: (a) without inverter; (b) with inverter.

61. D. R. Means and M. S. Ghausi, Inductorless filter design using active elements and piezo-electric resonators, *IEEE Trans. Circuit Theory*, **CT-19**, 247–253 (1972).
62. M. Müller, A very high-*Q RC* filter, *Mitteilungen AGEN*, Zürich, No. 12, August 1971, pp. 55–66.

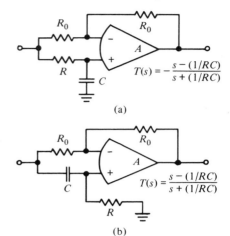

FIG. 3-64. First-order all-pass networks with single-ended amplifiers: (a) inverting; (b) noninverting.

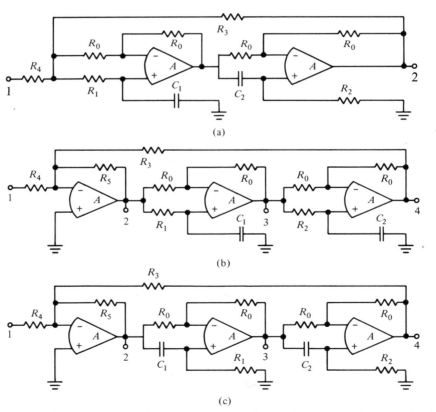

FIG. 3-65. Various configurations of the all-pass loop using single-ended operational amplifiers.

an inversion must be carried out in the loop (Fig. 3-63b). First-order all-pass sections with either polarity are readily obtained with single-ended operational amplifiers[63] (Fig. 3-64). Various realizations of Fig. 3-63a or b therefore immediately present themselves.[64] Some of these are shown in Fig. 3-65. Fig. 3-65a is attractive because it uses only two amplifiers. However, the pole frequency and Q are not as independently tunable as with the three-amplifier configurations. For ease of implementation, tuning, and computation, the configuration in Fig. 3-65b is superior; the capacitors (or crystal resonators, etc.) are grounded and share a common electrode. Furthermore, the low-pass structures connected to the noninverting terminals can be tuned to compensate for residual phase shift in the operational amplifiers.

Consider now the transfer function from terminal 1 to terminal 4 of the circuit in Fig. 3-65b. We obtain

$$T_{14}(s) = \frac{V_4}{V_1} = -\frac{R_5}{R_4} \frac{\left(\dfrac{s - \alpha_1}{s + \alpha_1}\right)\left(\dfrac{s - \alpha_2}{s + \alpha_2}\right)}{1 + \dfrac{R_5}{R_3}\left(\dfrac{s - \alpha_1}{s + \alpha_1}\right)\left(\dfrac{s - \alpha_2}{s + \alpha_2}\right)} \tag{3-205}$$

where $\alpha_1 = 1/R_1 C_1$ and $\alpha_2 = 1/R_2 C_2$.

This is a function of the form[65]

$$T(s) = -K_0 \frac{s^2 + \dfrac{\omega_0}{q_0} s + \omega_0^2}{s^2 + \dfrac{\omega_p}{q_p} s + \omega_p^2} \tag{3-206}$$

Multiplying out (3-205) and comparing coefficients with (3-206) we obtain

$$K_0 = \frac{R_5}{R_4} \cdot \frac{R_3}{R_3 + R_5} = \frac{R_3 \| R_5}{R_4} \tag{3-207a}$$

$$\omega_0 = \omega_p = \sqrt{\alpha_1 \alpha_2} \tag{3-207b}$$

$$q_0 = -\frac{\sqrt{\alpha_1 \alpha_2}}{\alpha_1 + \alpha_2} \tag{3-207c}$$

and

$$q_p = -\frac{1 + (R_5/R_3)}{1 - (R_5/R_3)} q_0 \tag{3-207d}$$

63. G. S. Moschytz and W. Thelen, Design of hybrid-integrated filter building blocks, *IEEE J. Solid-State Circuits*, **SC-5**, 99–107 (1970).
64. G. S. Moschytz, A high-Q, insensitive active *RC* network, similar to the Tarmi–Ghausi circuit, but using single-ended operational amplifiers, *Electronics Lett.* **8**, 458–459 (1972).
65. We do not use ω_z and q_z, corresponding to the general form (3-1), in the numerator here because the zeros of (3-206) do not yet coincide with the zeros of our general second-order function; they are limited to the real axis.

For the sake of circuit uniformity and ease in practical implementation, it is convenient to let $\alpha_1 = \alpha_2 = \alpha$. At the same time this provides a maximum q_p (and q_z) for a given ratio R_5/R_3. Rewriting (3-205) we therefore obtain

$$T_{14}(s) = -K_0 \frac{(s - \alpha)^2}{s^2 + \dfrac{\omega_p}{q_p} s + \omega_p^2} \tag{3-208}$$

Likewise, calculating the voltage-transfer functions from terminals 1 to 3 and 1 to 2, respectively, in the network of Fig. 3-65b, we obtain

$$T_{13}(s) = K_0 \frac{s^2 - \alpha^2}{s^2 + \dfrac{\omega_p}{q_p} s + \omega_p^2} \tag{3-209}$$

and

$$T_{12}(s) = -K_0 \frac{(s + \alpha)^2}{s^2 + \dfrac{\omega_p}{q_p} s + \omega_p^2} \tag{3-210}$$

where

$$K_0 = \frac{\gamma_4}{1 + \gamma_3} = \frac{R_3 \| R_5}{R_4} \tag{3-211a}$$

$$\omega_p = \alpha = \frac{1}{RC} \tag{3-211b}$$

$$q_p = \frac{1}{2} \cdot \frac{1 + \gamma_3}{1 - \gamma_3} \tag{3-211c}$$

and

$$\gamma_3 = \frac{R_5}{R_3}; \qquad \gamma_4 = \frac{R_5}{R_4} \tag{3-211d}$$

For high q_p we have

$$\gamma_3 = \frac{R_5}{R_3} = \frac{1 + \dfrac{q_0}{q_p}}{1 - \dfrac{q_0}{q_p}} \approx \left. 1 + 2 \frac{q_0}{q_p} \right|_{\alpha_1 = \alpha_2 = \alpha} = 1 - \frac{1}{q_p} \tag{3-212}$$

The resistor ratio γ_4 depends on the specified constant K_0.

Notice that the zeros of the three transfer functions $T_{12}(s)$, $T_{13}(s)$, and $T_{14}(s)$ lie either on the real or the imaginary axis; thus the all-pass loop by itself does not provide the general second-order function (3-1) whose zeros may be

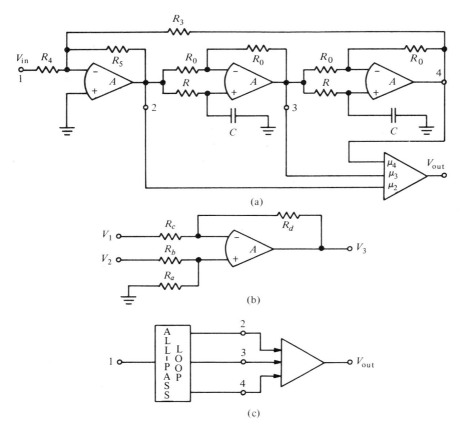

FIG. 3-66. The all-pass-loop building block using single-ended amplifiers: (a) circuit diagram; (b) simple realization of the summing amplifier; (c) simplified symbolic representation of the buiding block.

arbitrarily chosen in the s-plane. However, using a summing amplifier to add the three functions (appropriately weighted by the coefficients μ_2, μ_3, and μ_4) the general second-order building block shown in Fig. 3-66a is obtained. The corresponding transfer function is

$$T(s) = K_0 \frac{s^2(-\mu_2 + \mu_3 - \mu_4) + 2\alpha s(-\mu_2 + \mu_4) - \alpha^2(\mu_2 + \mu_3 + \mu_4)}{s^2 + \frac{\omega_p}{q_p}s + \omega_p^2} \qquad (3\text{-}213)$$

Depending on the polarity and value of the individual coefficients μ_2, μ_3, and μ_4, a zero pair can now be obtained anywhere in the s-plane. A simple realization of the summing amplifier is shown in Fig. 3-66b. It is the single-ended-output equivalent of the differential-output summing amplifier in Fig.

TABLE 3–10. THE ALL–PASS–LOOP BUILDING BLOCK

The All-pass-loop

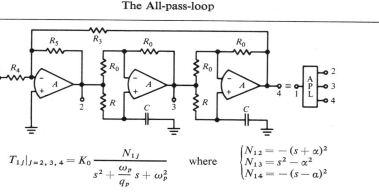

$$T_{1J}|_{J=2,\,3,\,4} = K_0 \frac{N_{1J}}{s^2 + \dfrac{\omega_p}{q_p} s + \omega_p^2} \qquad \text{where} \qquad \begin{cases} N_{12} = -(s+\alpha)^2 \\ N_{13} = s^2 - \alpha^2 \\ N_{14} = -(s-a)^2 \end{cases}$$

$$\omega_p = \alpha = \frac{1}{RC}; \qquad q_p = \frac{1}{2}\frac{1+\gamma_3}{1-\gamma_3}; \qquad K_0 = \frac{\gamma_4}{1+\gamma_3} = \frac{R_3\|R_5}{R_4}; \qquad \gamma_3 = \frac{R_5}{R_3}; \qquad \gamma_4 = \frac{R_5}{R_4}$$

The all-pass-loop building block

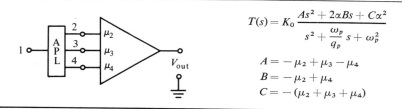

$$T(s) = K_0 \frac{As^2 + 2\alpha Bs + C\alpha^2}{s^2 + \dfrac{\omega_p}{q_p} s + \omega_p^2}$$

$$A = -\mu_2 + \mu_3 - \mu_4$$
$$B = -\mu_2 + \mu_4$$
$$C = -(\mu_2 + \mu_3 + \mu_4)$$

3-61, and can be derived from the latter by grounding the $-V_3$ output terminal. The output signal is given by

$$V_3 = -\frac{R_d}{R_c} V_1 + \frac{R_c}{R_a + R_b} \frac{R_d}{R_c} V_2 \tag{3-214}$$

Using this summing amplifier and the all-pass loop we obtain an all-purpose active-filter building block whose design equations are summarized in Table 3-10. Here we use the symbolic representation of the allpass loop shown in Figure 3-66c. With this building block we can now obtain any second-order function merely by interconnecting appropriately (and with the proper weighting) between loop and summing amplifier. The resulting network configurations together with the corresponding design equations for some typical second-order network types are given in Table 3-11.

Since the circuit of Fig. 3-65c is the $RC : CR$ dual of Fig. 3-65b, it can also be used as the all-pass loop of a second-order building block. The resulting configurations and design equations are similar to those just discussed. However, this network may be more sensitive to noise than the one shown in

Fig. 3-65b because of the capacitive inputs to the amplifiers which form high-pass or differentiator-type structures.

As expected, the expressions derived for the circuit of Fig. 3-66a closely resemble those derived for the circuit of Fig. 3-62. As a consequence, the favorable sensitivity behavior also applies here. So does the excellent high-frequency capability, since the gain of the individual amplifiers is required to be low, resulting in a wide bandwidth and stable gain. Thus we have here a filter building block providing stable high pole Q's and arbitrary zeros over a frequency band extending over several megacycles. It is no wonder then that we will return to this network in Chapter 7, where it will be examined in more detail with respect to its high-frequency capabilities. Because these can be shown to be very promising, the network is proposed as a useful high-frequency building block.

A Brief Assessment of the Analog-Computer Approach to Building-Block Design   It was shown earlier in this chapter (Section 3.2) that as the pole Q of a second-order network increases, the number of amplifiers per pole pair among which the gain is to be distributed must also be increased, in order to maintain gain stability and with it Q stability. The analog-computer methods described above have been shown to provide useful multi-amplifier building blocks for high-Q applications. Of course, they can be used for low-Q applications as well (say $Q < 25$), but there they compete, and not necessarily successfully, with various single-amplifier circuits that are equal in performance and cheaper in cost. After all, we must not forget that, beside the cost of the amplifiers themselves, the cost of supplying power to them may be appreciable, particularly if three to four amplifiers are required for each second-order section. If we add to this the need to control the heat generated by the dissipated power of the amplifiers, which may be considerable, it becomes clear that there is much to be gained by using single-amplifier active filters wherever the pole-Q's are low enough to permit this.

A recent survey[66] has shown that a large number of the filtering functions needed in modern communication systems can be realized with pole Q's low enough to be generated by single-amplifier building blocks. In such cases, for the reasons mentioned above, single-amplifier realizations are generally preferable to their multi-amplifier equivalents. Among the latter, we include the analog-computer circuits based on positive and negative feedback, both of which may require up to four amplifiers per second-order section.

From the preceding discussion it follows that with the circuits described so far we require a single-amplifier building block for all those applications in which the pole Q's are low enough to justify their use, and an analog-computer type circuit for the high-Q cases. In other words, we require the development of two entirely different building-block types realizing the same filter functions

---

66. G. B. Thomas, internal Bell Telephone Laboratories memorandum.

TABLE 3-11. THE ALL-PASS-LOOP BUILDING BLOCK: SOME TYPICAL NETWORK CONFIGURATIONS

Function	Circuit	Design equations
Low-pass $$T(s) = K \dfrac{\omega_p^2}{s^2 + \dfrac{\omega_p}{q_p} s + \omega_p^2}$$		$\mu_2 = \mu_4 = -\mu$ $\mu_3 = -2\mu$ $K = 4\mu K_0$
High-pass $$T(s) = K \dfrac{s^2}{s^2 + \dfrac{\omega_p}{q_p} s + \omega_p^2}$$		$\mu_2 = \mu_4 = -\mu$ $\mu_3 = 2\mu$ $K = 4\mu K_0$

TABLE 3.11—cont.

Function	Circuit	Design equations
Bandpass $T(s) = K \dfrac{\omega_p s}{s^2 + \dfrac{\omega_p}{q_p} s + \omega_p^2}$		$\mu_2 = -\mu_4 = -\mu$ $\mu_3 = 0$ $K = 4\mu K_0$
FRN $T(s) = K \dfrac{s^2 + \omega_z^2}{s^2 + \dfrac{\omega_p}{q_p} s + \omega_p^2}$		$\mu_2 = \mu_4 = -\mu$ $\mu_3 = 0$ $K = 2\mu K_0$

but covering different ranges of pole Q. The consequent decrease in production quantities—and increase in development costs—will only be acceptable in those cases in which the demand for both types of building blocks is sufficiently high. In other cases, an approach to building-block design is necessary that uses a *single*-amplifier building block not only for low-Q applications but, in multi-building-block structures, for high-Q applications as well. In the latter case, pairs of these same building blocks must be combinable into multi-amplifier structures providing stable, second-order, high-Q functions. Such an approach will be discussed in the next section. We should point out, though, that the realization of low- and high-Q circuits using single-amplifier building-blocks, either as separate entities or in pairs, does not eliminate the need for the analog-computer circuits in those applications in which only the features peculiar to them are required. Thus, the biquad may well be used where variable (e.g., switchable) frequency characteristics are required or where more than one function (e.g., low-pass and bandpass) with identical poles are required at the same time. Similarly, as we have already pointed out, the positive-feedback, analog-computer circuits may very well provide stable high Q's at frequencies that are far too high for other circuit types. Whether the passive components provide stable enough frequencies to warrant the high pole Q's is a separate question that will be dealt with in more detail in Chapter 4. Certainly, the possibility of using crystals in such circuits has already been mentioned, in which case the question is no longer relevant.

3.3.2 Filter Design Using FEN Building Blocks

The approach to be discussed here[67] is based on the FEN decomposition described in Section 3.1.1. We shall therefore be concerned with circuit realizations of the functions $T_L(s)$ and $T_H(s)$ as defined by (3-24) and (3-25); however, we shall assume that the specified pole Q's are higher than those attainable with the single-amplifier FEN dealt with in the discussion of twin-T feedback networks in Section 3.1.1.

The High-Selectivity FEN   Consider the circuit shown in Fig. 3-67. It consists of an inverting operational amplifier with open-loop gain A in the forward path and an active null network in cascade with a voltage amplifier (gain β) in the feedback path. The null network has the transfer function

$$t(s) = \frac{s^2 + \omega_p^2}{s^2 + \dfrac{\omega_p}{q_L}s + \omega_p^2} \qquad (3\text{-}215)$$

67. Based on G. S. Moschytz, Miniaturized filter building blocks using frequency emphasizing networks, *Proc. Nat. Electronics Conf.*, **23**, 364–369 (1967); also Active *RC* filter building blocks using frequency emphasizing networks, *IEEE J. Solid-State Circuits*, **SC-2**, 59–62 (1967).

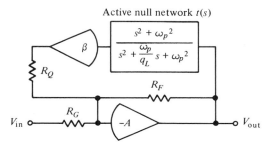

FIG. 3-67. The FEN: basic configuration.

Since it is assumed to be active, $q_L > 0.5$. Assuming that the two operational amplifiers have infinite input and zero output impedances we obtain the following transfer function:

$$T_H(s) = K_H \frac{s^2 + \dfrac{\omega_p}{q_L} s + \omega_p^2}{s^2 + \dfrac{\omega_p}{q_p} s + \omega_p^2} \qquad (3\text{-}216)$$

where

$$K_H = \frac{R_F}{R_G} \cdot \frac{q_L}{q_p} \qquad (3\text{-}217a)$$

and

$$q_p = \left(1 + \frac{R_F}{R_Q} \beta\right) q_L \qquad (3\text{-}217b)$$

ω_p and q_L being determined by the null network.

The subscript H is used here because the transfer function (3-216) is identical with (3-25b), the correcting function occurring in the decomposition for pole–zero cancellation. We recall that the corresponding frequency characteristic is that of a frequency-emphasizing network (FEN), as shown in Fig. 3-17a. In the following discussion, we shall examine separately the active null network used in the feedback loop of the FEN before we go on to discuss the FEN as a whole.

The Active Null Network   In selecting an active null network for the FEN we must take note of the fact that the FEN circuit in Fig. 3-67 incorporates two amplifiers in a loop. Consequently the precautions for stability that were discussed in Section 3.2.2 must be observed here; i.e., one of the amplifiers must have close-to-unity gain and a wider bandwidth than the other. Since the amplifier β can be used to provide the feedback for the active null network, we require that the active null network, including the amplifier, have close-to-unity forward gain. Another stipulation that we shall impose on the null network is that it, by itself, be realizable as an active-filter building block. From (3-217b) we see that, for a maximum q_p of, say, 500,[68] it will be quite adequate for q_L to be between 5 and 10, considering that β should not exceed 2 and that the resistor ratio R_F/R_Q can be selected to make up the rest. Thus, the active null network need not incorporate more than one amplifier and can, in fact, be selected from one of the building-block types described in Section 3.1. Of these, only the class 4 building blocks have close-to-unity gain, which brings us to the FRN circuits discussed in Section 3.1.2. Here we have the choice of two different null networks. Using the standard FRN (see Fig. 3-29a) we obtain the FEN circuit shown in Fig. 3-68a; with the split-feedback FRN we obtain the FEN in Fig. 3-68b.

Since the object of expanding to the FEN circuit is to provide stable high-Q poles, the criterion by which to select the FRN type must be which of the two provides a more stable q_L. As indicated earlier, a useful figure-of-merit for the Q stability in hybrid-integrated single-amplifier circuits is the gain–sensitivity product Γ, where

$$\Gamma = \beta \cdot S_\beta^q \tag{3-218}$$

This will be explained in Chapter 4. For the present, let us calculate Γ for the two FRN types.

THE SPLIT-FEEDBACK FRN

Starting out with the split-feedback (SF) type, we found from (3-59) that the gain–sensitivity product Γ_{SF} is minimum when the circuit has no loading network, i.e., $r = c = 0$. The circuit is shown this way in Fig. 3-68b. From (3-215) it follows that the null network must have $\omega_z = \omega_p$, and therefore, from Table 3-6,

$$q_{L\,SF}\bigg|_{r=c=0} = \hat{q}\,\frac{\alpha(1 - \lambda)}{\alpha(1 - \lambda) - \beta_{SF}\,\hat{q}_T} \tag{3-219}$$

Solving for β_{SF} we have

$$\beta_{SF} = \frac{\alpha(1 - \lambda)}{q_L}\left(\frac{q_L}{\hat{q}} - 1\right)\left(\frac{q_L}{\hat{q}_T} - \frac{\hat{q}}{\hat{q}_T}\right) \tag{3-220}$$

and

$$\Gamma_{SF} = \frac{\alpha(1 - \lambda)}{q_L}\left(\frac{q_L}{\hat{q}} - 1\right)\left(\frac{q_L}{\hat{q}_T} - \frac{\hat{q}}{\hat{q}_T}\right) \tag{3-221}$$

---

68. As pointed out in Section 1.5.2 it is questionable whether pole Q's of this magnitude are reasonable, since the frequency stability then required for a 10% (i.e., 1 dB) accuracy in the amplitude response is on the order of 0.002%.

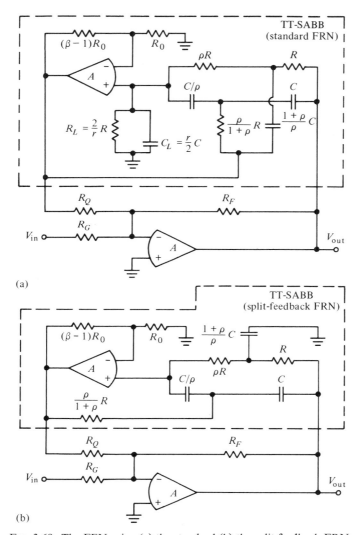

(a)

(b)

FIG. 3-68. The FEN using (a) the standard (b) the split-feedback FRN.

Since the loading network is zero (i.e., $r = c = 0$), we have $\hat{q} = \hat{q}_T$; further-more, assuming a potentially symmetrical[69] twin-T as in Fig. 3-68b, we have $\alpha = 1$, $\lambda = 1/(1 + \rho)$, and $\hat{q} = \rho/2(1 + \rho)$. Thus (3-221) becomes

$$\Gamma_{SF} = 2\left[\frac{q_L}{\hat{q}} + \frac{\hat{q}}{q_L} - 2\right] \tag{3-222a}$$

69. See Chapter 8, Section 8.1.2 of *Linear Integrated Networks: Fundamentals*.

For $q_L > 5$ we can write

$$\Gamma_{SF} \approx 2\left(\frac{q_L}{\hat{q}} - 2\right) \tag{3-222b}$$

For the symmetrical case in which $\rho = 1$, $\Gamma_{SF} \approx 4(2q_L - 1)$. Observe from (3-222a) that Γ_{SF} becomes smaller with increasing $\hat{q}$ and therefore with increasing ρ. (This is theoretically true only until $\hat{q}$ has reached the value q_L, which, of course, is physically impossible, since $q_L > 0.5$). In the limit, for $\rho \gg 1$, Γ_{SF} is bounded at the lower end by

$$\Gamma_{SF\,min} > 4(q_L - 1) \tag{3-223}$$

THE STANDARD FRN

Let us now compare Γ_{SF} with the gain–sensitivity product Γ_S of the standard FRN. Since we again require $\omega_z = \omega_p$ for the FEN, it follows that $r = c$. Thus we have, from Table 3-6,

$$q_{L\,S}|_{r=c} = \hat{q}_T \frac{1+r}{1 + \hat{q}_T r\left(\frac{1}{\alpha} + \alpha\right) - \beta_S} \tag{3-224}$$

and

$$\beta_S = \frac{\hat{q}_T}{q_L}(1+r)\left(\frac{q_L}{\hat{q}} - 1\right) \tag{3-225}$$

where

$$\hat{q} = q_L|_{\beta=0} = \hat{q}_T \frac{1+r}{1 + \hat{q}_T r\left(\frac{1}{\alpha} + \alpha\right)} \tag{3-226}$$

and $\hat{q}_T$ is given by (3-51h). With (3-57) we have

$$\Gamma_S = \frac{\hat{q}_T}{q_L}(1+r)\left(\frac{q_L}{\hat{q}} - 1\right)^2 \tag{3-227}$$

Assuming, again, a potentially symmetrical twin-T we have $\alpha = 1$, $\lambda = 1/(1 + \rho)$ and $\hat{q}_T = \rho/2(1 + \rho)$. Thus (3-227) can be rewritten as

$$\Gamma_S = \frac{1}{q_L \hat{q}_T} \frac{[\hat{q}_T r(2q_L - 1) + q_L - \hat{q}_T]^2}{1+r} \tag{3-228}$$

The object of our calculations is now to find the r value for which Γ_S is minimum and to compare the resulting $\Gamma_{S\,min}$ with $\Gamma_{SF\,min}$ as given by (3-223). The circuit with the smaller Γ value is the one to be used in the FEN.

For a given q_L, Γ_S as given by (3-228) depends on two variables, r and $\hat{q}_T$. Taking the derivative of this function with respect to $\hat{q}_T$ and r and solving the resulting two equations, we find a relative minimum equal to $4(2q_L - 1)$; it occurs for $\hat{q}_T = 0$ and $r = \infty$. No other relative minimum exists for positive and real values of r and $\hat{q}_T$. However, we must still examine Γ_S along the four boundaries defined by $r = 0$, $r = \infty$, $\hat{q}_T = 0$, and $\hat{q}_T = 0.5$. These boundaries are determined by the physical realizability of the network (e.g., $\hat{q}_T < 0.5$). We find that, at the intersection of the two boundaries $r = 0$ and $\hat{q}_T = 0.5$, Γ has a minimum:

$$\Gamma_{S\,min}\bigg|_{\substack{r=0 \\ \hat{q}_T = 0.5}} = \frac{(2q_L - 1)^2}{2q_L} \tag{3-229}$$

This minimum is less than the relative minimum obtained above for $\hat{q}_T = 0$ and $r = \infty$. Within the given "realizability boundaries," which are shown schematically in Fig. 3-69, (3-229) therefore represents an absolute minimum. For $q_L > 5$ we find the lower bound of (3-229):

$$\Gamma_{S\,min} > 2(q_L - 1) \tag{3-230}$$

Comparing with the gain–sensitivity product of the split-feedback FRN as given by (3-223) we find

$$\Gamma_{S\,min} \approx \tfrac{1}{2}\Gamma_{SF\,min} \tag{3-231}$$

Thus we conclude that the standard FRN is to be preferred for use in the FEN (or on its own) over the split-feedback FRN—provided that, in the standard FRN, r can be made to approach zero and $\hat{q}_T$ to approach 0.5. The gain–sensitivity product of the standard FRN is then approximately half that of the split-feedback FRN.

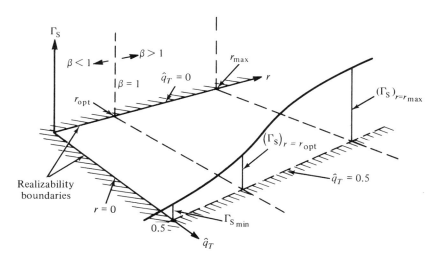

FIG. 3-69. The FEN, optimized for minimum gain-sensitivity product of the feedback network.

We must now ask ourselves how closely the minimum Γ given by (3-230) can be realized in practice. The resistor ratio r can readily be made zero simply by eliminating the loading network altogether. The twin-T pole Q, $\hat{q}_T$, can be made to approach the 0.5 limit arbitrarily closely by using a potentially symmetrical twin-T, where

$$\hat{q}_T = \frac{1}{2} \frac{\rho}{1 + \rho} \tag{3-232}$$

Already for $\rho = 10$, $\hat{q}_T$ is equal to 0.455. One problem does arise, however, as r approaches zero; the amplifier gain resulting from (3-225) takes on the value

$$\beta_S \bigg|_{r=0} = 1 - \frac{\hat{q}_T}{q_L} \tag{3-233}$$

which is less than unity. We recall that noninverting gains of less than unity cannot be obtained with an operational amplifier.

One way of overcoming this problem is to use the scheme discussed in Section 3.1.2 (see Fig. 3-32). There the resistor loading the twin-T is broken up into a voltage divider. The resulting attenuation permits the amplifier gain to be increased—if need be beyond unity. This solution tends to complicate the circuit, however, in particular as far as tuning is concerned. Since it also involves a high twin-T loading resistor, the substrate area required for a film realization may be high. The following alternative solution is therefore preferred (and will be assumed from here onwards). Here we establish the value of r for which β equals unity and select r equal to or greater than that value. Designating this lower limit by r_{opt} we find

$$r \geq r_{opt} = \frac{1}{2q_L - 1} \bigg|_{q_L \gg 1} \approx \frac{1}{2q_L} \tag{3-234}$$

and the corresponding gain–sensitivity product

$$\Gamma_S \bigg|_{r=r_{opt}} = \frac{2q_L - 1}{2\hat{q}_T} \tag{3-235}$$

Note that, as $\hat{q}_T$ approaches 0.5, this value is still lower than the lower bound of Γ for the split-feedback FRN as given by (3-223). For a potentially symmetrical twin-T we obtain, with (3-232),

$$\Gamma_S \bigg|_{r=r_{opt}} = \frac{1 + \rho}{\rho} (2q_L - 1) \tag{3-236}$$

and for $\rho = 10$ we obtain $2.2q_L - 1.1$, which is close to the lower bound given by (3-230). Incidentally, if $\hat{q}_T$ is not close to 0.5 (in the case of a symmetrical twin-T, for example, $\hat{q}_T$ equals 0.25) then (3-235) must be compared with (3-222a) of the split-feedback FRN; clearly the standard FRN still has a smaller Γ and is the preferred circuit.

It can be shown that, independently of r, Γ_S decreases monotonically with increasing $\hat{q}_T$. Similarly, along the boundary $\hat{q}_T = 0.5$, Γ_S decreases monotonically with decreasing r, as long as r is below an upper limit $r_{\max}$:

$$r_{\max} = \frac{2q_L + 1}{2q_L - 1}\bigg|_{q_L \gg 1} \approx 1 + \frac{1}{2q_L} \tag{3-237}$$

Hence, as can readily be shown,

$$\Gamma_S(r = 0) < \Gamma_S(r = r_{opt}) < \Gamma_S(r = r_{\max}) \tag{3-238}$$

This is illustrated qualitatively in Fig. 3-69.

In summary, in order to minimize Γ_S, $\hat{q}_T$ should be made as large, and r as small, as possible. Using the voltage-divider scheme of Fig. 3-32, r can be chosen as close to zero as the maximum resistor spread (specified by the resistor technology used) will allow. Alternatively, restricting ourselves to gains greater than or equal to unity, r may not be less than r_{opt}—as given by (3-234). In either case, however, the minimum Γ_S attainable is less, by approximately a factor of two, then $\Gamma_{SF\,min}$ as obtained for the split-feedback null network. The standard FRN is therefore used as the feedback null network in the FEN; it is also the preferred circuit when a single-amplifier null network is required. Because its gain is close to or equal to unity when used in the feedback network of the FEN, the maximum amplifier bandwidth can be used and optimum conditions for frequency stability are ensured. As we shall see in Chapter 6, it is sometimes advantageous to let the amplifier gain be slightly higher than unity for tuning purposes (unity-gain amplifiers are rigid and deprive one of a useful tuning parameter). This will have no detrimental effect on the FEN, however, since the plot of Γ_S in the vicinity of r_{opt} is very flat (see Fig. 3-69) and the amplifier bandwidth will hardly be reduced.

The HSFEN and its Input Network   The HSFEN shown in Fig. 3-70 uses the standard FRN, optimized for unity gain according to our discussion above, in the feedback loop. It is because this FEN configuration is used for high-Q applications that it is referred to as a "high-selectivity" FEN or HSFEN, in contrast to the "medium-selectivity" FEN (or MSFEN) to be discussed in the following section. The essential difference between the two is that the former has a pair of complex zeros (generated by an *active* feedback loop) whereas the zeros of the latter are negative real (generated by a passive feedback network in cascade with a buffer amplifier).

Having optimized the null network in the feedback path of the FEN for a minimum gain–sensitivity product, we should now do the same thing for the overall HSFEN. However, this involves a detailed discussion of FEN design, which we shall undertake in Chapter 6. Here we are more concerned with the general network topology of filter building blocks. Thus the preceding detailed discussion on FRN optimization was included here in order to establish which FRN topology—split-feedback or standard—to use in the HSFEN.

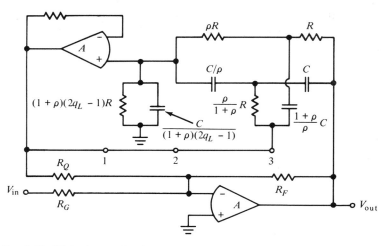

Fig. 3-70. High-Q second-order network using TT-SABBs for HSFEN design.

Having done so, we shall now briefly outline the implementation of the HSFEN (and MSFEN), leaving the additional design details and optimization considerations for Chapter 6.

The HSFEN may be tuned independently for frequency and Q as long as the parasitic phase shift of the amplifiers is negligible. The center frequency ω_p then depends only on the null frequency of the twin-T and, with a symmetrical or potentially symmetrical twin-T, is given by

$$\omega_p = \frac{1}{RC} \tag{3-239}$$

A value for q_L, the pole Q of the active null network in the feedback loop, must then be selected. In Chapter 6, we shall show that an optimum q_L exists for which the Q stability of the overall HSFEN is optimized. In practice, however, it is useful to decide on a standardized value for q_L; $q_L = 5$ has, for example, been found to be a suitable value (and, as we shall see in Chapter 6, not far from the optimum value), in which case

$$r_{\mathrm{opt}}\bigg|_{q_L=5} = \frac{1}{2q_L - 1} = \frac{1}{9} \tag{3-240}$$

and, with (3-51e),

$$R_L = R_s/r = 18R$$

and

$$C_L = rC_s = C/18 \tag{3-241}$$

Also, q_p can be tuned with R_Q according to (3-217b) whether β is unity or not, and, finally, K_H can be adjusted for with R_G (see (3-217a)).

The important feature of the HSFEN is that it can be realized using the single-amplifier, low-Q building block described in Section 3.1.2. The resistors R_G, R_Q, and R_F as well as the μ amplifier can remain separate components or, if the quantities warrant it, provision for their incorporation on the single-amplifier building block may be made. More will be said about these possibilities in Chapter 5. Here, we must still concern ourselves with the realization of the low-Q input network, whose transfer function, according to the FEN decomposition, must have the form of (3-25a):

$$T_L(s) = \frac{N(s)}{s^2 + \dfrac{\omega_p}{q_L} s + \omega_p^2} \tag{3-242}$$

where $N(s)$ is the numerator of the desired function $T(s) = N(s)/D(s)$. For pole–zero cancellation, the denominator of $T_L(s)$ must cancel the numerator of $T_H(s)$, i.e., of the HSFEN. The poles of the HSFE N in turn realize the denominator $D(s)$ of $T(s)$. Since the poles (i.e., ω_p and q_L) in (3-242) must be the same as those of the active FRN used in the HSFEN, $T_L(s)$ can be realized by the same building-block type as the FRN. A complete high-Q second-order network therefore has the form shown in Fig. 3-71. It consists of two twin-T single-amplifier building-blocks (TT-SABBs), the summing amplifier μ, and the three resistors R_G, R_F, and R_Q. Notice that this high-Q configuration uses three amplifiers much as the previously discussed high-Q circuits do, although, with finite zeros, the latter actually require four. The input network realizing $T_L(s)$ can be selected from Table 3-5 and can be obtained from the corresponding general-purpose circuit, whose diagram is shown in Fig. 3-27.

The Medium-Selectivity FEN   If we are interested in minimizing the number of amplifiers per second-order network, then the increase from one to three amplifiers, as we go from low to high pole Q's, may be unnecessarily

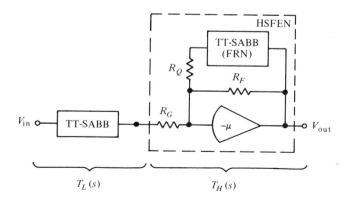

FIG. 3-71. Complete high-Q second-order network.

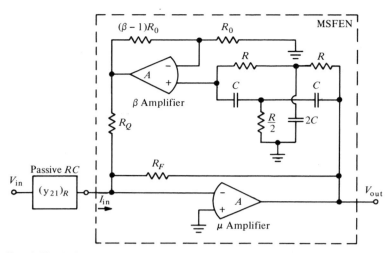

FIG. 3-72. Active network comprising passive RC input network and MSFEN.

large. We shall demonstrate here that, while using the same single-amplifier building-blocks as above, only two amplifiers are necessary when the pole Q's are only moderately high (i.e., above the low-Q range provided by the circuits with one amplifier, but not necessarily high enough to warrant three). A simple modification of the FEN circuit shown in Fig. 3-70 provides such a scheme. Opening up the feedback path between terminals 1 and 2 and grounding terminal 3, we obtain the medium-selectivity FEN shown in Fig. 3-72. Notice that the feedback circuit is now simply a passive twin-T in cascade with a buffer amplifier whose gain is β. Instead of (3-215) we now have the transfer function of a passive RC null network namely:

$$\hat{t}(s) = \frac{s^2 + \omega_p^2}{s^2 + \dfrac{\omega_p}{\hat{q}} s + \omega_p^2} \tag{3-243}$$

It is convenient to characterize the resulting medium-selectivity FEN (MSFEN) by its transfer impedance, which is now

$$(z_{21})_M = K_M \frac{s^2 + \dfrac{\omega_p}{\hat{q}} s + \omega_p^2}{s^2 + \dfrac{\omega_p}{q_p} s + \omega_p^2} \tag{3-244}$$

where we use the subscript M instead of H to differentiate between medium and high selectivity. We now have

$$K_M = R_F \cdot \frac{\hat{q}_p}{q_p} \tag{3-245a}$$

and

$$q_p = \left(1 + \beta \frac{R_F}{R_Q}\right)\hat{q} \qquad (3\text{-}245b)$$

The twin-T here requires no loading network, so that $\hat{q} = \hat{q}_T$ is given, for the general, balanced, twin-T, by (3-51h). For the symmetrical twin-T, as shown in Fig. 3-72, we have

$$\hat{q}\bigg|_{\substack{\lambda=\eta=\frac{1}{2}\\ \alpha=1}} = \frac{\alpha(1-\lambda)(1-\eta)}{\alpha^2(1-\lambda)+(1-\eta)} = \frac{1}{4} \qquad (3\text{-}246a)$$

and

$$\omega_p = 1/RC \qquad (3\text{-}246b)$$

Since the MSFEN of Fig. 3-72 is characterized by its transfer impedance, it is sufficient for the input network to be given by its transfer admittance. To cancel the zeros in (3-244) it must have the form

$$(y_{21})_R = \frac{N(s)}{s^2 + \dfrac{\omega_p}{\hat{q}}s + \omega_p^2} \qquad (3\text{-}247)$$

Clearly, it is realizable by a second-order RC network (indicated by the subscript R). A representative list of second-order $(y_{21})_R$ functions and their realizations are given in Table 3-12. The overall transfer function is, then,

$$T(s) = (y_{21})_R(z_{21})_M \qquad (3\text{-}248)$$

It is realized by a circuit configuration using only two operational amplifiers, namely the forward or μ amplifier and the buffer amplifier β in the feedback path. Because of the virtual ground at the inverting terminal of the μ amplifier the input network providing $(y_{21})_R$ can be fed directly into the MSFEN without the need for an additional buffer amplifier.

A comparison of the single-amplifier FEN circuits discussed in Section 3.1.1 with the MSFEN shown in Fig. 3-72 shows that the only basic difference between the two is the additional β amplifier in the MSFEN.[70] This amplifier allows the q_p-adjustment to be carried out with R_Q according to (3.245b), independently of the twin-T component values. Thus the MSFEN tuning procedure is, if anything, even simpler than that for the HSFEN in that ω_p can be set first by the null frequency of the twin-T and q_p then adjusted for by R_Q. Provided that ω_p is below the frequency range in which the amplifier

70. Thus, just as with the single-amplifier FEN discussed in Section 3.1.1, here we can also connect the passive RC input network to the *noninverting* instead of to the inverting input terminal. Naturally the input RC network must then realize a transfer rather than an admittance function, because the input network is essentially operating in an unloaded state. The necessary modification of the admittance networks shown in Table 3-12 is minimal, as a comparison with the corresponding RC networks providing voltage-transfer functions, given in Table 3-4, readily shows.

TABLE 3-12. TRANSFER ADMITTANCE OF SOME REPRESENTATIVE SECOND-ORDER RC NETWORKS

Transfer admittance	RC network	Parameter-component relations
(1) $K_R \cdot \dfrac{1}{s^2 + \dfrac{\omega_p}{\hat{q}} s + \omega_p^2}$		$K_R = \dfrac{1}{R_1 R_2 R_3 C_1 C_2}$ $\omega_p^2 = \dfrac{R_1 + R_2 + R_3}{R_1 R_2 R_3 C_1 C_2}$ $\hat{q} = \dfrac{\sqrt{R_1 R_2 R_3 (R_1 + R_2 + R_3) C_1 C_2}}{R_1(R_2 + R_3)C_1 + R_3(R_1 + R_2)C_2}$
(2) $K_R \cdot \dfrac{s^2}{s^2 + \dfrac{\omega_p}{\hat{q}} s + \omega_p^2}$		$K_R = \dfrac{1}{R_2}; \quad \omega_p^2 = \dfrac{1}{R_1 R_2 C_1 C_2}$ $\hat{q} = \dfrac{\sqrt{R_1 R_2 C_1 C_2}}{R_1(C_1 + C_2) + R_2 C_2}$
		$K_R = \dfrac{1}{R_1}; \quad \omega_p^2 = \dfrac{1}{R_1 R_2 C_1 C_2}$ $\hat{q} = \dfrac{\sqrt{R_1 R_2 C_1 C_2}}{(R_1 + R_2)C_1 + R_2 C_2}$

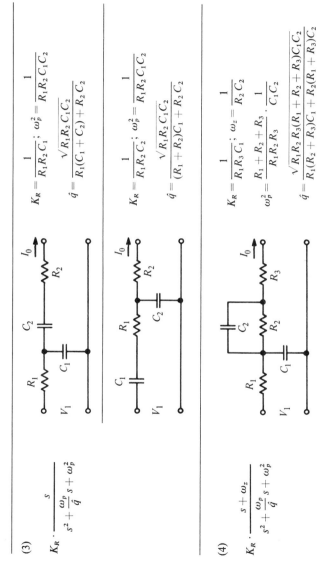

(3)

$$K_R \cdot \frac{s}{s^2 + \dfrac{\omega_p}{\hat{q}} s + \omega_p^2}$$

$$K_R = \frac{1}{R_1 R_2 C_1}; \quad \omega_p^2 = \frac{1}{R_1 R_2 C_1 C_2}$$

$$\hat{q} = \frac{\sqrt{R_1 R_2 C_1 C_2}}{R_1(C_1 + C_2) + R_2 C_2}$$

$$K_R = \frac{1}{R_1 R_2 C_2}; \quad \omega_p^2 = \frac{1}{R_1 R_2 C_1 C_2}$$

$$\hat{q} = \frac{\sqrt{R_1 R_2 C_1 C_2}}{(R_1 + R_2)C_1 + R_2 C_2}$$

(4)

$$K_R \cdot \frac{s + \omega_z}{s^2 + \dfrac{\omega_p}{\hat{q}} s + \omega_p^2}$$

$$K_R = \frac{1}{R_1 R_3 C_1}; \quad \omega_z = \frac{1}{R_2 C_2}$$

$$\omega_p^2 = \frac{R_1 + R_2 + R_3}{R_1 R_2 R_3} \cdot \frac{1}{C_1 C_2}$$

$$\hat{q} = \frac{\sqrt{R_1 R_2 R_3(R_1 + R_2 + R_3)C_1 C_2}}{R_1(R_2 + R_3)C_1 + R_2(R_1 + R_3)C_2}$$

(continued)

TABLE 3-12. Continued

Transfer admittance	RC network	Parameter-component relations

(5)

$$K_R \cdot \frac{s^2 + \dfrac{\omega_z}{q_z} s + \omega_z^2}{s^2 + \dfrac{\omega_p}{\hat{q}} s + \omega_p^2}$$

$K_R = \dfrac{1}{R_a}$; $\quad \omega_z^2 = \dfrac{1}{R_p R_q C_3 C_4}$

$q_z = \dfrac{\sqrt{\left(1 + \dfrac{R_4}{R_1 + R_2}\right) R_p R_4 C_3 C_4}}{R_p C_3 + R_4 C_4}$

$\omega_p^2 = \dfrac{1}{R_p R_q C_3 C_4}\left(1 + \dfrac{R_q}{R_m}\right)$

$\hat{q} = \dfrac{\sqrt{\left(1 + \dfrac{R_4}{R_1 + R_2}\right)\left(1 + \dfrac{R_q}{R_m}\right) R_p R_4 C_3 C_4}}{R_4 C_4 + R_p C_3\left[1 + \dfrac{R_4}{R_1} + \dfrac{R_4}{R_m}\right]}$

where $R_p = \dfrac{R_1 R_2}{R_1 + R_2}$

$R_m = \dfrac{R_a R_b}{R_a + R_b}$; $\quad R_q = \dfrac{R_4(R_1 + R_2)}{R_1 + R_2 + R_4}$

$K_R = \dfrac{1}{R_a}$; $\quad \omega_z^2 = \dfrac{1}{R_3 R_4 C_p C_q}$

$q_z = \dfrac{\sqrt{R_3 R_4 C_p C_q}}{R_3 C_p + R_4 C_4}$, $\quad \omega_p^2 = \dfrac{1 + R_4/R_m}{R_3 R_4 C_p C_q}$

$\hat{q} = \dfrac{\sqrt{(1 + R_4/R_m)R_3 R_4 C_p C_q}}{(1 + R_4/R_m)R_3 C_p + R_4(C_1 + C_4)}$

where $C_p = C_1 + C_2$

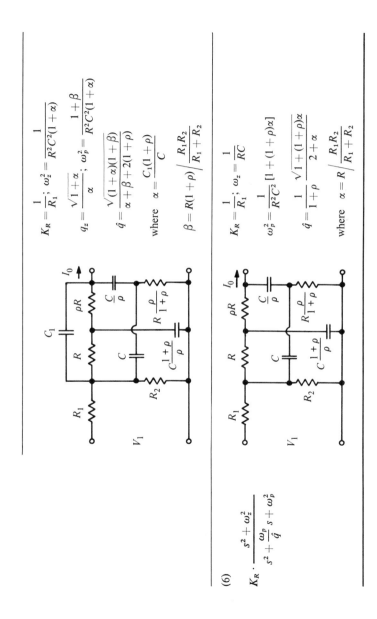

$$K_R = \frac{1}{R_1}; \quad \omega_z^2 = \frac{1}{R^2C^2(1+\alpha)}$$

$$q_z = \frac{\sqrt{1+\alpha}}{\alpha}; \quad \omega_p^2 = \frac{1+\beta}{R^2C^2(1+\alpha)}$$

$$\hat{q} = \frac{\sqrt{(1+\alpha)(1+\beta)}}{\alpha+\beta+2(1+\rho)}$$

where $\alpha = \dfrac{C_1(1+\rho)}{C}$

$$\beta = R(1+\rho)\sqrt{\frac{R_1R_2}{R_1+R_2}}$$

$$K_R = \frac{1}{R_1}; \quad \omega_z = \frac{1}{RC}$$

$$\omega_p^2 = \frac{1}{R^2C^2}[1+(1+\rho)\alpha]$$

$$\hat{q} = \frac{1}{1+\rho}\frac{\sqrt{1+(1+\rho)\alpha}}{2+\alpha}$$

where $\alpha = R\sqrt{\dfrac{R_1R_2}{R_1+R_2}}$

(6)

$$K_R \cdot \frac{s^2+\omega_z^2}{s^2+\dfrac{\omega_p}{\hat{q}}s+\omega_p^2}$$

gain becomes frequency dependent, the q_p-adjustment will have no effect on ω_p. If it does, a correction must be undertaken. This and other practical considerations pertaining to FEN design will be discussed in Chapter 6. Unless there is an input resistor R_G, K_M must be set by the selection of R_F according to (3-245a). If K_M is required accurately, one of the series resistors in $(y_{21})_R$ can be used for its fine adjustment.

The price to be paid for the independent tuning capabilities of the MSFEN is the buffer amplifier β. However, inspection of Fig. 3-72 shows that the feedback circuit comprising the twin-T and β is directly obtainable from the same single-amplifier building blocks used independently for low-Q applications and in the feedback path of the HSFEN. We have not, therefore violated the building-block concept developed in this chapter. Furthermore, at low frequencies the gain of the β amplifier may be larger than unity, in which case for a given q_p the gain of the μ amplifier, i.e., R_F/R_Q, can be reduced accordingly. In this way an improved gain distribution between β and μ can be obtained for maximum q_p stability. Also, the passive RC input network $(y_{21})_R$ can be realized by the TT-SABB, where the operational amplifier can either be left off the circuit, or disconnected to economize on power.

We shall postpone any further discussion of filter design using FENs until Chapter 6. There we shall describe in detail the application of this approach to hybrid-integrated filter design. Since the sensitivity and nonideal characteristics of these circuits are closely related to the technology used for their implementation, we shall also leave these topics for that chapter. This enables us to discuss the interrelationship between technology and realization in general, as seen from the perspective of hybrid integrated circuits, beforehand, in Chapter 5.

3.4 A NETWORK TOPOLOGY UTILIZING IDENTICAL BUILDING BLOCKS

So far in this chapter we have discussed the design of general-purpose filter building blocks which, when modified appropriately, provide every possible second-order network function. The incentive for this approach to network design was to reduce the unlimited variety of possible nth-order networks to a small family of basic building blocks with which any network of higher order can be reconstructed in integrated form. In following this path we were emulating the approach used to integrate digital systems; as we know, such systems can be broken down into a small number of basic functions such as AND–OR gates, shift registers, and the like. Just as it has been argued—and demonstrated—that most digital systems could actually be designed using exactly one basic function, namely the EXCLUSIVE–OR gate, so we shall show in the following, that in the field of linear-network design we can also reduce the number of necessary basic second-order functions drastically, and in many cases to a single one. This is possible if we leave the commonly used inter-connection of second-order networks by cascade and look for another inter-connection topology instead. The benefits of designing general linear networks with only one or two basic types of second-order networks are, of course, considerable when the networks are to be realized in integrated form. As we would expect, however, there are also disadvantages inherent in such an approach and the decision on which direction to take will depend on the technology available and the application at hand.

3.4.1 The Parallel Connection of Second-Order Networks

The Partial-Fraction Decomposition   It was pointed out in Chapter 1 (Section 1.3) that an nth-order transfer function can be decomposed either into a product or a sum of second- and third-order functions. The former results in a cascade connection, the latter in a parallel connection of second- and third-order networks.[71] Thus, having obtained our transfer function

$$T(s) = \frac{b_m s^m + b_{m-1} s^{m-1} + \cdots + b_0}{a_n s^n + a_{n-1} s^{n-1} + \cdots + a_0} \tag{3-249}$$

as a solution to the approximation problem (see step 1, Section 1.1) we can decompose it (step 2) either into a *product*

$$T(s) = \prod_{j=1}^{n/2} T_j(s) = \prod_{j=1}^{n/2} K_j \frac{s^2 + \dfrac{\omega_{zj}}{q_{zj}} s + \omega_{zj}^2}{s^2 + \dfrac{\omega_{pj}}{q_{pj}} s + \omega_{pj}^2} \tag{3-250}$$

or into a *sum*

$$T(s)\bigg|_{m<n} = \sum_{j=1}^{n/2} T_j'(s) = \sum_{j=1}^{n/2} \alpha_j \frac{N_j(s)}{s^2 + \dfrac{\omega_{pj}}{q_{pj}} s + \omega_{pj}^2} \tag{3-251}$$

of second-order functions. The poles of the sum are the same as those of the product. Consequently, if the order of the numerator is at least less by one than the order of the denominator, then (3-251) can be interpreted as the partial-fraction expansion of $T(s)$ and, as such, can be readily obtained. The individual second-order functions then *all* have the form

$$T_j'(s) = \alpha_j \frac{s + \omega_{zj}}{s^2 + \dfrac{\omega_{pj}}{q_{pj}} s + \omega_{pj}^2} \tag{3-252}$$

---

71. For brevity we shall omit reference to third-order networks in what follows. The term parallel describes the general topology of the networks; it should not be understood in the familiar sense of a parallel interconnection of two-ports.

This is the resonator function, as given in Table 1-3, entry 8. Note that we have no choice here in combining the pole pairs with the negative real zeros ω_{z_j}; the pairing results directly from the partial-fraction expansion of $T(s)$. The significance of this decomposition is in the fact that any *arbitrary* function $T(s)$, for which $m + 1 \leq n$, can be realized by a parallel connection of second-order networks, all possessing a resonator frequency response. This, of course, is precisely the kind of uniformity in network design that is desirable in hybrid integrated circuit fabrication.

With the decomposition (3-251), $T(s)$ can be realized by the parallel configuration[72] shown in Fig. 3-73. Each resonator $T_j''(s)$ is in cascade with a real multiplier α_j which must either amplify or attenuate the signal before being added or subtracted by the differential amplifier A. As long as the $|\alpha_j|$ are less than unity, resistor attenuators can be used for their realization; when the $|\alpha_j|$ are greater than unity, the necessary gain is obtainable from the amplifier A.

If the order of numerator and denominator of $T(s)$ are equal, i.e., $m = n$

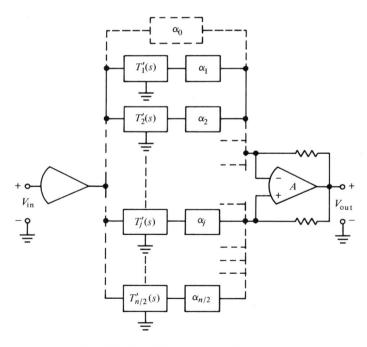

FIG. 3-73. Parallel network configuration.

72. J. F. O'Neill and M. S. Ghausi, Design of frequency selective networks using resonators, Proc. Midwest Symp. Circuit Theory, Purdue Univ., May 18–19, 1967.

in (3-249), then the corresponding partial-fraction expansion will have the form

$$T(s)\bigg|_{m=n} = \alpha_0 + \sum_{j=1}^{n/2} \alpha_j \frac{s + \omega_{z_j}}{s^2 + \frac{\omega_{p_j}}{q_{p_j}} s + \omega_{p_j}^2} \qquad [3\text{-}253]$$

where α_0 is a real constant. In a practical realization, α_0 is readily incorporated as shown in Fig. 3-73. In the decomposition of (3-253) each function $T_j'(s)$ has two unknowns, α_j and ω_{z_j}, associated with it. Thus, in an nth-order network we have a total of n parameters that must be determined, and if $m = n$ the number is $n + 1$. In some cases, it may be preferable for practical reasons to use a function type other than the resonator to perform the expansion of $T(s)$. Since each function is required to supply two unknowns, there are basically two function types, in addition to the resonator, that are appropriate, namely the FRN function

$$T_j'(s) = \alpha_j \frac{s^2 + \omega_{z_j}^2}{s^2 + \frac{\omega_{p_j}}{q_{p_j}} s + \omega_{p_j}^2} \qquad (3\text{-}254)$$

and the FEN function

$$T_j(s) = \alpha_j \frac{s^2 + \frac{\omega_{p_j}}{q_{z_j}} s + \omega_{p_j}^2}{s^2 + \frac{\omega_{p_j}}{q_{p_j}} s + \omega_{p_j}^2} \qquad (3\text{-}255)$$

Naturally, a general FRN function which includes a linear s term can be used instead of (3-254). The FRN function then has the form of (3-255) with $q_{z_j} > q_{p_j}$; in the FEN case, of course, $q_{z_j} < q_{p_j}$. However, the expansion of $T(s)$ into a sum of FRN or FEN functions (or a combination of the two) is not as readily available as a partial-fraction expansion, since computer programs exist only for the latter. On the other hand, if high-Q poles are involved, the multiple-amplifier FEN described earlier might be more convenient for this approach than a single-amplifier resonator.

Practical Considerations Pertaining to the Partial-Fraction Decomposition   Various considerations must be taken into account when designing a network based on the partial-fraction decomposition. Assuming that the pole Q's of $T(s)$ are low enough (say, less than 30) the class 4 resonator shown in Fig. 3-74 can readily be used to provide the individual $T_j'(s)$ functions. The transfer function of this circuit is

$$\frac{V_{\text{out}}}{V_{\text{in}}} = k \frac{s + \omega_z}{s^2 + \frac{\omega_p}{q_p} s + \omega_p^2} \qquad (3\text{-}256)$$

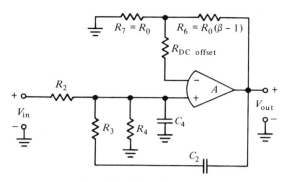

FIG. 3-74. Class-4 resonator.

where

$$k = \frac{\beta}{R_2 C_4} \tag{3-257a}$$

$$\omega_z = \frac{1}{R_3 C_2} \tag{3-257b}$$

$$\omega_p = \sqrt{\frac{R_2 + R_4}{R_2 R_3 R_4 C_2 C_4}} \tag{3-257c}$$

$$q_p = \frac{\sqrt{(R_2 + R_4) R_2 R_3 R_4 C_2 C_4}}{R_2 R_3 C_2 + R_3 R_4 C_2 + R_2 R_4 C_4 + R_2 R_4 C_2 (1 - \beta)} \tag{3-257d}$$

Note that the zero ω_z attainable with this circuit is negative real. However, when performing the partial-fraction expansion, the real zeros ω_{z_j} appearing in (3-253) may be either positive or negative. A positive zero can be obtained by cascading $T_j'(s)$ with a first-order all-pass function as follows:

$$T_j'(s) = \frac{s - \omega_{z_j}}{s^2 + \frac{\omega_{p_j}}{q_{p_j}} s + \omega_{p_j}^2}$$

$$= \frac{s + \omega_{z_j}}{s^2 + \frac{\omega_{p_j}}{q_{p_j}} s + \omega_{p_j}^2} \cdot \frac{s - \omega_{z_j}}{s + \omega_{z_j}} \tag{3-258}$$

A first-order all-pass network can be obtained by either of the two circuits shown in Fig. 3-64. If the order of $T(s)$ is odd, at least one of the T_j' functions will be required to provide only a negative real pole. This can be obtained by an RC buffer configuration, as shown in Fig. 3-75 for which we have

$$T_j'(s) = \frac{\beta}{RC} \frac{1}{s + \frac{1}{RC}} \tag{3-259}$$

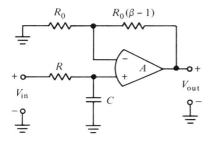

FIG. 3-75. Single-pole realization: RC buffer circuit.

Since full freedom is desirable in the choice of positive and negative α_j values (whose magnitudes may be larger or smaller than unity) the differential-amplifier configuration in Fig. 3-73 is impractical. Using a double-inversion scheme with resistive-T pads for the α_j, this problem can readily be overcome at the cost of an additional amplifier. The general circuit then has the form shown in Fig. 3-76. The actual α_j values adjusted for by the T pads must take into account the k values of the resonators (see (3-256) and (3-257a)) as well as the constant K of the overall function $T(s)$. Thus the actual value α_{act} of a T pad will be

$$(\alpha_j)_{act} = \frac{\alpha_j K}{k_j} \qquad [3\text{-}260]$$

Using a resistive T, α_j can be accurately adjusted for in either direction, even if film resistors are used.

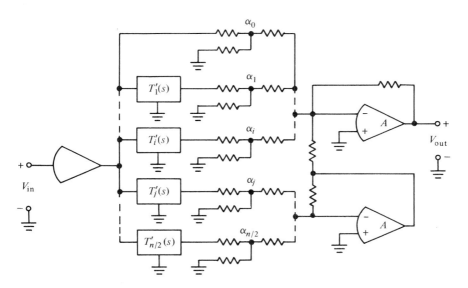

FIG. 3-76. Improved parallel network configuration.

Even with the advantage of bidirectional α_j adjustments, tuning a high-order network with the topology of Fig. 3-76 may cause some problems. First, the individual resonators must be precisely tuned; this can be done by monitoring the output of each resonator. Assuming that the whole circuit is in hybrid integrated form, the α_j T pads must then be individually adjusted while the whole circuit remains interconnected. With monitoring at the output (terminal 2) this can be done if the output of all but the α_j pad in question is shunted to ground with a sufficiently large capacitance. This may be difficult at low frequencies.[73] At high frequencies the virtual grounds of the two summing amplifiers may no longer be sufficiently ideal to provide perfect isolation between the signal paths joined together at the summing point.

Another problem that may be encountered in the parallel topology of Fig. 3-76 is limited dynamic range. If, as may well be the case, most or all of the T'_j networks have reasonably high gain within a confined frequency range (i.e., the pass band) then, to keep one or both of the summing amplifiers from being overdriven, the input signal will be limited accordingly. No optimization of pole–zero combinations, network sequence or gain distribution is possible here, and the dynamic range will depend on the particular function being designed and on the coefficients of its partial-fraction expansion. If we add to these problems the fact that the number of amplifiers and resistors may be considerably larger than in the cascade equivalent (remember that first-order all-pass networks may be required in series with the resonators) then we will understand why the prospect of designing our networks with only one kind of network function is not as attractive as it may at first have seemed. Conversely, the penalty for requiring somewhat more than one or two different function types, i.e., to design our networks by cascade, may well seem worthwhile.

---

73. Alternatively, the individual second-order T_j networks may be separate hybrid interconnected circuits externally connected to a hybrid integrated circuit containing the two summing amplifiers and the α_j T pads. In this case tuning of the α_j pads is straightforward and requires no shunting capacitors.

PART II

FUNDAMENTALS OF HYBRID
INTEGRATED NETWORK DESIGN

4

NETWORK DESIGN USING HYBRID INTEGRATED CIRCUITS

INTRODUCTION

The large variety of basic active networks discussed in the previous two chapters may well have induced a rather helpless feeling in any reader trying to select a filter scheme for a particular application. Qualitative statements and generalizations with regard to any particular synthesis method were purposely refrained from, however, because they can validly be made only within the framework of the specific component and device technology for which that method is intended.

The idea of tailoring a circuit-design or network-synthesis approach to a particular component technology may, at first, seem strange to the reader. Since the basic characteristics of a large variety of available passive components (e.g., resistors, capacitors, and inductors) did not change significantly for decades, the circuit designer has become accustomed to selecting components according to the design approach of his choice rather than to proceeding in the reverse direction. The flexibility afforded by discrete components also enabled him to make belated changes in his design, if the need arose. With the advent of mass-produced integrated circuits that can only be manufactured economically by using a minimum number of standardized technological processes, this convenience to the circuit designer no longer exists. The efficiency of a given design, both in terms of performance and cost, now depends directly on the degree to which it can be realized with commonly used integrated-circuit and component types. This was demonstrated effectively in the development of low-cost commercial operational amplifiers. These only became successful once linear silicon integrated circuit design was treated as a new discipline, obeying rules that differ significantly from those used for discrete-component circuit design.[1] Similarly, during the transition from vacuum-tube to transistor circuits, it rapidly became apparent that transistor circuits had to be considered in their own right and not as direct replacements for equivalent vacuum-tube circuits. Numerous other examples could be cited to demonstrate that the success of any circuit-design approach depends directly on how closely it is related to the technology intended for its implementation. Proven circuit designs of an established technology should not be forced onto new, emerging technologies, but should be appropriately modified and adapted to them, in order to fully benefit from the characteristics those new technologies afford.

In this section, hybrid integrated circuit (HIC) technology, i.e., the combination of film resistors and capacitors with silicon integrated active devices, has been singled out as the technology most suited to the design of integrated linear active networks. The passive components may be thick, or thin film, although in general, performance will be superior with the latter type. Incidentally, the analytical simplifications resulting from the condition of closely tracking resistor and capacitor values will also be more accurate with thin-film implementation.

In this chapter, we shall examine those aspects of active network design that are peculiar to hybrid integrated circuit implementation. First, we shall review the general characteristics of the networks discussed in Chapter 2 with respect to their suitability for HIC implementation. It will be shown that certain filter types are more suited to the characteristics of this technology than others. Those others may, for example, be more suited to discrete-component realization. Likewise, one criterion of goodness, or figure-of-merit, for the networks described will be shown to be preferable to another, depending on the technology for which the networks are intended. We shall then go on to the problems associated with the sensitivity and stability of hybrid integrated networks and also discuss some of the possible parasitic effects that may be encountered. Finally, we shall examine in some detail the adjustment and fine-tuning of HIC networks and consider methods of evaluating the initial tolerances of a tuned nth-order network.

4.1 THE STABILITY OF LINEAR HYBRID INTEGRATED NETWORKS

It will be recalled[2] that the relative change in a pole p can be accurately described by the relative change in the pole frequency ω_p and the pole Q as follows:

1. R. J. Widlar, Design techniques for monolithic operational amplifiers, *IEEE J. Solid-State Circuits*, **SC-4**, 184–191 (1969).
2. See Chapter 4, equation (4-129a) of *Linear Integrated Networks: Fundamentals*.

$$\frac{\Delta p}{p} = \frac{\Delta \omega_p}{\omega_p} - j \frac{\Delta Q/Q}{\sqrt{4Q^2 - 1}} \qquad [4\text{-}1]$$

For the single-amplifier circuit types used to design filter building blocks, the pole frequency ω_p depends on passive elements only. As a consequence, the frequency stability $\Delta \omega_p/\omega_p$ is independent of the active circuit configuration used*; it depends primarily on the quality of the resistors and capacitors. By contrast, the pole Q depends directly on the gain β of the active device and, in fact, does so exclusively if the passive components track perfectly. Thus, it is the Q stability, more than any other parameter, that best characterizes the performance of an active network and the suitability of a synthesis scheme to a given technological realization. Therefore, in deciding the merits of any one of the four network classes discussed in Chapter 2 with respect to its hybrid integrated circuit realizability, it will suffice to examine the stability of the resulting pole Q in each class.

The total variation of Q with respect to all r resistors R_i, and c capacitors C_j and the active gain device β can be written as

$$\frac{\Delta Q}{Q} = \sum_{i=1}^{r} V_{R_i}^Q + \sum_{j=1}^{c} V_{C_j}^Q + V_\beta^Q \qquad (4\text{-}2)$$

where

$$V_x^Q = S_x^Q \frac{\Delta x}{x} \qquad (4\text{-}3)$$

The resistor variations $\Delta R_i/R_i$ comprise two parts:

$$\frac{\Delta R_i}{R_i} = \delta_r + \varepsilon_{r_i} \qquad (4\text{-}4)$$

where δ_r accounts for the variation *common* to all the resistors (due to their tracking), and ε_{r_i} accounts for the nontracking, or random variation. The capacitor variations can be broken up in the same way:

$$\frac{\Delta C_j}{C_j} = \delta_c + \varepsilon_{c_j} \qquad (4\text{-}5)$$

Substituting (4-4) and (4-5) into (4-2) we have

$$\frac{\Delta Q}{Q} = \sum_{i=1}^{r} S_{R_i}^Q (\delta_r + \varepsilon_{r_i}) + \sum_{j=1}^{c} S_{C_j}^Q (\delta_c + \varepsilon_{c_j}) + S_\beta^Q \frac{\Delta \beta}{\beta} \qquad (4\text{-}6)$$

---

* We shall see later in this chapter (Section 4.2.3) that this is true only to a first approximation. The limited gain-bandwidth product of the operational amplifier does, in fact, affect the various network classes differently.

Due to the homogeneity of Q to order zero, we have[3]

$$\sum_{i=1}^{r} S_{R_i}^{Q} = \sum_{j=1}^{c} S_{C_j}^{Q} = 0 \tag{4-7}$$

Therefore (4-6) simplifies to

$$\frac{\Delta Q}{Q} = \sum_{i=1}^{r} S_{R_i}^{Q} \varepsilon_{r_i} + \sum_{j=1}^{c} S_{C_j}^{Q} \varepsilon_{c_j} + S_{\beta}^{Q} \frac{\Delta \beta}{\beta} \tag{4-8}$$

Assuming the maximum variations of ε_{r_i} and ε_{c_j} to lie within the limits

$$-\varepsilon_r \leq \varepsilon_{r_i} \leq \varepsilon_r \tag{4-9a}$$

and

$$-\varepsilon_c \leq \varepsilon_{c_j} \leq \varepsilon_c \tag{4-9b}$$

we have a maximum Q variation of

$$\left. \frac{\Delta Q}{Q} \right|_{max} = \varepsilon_r \sum_{i=1}^{r} |S_{R_i}^{Q}| + \varepsilon_c \sum_{j=1}^{c} |S_{C_j}^{Q}| + S_{\beta}^{Q} \frac{\Delta \beta}{\beta} \tag{4-10}$$

Equation (4-10) will be used later to compare the four network classes discussed in Chapter 2 with respect to hybrid integrated realizability.[4] First however we must examine the Q sensitivity of these network classes.

4.1.1 Q Sensitivity

It will be recalled from Chapter 2 that the pole Q of a class 1 or 2 negative feedback network is obtained from $\hat{Q}$ (the Q of the passive RC network in the feedback loop) by an expression of the form

$$Q = \hat{Q}\sqrt{1 + k\beta} \tag{4-11}$$

where k and $\hat{Q}$ are functions of the passive RC network only and β is the gain of the active element. The Q expression for class 3 networks is somewhat more complicated, since it depends also on the zeros of the passive band-rejection network in the feedback loop (i.e., on ω_z and q_z). The basic form may be given as follows [see Chapter 2, (2-20)]:

$$Q = \hat{Q} \frac{1 + k\beta}{1 \pm k\beta v} \tag{4-12}$$

For the purely negative feedback case (e.g., bridge-T in the feedback loop), $v > 0$. When $v \leq 0$ (e.g., twin-T in the feedback loop), (4-12) takes on the form of a network combining negative *and* positive feedback (i.e. negative sign in the denominator).

---

3. See Chapter 4, Section 4.2.4 of *Linear Integrated Networks: Fundamentals*.
4. This comparison follows the approach taken by R. W. Daniels in an unpublished Bell Telephone Laboratories memorandum.

TABLE 4-1. SENSITIVITY RELATIONS*

1.	$S_x^x = 1$	14.	$S_x^{1/y} = -S_x^y$
2.†	$S_x^{cx} = 1$	15.	$S_{1/x}^y = -S_x^y$
3.	$S_x^{cx^n} = n$	16.	$S_x^{y+c} = \dfrac{y}{y+c} S_x^y$
4.‡	$S_x^{y^n} = n S_x^y$		
5.	$S_{x^n}^y = \dfrac{1}{n} S_x^y$	17.	$S_x^{u+v+\cdots} = \dfrac{1}{u+v+\cdots}(u S_x^u + v S_x^v + \cdots)$
6.	$S_x^{cy} = S_x^y$	18.	$S_{cx}^y = S_x^y$
7.	$S_x^y = S_{u_1}^y \cdot S_x^{u_1} + S_{u_2}^y \cdot S_x^{u_2} + \cdots$		
	where $y = y(u_1, u_2, \ldots, u_n)$	19.	$S_{u/v}^y = \dfrac{1}{2}(S_u^y - S_v^y) = S_u^{\sqrt{y}} - S_v^{\sqrt{y}}$
8.	$S_x^y = S_x^{|y|} + j\phi_y S_x^{\phi_y}$	20.	$S_x^{e^y} = y S_x^y$
9.	$S_x^{|y|} = Re S_x^y$		
10.	$S_x^{y^*} = (S_x^y)^*$	21.	$S_x^{\ln y} = \dfrac{1}{\ln y} \cdot S_x^y$
11.	$S_x^{\phi_y} = \dfrac{1}{\phi_y} Im S_x^y$	22.	$S_x^{\sin y} = y \cot y \cdot S_x^y$
		23.	$S_x^{\cos y} = y \tan S_x^y$
12.	$S_x^{u \cdot v \cdots} = S_x^u + S_x^v + \cdots$	24.	$S_x^{\sinh y} = y \coth y \, S_x^y$
13.	$S_x^{u/v} = S_x^u - S_x^v$	25.	$S_x^{\coth y} = y \tanh y \, S_x^y$

† c and n are constants.
‡ y, u, and v are single-valued differentiable functions of x; also $y = |y| \cdot e^{j\phi_y}$.
* Many of these relations have been adapted from J. Gorski-Popiel, Classical sensitivity—A collection of formulas, *IEEE Trans. Circuit Theory*, CT **10**, 300–302 (1963); also P. R. Geffe, *Active Filters*, Westinghouse Research and Development Rept. ECOM-0363–4.

Class 1 and 2 Networks To obtain the Q sensitivity for any of our four network classes we can use the sensitivity formulas listed in Table 4-1. For example, to calculate $S_{R_i}^Q$ we proceed as follows. From (4-11) we have

$$S_{R_i}^Q = S_{R_i}^{\hat{Q}(1+k\beta)^{1/2}}$$

where $\hat{Q}$ and k are single-valued, differentiable functions of R_i.

With formula 11 of Table 4-1 we have

$$S_{R_i}^{\hat{Q}(1+k\beta)^{1/2}} = S_{R_i}^{\hat{Q}} + S_{R_i}^{(1+k\beta)^{1/2}}$$

with formula 4:

$$= S_{R_i}^{\hat{Q}} + \tfrac{1}{2} S_{R_i}^{(1+k\beta)}$$

with formula 15:

$$= S_{R_i}^{\hat{Q}} + \frac{1}{2} \frac{k\beta}{1 + k\beta} S_{R_i}^{k\beta}$$

and with formula 6:

$$= S_{R_i}^{\hat{Q}} + \frac{1}{2} \frac{k\beta}{1 + k\beta} S_{R_i}^k \qquad (4\text{-}13)$$

Thus, from the last line, we have for high-Q values (i.e., $Q \gg \hat{Q}$ and therefore $k\beta \gg 1$)

$$S_{R_i}^Q \bigg|_{Q \gg \hat{Q}} \approx S_{R_i}^{\hat{Q}} + \frac{1}{2} S_{R_i}^k \qquad \text{[4-14a]}$$

In precisely the same way we obtain

$$S_{C_j}^Q \bigg|_{Q \gg \hat{Q}} \approx S_{C_j}^{\hat{Q}} + \frac{1}{2} S_{C_j}^k \qquad \text{[4-14b]}$$

and

$$S_{\beta}^Q \bigg|_{Q \gg \hat{Q}} = \frac{1}{2} \frac{k\beta}{1 + k\beta} \approx \frac{1}{2} \qquad \text{[4-14c]}$$

Notice that for $Q = \hat{Q}(1 + k\beta)$ the three expressions in (4-14) remain the same except that the factor $1/2$ is replaced by unity.

Class 3 Networks   Proceeding in precisely the same way as above, we can readily obtain the Q sensitivities of a class 3 network. Since we are examining the effects of negative feedback on Q stability here we shall assume that $v > 0$ and therefore restrict ourselves to the case in which the sign in the denominator of (4-12) is positive. When this sign is negative (corresponding to positive feedback), it will influence the Q stability in the manner discussed under the effects of positive feedback.

To calculate the Q sensitivities of class 3 networks we start out from (4-12). Since $\hat{Q}$, k, and v are single-valued and differentiable functions of R_i, the Q sensitivity can be decomposed as follows (see Table 4-1, formula 7):

$$S_{R_i}^Q = S_{\hat{Q}}^Q S_{R_i}^{\hat{Q}} + S_k^Q S_{R_i}^k + S_v^Q S_{R_i}^v \qquad (4-15)$$

Thus we obtain

$$S_{R_i}^Q = S_{R_i}^{\hat{Q}} + \frac{k\beta}{1 + k\beta} \left(\frac{1 - v}{1 + k\beta v} \right) S_{R_i}^k - \frac{k\beta v}{1 + k\beta v} S_{R_i}^v \qquad (4-16)$$

An equivalent expression is obtained for the capacitors C_j. The sensitivity with respect to β is

$$S_{\beta}^Q = \frac{k\beta}{1 + k\beta} \left(\frac{1 - v}{1 + k\beta v} \right) = \frac{(Q/\hat{Q}) - 1}{(Q/\hat{Q})(1 - v)} \left(1 - \frac{Q}{\hat{Q}} v \right) \qquad (4-17)$$

Now, we recall that

$$v = \frac{\hat{Q}}{q_z} \qquad (4-18)$$

where $\hat{Q}$ is the pole Q and q_z the zero Q of the null network in the feedback loop. Then, for $Q \gg 1$, in which case $k\beta \gg 1$, (4-16) and (4-17) simplify to

$$S_{R_i}^Q \bigg|_{Q \gg 1} \approx S_{R_i}^{\hat{Q}} + \frac{1 - (Q/q_z)}{1 - (\hat{Q}/q_z)} S_{R_i}^k - S_{R_i}^v \qquad [4\text{-}19a]$$

$$S_{C_j}^Q \bigg|_{Q \gg 1} \approx S_{C_j}^{\hat{Q}} + \frac{1 - (Q/q_z)}{1 - (\hat{Q}/q_z)} S_{C_j}^k - S_{C_j}^v \qquad [4\text{-}19b]$$

and

$$S_\beta^Q \bigg|_{Q \gg 1} \approx \frac{1 - Q/q_z}{1 - \hat{Q}/q_z} \qquad [4\text{-}20]$$

Since $\hat{Q}/q_z$ is generally much smaller than unity, the middle terms in (4-19a) and (4-19b) as well as the expression given by (4-20) approach zero as Q approaches q_z.

Class 4 Networks   Turning our attention now to positive feedback, we recall from Chapter 2 (Section 2.1.2) that Q as a function of $\hat{Q}$ and β generally has the form:

$$Q = \frac{\hat{Q}}{1 - k\beta} \qquad (4\text{-}21a)$$

where

$$k = \hat{Q} \frac{\omega_k}{\omega_0} \qquad (4\text{-}21b)$$

ω_k and ω_0 were defined in Chapter 2; they are the constant coefficient and the natural frequency, respectively, of the RC feedback network. Both $\hat{Q}$ and k are therefore determined by the passive network in the positive-feedback loop; β is the gain of the active element. With the sensitivity formulas of Table 4-1 we obtain the Q sensitivity expressions pertaining to (4-21a) by inspection:

$$S_{R_i}^Q = S_{R_i}^{\hat{Q}} + \frac{k\beta}{1 - k\beta} S_{R_i}^k \qquad (4\text{-}22)$$

This can be rewritten as

$$S_{R_i}^Q = S_{R_i}^{\hat{Q}} + \left(\frac{Q}{\hat{Q}} - 1\right) S_{R_i}^k \qquad (4\text{-}23)$$

For high Q values, $k\beta$ approaches unity (see (4-21a)) and we have

$$S_{R_i}^Q \bigg|_{Q \gg \hat{Q}} \approx \frac{Q}{\hat{Q}} S_{R_i}^k \qquad [4\text{-}24a]$$

Similarly, we obtain:

$$S_{C_j}^Q\bigg|_{Q \gg \hat{Q}} \approx \frac{Q}{\hat{Q}} S_{C_j}^k \qquad \text{[4-24b]}$$

The Q sensitivity to β then follows:

$$S_\beta^Q = k\beta \frac{Q}{\hat{Q}} = \frac{Q}{\hat{Q}} - 1 \approx \frac{Q}{\hat{Q}} \qquad \text{[4-25]}$$

Comparing the sensitivities corresponding to negative feedback (classes 1 to 3) with those for positive feedback (class 4), it is clear that the former are considerably smaller than the latter. This is particularly true for high Q's, since the positive-feedback sensitivities are proportional to Q. This comes about because high Q are obtained in negative-feedback systems by multiplying the low-valued $\hat{Q}$ by a large factor $k\beta$; in positive feedback, high Q's are obtained by dividing $\hat{Q}$ by the difference of two numbers which are approximately equal to unity. In this way a quantity much smaller than unity is obtained in the denominator. As is well known, the subtraction of two large numbers to produce a small one is a sensitive process. Thus, based on a comparison of sensitivities alone, negative-feedback networks are clearly preferable to those using positive feedback.

This conclusion has very much less practical significance than might, at first, be expected. In fact, one does well to beware of comparing active networks on the basis of sensitivity when it is on the basis of actual network-parameter variations, resulting from component changes, that they will succeed or fail to perform satisfactorily in the field. We are, for example, interested in the actual *change* in Q, namely $\Delta Q/Q$, resulting from worst-case temperature drift and aging of the components of, say, a selective bandpass filter; we are not really, or only indirectly, interested in the Q *sensitivity* to these components. Thus, having derived the sensitivity functions, we must return to the Q variation defined by (4-10) in order to examine how it is affected by the sensitivity functions in conjunction with the variations ε_r, ε_c, and $\Delta\beta/\beta$. Before we do so, however, we shall define the following important quantity.

4.1.2 The Gain–Sensitivity Product[5]

It will be apparent shortly, when calculating the Q stability of active networks realized in hybrid integrated form, that no matter how good the quality of the passive components is (in particular their tracking capabilities), the variation of Q with β, or V_β^Q, must still be contended with. We shall therefore now examine this term in more detail.

5. G. S. Moschytz, Gain–sensitivity product—A figure of merit for hybrid integrated filters using single operational amplifiers, *IEEE J. Solid-State Circuits*, **SC-6**, 103–110 (1971).

Let us begin by considering the gain variation $\Delta\beta/\beta$. Since β is assumed to be the closed-loop gain of an operational amplifier we have

$$\frac{\Delta\beta}{\beta} = V_A^\beta + V_{R_{in}}^\beta + V_{R_{out}}^\beta + V_{R_F/R_G}^\beta \qquad (4\text{-}26)$$

where A, R_{in}, R_{out} and R_F/R_G are the open-loop gain, input resistance, output resistance, and ratio of external feedback resistors of the operational amplifiers, respectively. It can be shown[6] that the largest variation in (4-26) is due to the drift and changes in the open-loop gain A; therefore[7]

$$\frac{\Delta\beta}{\beta} \approx V_A^\beta = S_A^\beta \cdot \frac{\Delta A}{A} \qquad (4\text{-}27)$$

S_A^β is inversely proportional to the loop gain (LG) of the operational amplifier. The interdependence between loop gain, open-loop gain, and closed-loop gain, is qualitatively shown in Fig. 4-1. Accordingly, S_A^β can be expressed in terms of β and A:

$$S_A^\beta \approx \frac{1}{LG} = \frac{\beta}{A} \qquad (4\text{-}28)$$

The variation of Q with respect to β then follows as

$$V_\beta^Q = S_\beta^Q V_A^\beta = S_\beta^Q S_A^\beta \frac{\Delta A}{A} = [\beta S_\beta^Q] \frac{\Delta A}{A^2} \qquad [4\text{-}29]$$

Thus, for the third term of (4-2) we have

$$V_\beta^Q = \Gamma \cdot \frac{\Delta A}{A^2} \qquad [4\text{-}30a]$$

where

$$\Gamma = \beta S_\beta^Q \qquad [4\text{-}30b]$$

The Q variation as expressed by (4-30) consists of two separate and independent quantities: (i) $\Delta A/A^2$, which depends only on the specifications of

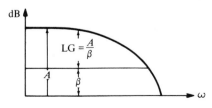

dB

$LG = \dfrac{A}{\beta}$

A

β

ω

FIG. 4-1. Qualitative relationship between A, β, and the loop gain (LG).

6. See Chapter 7, Section 7.2.4 of *Linear Integrated Networks: Fundamentals.*
7. Beside the open-loop gain, the next largest variable in (4-26) is $V_{R_{in}}^\beta$. However, $\Delta A/A$ can be assumed to include both variations in open-loop gain and input impedance, since the variation $\Delta R_{in}/R_{in}$ is also inversely proportional to the loop gain, i.e., $\Delta\beta/\beta \approx [(\Delta A/A) + (\Delta R_{in}/R_{in})]/LG$.

the operational amplifier used (e.g., temperature and aging characteristics, frequency-compensated open-loop gain); and (ii) the gain–sensitivity product (GSP) Γ, which depends only on the method used to realize the active network. As we shall see in what follows, Γ represents a figure-of-merit for hybrid integrated circuits (for which ε_r, $\varepsilon_c \ll 1$) and may be used to compare different hybrid integrated network realizations independently of the operational amplifier. In comparing V_β^Q for various networks, the same amplifier— and with it the same quantity $\Delta A/A^2$—is assumed for each. The difference found in V_β^Q is then a result of the difference in the gain-sensitivity product associated with the networks in question.

It should be clear that we can define the gain-sensitivity product with respect to any parameter F associated with an active network incorporating an operational amplifier. Thus, generalizing the derivation of (4-29) we have:

$$V_\beta^F = S_\beta^F \cdot V_A^\beta = [\beta S_\beta^F] \cdot \frac{\Delta A}{A^2} = \Gamma_\beta^F \cdot \frac{\Delta A}{A^2} \qquad [4\text{-}31]$$

where F may for example be the amplitude or phase at a given frequency, or the pole frequency ω_p. We shall see, however, that in characterizing the (pole) stability of hybrid integrated networks, the GSP-term Γ_β^Q plays an exclusive role (also in characterizing the stability of the pole frequency ω_p), so that we shall continue, for convenience, to refer to Γ_β^Q simply as the gain-sensitivity product Γ.

Let us, for example, now find the gain-sensitivity product for our four network classes.

Class 1 and 2 Networks To obtain Γ for a given network class we must first find the gain β necessary to obtain a given Q. From (4-11) we have, for class 1 and 2 networks,

$$\beta_N = \frac{1}{k}\left[\left(\frac{Q}{\hat{Q}}\right)^2 - 1\right] \approx \frac{1}{k}\left(\frac{Q}{\hat{Q}}\right)^2 \qquad (4\text{-}32)$$

and, with (4-14c),

$$\Gamma_N\big|_{\text{class 1, 2}} \approx \frac{1}{2k}\left(\frac{Q}{\hat{Q}}\right)^2 \qquad [4\text{-}33]$$

Thus

$$V_{\beta_N}^Q\bigg|_{\text{class 1, 2}} \approx \frac{1}{2k}\left(\frac{Q}{\hat{Q}}\right)^2 \frac{\Delta A}{A^2} \qquad [4\text{-}34]$$

Class 3 Networks Here we have, from (4-12),

$$\beta_N = \frac{1}{k}\frac{(Q/\hat{Q}) - 1}{1 \mp (vQ/\hat{Q})} \qquad (4\text{-}35)$$

Often a bridged-T network is used in the feedback loop resulting in the negative sign in the denominator. If a twin-T is used, either the positive (LHP zeros) or the negative sign (RHP zeros) may occur. RHP zeros affect the network in the same way that positive feedback does—e.g., as in a class 3 network incorporating a bridged-T *and* dual feedback (see Chapter 2, Section 2.4.1). Since positive feedback will be discussed separately later, we shall confine ourselves to negative feedback here (i.e., to the positive sign in the denominator of (4-35)). Nevertheless, even with LHP zeros only, we must still consider the case of a bridged-T and a twin-T in the feedback loop of a class 3 network separately, since the outcome, with respect to the gain–sensitivity product, is significantly different.

The general form of Γ_N for class 3 networks is readily obtained from (4-17) and (4-35) as

$$\Gamma_N \big|_{\text{class 3}} = \frac{1}{k} \frac{[(Q/\hat{Q}) - 1]^2}{(Q/\hat{Q})(1 - v)} \tag{4-36}$$

Since $v = \hat{Q}/q_z \ll 1$, we obtain, for high Q values,

$$(\Gamma_N \big|_{Q \gg 1})_{\text{class 3}} \approx \frac{1}{k} \frac{Q}{\hat{Q}} \tag{4-37}$$

where k is the multiplicative constant associated with the feedback network (i.e., bridged- or twin-T). Comparing (4-37) with (4-33) we note that, where Γ_N for class 1 and 2 networks is proportional to Q^2, it is only proportional to Q for the general class 3 network. This is a considerable improvement.

As will now be shown, (4-37) is not valid if the feedback-network is a *bridged-T*. We recall from Chapter 2, Section 2.3.3, that $\hat{Q}$ and q_z are *not* independent of each other in a bridged-T. We found there that the two are related by the expression

$$\hat{Q} = \frac{q_z}{1 + q_z^2\left(1 + \dfrac{C_2}{C_1}\right)}\Bigg|_{q_z \gg 1} \approx \frac{1}{\gamma q_z} \tag{4-38}$$

where $\gamma = 1 + (C_2/C_1)$ and C_1, C_2 are the capacitors of the bridged-T. The practical consequences of this interdependence in terms of operating range, component spread, and so on were discussed in Chapter 2. The consequences with respect to the gain–sensitivity product of the circuit follow here. If we substitute (4-38) into (4-36), we obtain, with (4-18),

$$(\Gamma_N)_{\substack{\text{class 3} \\ \text{(bridged-}T)}} = \frac{q_z}{kQ} \frac{(\gamma Q q_z - 1)^2}{\gamma q_z^2 - 1} \tag{4-39}$$

For high Q's we have

$$Q \approx q_z \tag{4-40}$$

and (4-39) simplifies to

$$\left(\Gamma_N\bigg|_{Q \approx q_z}\right)_{\substack{\text{class 3} \\ \text{(bridged-}T)}} \approx \frac{1}{k}(\gamma Q^2 - 1)\bigg|_{Q \gg 1} \approx \frac{\gamma}{k}Q^2 \qquad [4\text{-}41]$$

Thus, using the bridged-T, the gain–sensitivity product is proportional to Q^2, as we found it to be for the other two negative-feedback classes.

The foregoing discussion of class 3 network types permits the following conclusions to be drawn:

1. Using a twin-T in the feedback loop, a class 3 network is superior to other negative-feedback types. The gain–sensitivity product, given by (4-37), is proportional to Q. By maximizing $\hat{Q}$ and k, both of which are parameters of the twin-T, Γ_N can be minimized. For an unloaded twin-T $k = 1$; if it is also potentially symmetric,[8] the $\hat{Q}$ can be selected close to its maximum of 0.5. Thus, the lower bound on Γ_N is

$$(\Gamma_N)_{\substack{\text{class 3} \\ \text{(twin-}T)}} \geq 2Q \qquad [4\text{-}42]$$

2. Using a bridged-T in the feedback-loop of a class 3 network, Γ_N given by (4-41) is comparable to that attained with other negative feedback types; it is proportional to Q^2. For an unloaded bridged-T, $k = 1$ and (4-41) becomes

$$(\Gamma_N)_{\substack{\text{class 3} \\ \text{(bridged-}T)}} \approx \left(1 + \frac{C_2}{C_1}\right)Q^2 \qquad [4\text{-}43]$$

By minimizing the capacitive ratio C_2/C_1 within the bounds permitted by the hybrid technology used, Γ_N can be minimized accordingly. The lower bound is therefore given by

$$(\Gamma_N)_{\substack{\text{class 3} \\ \text{(bridged-}T)}} \geq Q^2 \qquad [4\text{-}44]$$

By contrast, the lower bound on Γ_N for class 1 and 2 networks is

$$(\Gamma_N)_{\text{class 1,2}} \geq 2Q^2 \qquad [4\text{-}45]$$

This is obtained from (4-33) with $k = 1$[9] and $\hat{Q}_{\text{max}} = 0.5$ and corresponds to a class-3 network with a symmetrical bridged-T (i.e., $C_1 = C_2$).

---

8. See Chapter 3, Section 3.5 of *Linear Integrated Networks: Fundamentals.*
9. In all cases, k, the multiplicand associated with the passive RC feedback network, has been assumed to have a maximum value of unity. Whereas this is unquestionably true for the unloaded bridged- and twin-T networks, it is not necessarily true for the RC ladder networks used in the feedback paths of class 1 and 2 networks. More specifically, k and $\hat{Q}$ are then not necessarily independent, so that $k = 1$ and $\hat{Q} \lesssim 0.5$ may be unobtainable simultaneously. In this case a lower bound of $4Q^2$ in (4-45) may be more accurate. More will be said about this in conjunction with positive-feedback networks in Chapter 6.

Class 4 Networks   For class 4, or positive-feedback, networks we have, from (4-21a),

$$\beta_P = \frac{1}{k}\left(1 - \frac{\hat{Q}}{Q}\right) \tag{4-46}$$

and therefore, with (4-25) and $Q \gg 1$

$$\Gamma_P\bigg|_{\text{class 4}} = \frac{1}{k}\frac{Q}{\hat{Q}}\left(1 - \frac{\hat{Q}}{Q}\right)^2\bigg|_{Q \gg 1} \approx \frac{1}{k}\frac{Q}{\hat{Q}} \tag{4-47}$$

Thus

$$V_{\beta_P}^Q\bigg|_{\text{class 4}} \approx \frac{1}{k}\frac{Q}{\hat{Q}}\frac{\Delta A}{A^2} \tag{4-48}$$

Notice that Γ_P is proportional to Q. Only a class 3 network with a twin-T in the feedback path has an equally small gain–sensitivity product; in fact, (4-47) and (4-37) have exactly the same form. The lower bound is also the same, assuming that $k_{\text{max}} = 1$ and $\hat{Q}_{\text{max}} = 0.5$:

$$(\Gamma_P)_{\text{class 4}} \geq 2Q \tag{4-49}$$

In the course of this and the following chapters we shall have occasion to come back to the expressions derived here whenever we are concerned (as we are in what follows) with the Q stability of a network or of a network class.

4.1.3 Q Stability

Returning to (4-10), in order to compare the overall Q variation obtainable with our four network classes we shall, for the time being, designate as "negative-feedback" networks (with subscript N), those of classes 1 and 2 as well as those of class 3 with a bridged-T in the feedback path. Γ_N for all these networks is proportional to Q^2. These negative-feedback networks will be compared with "positive-feedback" networks (with subscript P), meaning those of class 4, with a view to their use in hybrid integrated circuits. Subsequently, we shall consider the class 3 network using a twin-T in the feedback path with the same objective in mind.

Substituting (4-14) and (4-34) into (4-10) we have, for *negative-feedback* networks,

$$\left(\frac{\Delta Q}{Q}\bigg|_{\text{max}}\right)_N = \varepsilon_r \sum_{i=1}^r \left| S_{R_i}^{\hat{Q}} + \frac{1}{2} S_{R_i}^k \right| + \varepsilon_c \sum_{j=1}^c \left| S_{C_j}^{\hat{Q}} + \frac{1}{2} S_{C_j}^k \right| + \frac{1}{2k}\left(\frac{Q}{\hat{Q}}\right)^2\frac{\Delta A}{A^2} \tag{4-50}$$

For *class 3 networks*, (4-41) may be used instead of (4-34); however, since both expressions are proportional to Q^2 the difference is insignificant.

Furthermore (4-19a) and (4-19b) should, strictly speaking, be used instead of (4-14a) and (4-14b). However, in either case, the sensitivities to the passive components are low. Thus the Q variation of class 1, 2, and 3 networks does not warrant separate treatment (assuming, that is, that a bridged-T is used in the class 3 network).

Substituting (4-24) and (4-48) into (4-10) we obtain, for *positive-feedback* networks,

$$\left(\frac{\Delta Q}{Q}\bigg|_{\max}\right)_P = \left(\varepsilon_r \sum_{i=1}^{r}\left|S_{R_i}^k\right| + \varepsilon_c \sum_{j=1}^{c}\left|S_{C_j}^k\right|\right)\frac{Q}{\hat{Q}} + \frac{1}{k}\frac{Q}{\hat{Q}}\frac{\Delta A}{A^2} \qquad [4\text{-}51]$$

A comparison of (4-50) with (4-51) shows up some interesting differences. The Q variation caused by the uncorrelated changes in passive elements is large for positive feedback, since it is proportional to $Q/\hat{Q}$; it is small, and independent of Q, for negative feedback. On the other hand, the Q variation due to changes in the active device is proportional to Q^2 for negative feedback and only proportional to Q for positive feedback. The following general statement may therefore be made:

To obtain a given pole Q with a network containing an active device A, positive feedback should be used if closely tracking (i.e., $\varepsilon_r \ll 1$, $\varepsilon_c \ll 1$) passive components are available; with poorly tracking passive components, negative feedback provides a more stable network.

This statement is no more than qualitative in that it leaves unanswered the question of how close the passive-component tracking must be for the network to qualify for positive feedback. Certainly it leaves no doubt as to which feedback type to use in the case of *ideal* hybrid integrated networks, which are characterized by passive components with perfect tracking. This will briefly be demonstrated below. After that we shall return to the question of the degree of component tracking necessary to qualify for positive feedback when using nonideal hybrid integrated networks.

Ideal Hybrid Integrated Networks   For *ideal* hybrid integrated circuits the passive components track perfectly, i.e., $\varepsilon_R \approx \varepsilon_C \approx 0$. We then have (see (4-10) and (4-30))

$$\left(\frac{\Delta Q}{Q}\bigg|_{\max}\right)\bigg|_{\varepsilon_r = \varepsilon_c = 0} = V_\beta^Q = \beta S_\beta^Q \frac{\Delta A}{A^2} = \Gamma\left(\frac{\Delta A}{A^2}\right) \qquad (4\text{-}52)$$

With (4-50) and (4-51), respectively, we then obtain

$$\Gamma_N = \frac{1}{2k}\left(\frac{Q}{\hat{Q}}\right)^2 \qquad [4\text{-}53a]$$

and

$$\Gamma_P = \frac{1}{k}\frac{Q}{\hat{Q}} \qquad [4\text{-}53b]$$

We see here, that Γ represents a figure-of-merit for hybrid integrated networks in that it gives a measure for the Q stability attainable with such networks, independently of the operational amplifiers used. That Γ provides a more meaningful measure of Q stability than Q sensitivity alone, and that, in fact, comparisons based on the latter can be quite misleading will be illustrated by the following examples.

Consider the four circuits that were listed in Table 2-2, Chapter 2. Each was shown to be representative of one of the four basic network classes discussed in Chapter 2. As was shown in Table 2-2, based on sensitivity alone the class 1 and 2 networks are by far the best; the class 4 network is the worst, since the Q sensitivity is proportional to Q. Based on the gain–sensitivity product, however, the conclusions to be drawn are exactly reversed. That this is so is shown by the calculations listed in Table 4-2.[10] Here Γ and $\Delta Q/Q$ are given for the four networks in Table 2-2 for Q's of 10, 50, and 100, respectively. The gain–sensitivity product clearly predicts that the positive-feedback circuits (class 4), when realized in hybrid integrated form, are superior to the other three types; a fact that sensitivity comparisons invariably obscure.

Two cases, corresponding to two different operational amplifiers, are calculated in Table 4-2. Opamp 1 corresponds to a high-gain low-frequency type (e.g., a Fairchild μF 709 or μF 741) whose open-loop gain may typically vary by 40 %. Opamp 2 is more characteristic of a medium-gain high-frequency type (e.g., Fairchild μF 702 or RCA 3015) whose open-loop gain at higher frequencies varies by only 20 %.[11] Note that the less stable, high-gain Opamp is superior in all cases. The examples given for class 1 and 2 networks cannot be used up to Q's of 50 or 100, since the required closed-loop gain β is higher than the open-loop gain of either amplifier. The class 3 network can be used in all cases, but the Q variation becomes prohibitive already at a Q of 50. The class 4 network performs well, even at a Q of 100 if the open-loop gain is high enough, but rapidly becomes marginal with decreasing open-loop gain.

The conclusions to be drawn from Table 4-2 can be summarized as follows:

1. The Q stability of ideal hybrid integrated second-order active filter sections using one operational amplifier can be accurately characterized by the gain–sensitivity product Γ.

2. Such filter sections are useful only for low-to-medium Q's, depending on the available loop gain A/β of the amplifier. Since A/β decreases with frequency, medium Q's (50 to 100) of high stability can only be obtained at low frequencies (e.g., less than 20 kHz) with high-gain, low-frequency operational amplifiers.

10. G. S. Moschytz, op. cit.
11. This is due to frequency compensation by capacitive feedback around the high-gain stage of the amplifier; see Chapter 7, Section 7.2.5 of *Linear Integrated Networks: Fundamentals*.

TABLE 4–2. COMPARISON OF Q STABILITIES FOR FILTER EXAMPLES GIVEN IN TABLE 2-2 (CHAPTER 2)

| Class and filter type | $Q = 10$ | | | $Q = 50$ | | | $Q = 100$ | | |
| | | $\Delta Q/Q$ [%] | | | $\Delta Q/Q$ [%] | | | $\Delta Q/Q$ [%] | |
	Γ	Opamp 1	Opamp 2	Γ	Opamp 1	Opamp 2	Γ	Opamp 1	Opamp 2
1. Low-pass filter	1200	5	25	$3 \cdot 10^4$	$\beta > A$	$\beta > A$	$12 \cdot 10^4$	$\beta > A$	$\beta > A$
2. Bandpass filter	400	2	9	10^4	$\beta > A$	$\beta > A$	$4 \cdot 10^4$	$\beta > A$	$\beta > A$
3. Low-pass filter	300	1.2	6	$0.75 \cdot 10^4$	30	(150)	$3 \cdot 10^4$	(120)	(600)
4. Bandpass filter	84	0.35	1.7	$0.045 \cdot 10^4$	1.8	9.25	$0.092 \cdot 10^4$	3.7	18.5

Opamp 1: $A = 80$ dB; $\Delta A/A = 40\%$.   Opamp 2: $A = 60$ dB; $\Delta A/A = 20\%$.

3. Positive-feedback circuits in ideal hybrid integrated form are superior to negative-feedback (high-gain) circuits in spite of their high Q sensitivity. Due to the low closed-loop gains β required, they can be used up to relatively high frequencies to provide medium Q's, since the available loop gain A/β remains sufficiently large over a wide frequency range.

These conclusions are based on our discussion of ideal hybrid integrated circuits, in which the passive components are assumed to track perfectly. Naturally, we must now ask ourselves to what extent these conclusions are valid under more realistic circumstances, i.e., when ε_r and ε_c in (4-50) and (4-51) are not zero; this is the case with nonideal hybrid integrated networks.

Nonideal Hybrid Integrated Networks   To deal with the case of nonideal hybrid integrated networks based on negative feedback it is convenient to rewrite (4-50) as follows:

$$\left[\left(\frac{\Delta Q}{Q}\right)_{\max}\right]_N = N_R\,\varepsilon_r + N_C\,\varepsilon_c + N_\beta\,Q^2\left(\frac{\Delta A}{A^2}\right) \qquad [4\text{-}54]$$

where N_R, N_C, and N_β are the coefficients of the resistive, capacitive, and gain drift components affecting the Q variation in a negative-feedback system. These coefficients depend directly on the *synthesis method* used to obtain the network function $T(s)$. By contrast ε_r, ε_c, and $\Delta A/A^2$ characterize the *component technology* used to realize $T(s)$. Thin-film resistors and capacitors, for example, will track, if not perfectly, at least very closely; therefore ε_r and ε_c will be very small. Similarly, $\Delta A/A^2$ will depend on the open-loop stability, gain, and frequency response of the operational amplifier used. Q, of course, is given as the pole Q of the specified network function $T(s)$.

The positive-feedback maximum Q variation given by (4-51) can also be rewritten to separate the effects of the passive and active component technology from the synthesis method used:

$$\left[\left(\frac{\Delta Q}{Q}\right)_{\max}\right]_P = Q\left[P_R\,\varepsilon_r + P_C\,\varepsilon_c + P_\beta\left(\frac{\Delta A}{A^2}\right)\right] \qquad [4\text{-}55]$$

The whole expression here is proportional to Q, whereas in (4-54) the gain-variation term is proportional to Q^2. P_R, P_C, and P_β are the positive feedback counterparts of N_R, N_C, and N_β in (4-54). They result directly from a comparison of the coefficients in (4-51) and (4-55).

To obtain some insight into the effect of nonperfect component tracking on positive and negative feedback systems, we can plot the worst-case Q variation as a function of Q as shown qualitatively in Fig. 4-2. The Q variation increases linearly with positive feedback, quadratically with negative feedback. Let us assume that the two curves intersect, designating the corresponding Q values Q_1 and Q_2. In order to minimize the Q variation, positive feedback should be used for Q values below Q_1 and above Q_2, and negative

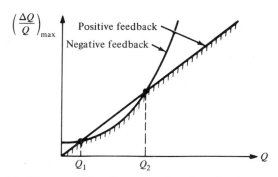

FIG. 4-2. Worst-case Q variation as a function of Q (qualitative).

feedback in between the two. In order to tell which feedback type is preferable, we must know how far apart Q_1 and Q_2 are from one another, and also in which Q range they lie. As we shall see, the location of Q_1 and Q_2 depends directly on the technology used, since this determines the relative location and the shape of the two curves in Fig. 4-2.

In order to evaluate the parameters affecting Q_1 and Q_2 we equate (4-54) with (4-55) and obtain

$$Q^2 - \frac{P_R \varepsilon_r + P_C \varepsilon_c + P_\beta(\Delta A/A^2)}{N_\beta(\Delta A/A^2)} Q + \frac{N_R \varepsilon_r + N_C \varepsilon_c}{N_\beta(\Delta A/A^2)} = 0 \qquad (4\text{-}56)$$

Letting

$$P = \frac{P_R \varepsilon_r + P_C \varepsilon_c + P_\beta(\Delta A/A^2)}{N_\beta(\Delta A/A^2)} \qquad (4\text{-}57a)$$

and

$$N = \frac{N_R \varepsilon_r + N_C \varepsilon_c}{N_\beta(\Delta A/A^2)} \qquad (4\text{-}57b)$$

and solving for Q we obtain

$$Q_1 = \frac{P}{2}[1 - \sqrt{1 - 4(N/P^2)}] \qquad (4\text{-}58a)$$

and

$$Q_2 = \frac{P}{2}[1 + \sqrt{1 - 4(N/P^2)}] \qquad (4\text{-}58b)$$

It is useful, at this point, to introduce the following Q variations:

$$(V_\varepsilon^Q)_N = N_R \varepsilon_r + N_C \varepsilon_c \qquad (4\text{-}59a)$$

$$(V_\beta^Q)_N = N_\beta \left(\frac{\Delta A}{A^2}\right) \qquad (4\text{-}59b)$$

$$(V_{\varepsilon,\beta}^{Q})_P = P_R\varepsilon_r + P_C\varepsilon_c + P_\beta\left(\frac{\Delta A}{A^2}\right) \tag{4-59c}$$

Then

$$P = \frac{(V_{\varepsilon,\ \beta}^{Q})_P}{(V_\beta^{Q})_N} \tag{4-60a}$$

$$N = \frac{(V_\varepsilon^{Q})_N}{(V_\beta^{Q})_N} \tag{4-60b}$$

and

$$4\,\frac{N}{P^2} = 4\,\frac{(V_\varepsilon^{Q})_N(V_\beta^{Q})_N}{[(V_{\varepsilon,\beta}^{Q})_P]^2}$$

$$= 4\,\frac{(N_R\varepsilon_r + N_C\varepsilon_c)N_\beta(\Delta A/A^2)}{[P_R\varepsilon_r + P_C\varepsilon_c + P_\beta(\Delta A/A^2)]^2} \tag{4-61}$$

From (4-58a) and (4-58b) it now follows that as the quantity $4N/P^2$ varies from zero to unity the two roots of (4-56), Q_1 and Q_2, vary as shown in Fig. 4-3, so that

$$0 \le Q_1 \le \frac{P}{2} \tag{4-62a}$$

and

$$\frac{P}{2} \le Q_2 \le P \tag{4-62b}$$

From Fig. 4-2 we can state generally that in order to obtain a small change $(\Delta Q/Q)_{max}$, negative feedback is preferable for Q values lower than Q_2, positive feedback for Q values above Q_2. The Q range between zero and Q_1

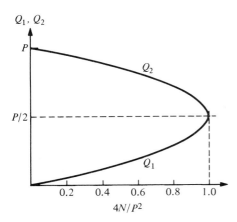

FIG. 4-3. Variation of Q_1 and Q_2 with $4N/P^2$.

is generally small, *and pertains to low* Q *values*; it is therefore negligible, i.e., need not be examined separately. Thus, we are interested in the maximum Q_2 value that can occur in a given circuit. From (4-62b) this is given by

$$Q_{2\,max} = P = \frac{P_R \varepsilon_r + P_C \varepsilon_c + P_B\left(\dfrac{\Delta A}{A^2}\right)}{N_\beta\left(\dfrac{\Delta A}{A^2}\right)} \qquad [4\text{-}63]$$

The actual Q_2 value occurring in the circuit will be all the closer to this maximum value the closer $4N/P^2$ is to zero. Incidentally, a glance at (4-63) shows that $Q_{2\,max}$ will be all the smaller the closer ε_r and ε_c are to zero (i.e., perfect tracking). In fact, for this special case,

$$Q_{2\,max}\Big|_{\varepsilon_r = \varepsilon_c = 0} = \frac{P_\beta}{N_\beta} = 2\frac{\hat{Q}}{Q} < \frac{1}{Q} \qquad (4\text{-}64a)$$

and

$$4\,\frac{N}{P^2}\bigg|_{\varepsilon_r = \varepsilon_c = 0} = 0 \qquad (4\text{-}64b)$$

Thus, for $\varepsilon_r = \varepsilon_c = 0$, $Q_{2\,max}$ is less than unity and positive feedback is preferable for all Q's. This, of course, coincides with the conclusions already reached in our discussion on ideal hybrid integrated networks.

In the practical case, with nonideal components, we must compare actual circuits and use ε_r, ε_c and $\Delta A/A^2$ values that are typical of hybrid integrated circuits. Consider, for example, the positive-feedback (class 4) network in Fig. 4-4. By straightforward analysis we obtain

$$Q = \frac{[(G_1 + \bar{G}_1)G_2 C_1 C_2]^{1/2}}{(G_1 + \bar{G}_1 + G_2)C_2 + G_2 C_1 - \beta G_1 C_2} \qquad (4\text{-}65)$$

By comparison with (4-21a) we have

$$\hat{Q} = Q(\beta = 0) = \frac{[(G_1 + \bar{G}_1)G_2 C_1 C_2]^{1/2}}{(G_1 + \bar{G}_1 + G_2)C_2 + G_2 C_1} \qquad (4\text{-}66a)$$

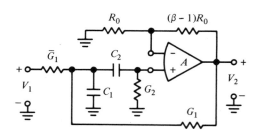

FIG. 4-4. Class 4 bandpass network.

and from (4-21b)

$$k = \frac{G_1 C_2}{(G_1 + \bar{G}_1 + G_2)C_2 + G_2 C_1} \tag{4-66b}$$

To obtain $(\Delta Q/Q)_{max}$ we require the following sensitivities (see (4-51)):

$$S_{G_1}^k = 1 - \frac{G_1 C_2}{(G_1 + \bar{G}_1 + G_2)C_2 + G_2 C_1}$$

$$= \frac{(\bar{G}_1 + G_2)C_2 + C_2 C_1}{D} \tag{4-67a}$$

$$S_{\bar{G}_1}^k = - \frac{\bar{G}_1 C_2}{D} \tag{4-67b}$$

$$S_{G_2}^k = - \frac{G_2(C_1 + C_2)}{D} \tag{4-67c}$$

$$S_{C_1}^k = - \frac{G_2 C_1}{D} \tag{4-67d}$$

$$S_{C_2}^k = \frac{G_2 C_1}{D} \tag{4-67e}$$

where

$$D = (G_1 + \bar{G}_1 + G_2)C_2 + G_2 C_1 \tag{4-67f}$$

Furthermore, solving (4-65) for β,

$$\beta = \frac{1}{G_1 C_2} \left[C_2(G_1 + \bar{G}_1 + G_2) + G_2 C_1 - \frac{\sqrt{G_2(G_1 + \bar{G}_1)C_1 C_2}}{Q} \right] \tag{4-68}$$

Note that[12]

$$\sum_{i=1}^{3} S_{G_i}^k = \sum_{j=1}^{2} S_{C_j}^k = 0 \tag{4-69}$$

With the component values assumed in Chapter 2, Table 2-2, namely,

$$\begin{aligned} G_1 &= \bar{G}_1 = 5 \\ G_2 &= 0.1 \\ C_1 &= 10 \\ C_2 &= 0.1 \\ \beta &= 4.02 - (2/Q) \end{aligned} \tag{4-70}$$

---

12. This must be the case, since $\Sigma S_{G_i} Q = 0 = \Sigma \{ S_{R_i} \hat{Q} + [k\beta/(1 - k\beta)] S_{R_i}{}^k \}$ and $\Sigma S_{R_i} \hat{Q} = 0$. Consequently $[k\beta/(1 - k\beta)] \Sigma S_{R_i}{}^k = 0$, from which (4-69) follows.

we obtain

$$S_{G_1}^k = \frac{1.51}{2.01}$$

$$S_{G_1}^k = -\frac{0.5}{2.01}$$

$$S_{G_2}^k = -\frac{1.01}{2.01}$$

$$S_{C_1}^k = -\frac{1}{2.01} \qquad (4\text{-}71)$$

$$S_{C_2}^k = \frac{1}{2.01}$$

$$\hat{Q} = \frac{1}{2.01}$$

$$k = \frac{0.5}{2.01}$$

Consequently

$$P_R = \frac{1}{\hat{Q}} \sum_{i=1}^{3} |S_{R_i}^k| = 3.02 \qquad (4.72a)$$

$$P_C = \frac{1}{\hat{Q}} \sum_{j=1}^{2} |S_{C_j}^k| = 2 \qquad (4\text{-}72b)$$

$$P_\beta = \frac{1}{k\hat{Q}} = 8.08 \qquad (4\text{-}72c)$$

and

$$\left(\frac{\Delta Q}{Q}\bigg|_{\max}\right)_P = Q\left(3.02\varepsilon_r + 2\varepsilon_c + 8.08\frac{\Delta A}{A^2}\right) \qquad (4\text{-}73)$$

Let us now assume an amplifier with $\Delta A/A = \pm 50\%$ and $A = 60$ dB; thin-film resistors tracking to within ± 5 ppm/°C and thin-film capacitors tracking to within ± 10 ppm/°C. Over a 60°C temperature range $\varepsilon_r = 0.03\%$ and $\varepsilon_c = 0.06\%$. Thus, with (4-73),

$$\left(\frac{\Delta Q}{Q}\bigg|_{\max}\right)_P = Q[0.09 + 0.12 + 0.4]\% = \pm 0.61 Q\% \qquad (4\text{-}74)$$

Note that for the tracking and opamp variations assumed, two-thirds of the resulting Q variation is due to the amplifier, the remaining one-third due to the passive components.

$\Big($Equation (4-73) is based on (4-51), which uses the approximations (4-24a) to (4-25). It may be of interest to compare this result with the more exact expression for $(\Delta Q/Q)_{max}$ given by (4-10). Using the formulas of Table 4-1, we obtain, from (4-65) and (4-70),

$$S_{\bar{G}_1}^Q = \frac{1}{2}\frac{\bar{G}_1}{G_1 + \bar{G}_1} - Q\frac{\bar{G}_1}{\sqrt{G_2(G_1 + \bar{G}_1)}}\sqrt{\frac{C_2}{C_1}} = \frac{1}{2}\left(\frac{1}{2} - Q\right) \quad (4\text{-}75a)$$

$$S_{\bar{G}_1}^Q = \frac{1}{2}\frac{G_1}{G_1 + \bar{G}_1} - Q\frac{G_1}{\sqrt{G_2(G_1 + \bar{G}_1)}}\sqrt{\frac{C_2}{C_1}}(1 - \beta) = -\frac{1}{2}\left(\frac{3}{2} - 3.02Q\right) \quad (4\text{-}75b)$$

$$S_{G_2}^Q = \frac{1}{2} - Q\sqrt{\frac{G_2}{G_1 + \bar{G}_1}}\left(\sqrt{\frac{C_1}{C_2}} + \sqrt{\frac{C_2}{C_1}}\right) = \frac{1}{2} - 1.01Q \quad (4\text{-}75c)$$

$$S_{C_1}^Q = \frac{1}{2} - Q\sqrt{\frac{G_2}{G_1 + \bar{G}_1}}\sqrt{\frac{C_1}{C_2}} = \frac{1}{2} - Q \quad (4\text{-}75d)$$

$$S_{C_2}^Q = \frac{1}{2} - Q\frac{G_1 + \bar{G}_1 + G_2 - \beta G_1}{\sqrt{G_2(G_1 + \bar{G}_1)}}\sqrt{\frac{C_2}{C_1}} = -\left(\frac{1}{2} - Q\right) \quad (4\text{-}75e)$$

$$S_\beta^Q = \frac{Q}{\hat{Q}} - 1 = 2.01Q - 1 \quad (4\text{-}75f)$$

and, with (4-68),

$$\beta = 3.82 \quad (4\text{-}75g)$$

With (4-10) we obtain

$$\left(\frac{\Delta Q}{Q}\Big|_{max}\right)_P = \left(3Q - \frac{3}{2}\right)\varepsilon_r + (2Q - 1)\varepsilon_c + (2Q - 1)\frac{\Delta\beta}{\beta} \quad (4\text{-}76)$$

For large Q this can readily be approximated by

$$\left(\frac{\Delta Q}{Q}\Big|_{max}\right)_P \approx Q\left[3\varepsilon_r + 2\varepsilon_c + 2\beta\frac{\Delta A}{A^2}\right] \quad (4\text{-}77)$$

which is practically the same as our derivation based on (4-51), namely (4-73).$\Big)$

By way of comparison let us now consider the negative-feedback (class 1) network shown in Fig. 4-5. Straightforward analysis gives us the pole Q in the form of (4-11), namely,

$$Q = \hat{Q}\sqrt{1 + k\beta} \quad (4\text{-}78)$$

where

$$\hat{Q} = \frac{\{C_1C_2[G_2(\bar{G}_1 + G_1) + G_3(\bar{G}_1 + G_1 + G_2)]\}^{1/2}}{C_2(\bar{G}_1 + G_1 + G_2) + C_1(G_2 + G_3)} \quad (4\text{-}79a)$$

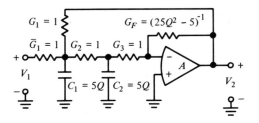

FIG. 4-5. Class 1 low-pass network.

and

$$k = \frac{G_1 G_2}{G_2(\bar{G}_1 + G_1) + G_3(\bar{G}_1 + G_1 + G_2)} \tag{4-79b}$$

Calculating the Q sensitivities for the component values used in Chapter 2, Table 2-2, namely,

$$G_1 = \bar{G}_1 = G_2 = G_3 = 1$$
$$C_1 = C_2 = 5Q$$
$$\beta = -(25Q^2 - 5) \tag{4-80}$$

we obtain

$$S_{\bar{G}_1}^Q = -0.2$$
$$S_{G_1}^Q = 0.3$$
$$s_{G_2}^Q = 0.1$$
$$S_{G_3}^Q = -0.2$$
$$S_{G_F}^Q = -\tfrac{1}{2}$$
$$S_{C_1}^Q = 0.1$$
$$S_{C_2}^Q = -0.1$$
$$S_\beta^Q \approx \tfrac{1}{2} \tag{4-81}$$

Furthermore

$$\hat{Q} = 1/\sqrt{5}$$
$$k = \tfrac{1}{5}$$

Thus

$$N_R = \sum_{i=1}^{5} |S_{R_i}^Q| \approx \sum_{i=1}^{5} \left| S_{R_i}^{\hat{Q}} + \frac{1}{2} S_{R_i}^k \right| \approx 1.3 \tag{4-82a}$$

$$N_C = \sum_{j=1}^{2} |S_{C_j}^Q| \approx \sum_{j=1}^{2} \left| S_{C_j}^{\hat{Q}} + \frac{1}{2} S_{C_j}^k \right| \approx 0.2 \tag{4-82b}$$

$$N_\beta = \frac{1}{2k\hat{Q}^2} = \frac{25}{2} \tag{4-82c}$$

and

$$\left(\frac{\Delta Q}{Q}\bigg|_{max}\right)_N = 1.3\varepsilon_r + 0.2\varepsilon_c + \frac{25}{2}\,Q^2\,\frac{\Delta A}{A^2} \qquad (4\text{-}83)$$

With the same assumptions for the opamp and the thin-film components as were made to obtain (4-74), we have

$$\left(\frac{\Delta Q}{Q}\bigg|_{max}\right)_N = [(1.3)(0.03) + (0.2)(0.06) + 0.6Q^2]\% \approx \pm 0.6Q^2\,\% \qquad (4\text{-}84)$$

Note that the contribution from the passive components is negligibly small here; it is the variation of the gain element, or more accurately the gain–sensitivity product, that determines the Q variation.

In order to find the maximum Q below which negative, or above which positive, feedback should be used, we calculate $Q_{2\,max}$ from (4-63) and obtain, from (4-74) and (4-84),

$$Q_{2\,max} = \frac{0.61}{0.6} \approx 1 \qquad (4\text{-}85)$$

Thus, for any Q above unity, the worst-case Q variation will be smaller with positive-feedback than with negative-feedback networks. For $Q = 10$, for example, the negative-feedback circuit will have a Q variation of $\pm 60\%$, whereas the positive-feedback circuit will be ten times less. For unity Q the variation is the same for both.

What if we use the same opamp as above, but now assume discrete passive components where, say, $\varepsilon_r = \varepsilon_c = 2\%$? Over $60°C$ we obtain for the first example

$$\left(\frac{\Delta Q}{Q}\bigg|_{max}\right)_P = Q[(3.02)(2) + 2(2) + 0.4]\% = 10.44Q\,\% \qquad (4\text{-}86)$$

and for the second

$$\left(\frac{\Delta Q}{Q}\bigg|_{max}\right)_N = [(1.3)(2) + (2)(0.2) + 0.6Q^2] = (3 + 0.6Q^2)\,\% \qquad (4\text{-}87)$$

In this case we have with (4-63)

$$Q_{2\,max} = \frac{10.44}{0.6} \approx 17.5 \qquad (4\text{-}88)$$

and, accordingly, for $Q = 10$, $[(\Delta Q/Q)_{max}]_P$ equals 104.4% whereas $[(\Delta Q/Q)_{max}]_N$ is only 63%. Thus, as the drift of the passive components increases, the effect is considerable on the Q variation of the positive-feedback networks, but only very slight for those with negative feedback.

The discussion above permits us to add a more specific statement to the three points made earlier. *The limit value Q_{2max} determines whether positive or negative feedback in single amplifier circuits is preferable for any given* *active-filter application. $Q_{2\max}$ is dependent on the technology used for the active-filter design.* For *hybrid integrated circuits* with closely tracking passive components, $Q_{2\max}$ may be on the order of unity; single-amplifier circuits with positive feedback will then be preferable. Using discrete passive components with wide tolerances and no tracking, $Q_{2\max}$ may be on the order of 20; then single-amplifier circuits using negative feedback should be used. Furthermore, it should be emphasized that with nonideal, but closely tracking, passive components, i.e., with practical hybrid integrated circuitry, it is generally the gain–sensitivity product that primarily determines the Q variation of a circuit. In the case of negative-feedback circuits this is always the case, no matter what the technology used, since the Q variations attributable to passive-component drift are negligible.[13]

The comparison between negative- and positive-feedback networks above dealt with class 1 and 2 networks, as well as class 3 networks with a bridged-T in the feedback path on the one hand, and class 4 networks on the other. It remains for us to consider a class 3 network with a twin-T in the feedback path. From a gain–sensitivity point of view, it has already been shown that there is no difference between it and a class 4 network (compare (4-42) and (4-49)). Indeed, from a Q-variation standpoint, the two network types will differ little, particularly when implemented in hybrid integrated form. In discrete form the twin-T network might at first seem preferable, were it not for the problem of tuning, and maintaining, a null with a discrete-component twin-T. This problem may well offset the problem of higher Q variation encountered by the positive-feedback network. In the case of hybrid integrated circuit implementation, the decisive factor may be an entirely different one. Using the class 3 network for building-block design, we found, in Chapter 3, that the single-amplifier FEN scheme was a very useful one (see Section 3.1.1). However, it will be recalled that a passive RC input network is required beside the twin-T feedback network in the FEN. The resulting large number of resistors and, in particular, capacitors, may lead to a large substrate size per second-order network and consequently high costs. The latter will be increased still more by the actual cost of the thin-film or chip capacitors, both of which are expensive when required with controlled temperature coefficients, low aging properties, and low losses. By contrast, a second-order, single-amplifier building-block, based on positive feedback as described in Section 3.1.2 of Chapter 3, is canonic,[14] at least with respect to the capacitor count, for all all-pole functions, and requires only one twin-T for the realization of finite zeros (i.e., the FRN).

13. Thus, negative-feedback networks have relatively little to gain, in terms of Q stability, by going from discrete to hybrid integrated circuitry; by contrast, positive-feedback networks benefit appreciably and actually become superior to their negative-feedback counterparts in the process.
14. A network using the minimum number of elements is often called *canonic*. To realize a second-order function a minimum of two resistors and two capacitors is required. For more on canonic networks see, for example, L. Weinberg, *Network Analysis and Synthesis* (New York: McGraw-Hill Book Co., 1962), p. 400.

4.1.4 Frequency Stability

The Frequency is Gain-Dependent   Although the networks used for building-block design are such that the frequency stability is independent, at least to the first order, of the active device, we shall, for the sake of completeness, first briefly examine the frequency stability of those circuits for which this independence is not valid. Referring to (4-1) and assuming a single active device β, we have the pole-frequency variation

$$\frac{\Delta\omega_p}{\omega_p} = \sum_{i=1}^{r} V_{R_i}^{\omega_p} + \sum_{j=1}^{c} V_{C_j}^{\omega_p} + V_{\beta}^{\omega_p} \qquad (4\text{-}89)$$

For simplicity, and with no loss in generality, we assume perfectly tracking hybrid integrated components, in which case (4-89) simplifies to[15]

$$\frac{\Delta\omega_p}{\omega_p} = -\left(\frac{\Delta R}{R} + \frac{\Delta C}{C}\right) + V_{\beta}^{\omega_p} \qquad [4\text{-}90]$$

The first term in this expression can be minimized by selecting resistors and capacitors with equal and opposite temperature coefficients. If ω_p were independent of β, the second term in (4-90) would be equal to zero. Since, however, $V_{\beta}^{\omega_p} \neq 0$, this term accounts for a radial shift of the pole p because, as we know, $V_{\beta}^{\omega_p}$ must be real. Referring to Fig. 4-6, it depends on the root locus[16] of p with respect to β, and on the value of Q (i.e., high or low) of the pole being considered, whether $V_{\beta}^{\omega_p}$ is negligible or not. In the case shown in Fig. 4-6a where the pole displacement dp is parallel to the $j\omega$ axis, $V_{\beta}^{\omega_p}$ will be relatively large. In Fig. 4-6b, where dp is almost horizontal, $V_{\beta}^{\omega_p}$ will

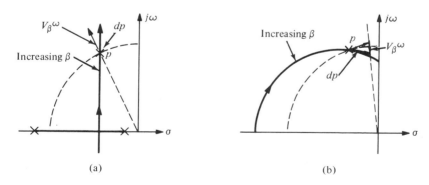

FIG. 4-6. Root loci of p with respect to β for a network with a single active device: (a) dp is parallel to the $j\omega$ axis; (b) dp is nearly horizontal.

15. See, for example, Chapter 4, equation (4-145) of *Linear Integrated Networks: Fundamentals*.
16. Remember that the pole displacement dp due to β is tangential to the root locus with respect to β.

be small. Thus, in those cases where $V_\beta^{\omega_p}$ is not exactly equal to zero, it depends entirely on the method of pole decomposition used (i.e., on the resulting root locus) and, of course, on the stability of β, whether it need be considered or not.

The Frequency Is Not Gain-Dependent   When ω_p is independent of β, i.e., $V_\beta^{\omega_p} = 0$, then the root locus with respect to β is a circle about the origin with radius ω_p. This is the most common case in practice. Then

$$\frac{\Delta\omega_p}{\omega_p} = - \left(\frac{\Delta R}{R} + \frac{\Delta C}{C}\right)$$ [4-91]

where all resistors are assumed to drift by $\Delta R/R$ and all capacitors by $\Delta C/C$.

Actually, there are numerous factors that can influence the drift of the components, and thereby the frequency stability, of a hybrid integrated network. The most important factors are:

1. Temperature (i.e., TCR and TCC)
2. Aging and humidity (A&H)
3. Initial frequency accuracy (I)

Thus, the total variation of frequency can be broken down into the following components:

$$\frac{\Delta f}{f} = F\left[\left(\frac{\Delta f}{f}\right)_{TCR}, \left(\frac{\Delta f}{f}\right)_{TCC}, \left(\frac{\Delta f}{f}\right)_{A\&H}, \left(\frac{\Delta f}{f}\right)_{I}\right]$$ (4-92)

The term $(\Delta f/f)_{TCR}$ is the frequency drift resulting from the temperature drift of the resistors. From (4-91) it follows that the frequency will drift by the same amount as the resistors and capacitors (although with the opposite polarity). Using nichrome thin film for the resistors, for example, it is possible to obtain a temperature coefficient (i.e., TCR) lower than ± 50 ppm/°C over a wide temperature range. We then obtain, typically,

$$\left(\frac{\Delta f}{f}\right)_{TCR} = \pm 50 \text{ ppm/°C}\bigg|_{\Delta T = 50°C} = \pm 0.25\%$$

The term $(\Delta f/f)_{TCC}$ is the frequency drift resulting from the temperature drift of the capacitors. To stay with the example given above, the capacitors used with nichrome thin-film resistors are usually NPO, multilayer, ceramic-chip types whose temperature coefficient (i.e., TCC) is typically ± 30 ppm/°C. Then

$$\left(\frac{\Delta f}{f}\right)_{TCC} = \pm 30 \text{ ppm/°C}\bigg|_{\Delta T = 50°C} = \pm 0.15\%$$

The term $(\Delta f/f)_{A\&H}$ is the frequency drift due to the effects of aging and humidity on the resistors and capacitors. For high-quality networks, in which frequency precision is critical, aging effects are minimized by initial accelerated aging and/or passivation of the components; humidity effects are minimized by assembling the individual components, or the entire networks, in hermetically sealed packages. Assuming all measures for long-term stability have been taken, values of 0.1 % are often quoted individually for resistors and capacitors, for time spans up to 20 years. For an expected life of, say, 5 years it may be realistic to assume 0.1 % for the combined aging effect of the resistors *and* capacitors.[17] Assuming a typical drift of $\pm 0.1 \%$ due to the aging of the resistors and the same amount due to the capacitors (whereby we include the effects of humidity) we obtain

$$\left(\frac{\Delta f}{f}\right)_{A\&H} = \pm 0.2\%$$

The term $(\Delta f/f)_I$ is the initial frequency precision attainable, be it by network tuning or otherwise. A glance at the typical tuning errors due to the factors enumerated above shows that, in order to be acceptable in a high-precision network, the tuning error should lie within 0.1 % and 0.25 %. With anything larger, the total error rapidly becomes excessive and the network incompatible with applications requiring a high degree of frequency precision. That tuning a network to this accuracy is no trivial matter will become apparent in Section 4.3. Nevertheless, it is reasonable to assume that the frequency error due to tuning is typically no larger than

$$\left(\frac{\Delta f}{f}\right)_I = \pm 0.2\%$$

Having obtained the individual frequency errors in (4-92) we must find the overall frequency error $\Delta f/f$ related to them. We can, of course, assume a worst-case frequency error such that

$$\left(\frac{\Delta f}{f}\right)_{\text{worst case}} = \pm \left[\left(\frac{\Delta f}{f}\right)_{TCR} + \left(\frac{\Delta f}{f}\right)_{TCC} + \left(\frac{\Delta f}{f}\right)_{A\&H} + \left(\frac{\Delta f}{f}\right)_{I}\right] \quad (4\text{-}93)$$

For the typical values assumed above we then obtain

$$\left(\frac{\Delta f}{f}\right)_{\text{worst case}} = \pm(0.25 + 0.15 + 0.2 + 0.2)\% = \pm 0.8\%$$

Whether this worst-case value is too pessimistic or not depends to a large extent on the statistical distributions of the individual errors. This is a subject that goes beyond the scope of this book. Suffice it to say that the root-

---

17. This is all the more realistic in those cases (e.g., tantalum thin-film resistors and capacitors) where the effects of aging of the two component types cancel each other out, at least partially.

mean-square error may often be considered a more realistic indication of the actual frequency error of a network, in which case we have[18]

$$\left(\frac{\Delta f}{f}\right)_{\text{RMS}} = \pm\frac{1}{2}\left[\left(\frac{\Delta f}{f}\right)^2_{\text{TCR}} + \left(\frac{\Delta f}{f}\right)^2_{\text{TCC}} + \left(\frac{\Delta f}{f}\right)^2_{\text{A\&H}} + \left(\frac{\Delta f}{f}\right)^2_{\text{I}}\right]^{1/2} \quad (4\text{-}94)$$

For the typical values given above we then have

$$\left(\frac{\Delta f}{f}\right)_{\text{RMS}} = \pm\tfrac{1}{2}\sqrt{0.25^2 + 0.15^2 + 0.2^2 + 0.2^2}\,\% = \pm 0.2\%$$

The large difference between the RMS and the worst-case values in our example above demonstrates well how careful one must be to use the appropriate method, which should mean the *realistic* method, of specifying a circuit, and how misleading the results may otherwise be. The RMS frequency error being one quarter of the worst-case error may make all the difference between acceptance and rejection of a network, or even of an entire technology. We can do no more here than to draw the reader's attention to the drastic consequences that the method of network specification may have, and to emphasize to him the importance of this sphere of the network designer's activity.

4.1.5 Pole Stability

Having discussed the stability of the pole Q and pole frequency of a (second-order) hybrid integrated network, it is of interest to consider ways of stabilizing both these factors simultaneously, i.e., of stabilizing the corresponding pole itself. This is perhaps the most direct way of approaching network stabilization, since we have shown elsewhere that it is primarily the stability and sensitivity of the dominant pole pair that determines the stability of a given network function. In this section we shall therefore consider the pole variations of a hybrid integrated network; naturally the same considerations apply to its zeros as well.

As we know,[19] we can write the pole variation of a hybrid integrated RC active network with closely tracking passive components, and with the effects of the active elements lumped into those of a single equivalent active device G, as follows:

$$\frac{dp}{p} = -\left(\frac{dR}{R} + \frac{dC}{C}\right) + S^p_G \frac{dG}{G} \quad (4\text{-}95)$$

---

18. A similar, detailed calculation of the frequency stability of typical tantalum thin-film components is given in Chapter 6, Section 6.1.4 of *Linear Integrated Networks: Fundamentals*. Note that there both the median value and the distribution about the median are taken into account for the individual terms of (4-92). Nevertheless, the resulting frequency stability obtained is comparable with the RMS value obtained here.
19. See Chapter 4, equation (4.120) of *Linear Integrated Networks: Fundamentals*.

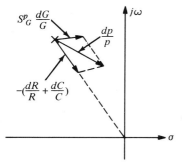

FIG. 4-7. Pole variation of a hybrid integrated RC active network with closely tracking passive components.

This corresponds to the vector representation shown in Fig. 4-7. For the pole sensitivity to be zero the following condition must be satisfied

$$S_G^p \frac{dG}{G} - \left(\frac{dR}{R} + \frac{dC}{C}\right) = 0 \qquad [4\text{-}96]$$

With (4-1) this can be written as

$$V_G^{\omega_p} - j \frac{V_G^Q}{\sqrt{4Q^2 - 1}} = \frac{dR}{R} + \frac{dC}{C} \qquad [4\text{-}97]$$

These expressions provide several means of desensitizing an active RC network to variations in network elements. In fact, the main differences in some of the numerous active synthesis techniques available involve the approach taken to satisfy (4-96).

There are three principal methods of desensitizing the dominant poles of an active RC network by realizing (4-96). These are briefly discussed below.[20]

Method 1: Compensation of Passive Sensitivity by Active Sensitivity
Pole desensitization is here achieved by designing the network in such a way that the passive sensitivity component is compensated by the active component, i.e.,

$$S_G^p = \frac{(dR/R) + (dC/C)}{dG/G} \qquad (4\text{-}98)$$

This is illustrated in Fig. 4-8a. The transmission poles are here dependent on both passive and active network elements. Pole-drift compensation is achieved by a combination of multiple forward transmission paths in single or multiloop feedback structures. These configurations result when a network is designed to obtain a specified transmission function and a desired

20. G. S. Moschytz, The operational amplifier in linear active networks, *IEEE Spectrum*, **7**(1), 42–50 (January 1970).

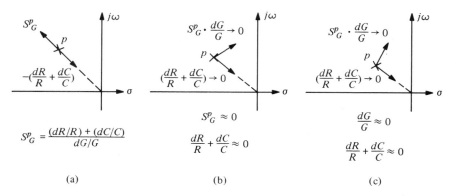

Fig. 4-8. Desensitizing of the dominant poles of an active RC network: (a) method 1; (b) method 2; (c) method 3.

active pole-sensitivity component at the same time. The method has been suggested for semiconductor integrated frequency-selective networks.[21] An intricate feedback scheme was used for the compensation of the drift properties of certain semiconductor integrated resistors and capacitors. The method is somewhat complicated and not very general. The chosen feedback configuration depends both on the properties of the semiconductor material being used and on the location of the particular pole being desensitized. It may well be, however, that this complexity is the price that must be paid when designing all-semiconductor networks of low sensitivity.

The requirement imposed by (4-98) is quite restrictive. It severely limits the choice of network configurations which must be capable of satisfying it, and providing the characteristics of a specified network function, at the same time. Thus this method is not suitable for network synthesis with operational amplifiers, since the gain characteristics of the amplifiers must be individually controllable or variable with ambient variations—which is rarely the case in practice.

Method 2: Zero Active Sensitivity and Minimum Passive Sensitivity
Here the critical transmission frequencies (i.e., dominant poles) depend only on the passive network elements; the effect of the active elements is negligible. The poles can then be desensitized like those of a purely passive RC network —i.e., by using resistors and capacitors with uniformly equal but opposite temperature coefficients and aging characteristics. Thus

$$S_G^p = S_G^\omega - (4Q^2 - 1)^{-1/2} \cdot S_G^Q \approx 0 \qquad (4\text{-}99a)$$

21. A. Gaash, R. S. Pepper, and D. O. Pederson, Design of integrable desensitized frequency selective amplifiers, *ISSCC Digest Tech. Papers*, February 1966, p. 34.

and

$$\frac{dR}{R} + \frac{dC}{C} \approx 0 \qquad \text{(4-99b)}$$

This is shown qualitatively in Fig. 4-8b.

Typical networks that permit the realization of (4-99) are high-gain negative-feedback types (e.g., class 3) and positive feedback configurations with unity forward gain. In both cases, the pole sensitivity to G can be made arbitrarily small by the use of amplifiers with sufficiently high loop gain. Obviously, the operational amplifier is ideal for this approach. It is, therefore, no coincidence that some of the oldest and best-known methods of active RC filter synthesis used high-gain amplifiers (as the closest available alternative to operational amplifiers) to obtain stable network characteristics in this way.[22]

Method 2 can be very effective in decreasing sensitivity. However, it deprives the designer of the added versatility afforded by the active network parameter, since he is limited in his choice of network configurations to those that are capable of satisfying (4-99a). This limitation is similar to, but by no means as severe as, that specified in method 1. Thus, even here the scope of realizable transmission functions is limited. Furthermore, high-Q networks require a wide spread of resistive or capacitive component values— incompatible both with thin-film and semiconductor integrated-circuit processing techniques.

Method 3: Minimum Active and Passive Sensitivity   Here, as in method 1, the critical transmission frequencies are permitted to depend both on passive and active network elements. However, the active and passive pole displacements are minimized independently, as in method 2. The passive pole displacement is compensated in the usual way, by the use of resistors and capacitors with uniformly equal but opposite drift properties. However, in contrast to the two preceding methods, the active pole displacement is not minimized by placing any constraint on the corresponding pole *sensitivity* but by minimizing drift *in the active element itself.* Thus, in this case (see Fig. 4-8c) the following expressions must apply:

$$\frac{dG}{G} \approx 0 \qquad \text{(4-100a)}$$

and

$$\frac{dR}{R} + \frac{dC}{C} \approx 0 \qquad \text{(4-100b)}$$

22. H. H. Scott, A new type of selective circuit and some applications, *Proc. IRE*, **26**, 226–235 (1938).

The active element can be stabilized (i.e., (4-100a) can be satisfied) by a local negative-feedback network consisting of passive components with tight tracking properties. Since this stabilizing process is much more effective when the available open-loop gain of the active element is as high as possible, it is clear that operational amplifiers are ideal for its implementation. However, since the available closed-loop gain of an amplifier is greatly reduced by individual feedback, more than one amplifier is often required in high-Q applications.

Method 3 has been found very effective in the design of highly selective hybrid integrated networks in which monolithic operational amplifiers have been combined with high-precision tantalum thin-film resistors and capacitors.[23] The temperature coefficients of the tantalum thin-film resistors and capacitors of the frequency-determining networks can be matched closely over a given temperature range, and the gain of the semiconductor amplifiers can be stabilized by tantalum thin-film resistors that track closely. As will be discussed in more detail in Chapter 6, this hybrid integrated technology lends itself particularly well to the method of pole desensitization under discussion.

To summarize, of the three methods of pole desensitization discussed, method 3 is the most practical and affords the greatest design flexibility, assuming that high-quality RC components with compensating drift characteristics are available. *However, both methods 2 and 3 can only be as successful as the quality of the available passive components permits.* With both methods, desensitizing becomes a *technological* as opposed to a circuit-theoretical problem. This puts network sensitivity back into the domain it was in previously, with passive LCR networks.

4.2 PARASITIC EFFECTS IN HYBRID INTEGRATED NETWORKS

In this section we shall point out some of the parasitics that are peculiar to hybrid integrated circuits. Such parasitics must be given full attention because, if overlooked, they can alter the performance of the hybrid integrated equivalent of a discrete-component circuit drastically. We shall, however, exclude those parasitic effects that hybrid integrated circuits have in common with more conventional circuits, such as circuits mounted on printed-wiring boards, since these are well known. Thus, for example, most circuit designers well appreciate the importance of keeping a critical signal path feeding into a low-impedance point as short as possible, particularly if the sheet resistance of the conductor film material is not negligibly low.

23. G. S. Moschytz, FEN-filter design using tantalum and silicon integrated circuits, *Proc. IEEE,* **58**, 550–566 (1970).

4.2.1 The Effects of Parasitic Capacitance

One significant difference between a hybrid integrated circuit and an equivalent conventional circuit mounted on a printed-wiring board is the fact that the dielectric constant of the substrate used in the former may be relatively large. If we recall that tantalum oxide, the dielectric material used in a tantalum thin-film capacitor, typically has a dielectric constant of 22, then it may come as something of a surprise to find that the dielectric constant of a glass substrate may be between 4 and 7, that of a ceramic substrate as high as 11.[24] More typically, the dielectric constant of a commonly used 96% alumina ceramic substrate may be 9, that of 7059 Corning Glass, 7. The implication of these numbers is clear. If conductor paths were deposited on either side of a ceramic substrate a usable, if low-valued, capacitor would result.[25] But even side by side, and over reasonably long lengths, two parallel conductor paths may pick up several, or even tens, of parasitic picofarads that may, in various applications, be highly detrimental to circuit performance. They will be so particularly in those circuits in which high-frequency and high-gain devices are used. Consequently, circuits comprising thin-film components and silicon integrated operational amplifiers are particularly vulnerable to parasitics of this kind. Since this is the circuit category with which we are most concerned here, let us pursue this point a little further.

Parasitic capacitance can affect an operational amplifier most critically in the following ways:

1. *Parasitic capacitance between the input terminals.* This case is most critical with the amplifier in the noninverting mode (see Fig. 4-9a). It can be shown[26] that with C_i equal to only a few picofarads a circuit may become unstable. To avoid this, the conductor paths attached to the amplifier input terminals should be spread apart on a substrate or should, at least, not run parallel to each other over any distance.

2. *Parasitic capacitance between output and inverting input terminals* (see Fig. 4-9b). Instead of obtaining the ideal gain $G_1 = -R_F/R_G$ in the inverting mode, the parasitic capacitance C_F introduces a pole in the gain, namely

$$G_1 \approx -\frac{R_F}{R_G} \frac{1/R_F C_F}{s + (1/R_F C_F)} \tag{4-101}$$

24. R. W. Berry et al., *Thin Film Technology*, Chapter 9, pp. 345ff. Bell Telephone Laboratory Series, Van Nostrand Reinhold, New York, N.Y., 1968.
25. Naturally, the resulting capacitance will still be orders of magnitude less than that of a thin-film capacitor even though the dielectric constant of the latter may be only twice as large. Recall that the capacitance density of a planar capacitor is $C/A = (0.0885/d)\,\varepsilon_r\,[\text{pF/cm}^2]$. For a thin-film capacitor, d (the dielectric thickness) is typically 3000–4000 Å; in the case of the glass or ceramic substrate d will be between 20 and 40 mils.
26. See Chapter 7, Section 7.2.7 of *Linear Integrated Networks: Fundamentals*.

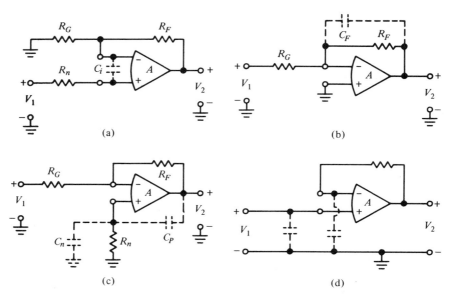

FIG. 4-9. Parasitic capacitances of an operational amplifier: (a) between the input terminals; (b) between the output and inverting input terminal; (c) between the output and the noninverting input terminal; (d) between the input terminals and ground.

The main cause for a parasitic capacitance shunting a film resistor is the meander pattern of the resistor. Of course, in the configuration of Fig. 4-9b, R_G will also have a meander pattern and also cause a shunt capacitance C_G. However, if G_I is large, $R_F C_F$ will be larger than $R_G C_G$, so that a cancellation of the pole $-1/R_F C_F$ will be unlikely. Some of the considerations to be kept in mind, when laying out a film resistor for minimum shunt capacitance, are:[27]

a. The shunt capacitance is minimized by using a thin substrate with a low dielectric constant.

b. The meandering resistor pattern has more shunting capacitance than the single-line pattern. Furthermore, the meandering pattern is usually needed for large resistances where the shunting effect is more critical. Many short meanders will give less shunting capacitance than a few long meanders, since the capacitances between meanders are in series.

c. Decreasing the size of the resistor pattern can decrease the shunting capacitance. It will not change the resistor value, provided the number of squares remains the same.

27. R. W. Berry et al., op. cit., Chapter 11, pp. 467ff.

d. A higher sheet resistance will mean a shorter path and a narrower pattern for the same resistor value; hence, there will be less shunting capacitance. For a given pattern, however, increasing the sheet resistance increases the resistance of the path without changing the capacitance. In this case, the reactance of the shunting capacitance forms a larger part of the total impedance of the resistor.

e. Decreasing the space-width to line-width ratio tends to *increase* the shunting capacitance.

f. The high-frequency performance of a meander resistor is limited either by the series inductance or by the shunting capacitance, depending to a large extent on the value of the resistor. Typically, resistors less than 1000 Ω will be limited at high frequencies by the parasitic inductance, above 1000 Ω by the shunt capacitance. Nevertheless, with adequate care, film resistors on the order of 1000 Ω may be designed and laid out such that their impedance is purely resistive up to, and beyond, 100 MHz.

3. *Parasitic capacitance between ouput and noninverting input terminals* (see Fig. 4-9c). Here the parasitic capacitance C_P feeds the output signal back to the noninverting input terminal. The resulting positive-feedback configuration is potentially unstable. One way of avoiding this condition is to arrange the layout such that the signal path associated with the output terminal is as far as possible from the noninverting input terminal. Another is to use a circuit requiring a large capacitor C_n in parallel with the noninverting source resistor R_n such that the signal in the parasitic positive-feedback path is attenuated effectively.

4. *Parasitic capacitance between the input terminals and ground.* This case is most critical with the amplifier both in the differential-input, and the non-inverting voltage-follower, modes (see Fig. 4-9d). In either case the signal must be equal at both input terminals (common-mode signal). If the parasitic capacitance to ground is different at these two terminals, the common-mode rejection is reduced. Here again the input signal paths should therefore be laid out to be as short and as far from ground—or other paths—as possible, in order to minimize the parasitic capacitance.

4.2.2 The Parasitics of Thin–Film Capacitors

Thin-film capacitors and their characteristics are discussed in some detail in Chapter 6 of *Linear Integrated Networks: Fundamentals.* Here we return to those aspects of thin-film capacitors, their frequency dependence and losses, that have a direct bearing on the performance and the tunability of a hybrid integrated circuit.

Frequency Dependence   It can be shown[28] that because of the distributed nature of a thin-film capacitor, its effective capacitance decreases with increasing frequency. In fact, it has been found,[29] both in theory and experiment, that the frequency dependence within a certain range follows an exponential function of the form

$$C(\omega) = C_0 e^{-\omega \tau_c} \tag{4-102}$$

In other words, a semilogarithmic representation of (4-102) is linear with slope τ_c, as shown qualitatively in Fig 4-10. Measurements have shown that when plotted on semilog paper, the slope τ_c is independent of the capacitance value over a frequency range extending from below 1 kHz up to approximately 20 kHz. Consequently, C_0 and τ_c can be obtained by measuring $C(\omega)$ at two frequencies ω_1 and ω_2; any capacitance value within the "linear" frequency range can then be obtained by interpolation. At higher frequencies $C(\omega)$ must be measured at the frequency at which the capacitor is to be used (e.g., the pole frequency of the corresponding RC network). Naturally, even below 20 kHz, the capacitance is obtained most accurately if measured at the actual frequency of operation.

Typically, a tantalum thin-film capacitor may decrease in value by about 0.3 % from 1 kHz to 10 kHz. If $C(\omega_1)$ and $C(\omega_2)$ are measured to within 0.1 % accuracy, interpolations within the frequency range of linearity will be comparably accurate.

The Losses of Thin-Film Capacitors   The impedance of a thin-film capacitor can be represented by[30]

$$Z = R + \frac{\tan \delta'}{\omega C(\omega)} + \frac{1}{j\omega C(\omega)} \tag{4-103}$$

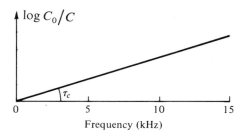

FIG. 4-10. Qualitative semilog plot of the frequency dependent decrease in capacitance of a thin-film capacitor.

28. L. Young, *Anodic Oxide Films* (London and New York: Academic Press, 1961) pp. 159ff.
29. A. R. Morley, The effect of the distributed resistance of the electrodes of thin film capacitors, *Microelectronics and Reliability*, Pergamon Press, Vol. 9, pp. 189–191.
30. See Chapter 6, Section 6.1.3 of *Linear Integrated Networks: Fundamentals*.

where R is the ohmic series resistance caused by the electrodes and the lead-in paths, and tan δ' is the dielectric loss factor of the dielectric material; tan δ' is frequency independent. $C(\omega)$ is the actual (frequency dependent) capacitor value. We can now define an overall loss factor

$$\tan \delta(\omega) = \frac{\text{Re } Z(j\omega)}{\text{Im } Z(j\omega)} = \tan \delta' + \omega RC(\omega) \qquad [4\text{-}104]$$

which, because of the frequency dependence of C, is now no longer linearly dependent on frequency, but dependent in a more complex way. Assuming that $C(\omega)$ is known, then the unknown quantities tan δ' and R can be determined by measuring tan δ at two different frequencies ω_1 and ω_2. From (4-104) we then obtain:

$$\tan \delta' = \frac{\omega_1 C(\omega_1) \tan \delta(\omega_2) - \omega_2 C(\omega_2) \tan \delta(\omega_1)}{\omega_1 C(\omega_1) - \omega_2 C(\omega_2)} \qquad (4\text{-}105\text{a})$$

and

$$R = \frac{\tan \delta(\omega_1) - \tan \delta(\omega_2)}{\omega_1 C(\omega_1) - \omega_2 C(\omega_2)} \qquad (4\text{-}105\text{b})$$

For tantalum thin-film capacitors ranging in value from 1000 to 20,000 pF, the resistor R will typically be in the order of 10 Ω. At low frequencies, say below 3 kHz, tan δ' will therefore be the dominant term in (4-104). This is seen in the set of typical curves for various capacitor values shown in Fig. 4-11.

In the analysis of RC networks with lossy capacitors it is useful to introduce an equivalent capacitor C^* that includes the loss factor tan δ. We

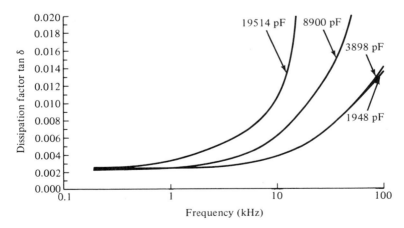

FIG. 4-11. Plot of dissipation factor vs. frequency for various capacitance values of tantalum thin-film capacitors.

obtain this equivalent capacitor $C^*(\omega)$ in terms of the actual capacitor $C(\omega)$ by inspection of (4-103):

$$C^*(\omega) = \frac{C(\omega)}{1 + j(\tan \delta' + \omega R C(\omega))} \tag{4-106}$$

With (4-104) this becomes

$$C^*(\omega) = \frac{C(\omega)}{1 + j \tan \delta} \tag{4-107}$$

Since $\tan \delta \ll 1$ (see Fig. 4-11), (4-107) can be rewritten as

$$C^*(\omega) = \frac{C(\omega)}{1 + j\delta} \tag{4-108}$$

At a given frequency, or over a limited frequency range, C^* and C may be assumed to be independent of frequency. Expanding (4-108) in a Taylor series and retaining, at most, second-order terms we then have

$$C^* = C(1 - j\delta - \delta^2) \tag{4-109}$$

Note that C^* is complex and will therefore affect both amplitude and phase of the corresponding network function. At low frequencies (e.g., below 3 kHz) $\tan \delta'$ is the dominant term in (4-104) and δ may be replaced by δ' in (4-109). Very often only the linear term in (4-109) need be considered, since $\delta' < \delta$ and $\delta \ll 1$. Either way, the effect of the capacitance loss on the amplitude or frequency response is obtained by substituting C^* for C in the transfer function:

$$T(R_i, C_i, j\omega) \rightarrow T(R_i, C_i^*, j\omega) \tag{4-110}$$

where C^*, given by (4-109), is complex.

Another way of gaining insight into the influence of capacitor losses on the performance of a network is to examine their effects on the critical frequencies of the network in the s-plane. The dissipation factor $\tan \delta$ of a capacitor can be written as

$$\tan \delta = \omega r C \tag{4-111a}$$

or as

$$\tan \delta = \frac{g}{\omega C} \tag{4-111b}$$

where r is the equivalent resistor in series, and g the equivalent conductance in parallel with the capacitor. Expressed in terms of $\tan \delta'$ and R of (4-103) we have

$$r(\omega) = R + \frac{\tan \delta'}{\omega C} \tag{4-112a}$$

and since $g = \omega C(\tan \delta)$ we have, with (4-104),

$$g(\omega) = \omega^2 C^2 \left(R + \frac{\tan \delta'}{\omega C} \right) \tag{4-112b}$$

Note that both r and g are functions of frequency. Measuring either value at a particular frequency we are free to characterize the capacitor loss by r or g. Let us do so by the latter. With (4-111b), we obtain for the admittance Y of the lossy capacitor at the frequency ω_0

$$Y = sC + g = C(s + \omega_0 \tan \delta) \tag{4-113}$$

Hence, we can consider the effect of lossy capacitors on a transfer function as that of translation in the s-plane by $\omega_0 \tan \delta$:

$$s \rightarrow s + \omega_0 \tan \delta \tag{4-114}$$

Consequently, all critical frequencies in the s-plane will be shifted to the left by an amount $\omega_0 \tan \delta$ as indicated in Fig. 4-12. Note that we have simplified the situation by assuming equal loss for all capacitors of the network. In practice this will not be the case; however, as will be shown in what follows, taking an average value for $\tan \delta$, some useful insight can be obtained nevertheless as to the overall effect that lossy capacitors will have on the transfer function of a given (active) RC network.

The Effect of Capacitance Losses on a Pole Pair   Let us, for example, consider the effect of lossy capacitors on the Q and undamped natural frequency of a complex conjugate pole pair. We have, for a second-order polynomial $D(s)$,

$$D(s) = (s - p)(s - p^*) = s^2 + \frac{\omega_p}{q_p} s + \omega_p^2 \tag{4-115a}$$

where

$$p = -\sigma + j\omega_c \tag{4-115b}$$

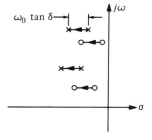

FIG. 4-12. The effect of lossy capacitors on the critical frequencies of a transfer function.

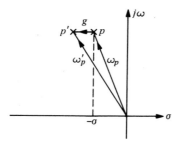

FIG. 4-13. The shifting of a pole due to a lossy capacitor ($g = \omega_p \tan \delta$).

After shifting p by the amount g (see Fig. 4-13), where

$$g = \omega_p \tan \delta \tag{4-116}$$

we obtain the new pole pair p', $(p')^*$ and the new polynomial

$$D'(s) = [(s+g) - p][(s+g) - p^*] = (s - p')[s - (p')^*]$$

$$= s^2 + \frac{\omega'_p}{q'_p} s + (\omega'_p)^2 \tag{4-117}$$

Since

$$D'(s) = D(s+g) = (s+g)^2 + \frac{\omega_p}{q_p}(s+g) + \omega_p^2 \tag{4-118}$$

we can solve for ω'_p and q'_p and obtain

$$\omega'_p = \omega_p \left[1 + \left(\frac{g}{\omega_p}\right)^2 + \frac{1}{2q_p^2} \cdot \frac{g}{\sigma} \right]^{1/2} \tag{4-119a}$$

and

$$q'_p = q_p \frac{\left[1 + \left(\frac{g}{\omega_p}\right)^2 + \frac{1}{2q_p^2}\frac{g}{\sigma} \right]^{1/2}}{1 + \frac{g}{\sigma}} \tag{4-119b}$$

Since in general $g \ll \omega_p$, we obtain for $q_p \gg 1$

$$\omega'_p \approx \omega_p \tag{4-120a}$$

and

$$q'_p \approx \frac{q_p}{1 + \frac{g}{\sigma}} \tag{4-120b}$$

Thus, to first order, the pole frequency remains unaffected by the capacitance loss g, whereas the pole Q, q_p, decreases in inverse proportion to the ratio g/σ. For medium-valued q_p, $g \ll \sigma$ and $q'_p \approx q_p[1 - (g/\sigma)]$.

The Effect of Capacitance Losses on a Twin-T Null   Let us now examine the effect of the capacitance loss g on an *RC* circuit, namely on the twin-*T*. In the balanced state, i.e., with $j\omega$-axis zeros, the transfer functions are

$$T(s) = \frac{s^2 + \omega_N^2}{s^2 + \dfrac{\omega_N}{\hat{q}} s + \omega_N^2} \tag{4-121}$$

and

$$T'(s) = T(s + g) = \frac{s^2 + 2gs + (\omega_N^2 + g^2)}{s^2 + \left(2g + \dfrac{\omega_N}{\hat{q}}\right)s + \left(\omega_N^2 + g^2 + \dfrac{\omega_N g}{\hat{q}}\right)} \tag{4-122}$$

Since $g \ll 1$, the frequency of minimum transmission can still be assumed to be at ω_N. Thus, with $2\hat{\sigma} = \omega_N/\hat{q}$,

$$T(j\omega_N + g) = \frac{j2g\omega_N + g^2}{2(g + \hat{\sigma})j\omega_N + g^2 + g\hat{\sigma}}$$

$$\approx \frac{g}{g + \hat{\sigma}} \approx \frac{g}{\hat{\sigma}} = \frac{2\hat{q}g}{\omega_N} \tag{4-123a}$$

With $q_z = \omega_N/2g$ this can also be written as

$$T(j\omega_N + g) \approx \hat{q}/q_z \tag{4-123b}$$

For a symmetrical twin-*T*, $\hat{q} = \tfrac{1}{4}$ and we obtain, with (4-116),

$$T(j\omega_N + g) = \frac{\tan \delta}{2} \tag{4-124}$$

Furthermore, a change $\Delta(\tan \delta)$ causes the following change in null depth:

$$\Delta T(j\omega_N) = T(j\omega_N + \Delta g) - T(j\omega_N) \approx \frac{\Delta(\tan \delta)}{2} \tag{4-125}$$

According to (4-124), the achievable twin-*T* null depth is proportional to the loss factor of the capacitors used in it. This is less significant for the initial null depth, since the effects of tan δ can virtually be tuned out by appropriate corrections in the resistor values. However, the twin-*T* having been tuned, its null depth will vary with temperature and other ambient effects proportionately with the variations of tan δ as given by (4-125). To reduce these variations two alternatives exist:

1. Use of capacitors with sufficiently low losses so that a large variation in tan δ remains insignificant. Assume, for example, an initial null depth of -60 dB (i.e., $|T(j\omega_N)| = 0.001$) and an average capacitor loss tan $\delta = 0.0002$. Then, according to (4-125), a 100 % increase of tan δ will cause the null depth to increase by only 1 dB, namely to 0.0011.

2. If capacitors with relatively high losses are used, the loss variations due to ambient influences must be small and well controlled. If, again, we assume an initial null depth of -60 dB but now an average capacitor loss $\tan \delta = 0.004$, then a 100% increase in $\tan \delta$ would correspond to an increase in null depth by a factor of three, or of appoximately 10 dB. In this case, to limit the increase in null depth to 1 dB we could tolerate an increase in $\tan \delta$ of only 5%.

4.2.3 The Finite Gain–Bandwidth Product of Operational Amplifiers

The loop gain of an operational amplifier that has been frequency-compensated for a rate of closure of 6 dB/oct (see Fig. 4-14) can be expressed by[31]

$$AB(s) = AB \frac{\Omega}{s + \Omega} \tag{4-126a}$$

where (see Fig. 4-9)

$$B = \frac{R_G}{R_F + R_G} \tag{4-126b}$$

The corresponding closed-loop gain follows as

$$\beta(s) = \alpha \cdot \frac{AB\Omega}{s + \Omega(1 + AB)} = \alpha \frac{AB\Omega}{s + \omega_\alpha} \tag{4-127}$$

where

$$\omega_\alpha = \Omega(1 + AB) \tag{4-128}$$

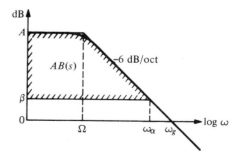

FIG. 4-14. Graphical representation of the loop gain of a frequency-compensated operational amplifier (rate of closure: 6 dB/oct).

31. To avoid confusion we use B for the feedback factor in this chapter instead of the more commonly used β since the latter has been used to denote the closed-loop gain of an operational amplifier in this and previous chapters. Furthermore α is the ideal closed-loop gain of an inverting (α_I) or noninverting (α_N) amplifier. Thus, according to the designations used in this and previous chapters, for an ideal amplifier the closed-loop gain β is equal to α_I or α_N as the case may be. For the justification of equations (4-126) see Chapter 7, Section 7.2.3 of *Linear Integrated Networks: Fundamentals*.

Table 4-3 summarizes the specific expressions obtained for an operational amplifier in the open-loop, the inverting and the noninverting modes, respectively. To guarantee sufficient loop gain the frequency range of operation must be well below ω_α. Then $|AB(j\omega)| \gg 1$ and, since $\alpha \approx 1/B$, (4-127) can be approximated by

$$\beta(s) \approx \alpha \frac{(A\Omega)B}{s + (A\Omega)B} = \alpha \cdot \frac{\omega_\alpha}{s + \omega_\alpha} = \alpha \frac{\omega_g B}{s + \omega_g B} \approx \frac{\omega_g}{s + \omega_\alpha} \qquad [4\text{-}129a]$$

and

$$\omega_\alpha \approx AB\Omega \qquad [4\text{-}129b]$$

TABLE 4-3. THE FREQUENCY-COMPENSATED OPERATIONAL AMPLIFIER IN TYPICAL MODES OF OPERATION

1. Open-Loop Mode:

$$\frac{V_0}{V_i} = -A(s) = -A_0 \frac{\Omega}{s+\Omega} = -A_0 \frac{\dfrac{\omega_g}{A_0}}{s + \dfrac{\omega_g}{A_0}}$$

where $\qquad \omega_g = A_0 \Omega$

2. Inverting Mode

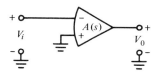

$$\frac{V_0}{V_i} = \beta_I(s) = -\alpha_I \frac{AB\Omega}{s + \Omega(1 + AB)}$$

$$= -\frac{\alpha_I}{1 + \alpha_I} \cdot \frac{\omega_g}{s + \omega_\alpha} = -\alpha_I \frac{\dfrac{\omega_g}{1 + \alpha_I}}{s + \dfrac{\omega_g}{1 + \alpha_I}}$$

where

$$\alpha_I = \frac{R_F}{R_G}; \qquad B = \frac{R_G}{R_F + R_G}; \qquad \omega_g = A\Omega; \qquad \omega_\alpha = \Omega(1 + AB) \approx \frac{\omega_g}{1 + \alpha_I}$$

3. Noninverting Mode

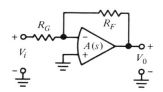

$$\frac{V_0}{V_i} = \beta_N(s) = \alpha_N \cdot \frac{AB\Omega}{s + \Omega(1 + AB)}$$

$$= \frac{\omega_g}{s + \omega_\alpha} = \alpha_N \frac{\dfrac{\omega_g}{\alpha_N}}{s + \dfrac{\omega_g}{\alpha_N}}$$

where

$$\alpha_N = \frac{R_F + R_G}{R_G}; \qquad B = \frac{R_G}{R_F + R_G}; \qquad \omega_g = A\Omega; \qquad \omega_\alpha = \Omega(1 + AB) \approx \omega_g/\alpha_N; \qquad \alpha_N = 1 + \alpha_I$$

where ω_g, the *gain–bandwidth product* of the operational amplifier, is defined by

$$\omega_g = A\Omega \approx \frac{\omega_\alpha}{B}$$

[4-130]

For the compensation scheme assumed here, ω_g corresponds to the unity-gain bandwidth of the amplifier (see Fig. 4-14); at any frequency ω beyond Ω, the magnitude of the amplifier gain is approximated by ω_g/ω.

In order to find the effect of the finite gain–bandwidth product on an active network, we must replace the frequency-independent gain β by $\beta(s)$ in the transfer function of the network. Considering only second-order networks, and the fact that we are interested only in networks in which the pole (and zero) frequencies are independent of gain, this amounts to replacing β by $\beta(s)$ in the expressions for Q and for the constant factor K; thus

$$T(s) = K(\beta, s)\,\frac{N(\beta, s)}{D(\beta, s)} = K(\beta, s) \cdot \frac{s^2 + \dfrac{\omega_z}{q_z(\beta, s)}\,s + \omega_z^2}{s^2 + \dfrac{\omega_p}{q_p(\beta, s)}\,s + \omega_p^2}$$

(4-131)

Let us examine the effect of $\beta(s)$ on q_p, q_z, and K separately.

The Effect of the Finite Gain–Bandwidth Product on the Poles As we know, the relationship between q_p and β differs according to the network class considered. We ought therefore to examine this relationship individually for each class, thereby obtaining the influence of $\beta(s)$ on the poles of the corresponding transfer functions. However, since the pole frequencies ω_p of class 1 and class 2 networks are not independent of gain, we shall not examine these two classes but go directly to class 3 networks.

Negative-Feedback: Class 3 Networks With $v = \hat{q}/q_z$,[32] and assuming open-loop zeros in the LHP, we have from (4-12)

$$q_p = q_z\,\frac{1 + k\beta}{\dfrac{q_z}{\hat{q}} + k\beta}\Bigg|_{k\beta \gg 1} \approx \frac{q_z}{1 + \dfrac{q_z/\hat{q}}{k\beta}}$$

(4-132)

Since this circuit type is used with the operational amplifier in the (frequency-compensated) open-loop mode, we must here replace β by

$$A(s) = \frac{A_0\,\Omega}{s + \Omega} = \frac{\omega_g}{s + \Omega}$$

(4-133)

---

32. We must differentiate here between $q_z(\beta, s)$, the Q of the zeros of $T(s)$ in (4-131), and q_z, the Q of the zeros of the passive RC feedback network. In class 3 networks the latter is a passive RC frequency-rejection network and has complex conjugate zeros.

In (4-132) we then have

$$q_p(s) = \frac{q_z}{1 + \dfrac{q_z(s + \Omega)}{\hat{q}k\omega_g}} \tag{4-134}$$

Substituting (4-134) into the denominator of (4-131) and solving for the new pole frequency and Q, namely $\omega_p + \Delta\omega_p$ and $q_p + \Delta q_p$ we obtain[33]

$$(\omega_p + \Delta\omega_p) = \frac{\omega_p}{\left(1 + \dfrac{1}{\hat{q}}\dfrac{\omega_p}{\omega_g}\right)^{1/2}} \approx \omega_p'\left(1 - \frac{1}{2\hat{q}}\frac{\omega_p}{\omega_g}\right) \tag{4-135a}$$

and, with (4-132)

$$(q_p + \Delta q_p) = q_z\frac{\left(1 + \dfrac{1}{\hat{q}}\dfrac{\omega_p}{\omega_g}\right)^{1/2}}{1 + \dfrac{\Omega}{\omega_g}\dfrac{q_z}{\hat{q}}} = q_p\left(1 + \frac{1}{\hat{q}}\frac{\omega_p}{\omega_g}\right)^{1/2}$$

$$\approx q_p\left(1 + \frac{1}{2\hat{q}}\frac{\omega_p}{\omega_g}\right) \tag{4-135b}$$

Thus

$$-\frac{\Delta\omega_p}{\omega_p} = \frac{\Delta q_p}{q_p} \approx \frac{1}{2\hat{q}}\frac{\omega_p}{\omega_g} \tag{4-136a}$$

and, since $\hat{q} < 0.5$,

$$-\left(\frac{\Delta\omega_p}{\omega_p}\right)_{min} = \left(\frac{\Delta q_p}{q_p}\right)_{min} \approx \frac{\omega_p}{\omega_g} \tag{4-136b}$$

It follows, then, that the effect of the finite gain–bandwidth product on a class 3 single-amplifier network is to increase the pole Q, q_p, and decrease the pole frequency ω_p by an amount proportional to ω_p/ω_g. With a unity-gain bandwidth of the frequency-compensated amplifier equal to 1 MHz, for example, the maximum pole frequency should be less than 10 kHz if $\Delta\omega_p/\omega_p$ and $\Delta q_p/q_p$ are not to exceed 1%.

In the case of a bridged-T, where $\hat{q} = (q_z\gamma)^{-1}$ (see (4-38), we have

$$-\frac{\Delta\omega_p}{\omega_p} = \frac{\Delta q_p}{q_p} \approx \frac{1}{2}q_z\gamma\frac{\omega_p}{\omega_g} \tag{4-137a}$$

where $\gamma = 1 + (C_2/C_1)$ (or $\gamma = 1 + (R_1/R_2)$ if the dual bridged-T is used). Since $\gamma > 1$ and, typically, may equal 2, we have

$$-\left(\frac{\Delta\omega_p}{\omega_p}\right)_{min} = \left(\frac{\Delta q_p}{q_p}\right)_{min} \approx \frac{1}{2}q_z\frac{\omega_p}{\omega_g} \tag{4-137b}$$

---

33. Note that k has been set equal to unity. This is generally accurate in class 3 networks when the feedback network is an unloaded bridged- or twin-T. Furthermore $k = 1$ corresponds to the case of minimum loop gain, as discussed in more detail in Chapter 6.

Thus, the error in ω_p and q_p is between $q_z/2$ and q_z times larger using a bridged-T network than, say, using a twin-T in the feedback network (in which case (4-136a) and (4-136b) are valid). This difference may be substantial, since $q_z \geq q_p$. Thus, again referring to a unity-gain bandwidth of 1 MHz, for $q_p = 10$ the maximum pole frequency for a 1% error should be on the order of 1 kHz.

It should be pointed out that if the approximation $k\beta \gg 1$, which is assumed in (4-132), is not valid, then we have, instead of (4-134),

$$q_p(s) = q_z \frac{s + \Omega + k\omega_g}{\dfrac{q_z}{\hat{q}}(s + \Omega) + k\omega_g} \qquad (4\text{-}138)$$

Substituting this expression into the denominator of (4-131), we obtain a third-order polynomial with a negative real pole approximately given by

$$p_1 \approx -\omega_g\left(1 + \frac{\omega_p}{\omega_g}\frac{1}{\hat{q}}\right) \qquad (4\text{-}139)$$

With p_1, the resulting errors in q_p and ω_p of the complex conjugate pole pair are still basically given by (4-136) except that a negative term, proportional to $(\omega_p/\omega_g)^2 q_p/\hat{q}$ is now added to $\Delta q_p/q_p$. Since only small frequency errors (still given by (4-136a) even when (4-138) is used) can generally be tolerated, ω_p/ω_g must be small and terms proportional to $(\omega_p/\omega_g)^2$ will in most cases be negligible.

Positive Feedback: Class 4 Networks   From (4-21a) we have

$$q_p(s) = \frac{\hat{q}}{1 - k\beta(s)} \qquad (4\text{-}140)$$

Since $\beta(s)$ is the gain of a *noninverting* operational amplifier, it follows from (4-129) that

$$\beta = \beta(s = 0) \approx \alpha_N \qquad (4\text{-}141\text{a})$$

and

$$\beta(s) = \beta\frac{\omega_g B}{s + \omega_g B} = \frac{\omega_g}{s + \omega_\alpha} = \frac{\omega_g}{s + \dfrac{\omega_g}{\beta}} \qquad (4\text{-}141\text{b})$$

where β is the DC gain and B the feedback factor of the amplifier; thus $\beta = 1/B$. Since the calculation of the new pole pair due to the frequency-dependent opamp gain is more complex here, we shall go through it in somewhat more detail than in the negative feedback case.

Let the new denominator of $T(s)$ have the form

$$D_0(s) = s^2 + \frac{\omega_0}{q_0} s + \omega_0^2 \qquad (4\text{-}142)$$

Inserting (4-140) with (4-141b) into the denominator $D(s)$ of (4-131), comparing the result with (4-142), and solving for ω_0 and q_0 then gives us the new pole frequency and Q.

Inserting (4-140) into $D(\beta, s)$ we obtain the third-order polynomial[34]

$$D(\beta, s) = a_3 s^3 + a_2 s^2 + a_1 s + a_0 \qquad (4\text{-}143)$$

where

$$a_3 = \frac{\beta}{\omega_g} \qquad (4\text{-}144a)$$

$$a_2 = 1 + \frac{\omega_p}{\omega_g} \frac{\beta}{\hat{q}} \qquad (4\text{-}144b)$$

$$a_1 = \frac{\omega_p}{q_p} \left(1 + \frac{\omega_p}{\omega_g} \beta q_p \right) \qquad (4\text{-}144c)$$

$$a_0 = \omega_p^2 \qquad (4\text{-}144d)$$

To obtain the roots of (4-143) we reason as follows. The third pole in $D(s)$, which we shall call p_3, is caused by the pole ω_α of $\beta(s)$. But ω_α, and with it ω_g, the gain–bandwidth product of our operational amplifier, must of necessity be much larger than our pole frequency ω_p. If it were not so, the available loop gain of our amplifier would be insufficient to guarantee stable Q, since in the term $\Delta q_p / q_p = (\beta S_\beta^q) \, \Delta A / A^2$, we would have $\beta \approx A$. Consequently $\Delta q_p / q_p$ would be directly proportional to the open-loop gain variations $\Delta A / A$. Designating the other two poles by p_1 and p_2 we can therefore say that

$$p_3 \gg p_1, p_2 \qquad (4\text{-}145)$$

Writing $D(\beta, s)$ in the form

$$D(\beta, s) = (s - p_1)(s - p_2)(s - p_3) \qquad (4\text{-}146)$$

multiplying out, and comparing the coefficients with (4-143) it follows that

$$\frac{a_2}{a_3} = -(p_1 + p_2 + p_3) \qquad (4\text{-}147a)$$

$$\frac{a_1}{a_3} = (p_1 p_2 + p_1 p_3 + p_2 p_3) \qquad (4\text{-}147b)$$

$$\frac{a_0}{a_3} = -p_1 p_2 p_3 \qquad (4\text{-}147c)$$

---

34. Notice that in the negative-feedback case the negative real pole could be computed separately and no third-order polynomial needed to be solved.

With (4-145) we obtain, from (4-147a),[35]

$$p_3 \approx -\frac{a_2}{a_3} = -\omega_g\left(\frac{1}{\beta} + \frac{\omega_p}{\omega_g}\frac{1}{\hat{q}}\right) \tag{4-148}$$

Having obtained p_3, we can now calculate ω_0 and q_0. We have

$$(s - p_3)\left(s^2 + \frac{\omega_0}{q_0}s + \omega_0^2\right) = s^3 + \frac{a_2}{a_3}s^2 + \frac{a_1}{a_3}s + \frac{a_0}{a_3} \tag{4-149}$$

Thus

$$\omega_0 = \omega_p + \Delta\omega_p = \left(-\frac{1}{p_3}\frac{a_0}{a_3}\right)^{1/2} = \frac{\omega_p}{\left(1 + \frac{\omega_p}{\omega_g}\frac{\beta}{\hat{q}}\right)^{1/2}} \tag{4-150}$$

and

$$\frac{\Delta\omega_p}{\omega_p} \approx -\frac{1}{2}\frac{\omega_p}{\omega_g}\frac{\beta}{\hat{q}} \tag{4-151}$$

Similarly

$$\omega_0^2 - \frac{\omega_0}{q_0}p_3 = \frac{a_1}{a_3} \tag{4-152}$$

Solving for q_0 we obtain

$$q_0 = q_p\frac{\omega_0}{\omega_p}\frac{1 + \dfrac{\omega_p}{\omega_g}\dfrac{\beta}{\hat{q}}}{1 + \beta q_p\dfrac{\omega_p^2 - \omega_0^2}{\omega_p\omega_g}} \tag{4-153}$$

and with (4-150)

$$q_0 = q_p + \Delta q_p \approx q_p\frac{\left(1 + \dfrac{\omega_p}{\omega_g}\dfrac{\beta}{\hat{q}}\right)\left(1 + \dfrac{\Delta\omega_p}{\omega_p}\right)}{1 - 2\beta q_p\dfrac{\omega_p}{\omega_g}\left(\dfrac{\Delta\omega_p}{\omega_p}\right)} \tag{4-154}$$

With (4-151) we obtain

$$\frac{\Delta q_p}{q_p} \approx \frac{1}{2}\frac{\omega_p}{\omega_g}\frac{\beta}{\hat{q}} - \left(\beta\frac{\omega_p}{\omega_g}\right)^2\frac{q_p}{\hat{q}} - \frac{1}{2}\left(\frac{\omega_p}{\omega_g}\frac{\beta}{\hat{q}}\right)^2 \tag{4-155a}$$

To first order we therefore find

$$-\frac{\Delta\omega_p}{\omega_p} = \frac{\Delta q_p}{q_p} \approx \frac{\beta}{2\hat{q}}\frac{\omega_p}{\omega_g} \tag{4-155b}$$

35. If, in addition to (4-145), $p_2 \gg p_1$, we also have $p_2 \approx -a_1/a_2$ and $p_1 \approx -a_0/a_1$.

Thus the frequency and Q errors are β times larger than for the negative-feedback case *using a twin-T* in the feedback loop (see (4-136a)). Since the closed-loop gain β is generally close to unity, the errors will, in fact, be almost the same.

To summarize, so far we have found that the pole-frequency and Q errors due to a finite gain-bandwidth product ω_g are essentially the same for negative feedback (class 3) twin-T networks as for positive feedback networks; they are approximately equal in magnitude to the ratio ω_p/ω_g. By contrast, for class 3 bridged-T networks the errors are approximately q_p times larger.

To examine the effects of finite ω_g on some of the building blocks discussed in Chapter 3 the same methods as those used above can be used. We find that, in general, the pole frequency is *decreased*, the pole Q *increased*, by an amount proportional to ω_p/ω_g. As discussed in Chapter 3 (Section 3.3.1), in the case of the biquad, the error in q_p can actually become infinite, i.e., for a certain ratio ω_p/ω_g the circuit oscillates. The reason for the high sensitivity of this circuit to finite ω_g is that the undesirable phase shift causing the error is increased by the number of amplifiers in the feedback loop.[36]

Up to this point we have calculated the initial error in q_p and ω_p that ensues as a result of the finite gain-bandwith product of the operational amplifier used in a second-order network. In high precision networks those errors must be corrected for; this is done by filter tuning as discussed under section 4.3. More important than this initial error, that can be corrected for, is the variation of q_p and ω_p with variations of ω_g *during operation of the circuit* (due to ambient effects such as temperature, aging etc.), since any subsequent correction is generally impractical. As we know, the response of a network is particularly sensitive to variations of the pole frequency ω_p. It can be shown[37] that for circuits for which $V_\beta^{\omega_p}$ (see (4-89)) is very small, or equal to zero, the variation $\Delta\omega_p/\omega_p$ with respect to variations of the gain-bandwith product $\Delta\omega_g/\omega_g$ has the form:

$$\frac{\Delta\omega_p}{\omega_p} = -\frac{1}{2q_p}(\beta S_\beta^{q_p})\frac{\omega_p}{\omega_g} \cdot \left(\frac{\Delta\omega_g}{\omega_g}\right)$$

$$= -\frac{1}{2q_p}\Gamma\frac{\omega_p}{\omega_g}\left(\frac{\Delta\omega_g}{\omega_g}\right) \qquad [4\text{-}156]$$

Note that $\Delta\omega_p/\omega_p$ is inversely proportional to the specified pole Q, q_p and, as in the expressions derived above, to the ratio ω_p/ω_g. Interestingly enough it is also proportional to the gain-sensitivity product Γ (see Section 4.1.2). In fact this is the only term in (4-156) which is dependent on the circuit configuration used. Thus in selecting a circuit with minimum Γ in order to

36. E. A. Faulkner, and J. B. Grimbleby, Active filters and gain-bandwidth product, Electronic Lett,. **6**, 549-550 (1970); also, The effect of amplifier gain-bandwidth product on the performance of active filters, *The Radio and Electronic Engineer*, **43**, 547–552 (1973).
37. A. S. Sedra, and J. L. Espinoza, Sensitivity and frequency limitations of biquadratic active filters, *Proc. 1974 IEEE Int. Symp. on Circuits and Systems*, pp. 645–650, April 1974.

minimize q_p variations with respect to variations in gain, *we are simultaneously ensuring that the variations of ω_p with respect to variations of the gain-bandwidth product are minimal.* We therefore obtain a circuit whose pole pair is optimally desensitized with respect to variations in opamp gain and gain-bandwidth product. This recognition further validates the use of Γ as a figure-of-merit for hybrid integrated networks.

The Effect of the Finite Gain–Bandwith Product on the Zeros In establishing the effect of finite ω_g on the zeros of $T(s)$, we can be brief. We recall that the difference between type I and type II networks is that, for the former, $A(s)$ or $B(s)$ is zero in the bilinear representation

$$T(s) = \frac{A(s) + xB(s)}{U(s) + xV(s)} \tag{4-157}$$

Assuming that x is the gain of the operational amplifier it therefore follows that:

The zeros of a type I network are independent of the nonideal properties of the amplifiers in the feedback loop generating its poles.

In the case of type II networks, the gain $\beta(s)$ will generally be associated both with the Q of the zeros, $q_z(\beta, s)$[38] and with the undamped natural frequency ω_z. In that case analyses similar to those carried out above will be required. Here, though, the effect of $\beta(s)$ on $q_z(\beta, s)$ and ω_z will not depend on the type of feedback used to generate the poles; rather it will depend on the input characteristics of the operational amplifier when used in the differential mode.

Consider, for example, the transfer function of the frequency-rejection network (FRN) shown in Fig. 4-15, derived from the twin-T single-amplifier

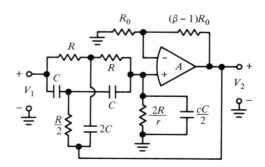

FIG. 4-15. An FRN derived from the TT-SABB.

38. The terminology indicates that we are referring to the q_z of the active network as in (4-131).

building block. For a balanced twin-T, the transfer function is

$$T(s) = \frac{V_2}{V_1} = \frac{\beta(s)}{1 + c} \cdot \frac{s^2 + \dfrac{1}{RC}}{s^2 + \dfrac{4}{RC}\left[\dfrac{1 + (r + c)/4 - \beta(s)}{1 + c}\right]s + \left(\dfrac{1 + r}{1 + c}\right)\dfrac{1}{RC}} \tag{4-158}$$

Clearly, the zeros are unaffected by the characteristics of $\beta(s)$. Consider now the FRN derived from the biquad, as shown in Fig 4-16. With

$$V_1 = K_1(s) \frac{s}{s^2 + \dfrac{\omega_p}{q_p}s + \omega_p^2} V_{in} \tag{4-159a}$$

and

$$V_3 = \frac{K_3(s)}{s^2 + \dfrac{\omega_p}{q_p}s + \omega_p^2} V_{in} \tag{4-159b}$$

and

$$V_0 = \frac{R_2}{R_1 + R_2} V_{in} = K_0 V_{in} \tag{4-159c}$$

we obtain the transfer function

$$T(s) = [aV_0 - (bV_1 + cV_3)]\frac{1}{V_{in}} \tag{4-160}$$

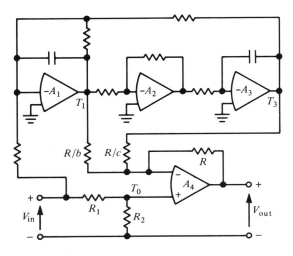

FIG. 4-16. An FRN derived from the biquad.

With equations (4-159) this becomes

$$T(s) = \frac{V_{out}}{V_{in}} = aK_0 \frac{s^2 + \left\{\frac{\omega_p}{q_p} - \frac{bK_1(s)}{aK_0}\right\}s + \left\{\omega_p^2 - \frac{cK_3(s)}{aK_0}\right\}}{s^2 + \frac{\omega_p}{q_p}s + \omega_p^2} \qquad (4\text{-}161)$$

where $a = 1 + b + c$ and $K_1(s)$ and $K_3(s)$ are amplifier-dependent gain functions. The dependence of the zeros on the gain characteristics is evident by inspection. A similar dependence of FRN zeros on amplifier characteristics would be found using the single-amplifier biquad (see Chapter 3, Section 3.1.1). By contrast, the single-amplifier and multiple-amplifier circuits using FENs (see Chapter 3, Sections 3.1.1 and 3.3.2), being type I, provide finite zeros that depend on the passive components only.

The Effect of the Finite Gain–Bandwidth Product on the Constant K
In most of the second-order filter building blocks considered for practical use, the gain $\beta(s)$ is linearly related to $K(s)$ (now no longer a constant) by a multiplicative constant factor $\overline{K}$. Thus, for example, in (4-158), $\overline{K} = 1/(1 + c)$. With (4-129a) we then have

$$K(s) = \overline{K}\beta(s) = \overline{K}\frac{\omega_g}{s + \omega_\alpha} \qquad (4\text{-}162)$$

The pole $\omega_\alpha = B\omega_g$, where B is the feedback factor of the amplifier, will be smaller by the factor B than ω_g. Naturally the closer ω_α is to the pole frequency ω_p, the more it will influence the transfer function $T(s)$ by introducing a parasitic third pole in what is supposed to be a second-order function. B is inversely proportional to the closed-loop gain β required; for example, in the configuration of Fig. 4-15, $B = 1/\beta$. This, of course, is also evident from the representation in Fig. 4-14. Thus networks requiring low closed-loop gain, such as positive-feedback networks, will be less susceptible to this form of error in the transfer function than those requiring high gain.

4.3 THE TUNING OF HYBRID INTEGRATED NETWORKS

The previous section on parasitics in hybrid integrated networks (4.2) was intended to demonstrate some of the many unpredictable effects causing a circuit, based on equations associated with ideal components, to stray widely from the desired response. Quite apart from the parasitic effects mentioned, both passive components and active devices have tolerances on their initial values, all of which contribute to the deviation of a practical network from its nominal characteristic. Naturally, the component tolerances can be tightened arbitrarily, but the resulting increase in cost of the individual components is generally far beyond the reach of an economically feasible system. Thus, just as in discrete systems, it will generally be necessary to adjust or tune a processed circuit to value, before it meets the specifications for which it was designed. This tuning procedure, which will generally be the last in a series of manufacturing steps before encapsulation or packaging, will permit the tolerances specified in the preceding steps to be relaxed, thereby resulting in a cost reduction of the final circuit.

In this section we shall discuss some of the most important aspects related to the tuning of linear networks in general, and of those in hybrid integrated form in particular. We shall first discuss the relationship between component variations and network characteristics by way of the sensitivity matrix. We shall then assess various methods and criteria for the tuning of second-order networks, go on to discussing means of testing whether a tuned filter, consisting of cascaded building blocks, meets the specified initial tolerances, and, finally, consider some of the constraints imposed on the tuning procedure by hybrid integrated circuits.

4.3.1 The Root–Sensitivity Matrix[39]

In tuning any type of linear network it is of interest to know how a particular component x_i affects a particular pole (or zero) of the corresponding network function. Once we have obtained the pole (or zero) variation dp/p (dz/z) it is simple enough, in the case of the all-important second-order network, to associate the real and imaginary parts with measurable (and tunable) frequency and Q variations respectively.

The Dependence of Root Displacements on Coefficient Increments

Let us consider the nth order polynomial

$$P(s) = K \prod_{i=1}^{n} (s - p_i) = \sum_{j=0}^{n} a_j s^j \qquad (4\text{-}163)$$

where p_i are the roots, a_j the coefficients, of $P(s)$. We require the p_i to be distinct roots of $P(s)$, other than that they may be real or complex. In the latter case, of course, to each p_i there belongs a corresponding complex conjugate p_i^*. If, now, we let each coefficient a_j take on an incremental value Δa_j, then each root p_i will be perturbed by an amount Δp_i. Then we obtain, from (4-163)

$$(K + \Delta K) \prod_{i=1}^{n} [(s - p_i) - \Delta p_i] = \sum_{j=0}^{n} (a_j + \Delta a_j)s^j \qquad (4\text{-}164)$$

39. Much of the material in this section is based on the papers by L. P. Huelsman, Matrix analysis of network sensitivities, *Proc. Nat. Electronics Conf.* **19**, 1–5 (1963); and G. Martinelli, On the matrix analysis of network sensitivities, *Proc. IEEE* **54**, 72–73 (1966).

With no loss in generalization we can assume that $K \equiv a_n \equiv 1$, in which case $\Delta a_n = \Delta K = 0$. Then (4-164) can be rewritten as

$$P(s) - \sum_{i=1}^{n} Q_i(s) \cdot \Delta p_i + \varepsilon(s) = P(s) + \sum_{j=0}^{n-1} \Delta a_j s^j \qquad (4\text{-}165)$$

where

$$Q_i(s) = \frac{P(s)}{s - p_i} \qquad (4\text{-}166)$$

and $\varepsilon(s)$ is a polynomial involving higher than first-order terms of Δp_i. Neglecting $\varepsilon(s)$, (4-165) becomes

$$-\sum_{i=1}^{n} Q_i(s) \Delta p_i = \sum_{j=0}^{n-1} \Delta a_j s^j \qquad (4\text{-}167)$$

If, now, we let $s = p_k$, then all terms in the summation on the left of (4-167) will be zero, except the kth one. Thus, we obtain the perturbation of the kth root Δp_k, caused by incremental changes of *all* the coefficients a_j, from (4-167), as

$$\Delta p_k = -\frac{1}{Q_k(p_k)} \sum_{j=0}^{n-1} \Delta a_j (p_k)^j \qquad (4\text{-}168)$$

The perturbation Δp_k resulting from the variation of only the jth coefficient a_j is then

$$\Delta p_k = -\frac{\Delta a_j (p_k)^j}{Q_k(p_k)} \qquad (4\text{-}169)$$

Assuming infinitesimal increments, we can now define the sensitivity of the kth root to variations of the jth coefficient as

$$q_{kj} = \frac{dp_k}{da_j} = -\frac{(p_k)^j}{Q_k(p_k)} \qquad (4\text{-}170)$$

and obtain a matrix of the form

$$
\begin{bmatrix} dp_1 \\ dp_2 \\ \vdots \\ dp_n \end{bmatrix}
=
\begin{bmatrix}
-\dfrac{1}{Q_1(p_1)} & -\dfrac{p_1}{Q_1(p_1)} & -\dfrac{p_1^2}{Q_1(p_1)} & \cdots & -\dfrac{(p_1)^{n-1}}{Q_1(p_1)} \\[2ex]
-\dfrac{1}{Q_2(p_2)} & -\dfrac{p_2}{Q_2(p_2)} & -\dfrac{p_2^2}{Q_2(p_2)} & \cdots & -\dfrac{(p_2)^{n-1}}{Q_2(p_2)} \\[2ex]
\vdots & \vdots & \vdots & & \vdots \\[1ex]
-\dfrac{1}{Q_n(p_n)} & -\dfrac{p_n}{Q_n(p_n)} & -\dfrac{p_n^2}{Q_n(p_n)} & \cdots & -\dfrac{(p_n)^{n-1}}{Q_n(p_n)}
\end{bmatrix}
\begin{bmatrix} da_0 \\ da_1 \\ \vdots \\ da_{n-1} \end{bmatrix}
$$

$$(4\text{-}171)$$

Thus we have an n-row column vector $[\Delta p]$ with elements dp_i, an n-row column vector $[\Delta a]$ with elements da_j, and an $n \times n$ matrix $[Q]$ whose elements are defined by (4-170). This can be written in the matrix relationship

$$[\Delta p] = [Q][\Delta a] \qquad [4\text{-}172]$$

As an example, consider the third-degree polynomial

$$P_3(s) = s^3 + a_2 s^2 + a_1 s + a_0 = (s - p_1)(s - p_2)(s - p_3) \qquad (4\text{-}173)$$

The terms $Q_i(s)$ follow from (4-166) as

$$Q_1(s) = (s - p_2)(s - p_3)$$
$$Q_2(s) = (s - p_1)(s - p_3)$$
$$Q_3(s) = (s - p_1)(s - p_2)$$

and from (4-168) we obtain, for $k = 1$,

$$\Delta p_1 = -\frac{1}{(p_1 - p_2)(p_1 - p_3)}(\Delta a_0 + \Delta a_1 p_1 + \Delta a_2 p_1^2) \qquad (4\text{-}174)$$

Proceeding in this way for $k = 2$ and 3, we obtain the matrix

$$
\begin{bmatrix} dp_1 \\ dp_2 \\ dp_3 \end{bmatrix} =
\begin{bmatrix}
-\dfrac{1}{(p_1 - p_2)(p_1 - p_3)} & -\dfrac{p_1}{(p_1 - p_2)(p_1 - p_3)} & -\dfrac{p_1^2}{(p_1 - p_2)(p_1 - p_3)} \\[2ex]
-\dfrac{1}{(p_2 - p_1)(p_2 - p_3)} & -\dfrac{p_2}{(p_2 - p_1)(p_2 - p_3)} & -\dfrac{p_2^2}{(p_2 - p_1)(p_2 - p_3)} \\[2ex]
-\dfrac{1}{(p_3 - p_1)(p_3 - p_2)} & -\dfrac{p_3}{(p_3 - p_1)(p_3 - p_2)} & -\dfrac{p_3^2}{(p_3 - p_1)(p_3 - p_2)}
\end{bmatrix}
\begin{bmatrix} da_0 \\ da_1 \\ da_2 \end{bmatrix}
$$

$$(4\text{-}175)$$

The matrix equation (4-172) relates the increments of the roots of $P(s)$ to incremental changes in its coefficients. Our objective, however, is to relate the root increments to the *component* variations which, in turn, cause the co-efficients to vary by the amounts Δa_j. Thus, we must now find a relationship between the coefficient increments Δa_j and the component variations Δx_r causing them.

The Dependence of Coefficient Increments on Component Variations
For this purpose, we assume that each coefficient a_j is a function of m components x_r:

$$a_j = f_j(x_1, x_2, \ldots x_r \ldots x_m) \tag{4-176}$$

We now assume that all the components x_r vary by some increment Δx_r; then there will be a corresponding variation Δa_j in the jth coefficient of $P(s)$. Retaining only first-order terms we obtain

$$\Delta a_j = \sum_{r=1}^{m} \frac{\partial}{\partial x_r} f_j(x_1, \ldots, x_m) \Delta x_r = \sum_{r=1}^{m} f_{jr} \Delta x_r \tag{4-177}$$

Since we have n coefficients a_j, there will be n equations relating the changes in the coefficients a_j to the component changes Δx_r. Assuming again that $a_n = 1$, we therefore have

$$\begin{bmatrix} \Delta a_0 \\ \Delta a_1 \\ \vdots \\ \Delta a_{n-1} \end{bmatrix} = \begin{bmatrix} f_{01} & f_{02} & \cdots & f_{0m} \\ f_{11} & f_{12} & \cdots & f_{1m} \\ \vdots & \vdots & & \vdots \\ f_{(n-1)1} & f_{(n-1)2} & \cdots & f_{(n-1)_m} \end{bmatrix} \begin{bmatrix} \Delta x_1 \\ \Delta x_2 \\ \vdots \\ \Delta x_m \end{bmatrix} \tag{4-178}$$

Thus we now have an n-row column vector $[\Delta a]$ whose elements are Δa_j, an m-row column vector $[\Delta x]$ whose elements are Δx_r, and an $n \times m$ matrix $[F]$ whose elements are defined by (4-177). We can therefore write in abbreviated form

$$[\Delta a] = [F][\Delta x] \tag{4-179}$$

The Dependence of Root Displacements on Component Variations
We can now combine (4-172) and (4-179) to obtain a relationship between the pole displacements and the component variations causing them, namely

$$[\Delta p] = [Q][F][\Delta x] \tag{4-180}$$

This expression is in terms of absolute displacements Δp and variations Δx. In practice, however, percentage component changes as well as relative pole displacements are more useful; we must therefore normalize the column vectors $[\Delta p]$ and $[\Delta x]$.

The percentage column vector $[(\Delta x/x)]$ is related to $[\Delta x]$ by a diagonal $m \times m$ normalizing matrix $[X]$ as follows:

$$\begin{bmatrix} \Delta x_1 \\ \vdots \\ \Delta x_r \\ \vdots \\ \Delta x_m \end{bmatrix} = \begin{bmatrix} x_1 & & & & 0 \\ & \ddots & & & \\ & & x_r & & \\ & & & \ddots & \\ 0 & & & & x_m \end{bmatrix} \begin{bmatrix} \Delta x_1/x_1 \\ \vdots \\ \Delta x_r/x_r \\ \vdots \\ \Delta x_m/x_m \end{bmatrix} \tag{4-181}$$

Thus

$$[\Delta x] = [X]\left[\frac{\Delta x}{x}\right]$$

(4-182)

Similarly, the relative pole variations are related to the $[\Delta p]$ column vector by the $n \times n$ diagonal normalizing matrix $[P]$ such that

$$\begin{bmatrix} \Delta p_1 \\ \vdots \\ \Delta p_k \\ \vdots \\ \Delta p_n \end{bmatrix} = \begin{bmatrix} p_1 & & & 0 \\ & \ddots & & \\ & & p_k & \\ & & & \ddots \\ 0 & & & p_n \end{bmatrix}\begin{bmatrix} \Delta p_1/p_1 \\ \vdots \\ \Delta p_k/p_k \\ \vdots \\ \Delta p_n/p_n \end{bmatrix}$$

(4-183)

Thus

$$[\Delta p] = [P]\left[\frac{\Delta p}{p}\right]$$

(4-184)

Substituting (4-182) and((4-184) into (4-180) we obtain

$$\left[\frac{\Delta p}{p}\right] = [P]^{-1}[Q][F][X]\left[\frac{\Delta x}{x}\right]$$

[4-185]

We can now define the sensitivity matrix

$$[S] = [P]^{-1}[Q][F][X]$$

[4-186]

which relates the relative pole displacements $\Delta p/p$ to the percentage change in the component values x_r. From (4-1) the individual terms s_{kr} of this root matrix are related to the corresponding root frequency and Q by

$$s_{kr} = S_{x_r}^{p_k} = S_{x_r}^{\omega p_k} - \frac{j}{\sqrt{4q_{p_k}^2 - 1}} S_{x_r}^{q_{p_k}}$$

[4-187]

Thus we obtain the frequency and Q sensitivities to variations in a component x_r from the real and imaginary parts of s_{kr} as follows:

$$S_{x_r}^{\omega p_k} = \text{Re } s_{kr}$$

[4-188a]

$$S_{x_r}^{q_{p_k}} = -\sqrt{4q_{p_k}^2 - 1} \text{ Im } s_{kr}$$

[4-188b]

EXAMPLE: As an example of the root sensitivity matrix, consider the denominator polynomial of the class 4 bandpass network shown in Fig. 4-17. The transfer function is given by

$$T(s) = \frac{V_2}{V_1} = \frac{\beta \dfrac{G_1}{C_1} s}{s^2 + \left[\dfrac{G_1 + G_2 + G_3}{C_1} + \dfrac{G_3}{C_2} - \beta\dfrac{G_2}{C_1}\right]s + \dfrac{G_3(G_1 + G_2)}{C_1 C_2}}$$

(4-189)

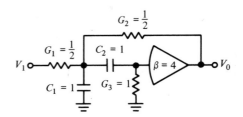

FIG. 4-17. A class 4 bandpass network.

and

$$a_0 = \frac{G_3(G_1 + G_2)}{C_1 C_2} \tag{4-190}$$

$$a_1 = \frac{G_1 + G_2 + G_3}{C_1} + \frac{G_3}{C_2} - \beta \frac{G_2}{C_1} \tag{4-191}$$

Calculating $[F]$ according to (4-177) and (4-178) we obtain

$$\begin{bmatrix} \Delta a_0 \\ \\ \Delta a_1 \end{bmatrix} = \begin{bmatrix} \dfrac{G_3}{C_1 C_2} & \dfrac{G_3}{C_1 C_2} & \dfrac{G_1 + G_2}{C_1 C_2} \\ \\ \dfrac{1}{C_1} & \dfrac{1}{C_1}(1 - \beta) & \dfrac{1}{C_1} + \dfrac{1}{C_2} \end{bmatrix}$$

$$\begin{bmatrix} -\dfrac{G_3(G_1 + G_2)}{(C_1)^2 C_2} & -\dfrac{G_3(G_1 + G_2)}{C_1 (C_2)^2} & 0 \\ \\ -\dfrac{(G_1 + G_2 + G_3) + \beta G_2}{(C_1)^2} & -\dfrac{G_3}{(C_2)^2} & -\dfrac{G_2}{C_1} \end{bmatrix} \begin{bmatrix} \Delta G_1 \\ \Delta G_2 \\ \Delta G_3 \\ \Delta C_1 \\ \Delta C_2 \\ \Delta \beta \end{bmatrix} \tag{4-192}$$

Calculating $[Q]$ from (4-170) and (4-171) we obtain

$$\begin{bmatrix} dp_1 \\ \\ dp_2 \end{bmatrix} = \begin{bmatrix} -\dfrac{1}{p_1 - p_2} & -\dfrac{p_1}{p_1 - p_2} \\ \\ -\dfrac{1}{p_2 - p_1} & -\dfrac{p_2}{p_2 - p_1} \end{bmatrix} \begin{bmatrix} da_0 \\ \\ da_1 \end{bmatrix} \tag{4-193}$$

and from (4-182) and (4-183)

$$[X] = \begin{bmatrix} G_1 & & & & & \\ & G_2 & & & 0 & \\ & & G_3 & & & \\ & & & C_1 & & \\ & 0 & & & C_2 & \\ & & & & & \beta \end{bmatrix} \tag{4-194}$$

and

$$[P] = \begin{bmatrix} p_1 & 0 \\ 0 & p_2 \end{bmatrix}; \quad [P]^{-1} = \frac{1}{p_1 p_2} \begin{bmatrix} p_2 & 0 \\ 0 & p_1 \end{bmatrix} \qquad (4\text{-}195)$$

For the nominal values given in Fig. 4-17, the denominator of (4-189) becomes $s^2 + s + 1$, i.e., $a_0 = a_1 = 1$ and $p_{1,2} = -\frac{1}{2} \pm \frac{1}{2}\sqrt{3}$. Furthermore, we obtain the matrices necessary to calculate $[S]$ as follows:

$$[P]^{-1} = -2 \begin{bmatrix} -\dfrac{1}{2} - j\dfrac{\sqrt{3}}{2} & 0 \\ 0 & -\dfrac{1}{2} + j\dfrac{\sqrt{3}}{2} \end{bmatrix} \qquad (4\text{-}196a)$$

$$[Q] = \begin{bmatrix} j\dfrac{\sqrt{3}}{3} & -\dfrac{1}{2}\left(1 + j\dfrac{\sqrt{3}}{3}\right) \\ -j\dfrac{\sqrt{3}}{3} & -\dfrac{1}{2}\left(1 - j\dfrac{\sqrt{3}}{3}\right) \end{bmatrix} \qquad (4\text{-}196b)$$

$$[F] = \begin{bmatrix} 1 & -3 & 2 & 0 & -1 & -\dfrac{1}{2} \\ 1 & 1 & 1 & -1 & -1 & 0 \end{bmatrix} \qquad (4\text{-}196c)$$

and

$$[X] = \begin{bmatrix} \dfrac{1}{2} & 1 & & & \\ & \dfrac{1}{2} & 1 & & 0 \\ & & 1 & & \\ & 0 & & 1 & \\ & & & & 4 \end{bmatrix} \qquad (4\text{-}196d)$$

Multiplying out, we obtain the desired result in the form of (4-185):

$$\begin{bmatrix} \dfrac{\Delta p_1}{p_1} \\ \dfrac{\Delta p_2}{p_2} \end{bmatrix} = \begin{bmatrix} \dfrac{1}{4} + j\dfrac{\sqrt{3}}{12} & \dfrac{1}{4} - j\dfrac{7\sqrt{3}}{12} & \dfrac{1}{2} + j\dfrac{\sqrt{3}}{2} \\ \dfrac{1}{4} - j\dfrac{\sqrt{3}}{12} & \dfrac{1}{4} + j\dfrac{7\sqrt{3}}{12} & \dfrac{1}{2} - j\dfrac{\sqrt{3}}{2} \end{bmatrix}$$

$$\begin{bmatrix} -\dfrac{1}{2} + j\dfrac{\sqrt{3}}{6} & -\dfrac{1}{2} - j\dfrac{\sqrt{3}}{6} & -j\dfrac{4\sqrt{3}}{6} \\ -\dfrac{1}{2} - j\dfrac{\sqrt{3}}{6} & -\dfrac{1}{2} + j\dfrac{\sqrt{3}}{6} & +j\dfrac{4\sqrt{3}}{6} \end{bmatrix} \begin{bmatrix} \Delta G_1/G_1 \\ \Delta G_2/G_2 \\ \Delta G_3/G_3 \\ \Delta C_1/C_1 \\ \Delta C_2/C_2 \\ \Delta \beta/\beta \end{bmatrix}$$

$$(4\text{-}197)$$

The frequency and Q sensitivities follow by inspection of the individual terms s_{kr}. For example, from (4-188a),

$$S_{G_1}^{\omega_{P_1}} = S_{G_2}^{\omega_{P_1}} = \frac{1}{2} S_{G_3}^{\omega_{P_1}} = \frac{1}{4} \tag{4-198a}$$

or, from (4-188b),

$$S_{\beta}^{q_{P_1}} = -\sqrt{4q_{P_1}^2 - 1} \left(-\frac{4\sqrt{3}}{6} \right)\Bigg|_{q_{P_1}=1} = 2 \tag{4-198b}$$

(4-198a) can be checked directly from (4-190) where we have, for example,

$$S_{G_1}^{\omega_P} = \frac{1}{2} S_{G_1}^{a_0} = \frac{1}{2} \frac{G_1}{G_1 + G_2} = \frac{1}{4}$$

Similarly, to check (4-198b), we know (see (4-25)) that:

$$S_{\beta}^{q_p} = \frac{q_p}{\hat{q}} - 1 = \left(\frac{q_p}{q_p(\beta = 0)} - 1 \right)_{q_p=1} = \frac{1}{1/3} - 1 = 2$$

In the same way as in (4-198), the ω_p and q_p sensitivities to any other components in the network can be immediately obtained.

To check the validity of the terms in $[S]$ it is useful to remember that

$$\sum S_{G_i}^{\omega_P} = -\sum S_{C_i}^{\omega_P} = 1 \tag{4-199a}$$

and

$$\sum S_{G_i}^{q_P} = \sum S_{C_i}^{q_P} = 0 \tag{4-199b}$$

Thus, for example, the sum of the real parts of the first three terms in the first row of (4-197) must equal unity, the sum of the corresponding imaginary parts must equal zero. Since a_0 or ω_p of (4-189) is independent of β, the real part of the last terms in the two rows of (4-197) must equal zero. Finally, since p_1 and p_2 are complex conjugate, the terms in the second row of (4-197) must be complex conjugate to those in the first row.

Real-Root and Complex-Root-Pair Displacements as Functions of Coefficient Increments or Component Variations   The root-sensitivity matrix (4-186) describes the relative displacements of the roots of a network polynomial $P(s)$ (i.e., the numerator or denominator) as a function of incremental changes of the components x_r. Thus it provides the information necessary to "tune" the poles or zeros of a network by indicating the effects which a slight adjustment of any component x_r will have on a given pole or zero. We will say more about the actual tuning process later.

In some cases, it may be preferable to expand the given polynomial $P(s)$ into its complex conjugate pairs and single roots instead of directly into the product of single roots as in (4-163). Assume that we have l pairs of complex conjugate pairs among the n roots of $P(s)$ and that the remaining roots are real. Then, instead of factoring $P(s)$ as in (4-163), we now have[40]

$$P(s) = \sum_{j=0}^{n} a_j s^j = K \prod_{i=1}^{l} (s^2 + b_{2i-1}s + b_{2i}) \prod_{i=2l+1}^{n} (s + b_i) \qquad (4\text{-}200)$$

where the indexing is such as to specify n real numbers b_i. The first l pairs b_{2i-1}, b_{2i} correspond to the 2σ and ω_p^2 terms of the complex conjugate root pairs given by the form $(s^2 + 2\sigma s + \omega_p^2)$; the remaining $n - 2l$ terms give the location of the negative real roots of $P(s)$. Instead of relating the perturbations Δa_j of the coefficients a_j to pole displacements, as in (4-172) we now relate them to variations Δb_i of the b_i terms in (4-200). Equating the corresponding expressions and retaining only first-order terms, we obtain n expressions of the kind

$$\Delta a_i = \sum_{j=1}^{n} d_{ij} \Delta b_j \qquad (4\text{-}201)$$

where the d_{ij} are functions of the coefficients b_i. Assuming $K = 1$ in (4-200), we therefore have a set of equations

$$\begin{bmatrix} \Delta a_0 \\ \Delta a_1 \\ \vdots \\ \Delta a_{n-1} \end{bmatrix} = \begin{bmatrix} d_{11} & \cdots & d_{1n} \\ \vdots & & \vdots \\ d_{n1} & \cdots & d_{nn} \end{bmatrix} \begin{bmatrix} \Delta b_1 \\ \vdots \\ \Delta b_n \end{bmatrix} \qquad (4\text{-}202)$$

which can be expressed in matrix form as follows:

$$[\Delta a] = [D][\Delta b] \qquad (4\text{-}203)$$

where $[\Delta a]$ is defined as in (4-178), $[D]$ is an $n \times n$ matrix with elements d_{ij} and $[\Delta b]$ is an n-row column vector with elements Δb_i. Instead of finding the relationship between $[(\Delta p/p)]$ and $[\Delta x]$ we are now interested in the relationship between $[\Delta b]$ and $[\Delta x]$. Solving for $[\Delta b]$ we obtain, from (4-203),

$$[\Delta b] = [D]^{-1}[\Delta a] \qquad (4\text{-}204)$$

and with (4-179) we obtain the desired expression,

$$[\Delta b] = [D]^{-1}[F][\Delta x] \qquad (4\text{-}205)$$

Normalizing $[\Delta x]$ as in (4-182), and also $[\Delta b]$ by introducing the $n \times n$ diagonal matrix $[B]$ such that

$$[\Delta b] = [B]\left[\frac{\Delta b}{b}\right] \qquad (4\text{-}206)$$

we obtain the matrix expression

$$\left[\frac{\Delta b}{b}\right] = [B]^{-1}[D]^{-1}[F][X]\left[\frac{\Delta x}{x}\right] \qquad [4\text{-}207]$$

40. L. P. Huelsman, *Theory and Design of Active RC Circuits*, (New York: McGraw-Hill Book Co., 1968), p. 34.

The four matrices on the right-hand side can again be combined into a "sensitivity matrix." Note that this sensitivity representation requires the matrix inversion $[D]^{-1}$, which can be quite complicated, particularly if the degree of $P(s)$ is higher than three. (The inversion of $[B]$, a diagonal matrix, is simpler to begin with, and can be avoided by using (4-206).) Furthermore, the resulting terms $\Delta b_i/b_i$ cannot be directly interpreted as physically measurable quantities (e.g., frequency or Q) as in the case of (4-185). On the other hand, the advantage of this representation lies in the fact that it does not require computations with complex numbers.

Consider, for example, our polynomial $P_3(s)$ in (4-173), which now is factored as follows:

$$P_3(s) = s^3 + a_2 s^2 + a_1 s + a_0 = (s^2 + b_1 s + b_2)(s + b_3)$$

Solving for the Δa_j, corresponding to perturbations Δb_i, we readily obtain the matrix $[D]$:

$$\begin{bmatrix} \Delta a_0 \\ \Delta a_1 \\ \Delta a_2 \end{bmatrix} = \begin{bmatrix} 0 & b_3 & b_2 \\ b_3 & 1 & b_1 \\ 1 & 0 & 1 \end{bmatrix} \begin{bmatrix} \Delta b_1 \\ \Delta b_2 \\ \Delta b_3 \end{bmatrix} \tag{4-208}$$

Inverting $[D]$, we obtain

$$[D]^{-1} = \frac{1}{b_1 b_3 - b_2 - b_3^2} \begin{bmatrix} 1 & -b_3 & b_1 b_3 - b_2 \\ b_1 - b_3 & -b_2 & b_2 b_3 \\ -1 & b_3 & -b_3^2 \end{bmatrix} \tag{4-209}$$

Assuming that the elements b_i are functions of the components x_r, the corresponding matrix $[F]$ can be obtained and, with the normalizing diagonal matrices $[B]^{-1}$ and $[X]$, the sensitivity matrix corresponding to (4-207) results.

Besides its use for tuning purposes, the process of obtaining the root-sensitivity matrix as given by (4-186), as well as the representation given by (4-207), readily enables us to acquire the root sensitivities of networks of higher than second order. This includes such networks as the general twin-T and the third-order networks discussed in Chapter 3, Section 3.1.2. It is left as an exercise for the reader to derive the third-order sensitivity expressions given in Chapter 1 (section 1.6) by the methods described above.

4.3.2 The General Sensitivity Matrix

The root-sensitivity matrix described above is useful for the purpose of positioning individual roots (poles or zeros) of a network function in the s-plane. Often, however, a more comprehensive measure of network performance is desirable, e.g., the amplitude or phase response over a given frequency range. In such a situation it may be rather cumbersome (although possible) to piece together the relevant tuning information from the pole and zero sensitivity matrix, and it may be useful to define a general sensitivity matrix[41] that describes the sensitivity of a desired network parameter F_j to an incremental change in a component x_r. It will have the form

$$
\begin{bmatrix} \Delta F_1/F_1 \\ \vdots \\ \Delta F_t/F_t \end{bmatrix} = \begin{bmatrix} \Sigma_{11} & \cdots & \Sigma_{1m} \\ \vdots & & \vdots \\ \Sigma_{t1} & \cdots & \Sigma_{tm} \end{bmatrix} \begin{bmatrix} \Delta x_1/x_1 \\ \vdots \\ \Delta x_m/x_m \end{bmatrix}
\tag{4-210}
$$

where

$$
\Sigma_{jr} = S_{x_r}^{F_j}
\tag{4-211}
$$

In abbreviated matrix form, we have

$$
\begin{bmatrix} \dfrac{\Delta F}{F} \end{bmatrix} = \begin{bmatrix} \Sigma \end{bmatrix} \begin{bmatrix} \dfrac{\Delta x}{x} \end{bmatrix}
\tag{4-212}
$$

The elements Σ_{jr} here represent real numbers, in contrast to the complex root displacements of the root-sensitivity matrix. Each number indicates how strongly a relative component change $\Delta x_r/x_r$ affects the associated relative parameter $\Delta F_j/F_j$. A strong dependence of F_j on a component x_r will result in a large number Σ_{jr}. A desired change in F_j can then be made by adjusting x_r. If, on the other hand, F_j is independent of x_r, then $\Sigma_{jr} = 0$ and x_r may be used to adjust another performance criterion F_k, without affecting F_j.

In the discussion of the characteristics of an ideal filter building block in the introduction to Chapter 3, the desirability was pointed out of having one adjustable resistor for each network parameter (remember, capacitors cannot be fine-tuned). Thus, for example, in a general second-order building block a separate resistor would, ideally, be available for the noninteractive adjustment of each of the five parameters K, ω_z, q_z, ω_p, and q_p. Let us now consider how the capability of noninteractive tuning is expressed in the corresponding sensitivity matrix. If two quantities, say, ω and q, are independently adjustable by R_ω and R_q, respectively, then

$$
S_{R_\omega}^{\omega} = k_1 \neq 0
\tag{4-213a}
$$

$$
S_{R_\omega}^{q} = 0
\tag{4-213b}
$$

$$
S_{R_q}^{\omega} = 0
\tag{4-213c}
$$

$$
S_{R_q}^{q} = k_2 \neq 0
\tag{4-213d}
$$

The corresponding sensitivity matrix according to (4-210) has the form

$$
\begin{bmatrix} \Delta\omega/\omega \\ \Delta q/q \end{bmatrix} = \begin{bmatrix} k_1 & 0 \\ 0 & k_2 \end{bmatrix} \begin{bmatrix} \Delta R_\omega/R_\omega \\ \Delta R_q/R_q \end{bmatrix}
\tag{4-214}
$$

---

41. Much of the material in this section is based on J. J. Golembeski and E. S. Mitchell, unpublished Bell Telephone Laboratories memorandum.

It therefore follows that:

the sensitivity matrix of a noninteractively tunable network is a diagonal matrix. The off-diagonal elements are all zero, since each circuit element controls only one specification; the sequence in which the elements are tuned is unimportant.

In practice the individual parameters of a network will rarely be tunable independently; the off-diagonal elements of the sensitivity matrix will not be zero. What very often does occur, though, is that, although each circuit element will influence more than one network parameter, a particular sequence of tuning steps exists which permits a single noniterative series of tuning operations to meet all performance specifications. Since it is only unilaterally interactive, such a tuning series progressively achieves the desired overall response in one pass. Each tuning step in the series leaves all previously tuned specifications F_j unchanged. Thus, in the kth step, F_k is adjusted to specification by x_k, whereby any so far untuned parameters $F_q(q > k)$ may also be affected, but all F_j $(j < k)$ remain unchanged. The corresponding sensitivity matrix can be represented by the form

$$\begin{bmatrix} \Delta F_1/F_1 \\ \vdots \\ \Delta F_n/F_n \end{bmatrix} = \begin{bmatrix} \Sigma_{11} & 0 & 0 & \cdots & 0 \\ \Sigma_{21} & \Sigma_{22} & 0 & \cdots & 0 \\ \Sigma_{31} & \Sigma_{32} & \Sigma_{33} & 0 & \cdots & 0 \\ \vdots & \vdots & & & \vdots \\ \Sigma_{n1} & \Sigma_{n2} & & \cdots & & \Sigma_{nn} \end{bmatrix} \begin{bmatrix} \Delta x_1/x_1 \\ \vdots \\ \Delta x_n/x_n \end{bmatrix} \quad (4\text{-}215)$$

It follows that:

The sensitivity matrix of a network that is tunable by a single, noniterative series of tuning steps is a triangular matrix. For each network parameter F_k there is a tuning element x_k that leaves all previously tuned parameters F_j $(j < k)$ unaffected. The tuning sequence is critical; it results directly from the sensitivity matrix after the latter has been arranged in triangular form.

In many cases even the triangular form of the sensitivity matrix does not exist; each component x_k affects each parameter F_k in whatever order the tuning steps are carried out. The tuning steps are interactive; the tuning procedure an iterative one. At best, in this situation, some sensitivities will be smaller than others. By arranging the corresponding sensitivity matrix such that the elements to the right of the diagonal are relatively small and decrease in value as the row progresses past the diagonal, a noniterative tuning sequence can be approximated for a given network. By selecting the tuning sequence according to the matrix rearranged in this way, the number of iterations necessary to overcome the interaction between components can at least be minimized.

EXAMPLE: AN ACTIVE BANDPASS NETWORK   As an example of the sensitivity matrix, consider once more the circuit in Fig. 4-17 and the corresponding transfer function given by (4-189). Let us assume that with this network we are to realize the function

$$T(s) = K \frac{s}{s^2 + \frac{\omega_p}{q_p} s + \omega_p^2} \tag{4-216}$$

where K, ω_p, and q_p are specified. By a comparison of coefficients between (4-189) and (4-216) we can, of course, readily obtain design equations for the network. However, to overcome the parasitic effects and component tolerances, some fine adjustment will very likely be necessary. In this case we need to know the relationships among K, ω_p, and q_p and the circuit components, in order to select three components with which the fine tuning can be carried out—noniteratively if possible. Since we are assuming hybrid integrated networks here, we constrain ourselves to tuning with resistors only. By inspection of the pole matrix (4-197) and utilization of the relationships given by (4-188a) and (4-188b) we can write down the ω_p and q_p sensitivities by inspection. The K sensitivities are also readily obtained, and the initial sensitivity matrix results as follows:

$$\begin{bmatrix} \Delta K/K \\ \Delta\omega_p/\omega_p \\ \Delta q_p/q_p \end{bmatrix} = \begin{bmatrix} 1 & 0 & 0 & 1 \\ 0.25 & 0.25 & 0.5 & 0 \\ -0.25 & 1.75 & -1.5 & 2 \end{bmatrix} \begin{bmatrix} \Delta G_1/G_1 \\ \Delta G_2/G_2 \\ \Delta G_3/G_3 \\ \Delta\beta/\beta \end{bmatrix} \tag{4-217}$$

Since we only require three tuning components for our three parameters, we can select those three of the four in (4-217) that most closely approximate a 3×3 triangular matrix. (Needless to say, there is no way of obtaining a diagonal matrix with the values in (4-217).) Since ω_p is influenced twice as much by G_3 than by G_1 or G_2, we retain G_3 in the matrix. Furthermore, since only G_1 of the three components G_1, G_2, and G_3 affects K, we retain it and obtain the final sensitivity matrix

$$\begin{bmatrix} \Delta K/K \\ \Delta\omega_p/\omega_p \\ \Delta q_p/q_p \end{bmatrix} = \begin{bmatrix} 1 & 0 & 1 \\ 0.25 & 0.5 & 0 \\ -0.25 & -1.5 & 2 \end{bmatrix} \begin{bmatrix} \Delta G_1/G_1 \\ \Delta G_3/G_3 \\ \Delta\beta/\beta \end{bmatrix} \tag{4-218}$$

Because of the dependence of K on β the matrix is not triangular. Nevertheless, the sequence of tuning steps is clearly evident and the consequences of nonperfect triangularity also clear. We first tune K with G_1, then ω_p with G_3. This second step is noninteractive with the first; K is unaffected by G_3. In our third step we adjust q_p with β; ω_p is not thereby affected, but K is, by half as much as q_p itself. If K is to be accurately realized, an iterative procedure must ensue, wherein all three components must be successively corrected for until the desired result is obtained. However, in this, as in most other cases, an error in K only affects the overall gain level of the frequency characteristic and the iterative tuning steps will be given up, leaving only the three-step procedure to carry out. The more important parameters ω_p and q_p can thereby be obtained arbitrarily accurately in a noniterative process, and the error in K will, at most, amount to a few percent. Should it amount to more, it can be compensated for by a flat gain stage in cascade with the network.

One further point should be made. Note that the ω_p sensitivities are all positive. (Taken with respect to resistors instead of conductors the sensitivities would, of course, all be negative.) This points to the physical reality that an increase in any resistor in the network (as well as any capacitor) invariably results in a decrease in a critical frequency. Since film resistors can generally be adjusted in only one (the increasing) direction, the impossibility of correcting for an overshoot in the frequency adjustment should immediately be clear. So should be the problems associated with very tight frequency tolerances; the tighter these tolerances are, the more careful one must be in adjusting for them, since once they have been exceeded there is no way of returning. The situation is quite different for the q_p adjustment. For one thing, there is no physical constraint imposing only unidirectional sensitivities. A glance at (4-217), for example, shows that an increase in G_2 increases q_p while an increase in G_3 decreases it. Furthermore β, the gain of a (noninverting) amplifier can be increased or decreased at will, purely by increasing resistors, since it is determined by a resistor ratio (see, e.g., Table 4-3).

EXAMPLE: THE TWIN-T Let us now consider the use of the sensitivity matrix in a different way by applying it to the general twin-T (Fig. 4-18a). To do so we can compute the zero sensitivity $\mathscr{S}_x^z = dz/(dx/x)$ of the general twin-T for $z = j\omega_N$, i.e., for the case that the twin-T is balanced for $j\omega$-axis

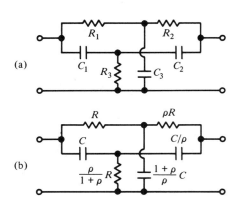

FIG. 4-18. The twin-T: (a) general case; (b) potentially symmetric case.

zeros.[42] In order to find dz/z, and from that $d\omega_z/\omega_z$ and dq_z/q_z, we must then divide the expressions for $\mathscr{S}_x^{j\omega_N}$ by $j\omega_N$. Restricting ourselves to variations with respect to resistive components only, we obtain

$$S_{R_1}^{j\omega_N} = -\frac{\alpha}{2(1+\alpha^2)}\left[\frac{1}{\alpha} + \alpha\lambda + j(1-\lambda)\right] \tag{4-219a}$$

$$S_{R_2}^{j\omega_N} = -\frac{\alpha}{2(1+\alpha^2)}\left[\frac{1}{\alpha} + \alpha(1-\lambda) + j\lambda\right] \tag{4-219b}$$

$$S_{R_3}^{j\omega_N} = -\frac{\alpha}{2(1+\alpha^2)}(\alpha - j) \tag{4-219c}$$

where, referring to Fig. 4-18a,

$$\lambda = \frac{R_1}{R_1 + R_2} \tag{4-220a}$$

and

$$\alpha = \frac{1}{C_1 + C_2}\sqrt{\frac{R_1 + R_2}{R_3}C_1 C_2} \tag{4-220b}$$

With (4-188a) and (4-188b) the corresponding sensitivity matrix with respect to $\Delta\omega_z/\omega_z$ and $\Delta q_z/q_z$ follows by inspection. Since the zero is in the vicinity of the $j\omega$ axis, $q_z \gg 1$ and $-\sqrt{4q_z^2 - 1} \approx -2q_z$. Thus

$$\begin{bmatrix} \dfrac{\Delta\omega_z}{\omega_z} \\[3mm] \dfrac{\Delta q_z}{q_z} \end{bmatrix} = -\frac{\alpha}{2(1+\alpha^2)}\begin{bmatrix} \dfrac{1}{\alpha} + \alpha\lambda & \dfrac{1}{\alpha} + \alpha(1-\lambda) & \alpha \\[3mm] -2q_z(1-\lambda) & -2\lambda q_z & 2q_z \end{bmatrix}\begin{bmatrix} \Delta R_1/R_1 \\[2mm] \Delta R_2/R_2 \\[2mm] \Delta R_3/R_3 \end{bmatrix} \tag{4-221}$$

In order to obtain a 2×2 triangular matrix permitting noniterative tuning of the twin-T, we note that $\Delta q_z/q_z$ becomes independent of R_2 when $\lambda \to 0$. Then (4-221) becomes

$$\begin{bmatrix} \dfrac{\Delta\omega_z}{\omega_z} \\[3mm] \dfrac{\Delta q_z}{q_z} \end{bmatrix} = \begin{bmatrix} -\dfrac{1}{2(1+\alpha^2)} & -\dfrac{1}{2} & \dfrac{-\alpha^2}{2(1+\alpha^2)} \\[3mm] \dfrac{q_z\alpha}{1+\alpha^2} & 0 & -\dfrac{q_z\alpha}{1+\alpha^2} \end{bmatrix}\begin{bmatrix} \dfrac{\Delta R_1}{R_1} \\[3mm] \dfrac{\Delta R_2}{R_2} \\[3mm] \dfrac{\Delta R_3}{R_3} \end{bmatrix} \tag{4-222}$$

42. See Chapter 8, Section 8.2 of *Linear Integrated Networks: Fundamentals*; also G. S. Moschytz, A general approach to twin-T design and its application to hybrid integrated linear active networks, *Bell Syst. Tech. J.* **49**, 1105–1149 (1970).

The variable q_z is equally dependent on R_1 and R_3 (with opposite signs). Since $\alpha \le 1$ for maximum pole Q of the twin-T,[43] $S_{R_3}^{\omega_z} < S_{R_1}^{\omega_z}$. Hence R_3 is to be preferred over R_1 for the tuning of q_z since it provides a smaller off-diagonal matrix element, improving noniterative tuning. Thus, rearranging (4-222), for an optimum two-step tuning sequence, we obtain

$$
\begin{bmatrix} \dfrac{\Delta q_z}{q_z} \\[3mm] \dfrac{\Delta \omega_z}{\omega_z} \end{bmatrix} = \begin{bmatrix} -\dfrac{\alpha q_z}{1+\alpha^2} & 0 \\[3mm] -\dfrac{\alpha^2}{2(1+\alpha^2)} & -\dfrac{1}{2} \end{bmatrix} \begin{bmatrix} \dfrac{\Delta R_3}{R_3} \\[3mm] \dfrac{\Delta R_2}{R_2} \end{bmatrix}
\tag{4-223}
$$

Thus the null depth is tuned first with R_3, the null frequency then tuned with R_2.

Because it is of particular practical interest, let us now examine the tuning sequence for the potentially symmetric twin-T (see Fig. 4-18b). From equations (4-219) we obtain the zero sensitivities

$$
S_{R_1}^{j\omega_N} = -\frac{1}{4}\left(\frac{2+\rho}{1+\rho} + j\frac{\rho}{1+\rho}\right)
\tag{4-224a}
$$

$$
S_{R_2}^{j\omega_N} = -\frac{1}{4}\left(\frac{1+2\rho}{1+\rho} + j\frac{1}{1+\rho}\right)
\tag{4-224b}
$$

$$
S_{R_3}^{j\omega_N} = -\frac{1}{4}(1-j)
\tag{4-224c}
$$

and the sensitivity matrix

$$
\begin{bmatrix} \dfrac{\Delta \omega_z}{\omega_z} \\[3mm] \dfrac{\Delta q_z}{q_z} \end{bmatrix} = \begin{bmatrix} -\dfrac{1}{4}\dfrac{2+\rho}{1+\rho} & -\dfrac{1}{4}\dfrac{1+2\rho}{1+\rho} & -\dfrac{1}{4} \\[3mm] \dfrac{q_z}{2}\dfrac{\rho}{1+\rho} & \dfrac{q_z}{2}\dfrac{1}{1+\rho} & -\dfrac{q_z}{2} \end{bmatrix} \begin{bmatrix} \dfrac{\Delta R_1}{R_1} \\[3mm] \dfrac{\Delta R_2}{R_2} \\[3mm] \dfrac{\Delta R_3}{R_3} \end{bmatrix}
\tag{4-225}
$$

Assuming $\rho = 10$ we obtain

$$
\begin{bmatrix} \dfrac{\Delta \omega_z}{\omega_z} \\[3mm] \dfrac{\Delta q_z}{q_z} \end{bmatrix} = \begin{bmatrix} -0.27 & -0.48 & -0.25 \\[3mm] 0.44q_z & 0.045q_z & -0.5q_z \end{bmatrix} \begin{bmatrix} \dfrac{\Delta R_1}{R_1} \\[3mm] \dfrac{\Delta R_2}{R_2} \\[3mm] \dfrac{\Delta R_3}{R_3} \end{bmatrix}
\tag{4-226}
$$

43. See Chapter 8, Fig. 8-15 of *Linear Integrated Networks: Fundamentals*.

We see that, here again, R_2 and R_3 are the preferable tuning components. As the rearranged sensitivity matrix below shows, the optimum tuning sequence is the same as for (4-223); q_z is trimmed first with R_3, followed by ω_z with R_2:

$$\begin{bmatrix} \Delta q_z/q_z \\ \Delta \omega_z/\omega_z \end{bmatrix} = \begin{bmatrix} -0.5q_z & 0.045q_z \\ -0.25 & -0.48 \end{bmatrix} \begin{bmatrix} \Delta R_3/R_3 \\ \Delta R_2/R_2 \end{bmatrix} \tag{4-227}$$

The twin-T example has illustrated how the sensitivity matrix can be used to derive not only an optimum tuning sequence for a network but also, to derive design equations for a network in order to improve its tunability. In the case of the twin-T with $\lambda \to 0$, improved tunability can be achieved practically at no cost at all (except for the spread between R_1 and R_2, which must be limited to an amount compatible with the technology used). Thus, it can be shown [44] that even as λ approaches zero a pole Q close to 0.5 can still be attained.

Simplifying the twin-T tuning process from a three-component to a two-component procedure by a corresponding modification of the sensitivity matrix is justified from the point of view of identifying the most expedient way of obtaining a specified ω_z and q_z. With film resistors, however, we are again confronted with the problem of irreversible resistor adjustments and the consequent loss of circuits due to overshooting of the specified tolerances. In such cases it is desirable to have additional components whose adjustment reverses the effects of the selected tuning components. This is possible if the sensitivity of a network parameter is opposite in sign with respect to variations of two different components. As mentioned above, the sensitivity of frequency (e.g., ω_z) with respect to resistor variations always has the same (negative) sign. In the case of q_z, however, the signs with respect to variations of R_1 and R_3 are opposite (see (4-222) and (4-225)). Thus, even though R_2 and R_3 are selected as the tuning components for the twin-T, one does well to keep in mind that any overshoot in R_3 can be corrected for by R_1. Indeed, since the sensitivities with respect to R_1 and R_3 not only have opposite signs but also approximately the same magnitude, the correction will proceed as readily as the initial tuning step itself.

4.4 THE TUNING OF SECOND-ORDER NETWORKS

The sensitivity matrices discussed above are useful in the discussion of network tuning in general. They provide the relationship between any given network parameter—used to characterize the network performance—and the components of the network. In practice, the problem of active-filter tuning is more specific; we are required to place the pole and zero pairs of

---

44. Chapter 8, Section 8.2 of *Linear Integrated Networks: Fundamentals*.

individual second-order networks accurately in the s-plane. Because of the isolation property, which is an inherent characteristic of usable active-filter building blocks, additional tuning should not, as a rule, be necessary once the building blocks are in cascade.

4.4.1 Functional and Deterministic Tuning

In discussing the methods of network tuning we shall distinguish between two basically different methods, namely between functional and deterministic tuning.

When a network is *functionally* tuned, certain parameters that sufficiently characterize the performance of the network are tuned for while the network is functional, i.e., in operation. The critical questions here are which parameters to select, how to measure them with the greatest accuracy, and how to convert the specified tolerances of the overall circuit into tolerances of the individual parameters to be tuned. It is because parameters characterizing the performance of the active network are being tuned that the network must be functional, i.e., assembled and powered as if for operation in the final system. As a consequence, any parasitics built into the network are automatically taken into account and "tuned out." Tuning a "powered" network generally presents no problems, provided that appropriate precautions are taken against accidentally altering the DC operation of the circuit during the tuning process or introducing temporary parasitics with the tuning apparatus. In general, functional tuning is iterative, particularly if the tuning steps are interactive. This can be time consuming in production; the number of necessary iterations will increase with the degree of tuning accuracy specified.

By *deterministic* tuning of a network, we understand the tuning, or trimming to value, of individual components of a network as predicted by a combination of comprehensive network equations (in which parasitic effects are taken into account) and component measurements. Since there are generally more components than equations, some additional optimizing constraints can be assumed, or else the undefined component values can be selected for convenience, economy, or the like. The undefined components are measured and their values entered into the equations. The solutions of the equations provide the values of the components to be tuned. Tuning is carried out "to value," so it makes no difference whether the network is assembled or not. Since the components to be tuned are invariably resistors, this method consists of "resistor trimming," in contrast to the tuning of network parameters (e.g., amplitude, phase, frequency) that occurs in functional tuning. Resistor trimming is a relatively simple operation and an essential one in any hybrid-integrated manufacturing plant. However, when applied to linear networks, what it saves in equipment sophistication it makes up for in computational complexity. Since the tuned circuit, once assembled, is expected to function as accurately as the functionally tuned circuit, all first-order, and often even second-order, parasitic effects must be taken into account in the equations. This may increase the degree and complexity of the equations drastically. Since even under ideal conditions the equations are nonlinear, the method, although rapid in execution (generally very few if any iterations are necessary), may require powerful computer programs for the solution of nonlinear equations of high degree. An on-line computer for the manufacturing process is therefore almost indispensable and automatic resistor- and capacitor-measuring and resistor-tuning equipment desirable. Clearly, deterministic tuning is the more efficient of the two methods, but the necessary expenditure and computational effort can generally be justified only by very high production volumes.

In practice it will very often be found useful to combine functional with deterministic tuning. In particular, the initial adjustments can be carried out by deterministic tuning, where the values are obtained from either the idealized network equations or those containing at most first-order parasitic effects. To overcome more subtle second- order parasitics, a fine-tuning step is then undertaken in which the circuit is functional (i.e., assembled and powered) and a fine adjustment of one or more of the critical parameters carried out.

4.4.2 Functional Tuning of Second–Order Networks

Our task here is to tune to value the five parameters of a general second-order network. We shall consider this task in three separate parts.

All-Pole Networks By all-pole networks we mean those comprising a complex conjugate pole pair, and zeros only at the origin or infinity (i.e., low-pass, high-pass, and bandpass networks). Consider the low-pass function

$$T(s) = \frac{K}{s^2 + \dfrac{\omega_p}{q_p} s + \omega_p^2} \tag{4-228}$$

For $s = j\omega$ the amplitude and phase response is as shown in Fig. 4-19. We could, of course, tune the circuit to the specified amplitude response. For high q_p values (e.g., > 5), the peak frequency corresponds (approximately) to ω_p; q_p corresponds (approximately) to the ratio of ω_p to the 3 dB bandwidth. However, the amplitude response in the vicinity of ω_p is flat; therefore an accurate ω_p adjustment is difficult. This is a serious disadvantage, since, as will be remembered from previous discussions (e.g., Section 4.1.4), above all the pole *frequency* should be adjusted as accurately as possible, often to within better than 0.2%.

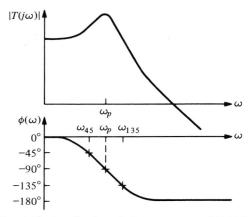

FIG. 4-19. Amplitude and phase response of (4-228).

A glance at the phase response in Fig. 4-19 shows that, in contrast to the amplitude, the phase has its *maximum slope* at ω_p and vicinity (i.e., covering the 3 dB frequency range). Thus, as will be shown next, the phase curve provides a far less ambiguous measure for the parameters ω_p and q_p than does the amplitude curve.[45] For $s = j\omega$ the response of $T(s)$ is

$$T(j\omega) = \frac{K}{\omega_p^2 - \omega^2 + j\omega \dfrac{\omega_p}{q_p}} \qquad (4\text{-}229)$$

and the phase is

$$\phi(\omega) = -\tan^{-1}\left[\frac{\omega\omega_p}{q_p(\omega_p^2 - \omega^2)}\right] \qquad (4\text{-}230)$$

A phase angle of $-90°$ is obtained for

$$\omega_{90} = \omega_p \qquad (4\text{-}231)$$

For the phase angles $-45°$ and $-135°$ the argument of (4-230) is ± 1. Solving for ω we obtain

$$\omega_{45} = \frac{\omega_p}{2q_p}[\sqrt{4q_p^2 + 1} - 1] \qquad (4\text{-}232a)$$

$$\omega_{135} = \frac{\omega_p}{2q_p}[\sqrt{4q_p^2 + 1} + 1] \qquad (4\text{-}232b)$$

45. K. Mossberg and D. Åkerberg, Accurate trimming of active *RC* filters by means of phase measurements, *Electronics Lett.* **5**, 520–521 (1969).

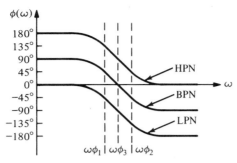

FIG. 4-20. Phase curves of general second-order networks. HPN, high-pass network; BPN, bandpass network; LPN, low-pass network.

Solving for q_p, we have, for the low-pass network (LP),

$$(q_p)_{LP} = \frac{\omega_p}{\omega_{135} - \omega_{45}} \tag{4-233}$$

Note that (4-231) and (4-233) hold for all q_p values, including those less than 0.5. Furthermore, for the bandpass and high-pass networks we must simply shift the phase curves by the constant amount introduced by their zeros (Fig. 4-20). This means adding 90° to the phase curve of the bandpass network, 180° to that of the high-pass network. Thus, (4-231) and (4-233) become, respectively,

$$\omega_p = \omega_{\phi_3} \tag{4-234a}$$

and

$$q_p = \frac{\omega_{\phi_3}}{\omega_{\phi_2} - \omega_{\phi_1}} \tag{4-234b}$$

where

low-pass network: $\phi_1 = -45°$; $\phi_2 = -135°$; $\phi_3 = -90°$
bandpass network: $\phi_1 = 45°$; $\phi_2 = -45°$; $\phi_3 = 0°$
high-pass network: $\phi_1 = 135°$: $\phi_2 = 45°$; $\phi_3 = +90°$

and

$$\omega_{\phi_1, \phi_2} = \frac{\omega_p}{2q_p} [\sqrt{4q_p^2 + 1} \mp 1] \tag{4-235}$$

The tuning procedure suggested above can be summarized by the following two steps:

1. ω_p Adjustment:
   At ω_p adjust the network for the phase shift ϕ_3.

2. q_p Adjustment:

Calculate ω_{ϕ_1} and ω_{ϕ_2} from (4-235) and adjust for ϕ_1 or ϕ_2 at the corresponding frequency.

The slope of $\phi(\omega)$, that is the delay $\tau(\omega)$, is obtained by taking the derivative of (4-230):

$$\frac{d\phi}{d\omega} = \tau(\omega) = -\frac{(\omega_p^2 + \omega^2)\omega_p q_p}{q_p^2(\omega_p^2 - \omega^2)^2 + \omega^2\omega_p^2} \tag{4-236}$$

Hence at ω_p

$$\frac{d\phi}{d\omega}\bigg|_{\omega=\omega_p} = \tau(\omega_p) = -\frac{2q_p}{\omega_p} \tag{4-237}$$

Thus, the larger q_p the steeper the phase slope, i.e., the larger the delay τ. This is very fortunate, since the accuracy of the frequency adjustment becomes all the more important the higher q_p is specified (since the amplitude error at the 3 dB frequencies is $q_p \cdot \Delta\omega_p/\omega_p$); as we see, with increasing q_p this accuracy also becomes all the more easy to attain.

We spoke above of a "two-step" tuning procedure in which ω_p and q_p are adjusted for. The reader will be aware, after our discussion on sensitivity matrices, that the procedure will then be truly "two-step" only if we have a noninteractive, i.e., a diagonal or triangular, sensitivity matrix. If we do not, then the two steps outlined above must be repeated iteratively until the desired accuracy is obtained.

In the q_p adjustment described above, we have the choice of adjusting for either ϕ_1 or ϕ_2 at the respective frequency. By definition ω_{ϕ_2} is the higher of the two frequencies; thus if the operational amplifier introduces any parasitic phase lag, this will be more apparent at ω_{ϕ_2} than at ω_{ϕ_1}. However, compensation of the parasitic phase $\Delta\phi°$ at ω_{ϕ_2}, by tuning only to $(\phi_2 - \Delta\phi)°$, may overcompensate the phase at ω_{ϕ_1}. Thus, which of the two frequencies to select for the q_p adjustment depends on which of the two, ω_{ϕ_1} or ω_{ϕ_2}, is in the more critical range of the filter band. Alternatively, an average value of parasitic phase can be compensated for at ω_p by tuning to $[\phi(\omega_p) - (\Delta\phi/2)]°$; q_p is then tuned by adjusting for ϕ_1 or ϕ_2 at ω_{ϕ_1} or ω_{ϕ_2}, respectively.

The tuning method described above entails the setting of an input sinusoidal signal generator to the required frequencies ω_p and, say, ω_{ϕ_2}, and then adjusting for ϕ_{ω_p} and ϕ_2, respectively. Naturally, for high tuning accuracy, the frequencies ω_p and ω_{ϕ_2} must be set accurately, a process that is time consuming with an oscillator of average accuracy, or costly using, say, a frequency synthesizer. To obviate the need for high-accuracy frequency setting, an oscillator of average accuracy (but high stability) combined with a high-accuracy frequency counter can be set to the *approximate* frequencies ω_p and ω_{ϕ_2}, say to ω_p' and ω_{ϕ_2}', and the corresponding phase ϕ_{ω_p}' and ϕ_2' calculated from (4-230). However, since $\tan^{-1}\phi(\omega_p')$ will be very large and inaccurate if $\omega'_p \approx \omega_p$, the two frequencies to use, in this case, will be ω'_{ϕ_1} and ω'_{ϕ_2}, as calculated from (4-235). For the three network types we then have:

low-pass network: $\phi'_1 = \phi(\omega'_{\phi_1});$ $\phi'_2 = \phi(\omega'_{\phi_2})$

bandpass network: $\phi'_1 = 90° + \phi(\omega'_{\phi_1});$ $\phi'_2 = 90° + \phi(\omega'_{\phi_2})$

high-pass network: $\phi'_1 = 180° + \phi(\omega'_{\phi_1});$ $\phi'_2 = 180° + \phi(\omega'_{\phi_2})$

The adjustments at ω'_{ϕ_1} and ω'_{ϕ_2} here take the place of the adjustments at ω_p and, say, ω_{ϕ_2}, and between them[46] tune the circuit for ω_p and q_p.

We have not mentioned the tuning of the constant coefficient K yet, partly because its accuracy is generally not very critical, partly because a noninteractive component is frequently unavailable for its adjustment. If ω_p and q_p can be tuned with relatively few iterative steps it will only rarely be worth increasing this number for the sake of an accurate K. If indeed high K accuracy is required, it can generally be attained by a gain or attenuator adjustment somewhere preceding or following the network. In any event, the K adjustment is a gain adjustment setting the overall level of the output signal; in most cases the initial or untuned K value, accurate to within a few percent, will be quite sufficient.

Networks with Finite Zeros   Here we consider minimum-phase networks of the general form

$$T(s) = K \frac{s^2 + \dfrac{\omega_z}{q_z} s + \omega_z^2}{s^2 + \dfrac{\omega_p}{q_p} s + \omega_p^2}$$ (4-238)

Besides the pole parameters ω_p and q_p we must now also adjust ω_z and q_z. The phase function is now

$$\phi(\omega) = \phi_z(\omega) - \phi_p(\omega)$$

$$= \tan^{-1}\left(\frac{\omega\omega_z}{q_z(\omega_z^2 - \omega^2)}\right) - \tan^{-1}\left(\frac{\omega\omega_p}{q_p(\omega_p^2 - \omega^2)}\right)$$ (4-239)

The contribution of the zeros has the same form as that of the poles but the opposite sign.

In those cases in which the zeros are realized by a summing operation, as in the biquad (e.g., Fig. 4-16), the zeros can be removed from the circuit by opening up the summing path. The poles are then tuned separately, as described above, and the phase measured at the two frequencies ω'_{ϕ_1} and ω'_{ϕ_2}.

46. Note, though, that in contrast to the ω_p and ω_{ϕ_2} adjustment, the two tuning steps are now interactive, since both ϕ_1' and ϕ_2' are functions of ω_p and q_p.

Reconnecting the summing circuitry, the zeros are then tuned at ω'_{ϕ_1} in such a way that

$$\phi(\omega'_{\phi_1}) = \tan^{-1}\left(\frac{\omega'_{\phi_1}\omega_z}{q_z[\omega_z^2 - (\omega'_{\phi_1})^2]}\right) - \tan^{-1}\left(\frac{\omega'_{\phi_1}\omega_p}{q_p[\omega_p^2 - (\omega'_{\phi_1})^2]}\right) \quad (4\text{-}240)$$

and at ω'_{ϕ_2} such that the equivalent phase value $\phi(\omega'_{\phi_2})$ is obtained.

In other circuits the feedback loop responsible for the pole generation can be measured and tuned separately from the circuitry realizing the zeros simply by measuring the network between appropriate terminals. This is especially simple in the case of class 3 networks where the feedback loop consists of an active frequency-rejection network and in the case of class 4 networks where the feedback loop has active bandpass character.

Consider, for example, the somewhat generalized class 4 network shown in Fig. 4-21. In regular operation, terminal 1 is grounded and the specified voltage transfer function is given by (see Chapter 2, equation (2-92))

$$T'(s) = \frac{N(s)}{D(s)} = \frac{V_3}{V_1'} \approx \frac{t'_{12}}{t_{32} - t'_{32}} = \frac{\beta t'_{12}}{1 - \beta t'_{32}} \quad (4\text{-}241)$$

where t'_{12} determines the zeros of $T'(s)$ and has the form

$$t'_{12} = \frac{V_2'}{V_1'}\bigg|_{V_3'=0} = \frac{N(s)}{s^2 + \dfrac{\omega_p}{\hat{q}}s + \omega_p^2} \quad (4\text{-}242)$$

t'_{32} is realized by a passive RC bandpass network with the form

$$t'_{32} = \frac{V_2'}{V_3'}\bigg|_{V_1'=0} = \frac{\omega_{32}s}{s^2 + \dfrac{\omega_p}{\hat{q}}s + \omega_p^2} \quad (4\text{-}243)$$

and

$$t_{32} = \frac{V_2}{V_3}\bigg|_{V_1=0} = \frac{1}{\beta} \quad (4\text{-}244)$$

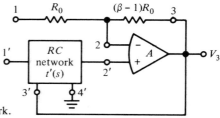

FIG. 4-21. Generalized class 4 network.

Consider now the transfer function between terminal 1 and the output, with terminal $1'$ grounded. We obtain (see Chapter 2, equation (2-91))

$$T(s) = \frac{V_3}{V_1} \approx -\frac{t_{12}}{t_{32} - t'_{32}} \tag{4-245}$$

where t_{32} and t'_{32} are given by (4-244) and (4-243), respectively, and

$$t_{12} = \left.\frac{V_2}{V_1}\right|_{V_3=0} = \frac{\beta - 1}{\beta} \tag{4-246}$$

Note that $T(s)$ is independent of t'_{12} and therefore independent of the zeros of the desired transfer function $T'(s)$. On the other hand the poles of $T(s)$ are precisely the poles of $T'(s)$. With (4-243) to (4-246) we obtain

$$T(s) = -(\beta - 1) \frac{s^2 + \dfrac{\omega_p}{\hat{q}} s + \omega_p^2}{s^2 + \dfrac{\omega_p}{q_p} s + \omega_p^2} \tag{4-247}$$

where (see (2-32b))

$$q_p = \frac{\hat{q}}{1 - \beta \dfrac{\omega_{32}}{\omega_p} \hat{q}} \tag{4-248}$$

The term q_p is, of course, the pole Q of the specified denominator $D(s)$ in (4-241), and $T(s)$ is the frequency response of an FEN, *irrespective of the zeros of $T'(s)$*. Thus the poles of the FEN, seen between terminals 1 and 3 of the general class 4 network, can be tuned first, followed by the separate tuning of the zeros.

The poles of (4-247) are tuned for as follows. From (4-239) we have

$$\phi(\omega) = \tan^{-1} \frac{\dfrac{\omega \omega_z}{q_z(\omega_z^2 - \omega^2)} - \dfrac{\omega \omega_p}{q_p(\omega_p^2 - \omega^2)}}{1 + \dfrac{1}{q_z q_p} \dfrac{\omega^2 \omega_z \omega_p}{(\omega_z^2 - \omega^2)(\omega_p^2 - \omega^2)}} \tag{4-249}$$

For $\omega_p = \omega_z$ (4-249) becomes

$$\phi(\omega) = \tan^{-1} \frac{\left(\dfrac{\omega \omega_p}{\omega_p^2 - \omega^2}\right)\left(\dfrac{q_p - q_z}{q_p q_z}\right)}{1 + \dfrac{1}{q_z q_p}\left(\dfrac{\omega \omega_p}{\omega_p^2 - \omega^2}\right)^2}$$

$$= \tan^{-1} \frac{\Omega\left(\dfrac{q_p - q_z}{q_p q_z}\right)}{\Omega^2 + \dfrac{1}{q_z q_p}} \tag{4-250}$$

where

$$\Omega = \left(\frac{\omega_p}{\omega} - \frac{\omega}{\omega_p}\right) \tag{4-251}$$

For $\Omega = 0$, i.e., $\omega = \omega_p$, we have

$$\phi(\omega_p) = 180° \tag{4-252}$$

When the denominator of (4-250) is zero, ϕ is $90°$ or $270°$. Solving for the corresponding frequencies, we obtain

$$\omega_{90,270} = \frac{\omega_p}{2\sqrt{q_p q_z}} \left(\sqrt{4q_p q_z + 1} \mp 1\right) \tag{4-253}$$

Thus to tune for ω_p and q_p we select two frequencies, e.g., ω_p and ω_{90}, and tune for the corresponding phase. To compute ω_{90} or ω_{270} we replace q_z by $\hat{q}$, which we can either compute from the nominal network elements (i.e., $q_p(\beta = 0) = \hat{q}$) or measure according to (4-243). The zeros are tuned after the circuit has been returned to its normal operating condition (terminal 1 grounded). It is preferable to tune the zeros using (4-241) instead of (4-242), since terminal 2′ is a high-impedance point and therefore sensitive to external measuring equipment, whereas the output impedance at terminal 3 is low.

Note that the expression for $\phi(\omega)$ in (4-250) is perfectly general for a function for which $\omega_p = \omega_z$. Thus depending on whether q_p is larger or smaller than q_z it describes the phase response of a symmetrical FEN or FRN, respectively.

In a frequency-rejection network (FRN) in which the zeros are close to, or on, the $j\omega$ axis (i.e., $q_z \rightarrow \infty$), the phase contribution of the zeros, $\phi_z(\omega)$, is restricted to a narrow frequency range around ω_z. This is shown in Fig. 4-22, where $\phi_z(\omega)$ is plotted for large q_z values. As a frame of reference for typical q_z values, one should bear in mind that the null depth of a symmetrical FRN (i.e., $\omega_z = \omega_p$) is proportional to q_p/q_z, so that for a -60 dB null and $q_p = 5$, q_z must be 5000. Even in the case of a passive twin-T with $\hat{q} = 0.25$, q_z must be 250. Naturally, for deeper null depths, q_z must be increased accordingly. Thus, referring to Fig. 4-22, it follows that ω_z need not be separated much from ω_p before the phase contribution of the zeros to the phase of the poles can be neglected. In such cases, the poles and zeros can be tuned separately and between the same input–output terminals instead of between two different terminal pairs, as was described above. Thus, for the range ω_z/ω_p less than 0.5 or greater than 2, the phase contribution of $\phi_z(\omega)$ on $\phi_p(\omega)$ in (4-239) can be neglected, and the poles can be tuned as in an all-pole network.[47] For the range of ω_z/ω_p between 0.9 and 1.1 the contribution of the zero phase to the pole phase is significant and the two will have to be tuned separately, as outlined above.

---

47. C. J. Steffen, unpublished Bell Telephone Laboratories memorandum.

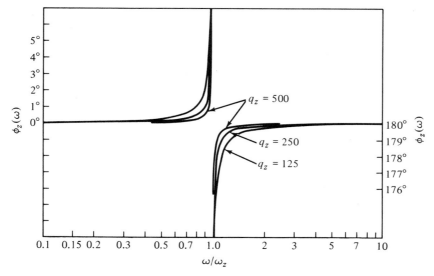

FIG. 4-22. Phase contribution of the zeros of an FRN as a function of ω/ω_z.

The fact that certain networks can be more easily tuned when their (finite) zeros are sufficiently far apart from the poles is one more criterion that should be kept in mind during the pole–zero pairing process discussed in Chapter 1 (Section 1.3.1). Thus, all things being equal, for tuning purposes it is usually preferable to pair poles with zeros that are as far apart as possible or at least have ω_z/ω_p ratios below 0.9 or above 1.1. Thus, using type I networks in applications having uncritical dynamic-range and signal-to-noise requirements, the pole–zero pairing may well proceed directly with a view to simplifying the tuning process. This means pairing the dominant or most critical poles with the zeros furthest away.

By considering the dual of Fig. 4-21 in the sense that we did in Chapter 2 (Fig. 2-39), a class 3 network can also be tuned separately for its poles and zeros by entering the network at appropriate terminals. Whether the lack of interaction between the pole and zero phase occurring for high q_z values can also be utilized there depends on the circuit type. It should be pointed out that in either network class this is possible only if a certain well defined part of the network realizes the zeros and additional components are then available to tune the poles. For example, in the class 4 FRN shown in Fig. 4-15, the twin-T can first be tuned for the zeros. *Subsequently,* and without any further adjustments of the twin-T, the twin-T loading network and the amplifier gain are used to tune the poles. These latter adjustments in turn do not affect the previously tuned zeros in any way (provided, of course, that ω_z and ω_p are sufficiently far apart). More will be said on this subject in Chapter 6.

Non-Minimum-Phase Networks   In the case of non-minimum-phase second-order networks we can restrict ourselves to all-pass functions, since any other function (quite apart from occurring only rarely) can always be expanded into the cascade of an all-pass network and a network with finite LHP zeros. This expansion is achieved by adding a phantom pole and zero pair in the left half s-plane:

$$
T(s) = \frac{s^2 - \dfrac{\omega_z}{q_z} s + \omega_z^2}{s^2 + \dfrac{\omega_p}{q_p} s + \omega_p^2}
$$

$$
= \frac{s^2 - \dfrac{\omega_z}{q_z} s + \omega_z^2}{s^2 + \dfrac{\omega_z}{q_z} s + \omega_z^2} \cdot \frac{s^2 + \dfrac{\omega_z}{q_z} s + \omega_z^2}{s^2 + \dfrac{\omega_p}{q_p} s + \omega_p^2}
\tag{4-254}
$$

This process is illustrated in Fig. 4-23.

Consider, then, the second-order all-pass function

$$
T(s) = \frac{s^2 - \dfrac{\omega_0}{q} s + \omega_0^2}{s^2 + \dfrac{\omega_0}{q} s + \omega_0^2}
\tag{4-255}
$$

Very often this function is characterized by a "stiffness" factor b instead of the root Q, where

$$
b = 2q
\tag{4-256}
$$

From (4-250) it follows that

$$
\phi(\omega) = -2 \cot^{-1}\left[q\left(\frac{\omega_0}{\omega} - \frac{\omega}{\omega_0}\right) \right]
\tag{4-257}
$$

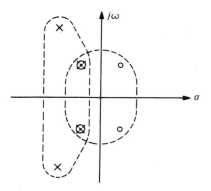

FIG. 4-23. Expansion of a non-minimum-phase second-order network function by adding a phantom pole-zero pair to obtain an all-pass function and a function with finite LHP zeros.

The delay or slope of the phase curve is then:

$$\tau(\omega) = \frac{d\phi(\omega)}{d\omega} = -\frac{2q}{\omega_0} \frac{1 + \left(\dfrac{\omega_0}{\omega}\right)^2}{1 + q^2\left(\dfrac{\omega}{\omega_0} - \dfrac{\omega_0}{\omega}\right)^2} \qquad (4\text{-}258)$$

At ω_0, we have

$$\tau(\omega_0) = -4q/\omega_0 \qquad (4\text{-}259a)$$

whereas the maximum delay occurs at a slightly lower frequency, namely at

$$\omega_{\tau_{max}} = \omega_0\left[\sqrt{4 - (1/q^2)} - 1\right] \qquad (4\text{-}259b)$$

The all-pass network is characterized by the two parameters ω_0 and q. We have from (4-257)

$$\phi(0) = 0°$$
$$\phi(\omega_0) = 180° \qquad (4\text{-}260)$$
$$\phi(\omega \rightarrow \infty) = 360°$$

From (4-253) the frequencies at which the phase is 90° and 270° are, respectively,

$$\omega_{90,270} = \frac{\omega_0}{2q}\left(\sqrt{4q^2 + 1} \mp 1\right) \qquad (4\text{-}261)$$

Thus the procedure for tuning a second-order all-pass network is as follows:

1. To set $\omega_z = \omega_p = \omega_0$, $\phi(\omega_0)$ is adjusted for 180°.
2. To set $-q_z = q_p = q$, the phase at ω_{90} or ω_{270} is adjusted for $-90°$ or $-270°$, respectively.

As in the tuning of the other functions discussed above we have merely indicated here which parameters (e.g., ω_z, ω_p, etc.) to tune and which measure (e.g., phase) to use in doing so. Naturally, each network realization must be examined individually to ascertain which components to use for the adjustment of each parameter, and the best sequence to follow. This will be demonstrated by a detailed example in Chapter 6. There we shall apply the tuning instructions which have been outlined in general here, to a specific family of active-filter building blocks.

Some networks may require a different measure for tuning a particular parameter than others. Consider, for example, the case of the all-pass network in which the tuning component x used for step 2 above moves the zeros away from the $j\omega$ axis while moving the poles closer to it (see Fig. 4-24). This will happen if q_z is inversely and q_p directly proportional to changes in the component x. In this case, the decrease in the phase contribution of the

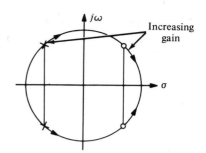

FIG. 4-24. Effect of step 2 of the tuning procedure for a second-order all-pass network.

zeros is approximately cancelled by the increase in the phase contribution of the poles; thus, in spite of high $S_x^{q_p}$ and $S_x^{q_z}$ values, the overall phase will remain practically constant while adjusting x. In such a case the amplitude is a better measure of the q adjustment than the phase. For $q_p = -q_z = q$, the amplitude at ω_0 must be equal to the amplitude at DC or any other frequency. Note, then, that whereas the phase measurements suggested above may in general be the most suitable indicators for the adjustment of network parameters, a given network realization may still require a tuning indicator other than phase. One therefore does well to ascertain individually the optimum tuning indicator every time a basically different network type is used.

4.4.3 Accuracy Considerations for Functional Tuning

If voltage gain is used to measure the frequency response of a network, then the accuracy of the voltmeter will directly determine the accuracy with which the frequency response can be tuned. We must now determine the accuracy with which the frequency response (e.g., the magnitude of the transfer function), and such parameters as root frequency and Q, can be tuned when phase is used as the tuning indicator.

It can be shown[48] that the tuning accuracy in the amplitude response of a second-order all-pole network is within $\pm 2 \cdot \Delta\phi \%$ or $\pm 0.2 \cdot \Delta\phi$ dB when trimmed by means of a phase meter with an error of $\pm \Delta\phi°$. This is independent of the pole Q. Thus, with $\Delta\phi$ given in degrees

$$\frac{\Delta |T|}{|T|} \leq 2 \cdot \Delta\phi \% \approx 0.2 \cdot \Delta\phi \text{ dB} \tag{4-262}$$

We may therefore conclude that a $1°$ *phase error corresponds to a* 0.2 dB *error in amplitude*. Phase meters with up to $0.1°$ phase accuracy are available at prices considerably lower than voltage meters with comparable accuracy (i.e., 0.02 dB). It follows that, wherever possible, parameter tuning by phase

48. K. Mossberg and D. Åkerberg, op. cit.

is to be preferred over tuning by amplitude. Strictly speaking this conclusion only applies to all-pole networks. However, since networks with finite zeros can generally be tuned for zeros and poles separately (in which case the pole tuning corresponds to the tuning of an all-pole network), our conclusion can be extended to most minimum-phase network types. With all-pass networks the situation is similar. Here it is primarily the phase (or delay) characteristic that is specified, so that the relationship between phase and amplitude is of secondary interest. The ripple in the (flat) amplitude response will, of course, be specified, but generally in less stringent terms than the phase. Thus, it will suffice to have an accurate phase meter for the phase specifications and a less accurate voltage meter for the amplitude.

Let us now examine the frequency accuracy that can be attained by phase measurements. In this respect the pole frequency ω_p is of most importance; it generally occurs in, or on the edge of, the pass band and, as we have seen earlier, should be tuned q_p times more accurately than q_p. From (4-237) we can solve directly for the frequency error at ω_p resulting from a phase error $\Delta\phi$ in degrees:

$$\frac{\Delta f_p}{f_p} [\%] = \frac{-\pi \cdot 100}{360 q_p} \Delta\phi \approx -\frac{\Delta\phi}{q_p} \qquad (4\text{-}263)$$

where $\Delta\phi$ is measured in degrees. This relationship has been plotted in Fig. 4-25 for different values of $\Delta\phi$. Note that the higher q_p, the smaller the

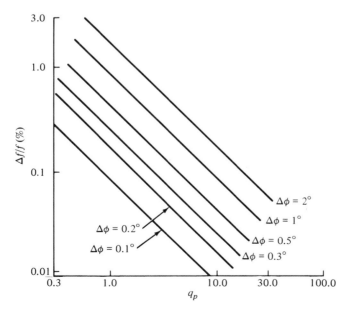

FIG. 4-25. Plot of frequency error vs. q_p for various phase errors.

frequency error resulting from a given phase error. Thus, for $q_p = 10$ and a phase accuracy of $0.1°$, the resulting frequency error is less than 0.01%; for $q_p = 1$ the frequency error is an order of magnitude larger. This is very fortunate, for, as we know, we require higher frequency precision with higher q_p values.[49]

Finally, we are interested in the attainable accuracy in q_p as a function of the available phase-measuring accuracy. From (4-230) we readily obtain

$$\frac{d\phi}{dq_p} = -\frac{\omega\omega_p(\omega_p^2 - \omega^2)}{q_p^2(\omega_p^2 - \omega^2)^2 + \omega^2\omega_p^2} \tag{4-264}$$

It is desirable to tune q_p at the frequency at which (4-264) has a maximum. It can be shown[50] that the frequencies of maximum $d\phi/dq_p$ are ω_{45} or ω_{135}, which are given by (4-232a) and (4-232b), respectively. At either of these frequencies, we have

$$\omega\omega_p = q_p(\omega_p^2 - \omega^2) \tag{4-265}$$

Thus, with (4-264),

$$\left.\frac{d\phi}{dq_p}\right|_{\substack{\omega_{45} \\ \omega_{135}}} = -\frac{1}{2q_p} \tag{4-266}$$

or

$$\left.\frac{\Delta q_p}{q_p}\right|_{\substack{\omega_{45} \\ \omega_{135}}} [\%] = -\frac{\pi \cdot 100}{90} \cdot \Delta\phi \approx -3.5\Delta\phi \tag{4-267}$$

where again $\Delta\phi$ is measured in degrees. This interesting and simple result shows us that at *the $-45°$ and $-135°$ frequencies the relative variation of q_p with changes in phase is independent of q_p*. Thus, no matter what the value of q_p is, a given error in phase will cause the same percentage error in q_p; with a $0.1°$-accurate phase meter we can tune any q_p value to within 0.35% accuracy at ω_{45} or ω_{135}. Since q_p is generally not required to be more than a few percent accurate (taking all ambient factors into account), a $0.1°$ phase accuracy will generally be more than sufficient.

4.4.4 Deterministic Tuning of Second–Order Networks

In this method, network equations are derived with parasitic effects taken into account, the values of the components to be trimmed are computed, and the components are subsequently trimmed accordingly. The tuning

---

49. On the other hand, for a specified frequency accuracy it is the pole with the *lowest* Q that determines the accuracy required of the phase meter used for tuning.
50. C. J. Steffen, unpublished Bell Telephone Laboratories memorandum.

process is not functional, the circuit is not in the operating mode (i.e., powered); each component, i.e., resistor, is trimmed to value separately. The equations must take all relevant parasitic effects into account to the extent that the response of the assembled circuit must be as accurate as if it had been tuned functionally.

The main parasitic effects that must be taken account of in the equations are:

1. The frequency-dependent gain of the operational amplifier.
2. Capacitor losses.
3. Frequency dependence of the (thin-film) capacitor values.
4. Parasitic capacitances on the circuit substrate and resistive losses encountered along conductance paths.

These parasitic effects have been discussed in detail in previous sections of this chapter. To demonstrate how they are incorporated into the design equations of a network we shall go through the considerations encountered in a specific design example.

Low-pass Network with Ideal Components   Consider the class 4 second-order low-pass circuit shown in Fig. 4-26. The transfer function of the ideal network is given by

$$T(s) = \frac{K}{s^2 + \dfrac{\omega_p}{q_p} s + \omega_p^2} \tag{4-268}$$

where

$$K = \frac{\beta}{R_1 R_2 C_3 C_4} \tag{4-269a}$$

$$\omega_p^2 = \frac{1 + \dfrac{R_1 + R_2}{R_4}}{R_1 R_2 C_3 C_4} \tag{4-269b}$$

$$\frac{\omega_p}{q_p} = \frac{R_1 + R_2}{R_1 R_2 C_3} + \frac{1}{R_4 C_4} + \frac{1 - \beta}{R_2 C_4} \tag{4-269c}$$

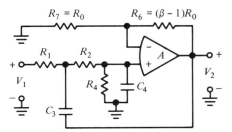

FIG. 4-26. A class 4 second-order low-pass circuit, whose ideal transfer function is given in (4-268).

It is easy to see that in the case of functional tuning, R_1 or R_2 can be used to tune K, R_4 to tune ω_p, and β to tune q_p. In this sequence ω_p and q_p can be tuned independently. K and q_p are not independent; they both depend on β. However S_β^K is unity, whereas $S_\beta^{q_p}$ is proportional to q_p. Thus the small changes of β necessary to adjust for q_p will hardly affect K.

Let us now follow the procedure necessary to tune our circuit by trimming appropriate components to value. We have three equations (see (4-269)) that must be satisfied by three components. Selecting the three resistors R_1, R_2, and R_4 for this purpose we can eliminate R_1 and R_4 and obtain[51] a fourth-degree equation in R_2:

$$R_2^4 - R_2^3 A + R_2^2 B - R_2 C + D = 0 \tag{4-270}$$

where

$$A = \frac{\omega_p \beta C_4}{q_p K}; \qquad B = \frac{2\beta}{KC_3 C_4} + \frac{\omega_p^2 \beta^2}{K^2} - \frac{\beta^2}{K}$$

$$C = \frac{\omega_p \beta^2}{q_p K^2 C_3}; \qquad D = \frac{\beta^2}{K^2 C_3^2 C_4^2} + \frac{\beta}{KC_3 C_4} - \frac{\beta^2}{K} \tag{4-271}$$

The components not solved for, i.e., those appearing in the coefficients A, B, C, and D above, must be measured as accurately as possible. More will be said about the accuracy of this method later. Solving for R_2 from (4-270) we can then calculate R_1 and R_4 from the expressions

$$R_1 = \frac{\beta}{KC_3 C_4} \frac{1}{R_2} \tag{4-272}$$

and

$$R_4 = \frac{R_1 + R_2}{\dfrac{\beta}{K} \omega_p^2 - 1} \tag{4-273}$$

So much for the deterministic tuning of our low-pass network when *ideal* components are assumed. When the parasitics are included the equations become considerably more complicated. In the following discussion we shall outline the tuning procedure to be followed in the nonideal case.

Low-pass Network with Nonideal Components   First of all we have to take into account the frequency-dependent gain of the operational amplifier, i.e., the finite, unity-gain bandwidth ω_g. Since our low-pass network is class 4, we can express the frequency-dependent gain by (4-141b), i.e. (see Fig. 4-14),

$$\beta(s) \approx \frac{A\Omega}{s + \dfrac{A\Omega}{\beta}} = \frac{\omega_g}{s + \omega_\alpha} \tag{4-274}$$

51. E. Lueder and G. Malek, unpublished Bell Telephone Laboratories memoranda.

Substituting $\beta(s)$ for β in the design equations (4-269) we obtain a new transfer function of the form

$$T'(s) = \frac{\beta}{(a_2 s^2 + a_1 s + a_0) + s'(a_2' s^2 + a_1' s + a_0')} \tag{4-275}$$

where

$$s' = \frac{s + \Omega}{\omega_\alpha} \approx \frac{s}{\omega_\alpha} \tag{4-276}$$

The coefficients a_i, $i = 0, 1, 2$, pertain to the network with ideal components and can be obtained by inspection of (4-268) to (4-269c); in the nonideal case, the coefficients a_i' are added to the transfer function in the manner shown. Both the coefficients a_i and a_i' are functions of all, or most, of the components of the circuit. Note that the introduction of $\beta(s)$ makes $T(s)$ a third-order network; it is now our task to tune it in such a way that, at least in the frequency range of interest, it behaves like a second-order network whose transfer function is given by (4-268). This can generally be achieved extremely well below, say, 10 kHz because with most of the available general-purpose operational amplifiers ω_g is at least two orders of magnitude larger.

First, however, we must still include the capacitor losses in our equations. This is done by substituting the complex capacitor values C_i^* for the nominal C_i values in the equations. From (4-109) we have

$$C_i^* = C_i(1 - j\delta_i - \delta_i^2) \tag{4-277}$$

where the δ_i correspond to the capacitor losses as explained in Section 4.2.2. For $s = j\omega$, the transfer function now has the form

$$T'(j\omega) = \frac{\beta}{A_3 \omega^3 + A_2 \omega^2 + A_1 \omega + A_0 + j(B_3 \omega^3 + B_2 \omega^2 + B_1 \omega)} \tag{4-278}$$

where each A_i and B_i is a function f_{A_i} and f_{B_i}, respectively, of the circuit components:

$$\begin{aligned} A_i (i = 0, \ldots, 3) &= f_{A_i}(R_1, R_2, R_4, C_3, C_4, \beta, \omega_\alpha, \delta_3, \delta_4) \\ B_i (i = 1, \ldots, 3) &= f_{B_i}(R_1, R_2, R_4, C_3, C_4, \beta, \omega_\alpha, \delta_3, \delta_4) \end{aligned} \tag{4-279}$$

The corresponding phase response is

$$\phi'(\omega) = -\tan^{-1} \frac{B_3 \omega^3 + B_2 \omega^2 + B_1 \omega}{A_3 \omega^3 + A_2 \omega^2 + A_1 \omega + A_0} \tag{4-280}$$

Obtaining the Tuning Equations   We must now select three conditions, corresponding to the given parameters K, ω_p, and q_p, that will enable us to approximate the specified second-order low-pass function $T(s)$ using the third-order function $T'(s)$. We select conditions similar to those used to functionally tune the network:

1. The DC gain of the two functions should be equal:

$$T'(0) = T(0) = \frac{K}{\omega_p^2} \qquad (4\text{-}281)$$

Thus, from equations (4-269),

$$T'(0)\left(1 + \frac{R_1 + R_2}{R_4}\right) = \beta \qquad (4\text{-}283)$$

2. The phase ϕ at ω_p should be $-90°$. Thus, setting the denominator of (4-280) equal to zero, we obtain

$$A_3\,\omega_p^3 + A_2\,\omega_p^2 + A_1\omega_p + A_0 = 0 \qquad (4\text{-}283)$$

3. The phase ϕ_2 at ω_{ϕ_2} should be $-135°$. Thus, setting the argument in (4-280) equal to unity, we have

$$\omega_{\phi_2}^3(B_3 - A_3) + \omega_{\phi_2}^2(B_2 - A_2) + \omega_{\phi_2}(B_1 - A_1) - A_0 = 0 \qquad (4\text{-}284)$$

Note that step 1 determines K, step 2 determines ω_p, and step 3 determines q_p. In calculating ω_{ϕ_2}, the specified q_p value is used.

Equations (4-281)–(4-284) represent the tuning equations which must be solved for the three components that have been selected for tuning. In the case of the idealized circuit, the three resistors R_1, R_2, and R_4 were found to be the most suitable components; here the most suitable components are R_1, R_2, and the gain β (i.e., the resistor ratio $(R_6 + R_7)/R_7$).

To solve our tuning equations, we need to know the coefficients A_i and B_i of (4-279), which are implicit functions of our three unknown components as well as of all the other components and parasitic characteristics of the network. To obtain the values of those other components and of the parasitic characteristics, we proceed as follows:

1. *Measure R_4, C_3, and C_4.* The measurement of R_4 is straightforward; it requires no more than an accurate resistor bridge (the question of attainable accuracies will be discussed later). The measurement of C_3 and C_4 requires an accurate capacitance bridge. If multilayer ceramic-chip capacitors are used, their values may be measured at a standard frequency, say 1kHz. As discussed in Section 4.2.2, thin-film capacitors are frequency dependent; they should therefore be measured at the actual pole frequency ω_p, this being the network frequency that they determine. In general, high-accuracy, programmable capacitor bridges do not permit the choice of arbitrary test frequencies. Fortunately, because of the predictable frequency characteristic of thin-film capacitors, it is generally sufficient to measure each capacitor at two frequencies, say $\omega_1 = 2\pi \cdot 1 \text{ kHz}$ and $\omega_2 = 2\pi \cdot 10 \text{ kHz}$, and to inter-

polate or extrapolate the value for ω_p as the case may be. We then have, from (4-102),

$$\log C(\omega_p) = \log C_0 + \omega_p \tau_c \qquad (4\text{-}285)$$

where

$$\log C_0 = \frac{\omega_2 \log C_1 - \omega_1 \log C_2}{\omega_2 - \omega_1} \qquad (4\text{-}286a)$$

$$\tau_c = \frac{\log (C_2/C_1)}{\omega_2 - \omega_1} \qquad (4\text{-}286b)$$

and $C_i = C(\omega_i)$.

If a capacitor measurement at two different frequencies cannot be made (e.g., if the bridge provides only one measuring frequency) then it is often sufficient to measure the frequency dependence of a number of representative film capacitors and to plot them on semilog paper, as shown in Fig. 4-10. It has been shown that over a relatively wide frequency and capacitance range the plot will be linear and the slope constant. Thus, based on the representative measurements, an average slope can be obtained which in turn provides the value τ_c. We then need to measure $C(\omega)$ at only one frequency, say ω_1, and obtain $C(\omega_p)$ from $C_1 = C(\omega_1)$ and τ_c:

$$\log C(\omega_p) = \log C_1 + \tau_c(\omega_p - \omega_1) \qquad (4\text{-}287)$$

One further problem may be encountered in measuring the capacitors. If the chip capacitors are already mounted on a substrate, or if thin-film capacitors are deposited on the same substrate as the resistors, the capacitor to be measured will not be isolated, i.e., additional components will be connected in parallel with it. The capacitor can still be measured accurately, using the so-called three-terminal capacitance measurement.[52] Referring to Fig. 4-27, for a given measurement all impedances forming a parallel path with the

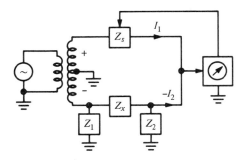

FIG. 4-27. Three-terminal capacitance-measurement circuit.

52. Thomas, H. E. and Clarke, C. A., *Handbook of Electronic Instruments and Measurement Techniques*, (Englewood Cliffs, N.J.: Prentice-Hall Inc., 1967).

unknown capacitor C_x are grounded at some point along the path. Two impedances to ground, Z_1 and Z_2, result, one at each terminal of the unknown. Z_1 becomes part of the source impedance of the signal generator, Z_2 becomes a part of the detector impedance. The major effect of these impedances is to reduce the sensitivity of the measuring set. Nevertheless, using this measuring system, good accuracy ($< 0.1\%$) can be obtained, even with high ratios (e.g., $10:1$) of the unknown impedance to the shunt impedances.[53]

2. *Measure the dissipation factors δ_3 and δ_4*. In the case of chip capacitors, the dissipation factors are very low, and below, say, 100 kHz they are practically independent of frequency. In general, the dissipation factor can therefore be ignored. With thin-film capacitors this is not the case, as discussed in Section 4.2.2 (see Fig. 4-11). The losses are both relatively high and frequency dependent. It would therefore be desirable to measure tan δ at the pole frequency. Since this is not generally possible, as pointed out for measurement 1, a two-frequency measurement at ω_1 and ω_2 again permits us to find tan $\delta(\omega_p)$ by interpolation. We obtain

$$\tan \delta(\omega_p) = \tan \delta' + \omega_p RC(\omega_p) \tag{4-288}$$

where

$$\tan \delta' = \frac{\tan \delta_1 + \tan \delta_2}{2} + \frac{\tan \delta_2 - \tan \delta_1}{2} \cdot \frac{\omega_1 C_1 + \omega_2 C_2}{\omega_1 C_1 - \omega_2 C_2} \tag{4-289a}$$

and

$$R = \frac{\tan \delta_2 - \tan \delta_1}{\omega_2 C_2 - \omega_1 C_1} \tag{4-289b}$$

Here $C_i = C(\omega_i)$ and tan $\delta_i = \tan \delta(\omega_i)$, $i = 1, 2$. In general tan $\delta_i \ll 1$ and can be replaced by δ_i. Naturally, the dissipation measurements δ_1 and δ_2 can be carried out at the same time as the capacitance measurements C_1 and C_2 in the preceding step.

In the event that measurements cannot be carried out at two different frequencies with a bridge, it is often possible to measure tan δ at a frequency low enough that

$$\tan \delta \approx \tan \delta' \tag{4-290}$$

The losses represented by R are primarily due to the relatively high sheet resistivity of the lower electrode of film capacitors (e.g., 5 $\Omega/\square$). This sheet resistivity is relatively well controlled. Given the geometry of the base electrodes of the thin-film capacitors, R can therefore be calculated relatively accurately for each capacitor and used in (4-288), together with (4-290), to compute tan $\delta(\omega_p) \approx \delta(\omega_p)$. In some cases, both ends of the base electrode of one or more of the capacitors on a substrate may be accessible (e.g.,

---

53. General Radio—1683 Automatic Capacitance Bridge Manual; also Teradyne-357 Capacitance Bridge.

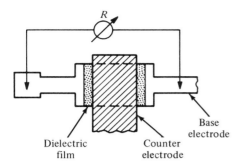

Dielectric   Counter   Base
film   electrode   electrode

Fig. 4-28. Capacitor on which resistance measurement may be made directly.

Fig. 4-28). In this case a resistance measurement can be made, either to obtain an average value for R or to compute the sheet resistivity with which to calculate the R values for each capacitance on a substrate. Alternatively, a dummy capacitor with representative dimensions and accessible lower electrode terminals (e.g., Fig. 4-28) may be included on each substrate, permitting the ready determination of an average value of R.

3. *Determine the 3 dB cutoff frequency ω_α of the closed-loop operational-amplifier gain* (Fig. 4-14). We could, of course, measure ω_α directly. It corresponds to the frequency at which the closed-loop gain of the operational amplifier has dropped by 3 dB. However, this constitutes a high-frequency measurement, and not a very accurate one at that. A more comprehensive procedure, comprising a simpler (i.e., low-frequency) measurement and providing information on the phase shift, not only due to the amplifier, but also due to other parasitic capacitance on a substrate, is based on the following reasoning. Assuming first an ideal amplifier ($\omega_\alpha \to \infty$) and ideal components, we have a phase ϕ_λ at the frequency ω_λ; this phase corresponds to ideal conditions, namely $s' = 0$ and $C_i^* = C_i$. It can be directly obtained from the transfer function of the ideal network (4-268). If, now, we *measure* the phase of the network at ω_λ we will not obtain the phase ϕ_λ. Because of the finite ω_α, and because of the other parasitics on the substrate (including such factors as the frequency-dependent and finite input impedance of the amplifier, the losses and frequency dependence of the capacitors, etc.), we will measure a phase ϕ_λ' which will be larger than the theoretical phase ϕ_λ. Substituting ϕ_λ' and ω_λ in (4-280) we can solve for ω_α, obtaining an expression of the form

$$\omega_\alpha = \frac{A_3' \omega_\lambda^3 + A_2' \omega_\lambda^2 + A_1' \omega_\lambda}{B_2' \omega_\lambda^2 + B_1' \omega_\lambda + B_0'} \tag{4-291}$$

where

$$A_i' (i = 1, \ldots, 3) = f_{A_i}'(R_1, R_2, R_4, C_3, C_4, \beta, \delta_3, \delta_4, \tan \phi_\lambda')$$
$$B_i' (i = 0, \ldots, 2) = f_{B_i}'(R_1, R_2, R_4, C_3, C_4, \beta, \delta_3, \delta_4, \tan \phi_\lambda') \tag{4-292}$$

All the quantities on the right-hand side of (4-292) are known by measuring the corresponding components of the as yet untuned circuit.

The Pretuning Step   The tuned circuit is required to have a pole frequency ω_p; the untuned circuit will have a *higher* pole frequency ω'_p (remember that resistors can only be trimmed in the *increasing* direction, therefore ω'_p *must* be higher than ω_p). For best accuracy the circuit should be *pretuned* such that ω'_p does not exceed ω_p by much more than 5 %. Due to the complicated frequency dependence of the parasitic effects of a circuit, which are only approximated reasonably well by $\beta(s)$ (see (4-274)) over a restricted frequency range, the frequency ω_λ at which ϕ'_λ is to be measured should be selected close to ω'_p. We do *not* select $\omega_\lambda = \omega'_p$, however, since the corresponding ϕ'_λ would be approximately 90°, and tan ϕ'_λ infinitely large. This will cause computational difficulties. Typically, if the circuit is pretuned such that ω'_p is 5 % higher than ω_p, ω_λ may be chosen, say, 8 % higher than ω_p.

To pretune the circuit to ω'_p we use our three tuning equations (4-282)–(4-284). Of course ω_α, which occurs in these equations, is still unknown. (Remember, we are pretuning our circuit to $\omega_p \approx 1.05\omega_p$ so that we can measure ϕ'_λ at $\omega_\lambda \approx 1.08\omega_p$ and therewith compute ω_α from (4-291).) To solve the equations we must use an *approximate* value for ω_α which we shall call ω'_α. This is not hard to obtain. The unity-gain bandwidth ω_g of our amplifier will be (approximately) specified by the manufacturer (e.g., $2\pi \cdot 1.5$ MHz). With a gain β of 2 we therefore have

$$\omega'_\alpha \approx \frac{\omega_g}{\beta} = \frac{2\pi \cdot 1.5}{2} \text{ [MHz]} \approx 5 \cdot 10^6 \text{ rad/sec} \tag{4-293}$$

We can now solve our three equations (4-282)–(4-284) for the unknowns R_1, R_2, and β as follows. We eliminate β from (4-283) and (4-284) by substituting the value for β given by (4-282). We then eliminate the product $R_1 R_2$ from (4-283) and (4-284); this results in a linear equation in R_2. Solving for R_2 and inserting the result in (4-284) we obtain a third-degree equation in R_1. We now have three equations of the form

$$D'_3 R_1^3 + D'_2 R_1^2 + D'_1 R_1 + D'_0 = 0 \tag{4-294}$$

$$E'_2 R_1^2 + E'_1 R_1 + E'_0 = R_2 \tag{4-295}$$

and

$$\beta = T(0)\left(1 + \frac{R_1 + R_2}{R_4}\right) \tag{4-296}$$

where

$$D'_i \ (i = 1, \ldots, 3) = f'_{D_i}(R_4, C_3, C_4, \delta_3, \delta_4, \omega'_\alpha, \omega'_{\phi_2}, K, \omega'_p, q_p)$$

and $\qquad\qquad\qquad\qquad\qquad\qquad\qquad\qquad\qquad\qquad\qquad\qquad$ (4-297)

$$E'_i \ (i = 0, \ldots, 2) = f'_{E_i}(R_4, C_3, C_4, \delta_3, \delta_4, \omega'_\alpha, \omega'_{\phi_2}, K, \omega'_p, q_p)$$

Solving these equations we obtain preliminary values for R_1, R_2, and β. These are used to *pretune* the circuit to ω_p'. Note that to compute the coefficient D_i' and E_i', the nominal design values for q_p and R_4 are used; C_3, C_4, δ_3, and δ_4 are measured; ω_α' is estimated as in (4-293); $\omega_{\phi_2}' = 1.05\,\omega_{\phi_2}$, where ω_{ϕ_2} is the nominal frequency at which the phase of $T(j\omega)$ is $-135°$; K is given and so is $\omega_p' = 1.05\omega_p$. Note that q_p is not explicit in the coefficients D_i' and E_i'; however it is implicit in the expressions used to determine ω_{ϕ_2}.

Of the three solutions for R_1 obtained from (4-294), the one closest to the design value is used to pretune R_1 to, and R_2 is pretuned to the value obtained by (4-295). The value of $\beta = 1 + (R_6/R_7)$ is obtained from (4-296) and the resistors R_6 and R_7 trimmed accordingly, while R_4 is trimmed to its design value.[54]

The Final Tuning Step and q_p Correction   After the pretuning step, ϕ_λ' can be measured at ω_λ and ω_α obtained from (4-291). Having computed ω_α, the three components R_1, R_2, and β can be calculated once again, using the same equations (4-294)–(4-296), except that now the coefficients D_i and E_i have no primes. They are functions of

$$D_i \ (i = 1, \ldots, 3) = f_{D_i}(R_4, C_3, C_4, \delta_3, \delta_4, \omega_\alpha, \omega_{\phi_2}, K, \omega_p, q_p)$$
$$E_i \ (i = 0, \ldots, 2) = f_{E_i}(R_4, C_3, C_4, \delta_3, \delta_4, \omega_\alpha, \omega_{\phi_2}, K, \omega_p, q_p) \quad (4\text{-}298)$$

Here R_4 is measured, ω_α is computed from (4-291), and ω_p and ω_{ϕ_2} are given by $T(s)$. The other components are the same as in the pretuning step. The circuit is now tuned by trimming R_1, R_2, and β (i.e., the resistor R_6 or R_7) to the final values obtained here.

Since q_p is particularly sensitive to component inaccuracies it will most likely have the largest error after the final tuning step. It is therefore often desirable to make a final correction of q_p using β, which alone can be adjusted in increasing *and* decreasing directions. We must therefore derive the relationship between the variation $\Delta\beta$ and the error Δq_p it is needed to correct for. Since $\Delta\beta$ will be a very small number, we can go directly to the transfer function of the ideal network as given by (4-268) and (4-269c), thereby neglecting the effects of ω_α and the $\tan \delta_i$. Solving for β we have

$$\beta = f(q_p) = 1 + \frac{R_2}{R_4} + \frac{(R_1 + R_2)C_4}{R_1 C_3} - \frac{\omega_p}{q_p} R_2 C_4 \quad (4\text{-}299)$$

Measuring q_p we can therefore calculate the corresponding gain β_m:

$$\beta_m = f(q_{p\,\text{meas}}) \quad (4\text{-}300a)$$

54. To ensure network stability it is recommended to pretune β to a value slightly lower than calculated. Thus R_7 may be tuned to the design value and R_6 5% lower than the calculated value. Similarly, taking the higher pole frequency $\omega_p' = 1.05\omega_p$ into account, it may be advisable to tune R_4 in two steps: in the pretuning step to a value $R_4' = 0.95R_4$, and in the final tuning step to the design value R_4. To achieve high tuning accuracy the measured value of R_4' and R_4, respectively, must then be used in the pretuning and in the final-tuning equations (see (4-297) and (4-298) respectively).

Similarly, the *specified* q_p corresponds to a gain β_s:

$$\beta_s = f(q_{p \text{ spec}}) \tag{4-300b}$$

We then readily obtain

$$\Delta\beta = \beta_s - \beta_m = \omega_p R_2 C_4 \left(\frac{1}{q_{p \text{ meas}}} - \frac{1}{q_{p \text{ spec}}} \right) \tag{4-301}$$

$q_{p \text{ meas}}$ can be obtained by measuring ϕ_1 and ϕ_2 at ω_{ϕ_1} and ω_{ϕ_2}, respectively, where ϕ_1 is in the vicinity of $-45°$, ϕ_2 in the vicinity of $-135°$ (see (4-234b)). By not specifying $-45°$ and $-135°$, we do not have to search for the two frequencies ω_{45} and ω_{135} but can use any two frequencies ω_{ϕ_1} and ω_{ϕ_2} in their vicinity. With (4-230) we then obtain

$$q_{p \text{ meas}} = \frac{\omega_{\phi_1}}{\tan \phi_1} \frac{\sqrt{(\Omega_0 \omega_{\phi_1}^2 - \omega_{\phi_2}^2)(\Omega_0 - 1)}}{\omega_{\phi_1}^2 - \omega_{\phi_2}^2} \tag{4-302}$$

where

$$\Omega_0 = \frac{\omega_{\phi_2} \tan \phi_1}{\omega_{\phi_1} \tan \phi_2} \tag{4-303}$$

The advantage of using (4-302) instead of (4-233) to evaluate q_p is that the measured value of ω_p does not occur in the former equation. This is important since ω_p, a parameter that is itself adjusted for, already has a tuning error associated with it. Thus, with (4-302) the accuracy of the ω_p and the q_p adjustments can be evaluated independently of one another. For the same reason, we can now eliminate ω_p from (4-301) and obtain

$$\Delta\beta = \frac{q_p - q_m}{q_m^2} \frac{1}{R_1 C_3} \frac{\omega_{\phi_1}}{2q_m \cdot \tan \phi_1} (\sqrt{1 + 4q_m^2 \tan^2 \phi_1} - 1) \tag{4-304}$$

where $q_m = q_{p \text{ meas}}$ and $q_p = q_{p \text{ spec}}$.

Having found $\Delta\beta = f(\Delta q_p)$ we can immediately obtain the corresponding adjustments for R_6 or R_7 from

$$\beta = 1 + \frac{R_6}{R_7} \tag{4-305}$$

namely

if $\Delta\beta > 0$ then $\Delta R_6 = \Delta\beta R_7;$ $\Delta R_7 = 0$

if $\Delta\beta < 0$ then $\Delta R_7 = -\Delta\beta \dfrac{R_7^2}{R_6};$ $\Delta R_6 = 0$ $\qquad$ (4-306)

In adjusting β by $\Delta\beta$ for an accurate value of q_p, we affect the already tuned DC gain $T(0)$ (see (4-282)) as well as the already tuned frequencies ω_p and ω_{ϕ_2}. However $S_\beta^{q_p}$ is approximately q_p times larger than $S_\beta^{T(0)}$ and the frequencies ω_p and ω_{ϕ_2} only depend on β through second-order parasitic effects. Thus, the changes of $T(0)$, ω_p and ω_{ϕ_2} due to $\Delta\beta$ will be negligibly small.

Summary of the Tuning Procedure With the (functional) correction of q_p by $\Delta\beta$ the tuning procedure is at an end. Let us briefly recapitulate the main steps involved in the deterministic tuning of our low-pass network in Fig. 4-26.

1. Measure the capacitors and their dissipation factors. If thin-film capacitors are used, values obtained either by interpolation or direct measurement at the pole frequency ω_p should be used.

2. Pretune the network to a pole frequency $\omega_p' = 1.05\omega_p$; the corresponding $-135°$ frequency is $\omega_{\phi_2}' = 1.05\omega_{\phi_2}$. This step is carried out as follows: Using ω_p', ω_{ϕ_2}', ω_α', the design values for R_4, $T(0)$, and q_p and the measured data from step 1, the values R_1, R_2, and β are calculated from (4-294)–(4-296). As a result, all the resistors can be trimmed to value: R_1, R_2, R_6, R_7 to the calculated values and R_4 to the design value.[55] (Since $\beta = 1 + R_6/R_7$, one of the resistors, say R_7, can also be trimmed to its design value, R_6 to the value corresponding to the calculated β.)

3. Measure the phase ϕ_λ' at a frequency $\omega_\lambda = 1.08\omega_p$. Calculate ω_α from (4-291).

4. Calculate the final values of R_1, R_2, and β using the specified parameters ω_p, ω_{ϕ_2}, q_p, the measured[56] value of R_4, the measured data from step 1, and ω_α from step 3. Trim R_1, R_2, R_6, or R_7 accordingly.

5. Determine q_p from (4-302) and, if necessary, correct according to (4-306).

The deterministic tuning procedure outlined above for a low-pass network is similar to that used for any other network type. Although involving some rather formidable computational steps, the procedure, in practice, is relatively simple. As mentioned earlier, an on-line computer (e.g., minicomputer or even in many cases a desk calculator) is necessary to carry out the calculations accurately and rapidly enough for manufacture in a production line. The procedure then breaks down into (i) an initial measurement followed by a pretuning step, (ii) a second measuring step followed by a tuning step, and (iii) if necessary, a final touching-up step.

Simplifications of the Tuning Procedure The procedure outlined above is quite elaborate in that it permits relatively gross parasitic effects to be accurately accounted for. In very many cases the procedure can be simplified

55. As explained in the discussion of pretuning above it may be preferable to use a value $R_4' = 0.95R_4$ in the calculations here and consequently to pretune R_4 5% low. Similarly, to ensure stability R_6 may be pretuned 5% low as well.
56. In the pretuning step it may be sufficient to use the design value of R_4 or R_4'; in the final tuning step, in order to guarantee the utmost accuracy, it is preferable to use the measured value of R_4. In this way the tuning error incurred during the tuning of R_4 to its design value is taken into account in the computations.

considerably. At voice frequencies (say, below 3 kHz) neither the pretuning (step 2 above) nor the inclusion of ω_α (step 3 above) is required. Even at higher frequencies, the approximate value of $\omega'_\alpha = \omega_g/\beta$ may be quite sufficient, thereby obviating the need for steps 2 and 3. Instead of measuring $C(\omega_p)$ and $\tan \delta(\omega_p)$ in step 1, either an average value or an interpolated one will often be enough. Furthermore, if ω_p is above 3 kHz (in which case the measurement of ω_α may be advisable) it will generally be sufficient to assume average $\tan \delta$ values or even to ignore them completely. The error thereby incurred is compensated for by the comprehensive ω_α measurement, which, it will be remembered, takes not only the frequency-dependent gain into account, but all other parasitic effects (e.g., the capacitor losses) as well. Due to all of these parasitic effects, the phase of the network is increased from ϕ_λ to ϕ'_λ at the frequency ω_λ.

In its simpler form, the deterministic tuning procedure will consist of the following steps:

For voice- and low-frequency applications (e.g., $f_p < 3$ kHz):

1. Measure the capacitors and their dissipation factors. For frequencies below 1 kHz the frequency dependence of C and $\tan \delta$ can be neglected.
2. Calculate the *final* values of R_1, R_2, and β using the specified parameters $T(0)$, ω_p, ω_{ϕ_2}, q_p, the design value R_4, and the measured data from step 1.
3. Determine q_p from (4-302) and, if necessary, correct according to (4-306).

For frequencies higher than the voice band (e.g., $f_p > 3$ kHz) the five steps outlined earlier should be carried out, except that the $\tan \delta$ values may be averaged or left out altogether. Naturally the simpler the procedure, the simpler will be the computations and the computer aids required. In many cases, as in those involving the three steps enumerated above, the computer can readily be replaced by a desk calculator.

Networks Requiring Other Procedures   In the example above, we noticed that when introducing parasitic effects into our low-pass network the initially second-order transfer function became third-order, i.e., the fundamental nature of the transfer function changed. We were able to cope with this situation, by realizing that the increase in the order of the transfer function affects the response mainly outside the band of interest, i.e., at frequencies well beyond the actual operating frequency range of our network. It was to compensate for its residual influences within the band of interest that we had to take this second-to-third-order transformation into consideration in our set of tuning equations, since for high accuracy and a one- or two-step deterministic tuning procedure even minimal departures from the ideal must be accounted for.

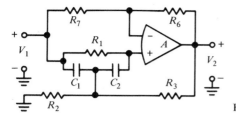

FIG. 4-29. A second-order all-pass network.

In contrast to this situation, there are cases in which the introduction of parasitics changes the nature of the transfer function significantly at all frequencies. In such cases we cannot merely modify the tuning equations of the original, ideal, network but must proceed in an altogether different manner. An example will help to illustrate such a case. Consider the second-order all-pass network shown in Fig. 4-29. As we know the desired transfer function of a second-order all-pass network has the form

$$T(s) = \frac{s^2 - \dfrac{\omega_0}{q} s + \omega_0^2}{s^2 + \dfrac{\omega_0}{q} s + \omega_0^2} \tag{4-307a}$$

and

$$T(j\omega) = \frac{(\omega_0^2 - \omega^2) - j\dfrac{\omega_0}{q}\omega}{(\omega_0^2 - \omega^2) + j\dfrac{\omega_0}{q}\omega} \tag{4-307b}$$

If, now, we analyze the circuit of Fig. 4-29, and include the dissipation factors $\tan \delta_1 \approx \delta_1$ and $\tan \delta_2 \approx \delta_2$ associated with the network capacitors C_1 and C_2, we obtain a transfer function of the form

$$T'(j\omega) = \frac{A_0 - jB_0}{A_0 - \delta A_1 + j(B_0 + A_1)} \tag{4-308}$$

where the coefficients A_i and B_i are functions of the circuit components and of the parameters of $T(s)$, and we assume $\delta_1 = \delta_2 = \delta$. Note that (4-308) can only realize the conditions of an all-pass function (i.e., real parts of numerator and denominator equal; imaginary parts equal but opposite in sign) in the ideal case, i.e., when $\delta = 0$. With the inclusion of nonideal capacitors the all-pass nature of the circuit is therefore immediately impaired. Of course, we must now ask ourselves to what degree this impairment takes place and whether it is not possible to approximate an all-pass function after all. Fortunately the δ values are very small (typically 0.004 or less) and a close approximation to the ideal is indeed possible.

Obtaining the Tuning Equations   In approximating (4-307) by (4-308) we must find a set of resistor values minimizing the inevitable errors in magnitude and phase that occur in our nonideal all-pass network. Since we have a limited number of variables, namely the resistors R_1, R_2, and R_3 and the gain $\beta = 1 + (R_6/R_7)$, we are limited in the number of points along the frequency characteristic at which we can minimise the error. Furthermore, we must decide whether the optimized points should apply to the magnitude or to the phase response since we do not have enough variables to optimize both. One approach to this problem is to use an optimization computer routine and to perturb the nominal resistor values until the phase, for example, is close enough to the desired response within the frequency band of interest. The approximation can be carried out by minimizing the error between the actual and desired response at various frequencies; the error can be minimized in a mean-square or Chebyshev sense. In the case of the all-pass, in which the phase generally has higher priority than the amplitude response, it may well turn out that optimizing the phase actually brings a satisfactory amplitude response with it.

Instead of simply selecting various points along the desired frequency response at which the errors between actual and desired values are to be minimized, it may be possible to select a set of criteria (as many as there are independent tuning parameters) whose accurate fulfillment approximates the response over the specified frequency range sufficiently well. This approach resembles the previous deterministic tuning method illustrated for the low-pass network, except that there the criteria chosen were not very critical, since their realization almost invariably guaranteed a satisfactory overall response. Here, however, it is a question of judiciously selecting a set of criteria that will at least satisfy the most important of the specified requirements while compromising on those of lesser significance. Such a set of constraints will depend on the circumstances involving the network and may therefore differ from one application to another. However, in the case of an all-pass network, it has been found[57] that the following set of constraints (or equations) would generally provide satisfactory all-pass characteristics (compared to the ideal), independently of the network used. They seem to pin down the most crucial aspects of the all-pass network accurately, relegating the deteriorating effects of the parasitics to aspects of the function that are of lesser concern. These constraints are:

1. The phase $\phi(\omega)$ pertaining to the transfer function $T(j\omega)$ should be $-180°$ at ω_0. Thus, from (4-308)

$$\tan \phi(\omega_0) = \tan \left. \frac{-A_1(A_0 + \delta B_0)}{A_0^2 + B_0^2 + A_1(B_0 - \delta A_0)} \right|_{\omega = \omega_0} = 0 \quad (4\text{-}309a)$$

57. E. Lueder, op. cit.

Consequently we require that

$$(A_0 + \delta B_0)_{\omega = \omega_0} = 0 \qquad (4\text{-}309\text{b})$$

2. The delay $\tau(\omega) = d\phi/d\omega$ should take on its specified value (see (4-259a)) at ω_0:

$$\left. \frac{d\phi(\omega)}{d\omega} \right|_{\omega = \omega_0} = -\frac{4q}{\omega_0} \qquad (4\text{-}310)$$

3. $|T(j\omega)|$ should be unity at ω_0:

$$|T(j\omega_0)| = \left| \frac{A_0 - jB_0}{A_0 - \delta A_1 + j(B_0 + A_1)} \right|_{\omega = \omega_0} = 1 \qquad (4\text{-}311\text{a})$$

or, neglecting second-degree terms in δ,

$$(2B_0 + A_1 - 2\delta A_0)_{\omega = \omega_0} = 0 \qquad (4\text{-}311\text{b})$$

4. At the frequency ω_{90} or ω_{270} as given by (4-261), $\phi(\omega_{90})$ or $\phi(\omega_{270})$ should be $-90°$ or $-270°$, respectively. Thus, for ω_{90}, we have, from (4-309a),

$$[A_0^2 + B_0^2 + A_1(B_0 - \delta A_0)]_{\omega = \omega_{90}} = 0 \qquad (4\text{-}312)$$

Solving the Tuning Equations Equations (4-309) to (4-312) represent a nonlinear system of equations for the resistors of the all-pass network. The solution of such equations is a separate problem in itself. Sometimes only some of the equations are interrelated in a nonlinear way, and an appropriate computer program may be found for their solution. As an initial solution, the design values pertaining to the ideal circuit may be used; they will generally not differ greatly from the final optimized values.

In some cases it may be possible to circumvent altogether the problem of solving a set of nonlinear equations. In the case of our all-pass network, for example, we obtain two highly nonlinear equations in the unknowns R_1 and β of the form

$$A_0 + A_1 R_1 + A_2 R_1^2 + A_3(1 - \beta)R_1^3 + A_4(1 - \beta)R_1^2 + A_5(1 - \beta)^2 R_1^4 = 0 \qquad (4\text{-}313)$$

and

$$B_0 + B_1 R_1 + B_2 R_1^2 + B_3(1 - \beta)R_1^3 + B_4(1 - \beta)R_1^2 + B_5(1 - \beta)^2 R_1^4 = 0 \qquad (4\text{-}314)$$

The remaining two unknowns, R_2 and R_3, result from linear equations of the form

$$R_3 = \frac{\beta R_1}{a_0 + a_1 R_1 + a_2(1 - \beta)R_1^2} \qquad (4\text{-}315)$$

and

$$R_2 = \frac{R_3}{b_0 R_1} \cdot \frac{1}{R_3 - (1/b_1 R_1^2)} \qquad (4\text{-}316)$$

where the coefficients a_i and b_i are functions of the known circuit components and the parameters of $T(s)$. The normal procedure for solving this set of four equations (which result from equations (4-309) to (4-312)) is first to solve the two nonlinear equations (4-313) and (4-314) for R_1 and β and then from (4-315) and (4-316) to compute R_3 and R_2, respectively. It has been found,[58] however, that at least as good an approximation of the desired all-pass response can be obtained by assuming the *design* value of R_1 (asssociated with the ideal network) as the *final* value for R_1, solving a linear combination of (4-313) and (4-314) for β (after substituting R_1, this entails no more than the solution of a quadratic equation in β) and then solving (4-315) and (4-316) for R_3 and R_2, respectively.

4.4.5 Accuracy Considerations for Deterministic Tuning

We shall examine here how accurate the measurements of capacitors and dissipation factors and the tuning (and measurement) of resistors must be to attain a specified accuracy of the frequency ω_p or ω_{ϕ_2}.

The accuracy of a frequency, determined by resistors and capacitors, is related directly to the accuracy with which the capacitors can be measured and the resistors trimmed. Assuming an equal, worst-case measuring error of $\Delta C/C$ for all the capacitors and an equal, worst-case trimming error of $\Delta R/R$ for each of the resistors, we obtain a frequency error

$$\frac{\Delta \omega}{\omega} = -\left(\frac{\Delta R}{R} + \frac{\Delta C}{C}\right) \qquad (4\text{-}317)$$

Thus, for example, with a 0.1%-accurate capacitance bridge and a capability of trimming and measuring resistors to within 0.05%, the worst-case frequency error will be 0.15%.

The effect of measurement inaccuracies in the capacitor dissipation factors is a little more difficult to evaluate. Here we must examine the equation from which the frequency-determining component is derived. Returning, for example, to our low-pass network (Fig. 4-26) the corresponding equation is given by (4-283). Neglecting higher than first-order terms, and assuming an equal dissipation factor δ for all the capacitors, the pole frequency ω_p can be shown to have the form:[59]

$$\omega_p' = \omega_p \left[1 + \frac{\delta}{2q_p}\left(1 + \frac{1}{4q_p}\right)\right] \qquad (4\text{-}318)$$

58. S. E. Sussmann, Unpublished Bell Telephone Laboratories memorandum.
59. E. Lueder, op. cit.

where ω_p is the pole frequency of the network with ideal, that is, lossless capacitors. Assuming a worst-case measuring error $\Delta\delta$, it then follows from (4-318) that the corresponding frequency error is

$$\frac{\Delta\omega'_p}{\omega'_p} = \frac{1}{2q_p}\left(1 + \frac{1}{4q_p}\right)\Delta\delta \qquad (4\text{-}319)$$

With a dissipation factor $\delta = 0.008$ measured to within 10% accuracy (i.e., $\Delta\delta/\delta = 0.1$) and $q_p = 2$ we have a frequency error $\Delta\omega'_p/\omega'_p = \frac{1}{4}(1 + \frac{1}{8})\cdot 0.08\%$ $\approx 0.02\%$. As q_p increases this error, of course, becomes smaller. With a frequency error due to resistor and capacitor inaccuracies of 0.15%, the total worst-case frequency error in our example (which may be considered typical) is on the order of 0.17%.

Although (4-318) results from an explicit network example, the resulting frequency error has a form typical for most network types. Thus, in general, a $\tan\delta$ measurement of 10% accuracy will result in a frequency error of less than 0.05%.

To find the error of the pole Q, q_p, resulting from an error of the gain β, we can proceed as follows. Assuming that a resistor R_β is used to tune q_p in the final tuning step mentioned earlier, we have

$$\frac{\Delta q_p}{q_p} = S^{q_p}_\beta S^\beta_{R_\beta} \frac{\Delta R_\beta}{R_\beta} \qquad (4\text{-}320)$$

From (4-305) it follows that R_β may, for example, be one of the two resistors R_6 or R_7 that determine β. Thus

$$S^\beta_{R_6} \frac{\Delta R_6}{R_6} = \left(1 - \frac{1}{\beta}\right)\frac{\Delta R_6}{R_6} \qquad (4\text{-}321a)$$

$$S^\beta_{R_7} \frac{\Delta R_7}{R_7} = -\left(1 - \frac{1}{\beta}\right)\frac{\Delta R_7}{R_7} \qquad (4\text{-}321b)$$

Furthermore, for the class 4 network of Fig. 4-26 we have $S^{q_p}_\beta = (q_p/\hat{q}) - 1$, so that (4-320) has the form

$$\frac{\Delta q_p}{q_p} = \left(1 - \frac{1}{\beta}\right)\left(\frac{q_p}{\hat{q}} - 1\right)\frac{\Delta R_\beta}{R_\beta} \qquad (4\text{-}322)$$

The error in q_p is proportional to q_p and to the error incurred in the tuning of R_β. For example, for $\beta = 2$, $q_p = 2$, $\hat{q} = 0.4$ and a resistor tuning accuracy of 0.05% we obtain $\Delta q_p/q_p = \frac{1}{2}[(2/0.4) - 1]0.05\% = 0.1\%$. For $q_p = 20$, $\Delta q_p/q_p$ will only be 1%, which is still well within the initial accuracy generally specified for q_p. Note, furthermore, that the tuning error in q_p due to the resistor error $\Delta R_\beta/R_\beta$ will be all the smaller, the closer the gain β is to unity.

4.5 TUNING TOLERANCES OF CASCADED nth-ORDER NETWORKS[60]

We have now reached the point in the tuning of our active filters, where each second-order section may be assumed to have been tuned to some accuracy either by one of the two methods discussed above or by a combination of the two. Thus, the five parameters K, ω_z, q_z, ω_p, and q_p of each of the general second-order sections have been tuned to within $\pm\Delta K/K$, $\pm\Delta\omega_z/\omega_z$, $\pm\Delta q_z/q_z$, $\pm\Delta\omega_p/\omega_p$, and $\pm\Delta q_p/q_p$ percent, respectively, and we must give some thought to checking the overall response of each section (as opposed to checking its individual parameters K, ω_z, q_z, etc.) as well as the response of the final cascaded nth-order filter as a whole. We are, after all, generally interested in the frequency response of the final filter which will be required to lie within a specified tolerance band, as shown in Fig. 4-30. The question that we must now answer is this: how can we deduce the worst-case tolerance band associated with the frequency response of a network from the known pole and zero displacements caused by the tuning errors $\pm\Delta\omega_z/\omega_z$, $\pm\Delta q_z/q_z$, $\pm\Delta\omega_p/\omega_p$ and $\pm\Delta q_p/q_p$? (The tuning error $\pm\Delta K/K$ only affects the overall level and can therefore be taken into account directly. In Fig. 4-30 the tolerance band is simply widened by the amount $\pm\Delta K/K$).

4.5.1 Second-Order Networks

Let us start out with a general second-order network. Because of tuning errors, each pole and zero may lie within a tolerance domain defined by

$$\omega_{z\,min} \leq \omega_z \leq \omega_{z\,max}; \qquad \omega_{p\,min} \leq \omega_p \leq \omega_{p\,max}$$
$$q_{z\,min} \leq q_z \leq q_{z\,max}; \qquad q_{p\,min} \leq q_p \leq q_{p\,max} \qquad (4\text{-}323)$$

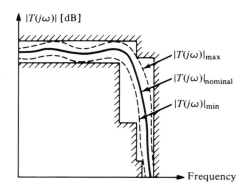

FIG. 4-30. Tolerance band of frequency response of an nth-order filter.

60. Much of this material is based on the paper by I. E. Berkovics and G. S. Moschytz, Analysis of worst-case gain for nth-order transfer functions realized by cascaded 2nd-order networks, Electronic Lett., 8, 199–200 (1972).

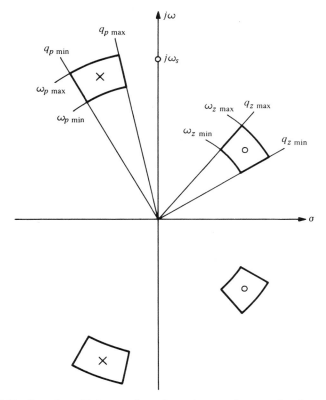

FIG. 4-31. Domains of tolerance for poles and zeros of a second-order network.

The corresponding domains in the *s*-plane are shown in Fig. 4-31. These domains are given; they result directly from the accuracy of our tuning method. Restating our problem with respect to these tolerance domains we ask, where within these domains must the poles and zeros lie so that the worst-case tolerance band of the frequency response is obtained? Restricting ourselves to the magnitude of the transfer function (i.e., amplitude response) we may drop the sign of the *q*'s, thereby assuming that all poles and zeros lie in the LHP.

Considering the magnitude, expressed in decibels of $T(j\omega)$ at a frequency ω_s, we obtain the following two equations for the extrema T_{min} and T_{max} of $|T(j\omega_s)|$:

$$T_{min}(\omega_s)\ [\text{dB}] = 20 \log |T(j\omega_s)|_{min}$$

$$= 10 \log \left[(\omega_z^2 - \omega_s^2)^2 + \frac{\omega_z^2 \omega_s^2}{q_z^2} \right]_{min}$$

$$- 10 \log \left[(\omega_p^2 - \omega_s^2)^2 + \frac{\omega_p^2 \omega_s^2}{q_p^2} \right]_{max} \qquad (4\text{-}324)$$

and

$$T_{max}(\omega_s) \, [\text{dB}] = 20 \log |T(j\omega_s)|_{max}$$

$$= 10 \log \left[(\omega_z^2 - \omega_s^2)^2 + \frac{\omega_z^2 \omega_s^2}{q_z^2} \right]_{max}$$

$$- 10 \log \left[(\omega_p^2 - \omega_s^2)^2 + \frac{\omega_p^2 \omega_s^2}{q_p^2} \right]_{min} \qquad (4\text{-}325)$$

In order to find T_{min} and T_{max} let us define the function

$$F(\omega_0, q_0) = (\omega_0^2 - \omega_s^2)^2 + \frac{\omega_0^2 \omega_s^2}{q_0^2} \qquad (4\text{-}326)$$

We now look for the values of ω_0 and q_0 (for a given ω_s) for which $F(\omega_0, q_0)$ has a maximum or a minimum. Clearly, these values will then lead us directly to the ω_0, q_0 combinations for which we obtain the extrema T_{max} and T_{min}.

It follows by inspection of (4-326) that, for a fixed ω_0, we have

$$F(q_0)|_{min} = F(q_{0\,max})$$

$$F(q_0)|_{max} = F(q_{0\,min}) \qquad (4\text{-}327)$$

Our next step is to examine the effect of variations of ω_0 on F for a fixed q_0. Let F have a minimum for $\omega_0 = \omega_{0m}$. We obtain ω_{0m} by setting the derivative of $F(\omega_0, q_0)$ with respect to ω_0 equal to zero. Then

$$\omega_{0m} = \omega_s \sqrt{1 - \frac{1}{2q_0^2}} \qquad (4\text{-}328)$$

By taking the second derivative of F with respect to ω_0, we find that F increases on both sides of ω_{0m}; hence $F(\omega_{0m})$ is an absolute minimum of $F(\omega_0)$. We now have

$$F(\omega_0, q_0)|_{min} = F(\omega_{0m}, q_{0\,max}) \qquad (4\text{-}329)$$

To obtain $F(\omega_0, q_0)_{max}$ we must examine the significance of ω_{0m} with respect to F more closely. First of all, a glance at (4-328) shows that ω_{0m} becomes imaginary for $q_0 < 1/\sqrt{2}$. Naturally, this is meaningless; the real minimum for $q_0 < 1/\sqrt{2}$ therefore occurs at the origin. For the more interesting case, when $q_0 \geq 1/\sqrt{2} = 0.707$, ω_{0m} is given by (4-328). The locus of $\omega_{0m} = \omega_{0m}(q_0)$ in the s-plane is a lemniscate situated symmetrically on the $j\omega$ axis, as shown in Fig. 4-32. As a function of the angle ϕ_0 (see Fig. 4-32) the lemniscate is given by

$$\omega_{0m}^2 = -\omega_s^2 \cos 2\phi_0 \qquad (4\text{-}330)$$

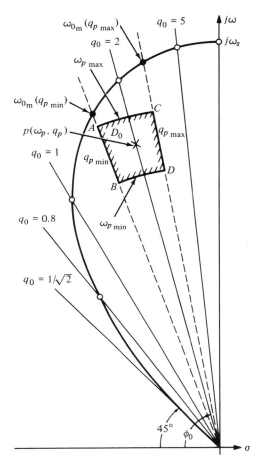

FIG. 4-32. The locus of $\omega_{0m} = \omega_{0m}(q_0)$ in the s-plane.

where the angle ϕ_0 is given by the frequency and Q of the original root pair, ω_0 and q_0. Thus

$$\phi_0 = \cos^{-1} \frac{\omega_0}{2q_0} \tag{4-331}$$

To understand the significance of the lemniscate in Fig. 4-32, consider the pole p in that figure and its corresponding pole domain D_0 given by the shaded area. This domain characterizes the accuracy of the process by which the pole could initially be tuned to its nominal location $p(\omega_p, q_p)$. Considering, now, that we have a second-order low-pass function, for example, whose poles are p and p^*, we are interested in knowing which deviation of p due to the tuning error we should assume within D_0, so as to find the maximum and minimum values of the frequency response at the frequency ω_s. In terms of $F(\omega_p, q_p)$ we can express the magnitude of our low-pass function as

$$|T(j\omega_s)| = -10 \log F(\omega_p, q_p) \text{ [dB]} \qquad (4\text{-}332)$$

From (4-327) it follows immediately that $T_{min}(\omega_s)$ must lie along the side AB corresponding to q_{min}; similarly $T_{max}(\omega_s)$ must lie along the side CD corresponding to q_{max}. To find the actual points along these two sides, where $T_{max}(\omega_s)$ and $T_{min}(\omega_s)$ are obtained, we reason as follows. Considering $T_{max}(\omega_s)$ first, we know that it lies along the side CD of our domain D_0. At the intersection of CD and the lemniscate we have the point $\omega_{0m}(q_{p\,max})$ at which F has a minimum, and therefore $|T(j\omega_s)|$ a maximum. The closest point in D_0 to this absolute minimum is the point C; thus $T_{max}(\omega_s)$ is obtained at the point C. Similar reasoning gives us the domain point for $T_{min}(\omega_s)$. Here we look for the point furthest away from the intersection $\omega_{0m}(q_{p\,min})$ along the stretch AB; it is, of course, the point B at which we obtain $T_{min}(\omega_s)$.

Referring to Fig. 4-33, we have so far found T_{max} and T_{min} at the frequency ω_s; in practice we shall require these extrema either over a whole frequency range or at least at a number of discrete test frequencies. For each additional frequency along the $j\omega$ axis we obtain a new lemniscate and must find the two points within, or on the bounds of D_0, relative to each new lemniscate, that give us T_{max} and T_{min}. In proceeding from low to high frequencies along the $j\omega$ axis we therefore generate an inflating lemniscate that is first below D_0, then traverses D_0, intersecting it at two sides in various ways and then leaving it behind, as shown qualitatively in Fig. 4-34. Depending on the location of the pole (or zero) pair, there are six relative positions of the lemniscate with respect to D_0, each of which results in a different pair of intersections along the periphery of D_0, corresponding to the values of T_{max} and T_{min}. These six relative positions with respect to D_0 are sketched in Fig. 4-35.

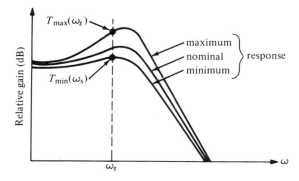

FIG. 4-33. Maximum, minimum, and nominal response curves for (4-332).

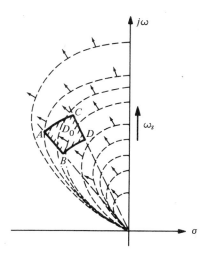

FIG. 4-34. Inflation of lemniscate locus of ω_{0m} as ω_s proceeds from low to high frequencies along the $j\omega$ axis.

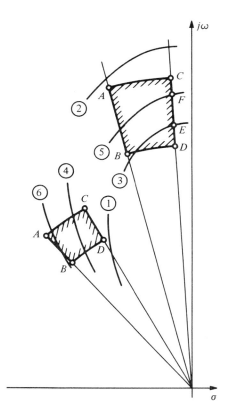

FIG. 4-35. The six possible relative positions of the lemniscate with respect to D_0 (see Figs. 4-32 and 4-34).

Referring to this figure, the extrema for each lemniscate–D_0 configuration are as follows:

$$\begin{aligned}
&\text{Case 1: } F = F_{min} \text{ in } D \qquad \text{Case 4: } F = F_{min} \text{ in } C \\
&\qquad\quad\; F = F_{max} \text{ in } A \qquad\qquad\quad\; F = F_{max} \text{ in } A \\
&\text{Case 2: } F = F_{min} \text{ in } C \qquad \text{Case 5: } F = F_{min} \text{ in } F \\
&\qquad\quad\; F = F_{max} \text{ in } B \qquad\qquad\quad\; F = F_{max} \text{ in } A \text{ or } B \\
&\text{Case 3: } F = F_{min} \text{ in } E \qquad \text{Case 6: } F = F_{min} \text{ in } C \\
&\qquad\quad\; F = F_{max} \text{ in } A \qquad\qquad\quad\; F = F_{max} \text{ in } A \text{ or } B
\end{aligned} \qquad (4\text{-}333)$$

Note that there is an ambiguity in the location of F_{max} for cases 5 and 6. This can be resolved if we observe that, for both cases, the following statements hold:

$$\begin{aligned}
&\text{if} \quad \omega_{p\,max}^2 + \omega_{p\,min}^2 > 2\omega_{0m}^2 \quad \text{then} \quad F = F_{max} \text{ in } A \\
&\text{if} \quad \omega_{p\,max}^2 + \omega_{p\,min}^2 < 2\omega_0^2 m \quad \text{then} \quad F = F_{max} \text{ in } B
\end{aligned} \qquad (4\text{-}334)$$

The same applies, of course, whether we are dealing with a pole or a zero.

4.5.2 nth-Order Networks

So far we have considered only the worst-case magnitude tolerances of a second-order function. Clearly, for an nth-order function we proceed as follows: to find the $T_{max}(\omega_s)$ and $T_{min}(\omega_s)$ values for the frequency response of the complete filter, we let ω_s traverse the frequency range of interest, i.e., we cause the lemniscate associated with each ω_s to expand from $\omega_{s\,min}$ to $\omega_{s\,max}$ (see Fig. 4-36). At discrete frequencies ω_s within this band we examine the relative position of each zero and pole domain with respect to the corresponding lemniscate and find the F_{max} and F_{min} values for each zero and pole at the ω_s in question. At that frequency ω_s the overall maximum and minimum magnitudes of the nth-order transfer function are then computed from

$$T_{max}(\omega_s) = \sum_i 10 \log F_{max}(\omega_{z_i}, q_{z_i}, \omega_s)$$

$$- \sum_j 10 \log F_{min}(\omega_{p_j}, q_{p_j}, \omega_s) \text{ [dB]} \qquad (4\text{-}335)$$

and

$$T_{min}(\omega_s) = \sum_i 10 \log F_{min}(\omega_{z_i}, q_{z_i}, \omega_s)$$

$$- \sum_j 10 \log F_{max}(\omega_{p_j}, q_{p_j}, \omega_s) \text{ [dB]} \qquad (4\text{-}336)$$

These two expressions determine the worst-case gain tolerance over a given frequency range. They therefore provide the test specifications for active

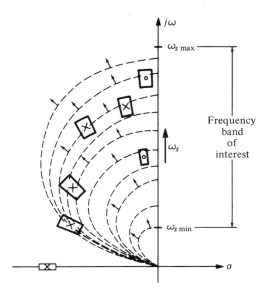

FIG. 4-36. Effect of letting ω_s traverse the frequency band of interest for an nth-order network.

filters consisting of cascaded building blocks, each of which has been tuned to a given combination of frequency and Q accuracy. Naturally, they also give the effect on the gain tolerance band of any other factors influencing the pole and zero domains D_0. For example, by expanding the domains D_0 to include ambient effects (e.g., temperature, aging, humidity etc.), the corresponding worst-case tolerance band of the filter can be directly obtained. Furthermore, statistical rather than worst-case bounds can be used for the domains, resulting in a corresponding statistical rather than worst-case tolerance band. The expressions may also be used in the reverse, e.g., to establish the tuning accuracy (i.e., the bounds on the domains D_0) necessary to provide filters with a specified initial tolerance band.

CHAPTER

5

THE DEVELOPMENT OF HYBRID
INTEGRATED FILTER BUILDING BLOCKS

INTRODUCTION

In the preceding chapters[1] we have attempted to provide the fundamentals, in terms of network analysis and synthesis as well as of component and device technology, necessary to design hybrid integrated linear networks. Having laid this groundwork we shall, in this and the following chapter, draw upon it to describe the practical realization of a family of active-filter building blocks for implementation in hybrid integrated form. We shall outline the various considerations leading to the selection of the circuit types described and, where possible, indicate alternative solutions that may be more suitable in different circumstances.

Before we go over to the realization of a family of viable filter building blocks we shall first briefly survey the inductorless filter field as a whole and indicate where the application of active filters seems to be warranted and where it either brings no advantages over other methods, or is quite simply not yet technically feasible. Having drawn bounds on useful active-filter applications we shall then attempt to draw bounds on the economical use of hybrid integrated circuit technology for the realization of active filters. Stated simply, we shall ask ourselves when it is economically feasible to use hybrid integrated circuit techniques for the implementation of active filters. Naturally there are no rigid answers to these questions, but there are guidelines, and we shall try to present them. Having done so, we shall describe the design in

1. We mean here the chapters of this book as well as those of the companion volume, *Linear Integrated Networks: Fundamentals.*

434

hybrid integrated form of one of the building-block schemes outlined in Chapter 3. The building blocks combine tantalum thin-film resistors and capacitors with beam-leaded operational-amplifier chips. Since the building blocks cannot be used to cover the whole frequency range over which active filters are desirable, other methods, complementing the building blocks at low and high frequencies, will be discussed in Chapter 7.

5.1 ACTIVE VS. OTHER INDUCTORLESS FILTERS

Before we begin to design active RC filters, we should know toward which frequency and Q ranges we are to direct our design efforts in order to avoid redundancy and overlap with other methods of inductorless filter design that may be more practical and economical for a given application. It is therefore appropriate, at this point, to briefly survey the various other analog types of inductorless filters and to establish where those of "active RC" origin have the most to offer and can most effectively be utilized. The following survey[2] will of necessity be brief, and will serve essentially as a basis of comparison with respect to active RC filters.

5.1.1 Electromechanical Filters

The frequency-sensitive element of the generalized electromechanical filter shown in Fig. 5-1 is a mechanical transmission device, or mechanical resonator, in which mass or moment of inertia, and elastic compliance or stiffness interact in the manner of an underdamped second-order system exhibiting the familiar phenomenon of resonance at a particular frequency. A direct analogy to electromagnetic resonance can be found in that there is a correspondence of mass to inductance, stiffness to capacitance and mechanical resonance to electrical resonance. The conversion of the energy within the mechanical resonator into electric energy and the coupling of the converted electric signals to an electric network are obtained by means of an electromechanical transducer, which is generally based on the piezoelectric, the magnetostrictive, or the electrostatic effect.

FIG. 5-1. Generalized electromechanical filter.

2. G. S. Moschytz, Inductorless filters: A survey, Parts I and II, *IEEE Spectrum*, August and September 1970.

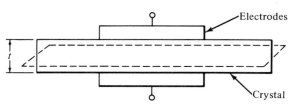

FIG. 5-2. Piezoelectric crystal resonator. Broken lines show shear deformation.

Broadly speaking there are two basic categories of electromechanical filters: those that provide mechanical resonance properties and the capability of energy conversion in one and the same device, and those that combine two separate materials or devices to perform these functions. The most important groups in the former category are *monolithic crystal* and *ceramic filters*; in the latter the most important group is that generally called *mechanical filters*.

Monolithic crystal filters[3]   These filters use coupled mechanical vibrations in a piezoelectric material to provide bandpass-type filter functions. A filter consists of a crystalline quartz wafer onto which pairs of metal electrodes are deposited. The operation of the filter is made possible by two factors:

1. The crystal is piezoelectric. It can transform electric energy into a mechanical form, specifically, into a transverse shear wave (see Fig. 5-2), and back again. Therefore, in addition to serving as resonator and interresonator coupling medium, it performs the transducing functions.

2. The metal electrodes lower the resonance frequency of the transverse shear wave in the plated region as compared with unplated quartz. As a result, the resonance created in the plated region does not extend into the areas without electrodes, but remains trapped under the electrodes, which are thin metal films.

In Fig. 5-3a a monolithic filter is shown in its simplest form. It consists of an input and an output resonator formed by a pair of electrodes deposited onto opposite faces of a quartz crystalline wafer. Vibration, induced by resonance between the plates, decays very rapidly outside the plated region; therefore, it is essentially trapped under the electrodes. Although the vibration is confined to the electrode area, the displacement decays exponentially in the surrounding region and this, mechanically or acoustically, couples adjacent resonators. The demensions and separation of the resonators determine the coupling, and therefore the bandwidth, transmission characteristics, and

3. R. A. Sykes, W. L. Smith, and W. J. Spencer, Monolithic crystal filters, 1967 *IEEE Intern. Conv. Record*, pt. 11, pp. 78–93.

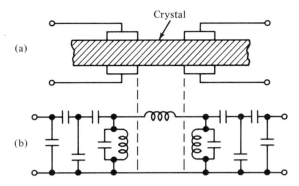

Fig. 5-3. The simplest form of a monolithic filter (a) and its equivalent electrical network (b).

terminal impedance of the monolithic filter. The equivalent electrical network is shown in Fig. 5-3b.

Table 5-1 presents a summary of the salient features of monolithic crystal filters as they are available today. The advantages of these filters from the standpoint of Q capability and frequency stability are evident; in fact, there is no comparable technique to match this performance. However, the frequencies are, and presumably will continue to be, limited to the megahertz range.

Functionally, the versatility of monolithic crystal filters is limited. They are inherently all-pole (more specifically, bandpass) filters with the number of transmission poles corresponding to the number of electrode pairs. As such, they permit the realization of any standard bandpass function (Chebyshev, Butterworth, etc.). Other filter functions (band elimination, transmission zeros, low pass, high pass, etc.) can be realized only with additional components, although a degree of functional variability can be obtained by splitting electrodes.

Initial filter adjustment and tuning can be carried out very accurately (e.g., 0.1-dB ripple in the pass band of, say, an eight-pole bandpass filter can be reproduced consistently). However, because of unwanted vibrations within the quartz plate, invariably there are additional pass bands at higher frequencies. The amount of interference with the required frequency response will depend on how close these spurious resonant modes are to the frequency band of interest.

The mechanical or acoustic coupling, which determines the bandwidth of the filter, depends principally on the electrode separation and, to a lesser extent, on the electrode mass and length. The bandwidth is controlled by electrode separation, mass, and area. It can be adjusted by placing stripes between the electrodes and either adding to these stripes by evaporation or removing them by laser. The center frequency, determined principally by

TABLE 5-1, CHARACTERISTICS OF SOME INDUCTORLESS FILTERS*

Type	Frequency range	Pole Q	$\Delta f/f$† (ppm/°C)	$\Delta Q/Q$† (%)	Functional versatility	Functional accuracy	Tunability	Signal dynamic range (dB)	Compatibility with HIC technology
Monolithic	5–150 MHz	1000–250,000	±1	0.1	Fair (bandpass and frequency-rejection networks)	Good (with exception of spurious modes)	Initial tuning good; system adjustment poor	40–80 (depending on proximity to spurious tones)	fair
Ceramic	0.1–10 MHz	30–1500	±100	0.2	Fair (requires additional components for most functions)	Good (with exception of spurious modes)	Initial tuning good; system adjustment poor	40–80 (depending on proximity to spurious tones)	fair
Mechanical	0.1 Hz–20 kHz	50–5000	±50	0.1	Fair (bandpass and frequency-rejection; numerous nonfilter functions possible)	Good	Good	60–80	poor
PLL	0.1 Hz–25 MHz	Up to several hundred	100	5	Poor (special purpose FM filter and discriminator)	Good	Good	40–60	good

* The characteristics listed may be mutually exclusive.
† Over temperature range from 0° to 60°C.

438

the plate thickness, is fine-tuned by the mass of electrodes. These operations can be tightly controlled in manufacture. However, once the filter is assembled into a system, it is very difficult to make any additional adjustments. Furthermore, since the entire resonator structure is coupled, it is impossible to measure the uncoupled frequency of any one resonator. Adjustments of resonators, therefore, must take this coupling into account, with the help of theoretical expressions derived for this purpose.

Compared with their conventional counterparts, monolithic crystal filters permit considerable size reduction, since no transformers, inductors, or other discrete components are needed. Further size reduction and compatibility with hybrid IC techniques also appear possible, since thin-film circuitry for passive components and beam-leaded semiconductors is now at a stage where these probably can be deposited on the same substrate as a monolithic crystal filter. In addition, once the necessary fixtures, masks, design data, etc., are obtained, filter costs become quite low, because only one quartz plate and one enclosure are needed for each filter.

Typical applications for monolithic crystal filters are in telephone carrier systems (frequency range, 2.6–19 MHz), point-to-point wire and radio transmission systems with single-sideband, double-sideband, or narrow-band frequency modulation, and broadband telephone multiplex systems. All of these systems, in addition to being in the megahertz frequency range, have stringent requirements on Q and frequency stability that are hard, if not impossible, to meet economically by other means.

Ceramic filters[4]   These filters combine the functions of the mechanical resonator and the electromechanical transducer shown in Fig. 5-1 (because of their combined mechanical resonance and piezoelectric[5] properties) in the same way that monolithic crystal filters do. In contrast to monolithic crystal filters, though, ceramic filters are generally used more in combinations of individual two- or three-electrode resonators of the kind shown in Figs. 5-4 and 5-5 than as monolithic structures. Two-electrode resonators may be interconnected with capacitors and/or amplifiers to provide high-order filter configurations.

The characteristics of ceramic filters, as determined by the ceramic resonators themselves, are summarized in Table 5-1. The lower frequency range and wider bandwidths (lower Q values) are apparent. In particular, the frequency range covered includes the intermediate frequencies (IF) of AM radios (455 kHz), audio television reception (4.5 MHz), and FM receivers (10.7 MHz). It is for this consumer market that ceramic filters hold the most promise, both as to performance and cost.

4. F. Sauerland and W. Blum, Ceramic IF filters for consumer products, *IEEE Spectrum*, **5**, 112–126 (November 1968).
5. Naturally, we refer here only to those high-Q piezoelectric ceramic materials that are used in ceramic filters.

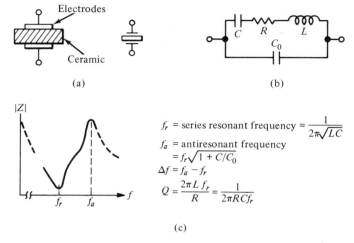

f_r = series resonant frequency = $\dfrac{1}{2\pi\sqrt{LC}}$

f_a = antiresonant frequency
$= f_r\sqrt{1 + C/C_0}$

$\Delta f = f_a - f_r$

$Q = \dfrac{2\pi L f_r}{R} = \dfrac{1}{2\pi R C f_r}$

(c)

FIG. 5-4. A two-electrode ceramic resonator (a), its equivalent network (b), and its reson-ance properties and electrical parameters (c).

The range over which the electrical parameters of ceramic resonators may be varied depends upon the ceramic material, processing, and resonator geometry. The quantities $\Delta f/f_r$ and C_0 (see Fig. 5-4) may be adjusted over a wide range by the manufacturing process. This makes the ceramic resonator a more versatile electric circuit element than, for example, the quartz resona-tor, whose equivalent circuit is the same as that of Fig. 5-4 but whose $\Delta f/f_r$ is a material constant of small and fixed value.

Compared with the average consumer-type IF transformer, ceramic mater-ials and filters have a better frequency–temperature stability and quality factor Q, the latter by an order of magnitude. A typical radial resonator with

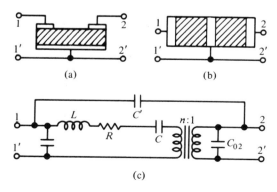

FIG. 5-5. Two types of three-electrode ceramic resonators (a and b) and their equivalent network (c).

$f_r = 455\,\text{kHz}$ has a diameter of 0.56 cm and a thickness of 0.038 cm. All equivalent circuits of the piezoelectric resonator are valid only in the vicinity of the operating frequency. For radial resonators, the circuit equivalent of Fig. 5-4 is accurate for frequencies up to approximately $1.5f_r$. At higher frequencies, other resonances (spurious modes) occur because of overtones of the radial mode and because of other vibrational modes.

Although ceramic filters, in their present form, cannot be integrated monolithically, hybrid integrated circuits have been made using thickness-mode ceramic IF filters. Also, radial-mode resonators become small enough at higher frequencies (>1 MHz) to be included in IC packages.

Mechanical filters[6]   These filters generally consist of a mechanical resonator, and rely on the electrostrictive, magnetostrictive, or piezoelectric effect of a *separate* transducer for the interaction between electrical and mechanical energy, particularly for a direct relation between electrical and mechanical resonances. Consider, for example, the H-shaped resonator shown in Fig. 5-6. It consists of two balanced masses (i.e., the bars of the H) connected by a flexible web. As the piezoelectric input transducer contracts and expands, the web flexes and the bars oscillate in opposite directions. The resonance signal can be picked off by a coil near a bar or by another piezoelectric transducer underneath the web.

Since the bars of the H-shaped resonator rotate counter to each other at their nodal points (i.e., points of least deflection), there is virtually no net transmission of energy through the common plane of the base. Thus, in contrast with most other mechanical resonators, where this is often a problem,

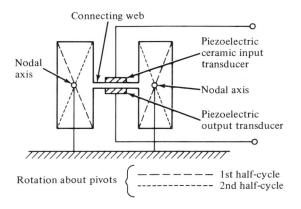

FIG. 5-6. An H-shaped mechanical resonator.

6. H. Baker and J. R. Cressey, H-shaped resonators signal upturn in tone telemetering, *Electronics*, 99–106, (October 2, 1967).

the resonant energy loss that stems from vibrations transmitted by the resonator to its mounting base is avoided. Since the Q of a resonator equals the ratio of conserved energy to dissipated energy, very high Q values can therefore be obtained. Another common problem of low-frequency mechanical resonators—their susceptibility to external shock and vibrations—is also greatly reduced. In the same way that transmission of vibrations from the resonator to the mount is prevented by the symmetrical push–pull configuration of the H-shaped resonator, external shocks are prevented from traveling from the mount to the resonator. This process is similar to the rejection of a common-mode signal at the input to a differential amplifier in that any vibrations originating at the base are common to both bars of the resonator and therefore cancel out.

The main characteristics of balanced (e.g., H-resonator-type) mechanical filters are summarized in Table 5-1. As shown, this representative mechanical filter type is a low-frequency device. The resonant frequencies range from a fraction of a hertz to 20 kHz; Q values can vary anywhere between 50 and 5000. If appropriate materials are used, dimensional variations due to temperature changes have little or no effect on the balance of the device, and the frequency stability of the filter can be very tightly controlled. The resonant frequency can be shifted 20–30% by balanced pairs of threaded tuning slugs inserted in threaded holes running the entire length of each bar. Low values of Q are obtained by setting the slugs at differing distances from the nodes of the bars. Consequently, this particular mechanical resonator can be accurately tuned initially, as well as readjusted after incorporation into a system.

The H-resonator is unconventional as far as mechanical filters go in that it does not use electrostrictive or magnetostrictive transducers; instead, it uses a piezoelectric transducer. This is possible because of its geometrical configuration. Consequently, it can deliver a substantial electric signal into a low-impedance load even though piezoelectric transducers are inherently high-impedance devices. In fact, if the transducer and load are properly adjusted, a voltage gain of as much as 2 : 1 can be obtained. The dynamic range of the H-resonator, which is determined by the signal capabilities of the piezoelectric transducer, is large. The output voltage, for example, can be maintained high enough to drive the gate of a silicon controlled rectifier directly. As with ceramic and monolithic circuits, multiple combinations of inputs and outputs can be achieved by coupling a number of separate transducers to the web of the H-shaped resonator. Since piezoelectric ceramic transducer material is polarized, transducer pairs can be placed so that their responses are in phase or 180° out of phase with each other. By suitably combining in- and out-of-phase transducers, a variety of circuit types—such as oscillators, FM discriminators, and bandpass and frequency-rejection filters—can be obtained.

Just as with other electromechanical filters, the functional versatility of mechanical filters is quite limited. Inherently these filters are resonators providing bandpass or frequency-rejection characteristics. Thus, to obtain, say, low-pass, high-pass, or elliptic filters with finite transmission zeros, auxiliary networks, which may be active RC in order to avoid inductors, must be used.

5.1.2 Frequency–Selective Networks Using the Phase–Locked Loop (PLL)

Referring to Fig. 5-7, the PLL provides frequency selectivity and filtering in the following way.[7] Assuming a number of signals present at the input to the PLL, the voltage-controlled oscillator (VCO) locks onto the signal closest to its own center frequency. The two signals (the incoming signal and the VCO signal), which are in phase to within a phase error $\Delta\phi(t)$, are passed through a phase comparator and low-pass filter. An error signal at the output, proportional to the phase difference $\Delta\phi(t)$ between the two signals, controls the frequency of the VCO. If necessary it can first be amplified by a DC amplifier. This error voltage can be considered as the output of a frequency discriminator, since it is a measure of the input frequency. At the same time, the PLL exercises a smoothing effect on the error voltage, equivalent to the effect of a second-order bandpass filter, with respect to noise (e.g., phase jitter) and interference from neighboring channels. In other words, if there are a number of signals at the input to the PLL, a filtered (and amplified) replica of the signal whose frequency is closest to that of the VCO will appear at the output of the VCO and, in the case of an FM input signal, a demodulated and filtered version of this signal will be available at the VCO's input.

The importance of the PLL for integrated circuit design is, first, that its individual functions can be designed in active RC form. Second, because of

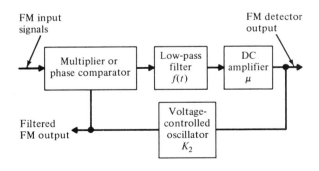

FIG. 5-7. Generalized scheme of the phase-locked loop (PLL).

7. G. S. Moschytz, Miniaturized RC filters using phase-locked loop, *Bell Syst. Tech. J.* **44**, 823–870 (1965).

the locking property of the PLL, which means that its frequency tracks or locks onto the input signal to be filtered and detected, the stability of the overall circuit is not as critical as that of other active filters. Thus, the PLL is perhaps the only filtering scheme that can be manufactured using exclusively monolithic integrated-circuit techniques.[8]

By its nature the PLL is limited to two basically different types of applications: (i) as a demodulator where it is used to follow phase or frequency modulation and may be thought of as a matched filter operating as a coherent detector, and (ii) to track a carrier or synchronizing signal which may vary in frequency with time. When used in the latter capacity the PLL may be thought of as a narrow-band filter for removing noise from a signal. In contrast to this limited functional versatility, the PLL can be used over a very wide frequency range merely by changing the center frequency of the VCO. It is, and will remain, a special-purpose circuit, which may, however, find its way into some of the same consumer communications systems for which numerous electromechanical filters are being developed. One possible application is a high-frequency (1–25 MHz) FM amplifier–demodulator circuit, which is designed as a monolithic replacement for the IF strip and the detector section of commercial FM receivers. In another application, the PLL has been used as a monolithic FM multiplex receiver circuit for industrial applications in a frequency range of 2.1 Hz to 300 kHz. The attractiveness of the PLL for these applications lies in the fact that it can be used as a combined channel filter and FM discriminator.

5.1.3 The Case for Active RC Filters

The capabilities of the analog inductorless filters briefly described above can be summarized by the Q vs. frequency representation of Fig. 5-8. Electromechanical filters provide very high pole Q's with a correspondingly high frequency stability (without which, as we know, the high pole Q's would be useless). They also cover a wide frequency range (with a marked gap between several kilohertz and several hundred kilohertz). However, none of them readily permit the realization of low to medium pole Q's, and all of them are limited in their functional versatility—typically, to bandpass and band-rejection frequency responses. Furthermore, the mechanical filters which cover the low-frequency end are incompatible with integrated circuit techniques and become ever larger in size with decreasing frequencies. The Q vs. frequency curve of typical LC filters, which is also shown, supplements the capabilities of the electromechanical filters. However, the obtainable Q values at high and low frequencies rapidly decrease and the inductors too become ever

8. A. B. Grebene and H. R. Camenzind, Frequency-selective integrated circuits using phase-locked techniques, *IEEE J. Solid-State Circuits*, SC-4, 216–225 (1969).

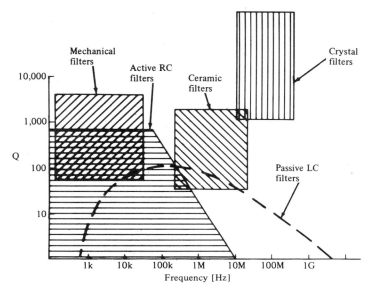

FIG. 5-8. Q vs. frequency plot for the analog inductorless filters described in Section 5.1.2. as well as for passive LC and active RC filters.

bulkier and more expensive as the frequency is decreased. Furthermore, as we know, the electromagnetic components in LC filters cannot be integrated.

From the representation in Fig. 5-8 together with Table 5-1, the case for active RC filters emerges rather clearly. For low-to-medium pole Q's, at low-to-very-low frequencies, no alternatives to active RC filters exist. In the past, filters with frequencies at several hundred hertz and below were avoided wherever possible. On the other hand, at those low frequencies, the required filtering functions will, as a rule, be only moderately complex. Relatively simple all-pole functions of low order and precision will generally be sufficient here, since the type of signals being transmitted and filtered will not require elaborate shaping and accurate equalization of the transmission medium.

By contrast, in the voice-frequency band, where filtering applications abound, the demands on the filter complexity may be considerable. Amplitude and delay equalization are common and the order of the networks can be high. Functional versatility is therefore an important consideration. Here, as in the low-frequency case, there is no viable alterative to active RC filters, if compatibility with hybrid integrated circuit techniques is desired.[9]

9. Clearly, digital filters have a place here, but space does not permit a discussion of this topic; hence we are restricting ourselves to *analog* inductorless filters. For a survey of digital filters in the realm of general inductorless filters we refer to the already cited paper, Inductorless filters: A survey, by G. S. Moschytz and the bibliography given therein, as well as to *Digital Signal Processing*; ed. by L. R. Rabiner and C. M. Rader, (New York: IEEE Press, 1972).

TABLE 5-2. SUGGESTED GUIDELINES FOR THE DESIGN OF ACTIVE *RC* NETWORKS

Frequency range	Maximum* pole Q	$\Delta f/f$	$\Delta Q/Q$	Remarks
Low frequencies (e.g., ≤ 100 Hz)	≤ 100	$\leq \dfrac{1}{Q}\dfrac{\Delta f}{BW}$ †	$\dfrac{\dfrac{\Delta f}{f} - \dfrac{\Delta BW}{BW}}{1 + \dfrac{\Delta BW}{BW}}$	Limited functional variety required: generally bandpass and band rejection networks. Frequency and Q stability characterized in terms of bandwidth (BW) and percentage bandwidth stability. For high stability and high Q's, combinations with mechanical resonators possible.
Medium frequency (e.g., 100 Hz–100 kHz)	≤ 500	$\leq \dfrac{\Delta\alpha[dB]}{8.68 Q}\bigg|_{f_{-3dB}}$ ‡	$[\beta \cdot S^q_\beta \cdot \dfrac{\Delta A}{A^2}]$ $= \Gamma\dfrac{\Delta A}{A^2}$	Unlimited functional variety and versatility required: e.g., minimum and nonminimum phase-shift networks, high-precision pulse-shaping networks. Stability characterized in terms of transmission characteristics (e.g., amplitude, phase, or delay requirements). Maximum Q depends on frequency stability attainable with *RC* networks.
High frequency >100 kHz	≤ 100	$\leq \dfrac{1}{Q}\dfrac{\Delta f}{BW}$	$\dfrac{\dfrac{\Delta f}{f} - \dfrac{\Delta BW}{BW}}{1 + \dfrac{\Delta BW}{BW}}$	Functional variety more limited than with "medium-frequency" filters. Supplement, rather than substitute, other analog inductorless filter methods. Combinations with ceramic and crystal resonators recommended for high-stability, high-Q applications. Used to extend the very limited functional variety of crystal filters.

* Depends on $\Delta f/f$ and $\Delta Q/Q$ requirements.
† BW $= -3$ dB bandwidth of second-order bandpass network or high-Q pole pair.
‡ $\Delta\alpha[dB]|_{f_{-3dB}} =$ amplitude variation in dB at -3 dB band-edge.

At high frequencies (e.g., above several hundred kHz) inductors are less bulky and *LC* filters less objectionable.[10] If high pole *Q*'s are required, ceramic and monolithic crystal filters can be used; however if a variety of filter characteristics other than those peculiar to electromechanical filters are to be realized, then the latter must be combined with *LC* or, for hybrid integrated realizability, with active *RC* networks. The combination of active *RC* networks with electromechanical devices can also be very useful at low frequencies if functional variety and very high *Q*'s are required simultaneously. The reason for this is that the frequency stability of the resistors and capacitors of the active *RC* networks cannot, in general, be made to equal the frequency-stability requirements of very high pole *Q*'s; conversely, the electromechanical resonators that do have the necessary frequency stability provide only a limited functional variety.

From the above discussion we can plot a desired *Q* vs. frequency curve for hybrid integrated active filters, which is such that they will be capable of supplementing the capabilities of other analog inductorless filtering methods. This curve is included in Fig. 5-8. We can also derive a table, similar to Table 5-1, but for low-, medium-, and high-frequency active *RC* filters whose characteristics satisfy the requirements indicated above (see Table 5-2). Note that the most critical requirements apply to the medium frequency filters, simply because at high frequencies alternatives are available, and at low frequencies the applications are fewer and, typically, less demanding. In what follows we shall therefore use medium-frequency active filters as a vehicle for the demonstration of the development necessary for the production of hybrid integrated networks. First, however, we shall take a brief look at some of the economic implications involved in the manufacture of hybrid integrated linear networks in order to derive some guidelines as to when the use of this technology is warranted and when it is not.

5.2 SOME CONSIDERATIONS ON THE ECONOMICS OF HYBRID INTEGRATED CIRCUITS

The key to the economic feasibility of hybrid integrated, as of all integrated circuit manufacture, is to have a sufficiently high production volume to absorb the high tooling and processing costs involved in these technologies. A typical cost per circuit vs. production volume curve is shown in Fig. 5-9.[11] A characteristic feature of this curve is that below some quantity, say 1000 circuits, the cost very rapidly increases, and above some quantity, say 100,000 circuits, the cost-curve flattens out. Naturally, the actual cost of each circuit will depend on the complexity of the circuits in question and on the technology used.

10. M. Grossman, Focus on fixed passive components, *Electronic Design*, **21** (12) 56–67 (1973).
11. *Integrated Circuits, Technical and Business Analysis*, Harvard University 1963, Integrated Associates, P.O. Box 131, Cambridge 40, Mass.

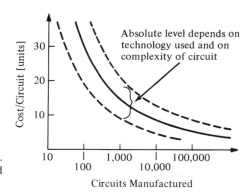

FIG. 5-9. A typical cost per circuit vs. production volume curve for a hybrid integrated circuit.

Also, the time span in which a given number of circuits is manufactured, that is, the time in which the invested manufacturing costs are recovered by sales, will be critical in determining the average cost of a circuit. Typically, the yearly circuit production is expected to be several thousand the first year, progressively growing to tens of thousands in following years, if the integrated circuit is to be economically more worthwhile than, say, its discrete counterpart.

In making a comparison between the cost of an integrated circuit and its discrete-component equivalent, more must be taken into account than the cost of circuit production alone. Peripheral costs affected by the technological realization, such as the number of necessary printed wiring boards and interconnecting sockets, the dissipated power, cooling requirements, etc., associated with each circuit realization must be included in any cost comparison. Since, in most cases, the production volume will be a decisive factor in this comparison, an integrated version of a circuit will only be warranted when either a given circuit is required in sufficiently large numbers or a group of different circuits can be derived from a common building block. The latter situation is common in the production of filter networks and accounts for the filter building blocks discussed in the next section. Either way, the circuit manufacture must be based on the use of batch-production techniques resulting in large enough production volumes at high enough yields to justify the cost of the manufacturing equipment.

5.3 A HYBRID INTEGRATED BUILDING BLOCK USED IN FEN FILTER DESIGN

In the preceding survey of inductorless filters we found that, because of the lack of appropriate substitutes, the greatest need for active *RC* networks is in the medium-frequency range; here also the demands on functional variety and precision are the most severe. We also found that for economical feasibility when the circuit is in hybrid integrated form production volumes must be high. Often the production volume can be considerably increased by designing building blocks that are either all-purpose, or can readily be modified and adapted to a large variety of applications. In what follows we describe the design of hybrid integrated filter building blocks that were developed to possess the latter qualifications.

The filter design method for which the building blocks are developed is based on the FEN decomposition, or pole–zero cancellation technique, given in Chapter 3. Whereas single-amplifier and multiple-amplifier methods of implementing this technique were both given there, our building blocks, realized with tantalum thin-film resistors and capacitors combined with monolithic silicon integrated (beam-leaded) operational amplifier chips are for the filter-design technique using the multiple-amplifier FEN.

The underlying objective upon which the concept of hybrid integrated filter building blocks using multiple-amplifier FENs is based is to provide a filter scheme with which low-Q, second-order networks can be realized with single-amplifier building blocks and, *using combinations of those same building blocks*, multiple-amplifier circuits for high-Q realizations can be obtained as well. Thus any low-Q building block is to be readily expandable to high-Q operation, simply by adding a second single-amplifier building block of the same type in a suitable structure. As we know from Chapter 3, Section 3.3.2, the single-amplifier building block most suitable for this purpose is the class 4 type (the twin-T single amplifier building block or TT-SABB) described in Section 3.1.2. In what follows, the hybrid integrated realization of this single-amplifier building block, which forms the core of active-filter design using multiple-amplifier FENs, will be described.[12]

5.3.1 The Twin-T Single-Amplifier Building Block (TT-SABB) Customized by Scribing

The Layout   When using an all-thin-film approach (i.e., using thin-film resistors *and* capacitors) to the development of linear networks, we must decide whether to deposit the film resistors and capacitors on the same substrate (single-substrate approach) or to design a separate resistor and capacitor substrate (double-substrate approach). The former results in a more compact assembly and, in spite of the more numerous and complex processing steps, is more economical,[13] provided the production volume is high enough. Thus, if one particular network type is required in very large numbers, the

12. D. G. Medill, G. S. Moschytz, and C. J. Steffen, Filter building blocks using tantalum and silicon integrated circuits, Proc. 1969 Electronic Components Conf., Washington, D.C.
13. W. H. Orr and J. J. Degan, Jr., Precision tantalum film RC circuits for communication systems, *Digest* of the 1972 IEEE Intercon, Paper 3H, 4; also, W. H. Orr et. al., Integrated tantalum film RC circuits, Proc. 20th Electronic Components Conf., 1970, pp. 602–612.

single-substrate approach may be expected to provide the most economical solution. Among other things, the circuit can then be optimized for minimum substrate area and a minimum number of crossovers, all of which help in reducing the cost of the final circuit. On the other hand, the multi-optional nature of our filter building blocks also makes the double-substrate approach appear attractive. For one thing, all crossovers can thereby be eliminated; for another, we thereby gain an exceptional circuit flexibility that would be hard to realize in any other way. However, it is not necessary for us here to describe both the single- and the double-substrate realization of the TT-SABB (although, in fact, both have been carried out), because the differences between the two are limited to processing details that go beyond the responsibilities of the network designer. In what follows we shall therefore confine ourselves to the development of a double-substrate building block with the knowledge that, at least from the network designer's point of view, the processing of a single-substrate building block is very similar.

The circuit diagram of the TT-SABB is shown in Fig. 5-10. With a few exceptions (of negligible significance) the diagram is identical to that shown in Chapter 3, Fig. 3-27. The circuit comprises all possible functional options, each of which is obtained by opening a predetermined set of circuit paths. The layouts of the resistor and capacitor substrates are shown in Fig. 5-11. The conductor patterns on both substrates include the circuits of every filter type provided by the building block. A desired filter circuit is obtained by open-circuiting the unwanted connections. The gain, Q, and frequency ranges of the building block are obtained by using tapped thin-film resistors and

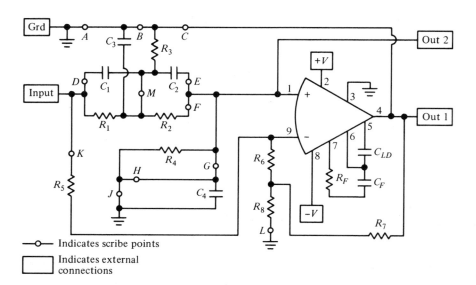

FIG. 5-10. Circuit diagram of the TT-SABB.

(a)

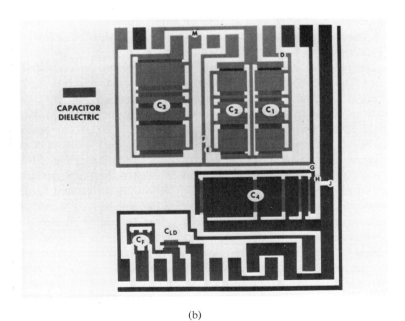

(b)

FIG. 5-11. Layout of the TT-SABB: (a) resistor substrate; (b) capacitor substrate.

capacitors.[14] A tapped thin-film resistor consists of resistor segments shunted by conducting paths. A tapped thin-film capacitor consists of finger-like counter electrodes deposited over a common base electrode. The elements (resistor segments or capacitor counter-electrode fingers) of a tapped component are combined in such a way as to realize a predetermined range of discrete component values by open-circuiting the conductor segments bridging the individual component elements. The element values are chosen in such a way that the increments between combinations are equal. One way of achieving this is by distributing the individual segments and finger electrodes in a binary series. The range of component values thereby obtained depends on the gain, Q, and frequency ranges required. These, in turn, are limited by the constraints of the technology used, such as by the maximum resistance and capacitance per substrate compatible with an acceptable substrate yield. In the case of tantalum technology these values are on the order of 0.5 MΩ to 1 MΩ and 0.05 μF to 0.1 μF, respectively.

We recall,[15] that in between the discrete values provided by a tapped resistor, any arbitrary value can be accurately obtained by anodization. The resistor value is thereby continuously increased by converting the upper layer of the resistive film into an insulating film of tantalum pentoxide. No suitable method of continuously adjusting thin-film capacitors is available; however, by using continuously variable resistors and discretely variable capacitors, highly accurate gain, Q, and frequency adjustments are made possible. In fact, the limiting factor in the tuning accuracy attainable is the instrumentation, rather than the anodization, accuracy.

Because resistors are adjustable only in the increasing direction and capacitors not fine-adjustable at all, the value of any resistor that is to be fine-tuned must initially be somewhat *lower* than its final value. By how much depends on the capacitor tolerances and on any other inaccuracies and parasitic effects that the resistor to be trimmed has to compensate for. Typically, with functional tuning, in which only a small number of resistors (one per parameter) are trimmed, the necessary tuning range may lie between 20% and 30%, meaning that the resistors concerned should be designed (and incorporated in the mask layout) at 15% below their nominal value. With deterministic tuning, in which all or most of the resistors are tuned, the layout value may be closer to the nominal value. Naturally, numerous effects may influence how far below the nominal value a mask resistor must be dimensioned. However, it should not be made lower in value than necessary, because extended anodization without the possibility of subsequent heat treatment deteriorates the long-term stability of a resistor and also takes unnecessary tuning time. A mask resistor has the value

$$R_{mask} = R_{nominal}[1 - (\rho_n/2)] \tag{5-1}$$

---

14. See Chapter 6, Figs. 6-10 and 6-11, of *Linear Integrated Networks: Fundamentals.*
15. See Chapter 6 of *Linear Integrated Networks: Fundamentals.*

where $\rho_n \cdot 100$ is the maximum (worst-case) resistor range in percent over which a resistor is expected to require continuous trimming, and R_{nominal} is the resistor value resulting from the design equations of the circuit. Like any other resistors, the segments of a tapped resistor must be laid out $\rho_n \cdot 50\%$ lower than their nominal values.

The only possible way of overcoming the unidirectional tunability of a resistor is by replacing it by a resistive-T network, as was described in Chapter 2. For convenience we remind the reader of this capability by referring to Fig. 5-12. The short-circuit transfer admittances of the two resistive two-ports are

$$(y_{21})_T = \frac{1}{R_a + R_b + (R_a R_b / R_c)} \tag{5-2a}$$

and

$$(y_{21})_R = 1/R \tag{5-2b}$$

If the two admittances are to be equal, we have

$$R = R_a + R_b + (R_a R_b / R_c) \tag{5-3}$$

This transformation can be used only where the short-circuit transfer admittance of a resistor R is required. When this is the case, it has several advantages. For one thing it permits a large resistor R to be replaced by a three-resistor combination of smaller total resistance; for another, it permits bidirectional tuning (i.e., increased value by trimming R_a or R_b; decreased value by trimming R_c). Finally, if required, the T circuit can be dimensioned to provide a relatively constant open-circuit driving impedance of $R_a + R_c$ (in spite of a certain amount of trimming). On the other hand, as was pointed out previously, caution is sometimes advised in its use: when connected in the feedback path of an operational amplifier it has the effect of decreasing the loop gain by the amount $R_c / (R_a + R_b)$. Furthermore, if used too liberally, it rapidly increases the complexity of a circuit.[16]

In general the network designer will not be directly concerned with those aspects of a circuit layout that are closely related to the film technology used. After having agreed on certain maximum and minimum component values, the device designer will determine the dimensions of the individual components, based on his knowledge of the optimum sheet resistance for a given film, the line widths compatible with his processing equipment, the desired

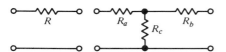

FIG. 5-12. Resistor and resistive-T.

---

16. The more components there are on a substrate, the more must be inspected (and rejected) after fabrication. Thus, the substrate yield decreases with component count, and consequently the substrate cost increases.

component stability, and so on. However, the network designer will be expected to give broad guidelines with respect to the layout topology so as to avoid the introduction of undesirable parasitic effects. For example, in the circuit of Fig. 5-10 the most critical parasitic effects are the following:

1. The conducting path from the output terminal 4 to the base of the twin-T (point B) should be as short as possible in order to maintain a low resistance in this path. An increase in resistance, say above 1 Ω, directly affects the null depth achievable with this circuit, much as lossy capacitors do (see Chapter 4): it shifts a zero on or close to the $j\omega$ axis to the left in the s-plane.

2. The conducting paths connected to the input terminals of the operational amplifier (terminals 1 and 9) should be spread apart as far as possible in order to eliminate any parasitic capacitance between these two paths. This parasitic capacitance may cause a severe decrease in the bandwidth of the amplifier, or even parasitic oscillations.[17]

3. The output terminal 4 should not be located close to the positive input terminal 1 of the amplifier, in order to prevent instability due to inadvertent positive feedback. This is particularly important for those circuit options in which C_4 is not present to shunt high-frequency feedback to ground.

Substrate and Film Materials Having decided on a two-substrate approach, a building block consists of a resistor substrate mounted on top of a capacitor substrate in sandwichlike fashion. A beam-leaded operational amplifier chip is bonded onto the resistor substrate in the area visible at the top right-hand corner of the resistor R_F in Fig. 5-11a. The resistor R_F, whose large area is due to its low value of 30 Ω, is required for the frequency stabilization of the operational amplifier; it is therefore used unmodified for every functional option and requires no taps. This resistor, as well as all the other passive components and interconnection paths, consist of tantalum thin film; the film is deposited on the two substrates in the manner described in detail in Chapter 6 of *Linear Integrated Networks: Fundamentals*. Gold ribbon or wire leads are bonded onto the pads lined up on either side of each substrate, connecting the two substrates together. Either these continue on from there to form interconnections with external points, or a lead frame is bonded onto the lower (capacitor) substrate for this purpose. Figure 5-13 shows both substrates generated from the mask layouts in Fig. 5-11 and the assembled combination (including the operational amplifier chip).

The resistors and the associated conductor pattern on the resistor substrate are generated from layers of tantalum and gold film deposited on an unglazed[18] (e.g., Corning 772) ceramic substrate, as indicated in the schematic

---

17. See Chapter 7, Section 7.2.7 of *Linear Integrated Networks: Fundamentals*.
18. For improved long-term stability it may be advisable to spot-glaze the resistor substrate, i.e., to generate glazed islands where the resistors are afterwards deposited, while keeping those areas unglazed where chip or lead bonding is to take place.

FIG. 5-13. TT-SABB and substrates: left, resistor substrate; center, capacitor substrate; right, assembled building block.

of Fig. 5-14a. The top and thickest layer is an evaporated gold film with a sheet resistance of less than 0.05 $\Omega/\square$; it adheres to the tantalum resistor film below because of the thin titanium (or titanium–palladium) layer sandwiched between the two. The tantalum nitride resistor film is doped with oxygen so that the resulting TCR cancels the TCC of the tantalum-film capacitors. Typically, the resistor film may be sputtered to a thickness corresponding to an initial sheet resistance of 35 $\Omega/\square$; after adjusting the resistors and thereby tuning the circuits by anodization, the sheet resistance is expected to be on the order of 50 $\Omega/\square$. We recall that the anodization processes (i.e., preanodization, thermal oxidization, final trim-anodization) also serve to stabilize and passivate the resistor exterior by growing an insulating tantalum-pentoxide film on its surface. If the average nominal resistance value is relatively high, resistor, and thereby substrate, area can be reduced by selecting a higher initial and final sheet resistance. Typically, this may be 50–70 $\Omega/\square$ and 100 $\Omega/\square$, respectively. Because of the power dissipated by the operational amplifier, it is advantageous to use a ceramic (rather than glass) substrate to maintain a uniform temperature distribution across the substrate. This improves both the tracking of resistor ratios on the substrate and the frequency-stability of the network due to TCR–TCC matching.

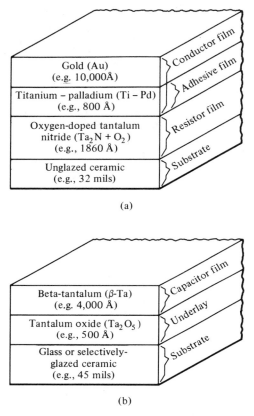

(a)

(b)

FIG. 5-14. Schematic diagrams of TT-SABB substrates: (a) resistor substrate; (b) capacitor substrate.

A schematic of the film layers associated with an unprocessed capacitor substrate is shown in Fig. 5-14b. The top layer is a beta-tantalum film sputtered on a glass (e.g., Corning 7059) or selectively-glazed ceramic substrate to a thickness corresponding to a sheet resistance of approximately 5.0 $\Omega/\square$. This sheet resistance must be held as low as possible in order to minimize the resistive (as opposed to the dielectric) loss of the capacitors. The advantage of using glass for the capacitor substrate is its excellent surface smoothness; the resulting uniform beta-tantalum film thickness provides high capacitor yields. On the other hand, because the brittleness of glass may result in excessive substrate breakage during production, glazed ceramic is often preferred, even though its surface smoothness is somewhat inferior to that of glass. In either case, glass or glaze must be coated with a thin layer of tantalum oxide to prevent undercutting during the etching processes involved in capacitor fabrication.

The interconnection pattern on both the resistor and capacitor substrates is generated from an evaporated gold film. In addition to having a low sheet resistivity ($0.05\ \Omega/\square$), the gold film is essential for the thermocompression bonding both of the operational amplifier beam-leaded chips and the gold leads used for the assembly of the circuits.

The substrate cross sections shown in Fig. 5-14 are to be understood qualitatively; naturally, certain variations will be undertaken in the actual production process, without, however, deviating significantly from the basic description given here. The same applies to the actual process steps that are followed to generate the individual components on the resistor and capacitor substrates. Typically, with minor variations, a three-mask process is used to etch the resistors and interconnecting paths out of the three film layers initially deposited on a ceramic substrate. The three masks are used as follows.

Mask 1: Using a *conductor-etch* mask the conductor pattern is generated. The gold film is etched away everywhere on the substrate except where it is needed for the conductor pattern. Because the etchant is to dissolve only the gold and not the tantalum film below, a "selective" etchant is used.

Mask 2: Using a *tantalum-etch* mask, the resistor patterns are generated by etching away the tantalum resistor film everywhere on the substrate, except where it is needed for the resistor meander patterns.

Mask 3: Using an *anodization* mask, windows are exposed in the photo-resist (which has previously been deposited on the substrate) to anodize the resistor patterns. Here the sheet resistivity of the tantalum film may, for example, be increased from $35\ \Omega/\square$ to $50\ \Omega/\square$. Depending on the required accuracy, the resistors are trimmed (i.e., anodized) individually or batch anodized.

A similar three-mask process is used to fabricate the capacitors and inter-connecting paths from the beta-tantalum film deposited on a glass or glazed ceramic substrate. Typically the process is as follows:

Mask 1: Using a *tantalum-etch* mask the beta-tantalum areas (i.e., base electrodes), on which the capacitor dielectric is to be formed, are generated.

Mask 2: The capacitor dielectric layers are now formed by anodization of the beta-tantalum base electrodes; the upper layer of the beta-tantalum is thereby converted into tantalum pentoxide, which here serves as the dielectric. Those tantalum areas that are not to be anodized are protected by a layer of hardened photoresist. The windows corresponding to the dielectric areas are exposed in the photoresist with an *anodization* mask. Typically the beta-tantalum may be anodized to 225 V, resulting in a capacitance density of $0.05\ \mu F/cm^2$.

Mask 3: Using a *conductor-etch* mask the counter electrodes and the conductor pattern are etched out of a dual gold and titanium (or nichrome–palladium) film layer which is evaporated onto the whole substrate after the dielectrics are formed. As with the resistors, the titanium or nichrome–palladium is required for adherance of the gold film to the tantalum. It is in this step that the capacitor values are determined, since these depend on the area of the dielectric covered by the counter electrodes.

Scribing and Circuit Adjustments   For any specified circuit option the corresponding interconnecting conductor paths must be opened and the unwanted taps on the resistors and capacitors (i.e., resistor segments and finger electrodes, respectively) separated in order to bring the poles and zeros of the network into the required frequency and Q range (coarse adjustment). For laboratory purposes and small production quantities, both these steps can be carried out manually using a diamond-tipped scribing tool. In assembly-line production, automatic means of open-circuiting the conductor paths are preferable. The use of a programmable x–y table and/or a deflectable laser beam has been found to have considerable potential for this type of operation, although abrasion machining and trimming has also been found suitable, at least with thick-film circuits.[19] Note that the building-block approach is indispensable if automatic tooling is to be used; each resistor and capacitor is then always at a predetermined location relative to the substrate, as are all the external leads (i.e., those connecting the circuit to the outside). Consequently the tooling jig and all other interface equipment between the circuit and the processing and test gear are always the same, irrespective of the actual network function to be realized by the circuit. It is one of the important features of the building block based on the adjustment method by scribing that a *single*, universal building block can be adapted with ease to any desired filter function and to any frequency within more than a decade, merely by scribing the appropriate tapped resistors, capacitors and conductor paths.

The fact that only resistors can be fine-tuned is no real disadvantage, since operating frequencies are determined by RC products and operational amplifier gains by resistor ratios. Fig. 5-15 shows how tantalum thin-film resistors may be experimentally fine-tuned by individual trim-anodization. By placing a solid electrolyte, which is mounted on a manual probe, onto the resistor to be adjusted, and applying an anodization voltage, the top layer of the resistor film is converted into an insulating tantalum pentoxide film. This film grows outward *and* inward, thereby decreasing the resistor cross section and providing (one-way) resistor vernier adjustment. Resistors can be increased in this way by up to 50 % (although, for high stability, the upper

19. M. L. Topfer, *Thick Film Microelectronics: Fabrication, Design and Applications* (New York: Van Nostrand-Reinhold, 1971).

FIG. 5-15. Fine tuning of tantalum thin-film resistors by individual trim-anodization.

limit should not exceed 30 %) and to within 0.01 % initial accuracy. The factor limiting this accuracy is the measuring resistor bridge, whose accuracy must then exceed 0.01 %. Also the resistor temperature coefficient may render such accuracy improbable. The TCR of oxygen-doped tantalum nitride resistors is increased from, say, -75 ppm/°C (undoped) to -200 ppm/°C in order to match the TCC of the capacitors ($+200$ ppm/°C); thus the resistor will change by 0.02 % with a 1°C change in temperature. Extreme resistor accuracy therefore calls for a temperature-controlled environment during the tuning process (to within less than 1°C); however the costs involved will justify such measures only in exceptional cases. Naturally, when the TCR is less, higher tuning accuracy becomes more readily feasible.

Resistor fine tuning (by any method) is an irreversible process and therefore calls for extreme care in order to avoid overshooting the desired value. Here again, automation and computer control can be used to advantage when the production volume is sufficiently large. High-precision tuning setups have been developed which automatically tune a given filter function by on-line computer control. Such set-ups can be used for a variety of linear networks including, for example, audio oscillators.[20]

20. F. H. Hintzman, Jr., Computer tuning of hybrid audio oscillators, *IEEE Spectrum*, Vol. 6, pp. 56–60 (February 1969).

If a laser is used for circuit scribing, it can also be used to fine-tune the resistors. This obviates the need for additional trim-anodization equipment.[21] Thus, as far as the execution of the two processes is concerned, scribing and trimming are virtually identical processes and can be carried out with the same manufacturing setup. Laser scribing and tuning promises various other advantages as well. The process can be controlled to extreme accuracy because of the narrow width of the laser beam (from 0.25 mil to 1 μm under well controlled conditions), and because the work piece can be viewed many times enlarged on the screen of a closed-circuit TV system during processing. The process is also rapid; the laser is capable of machining the work piece at speeds up to one inch per second, provided a suitable positioning table and measuring equipment are available. Above all, laser scribing and trimming is a noncontact process, leaving no residue on the resistors and therefore permitting the trimming results to be monitored and evaluated *during* the actual trimming process. By contrast, with solid-electrolyte anodization the probes shunting the resistors cause measuring errors, making it necessary to lift the probes off the resistors before each meter reading.

There may be one other important advantage associated with the laser trimming of resistors. We recall that tantalum film resistors should be heat-treated, i.e., baked at temperatures on the order of 250°C after anodization of any kind. This prevents drift due to thermal oxidization after anodization. The more the resistors have been (batch- or trim-) anodized, the greater the need for subsequent heat treatment. However, since heat treatment changes the value of a resistor, it cannot follow the trimming process of high-precision networks. Besides, when the *functional* tuning method is used, the completely assembled circuits may not withstand the baking process involved. Thus, trim anodizing invariably deteriorates resistor stability to a greater or lesser degree, depending on the extent of anodization required. Clearly, this source of deterioration of the resistor stability is avoided with laser trimming, if we assume that the laser trimming process itself does not deteriorate resistor stability in any way. This assumption is not universally accepted, however, and there are indications that the larger the "heat-affected" zone of the "active" area of a resistor (i.e., the area in which current flows), the poorer the resistor stability will be. Thus, the laser trimming of precision resistor networks involves special L and T cuts that are designed to minimize the heat-affected zone of the resistors to be trimmed.[22] Naturally the resistor layout is also influenced by these considerations.

Trimming and scribing by laser are identical processes used to achieve different ends. Thus, in an automated production line, both can be incorporated using a tape-controlled high-speed x–y table. For any desired network

21. M. I. Cohen, B. A. Unger, and J. F. Milkosky, Laser machining of thin films and integrated circuits, *Bell Syst. Tech. J.* **47**, 385–405 (1968).
22. See, for example, R. C. Headley, Laser trimming, an art that must be learned, *Electronics*, pp. 121–125 (June 21, 1973).

function, a corresponding tape can be used to program the high-precision x–y table.

When trimming functionally or deterministically, equipment must be present that is capable of monitoring either the corresponding network function or the resistor values. Naturally, the cost of the equipment involved in an automated setup can only be absorbed by a high production volume. For small numbers, or as a laboratory tool, manual anodization by solid electrolyte is more suitable; the equipment necessary is relatively simple and inexpensive, and the method itself easy to implement. In production-line use, however, it is less easily controllable and, for high precision, may be relatively slow.

Building-Block Assembly   Starting out with the two substrates shown in Fig. 5-13, the assembly of a building-block takes place as follows:

1. The desired filter function is obtained by open-circuiting the appropriate conducting paths.
2. Coarse frequency and Q adjustment is obtained by scribing open the unwanted resistor segments and capacitor counter-electrode fingers.
3. The operational amplifier is bonded to the resistor substrate.
4. The resistor substrate is attached to the top of the capacitor substrate with a silicone adhesive (e.g., Dow Corning RTV 3145).
5. The circuit is tuned to the desired frequency, Q, and gain by adjusting the corresponding resistors.

When functional tuning is being done, the last step is carried out while the circuit is in operation; when the deterministic method is used, the capacitors (and their losses) have to be measured and recorded before the resistor is lead-interconnected with the capacitor substrate. It is advantageous to measure the capacitors *after* step 4, so that any parasitics due to the RTV or the upper substrate can be taken into account. However, this implies that the terminals of each capacitor are accessible from the pads on the two sides of a substrate; then each capacitor can be measured separately.

5.3.2 The Twin-T Single-Amplifier Building Block Customized by Selective Film Removal

As long as the capabilities for scribing and coarse trimming by laser are not available, serious objections can be raised with respect to mechanical scribing as a process for high-volume production or even production at all. Since thin-film components, in particular capacitors, must be deposited on glass or at least glazed-ceramic substrates, damage may result from scribe marks propagating as cracks along the substrate surface. Conceivably, sand-blasting machines damage the surface less than a metal or diamond scribing tool.

However, machines of this kind, providing the accuracy in registration and resolution required for thin-film circuits with line widths of a few mils, are both expensive and difficult to develop. Absorbing the development costs of these tools may in turn increase the cost of the end product appreciably. Incidentally, the latter consideration is also valid for laser processing equipment. Another objection that concerns both mechanical and laser scribing is that both methods introduce a process into the manufacture of thin-film circuits that is incompatible with any of the other thin-film-type processes. The manufacturing procedure thereby becomes more complicated and time consuming, and, consequently, may be too expensive to compete effectively with more compatible thin-film production techniques.

It is to eliminate these problems that the "selective film-removal" rather than the scribing method may be used in the production of hybrid integrated filter building blocks. It accomplishes the same circuit versatility as the scribing method, but does so in a thin-film-compatible way. Where the scribing method opened all the unwanted paths, this method does not generate them in the first place. Using a conductor-etch mask the gold film layer that is initially deposited on top of the tantalum resistor film is removed selectively, leaving gold only on the conductor areas required for a particular function. Thus *each circuit function requires an individual conductor-etch mask* to delineate the corresponding circuit topology on the substrate. Resistor and capacitor scribing is eliminated essentially in the same way. To obtain a desired range of resistor values, a set of tantalum etch masks are used to selectively remove tantalum resistor film between the previously generated, gold-covered resistor terminals, leaving only the desired resistor patterns behind.[23]

The processing sequence for the resistor and capacitor substrate of a twin-T single-amplifier building block which is customized by selective film removal is shown in Fig. 5-16. In order to economize on masks, the conductor etch mask delineating the conductor paths and the mask that determines the tantalum resistor pattern can be combined into a tantalum-and-conductor etch mask. Referring to Fig. 5-16, the actual mask sequence for the resistor substrate is then as follows:

Mask 1: Window-etch mask (step 6) used to etch windows in the conductor film corresponding to the areas where the resistors of the circuit will be located. Naturally, each window is large enough to accommodate the *maximum* resistor value to be generated in it. Furthermore, this mask contains windows for every possible resistor, irrespective of which resistor combination is required for a given function.

Mask 2: A tantalum-and-conductor etch mask (step 7) is now used to delineate the conductor pattern corresponding to a particular function and

---

23. Examples of the high-, medium-, and low-valued versions of a resistor generated in this way are shown in Chapter 6, Fig. 6-10b, of *Linear Integrated Networks: Fundamentals.*

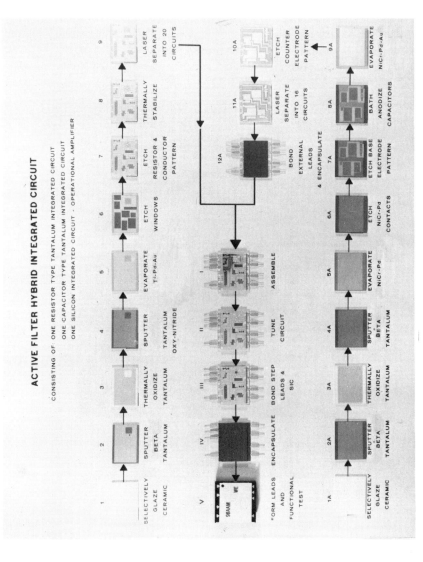

FIG. 5-16. The processing sequence for the resistor and capacitor substrates of a TT-SABB customized by selective film removal.

463

the meander paths corresponding to the resistor values required for that function. Clearly every different function, and every function with different poles and zeros, requires a separate tantalum-and-conductor etch mask.

Mask 3: An anodization mask may now be required to anodize the resistors to value.[24] However, in the process sequence of Fig. 5-16, only thermal stabilization of the resistors is carried out (step 8), which requires no mask of any kind. This simplification of the process is possible because each resistor is individually trim-anodized in the tuning step after the assembly of the two substrates (steps I and II).

Notice that masks 1 and 3 (if the latter is required) are common to all the functions and resistor values provided for in a building block, while a new

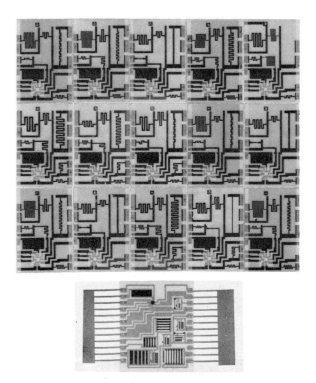

FIG. 5-17. The all-purpose resistor substrate (by itself) and 15 different substrates corresponding to various transfer functions (grouped).

24. Sometimes, to enhance resistor stability, the anodization is carried out in two steps separated by a thermal stabilization process (i.e., heat treatment). In this case two separate anodization masks are necessary, the second having slightly smaller windows than the first to prevent mask-to-substrate misalignment.

mask 2 is necessary each time a new function, or even a new resistor value is required. Add to this a glaze mask[25] used to selectively glaze the substrate in such a way that the chip and bonding areas remain unglazed (step 1, Fig. 5-16) and we have a total of three or four masks for the resistor substrate, one of which must be individually patterned for each new network function.

A selection of 15 resistor substrates corresponding to a variety of functions, is shown in Fig. 5-17. Also shown for comparison is the all-purpose resistor substrate which, when appropriately scribed, can replace the 15 "custom-designed" substrates (as well as an arbitrary number of others). Note that the scribable substrate is larger than the custom substrate, since the tapped resistors take up more area than conventional ones. This fact must be taken into account when comparing the costs of the two approaches.

Closely resembling the fabrication of film resistors, a wide range of capacitor values is obtained with the selective film-removal method, by varying the width of the counter electrode while maintaining a fixed area for the base electrode and dielectric. This fixed area corresponds to the maximum desired capacitor value.[26] It is obvious here that the conductor-etch mask delineating the conductor paths and the mask that determines the counter-electrode area can be combined into one, since conductor paths and counter electrodes are made of the same film material (e.g., gold). Referring to the process sequence shown in Fig. 5-16, a typical mask sequence for the capacitor substrate may therefore be as follows:

Mask 1: A *tantalum-etch* mask (step 7A) is used to generate the base-electrode pattern.

Mask 2: An *anodization-mask* (step 8A) is used to form the dielectric film on the base electrodes.

Mask 3: A *conductor-etch* mask (step 10A) is used to form the conductor paths and the counter electrodes.

Here again a glaze mask (step 1A) is necessary when glazed-ceramic rather than glass substrates are used. Masks 1 and 2 are common to all the functions and capacitor values provided by the building block. Mask 3 changes with every new function or capacitor value. An additional mask (step 6A) is required to etch the nichrome–palladium contacts deposited in step 5A. This measure is taken to ensure good contact between the gold terminal evaporated onto the tantalum base electrode in spite of the preceding bath anodization. Without this special NiCr–Pd deposition the contact between the base electrode and the gold conductor path leading away from it may be poor and the capacitor noisy. In all, then, we have five masks, one of which

25. This is also required if selectively glazed ceramic is used for the substrates in Fig. 5-13.
26. An example of this scheme was shown in Chapter 6, Fig. 6-11b of *Linear Integrated Networks: Fundamentals*.

occurs as many times as there are different capacitor values or interconnecting paths on a capacitor substrate.[27]

It should be noted that although we have referred to them as "custom designed," the substrates patterned by selective film removal still have not lost the uniformity vital to a building-block approach and thereby to batch-processing techniques. Every resistor and capacitor is always in the same location relative to the substrate (whether it is being used or not for a particular function). All external leads are positioned the same, as is the operational amplifier chip and the associated components required for frequency compensation. Thus the same tooling, processing, and test equipment is used for any function provided by a building block. This is evident from Fig. 5-18 where a resistor substrate, a capacitor substrate and a two-substrate assembly are shown at the top, and an encapsulated building block is shown below. Once encapsulated a building block only differs by its electrical characteristics; physically they all look alike. Thus there is no telling (other than by measurement) that the printed wiring board in Fig. 5-19 is a three-tap delay

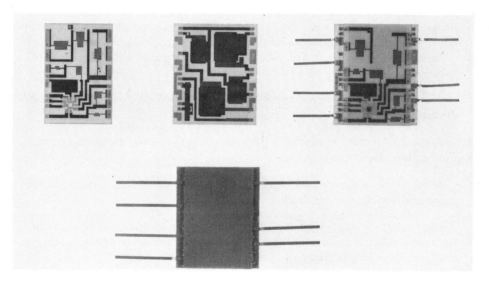

Fig. 5-18. Substrates and two-substrate assembly (top) and encapsulated building block (bottom).

27. Since the frequencies of the building blocks are determined by RC products, a single capacitor substrate can be combined with a number of resistor substrates to cover a given frequency range. The resistor substrates may also be used to determine the majority of the building block functions. Thus, in practice a filter set may require n different versions of resistor Mask 2 (step 7) but only n/4 or even less versions of capacitor Mask 3 (step 10A) to cover a given frequency range and set of filter functions. Naturally, the larger the permissible range of resistor trim anodization (typically 20% to 30%) the smaller n will be. The final set of filters then comprises n different building block types, any of which may, of course, occur in multiples.

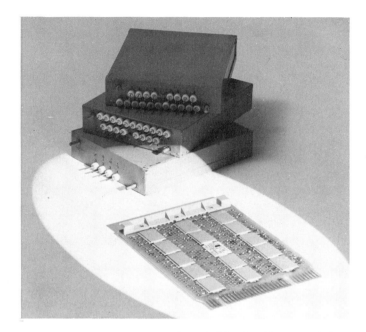

FIG. 5-19. Three-tap (hybrid integrated) active RC delay line (foreground) shown with the three *LC* filter cans it replaces.

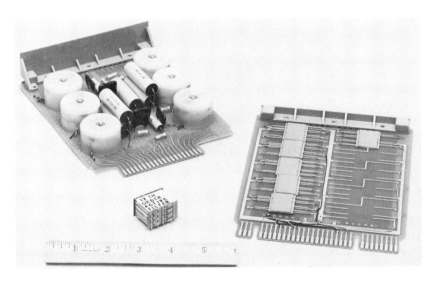

FIG. 5-20. A twelfth-order bandpass filter in three realizations: *LC* form (left background), six-building-block cube-type assembly (left foreground), and six-building-block printed wiring board (right).

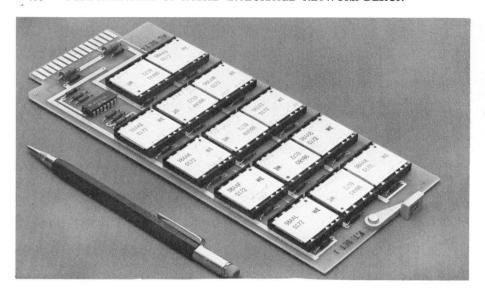

FIG. 5-21. Filter network assembled from building blocks in dual in-line packages.

line replacing the three LC filter cans shown in the background. The same applies to the twelfth-order bandpass filter shown in building-block and LC form in Fig. 5-20. The six building blocks are shown soldered on a printed wiring board and, in a far more efficient package from a space-savings point of view, in a cube-type assembly taking up less than one cubic inch in volume. In fact, the volume of the twelfth-order filter cube is comparable to that of one of the inductor coils making up the equivalent LC filter. Finally, in Fig. 5-21, the building blocks making up a filter network are shown in the form of dual in line packages. The printed writing board shown accommodates fifteen sections corresponding to filter networks with a total of fifteen complex conjugate pole pairs and, if required, an additional fifteen negative real poles. The space savings in any one of the forms shown is considerable, compared to the LC equivalent networks, and the building-block nature of the individual packages brings significant savings in equipment cost as well.

5.3.3 Other Hybrid Integrated Realizations

To find other types of hybrid integrated circuit realizations for the design of active-filter building blocks we have only to examine the most important combinations of resistor and capacitor materials available to us. These are summarized in Table 5-3. The first combination, tantalum oxynitride thin-film resistors and beta-tantalum thin-film capacitors, has been dealt with above. The remaining four combinations comprise film resistors and appli-quéed multilayer chip capacitors. The same considerations apply to these

TABLE 5-3. *RC* COMBINATIONS FOR HYBRID IC ACTIVE FILTERS

	Resistor			Matching capacitor		
Com-bination	Material	Typical TCR (ppm/°C)	Typical aging* (%)	Type	Typical TCR (ppm/°C)	Typical aging (%)
1	Tantalum oxinitride	-200 ± 15	0.1–0.5	Beta-tantalum thin film	200 ± 30	0.1–0.5
2	Nichrome	50 ± 30	0.1–0.5	Multilayer chip	$NPO \pm 30$	0.1
3	Thick film	± 50	0.1–1	Multilayer chip	$NPO \pm 30$	0.1
4	Tantalum nitride	-100 ± 20	0.1–0.5	Multilayer chip	$NPO \pm 30$	0.1
5	Tantalum nitride	-100 ± 20	0.1–0.5	Multilayer chip	$+100 \pm 30$	0.1

* Depending on life span.

combinations as to those discussed in relation to combination 1. Thus, a resistor substrate using either scribing or selective film removal may be used for any of the combinations 2 to 5. However, beside the operational amplifier, which may be either beam-leaded, wire-bonded, or soldered on to the substrate, provision must now be made for the capacitors. A scribable building block using tapped thin-film resistors, appliquéed capacitors, and a beam-leaded operational amplifier chip is shown complete in Fig. 5-22a and without the appliquéed components in Fig. 5-22b. Using the tantalum-resistor-film technology (combination 4 or 5) special provision must be made in order to be able to solder chip capacitors onto the conductor pads, since chip capacitors with bondable ribbon leads are uncommon.

The chip capacitors most frequently used in precision hybrid integrated networks are the NPO type, i.e., those whose TCC is centered at zero. TCCs of $+100$ ppm/°C, matching the TCR of tantalum nitride (combination 5) are also available, but in a smaller selection. Combination 2 is relatively common because the attainable frequency accuracy is good. To estimate the worst-case frequency error of any one of the combinations in Table 5-3, the individual drift components can be plotted additively. This is shown in Fig. 5-23a for combination 2, where we have assumed the TCR, TCC values given in Table 5-3, overall aging of -0.1%, and a tuning error of $\pm 0.1\%$. In Fig. 5-23b, combination 4 is plotted, where again we have assumed -0.1% aging and $\pm 0.1\%$ tuning error. From 10°C to 60°C, combination 2 (i.e., nichrome film resistors with NPO chip capacitors) has a total worst-case frequency drift of $+0.35\%$ and -0.6%, while combination 4 (tantalum nitride film resistors with NPO chip capacitors) has a total worst-case frequency drift of $\pm 0.5\%$. Thus combination 4 is only marginally inferior to combination 2.

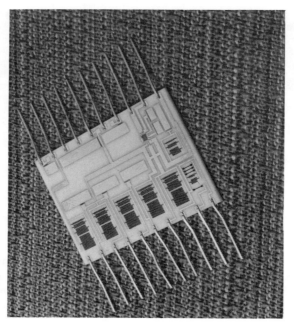

(a)

(b)

Fig. 5-22. (a) A scribable building block using tapped thin-film resistors, appliquéed capacitors, and a beam-leaded operational amplifier chip. (b) The same building block without the appliquéed components.

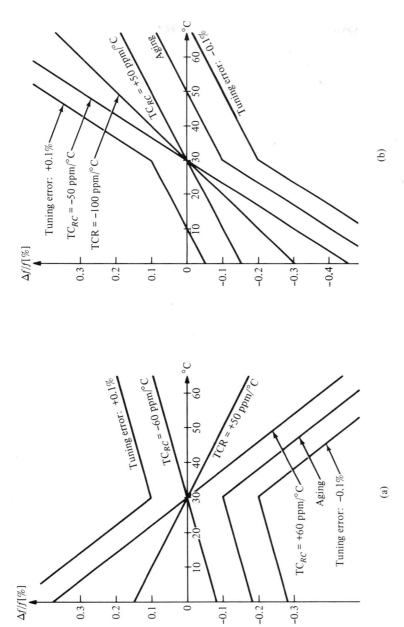

FIG. 5-23. Plots of individual drift components of (a) combination 2 (Table 5-3) and (b) combination 4 (Table 5-3).

471

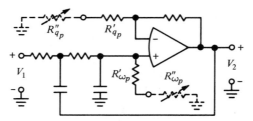

FIG. 5-24. Second-order low-pass network derivable from the TT-SABB whose principal parameters are tunable by means of external resistances.

A dilemma facing many systems designers is that they require small-size, hybrid integrated active RC networks on the one hand, but lack the facilities and the specialized capabilities necessary to implement this technology on the other. One way of overcoming this problem is to order whatever film components are necessary for the networks as individual substrates from a hybrid IC supplier, and to assemble the parts, as required, on the systems-designer's premises. However, the problem of final network adjustment remains. This requires both the specialized equipment and know-how generally available only to the hybrid-IC manufacturer and the network and systems understanding peculiar to the systems designer. A frequent and satisfactory solution to this problem is to design the circuit in hybrid IC form but to provide terminals for external tuning resistors with which the network designer may fine-tune his circuit without the need for hybrid IC capabilities. The tuning resistors may be either potentiometers or carefully selected discrete resistors. An example of a network implementing this concept is shown in Fig. 5-24. We have here a second-order low-pass network (obtainable from the building block in Fig. 5-10) whose principal parameters q_p and ω_p are tunable with the resistors R_{q_p} and R_{ω_p}, respectively. Both may be attached externally to the hybrid IC package as grounded potentiometers or discrete resistors. If the externally attached resistors are small enough, i.e., if the ratios R''_{q_p}/R'_{q_p} and $R''_{\omega_p}/R'_{\omega_p}$ are much less than unity, then the fact that the external resistors have different characteristics and track less well than the film resistors will matter little. In general the values of the externally attachable components need only be large enough to make up for the tolerances and parasitics inherent in the hybrid IC part of the network.

CHAPTER

6

DESIGNING HYBRID INTEGRATED FILTER BUILDING BLOCKS

INTRODUCTION

In the previous chapter we discussed some of the technological considerations involved in designing a hybrid integrated building block. As our example, we used the twin-T single-amplifier building block (TT-SABB) which has been found very suitable for hybrid integrated realization and which covers the medium-frequency range—at present the most important frequency range for linear integrated networks. Furthermore, combining two TT-SABBs appropriately with an additional operational amplifier according to the FEN decomposition, any high-Q second-order function is readily obtained. Thus not only the TT-SABB, but also the multiple-amplifier FEN configurations that can be realized with it, represent useful approaches to hybrid-integrated network design.

Using the FEN-design approach as an example, we shall, in this chapter, go into the practical details of hybrid integrated filter design. The design procedure outlined will cover the optimization of the TT-SABB for hybrid integrated implementation. In the course of this outline the problem areas most likely to be encountered by the designer will be discussed in detail. Foremost among these are parasitic effects and methods of network tuning that permit the elimination of those effects. Subsequently, some specific design examples involving the TT-SABB by itself as well as its incorporation in high-Q FEN configurations are presented.

473

6.1 DESIGNING THE TT–SABB FOR A MINIMUM GAIN–SENSITIVITY PRODUCT

As shown in Chapter 4, in the design of a hybrid IC filter network, an important consideration in selecting component values is the minimization of the gain–sensitivity product Γ. Here we shall examine in more detail the factors affecting Γ for the TT-SABB and, more generally, for class 4 single-amplifier networks. We shall also derive a lower limit Γ_{min} for these networks in order to provide a frame of reference against which to compare the actual Γ values achieved.

6.1.1 Minimizing Γ for Class 4 Networks[1]

To briefly summarize the pertinent expressions for class 4, (i.e., positive-feedback) networks, we found in Chapter 3 (equation (3-46)) that the pole Q can be written in the form

$$q_p = \frac{\kappa_1}{\kappa_2 - \kappa_3 \beta} \tag{6-1}$$

where κ_1, κ_2, and κ_3 are functions of the passive RC feedback network only and β is the closed-loop amplifier gain. Assuming close tracking of the resistors and capacitors, q_p is sensitive only to variations in β according to the relationship

$$S_\beta^{q_p} = \frac{\kappa_2}{\kappa_1} q_p - 1 = \kappa q_p - 1 \tag{6-2}$$

where

$$q_p(\beta = 0) = \hat{q} = \frac{1}{\kappa} = \frac{\kappa_1}{\kappa_2} \tag{6-3}$$

and $\hat{q}$ is the pole Q of the passive RC feedback network. From (6-3) it follows that

$$\kappa_{min} > 2 \tag{6-4}$$

Hence the lower limit on the Q sensitivity is:

$$[S_\beta^{q_p}]_{min} > 2q_p - 1 \tag{6-5}$$

Clearly, to minimize $S_\beta^{q_p}$ for any function $T(s)$, κ must be minimized, i.e., be as close to 2 as possible. This is achieved by an appropriate choice of the resistors and capacitors of the feedback network. Because κ is the inverse pole

---

1. G. S Moschytz, A universal low-Q active-filter building block suitable for hybrid integrated circuit implementation, *IEEE Trans. Circuit Theory*, **CT-20**, 37–47 (1973).

Q of the passive RC feedback network, minimizing κ reverts to the problem of maximizing $\hat{q}$. Since we are considering only second-order networks, $\hat{q}$ is related to the distance between the two poles of the RC feedback network on the negative-real axis. The maximum value of $\hat{q}$ (i.e., 0.5) is attained when the two poles coincide on a double pole. This condition can be approximated reasonably well by physically realizable networks; whereas a $\kappa = 1/\hat{q}$ value of 2 cannot be attained, a value of, say, 2.2 does not strain the physical realizability of a network.

A general method of approximating the maximum $\hat{q}$ value of a second-order RC network was briefly discussed in Chapter 3, Section 3.2.2. By impedance-mismatching the second half of the ladder so as to minimize its loading effect on the first ladder section, $\hat{q}$ can be made to approach 0.5. The second section is thereby impedance-scaled by an amount $\rho \gg 1$.

For the TT-SABB configurations listed in Table 3-5 of Chapter 3, κ is given in general terms as a function of the resistor ratios r_i and the capacitor ratios c_i. A special case is given for some of the circuits in the table where a value for one of the three component ratios has been assumed and κ minimized with respect to another. It is then further minimized with respect to the remaining third component ratio by letting the latter become very large or tend to zero, depending on which provides the smaller κ value.

Our real concern, however, is not to minimize $S_\beta^{q_p}$ (i.e., κ in the case of class 4 networks) but to minimize the gain–sensitivity product Γ. With (6-1) and (6-2) we obtain:

$$\Gamma = \beta S_\beta^{q_p} = \beta\kappa(q_p - \hat{q}) = \frac{\kappa_2}{\kappa_3}\frac{(q_p - \hat{q})^2}{q_p\hat{q}} \tag{6-6}$$

and

$$\Gamma|_{q_p \gg \hat{q}} \approx \beta\kappa q_p = \beta \frac{q_p}{q_p(\beta = 0)} \tag{6-7}$$

To obtain a lower bound on Γ from (6-6) it is useful to look more closely at the ratio κ_2/κ_3. For this purpose we recall from Chapter 2 that the feedback transfer function $t_{32}(s)$ of a class 4 network must have the form

$$t_{32} = k_{32} \cdot \frac{\omega_p s}{s^2 + \dfrac{\omega_p}{\hat{q}}s + \omega_p^2} \tag{6-8}$$

As we know, this is the transfer function of a passive RC second-order band-pass network. Thus, we can represent a general class 4 network in the block diagram form shown in Fig. 6-1. From (6-8) it follows that the peak value $\hat{t}_{32}$, which occurs at ω_p (see Fig. 6-2) is given by

$$\hat{t}_{32} = t_{32}(j\omega_p) = k_{32} \cdot \hat{q} \tag{6-9}$$

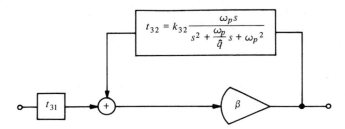

FIG. 6-1. Block diagram of general class 4 network.

Returning to the feedback loop in Fig. 6-1 and considering it at the frequency ω_p we obtain the configuration shown in Fig. 6-3. For oscillation to occur (i.e., $q_p \to \infty$) it follows by inspection that β must take on the value

$$\beta_{\text{osc}} = \frac{1}{\bar{t}_{32}} = \frac{1}{k_{32}\hat{q}} \tag{6-10}$$

However, from (6-1) it follows that the β required for infinite q_p is given by

$$\beta_{\text{osc}} = \kappa_2/\kappa_3 \tag{6-11}$$

Combining (6-10) and (6-11):

$$\frac{\kappa_2}{\kappa_3} = \frac{1}{\bar{t}_{32}} = \frac{1}{k_{32}\hat{q}} \tag{6-12}$$

and (6-6) becomes

$$\Gamma = \beta S_\beta^{q_p} = \frac{1}{\bar{t}_{32}} \frac{(q_p - \hat{q})^2}{q_p \hat{q}} = \frac{1}{k_{32}} \left(\frac{q_p - \hat{q}}{\hat{q}}\right)^2 \frac{1}{q_p} \tag{6-13}$$

For $q_p \gg \hat{q}$ this simplifies to

$$\Gamma = \frac{1}{\bar{t}_{32}} \left(\frac{q_p}{\hat{q}}\right) = \frac{1}{k_{32}} \frac{q_p}{\hat{q}^2} \tag{6-14}$$

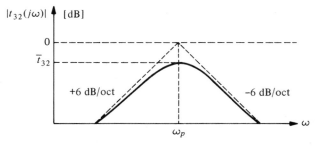

FIG. 6-2. Amplitude response of the feedback network t_{32}.

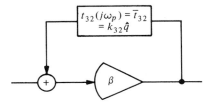

FIG. 6-3. Class 4 network at pole frequency ω_p.

The problem of minimizing the gain–sensitivity product Γ is now reduced to that of maximizing $\hat{q}$ and either $\bar{t}_{32}$ or k_{32}. Considering $\bar{t}_{32}$ first, it can be shown that the maximum transmission of a passive second-order RC network is approximately unity,[2] so that

$$0 < \bar{t}_{32} = |t_{32}(j\omega)|_{\max} < 1 \qquad (6\text{-}15)$$

Thus from (6-3) and (6-13)

$$\Gamma \geq \left(2q_p + \frac{1}{2q_p} - 2\right)_{q_p \gg 1} \approx 2(q_p - 1) \qquad (6\text{-}16)$$

A more accurate lower bound on Γ can be obtained by considering the individual coefficients k_{32} and $\hat{q}$. We know the limits of $\hat{q}$ to be zero and 0.5. Furthermore, since $t_{32}(s)$ is the transfer function of a 3-terminal RC network, k_{32} is limited by[3]

$$0 < k_{32} < 2 + \frac{1}{\hat{q}} \qquad (6\text{-}17)$$

in order that $0 < t_{32}(s) < 1$ for $0 < s < \infty$. With (6-13) we then have

$$\Gamma = \frac{(q_p - \hat{q})^2}{k_{32} q_p \hat{q}^2} \geq \frac{(q_p - \hat{q})^2}{\left(2 + \dfrac{1}{\hat{q}}\right) q_p \hat{q}^2} = \frac{(q_p - \hat{q})^2}{q_p \hat{q}(2\hat{q} + 1)} \qquad [6\text{-}18]$$

This expression decreases with increasing $\hat{q}$. Thus for maximum $\hat{q}$, namely 0.5, (6-18) implies that $\Gamma \geq (q_p - 0.5)^2/q_p$ or, for large q_p, $\Gamma \geq q_p - 1$. This conclusion is true for those cases in which $t_{32}(s)$ represents a general RC network. However, for those cases in which $t_{32}(s)$ is an RC ladder network, k_{32} and $\hat{q}$ are not independent of each other, as the following simple argument shows.

Consider the two frequency responses corresponding to two different second-order bandpass functions t_{32A} and t_{32B} shown in Fig. 6-4. The corner frequencies ω_{p_1} and ω_{p_2} of our first case t_{32A} are far apart, corresponding to a

2. The maximum of a general second-order RC network may actually exceed unity slightly, its upper bound being $\sqrt{4/3}$. For a ladder network the upper bound is unity. See A. D. Fialkow and I. Gerst, The maximum gain of an RC network, *Proc. IRE*, **41**, 392–395 (1953).
3. A. D. Fialkow, I. Gerst: "The transfer function of general two terminal-pair RC networks," *Quart. of Appl. Math*, **10**, 113–127, April, 1952.

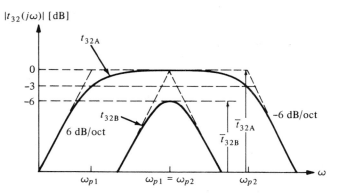

FIG. 6-4. Frequency response of two second-order bandpass functions t_{32A} and t_{32B}.

low value of $\hat{q}$, but permitting t_{32A} to reach its maximum value, $\bar{t}_{32A}$ equal to unity. By contrast, the corner frequencies of our second case, t_{32B}, coincide (i.e., $\omega_{p_1} = \omega_{p_2}$), corresponding to the maximum value of 0.5 for $\hat{q}$, and, at the same time, to an attenuation of 0.5 (i.e., -6 dB)[4] for $\bar{t}_{32B}$. Thus we note that $\hat{q}$ and $\bar{t}_{32}$ as well as k_{32} cannot have maximum values simultaneously. Adding a prime to the maximum $\bar{t}_{32}$ and k_{32} values compatible with maximum $\hat{q}$, we have

$$\bar{t}'_{32} = \bar{t}_{32}|_{\hat{q}=0.5} \leq 0.5 \tag{6-19}$$

Consequently, with (6-9),

$$k'_{32\,\text{max}} = k_{32\,\text{max}}|_{\hat{q}=0.5} = 2\bar{t}'_{32\,\text{max}} = 1 \tag{6-20}$$

From (6-13), the minimum gain–sensitivity product Γ then results as

$$\Gamma|_{k_{32}=1} = \frac{q_p}{\hat{q}^2} + \frac{1}{q_p} - \frac{2}{\hat{q}} = \kappa^2 q_p + \frac{1}{q_p} - 2\kappa \tag{6-21}$$

where $\kappa = 1/\hat{q}$. For $q_p \gg 1$ we have

$$\Gamma|_{k_{32}=1} \approx \kappa^2 q_p - 2\kappa \tag{6-22}$$

and the lower limit on Γ results as

$$\Gamma_{\text{min}} \geq 4(q_p - 1) \tag{6-23}$$

With (6-6) we have for the $\beta\kappa$ product

$$(\beta\kappa)_{\text{min}} \geq 4\left.\frac{q_p - 1}{q_p - \hat{q}}\right|_{q_p \gg 1} \approx 4\frac{q_p - 1}{q_p} \tag{6-24}$$

---

4. Assume that the bandpass network consists of first-order low- and high-pass sections, each with a pole at ω_p. The minimum attenuation of each section at ω_p is -3 dB, the minimum attenuation of the two in cascade is therefore -6 dB.

For $q_p \gg 1$ the lower bound on this product is 4; for medium and low values of q_p, $(\beta\kappa)_{min}$ may be less than 4. The lower bounds in (6-23) and (6-24) provide useful references with respect to which the gain–sensitivity product of any practical circuit can be measured.

6.1.2 All–pole Networks with Minimum Γ

Having found the lower bound on Γ, the question is now how to optimize an all-pole network so that its Γ is as close to the minimum as possible. (Networks with finite zeros will be dealt with separately later.) The most direct way of minimizing Γ is to minimize the $\beta\kappa$ product itself. For any given circuit, this can be done using a standard computer optimization program. Values for the circuit elements representing degrees of freedom are selected at random and the configuration with the minimum $\beta\kappa$ product is then found by a Monte Carlo routine. The resulting circuit will have minimum Q variation with respect to changes in the active element (i.e., the amplifier β). Furthermore, it can be shown that the configuration for which the $\beta\kappa$ product is a minimum differs only negligibly from the configuration for which the Q variation is minimum with respect to the active *and* the thin-film passive components. Thus, *minimizing the $\beta\kappa$ product optimizes the hybrid integrated circuit also if the small variations of the passive film components are taken into account.*

To demonstrate such a procedure we shall now optimize a representative all-pole, namely a class 4 bandpass network and its $RC : CR$ dual by finding the components minimizing the corresponding $\beta\kappa$ products. At the same time we shall observe whether the circuit configurations for which $\beta\kappa$ is a minimum are similar to those configurations for which the total Q-variation is a minimum, bearing in mind that hybrid integrated circuit performance (e.g., component tracking) is presumed. Finally, we shall compare the two bandpass circuits with respect to their respective achievable Q stability.

Computer Optimization   The bandpass network designated type B in table 3-5, and its design equations, are given in Table 6-1; also given is the corresponding $RC : CR$ dual bandpass network, designated type $\bar{\text{B}}$. In Table 6-2 the corresponding sensitivity expressions are listed. In optimizing the circuits for minimum Q variation we recall that q_p is a dimensionless quantity, i.e., it depends only on resistor and capacitor ratios. Thus, for the type B circuit, we define as our variables

$$x = \bar{G}_1/G_1$$
$$y = G_2/G_1$$
$$z = C_1/C_2$$

(6-25)

TABLE 6-1. TYPE B BANDPASS FILTER AND ITS DUAL TYPE $\bar{B}$

Type B	Type $\bar{B}$

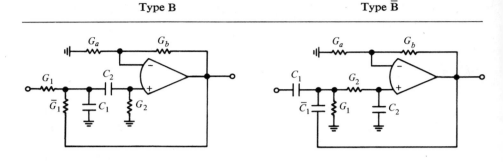

$$T(s) = K \frac{\omega_p s}{s^2 + \dfrac{\omega_p}{q_p} s + \omega_p^2}$$

Type B	Type $\bar{B}$
$K\omega_p = \beta \dfrac{G_1}{C_1}$	$K\omega_p = \beta \dfrac{G_2 C_1}{C_2(C_1 + \bar{C}_1)}$
$\omega_p^2 = \dfrac{G_2(G_1 + \bar{G}_1)}{C_1 C_2}$	$\omega_p^2 = \dfrac{G_1 G_2}{C_2(C_1 + \bar{C}_1)}$
$q_p = \dfrac{\sqrt{G_2(G_1 + \bar{G}_1)C_1 C_2}}{(G_1 + \bar{G}_1 + G_2)C_2 + G_2 C_1 - \beta \bar{G}_1 C_2}$	$q_p = \dfrac{\sqrt{C_2(C_1 + \bar{C}_1)G_1 G_2}}{G_2(C_1 + \bar{C}_1 + C_2) + G_1 C_2 - \beta G_2 \bar{C}_1}$
$= \dfrac{\sqrt{(1+x)yz}}{1 + x + y(1+z) - \beta x}$	$= \dfrac{\sqrt{(1+x)yz}}{1 + x + y(1+z) - \beta x}$

where

$$x = \frac{\bar{G}_1}{G_1}; \qquad y = \frac{G_2}{G_1}; \qquad z = \frac{C_1}{C_2} \qquad\qquad x = \frac{\bar{C}_1}{C_1}; \qquad y = \frac{C_2}{C_1}; \qquad z = \frac{G_1}{G_2}$$

and for the type $\bar{B}$ circuit, the variables

$$x = \bar{C}_1/C_1$$
$$y = C_2/C_1 \qquad\qquad (6\text{-}26)$$
$$z = G_1/G_2$$

A computer program can now be written to compute $\Delta Q/Q$ as a function of the component ratios x, y, and z for the two circuits. To determine the minimum q_p variation $(V^{q_p})_{\min}$ and $(\beta \kappa)_{\min}$ we proceed as follows:

1. Select x and y.
2. Compute z from the circuit equations.
3. Compute the corresponding sensitivities.

TABLE 6-2. SENSITIVITY EXPRESSIONS OF TYPE B AND $\bar{\text{B}}$ BANDPASS FILTERS

Sensitivity	Expression	
	Type B	Type $\bar{\text{B}}$
$S_{G_1}^{q_p}$	$\dfrac{1}{2}\dfrac{G_1}{G_1+\bar{G}_1} - \dfrac{G_1}{\sqrt{G_2(G_1+\bar{G}_1)}}\sqrt{\dfrac{C_2}{C_1}}\,q_p$	$\dfrac{1}{2} - \sqrt{\dfrac{C_2}{C_1+\bar{C}_1}}\sqrt{\dfrac{G_1}{G_2}}\,q_p$
$S_{\bar{G}_1}^{q_p}$	$\dfrac{1}{2}\dfrac{\bar{G}_1}{G_1+\bar{G}_1} - \dfrac{\bar{G}_1(1-\beta)}{\sqrt{G_2(G_1+\bar{G}_1)}}\sqrt{\dfrac{C_2}{C_1}}\,q_p$	—
$S_{G_2}^{q_p}$	$\dfrac{1}{2} - \sqrt{\dfrac{G_2}{G_1+\bar{G}_1}}\left[\sqrt{\dfrac{C_1}{C_2}}+\sqrt{\dfrac{C_2}{C_1}}\right]q_p$	$-\dfrac{1}{2} + \sqrt{\dfrac{C_2}{C_1+\bar{C}_1}}\sqrt{\dfrac{G_1}{G_2}}\,q_p$
$S_{C_1}^{q_p}$	$\dfrac{1}{2} - \sqrt{\dfrac{G_2}{G_1+\bar{G}_1}}\sqrt{\dfrac{C_1}{C_2}}\,q_p$	$\dfrac{1}{2}\dfrac{C_1}{C_1+\bar{C}_1} - \dfrac{C_1}{\sqrt{C_2(C_1+\bar{C}_1)}}\sqrt{\dfrac{G_2}{G_1}}\,q_p$
$S_{\bar{C}_1}^{q_p}$	—	$\dfrac{1}{2}\dfrac{\bar{C}_1}{C_1+\bar{C}_1} - \dfrac{\bar{C}_1(1-\beta)}{\sqrt{C_2(C_1+\bar{C}_1)}}\sqrt{\dfrac{G_2}{G_1}}\,q_p$
$S_{C_2}^{q_p}$	$-\dfrac{1}{2} + \sqrt{\dfrac{G_2}{G_1+\bar{G}_1}}\sqrt{\dfrac{C_1}{C_2}}\,q_p$	$\dfrac{1}{2} - \sqrt{\dfrac{C_2}{C_1+\bar{C}_1}}\left[\sqrt{\dfrac{G_1}{G_2}}+\sqrt{\dfrac{G_2}{G_1}}\right]q_p$
$S_{\beta}^{q_p}$	$\dfrac{\bar{G}_1 C_2\beta}{(G_1+\bar{G}_1+G_2)C_2 + G_2 C_1 - \beta\bar{G}_1 C_2}$	$\dfrac{G_2\bar{C}_1\beta}{G_2(C_1+\bar{C}_1+C_2)+G_1 C_2 - \beta G_2\bar{C}_1}$
$S_{Ga}^{q_p}=S_{\beta}^{q_p}S_{Ga}^{\beta}$	$\dfrac{G_a}{G_b}\dfrac{1}{\beta}\,S_{\beta}^{q_p}$	$\dfrac{G_a}{G_b}\dfrac{1}{\beta}\,S_{\beta}^{q_p}$
$S_{Gb}^{q_p}=S_{\beta}^{q_p}S_{Gb}^{\beta}$	$-\dfrac{G_a}{G_b}\dfrac{1}{\beta}\,S_{\beta}^{q_p}$	$-\dfrac{G_a}{\bar{G}_b}\dfrac{1}{\beta}\,S_{\beta}^{q_p}$

$$S_{A_0}^{q_p}=\frac{\beta}{A_0}\,S_{\beta}^{q_p}$$

4. Compute the total q_p variation for the type B circuit from the expression

$$\left.\frac{\Delta q_p}{q_p}\right|_{\text{type B}} = \left[\frac{1}{2}(S_{G_1}^{q_p} - S_{G_1}^{q_p}) + \frac{1}{2}(S_{G_2}^{q_p} - S_{G_1}^{q_p})\right]\frac{\Delta G}{G}$$

$$+ [S_{C_1}^{q_p} - S_{C_2}^{q_p}]\frac{\Delta C}{C} + S_{A_0}^{q_p}\frac{\Delta A_0}{A_0} + 2S_{Ga}^{q_p}\frac{\Delta G_a}{G_a} \qquad (6\text{-}27)$$

and for the type $\bar{\text{B}}$ circuit from the expression

$$\left.\frac{\Delta q_p}{q_p}\right|_{\text{type }\bar{\text{B}}} = (S_{G_1}^{q_p} - S_{G_2}^{q_p})\frac{\Delta G}{G} + \left[\frac{1}{2}(S_{C_1}^{q_p} - S_{C_1}^{q_p}) + \frac{1}{2}(S_{C_2}^{q_p} - S_{C_1}^{q_p})\right]\frac{\Delta C}{C}$$

$$+ S_{A_0}^{q_p}\frac{\Delta A_0}{A_0} + 2S_{Ga}^{q_p}\frac{\Delta G_a}{G_a} \qquad (6\text{-}28)$$

5. Compute the $\beta\kappa$ product. Since

$$\beta = \frac{1}{\kappa_3}\left(\kappa_2 - \frac{\kappa_1}{q_p}\right)$$ (6-29)

and

$$\kappa = \frac{\kappa_2}{\kappa_1}$$ (6-30)

where

$$q_p = \frac{\kappa_1}{\kappa_2 - \kappa_3\beta}$$ (6-31)

we have

$$\beta\kappa = \frac{\kappa_2}{\kappa_3}\left(\frac{\kappa_2}{\kappa_1} - \frac{1}{q_p}\right)$$ [6-32]

The corresponding κ_i values follow directly from the q_p expressions given in Table 6-1.
6. Compare the resulting q_p variation and the $(\beta\kappa)$ product with the corresponding previous values.
7. Save the smaller values until $(V^{q_p})_{min}$ and $(\beta\kappa)_{min}$ are obtained.

An example of an optimization study for the two bandpass networks is summarized in Table 6-3.[5] The following assumptions were made:

1. $K = 1$
2. $\Delta G/G = 0.03\%$. This corresponds to a tracking ratio of 5 ppm/°C for the film resistors on a substrate, over a temperature range of 60°C.
3. $\Delta C/C = 0.06\%$. This corresponds to a tracking ratio of 10 ppm/°C for the film or chip capacitors on a substrate, over a temperature range of 60°C.
4. $A_0 = 3000$ and $\Delta A_0/A_0 = 50\%$.

For the q_p values shown in Table 6-3, the ratios x and y were varied from 0.1 to 10 in steps of 0.1. The following conclusions can be drawn from the results listed in the table:

1. The amplifier variation $V_{A_0}^{q_p}$ represents the largest contribution to the Q variation V^{q_p}.

2. A comparison of the $(V^{q_p})_{min}$ values for the type B and B̄ circuits reveals only insignificant differences.

3. For q_p values larger than unity the computer-minimized $\beta\kappa$ values are relatively close to the $(\beta\kappa)_{min}$ values calculated from (6-24). To actually attain

5. R. M. Zeigler, unpublished Bell Telephone Laboratories Memorandum.

TABLE 6-3. COMPARISON OF TYPE B AND $\bar{B}$ BANDPASS CIRCUITS OPTIMIZED FOR MINIMUM $\beta\kappa$ AND V^{q_p}

q_p	Circuit type	x	y	z	β	$\beta\kappa$	$\beta\kappa_{min}$†	$V_G^{q_p}$	$V_C^{q_p}$	$V_{A0}^{q_p}$	V^{q_p}
1	$\bar{B}$	1.1	0.1	3.74	1.71	3.6		0.0189	0.0337	0.0854	0.1381M*
	B	1.1	0.1	3.74	1.71	3.6		0.0164	0.0379	0.0854	0.14M
	$\bar{B}$	1.4	0.1	2.79	1.37	3.28M*		0.0041	0.0379	0.2001	0.2422
	B	1.4	0.1	2.79	1.37	3.28M		0.0236	0.0383	0.2001	0.232
10	$\bar{B}$	2.1	0.1	3.78	2.10	4.63 ⎫	3.71	0.3770	0.5941	1.279	2.25M
	B	2.1	0.1	3.78	2.10	4.63 ⎭		0.2717	0.7541	1.279	2.31M
	$\bar{B}$	2.4	0.1	3.03	1.77	4.42M ⎫	3.70	0.2819	0.5055	1.768	2.556
	B	2.4	0.1	3.03	1.77	4.42M ⎭		0.2480	0.5638	1.768	2.58
20	$\bar{B}$	2.1	0.1	4.03	2.24	4.82 ⎫	3.86	0.8385	1.311	2.511	4.661M
	B	2.1	0.1	4.03	2.24	4.82 ⎭		0.5773	1.677	2.474	4.73M
	$\bar{B}$	2.5	0.1	2.97	1.76	4.48M ⎫	3.85	0.5728	1.029	3.716	5.317
	B	2.5	0.1	2.97	1.76	4.48M ⎭		0.5055	1.145	3.716	5.36
50	$\bar{B}$	2.2	0.1	3.74	2.12	4.70 ⎫	3.94	1.955	3.075	6.671	11.7M
	B	2.2	0.1	3.74	2.12	4.70 ⎭		1.400	3.911	6.671	11.98M
	$\bar{B}$	2.5	0.1	3.03	1.79	4.52M ⎫	3.94	1.507	2.585	9.009	13.1
	B	2.5	0.1	3.03	1.79	4.52M ⎭		1.253	3.014	9.009	13.28

* M = minimum value obtained by the computer optimization.
† Calculated from (6-24).

the calculated lower bound, the permissible ranges of x and y would presumably have to be extended. Since the minimum value of $y = 0.1$ is always required, this value, in particular, would have to be reduced; however this would increase the corresponding component spread more than is desirable with hybrid integrated circuits.

4. The V^{q_p} values corresponding to the computer-minimized $\beta\kappa$ values (the latter are followed by an M to indicate that they are minimum values) do not differ significantly from the $(V^{q_p})_{\min}$ values (also followed by an M). This supports the earlier statement that minimizing the $\beta\kappa$ product simultaneously provides a (hybrid integrated) circuit with minimum, or close-to-minimum, V^{q_p}.

5. In spite of its three capacitors, it follows from the values given in the table that the optimized type $\bar{\mathrm{B}}$ circuit actually has a *smaller* total capacitance than the corresponding type B circuit. Furthermore the Q variation of type B circuits is more sensitive to variations of the capacitors than that of type $\bar{\mathrm{B}}$ circuits, i.e.,

$$V_C^{q_p}\bigg|_{\text{type }\bar{\mathrm{B}}} < V_C^{q_p}\bigg|_{\text{type B}} \tag{6-33}$$

Analytical Optimization If a computer optimization program is not available, the expressions for Γ and the $\beta\kappa$ product derived above permit the following direct and simple procedure for the optimization of a circuit for minimum gain–sensitivity product.

From (6-6) we have

$$\beta\kappa = \frac{\Gamma}{q_p - \hat{q}}\bigg|_{q_p \gg \hat{q}} \approx \frac{\Gamma}{q_p} = \frac{\beta S_\beta^{q_p}}{q_p} \tag{6-34}$$

Hence, from (6-13),

$$\beta\kappa = \frac{1}{k_{32}\hat{q}^2}\left(1 - \frac{\hat{q}}{q_p}\right)^2\bigg|_{q_p \gg \hat{q}} \approx \frac{1}{k_{32}\hat{q}^2} \tag{6-35}$$

Thus to calculate the $\beta\kappa$ product we need only to know k_{32} and the pole Q of the RC feedback function $t_{32}(s)$; this is more convenient than calculating Γ, which is a function of q_p.

Some typical forms of the feedback network t_{32} are shown in Table 6-4, together with the corresponding k_{32} and $\hat{q}$ values. Consider, for example, the type B bandpass circuit shown. From (6-35) we obtain, for $q_p \gg \hat{q}$,

$$(\beta\kappa)_{\text{BP B}} \approx \left(1 + \frac{\bar{R}_1}{R_1}\right)\frac{\left(1 + \frac{R_2}{R_p} + \frac{C_1}{C_2}\right)^2}{\sqrt{\dfrac{R_2}{R_p}\dfrac{C_1}{C_2}}}\frac{R_p}{R_2} \tag{6-36}$$

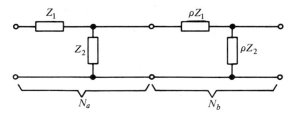

FIG. 6-5. Impedance-scaling the second L section of a four-element ladder network.

where, referring to the table, $R_p = R_1\bar{R}_1/(R_1 + \bar{R}_1)$. In minimizing $\beta\kappa$ for any of the circuits in which t_{32} is a ladder network it is generally simplest to unbalance the network by impedance scaling the second L section with respect to the first by a factor ρ, as shown in Fig. 6-5. With $\rho > 1$ we reduce the loading of the second L section on the first. Thus, with

$$R_p = R$$
$$C_1 = C$$
$$R_2 = \rho R \qquad\qquad (6\text{-}37)$$
$$C_2 = C/\rho$$

we obtain

$$(\beta\kappa)_{\text{BPB}} \approx \left(1 + \frac{\bar{R}_1}{R_1}\right)\left(2 + \frac{1}{\rho}\right)^2 \qquad (6\text{-}38)$$

A spread of 10 : 1 is quite reasonable with film resistors, so that with $\bar{R}_1/R_1 = 0.1$ and $\rho = 10$ we obtain $\beta\kappa = 4.85$. Comparing this with the high-$q_p\ \beta\kappa$ values in Table 6-3 we find that the minimization of $\beta\kappa$ according to (6-35) by impedance scaling gives values almost as small as those obtained by a computer optimization. Naturally the wider the resistor spread given by $\bar{R}_1/R_1$ and ρ, the closer to the lower bound of $(\beta\kappa)_{\text{min}}$ we can come. For the type $\bar{B}$ bandpass, the equivalent expressions are immediately obtained if the $RC : CR$ duality between the two circuits is taken into account.

   With (6-35) we can also compare the inherent capabilities of two circuits providing the same function. Consider, for example the type A and type B bandpass circuits shown in Table 6-4. Referring to the designations in the table we obtain for the type A circuit, using (6-35),

$$(\beta\kappa)_{\text{BP A}} \approx \left(1 + \frac{\bar{R}_1}{R_1}\right)\frac{\left(1 + \frac{R_p}{R_2} + \frac{C_2}{C_1}\right)^2}{\sqrt{\dfrac{R_p}{R_2}\dfrac{C_2}{C_1}}} \qquad (6\text{-}39)$$

Comparing this with $(\beta\kappa)_{\text{BPB}}$ given by (6-36) the similarity in form is apparent. However the additional multiplicand R_p/R_2 in (6-36) readily permits the

TABLE 6-4. TYPICAL FORMS OF THE FEEDBACK RC BANDPASS NETWORK t_{32}

Type	Circuit Diagram	t_{32}	k_{32}	$\hat{q}$
Low pass			$\sqrt{\dfrac{R_1 C_1}{R_2 C_2}}$	$\dfrac{\sqrt{R_1 C_1 R_2 C_2}}{R_1 C_1 + C_2 (R_1 + R_2)}$
High pass			$\sqrt{\dfrac{R_2 C_2}{R_1 C_1}}$	$\dfrac{\sqrt{R_1 C_1 R_2 C_2}}{R_1 C_1 + C_2 (R_1 + R_2)}$
Bandpass* type A			$\dfrac{R_1}{R_1 + \bar{R}_1} \sqrt{\dfrac{R_2 C_1}{R_p C_2}}$	$\dfrac{\sqrt{R_p C_1 R_2 C_2}}{R_p C_1 + R_2 (C_1 + C_2)}$

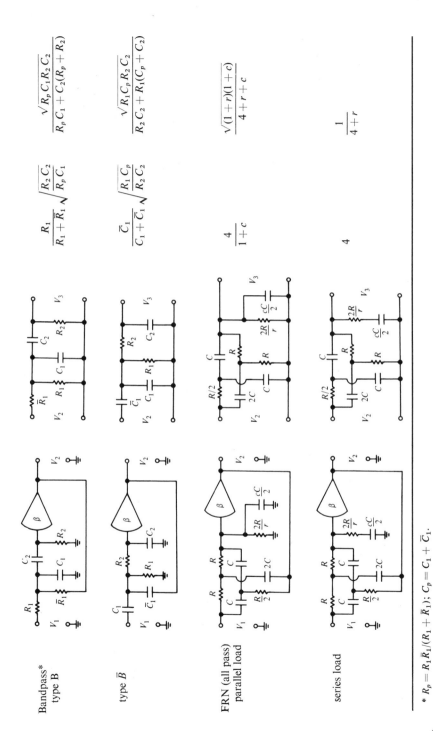

Bandpass*
type B

$$\frac{R_1}{R_1 + \bar{R}_1}\sqrt{\frac{R_2 C_2}{R_p C_1}}$$

$$\frac{\sqrt{R_p C_1 R_2 C_2}}{R_p C_1 + C_2(R_p + R_2)}$$

type $\bar{B}$

$$\frac{\bar{C}_1}{C_1 + \bar{C}_1}\sqrt{\frac{R_1 C_p}{R_2 C_2}}$$

$$\frac{\sqrt{R_1 C_p R_2 C_2}}{R_2 C_2 + R_1(C_p + C_2)}$$

FRN (all pass)
parallel load

$$\frac{4}{1 + c}$$

$$\frac{\sqrt{(1 + r)(1 + c)}}{4 + r + c}$$

series load

$$4$$

$$\frac{1}{4 + r}$$

* $R_p = R_1 \bar{R}_1/(R_1 + \bar{R}_1)$; $C_p = C_1 + \bar{C}_1$.

487

minimum value $(\beta\kappa)_{\min}$ given by (6-24) to be approached arbitrarily closely by impedance scaling. This possibility does not exist for the type A circuit. Thus the type B (and $\bar{B}$) bandpass circuits are preferable to the type A circuit. As for a preference between types B and $\bar{B}$, the former requires only two capacitors (but may have a larger total capacitance when optimized for minimum $\beta\kappa$). On the other hand one of the three capacitors of the latter shunts the noninverting input terminal of the amplifier to ground, which is useful in reducing positive-feedback parasitic high frequencies that may otherwise cause high-frequency oscillations in the circuit (see Chapter 4, Section 4.2.1).

Optimization of the remaining all-pole networks follows in precisely the same way. For example, with the k_{32} and $\hat{q}$ values in Table 6-4 and with (6-35) we obtain, for the low-pass network,

$$(\beta\kappa)_{\text{LP}} \approx \frac{\left(1 + \dfrac{C_1}{C_2} + \dfrac{R_2}{R_1}\right)^2}{\sqrt{\dfrac{R_2}{R_1}\dfrac{C_1}{C_2}}} \cdot \frac{C_2}{C_1} \tag{6-40}$$

This has exactly the same form as $\beta\kappa$ for the type B and $\bar{B}$ bandpass circuits (see (6-36)) and can be minimized by impedance-scaling accordingly. The same, of course, applies to the high-pass network (which is the $RC:CR$ dual of the low-pass network).

6.1.3 FRNs with Minimum Γ

Let us now consider the $\beta\kappa$ minimization of frequency-rejection networks (FRNs). From Table 6-4 we have, for the circuit using a symmetrical *parallel*-loaded twin-T,

$$(\beta\kappa)_{\text{FRN P}} = \frac{(4 + r + c)^2}{4(1 + r)} \tag{6-41a}$$

and, for the circuit using a *series*-loaded twin-T,

$$(\beta\kappa)_{\text{FRN S}}\big|_{r=c} = \frac{(4 + r)^2}{4} \tag{6-41b}$$

The series-loaded case can be used only when $\omega_p = \omega_z$ (i.e., $r = c$) which limits its use considerably. The $\beta\kappa$ product of the parallel-loaded case can be minimized by keeping r and c small; however, r and c cannot both be chosen arbitrarily since the ratio $(1 + r)/(1 + c)$ is prescribed by the ratio $(\omega_p/\omega_z)^2$.

We can obtain more degrees of freedom in order to minimize $\beta\kappa$ by considering the general twin-T whose zeros may be close to, but need not be on, the $j\omega$ axis. This network has already been discussed in Chapter 3 (Section 3.1.2); for convenience the corresponding diagram and design equations are summarized again in Table 6-5. We consider only the standard FRN here,

TABLE 6-5. THE STANDARD FRN

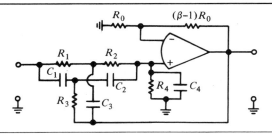

Transfer function:

$$T(s) = K \frac{\left(s^2 + \dfrac{\omega_z}{q_z} s + \omega_z^2\right)(s + \omega_1)}{\left(s^2 + \dfrac{\omega_p}{q_p'} s + \omega_p^2\right)(s + \omega_1')}$$

Abbreviations:

$$R_s = R_1 + R_2 \qquad C_s = \frac{C_1 C_2}{C_1 + C_2}$$

$$R_p = \frac{R_1 R_2}{R_1 + R_2} \qquad C_p = C_1 + C_2$$

$$\lambda = \frac{R_1}{R_1 + R_2} \qquad \eta = \frac{C_2}{C_1 + C_2}$$

$$r = \frac{R_s}{R_4} \qquad c = \frac{C_4}{C_s}$$

$$\alpha = \sqrt{\frac{R_s C_s}{R_3 C_p}} \qquad \omega_1 = \alpha \omega_p$$

$$\nu = \frac{\hat{q}_T}{q_z} \qquad \hat{q}_T = \hat{q}_{\text{4win-T}} = \frac{\alpha(1 - \lambda)(1 - \eta)}{\alpha^2(1 - \lambda) + (1 - \eta)}$$

Twin – T balance conditions:

$q_z \to \infty;$ $\nu = 0$	$q_z \gg 1;$ $\nu \neq 0$
I: $\omega_z^2 = \dfrac{1}{R_1 R_2 C_s C_3}$	I: $\omega_z^2 \approx \dfrac{1}{R_1 R_2 C_s C_3}$
II: $\omega_1 = \omega_1';$ $\dfrac{C_3}{R_3} = \dfrac{C_p}{R_p}$	II: $\dfrac{C_3}{R_3} \approx \dfrac{C_p}{R_p}$
	III: $\omega_1 = \omega_1';$ $R_1 C_1 = R_2 C_2$

(continued overleaf)

TABLE 6-5. THE STANDARD FRN (continued)

General design equations:

1. $\quad K = \dfrac{\beta}{1+c}$

2. $\quad \omega_z^2 = \dfrac{1}{R_1 R_2 C_s C_3} = \dfrac{1}{R_s R_3 C_1 C_2}$

3. $\quad \omega_p = \omega_z \sqrt{\dfrac{1+r}{1+c}}$

4.* $\quad q_p' = \dfrac{\alpha \sqrt{(1+r)(1+c)(1-\lambda)(1-\eta)}}{[\alpha^2(1-\lambda)+(1-\eta)][1-\beta(1 \mp v)]+(r+c\alpha^2)(1-\lambda)(1-\eta)}$

$\qquad = \dfrac{q_p}{1 \pm K v \dfrac{q_p}{\hat{q}_T} \cdot \dfrac{\omega_z}{\omega_p}} = \dfrac{\hat{q}}{1 - K(1 \mp v) \cdot \dfrac{\hat{q}}{\hat{q}_T} \cdot \dfrac{\omega_z}{\omega_p}}$

5. $\quad q_p = q_p'(v=0)$

6. $\quad \hat{q} = q_p'(\beta=0) = \dfrac{\alpha \sqrt{(1+r)(1+c)(1-\lambda)(1-\eta)}}{\alpha^2(1-\lambda)+(1-\eta)+(r+c\alpha^2)(1-\lambda)(1-\eta)}$

$\qquad = \hat{q}_T \dfrac{\sqrt{(1+r)(1+c)}}{1+\hat{q}_T\left(\dfrac{r}{\alpha}+c\alpha\right)}$

* Upper sign, LHP; lower sign RHP.

since it was shown in Chapter 3 that it is superior, from a gain–sensitivity point of view, to the split-feedback FRN. To obtain $\beta\kappa$ for the general FRN it is inconvenient to use (6-35), since we do not have the value k_{32}. We can, however, use (6-34). Designating the pole Q of an FRN with zeros off the $j\omega$ axis by q_p' (in contrast to q_p, corresponding to zeros on the $j\omega$ axis), we have

$$S_\beta^{q'} = \left(\frac{q_p'}{\hat{q}} - 1\right) \tag{6-42}$$

and, from Table 6-5, with

$$\beta = \left(\frac{1+c}{1 \mp v}\right)\left(\frac{\omega_p}{\omega_z}\right)\left(\frac{\hat{q}_T}{\hat{q}}\right)\left(1 - \frac{\hat{q}}{q_p'}\right) \tag{6-43}$$

we obtain

$$\Gamma = \beta S_\beta^{q_{p'}} = \left(\frac{1+c}{1 \mp v}\right)\left(\frac{\omega_p}{\omega_z}\right) \hat{q}_T\left(\frac{1}{\hat{q}} - \frac{1}{q_p'}\right)\left(\frac{q_p'}{\hat{q}} - 1\right) \tag{6-44}$$

and

$$\beta\kappa = \frac{\Gamma}{q_p'} = \left(\frac{1+c}{1 \mp v}\right)\left(\frac{\omega_p}{\omega_z}\right) \hat{q}_T\left(\frac{1}{\hat{q}} - \frac{1}{q_p'}\right)^2 \tag{6-45}$$

In (6-43) to (6-45) the negative sign preceding v is for twin-T zeros left of the $j\omega$ axis, the positive sign for RHP zeros. Naturally for $j\omega$-axis zeros, v equals zero and $q'_p = q_p$. Inspection of (6-45) shows that $\beta\kappa$ can be decreased by using RHP zeros (although they must remain close to the $j\omega$ axis if a given attenuation is to be attained at ω_z).

We can now come closer to solving the general FRN design problem by approximating (6-45) for the most important case, $q'_p \gg \hat{q}$. We obtain

$$\beta\kappa|_{q_{p}' \gg \hat{q}} \approx \left(\frac{1+c}{1 \mp v}\right)\left(\frac{\omega_p}{\omega_z}\right)\left(\frac{\hat{q}_T}{\hat{q}^2}\right) \qquad [6\text{-}46]$$

At first glance it would seem from (6-46) that ω_p should be smaller than ω_z for minimum $\beta\kappa$. This conclusion need not be correct, though, because the coefficients r and c which determine the ratio ω_p/ω_z are also contained elsewhere in the expression (including in $\hat{q}$). Besides, in most cases ω_p/ω_z will be determined by a pole–zero pairing routine such as the one minimizing in-band losses and distortion described in Chapter 1. In this case the ratio $(1 + r)/(1 + c)$ is also determined, but the absolute value of either r or c is still free to be chosen. To obtain some insight in how to go about this, we can eliminate c from (6-46) by making the substitution

$$c = (1 + r)\left(\frac{\omega_z}{\omega_p}\right)^2 - 1 \qquad (6\text{-}47)$$

In proceeding further we shall consider the most frequently used FRN, which comprises a potentially symmetrical twin-T. The corresponding design equations are summarized in Table 6-6. Furthermore, for simplicity, complex zeros on the $j\omega$ axis will again be assumed (i.e., $v = 0$); if $v \neq 0$, q_p is modified in exactly the same way as in Table 6-5. Referring to the expressions given in Table 6-6 and making the substitution (6-47), (6-46) then becomes

$$(\beta\kappa)\Big|_{q_p \gg \hat{q}} \approx \left(\frac{\omega_p}{\omega_z}\right) \frac{\left\{1 + \hat{q}_T(1 + r)\left[\left(\frac{\omega_z}{\omega_p}\right)^2 - \left(\frac{1-r}{1+r}\right)\right]\right\}^2}{\hat{q}_T(1 + r)} \qquad (6\text{-}48)$$

where

$$\hat{q}_T = \frac{1}{2}\frac{\rho}{1+\rho} \qquad (6\text{-}49)$$

Expression (6-48) is a multivariate function depending on the independent variables ω_p/ω_z, r, and $\hat{q}_T$ (or ρ). Assuming that the ratio ω_p/ω_z is given by a pole–zero pairing routine (e.g., for minimum in-band losses), we have $\beta\kappa$ as a function of the variables $\hat{q}_T$ and r. Taking the partial derivative of (6-48) with respect to these two variables, equating to zero and solving for $(\hat{q}_T)_{\text{opt}}$ and r_{opt} we find that the former is zero, the latter tends to infinity. Remember now, that we examined the general FRN in Chapter 3 (Section 3.3.2) for the special

TABLE 6-6. FRN USING SYMMETRICAL OR POTENTIALLY SYMMETRICAL TWIN-T

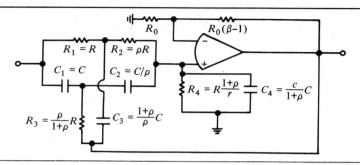

Transfer function:

$$T(s) = K \frac{s^2 + \omega_z^2}{s^2 + \dfrac{\omega_p}{q_p} s + \omega_p^2}$$

Design equations:

1. $\quad K = \dfrac{\beta}{1 + c}$

2. $\quad \omega_z = \dfrac{1}{RC}$

3. $\quad \omega_p = \omega_z \sqrt{\dfrac{1 + r}{1 + c}}$

4. $\quad q_p = \hat{q}_T \dfrac{\sqrt{(1 + r)(1 + c)}}{1 + (r + c)\hat{q}_T - \beta} = \dfrac{\hat{q}}{1 - K \dfrac{\hat{q}}{\hat{q}_T} \cdot \dfrac{\omega_z}{\omega_p}}$

5. $\quad \hat{q} = q_p(\beta = 0) = \hat{q}_T \dfrac{\sqrt{(1 + r)(1 + c)}}{1 + (r + c)\hat{q}_T}$

6. $\quad \hat{q}_T = \hat{q}_{\text{twin-}T} = \dfrac{1}{2} \dfrac{\rho}{1 + \rho}$

case that $\omega_p = \omega_z$. There we found that Γ (or $\beta\kappa$) not only has a minimum for $\hat{q}_T = 0$ and $r = \infty$, but another *lower* minimum for $\hat{q}_T = 0.5$ and $r = 0$. It can be shown that this absolute minimum (within the boundaries prescribed by circuit realizability) is valid also for the general case that $(\omega_p/\omega_z) \gtrsim 1$. We therefore stipulate:

The FRN with minimum Γ is obtained by using a twin-T whose pole Q, $\hat{q}_T$, is as large as possible and whose loading factor r is as small as possible.

To find the upper limit on $\hat{q}_T$ and the lower limit on r we refer to the circuit diagram of the FRN using a (potentially) symmetrical twin-T in Table 6-6. In order for r to be as small as possible, the loading resistor $R_4 = R(1 + \rho)/r$

must be large—in fact it will generally have the highest value in the circuit. But the highest resistor value in the circuit will be limited to a value R_{max} by the hybrid integrated circuit technology used to realize the circuit. Referring to the diagram in Table 6-6, the resistor ratio r is therefore given by

$$r = \frac{R}{R_4}(1 + \rho) = \frac{R}{R_{max}}(1 + \rho) \qquad (6\text{-}50)$$

With (6-49) we therefore have

$$r = \frac{R/R_{max}}{1 - 2\hat{q}_T} \qquad [6\text{-}51]$$

For a given resistor ratio R/R_{max}, r depends on $\hat{q}_T$—and, in fact, increases rapidly (to infinity) as $\hat{q}_T$ approaches its maximum of 0.5. It is therefore impossible for r and $\hat{q}_T$ to approach their optimum values of zero and 0.5, respectively, at the same time. A glance at Fig. 6-6, where $\hat{q}_T$ and the factor $(1 - 2\hat{q}_T)^{-1}$ multiplying the given resistor ratio R/R_{max} have been plotted as a function of the twin-T symmetry factor ρ, shows that a good compromise may be to select a symmetrical twin-T ($\rho = 1$) for which $\hat{q}_T = 0.25$ and $r = 2R/R_{max}$. For $\omega_p = \omega_z$ and $r = 1$, we then obtain $\beta\kappa = 4.5$ from (6-48). This compares well with the theoretical minimum obtained earlier. Note that β becomes smaller than unity for $\omega_p = \omega_z$ and $r < 1$. This is no problem if a voltage divider is used for R_4, as discussed in Chapter 3 (see Fig. 3-32). Then $\beta' = \beta R_4''/(R_4' + R_4'')$ where β may be equal to, or greater than, unity.

Design Steps for FRNs with Minimum Γ It is appropriate at this point to summarize the steps required to design an FRN for minimum $\beta\kappa$, or in other words for minimum gain–sensitivity product Γ. We assume that

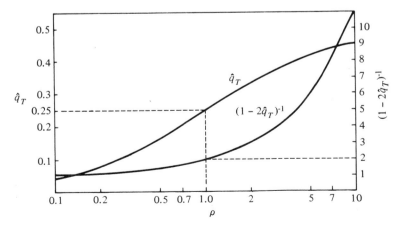

FIG. 6-6. $\hat{q}_T$ and $1/(1 - 2\hat{q}_T)$ as a function of twin-T symmetry factor ρ.

ω_z, ω_p, q_p, and K are given. However, in general we must accept the K value resulting from the design procedure and correct for any subsequent error in K in a neighboring stage (or, if necessary, with an additional amplifier or attenuator). Maximum resistance and capacitance values R_{max} and C_{max} are also given; they are determined by the hybrid IC technology used.

Referring to the designations used in Tables 6-5 and 6-6 we proceed as follows:

1. Assume a *symmetrical twin-T*, i.e., $\rho = 1$, and $\hat{q}_T = 0.25$.

2. The largest capacitor occurring in the circuit will be either C_3 or C_4. A simple procedure is to select $C_3 = C_{max}$. Should C_4 turn out to be larger than C_{max} in step 7 below, then we must return to the present step and decrease C_3 proportionately until $C_4 = C_{max}$. (Naturally the subsequent steps must then be retraced to accomodate the new C_3 value. More often than not, however, $C_3 \geq C_4$, so that this repetition will be unnecessary.)

3. $C_1 = C_2 = C = \frac{1}{2}C_{max}$.

4. $R_1 = R_2 = R = 2/\omega_z C_{max}$.

5. $R_3 = R_p C_3/C_p = R/2 = 1/\omega_z C_{max}$.

6. Select $R_4 = R_{max}$. Then $r = R_s/R_4 = 2R/R_{max} = 4/\omega_z R_{max} C_{max}$. However, r must not be chosen too small, i.e., R_4 not too large, otherwise C_4 will be negative in step 7 below. Thus

$$r > \left(\frac{\omega_p}{\omega_z}\right)^2 - 1 \tag{6-52a}$$

or

$$R_4 < \frac{R_s}{\left(\frac{\omega_p}{\omega_z}\right)^2 - 1} \tag{6-52b}$$

If $(\omega_p/\omega_z) < 1$, then these conditions simplify to

$$r \geq 0; \quad R_4 \leq \infty, \quad \text{i.e.,} \quad R_4 = R_{max}$$

7.

$$C_4 = C_s\left[(1 + r)\left(\frac{\omega_z}{\omega_p}\right)^2 - 1\right]$$

$$= \frac{C_{max}}{4}\left[\left(1 + \frac{4}{\omega_z R_{max} C_{max}}\right)\left(\frac{\omega_z}{\omega_p}\right)^2 - 1\right] \tag{6-53}$$

Note that for $r \ll 1$ and the common case that $(\omega_z/\omega_p) > 1$ we have

$$\frac{C_4}{C_s} \approx \left(\frac{\omega_z}{\omega_p}\right)^2 - 1 \tag{6-54}$$

i.e., *the capacitor ratio C_4/C_s determines the zero–pole separation.*

As mentioned in step 2, if C_4 here turns out to be larger than C_{max}, then we must return to step 2 and determine a smaller value for C_3 such that $C_4 = C_{max}$.

8. Calculate the amplifier gain as follows:

$$\beta = 1 + (r + c)\hat{q}_T - \frac{\hat{q}_T}{q_p}\sqrt{(1 + r)(1 + c)} = \hat{q}_T\left(\frac{1}{\hat{q}} - \frac{1}{q_p}\right)\left(\frac{\omega_p}{\omega_z}\right)(1 + c) \quad (6\text{-}55)$$

It may be that the resistor ratio R/R_{max} can be made small enough to allow a potentially symmetrical twin-T with $\rho > 1$ to be used in the FRN. Then $\hat{q}_T > 0.25$ and r will be sufficiently small to result in a $\beta\kappa$ value which is smaller than in the symmetrical twin-T case. As we know, the potentially symmetrical twin-T with $\rho > 1$ has the additional advantage of being easily tunable. For a *potentially symmetrical twin-T* the design steps are as follows:

1'. Select $C_1 = C_{max}$.

2'. $R_1 = R = 1/\omega_z C_{max}$.

3'. Select $R_4 = R_{max}$. Then $r = (1 + \rho)R/R_4 = (1 + \rho)/\omega_z R_{max} C_{max}$.

4'. Select ρ as large as possible but in such a way that r is still sufficiently small to guarantee a low $\beta\kappa$ value. In order for C_4 to be positive in step 9' it is necessary that

$$\rho > \left[\left(\frac{\omega_p}{\omega_z}\right)^2 - 1\right]\frac{R_4}{R} - 1 \quad (6\text{-}56)$$

or, for small ω_p/ω_z ratios, $\rho > 0$.

5'. $C_2 = C/\rho = C_{max}/\rho$. Naturally ρ must not be so large as to reduce C_2 below the minimum capacitance value technologically achievable. Similarly, there will be a minimum resistance value in step 2' below which R_1 should not be reduced.

6'. $R_2 = \rho R = \rho/\omega_z C_{max}$.

7'. $C_3 = C_{max}[1 + (1/\rho)] = C_1 + C_2$. C_3 here becomes larger than C_{max}. If $\rho \gg 1$, $C_3 \approx C_1$. If C_3 is too large, C_1 must be selected proportionately smaller than C_{max} in such a way that $C_3 = C_{max}$.

8'. $R_3 = \rho R/(1 + \rho) = \rho/(1 + \rho)\omega_z C_{max}$.

9'.

$$C_4 = C_s\left[(1 + r)\left(\frac{\omega_z}{\omega_p}\right)^2 - 1\right]$$

$$= \frac{C_{max}}{\rho + 1}\left[\left(1 + \frac{(1 + \rho)}{\omega_z R_{max} C_{max}}\right)\left(\frac{\omega_z}{\omega_p}\right)^2 - 1\right] \quad (6\text{-}57)$$

10'. Calculate the gain β as given in step 8 for the symmetrical twin-T.

Numerical Design Examples To demonstrate the design procedures outlined above, consider the following numerical example. The parameters given are

$$f_z = 5.5 \text{ kHz}; \qquad f_p = 3.0 \text{ kHz}; \qquad q_p = 20$$

$$R_{max} = 100 \text{ k}\Omega; \qquad C_{max} = 18,000 \text{ pF}$$

We assume $j\omega$-axis zeros; thus $v = 0$. Designing for a *symmetrical twin-T* and referring to the corresponding steps given above we have:

1. $\hat{q}_T = \frac{1}{4}$
2. $C_3 = 18,000 \text{ pF}$
3. $C_1 = C_2 = 9000 \text{ pF}$
4. $R_1 = R_2 = 3.2 \text{ k}\Omega$
5. $R_3 = 1.6 \text{ k}\Omega$
6. $R_4 = 100 \text{ k}\Omega; \qquad r = 6.4/100 = 0.064$
7. $C_4 = 4,500 [1.064 \cdot 3.35 - 1] \text{ pF} = 10,600 \text{ pF}$
   $c = C_4/C_s = 10,600/4,500 = 2.34$
8. $\beta = 1.55$

Then, from the expression for $\hat{q}$ in Table 6-6 we have $\hat{q} = 0.294$. Since $q_p \gg 1$ we can use (6-46) to calculate $\beta\kappa$. However, having calculated β and $\hat{q}$ it is still simpler to recall that $\beta\kappa = \beta/\hat{q}$, from which we obtain $\beta\kappa = 5.25$.

For the same example, using a *potentially symmetrical twin-T*, we proceed as follows while referring to the corresponding steps given above:

1'. $C_1 = 18,000 \text{ pF}$
2'. $R_1 = 1.6 \text{ k}\Omega$
3'. $R_4 = 100 \text{ k}\Omega; \qquad r = (1 + \rho)1.6/100 = (1 + \rho)0.016$
4'. $\rho = 10; \qquad r = 0.176$
5'. $C_2 = 1,800 \text{ pF}$
6'. $R_2 = 1.6 \text{ k}\Omega$
7'. $C_3 = 19,800 \text{ pF}$

Since the value of C_3 exceeds C_{max}, we must select $C_3 = 18,000$ pF. Retaining the twin-T capacitors C_1 and C_2 assumed in steps 1' and 5' above, as well as R_1 obtained in step 2', we then calculate R_3 and recalculate R_2 from the general twin-T conditions for a perfect null (see Table 6-5):

$$R_2 = \frac{1}{\omega_z^2 R_1 C_s C_3} \tag{6-58}$$

and

$$R_3 = \frac{C_3}{C_p} \cdot R_p \tag{6-59}$$

Thus, marking the correspondingly modified values with an asterisk, we obtain

$$R_2^* = 17.8 \text{ k}\Omega \text{ [from (6.58)]}; \qquad C_3^* = 18,000 \text{ pF}$$

$$R_3^* = 1.33 \text{ k}\Omega \text{ [from (6.59)]}$$

$$r^* = \frac{R_1 + R_2^*}{R_4} = 0.194$$

Using these modified values we continue with step 9' and obtain $C_4 = 4,900 \text{ pF}$. To calculate β from 10' we need to know $\hat{q}_T$ and $\hat{q}$. For $\hat{q}_T$ we cannot use the expression for a potentially symmetrical twin-T (i.e., $\rho/2(1 + \rho)$) because of the modification carried out above. Thus we must use the general expression given in Table 6-5:

$$\hat{q}_T = \frac{\alpha(1 - \lambda)(1 - \eta)}{\alpha^2(1 - \lambda) + (1 - \eta)} \tag{6-60}$$

where α, λ, and η are defined in the table. To obtain $\hat{q}$ we then use

$$\hat{q} = \hat{q}_T \frac{\sqrt{(1 + r)(1 + c)}}{1 + (r + c)\hat{q}_T} \tag{6-61}$$

With the numerical values given above we obtain $\hat{q}_T = 0.457$ and $\hat{q} = 0.386$. From (6-55) we then obtain $\beta = 2.46$ and $\beta\kappa = \beta/\hat{q} = 6.37$. Note that by using a potentially symmetrical twin-T with a scaling coefficient $\rho = 10$, the $\beta\kappa$ product has increased relatively little compared to the symmetrical case, i.e., from 5.25 to 6.37, even though the same $R_{\max}$ value was used for both cases. Thus if desirable for tuning or other purposes, the potentially symmetrical twin-T may well be used instead of the symmetrical one, even though the r value has increased from 0.064 to 0.194. The ideal minima given by (6-24) are 3.86 and 3.88 for the symmetrical and potentially symmetrical circuits, respectively. To approach these minimum values more closely, much larger spreads in the component values would be required.

When designing a hybrid integrated circuit with chip, rather than thin film, capacitors, the designer is not free to select arbitrary capacitor values but has only a series of discrete values at his disposal. This was not taken into consideration in the design steps outlined above. It can readily be done so, however, by first computing the component values according to the design steps given above, and subsequently, assuming available capacitor values closest to those calculated, recomputing the resistor values as follows:

1. Select suitable values for C_1, C_2, C_3, C_4, and R_1.
2. R_2 and R_3 follow from (6-58) and (6-59), respectively.

3. R_4 is obtained from

$$R_4 = \frac{R_s}{\left(\dfrac{\omega_p}{\omega_z}\right)^2 \left(1 + \dfrac{C_4}{C_s}\right) - 1}$$

4. β is obtained as in the design steps above.

6.1.4 All–Pass Networks with Minimum Γ

To optimize the all-pass circuit derived from the TT-SABB for minimum $\beta\kappa$ and thus for minimum Γ, we proceed exactly as outlined above. When the all-pass is a derivative of the FRN (i.e., uses an RC-loaded twin-T as described in Section 3.1.2), then the considerations valid for the FRN are also valid for the all-pass network. Similar considerations follow for the all-pass incorporating a bridged-T. They have already been discussed in Chapter 3, where the minimization of $\beta/\hat{q}$ (or, as referred to here, $\beta\kappa$) was dealt with following equation (3-107).

A final comment on the optimization of FRN and all-pass circuits. Note that the feedback networks $t_{32}(s)$ are not, in these two cases, ladder networks; therefore the lower bound on Γ is not given by (6-23), but, as mentioned earlier, by $\Gamma \geq (q_p - 0.5)^2/q_p$, which for large q_p simplifies to $\Gamma \geq q_p - 1$. The corresponding $\beta\kappa$ product follows from (6-34) as $\beta\kappa \geq (q_p - 0.5)^2/q_p(q_p - \hat{q})$. Thus, for large q_p, $\beta\kappa \geq (q_p - 1)/(q_p - \hat{q})$ and approaches unity. From the discussion above we see that with the symmetrical or potentially symmetrical twin-T we remain substantially above this minimum and must conclude that a more general configuration would be required to approach it more closely.

6.1.5 The Maximum Pole Q of the TT–SABB

Returning to the general TT-SABB networks as a whole, we note that, as a rule, a maximum pole-Q variation $(\Delta q_p/q_p)_{max}$ can be tolerated while a maximum open-loop gain variation $(\Delta A/A)_{max}$ is specified by the amplifier manufacturer. From

$$\frac{\Delta q_p}{q_p} = \beta \left(\frac{q_p}{\hat{q}} - 1\right) \frac{\Delta A}{A^2} \tag{6-62}$$

it therefore follows that

$$(q_p)_{max} \leq \frac{A}{\beta\kappa} \cdot \frac{(\Delta q_p/q_p)_{max}}{\left(\dfrac{\Delta A}{A}\right)_{max}} \tag{6-63}$$

For example, if $(\Delta q_p/q_p)_{\max} = 5\%$, $(\Delta A/A)_{\max}$ is given as 50%, the available loop gain $A/\beta = 1000$, and $\kappa = 2.5$, the maximum acceptable q_p is 40. Since the available loop gain decreases with frequency, the maximum acceptable q_p will decrease with frequency as well. It is clear from (6-63) that the higher the frequency of operation is, the more important minimizing the $\beta\kappa$ product becomes.

On the other hand, it must not be forgotten that sensitivity minimization is only one of the considerations that determines the choice of the component values. Frequently, to satisfy other requirements such as minimum substrate or (thin-film) component area, maximum circuit versatility, and the like, it may be worthwhile to accept a certain amount of sensitivity deterioration. Along these lines, even with a limited $R_{\max}$ value, the potentially symmetrical rather than the symmetrical twin-T may be used in an FRN in order to benefit from its superior tuning characteristics, provided the increase in the $\beta\kappa$ product remains acceptably small. Of course, if q_p is low—say, less than 5—then, the whole question of $\beta\kappa$ minimization (i.e., of q_p stabilization) becomes less critical.

6.2 TUNING THE TT–SABB

In Chapter 4 we discussed general methods of tuning second-order active networks. Here we shall illustrate how these methods are applied to the TT-SABB. We shall restrict ourselves to functional tuning, since the length and complexity of the equations required for deterministic tuning are beyond the scope of this book. Besides, given the appropriate computer aids, the equations can be derived by following the steps outlined in Section 4.4.4 in Chapter 4.

6.2.1 All–Pole Networks

There are three functions that fall into the all-pole network category: low-pass, high-pass, and bandpass. Only the poles need be tuned for these circuits; the zeros are always either at the origin or at infinity. The two pole parameters to be tuned are q_p and the frequency ω_p. In addition, there is the constant factor K. However, in general, the specified tolerance of the constant K is relatively loose and the deviation from the nominal design value, after tuning ω_p and q_p, is quite small. Hence, since it simplifies the tuning procedure considerably, it is preferable not to tune K at all. If need be, a single, overall correction for the errors in the K_i of a filter cascade can be made in one of the second-order stages, after the tuning procedure of the individual stages has been completed. This leaves ω_p and q_p as the only two parameters requiring tuning in the all-pole case, and we must find two components x_1 and x_2 with respect to which the corresponding sensitivity matrix is diagonal, i.e.,

$$\begin{bmatrix} \Delta\omega_p/\omega_p \\ \Delta q_p/q_p \end{bmatrix} = \begin{bmatrix} k_1 & 0 \\ 0 & k_2 \end{bmatrix} \begin{bmatrix} \Delta x_1/x_1 \\ \Delta x_2/x_2 \end{bmatrix} \tag{6-64}$$

or at least triangular, i.e.,

$$\begin{bmatrix} \Delta\omega_p/\omega_p \\ \Delta q_p/q_p \end{bmatrix} = \begin{bmatrix} k_1 & 0 \\ k_{21} & k_2 \end{bmatrix} \begin{bmatrix} \Delta x_1/x_1 \\ \Delta x_2/x_2 \end{bmatrix} \tag{6-65}$$

As discussed previously, in the TT-SABB (as well as in most other useful building blocks) ω_p depends only on the resistors and capacitors of the feedback network (i.e., $t_{32}(s)$); q_p, on the other hand, depends also on the gain β. Hence we must look for a suitable resistor R_{ω_p} with which to tune ω_p, while q_p is preferably tuned by β. Typically, therefore, the sensitivity matrix has the form

$$\begin{bmatrix} \Delta\omega_p/\omega_p \\ \Delta q_p/q_p \end{bmatrix} = \begin{bmatrix} k_1 & 0 \\ k_{21} & k_2 \end{bmatrix} \begin{bmatrix} \Delta R_{\omega_p}/R_{\omega_p} \\ \Delta\beta/\beta \end{bmatrix} \tag{6-66}$$

where

$$k_1 = S_{R_{\omega_p}}^{\omega_p} \tag{6-67a}$$

$$k_{21} = S_{R_{\omega_p}}^{q_p} \tag{6-67b}$$

and

$$k_2 = S_{\beta}^{q_p} \tag{6-67c}$$

R_{ω_p} should be selected so as to produce a large k_1 and a small k_{21}; i.e., the sensitivity of ω_p to R_{ω_p} should be large but the sensitivity of q_p to R_{ω_p} small. However, the magnitude of k_1 is limited to values less than unity; this follows directly from the fact that $\sum S_{Ri}^{\omega_p} = -1$ and that ω_p depends only on passive components. In fact, as a rule, the magnitude of k_1 will be less than $\frac{1}{2}$. Thus, in selecting a resistor for R_{ω_p} our primary concern is that it results in a k_1 value that is as large as possible. If k_1 is too small we require a large adjustment of R_{ω_p} to obtain a change in ω_p; this is undesirable from a technological standpoint (resistor trimming by a high percentage), as well as from the point of view of efficient production (lengthy trimming time). Our second consideration must be the relative magnitude of k_{21} with respect to k_1 and k_2. Very often k_{21} will actually be larger than k_1; furthermore, with increasing k_1 the ratio k_1/k_{21} may very likely decrease, instead of increasing as we would like it to. However, provided k_2 is large enough to effectively correct for q_p errors caused by R_{ω_p}, the relative magnitude of k_{21} with respect to k_1 does not really matter. *Thus, of the two ratios k_2/k_{21} and k_1/k_{21}, we are primarily interested in the former being as large as possible.* Fortunately, from the tuning point of view at least, k_2 is generally large. As we know, it is proportional to q_p; according to (6-5):

$$k_2 = \frac{q_p}{\hat{q}} - 1 = \frac{q_p}{q_p(\beta = 0)} - 1 > 2q_p - 1 \qquad (6\text{-}68)$$

and, in general, k_2/k_{21} will be larger than unity.

In selecting R_{ω_p} we therefore proceed as follows:

1. Compute all the sensitivities $(k_1)_{R_i}$ where R_i are the resistors of the RC network that determines ω_p. As a check for these computations remember that $\sum (k_1)_{R_i} = -1$. Those resistors R_j whose corresponding k_1 values are large are suitable candidates from which to select R_{ω_p}.

2. Compute the sensitivity ratios $(k_2/k_{21})_{R_j}$ for the resistors R_j selected in step 1. (If all the ratios $(k_2/k_{21})_{R_i}$ are computed, remember, as a useful check, that $\sum (k_{21})_{R_i} = 0$). The resistor R_j corresponding to the maximum ratio k_2/k_{21} is the optimum choice for R_{ω_p}.

In carrying out the computations outlined above it is useful to make up a sensitivity table in which the various sensitivities and sensitivity ratios can be easily examined and compared. Naturally the sensitivities in the table will depend on the actual component values selected for a given circuit. Ideally these result from a design optimized with respect to sensitivity or, in the case of hybrid integrated circuits, optimized for a minimum $\beta\kappa$ product.

Design Examples   Let us illustrate the foregoing discussion by the following design examples. Consider the type B bandpass circuit in Table 6-1 and the corresponding sensitivity expressions given in Table 6-2. Since it is more convenient to discuss tuning with respect to a resistor R than with respect to a conductance $G = 1/R$, remember that

$$S_R^F = -S_G^F \qquad (6\text{-}69)$$

We have three resistors to choose R_{ω_p} from: R_1, $\bar{R}_1$, and R_2. In making our choice we must examine the sensitivities $(k_1)_{R_i}$ and the ratios $(k_2/k_{21})_{R_i}$ for each. These may be readily derived from Tables 6-1 and 6-2, and are listed in general terms in Table 6-7 and for two specific circuit designs in Table 6-8. Circuit 1 corresponds to the configuration designed to have a low $\beta\kappa$ product by impedance scaling as prescribed by (6-37). Circuit 2 corresponds to a commonly used configuration in which all like components are equal. Both circuits are shown in Fig. 6-7. The price to be paid for the convenience of having equal resistors and capacitors is a $\beta\kappa$ product of 11.3 as compared to 4.85 for circuit 1. From Table 6-8 we see that in circuit 1, k_1 is almost equally large for $\bar{R}_1$ or R_2. Either one of these resistors could therefore serve well as R_{ω_p}. Since the ratio k_2/k_{21} is larger for R_2 than for $\bar{R}_1$, and this, as shown in

TABLE 6-7. SENSITIVITY TABLE FOR THE GENERAL TYPE B BANDPASS NETWORK

	R_1	$\bar{R}_1$	R_2
$(k_1)_{Ri} = S_{Ri}^{\omega p}$	$-\dfrac{1}{2}\dfrac{1}{1+x}$	$-\dfrac{1}{2}\dfrac{x}{1+x}$	$-\dfrac{1}{2}$
$(k_{21})_{Ri} = S_{Ri}^{qp}$	$\dfrac{q_p}{\sqrt{yz}(1+x)} - \dfrac{1}{2}\dfrac{1}{1+x}$	$\dfrac{1+\dfrac{x}{2}}{1+x} - q_p\,\dfrac{[1+y(1+z)]}{\sqrt{yz}(1+x)}$	$\dfrac{y(1+z)q_p}{\sqrt{yz}(1+x)} - \dfrac{1}{2}$
$k_2 = S_\beta^{qp}$		$\dfrac{q_p}{\hat{q}} - 1 = \dfrac{q_p}{q_p(\beta=0)} - 1 = q_p\left[\dfrac{1+x+y(1+z)}{\sqrt{yz}(1+x)}\right] - 1$	
$\left(\dfrac{k_2}{k_{21}}\right)_{Ri}$	$\dfrac{[1+x+y(1+z)]q_p - \sqrt{yz}(1+x)}{q_p - \dfrac{1}{2}\sqrt{\dfrac{yz}{(1+x)}}}$ For $q_p \gg 1$: $\approx 1+x+y(1+z)$	$\dfrac{[1+x+y(1+z)]q_p - \sqrt{yz}(1+x)}{\left(1+\dfrac{x}{2}\right)\sqrt{\dfrac{yz}{1+x}} - q_p[1+y(1+z)]}$ For $q_p \gg 1$: $\approx -\dfrac{1+x+y(1+z)}{1+y(1+z)}$	$\dfrac{[1+x+y(1+z)]q_p - \sqrt{yz}(1+x)}{[y(1+z)]q_p - \dfrac{1}{2}\sqrt{yz}(1+x)}$ For $q_p \gg 1$: $\approx \dfrac{1+x+y(1+z)}{y(1+z)}$

$x = R_1/\bar{R}_1$; $y = R_1/R_2$; $z = C_1/C_2$.

TABLE 6-8. SENSITIVITY TABLE FOR TWO TYPE B BANDPASS NETWORKS

	Circuit 1			Circuit 2		
	$x=10$; $y=1.1$; $z=10$; $R_2/R_p=10$; $\beta_\kappa \approx 4.85;^*$ $\beta \approx 2$			$x=1$; $y=1$; $z=1$; $R_2/R_p=2$; $\beta_\kappa \approx 11.3;^*$ $\beta \approx 4$		
$k_2 = S_\beta^{q_p}$	$2.1q_p - 1$			$2\sqrt{2}q_p - 1$		
	R_1	$\bar{R}_1$	R_2	R_1	$\bar{R}_1$	R_2
$(k_1)_{Ri} = S_{Ri}^{\omega_p}$	$-\dfrac{1}{22} = -0.0455$	$-\dfrac{1}{2.2} = -0.455$	$-\dfrac{1}{2}$	$-\dfrac{1}{4}$	$-\dfrac{1}{4}$	$-\dfrac{1}{2}$
$(k_{21})_{Ri} = S_{Ri}^{q_p}$	$\dfrac{1}{11}\left(q_p - \dfrac{1}{2}\right)$	$\dfrac{1}{11}(6 - 13.1q_p)$	$1.1q_p - \dfrac{1}{2}$	$\dfrac{q_p}{\sqrt{2}} - \dfrac{1}{4}$	$\dfrac{3}{4} - \dfrac{3}{\sqrt{2}}q_p$	$\sqrt{2}q_p - \dfrac{1}{2}$
$\left(\dfrac{k_2}{k_{21}}\right)_{Ri}$	$22\dfrac{2.1q_p - 1}{2q_p - 1}$	$-1.83\dfrac{2.1q_p - 1}{2.18q_p - 1}$	$2\cdot\dfrac{2.1q_p-1}{2.2q_p-1}$	4.0	$-\dfrac{4}{3}$	2.0
$q_p = 1$ $\quad k_2$	1.1			1.828		
$(k_{21})_{Ri}$	0.0455	-0.645	0.6	0.457	-1.37	0.914
$\left(\dfrac{k_2}{k_{21}}\right)_{Ri}$	24.2	-1.71	1.83	4.0	$-\dfrac{4}{3} = -1.33$	2.0
$q_p = 10$ $\quad k_2$	20			27.28		
$(k_{21})_{Ri}$	0.864	-11.36	10.5	6.82	-20.45	13.64
$\left(\dfrac{k_2}{k_{21}}\right)_{Ri}$	23.2	-1.76	1.91	4.0	$-\dfrac{4}{3} = -1.33$	2.0

$x = R_1/\bar{R}_1$; $y = R_1/R_2$; $z = C_1/C_2$; $R_2/R_p = (1+x)/y$; $R_p = R_1 \| \bar{R}_1$.
* Approximated by (6-35) for $q_p \gg q$.

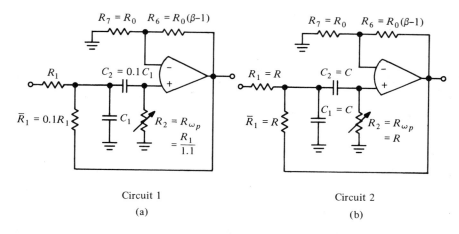

Circuit 1   Circuit 2

(a)   (b)

FIG. 6-7. Type B bandpass network: (a) optimized for low $\beta\kappa$ product; (b) conventional circuit (all like components equal).

the table, for a q_p of either unity or ten, R_2 is the best choice to make for R_{ω_p}. The same conclusion follows from the computations made for circuit 2.

As a second example consider the dual of the type B bandpass network, namely type $\bar{B}$, as given in Tables 6-1 and 6-2. Only two resistors, R_1 and R_2, are available for tuning here; the corresponding sensitivity tables, for the general circuit and for two specific circuits (see Fig. 6-8) are given in Tables 6-9 and 6-10, respectively. Note that $(k_1)_{R_1} = (k_1)_{R_2}$ and $(k_{21})_{R_1} = -(k_{21})_{R_2}$. As a consequence, it would seem that either R_1 or R_2 can be selected as R_{ω_p}. However, R_1 is the preferable choice for the following two reasons: First,

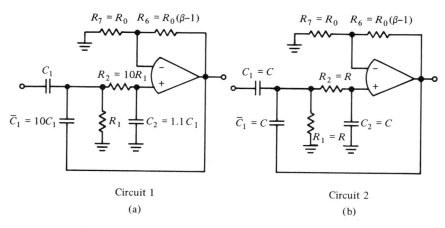

Circuit 1   Circuit 2

(a)   (b)

FIG. 6-8. Type-$\bar{B}$ bandpass network: (a) optimized for low $\beta\kappa$ product; (b) conventional circuit (all like components equal).

TABLE 6-9. SENSITIVITY TABLE FOR THE GENERAL TYPE $\bar{B}$ BANDPASS NETWORK

	R_1	R_2
$(k_1)_{Ri} = S_{Ri}^{\omega_p}$	$-\dfrac{1}{2}$	$-\dfrac{1}{2}$
$(k_{21})_{Ri} = S_{Ri}^{q_p}$	$\dfrac{yz}{1+x}q_p - \dfrac{1}{2}$	$\dfrac{1}{2} - \dfrac{yz}{1+x}q_p$
$k_2 = S_{\beta}^{q_p}$	$\dfrac{q_p}{\hat{q}} - 1 = \dfrac{q_p}{q_p(\beta=0)} - 1 = q_p\left[\dfrac{1+x+y(1+z)}{\sqrt{yz(1+x)}}\right] - 1$	
$\left(\dfrac{k_2}{k_{21}}\right)_{Ri}$	$\dfrac{[1+x+y(1+z)]q_p - \sqrt{yz(1+x)}}{q_p - \dfrac{1}{2}\sqrt{\dfrac{1+x}{yz}}}$ For $q_p \gg 1$: $\approx 1+x+y(1+z)$	$-\dfrac{[1+x+y(1+z)]q_p - \sqrt{yz(1+x)}}{q_p - \dfrac{1}{2}\sqrt{\dfrac{1+x}{yz}}}$ For $q_p \gg 1$: $\approx -[1+x+y(1+z)]$

$x = \bar{C}_1/C_1$; $y = C_2/C_1$; $z = R_2/R_1$

we see in Tables 6-9 and 6-10 that $S_{R_1}^{q_p} > 0$ and $S_{R_2}^{q_p} < 0$. Therefore initially leaving R_1 low in value *decreases* the initial q_p, whereas leaving R_2 low *increases* it. In the latter case the circuit may oscillate. Second, it may be advantageous that R_1 is grounded. Other than that, the tuning conditions are seen to be similar for the type B bandpass and its dual.

For tuning purposes, the sensitivity matrix for the general type B bandpass network now follows from Table 6-7:

$$
\begin{bmatrix} \dfrac{\Delta\omega_p}{\omega_p} \\[3mm] \dfrac{\Delta q_p}{q_p} \end{bmatrix} =
$$

$$
\begin{bmatrix} -\dfrac{1}{2} & 0 \\[4mm] q_p\dfrac{\dfrac{R_1}{R_2}\left(1+\dfrac{C_1}{C_2}\right)}{\sqrt{\dfrac{R_1 C_1}{R_2 C_2}\left(1+\dfrac{R_1}{R_1}\right)}} - 1 & q_p\dfrac{1+\dfrac{R_1}{\bar{R}_1}+\dfrac{R_1}{R_2}\left(1+\dfrac{C_1}{C_2}\right)}{\sqrt{\dfrac{R_1 C_1}{R_2 C_2}\left(1+\dfrac{R_1}{R_1}\right)}} - 1 \end{bmatrix}
\begin{bmatrix} \dfrac{\Delta R_2}{R_2} \\[3mm] \dfrac{\Delta\beta}{\beta} \end{bmatrix}
$$

$$(6\text{-}70)$$

TABLE 6-10. SENSITIVITY TABLE FOR TWO TYPE B̄ BANDPASS NETWORKS

	Circuit 1		Circuit 2	
	$x=10$; $y=1.1$; $z=10$; $C_p/C_2=10$; $\beta\kappa \approx 4.85$*		$x=1$; $y=1$; $z=1$; $C_p/C_2=2$; $\beta\kappa \approx 11.3$*	
$k_2 = S_\beta^{q_p}$	$2.1q_p - 1$		$2\sqrt{2}\,q_p - 1$	
	R_1	R_2	R_1	R_2
$(k_1)_{Ri} = S_{Ri}^{\omega p}$	$-\dfrac{1}{2}$	$-\dfrac{1}{2}$	$-\dfrac{1}{2}$	$-\dfrac{1}{2}$
$(k_{21})_{Ri} = S_{Ri}^{q_p}$	$q_p - \dfrac{1}{2}$	$\dfrac{1}{2} - q_p$	$\dfrac{1}{2}(\sqrt{2}\,q_p - 1)$	$-\dfrac{1}{2}(\sqrt{2}\,q_p - 1)$
$\left(\dfrac{k_2}{k_{21}}\right)_{Ri}$	$2\dfrac{2.1q_p - 1}{2q_p - 1}$	$-2\dfrac{2.1q_p - 1}{2q_p - 1}$	$2\dfrac{2\sqrt{2}\,q_p - 1}{\sqrt{2}\,q_p - 1}$	$-2\dfrac{2\sqrt{2}\,q_p - 1}{\sqrt{2}\,q_p - 1}$
$q_p = 1$ k_2	1.1		1.828	
$(k_{21})_{Ri}$	0.5	−0.5	0.207	−0.207
$\left(\dfrac{k_2}{k_{21}}\right)_{Ri}$	2.2	−2.2	8.82	−8.82
$q_p = 10$ k_2	20		27.28	
$(k_{21})_{Ri}$	9.5	−9.5	6.57	−6.57
$\left(\dfrac{k_2}{k_{21}}\right)_{Ri}$	2.1	−2.1	4.16	−4.16

$x = \bar{C}_1/C_1$; $y = C_2/C_1$; $z = R_2/R_1$; $C_p/C_2 = (1 + x)/y$; $C_p = C_1 + \bar{C}_1$.
* Approximated by (6-35) for $q_p \gg \hat{q}$.

Likewise, the sensitivity matrix for the type $\bar{B}$ bandpass follows from Table 6-9:

$$
\begin{bmatrix} \dfrac{\Delta\omega_p}{\omega_p} \\[2ex] \dfrac{\Delta q_p}{q_p} \end{bmatrix} = \begin{bmatrix} -\dfrac{1}{2} & 0 \\[3ex] q_p\sqrt{\dfrac{\dfrac{R_2 C_2}{R_1 C_1}}{1+\dfrac{\bar{C}_1}{C_1}}} - \dfrac{1}{2} & q_p\,\dfrac{1+\dfrac{\bar{C}_1}{C_1}+\dfrac{C_2}{C_1}\left(1+\dfrac{R_2}{R_1}\right)}{\sqrt{\dfrac{R_2 C_2}{R_1 C_1}\left(1+\dfrac{\bar{C}_1}{C_1}\right)}} - 1 \end{bmatrix} \begin{bmatrix} \dfrac{\Delta R_1}{R_1} \\[2ex] \dfrac{\Delta\beta}{\beta} \end{bmatrix}
$$

(6-71)

With $(k_1)_{R_{\omega_p}} = -\frac{1}{2}$ for either circuit, it follows that we must trim R_{ω_p} to within half the accuracy required for ω_p. For example, for a frequency accuracy of 0.1% a trimming accuracy of 0.2% is required for R_{ω_p}. Furthermore, we recall from the expressions derived in Chapter 4, Section 4.4.3, that the tuning of ω_p is best accomplished by monitoring the phase ϕ_{ω_p} at ω_p. For a bandpass network, $\phi_{\omega_p} = 0°$. With the relationship between ϕ_{ω_p} and ω_p (see (4-263)) we therefore obtain

$$
\Delta\phi_{\omega_p}\,[\text{degrees}] \approx -q_p\,\frac{\Delta\omega_p}{\omega_p}\,[\%] = -\frac{q_p}{2}\,\frac{\Delta R_{\omega_p}}{R_{\omega_p}}\,[\%]
$$

(6-72)

It follows that the phase ϕ_{ω_p} must be monitored to within 0.1 times q_p degrees accuracy in order to obtain a frequency accuracy of 0.1%.[6] For low q_p values the required phase accuracy will therefore be correspondingly close to $0.1°$.

To tune for a given value of q_p we found in Section 4.4.3 that the accuracy of q_p depends on the accuracy of the β-determining resistors R_β as follows (see (4-322)):

$$
\begin{aligned}
\frac{\Delta q_p}{q_p} &= \left(1 - \frac{1}{\beta}\right)\cdot S_\beta^{q_p}\,\frac{\Delta R_\beta}{R_\beta} \\[2ex]
&= \left(1 - \frac{1}{\beta}\right) k_2 \cdot \frac{\Delta R_\beta}{R_\beta}
\end{aligned}
$$

(6-73)

R_β will be either the resistor R_6 or R_7 shown in Figs. 6-7 and 6-8. For both type B and $\bar{B}$ circuits β is given by

$$
\beta = \frac{1}{x}\left[1 + x + y(1 + z) - \frac{\sqrt{yz(1 + x)}}{q_p}\right]
$$

(6-74)

---

6. In general the desired frequency-tuning error is given. From (6-72) it, together with the lowest q_p value occurring in a filter, then determines the accuracy of the phase meter required. The specified phase tolerance to which any given filter section must be tuned is, according to (6-72), directly proportional to the q_p of that section.

where x, y, and z are as given in Tables 6-7 and 6-9. Thus for the circuits designed for low $\beta\kappa$ (see circuit 1 in Figs. 6-7a and 6-8a) we have

$$\beta|_{\substack{x=z=10 \\ y=1.1}} = 2.31 - \frac{1.1}{q_p}. \tag{6-75a}$$

and for circuit 2 (Figs. 6-7b and 6-8b) we have

$$\beta|_{x=y=z=1} = 4 - \frac{\sqrt{2}}{q_p} \tag{6-75b}$$

Thus for circuit 1 we obtain

$$\frac{\Delta q_p}{q_p}\bigg|_{\substack{x=z=10 \\ y=1.1}} = \frac{1.21q_p - 1.1}{2.31q_p - 1.1}(2.1q_p - 1.1) \cdot \frac{\Delta R_\beta}{R_\beta}$$

$$\approx (1.21q_p - 1.1)\frac{\Delta R_\beta}{R_\beta} \tag{6-76}$$

For $q_p = 10$ we have

$$\frac{\Delta q_p}{q_p}\bigg|_{q_p=10} \approx 11\frac{\Delta R_\beta}{R_\beta} \tag{6-77}$$

i.e., to obtain a 1 % accuracy in q_p we require the capability of trimming R_β accurately to within 0.1 %. Similarly for circuit 2 we obtain

$$\frac{\Delta q_p}{q_p}\bigg|_{x=y=z=1} \left(\frac{3}{\sqrt{2}}q_p - 1\right)\frac{\Delta R_\beta}{R_\beta} \tag{6-78}$$

and for $q_p = 10$ and the same accuracy of 1 % we require R_β to be trimmed twice as accurately as for circuit 1, i.e., to within 0.05 %. A little reflection on the correlation between the $\beta\kappa$ product and the sensitivity of q_p to R_β should make clear this difference in the ease with which circuits 1 and 2 can be tuned. It should also demonstrate the desirability of minimizing $\beta\kappa$ from a tuning point of view.

We recall that q_p is also preferably tuned by monitoring the phase at the frequency ω_ϕ at which the phase has a predetermined value ϕ_{q_p}. At the frequency ω_ϕ the relationship between q_p and the phase error is given by (see (4-267))

$$\frac{\Delta q_p}{q_p}[\%] \approx -3.5\Delta\phi \text{ [degrees]} \tag{6-79}$$

so that a q_p accuracy of 1 % requires a phase reading accurate to within 0.3°. For a bandpass circuit either $-45°$ or $+45°$ can be used for ϕ_{q_p}.

Proceeding in the same way as we did for the bandpass circuits described above, we can find a suitable resistor to serve as R_{ω_p} for the remaining all-pole circuits. This has been done and the results summarized in Table 6-11. The

TABLE 6-11. TUNING THE TT-SABB: SOME ALL-POLE FUNCTIONS

	Low pass	High pass	Bandpass (type $\bar{B}$)
Circuit:			

Equations:

Low pass

$$T(s) = \frac{K}{s^2 + \frac{\omega_p}{q_p}s + \omega_p^2}; \qquad \omega_p^2 = \frac{R_1 + R_2 + R_3}{R_1 R_2 R_4 C_3 C_4}$$

$$\beta = 1 + \frac{R_6}{R_7} \qquad K = \frac{\beta}{R_1 R_2 C_3 C_4};$$

$$q_p = \frac{\sqrt{(R_1 + R_2 + R_4)(R_1 R_2 R_4 C_3 C_4)}}{(R_1 + R_2)R_4 C_4 + R_1 R_2 C_3 + R_1 R_4 C_3(1 - \beta)}$$

ϕ_{ω_p}* $-90°$

R_{ω_p} If $R_1 \gg R_2, R_4$, use R_4; If $R_1 \approx R_2 \approx R_4$, use R_1

ϕ_{q_p}* $-45°$

R_{q_p} If $|\phi| < 45°$, increase R_7; If $|\phi| > 45°$, increase R_6

High pass

$$T(s) = K\frac{s^2}{s^2 + \frac{\omega_p}{q_p}s + \omega_p^2}; \qquad \omega_p^2 = \frac{1 + \frac{C_4}{C_1} + \frac{C_4}{C_2}}{R_3 R_4 C_1 C_2}$$

$$K = \frac{\beta}{1 + \frac{C_4}{C_1} + \frac{C_4}{C_2}};$$

$$q_p = \frac{\sqrt{C_1 C_2 R_3 R_4}\sqrt{1 + \frac{C_4}{C_1} + \frac{C_4}{C_2}}}{R_3(C_1 + C_2) + R_4 C_4 + R_4 C_2(1 - \beta)}$$

ϕ_{ω_p} $+90°$

R_{ω_p} R_4

ϕ_{q_p} $+45°$

R_{q_p} If $\phi < 45°$, increase R_7; If $\phi > 45°$, increase R_6

Bandpass (type $\bar{B}$)

$$T(s) = K\frac{s}{s^2 + \frac{\omega_p}{q_p}s + \omega_p^2}; \qquad \omega_p^2 = \frac{1}{R_1 R_2 C_4(C_1 + C_3)}$$

$$K = \frac{\beta C_1}{R_2 C_4(C_1 + C_3)};$$

$$q_p = \frac{\sqrt{C_4 R_1 R_2(C_1 + C_3)}}{(R_1 + R_2)C_4 + R_1 C_1 + R_1 C_3(1 - \beta)}$$

ϕ_{ω_p} $0°$

R_{ω_p} R_1

ϕ_{q_p} $+45°$

R_{q_p} If $\phi < 45°$, increase R_6; If $\phi > 45°$, increase R_7

* As discussed in Chapter 4, Section 4.4.2, two arbitrary phases ϕ_1' and ϕ_2' may also be used for tuning.

gain β is used in all the circuits to tune for q_p. If the measured phase at ω_ϕ is less than ϕ_{q_p}, β and with it q_p must be increased, and vice versa. Thus at the frequency ω_ϕ, if the phase ϕ is less than the prescribed phase ϕ_{q_p}, R_6 must be increased, and if ϕ is greater than ϕ_{q_p}, R_7 must be increased, whereby the closed-loop amplifier gain $\beta = 1 + (R_6/R_7)$.

6.2.2 Networks with Finite Zeros: The FRN

Here we are concerned with the tuning of the TT-SABB to provide a function of the form

$$T(s) = K \frac{s^2 + \dfrac{\omega_z}{q_z} s + \omega_z^2}{s^2 + \dfrac{\omega_p}{q_p} s + \omega_p^2} \tag{6-80}$$

where, in general, $q_z \gg q_p$, i.e., the zeros are in the vicinity of, if not on, the $j\omega$ axis. The TT-SABB circuit providing this function, the FRN (frequency rejection network), is described in Tables 6-5 and 6-6. In Chapter 4 (Section 4.4.2) a procedure for the tuning of the poles and zeros of a function of the type (6-80) was discussed in general; here we shall indicate with which elements of the FRN such a procedure is to be carried out. In so doing we shall distinguish between an FRN with an infinite null ($q_z = \infty$) and an FRN with a finite null ($|q_z| < \infty$).[7] By the former we mean an FRN whose null is required to be as deep as possible without regard to the actual null depth, so long as it exceeds some limit value, typically -60 dB. Conversely an FRN with a finite null will be designed to a specified depth *and phase* at the null frequency. Thus, in the latter case the zeros are assumed to be at a well defined location in the vicinity of the $j\omega$ axis.

One of the advantages of the TT-SABB is that, in the FRN configuration, it permits the transmission zeros and poles to be tuned independently of one another. This is because the zeros are realized exclusively by the twin-T (irrespective of amplifier or other circuit characteristics).[8] The twin-T thereby resembles a secondary building block of its own which, after having been tuned for the zeros, is to be adjusted no more; the poles are subsequently tuned for by separate components. In order to tune the FRN, we must therefore first turn our attention to the tuning of the twin-T per se. In fact, since it is the twin-T which determines whether the FRN has a finite or infinite null,

---

7. The finite and infinite nulls referred to here should not be confused with the "finite zeros" referred to in the subtitle of this section. By networks with finite zeros we mean networks whose transfer functions have zeros in the (finite) s-plane rather than at the origin or at infinity; the latter are, of course, the all-pole networks discussed in the preceding section.
8. As such it is a type I circuit with the attendant advantage that pole–zero pairing of nth-order network functions can be carried out with a view to optimizing signal level and minimizing distortion, without regard for sensitivity minimization.

this aspect of the FRN depends entirely on how the twin-T is tuned. This topic will be dealt with in the following discussion.

Tuning the Passive Twin-T for Specified Zeros   When perfectly balanced for an infinite null, the transfer function of the general, unloaded twin-T (Fig. 6-9) is given by

$$T_N(s) = \frac{s^2 + \omega_N^2}{s^2 + \dfrac{\omega_N}{\hat{q}_T} s + \omega_N^2} \tag{6-81}$$

The transmission zeros z and z^* are on the $j\omega$ axis at the null frequency ω_N (Fig. 6-10), i.e., $z = j\omega_N$. The polar diagram of $T_N(j\omega)$ is a circle between zero and unity on the real axis (Fig. 6-11) and the corresponding frequency and phase characteristics are as shown in Fig. 6-12a. Note that there is ambiguity in the interpretation of the phase characteristic: characteristic 1 corresponds to the vector in the polar plot flipping counterclockwise by 180° when reaching ω_N, characteristic 2 corresponds to it flipping by 180° in the clockwise direction. There are two conditions that must be satisfied to obtain a perfect null at ω_N; they are (see Table 6-5)

$$\omega_N^2 = \frac{1}{R_1 R_2 C_s C_3} \tag{6-82}$$

and

$$\frac{R_3}{C_3} = \frac{C_p}{R_p} \tag{6-83}$$

In practice the three resistors R_i, will deviate from their nominal values by the amounts $\Delta R_i / R_i$, the capacitors C_j by the amounts $\Delta C_j / C_j$. Letting

$$\Delta R_i / R_i = \rho_i, \qquad \Delta C_j / C_j = \delta_j$$

we have

$$\begin{aligned} R_i' &= R_i(1 + \rho_i) & i &= 1, 2, 3 \\ C_j' &= C_j(1 + \delta_j) & j &= 1, 2, 3 \end{aligned} \tag{6-84}$$

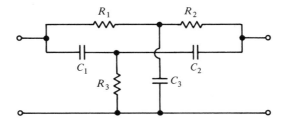

FIG. 6-9. General twin-T.

As a consequence of these component deviations, the null is shifted by some amount $\Delta\omega_N$ to the frequency $\omega_N' = \omega_N + \Delta\omega_N$. Furthermore, instead of being zero at ω_N', the response will merely have a minimum at that frequency, corresponding to the modified transfer function[9]

$$T_N'(s) = \frac{s^2 + \dfrac{\omega_N'}{q_z} s + \omega_N'^2}{s^2 + \dfrac{\omega_N'}{\hat{q}_T} s + \omega_N'^2} \tag{6-85}$$

Notice, that for $s = j\omega_N'$ this expression is real.

Seen in the s-plane, the deviations of the resistors and capacitors have caused the transmission zeros to shift off the $j\omega$ axis by some amount Δz. The zeros may now lie on either side of the $j\omega$ axis (RHP or LHP zeros), as shown in Fig. 6-10, and the corresponding polar plots will no longer pass through the origin (Fig. 6-11). With a no-longer-perfect null, the finite transmission at ω_N' results in a frequency response as shown in Fig. 6-12b. Depending on whether the zeros are in the left or right half-plane, the transmission at ω_N' is real and positive or negative, respectively. The ambiguity encountered in Fig. 6-12a has disappeared and *one can tell from the phase response in which plane the zeros are located.*

In order to tune the twin-T accurately, we must now examine how the increments ρ_i and δ_j affect the location of the zeros. Only then can we specify the resistor adjustments necessary to relocate the zeros as desired. Let us therefore first find the value that the twin-T transmission has taken on at the nominal null frequency ω_N because of the zero shift Δz. We can write

$$T(z + \Delta z)|_{z = j\omega_N} = [T(z) + \Delta T(z + \Delta z)]_{z = j\omega_N} = \Delta T(z + \Delta z)|_{z = j\omega_N} \tag{6-86}$$

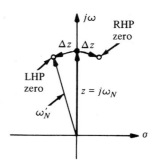

FIG. 6-10. Twin-T transmission zero z and zero shift Δz.

9. A simple consideration will convince the reader that the same frequency (i.e., ω_n') must still occur in the numerator and denominator of $T_n'(s)$: for zero and infinite frequency, the voltage transfer function of the *unloaded* twin-T must equal unity, whether its null is perfect or not.

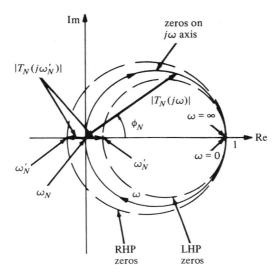

FIG. 6-11. Polar diagrams of twin-T for various transmission zeros.

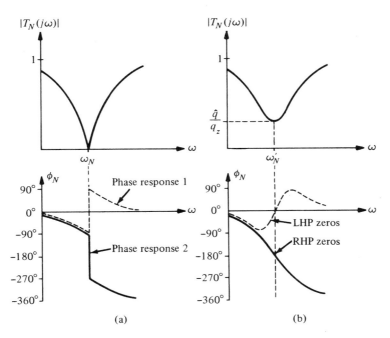

FIG. 6-12. Twin-T amplitude and phase response: (a) $j\omega$ axis zeros; (b) LHP and RHP zeros.

Recalling from Chapter 1, equation (1-148), that

$$S_z^{T(s)} = - \frac{z}{s - z} \tag{6-87}$$

we obtain[10]

$$\Delta T_N(s) = - \frac{\Delta z}{s - j\omega_N} T_N(s) \tag{6-88}$$

The zero displacement Δz is given by

$$\Delta z = \sum_{i=1}^{3} \mathcal{S}_{R_i}^{j\omega_N} \rho_i + \sum_{j=1}^{3} \mathcal{S}_{C_j}^{j\omega_N} \delta_j \tag{6-89}$$

The corresponding zero sensitivities for the twin-T can readily be calculated.[11] We obtain

$$\Delta z = \frac{\alpha \omega_N}{2(1 + \alpha^2)} [M(\rho_i, \delta_j) + jN(\rho_i, \delta_j)] \tag{6-90}$$

where

$$M(\rho_i, \delta_i) = (1 - \lambda)\rho_1 + \lambda\rho_2 - \rho_3 - (1 - \eta)\delta_1 - \eta\delta_2 + \delta_3 \tag{6-91a}$$

and

$$N(\rho_i, \delta_i) = - \left[\left(\frac{1}{\alpha} + \alpha\lambda\right)\rho_1 + \left(\frac{1}{\alpha} + \alpha - \alpha\lambda\right)\rho_2 + \alpha\rho_3 \right. $$
$$\left. + \left(\alpha + \frac{\eta}{\alpha}\right)\delta_1 + \left(\frac{1}{\alpha} + \alpha - \frac{\eta}{\alpha}\right)\delta_2 + \frac{\delta_3}{\alpha} \right] \tag{6-91b}$$

The twin-T parameters α, λ, and η are defined in Table 6-5. In the vicinity of the null frequency ω_N we have

$$\left. \frac{T_N(s)}{s - j\omega_N} \right|_{s=j\omega_N} = \left. \frac{s - z^*}{s^2 + \frac{\omega_N}{\hat{q}_T} s + \omega_N^2} \right|_{s=j\omega_N} = 2 \frac{\hat{q}_T}{\omega_N} \tag{6-92}$$

Thus, substituting (6-90) and (6-92) into (6-88), we obtain

$$\Delta T_N(j\omega_N) = - \frac{\alpha}{1 + \alpha^2} \hat{q}_T [M(\rho_i, \delta_j) + jN(\rho_i, \delta_j)] \tag{6-93}$$

---

10. This also follows directly from (1-46) and (1-51) as applied to $T_N(s)$ in the vicinity of the zero $z = j\omega_N$.
11. See Chapter 8, Table 8.2 of *Linear Integrated Networks: Fundamentals*.

For the symmetrical twin-T for which $\alpha = 1$, $\lambda = \eta = \frac{1}{2}$ and $\hat{q}_T = \frac{1}{4}$, we then have

$$\Delta T_N(j\omega_N)|_{\substack{\alpha=1 \\ \lambda=\eta=\frac{1}{2}}} = \frac{1}{8}\left[\rho_3 - \delta_3 + \frac{1}{2}(\delta_1 + \delta_2 - \rho_1 - \rho_2) \right.$$

$$\left. + j\left\{\frac{3}{2}(\rho_1 + \rho_2 + \delta_1 + \delta_2) + \rho_3 + \delta_3\right\}\right] \qquad (6\text{-}94)$$

For the potentially symmetrical twin-T, for which $\alpha = 1$, $\lambda = \eta = (1 + \rho)^{-1}$ $\hat{q}_T = 0.5\rho/(1 + \rho)$, where ρ is the scaling factor, we obtain

$$\Delta T_N(j\omega_N)|_{\substack{\alpha=1 \\ \lambda=\eta=(1+\rho)^{-1}}} = \frac{1}{4}\cdot\frac{\rho}{1+\rho}\left\{\rho_3 - \delta_3 + \frac{\rho}{1+\rho}(\delta_1 - \rho_1)\right.$$

$$\left. + \frac{1}{1+\rho}(\delta_2 - \rho_2) + j\left[\frac{2+\rho}{1+\rho}(\rho_1 + \delta_1) + \frac{1+2\rho}{1+\rho}(\rho_2 + \delta_2) + \rho_3 + \delta_3\right]\right\} \quad (6\text{-}95)$$

So far we have computed the increase of the twin-T transmission from the nominal value of zero to the value $\Delta T_N(j\omega_N)$, where ω_N is the desired null frequency. This transmission value, given by (6-93) for the general twin-T and by (6-94) and (6-95) for the symmetrical and potentially symmetrical twin-T, respectively, is caused by nonideal component values (i.e., initial tolerances and tuning errors ρ_i and δ_j). Note that instead of zero we now have a *complex* quantity at ω_N. Furthermore we know that $\Delta T_N(j\omega_N)$ is not the minimum transmission value; this has shifted to the frequency $\omega'_N = \omega_N + \Delta\omega_N$ at which $\Delta T_N(j\omega'_N)$ is real. We can therefore readily calculate ω'_N; it corresponds to the frequency at which the imaginary part of $\Delta T_N(j\omega_N)$ is zero.

To compute ω'_N, the frequency of the transmission minimum, we can write (see Fig. 6-13)

$$T_N(j\omega'_N) = T_N(j\omega_N + j\Delta\omega_N) = T_N(j\omega_N) - \Delta T_N(j\omega_N - j\Delta\omega_N)$$

$$= \Delta T_N(j\omega_N) - \Delta T_N(j\omega_N - j\Delta\omega_N) \qquad (6\text{-}96)$$

From (6-81) it readily follows that, to a first approximation,

$$\Delta T_N(j\omega_N - j\Delta\omega_N) \approx -j2\hat{q}_T\frac{\Delta\omega_N}{\omega_N} \qquad (6\text{-}97)$$

(see also (4-123a)). Thus, with (6-93) and (6-96) we obtain

$$T_N(j\omega'_N) = v = -\frac{\alpha}{1+\alpha^2}\hat{q}_T\left\{M(\rho_i, \delta_j) + j\left[N(\rho_i, \delta_j) - 2\left(\frac{1}{\alpha} + \alpha\right)\frac{\Delta\omega_N}{\omega_N}\right]\right\}$$

$$= -\frac{\alpha}{1+\alpha^2}\hat{q}_T M(\rho_i\delta_j) \qquad [6\text{-}98]$$

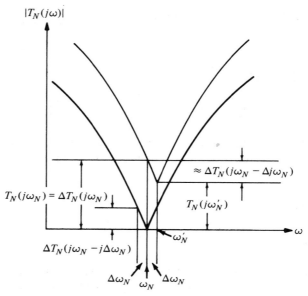

Fig. 6-13. Amplitude response of tuned and untuned twin-T.

where

$$v = \frac{\hat{q}_T}{q_z} \tag{6-99a}$$

and

$$q_z = -\left(\alpha + \frac{1}{\alpha}\right)\frac{1}{M(\rho_i, \delta_j)} \tag{6-99b}$$

The term v is the minimum value of $T_N(j\omega)$ as obtained at the frequency ω'_N; it has a positive or negative real value depending on whether the zeros are in the LHP or RHP, respectively. From (6-98), the new frequency of minimum transmission is shifted from the nominal value by the amount

$$\Delta\omega_N = \frac{1}{2}\frac{\alpha}{1 + \alpha^2}\, N(\rho_i, \delta_j)\omega_N \tag{6-100}$$

If, for example, all the resistor values are high by ρ_0 percent and the capacitors are at their nominal values, then, for the symmetrical twin-T,

$$\Delta\omega_N \Big|_{\substack{\alpha=1 \\ \lambda=\eta=\frac{1}{2} \\ \rho_i=\rho_0}} = -\rho_0 \cdot \omega_N \tag{6-101}$$

As we would expect, the twin-T has thereby been scaled down in frequency by the amount ρ_0 without affecting the null depth, since $M(\rho_1 = \rho_2 = \rho_3) = 0$.

Expressions (6-93) and (6-98) form the basis of our procedure for tuning the twin-T. As mentioned earlier we must distinguish between tuning the twin-T for finite and infinite nulls. As we shall see, however, once having shown how to tune for a finite null, the infinite-null case readily follows as a special case.

A finite null is determined by a pair of zeros off the $j\omega$ axis. The zeros are characterized by the twin-T null depth $v = \hat{q}_T/q_z$ and the zero[12] frequency ω_z. The polarity of v determines in which plane the zeros lie. Various methods of tuning a specified zero pair can now be derived from the discussion above. Among them, we can select a deterministic and a functional method as discussed in Chapter 4, Section 4.4.1.

Functionally Tuning the Twin-T

1. Assuming that an appropriate twin-T configuration has been selected (e.g., symmetrical or potentially symmetrical) calculate the corresponding twin-T parameters α, λ, and $\hat{q}_T$ (see Table 6-5).

2. Determine the nominal resistor and capacitor values R_1, R_2, R_3, C_1, C_2, and C_3 corresponding to the desired zero frequency ω_z, but scale down the resistors R_1, R_2 and R_3 by the factor ρ_0 so that

$$R_1' = R_1(1 - \rho_0); \qquad R_2' = R_2(1 - \rho_0); \qquad R_3' = R_3(1 - \rho_0) \quad (6\text{-}102)$$

where

$$\rho_0 > \delta_1 \left(\frac{C_2}{C_1 + C_2}\right) + \delta_2 \left(\frac{C_1}{C_1 + C_2}\right) + \delta_3 \quad (6\text{-}103)$$

and δ_j are the worst case tolerances of the capacitors C_j ($j = 1, 2, 3$). As a consequence the initial zero frequency of the untuned twin-T will be higher than ω_z by an amount approximately equal to ρ_0.

3. Find the frequency ω_z' at which the response of the untuned twin-T has a minimum. Measure this minimum, namely v', including its phase. If the phase is $0°$, the zeros of the untuned twin-T are in the LHP; if it is $180°$, they are in the RHP. Note that an accurate phase measurement is not required.

4. Calculate $\bar{R}_2'$ and $\bar{R}_3'$ corresponding to a perfect null ($v = 0$) at the frequency ω_z' from the expressions

$$\bar{R}_2' = \frac{1}{(\omega_z')^2 R_1' C_s C_3} \quad (6\text{-}104a)$$

and

$$\bar{R}_3' = \frac{C_3}{C_p} \frac{R_1' \bar{R}_2'}{R_1' + \bar{R}_2'} \quad (6\text{-}104b)$$

---

12. Since we are concerned now with the tuning of a network function of the type given by (6-80) we shall hereafter refer to a *zero* frequency ω_z rather than to a *null* frequency ω_N.

5. From (6-98) we obtain the values ρ_2 and ρ_3 necessary to obtain the required null v (where the desired polarity of v must be used):

$$\rho_2 = -\frac{\alpha}{\hat{q}_T}(v - v') \tag{6-105a}$$

$$\rho_3 = \frac{1}{\hat{q}_T}\left(1 - \frac{1}{\alpha} - \alpha\lambda\right)(v - v') \tag{6-105b}$$

Note that since the null v' was measured directly at the null frequency ω'_z, $\Delta\omega_z = 0$.

6. Trim R_2 and R_3 to the values

$$R''_2 = \bar{R}'_2(1 + \rho_2) \tag{6-106a}$$
$$R''_3 = \bar{R}'_3(1 + \rho_3) \tag{6-106b}$$

7. Measure the new null depth v'' at the frequency ω''_z. If v'' is not close enough to the specified value v, step 4 is repeated using ω''_z and obtaining $\bar{R}''_2$ and $\bar{R}''_3$. Then ρ'_2 and ρ'_3 are calculated as in step 5, where $v - v''$ is substituted for $v - v'$. Subsequently the new values $R'''_2 = \bar{R}''_2(1 + \rho'_2)$ and $R'''_3 = \bar{R}''_3(1 + \rho'_3)$ are adjusted for, and so on.

8. Having obtained the desired null depth v at some frequency, say ω'''_z, the three twin-T resistors are increased by the amount ω'''_z/ω_z, i.e., they are frequency-scaled so that the null depth v is shifted down to the desired frequency ω_z.

In general, not more than two or three iterations should be necessary, with this functional tuning method. Nevertheless, the deterministic method described next, being potentially a one-step procedure, will be preferable whenever the capability of high-accuracy resistance and capacitance measurements is available.

Deterministically Tuning the Twin-T

1'. As in step 1 above.

2'. Determine nominal resistor and capacitor values corresponding to the desired zero frequency ω_z. Select (or preadjust) R_1, C_1, C_2, and C_3 as close to their nominal values as is feasible.

3'. Select a frequency ω'_z which is sufficiently higher than ω_z to account for the worst-case initial tolerances δ_j of the capacitors $C_j(j = 1, 2, 3)$ and ρ_1 of R_1. Thus

$$\omega'_z = \omega_z\left(1 + \frac{\rho_0}{2}\right) \tag{6-107a}$$

where

$$\rho_0 > \rho_1 + \delta_1 \left(\frac{C_2}{C_1 + C_2}\right) + \delta_2 \left(\frac{C_1}{C_1 + C_2}\right) + \delta_3 \qquad (6\text{-}107b)$$

Note that in a potentially symmetrical twin-T in which $C_1 \gg C_2$, the capacitors C_2 and C_3 should have a lower tolerance than C_1.[13]

4′. In order for the untuned twin-T zero frequency to be in the vicinity of ω'_z, the initial (i.e., design) value for R_2 must be

$$R'_2 = R_2(1 - \rho_0) \qquad (6\text{-}108a)$$

The initial value for R_3 corresponding to a perfect null at ω'_z is $R_3[1 - \rho_0 R_1/(R_1 + R_2)]$. However, to guarantee a sufficiently large tuning range we also let

$$R'_3 = R_3(1 - \rho_0) \qquad (6\text{-}108b)$$

R_2 and R_3 are the nominal values corresponding to ω_z.

5′. Measure resistor R_1 and the capacitors C_1, C_2, and C_3 as accurately as possible. Since we really require $C_s = C_1 C_2/(C_1 + C_2)$ and $C_p = C_1 + C_2$, rather than the individual capacitance values, it is actually preferable to measure the series and parallel combinations of C_1 and C_2 directly. This may also alleviate the problem of measuring capacitors whose terminals are not accessible. If it does not, we refer to the discussion of this problem in Chapter 4 (Section 4.4.4).

6′. Calculate $\bar{R}'_2$ and $\bar{R}'_3$ required for a perfect null at $\omega'_z = \omega_z(1 + \rho_0/2)$ from[14]

$$\bar{R}'_2 = \frac{1}{(\omega'_z)^2 R_1 C_s C_3} \qquad (6\text{-}109a)$$

and

$$\bar{R}'_3 = \frac{C_3}{C_p} \frac{R_1 \bar{R}'_2}{R_1 + \bar{R}'_2} \qquad (6\text{-}109b)$$

7′. From (6-98) we can now calculate the values ρ_2 and ρ_3 necessary to obtain a required null v at ω_z (where the desired polarity of v must be used):

$$\rho_2 = 2 \frac{\omega'_z - \omega_z}{\omega'_z} - \frac{\alpha}{\hat{q}_T} v \qquad (6\text{-}110a)$$

$$\rho_3 = \frac{v}{\hat{q}_T}\left(1 + \frac{1}{\alpha} - \alpha\lambda\right) + 2\lambda \frac{\omega'_z - \omega_z}{\omega'_z} \qquad (6\text{-}110b)$$

13. Just how low the tolerances should be is an economic question. The price of chip capacitors, for example, increases rapidly as the tolerances are decreased from, say, 5% to 1%. On the other hand, it will be evident in what follows that the tuning time (number of iteration steps, etc.), and with it the cost of tuning a circuit, will increase as ρ_0 is increased.
14. The values R_1, C_s, and C_3 are, of course, the measured values obtained in step 5′.

8'. Trim R_2 and R_3 to the values

$$R_2'' = \bar{R}_2'(1 + \rho_2) \tag{6-111a}$$
$$R_3'' = \bar{R}_3'(1 + \rho_3) \tag{6-111b}$$

Step 8' is the only actual trimming step provided for in this deterministic procedure. Should the acquired null depth and frequency not be within the specified limits, it is quite likely that a correction will involve having to *decrease* either R_2 or R_3; but with film components this of course cannot be done. When high precision is required it is therefore often a good idea not to solve ρ_2 and ρ_3 in (6-110) directly for the final frequency ω_z but for an intermediate frequency several percent *higher* than ω_z. This permits a correction of null depth and frequency to be made after step 8'. The correction steps comprise the following:

9'. Measure the frequency ω_z'' and null depth v' obtained after trimming R_2 and R_3 in 8'.

10'. Repeat step 6', calculating $\bar{R}_2''$ and $\bar{R}_3''$ from (6-109), where ω_z'' is used in (6-109a) and $\bar{R}_2''$ in (6-109b).

11'. Calculate the new values ρ_2' and ρ_3' as follows:

$$\rho_2' = 2\,\frac{\omega_z'' - \omega_z}{\omega_z''} - \frac{\alpha}{\hat{q}_T}(v - v') \tag{6-112a}$$

$$\rho_3' = \frac{1}{\hat{q}_T}\left(1 + \frac{1}{\alpha} - \alpha\lambda\right)(v - v') + 2\lambda\,\frac{\omega_z'' - \omega_z}{\omega_z''} \tag{6-112b}$$

12'. Trim R_2 and R_3 to the values

$$R_2''' = \bar{R}_2''(1 + \rho_2') \tag{6-113a}$$
$$R_3''' = \bar{R}_3''(1 + \rho_3') \tag{6-113b}$$

Steps 9'–12' can of course be repeated, but it is unlikely that this will be necessary. No doubt, by adding these steps we have added functional measurements, thereby requiring both component *and* null depth and frequency measurements to be made. However, in so doing we have eliminated all but one iteration from the purely functional procedure described above, which should make this combination of measurements well worth while. Naturally the correction steps (i.e., 9'–12') can be omitted if the specified tolerances for the zeros are sufficiently loose.

In order to appreciate the accuracy that can actually be expected with the simple tuning procedure comprising steps 1'–8', let us look at a numerical example. Referring to Fig. 6-9, we assume a twin-T that has been designed to be as close to potentially symmetrical as the discrete (chip) capacitors used will allow. For an infinite null at 3.96 kHz, for example, we have the following component values: $R_1 = 2.23$ kΩ, $R_2 = 24.57$ kΩ, $R_3 = 1.86$ kΩ. $C_1 = 18,000$ pF,

$C_2 = 1,800$ pF, $C_3 = 18,000$ pF. Consequently, we obtain the following twin-T parameters: $\lambda = 0.083$, $\eta = 0.090$, $\alpha = 1.090$, and $\hat{q}_T = 0.455$. Using the deterministic tuning steps $1'$–$8'$ we now assume that R_1, C_1, C_2, and C_3 can be measured, and R_2 and R_3 trimmed, to within 0.1% accuracy. Thus we have $\rho_i = \delta_j = \pm 0.001$ ($i, j = 1, 2, 3$). The nominal R and C values given above correspond to an infinite null at the specified null frequency. We now wish to know, with the given errors in measurement and trimming, how far off nominal (in the worst case) the null frequency and the corresponding null depth can possibly be. For this we can use the expressions derived earlier. Instead of an infinite null, we now have the null depth v given by (6-98), i.e., $v = -\alpha(1 + \alpha^2)^{-1}\hat{q}_T M(\rho_i, \delta_j)$. With the numerical values given above we obtain $M(\rho_i, \delta_j) = 4 \cdot 10^{-3}$ and therefore $v \approx -0.9 \cdot 10^{-3}$, i.e., the null depth will still exceed -60 dB. From (6-100), the worst-case frequency shift of the null depth caused by our errors will be $\Delta\omega_N/\omega_N = 0.5\alpha(1 + \alpha^2)^{-1}$ $N(\rho_i, \delta_j)$. With the numerical values given we obtain $N(\rho_i, \delta_j) = -8 \cdot 10^{-3}$, and hence $\Delta\omega_N/\omega_N \approx 1.99 \cdot 10^{-3}$, i.e., the frequency will be better than 0.2% accurate. As an upper limit (i.e., worst case), both the nulldepth and the frequency accuracy attained are remarkably good. This example therefore shows us that, with facilities for component measuring and trimming to within 0.1%, the attainable null depth and frequency accuracy will be good enough for a large number of applications.[15]

Incidentally, the results obtained above could have been predicted rather accurately and simply as follows. Given a worst-case resistor and capacitor error of ρ_0 and δ_0 percent, the frequency will be off by $(\rho_0 + \delta_0)$ percent. To obtain the null depth we designate $\rho_0 = \delta_0 = \varepsilon_0$ and obtain from (6-91a): $|M(\rho_i \equiv \delta_j \equiv \varepsilon_0)|_{\max} = 2\varepsilon_0[(1 - \lambda) + (1 - \eta)] \approx 4\varepsilon_0$ since λ and η are very small for a potentially symmetrical twin-T. For the same reason $\hat{q}_T \approx 0.5$ and $\alpha \approx 1$, and we obtain, from (6-98), $v \approx \varepsilon_0$. Similarly, for a symmetrical twin-T, where $\lambda = \eta = 0.5$, $\hat{q}_T = 0.25$, and $\alpha = 1$, we obtain $v \approx \varepsilon_0/4$.

In both tuning methods described above, the importance of making the initial untuned null frequency ω_z' large enough with respect to ω_z should be evident. Only then can the calculated values ρ_2 and ρ_3 be guaranteed to be positive, meaning that they require *increases* in resistors R_2 and R_3. A comparison of (6-102) with (6-105) and (6-106), or of (6-108) with (6-110) and (6-111), should make this clear. Furthermore, to appreciate the interrelationships between ρ_2, ρ_3, v, and ω_z, the zero displacements shown graphically in Fig. 6-14 for the potentially symmetrical twin-T should be helpful.[16]

In selecting a pair of resistors to tune with, we have the choice of combining either R_1 or R_2 with R_3. However, for the symmetrical as well as for the

---

15. Note that the null of an FRN generally occurs *outside* the pass band. Thus the requirements on the frequency accuracy of the null are generally less tight than those for the network (pole) frequencies occurring *in* the pass band.

16. See also G. S. Moschytz, A general approach to twin-T design and its application to hybrid integrated linear active networks, *Bell Syst. Tech. J.* 1105–1149 (1970), and Chapter 8, Fig. 8-12 of *Linear Integrated Networks: Fundamentals* for the symmetrical twin-T.

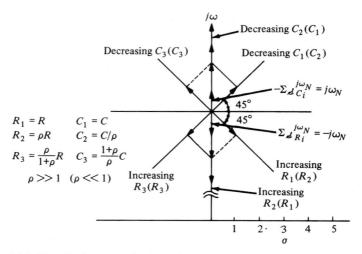

FIG. 6-14. Zero displacements in the s-plane for a potentially symmetrical twin-T.

potentially symmetrical twin-T, $\hat{q}_T$ increases with increasing R_2 (see Fig. 6-15) and decreases with increasing R_1. Since a large $\hat{q}_T$ is generally desirable, R_2 is mostly preferable over R_1.

Tuning the Twin-T for $j\omega$-Axis Zeros It should be clear to the reader that both tuning methods discussed above are readily adaptable to the twin-T with an infinite null merely by setting the specified value of v equal to zero in steps 5, 7′, and 11′. In general, the semideterministic procedure outlined above,

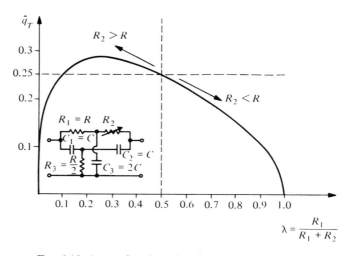

FIG. 6-15. $\hat{q}_T$ as a function of R_2 for a symmetrical twin-T.

comprising the steps $1'$–$12'$ will be most efficient if the specifications for the $j\omega$-axis zeros are tight (e.g., null depth below 70 or 80 dB, and ω_z to within 0.1%). Obviously the process is considerably simplified if the tolerances require no more than steps $1'$–$8'$.

In the case of the potentially symmetrical twin-T, the semideterministic tuning procedure becomes particularly simple when tuning for an infinite null. This follows from step $7'$ if we recall that, for the potentially symmetrical twin-T, λ approaches zero and $\alpha = 1$. Thus, solving for ρ_3 first, we have

$$\rho_3 \Big|_{\substack{\lambda \to 0 \\ \alpha = 1}} = \frac{v}{\hat{q}_T}\left(1 + \frac{1}{\alpha}\right) \qquad (6\text{-}114)$$

R_3 is then trimmed to $\bar{R}'_3(1 + \rho_3)$, whereby v is tuned to zero. Consequently

$$\rho_2 \Big|_{\substack{\lambda \to 0 \\ \alpha = 1}} = 2\,\frac{\omega'_z - \omega_z}{\omega'_z} \qquad (6\text{-}115)$$

permitting ω_z to be set *independently* by R_2. It will be recalled that this feature of the potentially symmetrical twin-T was demonstrated somewhat differently in Section 4.3.2 in Chapter 4. Furthermore it follows directly from Fig. 6-14 that we also have the capability of correcting for overshoot of R_3 by tuning R_1, because

$$\rho_1 \Big|_{\substack{\lambda \to 0 \\ \alpha = 1}} = -\frac{v}{\hat{q}_T}\left(1 + \frac{1}{\alpha}\right) \qquad (6\text{-}116)$$

In this case a simplified functional tuning procedure is useful, in which R_1 and R_3 are tuned alternatively until v is sufficiently small, and, as a last step, R_2 shifts the null down to the specified frequency ω_z. Naturally the initial frequency ω'_z must here again be sufficiently higher than ω_z to prevent overshoot. At the end of the following discussion on the tuning of the active twin-T we shall come back to the potentially symmetrical twin-T and show that a simple two-step tuning method presents itself there.

Tuning the Active Twin-T for Specified Zeros   So far we have discussed only the tuning of a passive twin-T for specified zeros. However in the practical case at hand, namely the tuning of an FRN (see Fig. 6-16) for specified zeros, the twin-T is imbedded within an active configuration and the tuning procedures mentioned above must be modified accordingly. Fortunately the modifications are relatively minor, as we shall see in the following discussion.

It is a simple matter to show that the zeros of the twin-T are independent of the loading network. The twin-T can readily be converted into an equivalent π network[17] as shown in Fig. 6-17. The voltage transfer function of this π

17. G. S. Moschytz, A general approach to twin-T design and its application to hybrid integrated linear active networks, loc. cit., p. 1141.

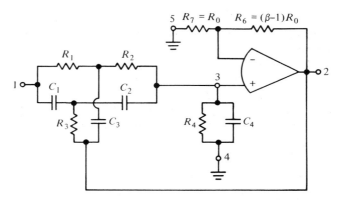

FIG. 6-16. A class 4 frequency-rejection network (FRN).

network is $V_2/V_1 = z_{21}/z_{11} = Z_c/Z_b$; by inspection the zeros are therefore independent of the loading network. In the FRN this means that the twin-T determines the zeros, independent of the loading network and the operational amplifier characteristics. It is therefore permissible to ground the output terminal of the FRN (terminal 2 in Fig. 6-16) and, monitoring the twin-T output at terminal 3, to tune the passive twin-T for the desired zeros exactly as outlined in the previous section. On "reactivating" the circuit, the zero locations of the overall circuit will correspond to those of the passive twin-T.[18]

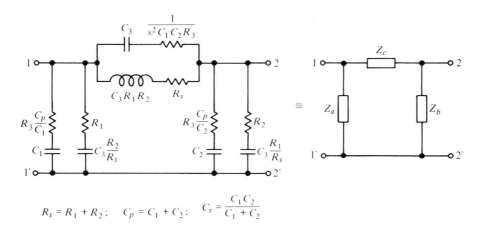

FIG. 6-17. Equivalent π network for the balanced twin-T.

18. The null depth in the two cases will be different. To understand this, assume that the active FRN requires $\omega_z = \omega_p$. Then in the activated network the null depth will be q_p/q_z whereas in the passive twin-T it is $\hat{q}_T/q_z$. Thus in the active case the null will be $q_p/\hat{q}$ *less* deep than in the passive twin-T. Nevertheless q_z, which pertains to the zeros only, will be the same in both cases.

Should it not be desirable to convert the FRN into a passive twin-T as above, then we must proceed differently, at least when functionally tuning the circuit. (Naturally the deterministic method—excluding the functional correction steps—can be applied here without any modifications.) Recall that the transfer function of the FRN can be written as follows (see Chapter 3, equation (3-70)):

$$T(s) = \frac{-\beta \dfrac{y_{21}}{y_{22}}}{1 - \beta \left(\dfrac{y_{21}}{y_{22}} + 1\right) + \dfrac{Y_L}{y_{22}}} \tag{6-117}$$

where y_{21} and y_{22} are the y parameters of the twin-T and Y_L is the admittance of the twin-T loading network. Since the transfer function of the twin-T is $T_N(s) = -y_{21}/y_{22}$, (6-117) can be written as

$$T(s) = H(s)T_N(s) \tag{6-118}$$

Thus, at any frequency, $T(j\omega)$ can be written as a complex constant $H(j\omega)$ times the transfer function of the twin-T.[19] By dividing the quantities measured on the active circuit by the appropriate constant, the tuning problem is reduced to that of tuning the passive twin-T as described above. However, at a transmission minimum v occurring at ω_z, the phase will be neither $0°$ nor $180°$ and we must now examine how we can infer the plane of the corresponding zeros from the polarity of the, now arbitrary, phase.

At the null frequency ω_z the general transfer function (6-80) will have the value

$$T(j\omega_z) = K \frac{\pm j\dfrac{\omega_z^2}{q_z}}{\omega_p^2 - \omega_z^2 + j\dfrac{\omega_p}{q_p}\omega_z} \tag{6-119}$$

The sign of the numerator determines the plane of the zeros (positive in the LHP and negative in the RHP) and, as we shall now show, also the phase of $T(j\omega_z)$. From (6-119), $T(j\omega_z)$ can be written in the form

$$T(j\omega_z) = \pm ja(b - jc) = |T(j\omega_z)|e^{j(\phi \pm 90°)} \tag{6-120}$$

where

$$a = \omega_z^2/q_z \tag{6-121a}$$

$$b = \frac{\omega_p^2 - \omega_z^2}{(\omega_p^2 - \omega_z^2)^2 + \left(\dfrac{\omega_p\omega_z}{q_p}\right)^2} \tag{6-121b}$$

19. Much of this and the following material on FRN tuning is based on unpublished material by C. J. Steffen, Bell Telephone laboratories.

$$c = \frac{\dfrac{\omega_p \omega_z}{q_p}}{(\omega_p^2 - \omega_z^2)^2 + \left(\dfrac{\omega_p \omega_z}{q_p}\right)^2} \tag{6-121c}$$

Notice that b is positive or negative depending on whether ω_p/ω_z is larger or smaller than unity. Thus, from the diagram shown in Fig. 6-18 it follows that the value of ϕ lies anywhere between $180°$ and $360°$, depending on ω_p/ω_z. (For $\omega_p = \omega_z$, ϕ equals $270°$.) Consequently the phase of $T(j\omega_z)$ is related to the plane of the zeros of $T(s)$ as follows:

$$\begin{aligned} -90° < \angle T(j\omega_z) < 90° &\to \text{LHP zeros} \\ 90° < \angle T(j\omega_z) < 270° &\to \text{RHP zeros} \end{aligned} \tag{6-122}$$

This correlation between the phase of $T(j\omega_z)$ and the location of the zeros in the s-plane can be put to good use in tuning the twin-T (active or passive) for an infinite null. We shall illustrate this with the potentially symmetrical passive twin-T for which the procedure becomes particularly simple to understand. It can be extended to any other active or passive configuration, resulting, however, in more than the two adjustment steps required for the potentially symmetrical twin-T. In the potentially symmetrical case we proceed as follows:

1. Use the nominal values for R_1, C_1, C_2, and C_3 corresponding to the desired infinite null frequency ω_z.

2. The initial values of R_2 and R_3 are $R_2' = R_2(1 - \rho_0)$ and $R_3' = R_3(1 - \rho_0)$ where ρ_0 is given by (6-107b) and takes the worst-case tolerances of R_1, C_1, C_2, and C_3 into account.

3. Because only R_2 and R_3 are less than nominal the zero z of the untuned twin-T is guaranteed to lie in the RHP as shown in Fig. 6-19a. Consequently

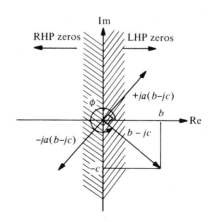

FIG. 6-18. Tuning the active twin-T.

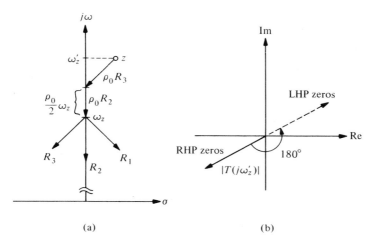

(a)   (b)

FIG. 6-19. Tuning the active twin-T: (a) zero shift; (b) phasor diagram.

the phase of $T(j\omega)$ in the vicinity of ω_z is anywhere between 90° and 270° (Fig. 6-19b).

4. At the frequency $\omega_z[1 + (\rho_0/2)]$ increase R_3 until the phase of the output signal changes by 180° (while the magnitude goes to zero) indicating that the zero z has been shifted to the $j\omega$ axis (Fig. 6-19a).

5. Monitoring the signal at ω_z, increase R_2 until the null obtained in step 4 has been shifted to ω_z.

Starting at some higher frequency than ω_z, and taking into account the relationship between component variation and zero displacement in the s-plane (e.g., Fig. 6-14), it should be clear how the procedure outlined in the steps above can be expanded into an iterative one. Every time a zero crosses the $j\omega$ axis the phase of the output signal changes by 180° while its magnitude goes through zero. Using a Lissajous figure (i.e., ellipse) to monitor the output signal level and phase, a $j\omega$ axis crossing then corresponds to a slanted ellipse folding into a horizontal line and reappearing slightly less slanted in the opposite direction. This process continues until the ellipse can be maintained as a horizontal line. In such a procedure all three resistors must be used for tuning. In order to avoid overshooting the frequency ω_z the initial frequency ω_z' should be high enough that, by the time the ellipse has been tuned into a horizontal line, the corresponding frequency is still higher than ω_z. Frequency-scaling the three resistors the null is then shifted down to ω_z.

One final point should be made. Whereas the *zeros* of a twin-T (whether active or passive) with a finite null ($q_z < \infty$) remain unaffected by a loading network, the actual *minimum* of the resulting transfer function of the type

(6-80), is theoretically no longer at the frequency ω_z as soon as $\omega_z \neq \omega_p$. It can be shown, however, that the deviation of the minimum from ω_z is generally negligibly small, so that we may, as a rule, equate the frequency of the measured null with the zero frequency ω_z. If this is not the case, we can obtain the actual null frequency by solving the general equation (1-30) given in Chapter 1, Section 1.3.2.

Tuning the FRN Poles   It was pointed out in Section 4.4.2 of Chapter 4 that as long as the pole and zero frequencies are sufficiently far apart from each other the poles of a network realizing a function of the type (6-80) can be tuned independently of the zeros in the same manner as they would be in an all-pole network. Of course, the separation between ω_p and ω_z necessary for this simplification to be valid depends on q_z (see Fig. 4-22). Fortunately, in practice it is valid far more often than not. We shall therefore consider this important case first.

Ignoring, the effects of the zeros, the poles of our FRN may be tuned in the same manner as those of an all-pole network. The only remaining network elements at our disposal are the resistor R_4 (Fig. 6-16) and the gain β, and it should be clear after our discussion on the tuning of all-pole networks that we shall use R_4 to tune ω_p, and the gain β to tune for q_p. Thus with the expressions given in Table 6-6 we readily obtain the necessary sensitivity matrix

$$\begin{bmatrix} \dfrac{\Delta\omega_p}{\omega_p} \\[2ex] \dfrac{\Delta q_p}{q_p} \end{bmatrix} = \begin{bmatrix} -\dfrac{1}{2}\dfrac{r}{1+r} & 0 \\[2ex] q_p\dfrac{r}{\sqrt{1+c}}-\dfrac{1}{2}\dfrac{r}{1+r} & \dfrac{q_p}{\hat{q}_T}-1 \end{bmatrix} \begin{bmatrix} \dfrac{\Delta R_4}{R_4} \\[2ex] \dfrac{\Delta\beta}{\beta} \end{bmatrix} \tag{6-123}$$

where r, c, and $\hat{q}_T$ are given in Table 6-6. Notice that from the point of view of frequency tuning it would be desirable to let $r = (R_1 + R_2)/R_4$ be as large, i.e., R_4 as small as possible. This conflicts with our design rules of Section 6.1.3, however, in which r was specified to be as small as possible in order to minimize the $\beta\kappa$ product. As a result a relatively large adjustment of R_4 may be required in order to obtain a given adjustment of ω_p. We shall come back to this problem shortly.

Assuming the phase of the zeros can be ignored, we do not yet know what kind of an all-pole network we can consider the FRN to be in order to tune its poles. However, if we assume that q_z is large enough to neglect the linear term in s, this becomes clear by inspection of our transfer function:

$$T(s) = K\frac{s^2 + \omega_z^2}{s^2 + \dfrac{\omega_p}{q_p}s + \omega_p^2} \tag{6-124}$$

When $\omega_p < \omega_z$ the function has a low-pass characteristic in the vicinity of ω_p, when $\omega_p > \omega_z$ the characteristic is high-pass. This is illustrated in Fig. 6-20.

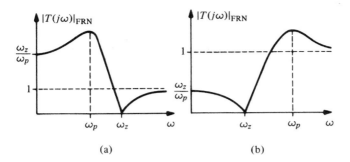

FIG. 6-20. FRN response: (a) low-pass, $\omega_p < \omega_z$; (b) high-pass, $\omega_p > \omega_z$.

With that, the pole-tuning procedure is straightforward and takes place as follows:

a. $\omega_p < \omega_z$.

   1. Adjust R_4 until $\phi_{\omega_p} = -90°$.

   2. At the frequency

$$\omega_\phi = \frac{\omega_p}{2q_p} \left[\sqrt{4q^2 + 1} - 1\right]$$

    adjust β for an output phase of $-45°$. (If the phase exceeds 45°, increase β, and vice versa).

b. $\omega_p > \omega_z$.

   1. Adjust R_4 for $\phi_{\omega_p} = -270°$.

   2. At the frequency

$$\omega_\phi = \frac{\omega_p}{2q_p} \left[\sqrt{4q_p^2 + 1} + 1\right]$$

    adjust β for an output phase of $-315°$. (If the phase exceeds 315° decrease β, and vice versa).

Note that, for obvious reasons, in both cases the frequency ω_ϕ furthest away from ω_z is used to tune q_p. Other than that the procedure is identical to that used with the corresponding all-pole networks.

We must now consider the case in which ω_z is so close to ω_p that the phase contribution of the zeros cannot be neglected at ω_p. One way of dealing with this situation is to compute the phase contribution of the zeros at the frequency ω_p and, if necessary, at ω_ϕ and to subtract these contributions from the total phase measured in steps 1 and 2 above. Another is to tune the poles by feeding the signal into the ungrounded terminal 5 (see Fig. 6-16) while grounding terminal 1. It will be recalled (Chapter 4, Section 4.4.2 and Fig. 4-21) that we thereby obtain an FEN-type frequency response whose poles we can tune independently of the previously tuned zeros (the latter by conventional means). The two tuning steps are then

1. Adjust R_4 for $\phi_{\omega_p} = -180°$.

2. At the frequency

$$\omega_\phi = \frac{\omega_p}{2q_p} [\sqrt{4q_p^2 + 1} - 1]$$

adjust β for an output phase of $-45°$ after having deducted the phase contribution of the zeros at ω_ϕ. (Increasing β will decrease the phase and vice versa).

When tuning the FRN (or any of the other circuits), the finite gain–bandwidth product of the operational amplifier may prevent the independent tuning of ω_p and q_p that we have assumed here. In other words the second term of the first row in (6-123) may not equal zero. This problem will be discussed in Section 6.3.

In some cases, notably when used as the active null network in the feedback loop of a multiamplifier FEN, ω_z is required to equal ω_p, whereas the pole Q need not be accurately adjusted for.[20] In such cases ω_p can be tuned to be equal to ω_z by adjusting R_4 until the magnitude of the FRN frequency response is equal at high and low frequencies.

Finally, let us consider by how much R_4 should initially be low in value so as to guarantee only increasing resistor adjustments when tuning for ω_p. From Table 6-6 we have

$$\frac{\omega_p}{\omega_z} = \sqrt{\frac{1 + r}{1 + c}} \tag{6-125}$$

It follows that

$$R_4 = \frac{R_s}{\left(\dfrac{\omega_p}{\omega_z}\right)^2 (1 + c) - 1} \tag{6-126}$$

Assuming the deterministic tuning method, in which R_1 is adjusted to its nominal value and R_2 subsequently adjusted to within ρ_2 percent in order to tune ω_z, we must now find ρ_4' such that the initial value of R_4 is $R_4(1 - \rho_4')$.[21]

---

20. As we shall see presently, the q_p adjustment is subsequently carried out in the overall FEN by adjusting the loop gain.
21. A prime is used here for ρ_4 because, as will be discussed in Section 6.3.1, a second term ρ_4'' must be considered, such that the overall factor $\rho_4 = \rho_4' + \rho_4''$ results.

By straightforward analysis we find the lower limit for ρ'_4 with worst-case tolerances δ_j of the capacitors C_j ($j = 1, 2, 4$) and with ρ_2 given by (6-107b) as

$$\rho'_4 > \frac{C_4}{C_s} \frac{\left[\delta_1\left(\dfrac{C_2}{C_1 + C_2}\right) + \delta_2\left(\dfrac{C_1}{C_1 + C_2}\right) + \delta_4\right]\left(\dfrac{\omega_p}{\omega_z}\right)^2}{\left(\dfrac{\omega_p}{\omega_z}\right)^2\left(1 + \dfrac{C_4}{C_s}\right) - 1} + \rho_2 \frac{R_2}{R_1 + R_2} \qquad (6\text{-}127)$$

As with ρ_2 we see that the tolerances determining ρ'_4 are primarily δ_4 and, depending on the ratio C_1/C_2, δ_1 or δ_2. In a potentially symmetrical twin-T, we have $C_1 \gg C_2$. In this case the tolerances of C_2 and C_4 should be as small as possible. Assuming for example a ratio $C_1/C_2 = 10$, $\delta_1 = 5\%$, and $\delta_2 = \delta_4 = 2\%$, the expression in square brackets in the numerator of the first term in (6-127) is approximately 4.5%. In a symmetrical twin-T, δ_1 should be as small as δ_2. Remember that, since δ_3 appears unattenuated in (6-107b), C_3 should also have a small tolerance. As mentioned earlier, these considerations are primarily of importance when using chip capacitors, since the price of these capacitors is strongly dependent on their tolerances.

6.2.3 Non–Minimum–Phase Networks : The All–Pass Network

As we have seen in the corresponding discussion in Chapter 4 (in Section 4.4.2), we can restrict ourselves here to the realization of a second-order all-pass function,

$$T(s) = K \frac{s^2 - \dfrac{\omega_0}{q}s + \omega_0^2}{s^2 + \dfrac{\omega_0}{q}s + \omega_0^2} \qquad (6\text{-}128)$$

As we know from Section 3.1.2, the TT-SABB provides two possible configurations to realize this function: the twin-T all-pass (Fig. 6-21a) and the bridged-T all-pass (Fig. 6-21b). We shall describe a functional procedure for tuning each.

The Twin-T All-Pass Network We recall that the twin-T all-pass network is obtained by connecting terminal 5 to terminal 1 in the FRN of Fig. 6-16. Consequently the all-pass is first connected as an FRN, with terminal 5 not necessarily grounded but left disconnected, corresponding to $\beta = 1$. This simplifies the zero-tuning process which is carried out according to one of the methods discussed in the previous section. The FRN is thereby tuned for an

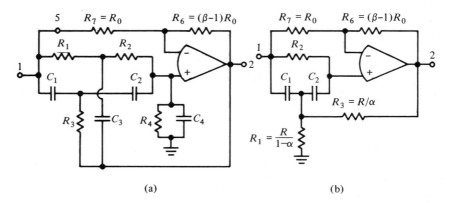

FIG. 6-21. Class 4 all-pass networks: (a) the twin-T all-pass; (b) the bridged-T all-pass.

infinite null[22] at ω_0. Subsequently, reconnected as an all-pass, the amplitude at ω_0 and at a low frequency (say, $\omega_0/100$) is compared. With β, the amplitude at ω_0 is then adjusted so that it equals the amplitude at $\omega_0/100$. Increasing β increases the amplitude at ω_0 and vice versa. Note that, as explained in Section 4.4.2, amplitude rather than phase is used here to tune q.

The Bridged-T All-pass Network   Since it is the more useful of the two all-pass networks (easier tuning, fewer components) we shall go into some-what more detail in discussing the functional tuning of the bridged-T all-pass network.

Since the poles and zeros of any single-amplifier all-pass network are closely related to each other, individual pole and zero tuning is generally impractical (although not impossible[23]). It is then desirable to follow a tuning strategy that simultaneously shifts both the pole and the zero pair into the specified positions in the s-plane. Such a strategy, consisting of four steps, is summarized in Table 6-12. In the first step the pole and zero pair are shifted onto the *same* circle, in the second they are both simultaneously shifted onto the *specified* circle, in the third they are both shifted along the circle till they are *symmetrical* with respect to the $j\omega$ axis, and in the fourth they are simul-taneously shifted along the circle until they reach the specified symmetrical positions. The desired sensitivity conditions and suitable tuning parameters are also given in the table; naturally it depends on the circuit whether the tuning conditions can be satisfied or not.

22. An infinite null is not really required here as long as the null frequency equals ω_0. However, in order to simplify the tuning instructions and to avoid ambiguity, we specify an infinite rather than a finite null, meaning that the null should be as deep as possible.
23. D. Hilberman, Input and ground as complements in active filters, *IEEE Trans. Circuit Theory*, **CT-20**, 540–547 (1973); also, G. E. Roberts, On tuning the group delay of an active *RC* all-pass resonator, *IEEE Trans. Circuit Theory*, **CT-20**, 172–173 (1973).

Let us now consider our bridged-T all-pass network and examine to what extent the four-step tuning procedure can be applied to it. Referring to Fig. 6-21b, in which, incidentally, the component designations indicate how the circuit may be derived from the general TT-SABB used in Fig. 6-21a, we have

$$\omega_{0z} = \omega_{0p} = \omega_0 = \frac{1}{\sqrt{RR_2\,C_1C_2}} \tag{6-129a}$$

$$q_{0z} = q = \frac{r_b c_b}{1 + c_b^2 + (1 - \beta)r_b^2} \tag{6-129b}$$

$$q_{0p} = q = \frac{r_b c_b}{1 + c_b^2 + (1 - \alpha\beta)r_b^2} \tag{6-129c}$$

where q_{0z} and q_{0p} are the zero and pole Q, respectively, and

$$r_b = \sqrt{R_2/R} \tag{6-130a}$$

$$c_b = \sqrt{C_1/C_2} \tag{6-130b}$$

and

$$\alpha = \frac{2}{\beta \dfrac{r_b}{c_b} \cdot \hat{q} - 1} \tag{6-130c}$$

Note that, as a resistor voltage divider, $0 < \alpha < 1$.

With the expressions given by (6-129) we must now examine to what extent we can apply the four-step tuning procedure of Table 6-12. Note that, with (6-129a), ω_{0z} is automatically equal to ω_{0p}, so that step 1 does not appear to be necessary. This conclusion is only true if parasitic effects such as operational amplifier phase shift do not interfere with the desired response. Correcting error sources of this kind are discussed in the following Section (6.3). There (6.3.3) we shall discuss means of improving the opamp frequency compensation so that the phase shift of high-gain low-frequency amplifiers is negligible, at least over the voice band. For higher frequencies, wide-band amplifiers must be used. Whether step 1 is necessary or not, the form of (6-129a) does not provide two separate components x_z and x_p permitting independent tuning of ω_{0z} and ω_{0p}. Hence if step 1 should be necessary the condition that $\omega_{0z} = \omega_{0p} = \omega_0$ can be satisfied approximately, using the type of tuning procedure outlined in Chapter 4, Section 4.4.4 (Networks Requiring Other Procedures).

To examine whether the remaining three steps in our tuning procedure can be realized, we must derive the sensitivities of the all-pass parameters to the network components. These are listed in Table 6-13. Since $\alpha < 1$ we note by inspection that the resistor R_2 (or conductance G_2) is the obvious component to be used as x_ω in step 2, the sensitivity $S_{G_2}^{\omega_0}$ being the largest of the three

TABLE 6-12. A FOUR-STEP TUNING STRATEGY FOR ALL-PASS NETWORKS

	Shift in the s-plane	Parameter measured	Amplitude response	Required sensitivity conditions
Step 1 $\omega_{0z} = \omega_{0p}$ $= \omega'_0$		Amplitude		Availability of two components x_z, x_p such that $S^{\omega_{0z}}_{x_z}$, $S^{\omega_{0p}}_{x_p}$ exist independently
Step 2 $\omega'_0 = \omega_0$		Phase, for $-180°$ at ω_0		Availability of component x_ω such that $S^{\omega_{0p}}_{x_\omega} = S^{\omega_{0z}}_{x_\omega} = S^{\omega_0}_{x_\omega}$

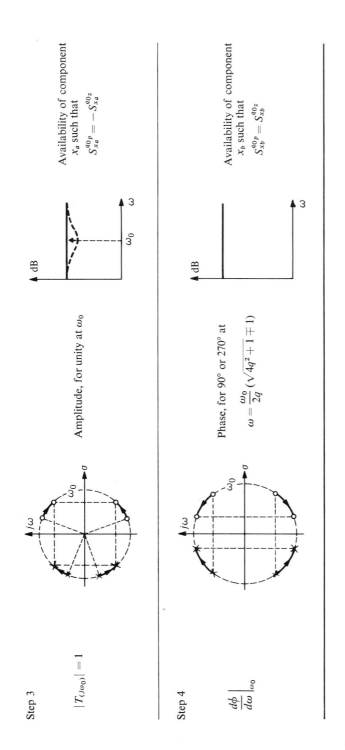

Step 3

$$|T_{(j\omega_0)}| = 1$$

Amplitude, for unity at ω_0

Availability of component
x_a such that
$$S_{x_a}^{q_0 p} = - S_{x_a}^{q_0 z}$$

Step 4

$$\left.\frac{d\phi}{d\omega}\right|_{\omega_0}$$

Phase, for 90° or 270° at
$$\omega = \frac{\omega_0}{2q}(\sqrt{4q^2 + 1} \mp 1)$$

Availability of component
x_b such that
$$S_{x_b}^{q_0 p} = S_{x_b}^{q_0 z}$$

535

TABLE 6-13. SENSITIVITY EXPRESSIONS FOR BRIDGED-T ALL-PASS NETWORK

S_{G_i}	G_1	G_2	G_3	G_6	G_7
$S_{G_i}^F$	$\frac{1}{2}(1-\alpha)$ $=\dfrac{r_b c_b}{r_b c_b + q(1+c_b^2+r_b^2)}$	$\frac{1}{2}$	$\frac{1}{2}\alpha =$	0	0
$S_{G_i}^{\omega_0 z}$	$-\dfrac{r_b c_b + 2q(1+c_b^2)}{r_b c_b + q(1+c_b^2+r_b^2)}$	$\frac{1}{2}+\dfrac{q(1+c_b^2)}{r_b c_b}$	$-\frac{1}{2}-\left[\dfrac{q(1+c_b^2)}{r_b c_b}\right]$ $-\dfrac{r_b c_b + 2q(1+c_b^2)}{r_b c_b + q(1+c_b^2+r_b^2)}$	$1+\dfrac{q(1+c_b^2)}{r_b c_b}$	$-\left[1+\dfrac{q(1+c_b^2)}{r_b c_b}\right]$
$S_{G_i}^{\omega_0 p}$	$\dfrac{r_b c_b - 2qr_b^2}{r_b c_b + q(1+c_b^2+r_b^2)}$	$\frac{1}{2}-\dfrac{q(1+c_b^2)}{r_b c_b}$	$-\frac{1}{2}+\dfrac{q(1+c_b^2)}{r_b c_b}$ $-\dfrac{r_b c_b - 2qr_b^2}{r_b c_b + q(1+c_b^2+r_b^2)}$	$1-\dfrac{q(1+c_b^2)}{r_b c_b}$ $-\dfrac{2qr_b^2}{r_b c_b + q(1+c_b^2+r_b^2)}$	$-\left[1-\dfrac{q(1+c_b^2)}{r_b c_b}\right.$ $\left.-\dfrac{2qr_b^2}{r_b c_b + q(1+c_b^2+r_b^2)}\right]$

sensitivities available. Hence this tuning step consists of adjusting the resistor R_2 (i.e., *increasing* it in the case of a hybrid integrated circuit) until the output phase, compared to that of the input, is $180°$ at the specified frequency ω_0.

Continuing with step 3 we find that we can use the resistor R_6 as the component x_a so that $S_{x_a}^{qop} = -S_{x_a}^{qoz}$. This follows immediately if we examine the condition necessary for step 4. There we find that R_1 (or G_1) can be used as x_b and that in order for $S_{G_1}^{qop} = S_{G_1}^{qoz}$ we must let

$$2qr_b^2 = r_b c_b + q(1 + c_b^2 + r_b^2) \tag{6-131a}$$

At the same time, to ensure that $S_{G_1}^{\omega_0}$ remains small (compared to 0.5) we require with (6-131a) that

$$\frac{r_b c_b}{r_b c_b + q(1 + c_b^2 + r_b^2)} = \left(1 + \sqrt{1 + 4q^2 \frac{1 + c_b^2}{c_b^2}}\right)^{-1} \ll 1 \tag{6-131b}$$

or, in other words, that

$$c_b^2 \ll 1 \tag{6-131c}$$

Returning to step 3, it follows with (6-131a) that $S_{G6}^{qoz} = 1 + q\,(1 + c_b^2)/r_b c_b$ and that $S_{G6}^{qop} = -q(1 + c_b^2)/r_b c_b$, or, with $q \gg 1$, that $S_{G6}^{qoz} \approx -S_{G6}^{qop}$, which is the tuning condition required. In summary, then, the tuning procedure for the bridged-T all-pass network is:

Step 1: —
Step 2: Adjust R_2 for $-180°$ phase at ω_0
Step 3: Adjust R_6 for unity gain at ω_0
Step 4: Adjust R_1 for $90°$ or $270°$ phase at ω_{90} or ω_{270} respectively, where

$$\omega_{90,270} = \frac{\omega_0}{2q} \left[\sqrt{4q^2 + 1} \mp 1\right]$$

Which of the two frequencies is used in step 4 depends on which of the two is supposed to provide a higher phase accuracy. This, in turn, will depend on which of the two is located in a more critical part of the pass band.

To provide an example of this procedure consider a circuit for which $q = 2$. To satisfy (6-131c) we select $c_b^2 = 0.1$ and, solving (6-131a) for r_b, obtain $r_b^2 = 1.28$. The resulting sensitivities are shown in Table 6-14a. By contrast let us derive the sensitivities for a more commonly used circuit, namely one with

$$r_b = 2$$

and

$$c_b = 1$$

TABLE 6-14. SENSITIVITIES FOR TWO BRIDGED-T ALL–PASS NETWORKS WITH $q=2$

$S_{G_i}^F$			G_i		
	G_1	G_2	G_3	G_6	G_7
		a. $c_b^2 = 0.1$; $r_b^2 = 1.28$			
ω_0	0.0698	0.5	0.4302	0	0
q_{0z}	−0.93	6.66	−5.63	7.16	−7.16
q_{0p}	−0.93	−5.66	−6.59	−6.16	6.16
		b. $c_b^2 = 1$; $r_b^2 = 4$			
ω_0	0.143	0.5	0.357	0	0
q_{0z}	−0.714	2.5	−1.786	3	−3
q_{0p}	−1	−1.5	2.5	−2.14	2.14

where again we specify that $q = 2$. Then

$$\beta|_{r_b = 2c_b = 2} = \frac{q(2 + r_b^2) + r_b}{qr_b^2} \tag{6-132a}$$

and

$$\alpha\Big|_{r_b = 2c_b = 2} = \frac{q(2 + r_b^2) - r_b}{q(2 + r_b^2) + r_b} \tag{6-132b}$$

The corresponding sensitivities are listed in Table 6-14b. Note that in the latter case, which is inferior from a tuning-strategy point of view, but more convenient in terms of component spread, the three-step procedure outlined above is still quite workable.

6.3 CORRECTING ERROR SOURCES IN THE TT–SABB

The most serious source of error in the TT–SABB is the finite gain–bandwidth product ω_g of the operational amplifier. The effects of finite ω_g on ω_p and q_p of a positive-feedback network were discussed in general terms in Section 4.2.3 of Chapter 4. We found that the errors $\Delta\omega_p/\omega_p$ and $\Delta q_p/q_p$ depend approximately on the ratio ω_p/ω_g as follows:

$$-\frac{\Delta\omega_p}{\omega_p} = \frac{\Delta q_p}{q_p} \approx \frac{\beta}{2\hat{q}}\frac{\omega_p}{\omega_g} = \frac{1}{2}\beta\kappa\frac{\omega_p}{\omega_g} \tag{6-133}$$

That is, the pole frequency is decreased and the pole Q increased approximately by the same amount. Note that these errors are proportional to $\beta\kappa$ and therefore to the gain-sensitivity product Γ. (Remember that $\beta\kappa \approx \Gamma/q_p$).

Thus minimizing Γ as outlined in Section 6.1 will also minimize these errors. We shall now examine the consequences of these errors—and methods of eliminating them—for the TT-SABB.

6.3.1 The Effect of the Finite Gain–Bandwidth Product on Frequency Tuning

We have assumed throughout the previous section that the pole frequency ω_p is independent of β. In terms of the sensitivity matrix (e.g., (6-66)) this has meant that $k_{12} = S_\beta^{\omega_p} = 0$. Whereas ideally this is true, (6-133) demonstrates that, to first order, ω_p is proportional to β, or that

$$k_{12} = S_\beta^{\omega_p} = -\frac{1}{2}\beta\kappa \cdot \frac{\omega_p}{\omega_g} \qquad (6\text{-}134)$$

Of course, the larger ω_g is compared to ω_p, the smaller this sensitivity will be. Nevertheless, with realistic values for the ratio ω_p/ω_g and with $\beta\kappa/2$ anywhere between unity and five or more, the already-tuned pole frequency ω_p may vary slightly when β is adjusted to tune q_p instead of, as previously supposed, remaining unaffected by this adjustment. Since we may, typically, require a frequency tuning accuracy of 0.1% to 0.2%, the slight shift in ω_p due to the q_p adjustment may be unacceptably large. Now, if β happens to be decreased while tuning q_p, ω_p will increase and can readily be corrected for by increasing the appropriate resistor R_{ω_p}. However, if β is increased in the q_p tuning step, R_{ω_p} must be decreased to correct for the error in ω_p which, as we know, cannot be done with a film resistor. One method of coping with this problem is to ensure that the β of the untuned circuit is always high, so that in the q_p tuning step ω_p will invariably be increased, allowing for a subsequent correction in R_{ω_p}. However, higher β results in higher q_p and the risk of initial instability. A better solution is to first adjust R_{ω_p} to a frequency slightly higher, say by one percent, than ω_p. The q_p adjustment will then shift this frequency slightly (either high or low but not by as much as one percent), allowing, in all cases, for a subsequent increase in R_{ω_p} in order to shift the pole frequency down to ω_p. A fine adjustment of q_p may be necessary after this (k_{21} is generally much larger than k_{12}) but, as a rule, this will be small enough to leave ω_p practically unaffected.

Quite apart from the small change in ω_p caused by what will generally be a very slight adjustment of β during the q_p tuning process, ω_p will already be low before the tuning process begins, by the amount given by (6-133). To permit correction of the ω_p due to this initial error caused by the limited amplifier bandwidth, R_{ω_p} must at first be still lower than indicated for example by ρ_0 in (6-107b) or by ρ_4' in (6-127). In principle this is no problem. The sensitivity of ω_p to R_{ω_p} is typically on the order of 0.5, so that R_{ω_p} must simply be low by about twice the maximum anticipated error $\Delta\omega_p/\omega_p$.

The problem is somewhat aggravated in the case of the FRN because k_1, the sensitivity of ω_p to R_4 (which is used as R_{ω_p} in the FRN) is small (see (6-123), and remember that r is selected as small as possible). Consequently, the error in ω_p may require a relatively large correction in R_4. Take, for example, the following typical numerical example: $f_p = 3$ kHz, $f_g = 1$ MHz, $\beta = 2, \hat{q} = 0.3$. From (6-133) it follows that ω_p will be low by approximately 1%. To correct for this in an FRN, R_4 must be decreased (or rather should by initially low) by an amount at least equal to ρ_4'', where $R_4(1 - \rho_4'')$. Assume that $r = R_s/R_4 = 0.25$. Then, from (6-123)

$$\rho_4'' > \frac{2(1 + r)}{r} \frac{\Delta\omega_p}{\omega_p} \qquad (6\text{-}135)$$

With our numerical example this is in the order of 10%. Since ρ_4'' grows rapidly with decreasing r, it must be taken into consideration when specifying the initial untrimmed value of R_4 (i.e., when it is deposited in film). Taking the overall amount by which R_4 must be initially low we have

$$\rho_4 = \frac{\Delta R_4}{R_4} > \rho_4' + \rho_4'' \qquad (6\text{-}136)$$

where ρ_4' and ρ_4'' are given by (6-127) and (6-135), respectively. Remember that the upper limit on ρ_4 will depend on the trimming range permitted by the resistor-film technology used. It may be that ρ_4 as given by (6-136) exceeds this limit, in which case capacitors with tighter tolerances must be used to reduce ρ_4' and some means may be necessary for reducing ρ_4''. The latter will be discussed later.

6.3.2 The Effect of Finite Output Impedance on TT–SABB Stability

Up to this time we have assumed that the output impedance of the amplifier in the TT-SABB is equal to zero. Let us now consider what effect a *finite* output impedance may have on the general circuit. For this purpose we consider the return difference of the general class 4 signal-flow graph corresponding to the block diagram in Fig. 6-1. It is shown in Fig. 6-22. The feedback function $t_{32}(s)$ has a bandpass characteristic and is given by (6-8). Thus, the return difference with respect to the gain β is given by

$$F_\beta(s) = 1 - \beta t_{32}(s) = 1 - \beta k_{32} \frac{\omega_p s}{s^2 + \dfrac{\omega_p}{\hat{q}} s + \omega_p^2} \qquad (6\text{-}137)$$

The stability of the network is most critical at the pole frequency ω_p where the *return ratio* $\beta t_{32}(j\omega_p)$ has maximum value (see (6-9)):

$$\beta t_{32}(j\omega_p) = \beta \bar{t}_{32} = \beta k_{32}\hat{q} \qquad (6\text{-}138)$$

Notice that $\bar{t}_{32}$ occurs at the frequency for which $t_{32}(j\omega)$ is real.

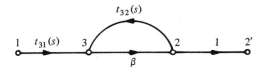

FIG. 6-22. Signal-flow graph of a general class 4 network.

Expression (6-138) applies to an ideal amplifier with frequency-independent gain β. If β is nonideal, the frequency of maximum return ratio will shift to the pole frequency $\omega_p + \Delta\omega_p$, where $\Delta\omega_p$ is given by (6-133). If this frequency dependence (or any other parasitic effect) increases the return ratio beyond its nominal value, the pole Q, q_p, will increase accordingly. In the limit, as the return ratio approaches and then equals unity, the network becomes unstable. This may happen if β is increased to β_{osc} as given by (6-10) but also if $\bar{t}_{32} = k_{32}\hat{q}$ is increased. Now if $\beta(s)$ should be higher than nominal due to parasitic effects, this will become apparent during the tuning process and can readily be corrected for. However, the situation is different with respect to $\bar{t}_{32}$. As we shall see in the following, the output impedance of a network may cause unexpected increases in this term.

Consider the general class 4 circuit redrawn as an active ladder network in Fig. 6-23a. As shown, the circuit is fed from an ideal voltage source, an assumption that we have tacitly made for all our networks so far. The basis for this assumption is that when these networks are in cascade one is connected to another and each has a negligibly small output impedance. In terms of Fig. 6-23a the feedback function is

$$t_{32}(s) = \frac{Z_1 Z_4}{Z_1(Z_2 + Z_3 + Z_4) + Z_2(Z_3 + Z_4)} \tag{6-139}$$

The magnitude of $t_{32}(j\omega)$ will increase with Z_1 from zero to the maximum value $Z_4/(Z_2 + Z_3 + Z_4)$. The same applies to the maximum value $\bar{t}_{32}$, which is real. Thus, for any nominal value of $\bar{t}_{32}$ corresponding to a nominal value of Z_1, an increase in Z_1 will increase $\bar{t}_{32}$ and deteriorate the stability of the network. This becomes immediately clear if we take the product $\bar{t}_{32} = k_{32}\hat{q}$ for the networks shown in Table 6-4. For the bandpass circuit type B, for example, we obtain

$$\bar{t}_{32}|_{\text{Bandpass B}} = k_{32}\hat{q} = \frac{R_1}{R_1 + R_2} \cdot \frac{R_2 C_2}{R_p(C_1 + C_2) + R_2 C_2} \tag{6-140}$$

Here $\bar{t}_{32}$ increases with R_1 from zero to its maximum $R_2 C_2/[\bar{R}_1(C_1 + C_2) + R_2 C_2]$. Similar expressions are obtained for the other circuits.

Since Z_1 is the network branch attached to the input terminal of the network, any source impedance R_s of the input voltage source V_1 will have a deteriorating effect on the stability of the network (see Fig. 6-23b). Increasing

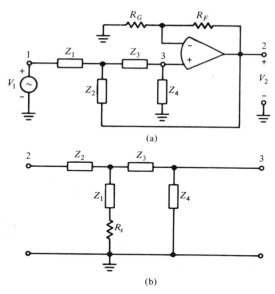

FIG. 6-23. General class-4 network: (a) as active ladder network; (b) the passive feedback network.

R_s until $i_{32} = 1/\beta$ will cause the network to oscillate. *The class 4 network is inherently open-circuit unstable.*[24]

Any increase of R_s above its anticipated nominal value will increase q_p above nominal and deteriorate the stability of the network. At the same time ω_p, which is determined by the components of the ladder network, will decrease. If R_s is known and well controlled it can be accounted for in the design of the network and during the tuning process without any deleterious effects. The R_s value for which the network was designed must then be specified so that it can be incorporated both into the tuning setup and into the final system. However, we shall see below that the output impedance of a class 4 network may take on relatively large, and what is worse, frequency-dependent, values. If such a network is cascaded with a second class 4 network, the second network may actually go into oscillation (if it is high-Q to start with) unless special measures, to be discussed later, are undertaken. First, however, let us consider other effects on a class 4 network, caused by an increase in its output impedance.

The preceding discussion concerned the effect of the output impedance of a signal source on a class 4 network. It was understood that this source impedance will most often be the output impedance of another class 4 network, since networks of this kind are generally connected in cascade. We shall now

24. This does not mean that *all* class 4 networks will necessarily oscillate when the input terminal is open, since i_{32} will not thereby necessarily increase to the value $1/\beta$.

examine the effect of the output impedance of a class 4 network *on itself*, i.e., on its own operation. In so doing, we shall resign on any analysis, since our qualitative examination is straightforward and illustrates the simple course to be taken, should a quantitative analysis be required.

In principle a finite output impedance of the amplifier in the circuit of Fig. 6-23a has two consequences.

1. It will increase the effective value of R_F and with it the closed-loop gain $\beta = 1 + R_F/R_G$; as a result q_p is increased. However, any increase in q_p, whether due to a constant or even a frequency-dependent increase in R_F, is of no consequence, since it can be corrected for in the tuning process. Thus, if deterministic tuning is to be used, either the finite output impedance must be included in the tuning equations or a functional correction of q_p will be required. As we have seen, the latter is to be recommended anyway, because of the high sensitivity of q_p to variations in loop gain.

2. It will increase the impedance of Z_2 and thereby *decrease* the factor $\bar{i}_{32}$ and with it q_p. In this respect it counteracts the effect of an increased source impedance at the input to the network. Like the latter, however, it will decrease the pole frequency ω_p. In actual fact, however, neither effect will take place if taken into account during the tuning process.

We see, then, that no matter what the effect is of an increase in the output impedance, it can generally be compensated for in the tuning process. In this respect, this case is simpler to cope with than the increase in the external source impedance. In the latter case, the increase depends on a separate circuit whose output impedance may not be known during the tuning of a given circuit. The case discussed here is particularly simple if Z_2 is a resistor. However, one case in which the output impedance may be troublesome is in the FRN (Fig. 6-16). A finite output impedance has the effect of adding a resistance in series with R_3 and C_3. This results in a degradation of the twin-T null depth. The only way of avoiding this, if the degradation is unacceptable, is to reduce the output impedance by one of the methods described later.

This brings us to the question of how large the output impedance of a class 4 network may be expected to be in the first place. We are so familiar with the very low output impedances of operational amplifiers that it may at first seem unnecessary to discuss their effects here in any detail. In the following we shall see that this, unfortunately, is not the case.

From (4-126) we find that the output impedance of an operational amplifier (inverting or noninverting) may be approximated by

$$Z_0(s) = \frac{R_0}{AB(s)} = \frac{\beta}{\omega_g}(s + \Omega)R_0$$

$$= \frac{\beta}{\omega_g}\left(s + \frac{\omega_g}{A_0}\right)R_0 \qquad (6\text{-}141)$$

where R_0 is the output resistance of the operational amplifier in the open-loop mode, $AB(s)$ is the loop gain (see Fig. 4-14), A_0 the DC open-loop gain, ω_g the gain–bandwidth product, Ω the 3 dB frequency of the open-loop gain, and β the DC closed-loop gain. The magnitude of $Z_0(j\omega)$ then follows as

$$|Z_0(j\omega)| = \beta \frac{\omega}{\omega_g} \sqrt{1 + \left(\frac{\Omega}{\omega}\right)^2} \, R_0 \qquad (6\text{-}142)$$

For $\omega \gg \Omega$, which is very often valid, we have

$$|Z_0(j\omega)| \approx \beta \frac{\omega}{\omega_g} R_0 \qquad (6\text{-}143)$$

For a high-performance, voiceband operational amplifier (e.g., the Fairchild μF 741), typical values are $R_0 = 75 \, \Omega$ and $f_g = 1$ MHz. With $\beta = 1.5$ and at a frequency of, say, 3 kHz, we have $Z_0 \approx 0.3 \, \Omega$. Certainly, by itself, this is negligible either as the source impedance or as the output impedance of a network stage. However, (6-142) and (6-143) take into account only the negative-feedback loop (e.g., R_F and R_G in Fig. 6-23a) of the class 4 circuit type from which the TT-SABB is derived. We must now examine what additional effect the positive-feedback loop of the TT-SABB has on the output impedance. This can readily be done if we recall the Blackman impedance relation for feedback networks.[25] According to this relation the impedance $Z(s)$ shunting the terminals of a feedback loop may be written as

$$Z(s) = \frac{Z_0}{F_x(s)} \qquad (6\text{-}144)$$

where Z_0 is the impedance without feedback and $F_x(s)$ is the return difference of the feedback loop with respect to the component x. In our case, x is the closed-loop gain β of our amplifier and $F_\beta(s)$ is given by (6-137). Thus

$$Z(s) = \frac{Z_0}{F_\beta(s)} = Z_0 \frac{s^2 + \dfrac{\omega_p}{\hat{q}} s + \omega_p^2}{s^2 + \dfrac{\omega_p}{q_p} s + \omega_p^2} \qquad (6\text{-}145)$$

where

$$q_p = \frac{\hat{q}}{1 - \beta k_{32}\hat{q}} \qquad (6\text{-}146)$$

as already encountered in earlier chapters. Assuming that Z_0 is real, $Z(s)$ will have an FEN-type characteristic as shown qualitatively in Fig. 6-24. The output impedance will have a maximum value at the pole frequency ω_p:

$$Z_{max} = Z(j\omega_p) = Z_0 \cdot \frac{q_p}{\hat{q}} \qquad (6\text{-}147)$$

25. See Chapter 2, Section 2.2.6 of *Linear Integrated Networks: Fundamentals*.

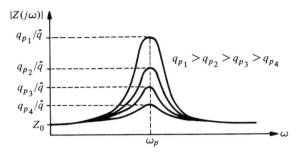

FIG. 6-24. Frequency response of $Z(s)$ for various ratios $q_{p_i}/\hat{q}$.

Z_0 is not a constant impedance but itself given by (6-143); thus we obtain

$$Z_{max} = Z(j\omega_p) \approx \beta \left(\frac{\omega_p}{\omega_g}\right) \left(\frac{q_p}{\hat{q}}\right) R_0 \qquad (6\text{-}148)$$

or, since $\kappa = 1/\hat{q}$,

$$Z_{max} = (\beta\kappa)(\omega_p q_p) \left(\frac{R_0}{\omega_g}\right) \qquad [6\text{-}149]$$

The first term in this expression is the $\beta\kappa$ product; it is determined by the circuit used and the design values selected. For optimum Q stability the $\beta\kappa$ product is minimized. The second-term, the ωQ product, is determined by the network, or system, specifications. Finally, the third term, R_0/ω_g, is determined exclusively by the characteristics of the operational amplifier used, and assumes that the amplifier is frequency compensated to have a -6 dB/octave roll-off. In terms of the gain–sensitivity product (6-149) can be written as

$$Z_{max} = \Gamma \frac{\omega_p}{\omega_g} R_0 \qquad [6\text{-}150]$$

Assuming, as above, the typical values $R_0 = 75\ \Omega$ and $f_g = 1$ MHz, as well as $\beta\kappa = 4$, $q_p = 20$, and $f_p = 3$ kHz, we obtain, from (6-149), $Z_{max} \approx 20\ \Omega$. As the source impedance of a stage that has been tuned for a source impedance of close to zero ohm, this may well be large enough, if not to cause stability problems, at least to increase the q_p of that stage by an unacceptable amount.[26] One way of avoiding this problem is to set an upper limit on q_p according to an acceptable value of Z_{max}:

$$(q_p)_{max} < \frac{Z_{max}}{\beta\kappa R_0} \cdot \frac{\omega_g}{\omega_p} \qquad (6\text{-}151)$$

26. In general the individual second-order circuits will be cascaded *after* they have been tuned *and encapsulated*, so that a subsequent correction of the second and following stages will not be feasible.

Of course, if the input impedance of the network (see Z_1 in Fig. 6-23a) is sufficiently high, then the tolerable value for Z_{max} is accordingly high and the upper bound on q_p will be determined by other factors. Thus, for example, in voiceband circuits, Z_1 may typically be on the order of 20 kΩ, so that a Z_{max} value of even 20 Ω will only account for 0.1 % of Z_1; this will very likely be negligible. If, however, the input impedance to the network Z_1 is low and Z_{max} is, say, 5 Ω, then the resulting limit on q_p will be unreasonably low, i.e., well below the q_p capabilities normally associated with circuits of this kind.

Another way of coping with the problem presented above is either to prevent two high-Q networks from following each other in a cascade or, if this is not possible, to isolate them from one another by using buffer amplifiers (e.g., voltage followers or emitter followers), as shown in Fig. 6-25a. Naturally, the output impedance of stage 1 will be most harmful to stage 2 if $\omega_{p_1} = \omega_{p_2}$.[27] If the two frequencies are sufficiently far apart the buffer amplifier may not be required. Likewise, even if ω_{p_1} is close to ω_{p_2}, the buffer will be unnecessary if either q_{p_1} or q_{p_2} is low. Thus the need for a buffer amplifier must be examined individually and will generally occur only when two stages, both with high q_p values, are in cacade. The same principle of reducing the effective output impedance with a buffer stage can be used *within* a circuit (Fig. 6-25b). If at all, this may be necessary with an FRN as mentioned above, and then only if a high q_p is required simultaneously with a deep null depth.

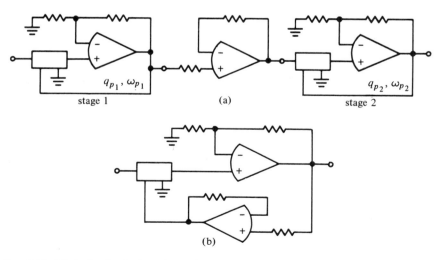

FIG. 6-25. Methods of reducing effects of output impedance $Z(s)$: (a) buffer in filter cascade; (b) buffer within second-order section.

27. The reason for this should be clear. The output impedance of the *driving* section will have its maximum value Z_{max} (as given by (6.149)) at its pole frequency, say ω_{p1}. The *driven* section will be most prone to instability when $t_{32}(j\omega_p)$ reaches its maximum value $\hat{t}_{32}$; this occurs at its pole frequency ω_{p2} (see (6.138)). Thus for maximum stability, ω_{p1} should be as far apart from ω_{p2} as possible.

6.3.3 Improving the Operational Amplifier Frequency Compensation

The inclusion of buffer amplifiers in a filter network will be acceptable only if no other method of reducing the output impedance of the amplifiers is available. Quite apart from increasing the cost of the filters directly, the buffer amplifiers also dissipate additional power, generate additional heat, and require additional space. A simple and often surprisingly effective method of reducing the output impedance of a filter building block is to better utilize the gain available from its amplifiers by a judicious method of frequency compensation. In (6-141) a 6 dB/octave roll-off of the amplifier is assumed. Hence at any frequency ω, the value of the output resistance R_0 is divided by the loop gain $AB(\omega)$, as shown by the solid lines in Fig. 6-26. As the frequency of operation is extended beyond Ω, $AB(\omega)$ decreases and the output resistance increases accordingly.

As we know, in applications requiring low voltage gain (e.g., from 0 to 6 dB gain) it is sufficient for the rate of closure between open- and closed-loop gain of an operational amplifier to be 6 dB/octave. At frequencies somewhat below the corresponding intersection (see ω_α in Fig. 6-26) the roll-off may be -12 dB/oct, whereby the loop gain is increased substantially while the circuit remains stable. One method of achieving this is to connect an RC T-network (hence "T compensation") in the feedback path of that particular transitor stage in an amplifier that provides most of the open-loop gain.[28] Referring to the schematic representation of an operational amplifier in Fig. 6-27, most of the open-loop gain is generally concentrated in a common-emitter transistor

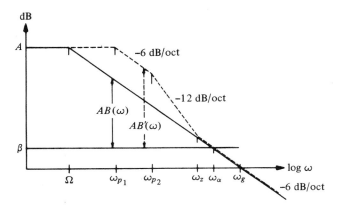

Fig. 6-26. Frequency response of compensated and uncompensated operational amplifiers.

28. See Chapter 7, Section 7.2.5 (Figs. 7-30 and 7-33) of *Linear Integrated Networks: Fundamentals*.

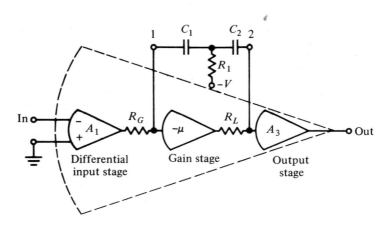

FIG. 6-27. Schematic representation of operational amplifier for frequency compensation.

stage located in between a differential input stage (A_1) and a driver output stage (A_3). The amplifier is normally frequency compensated by an external capacitor C connected between the terminals 1 and 2. With the gain μ of this stage the Miller capacitance μC is then effective in causing the -6 dB/octave roll-off while requiring only a small capacitance C (typically 30 pF). With the external frequency-compensating T network consisting of R_1, C_1, and C_2 and with the internal resistance R_G and R_L (the base and load resistance, respectively, of the transistor gain stage) the open-loop transfer function of the amplifier can be approximated by assuming that the gain stages A_1, μ, and A_3 have very high input and very low output impedances. We then obtain

$$T(s) = \frac{V_{\text{out}}}{V_{\text{in}}} = -A_0 \cdot \frac{s + \omega_z}{(s + \omega_{p_1})(s + \omega_{p_2})} \tag{6-152}$$

where

$$A_0 \approx \frac{A_1 A_3}{R_G C_s} \tag{6-153a}$$

$$\omega_z \approx \frac{1}{R_p C_p} \tag{6-153b}$$

$$\omega_{p_1} \approx \frac{1}{\mu \dfrac{R_G}{R_L} R_p C_1} \tag{6-153c}$$

$$\omega_{p2} \approx \frac{1}{R_L C_2} \tag{6-153d}$$

and

$$R_p = \frac{R_1 R_L}{R_1 + R_L}; \qquad C_s = \frac{C_1 C_2}{C_1 + C_2}; \qquad C_p = C_1 + C_2 \tag{6-154}$$

The corresponding open-loop characteristic is indicated by the broken lines in Fig. 6-26. R_G and R_L can be measured by connecting a known capacitor alternately from ground to terminal 1, and then to terminal 2, and measuring the corresponding 3 dB cutoff frequency. Similarly the gain μ can be measured by feeding a signal from terminal 1 to terminal 2 while stabilizing the amplifier with a capacitor from the output to the input terminal. Typically, in a voiceband amplifier (e.g., Fairchild μA 748), R_G and R_L may be on the order of 100 kΩ. With a closed-loop gain of 10 dB, typical component values for the T compensating network will be $C_1 = 27$ pF, $C_2 = 18$ pF, and $R_1 = 80$ kΩ. For lower closed-loop gains C_1 and C_2 may both be increased, say to 30 pF, while R_1 remains at 80 kΩ.[29] Using a T-compensated amplifier in the TT-SABB, the additional loop gain obtained ($AB'(\omega)$ in Fig. 6-26) can reduce the output impedance sufficiently to eliminate any stability problems that may otherwise arise when cascading high-Q sections. With the Fairchild μA 748 amplifier, for example, an additional 40 to 70 dB of loop gain can be obtained over several kilohertz by replacing the commonly used 30pF feedback capacitor by the RC T-network. Needless to say, the increased loop gain also increases the input impedance and decreases the parasitic phase shift of the amplifier considerably. Notice that the zero frequency ω_z in Fig. 6-26 may be controlled independently by R_1. By contrast, C_1 and C_2 affect one of the poles and the zero simultaneously. Furthermore, R_1 is connected to the negative supply voltage; this serves to suppress power-supply noise at the output terminal of the amplifier.

6.4 DESIGNING FEN'S WITH THE TT–SABB

We have pointed out repeatedly that the maximum pole Q attainable with single-amplifier circuits is limited. The upper limit depends on the amplifiers available and on the pole frequencies specified; presently it is on the order of

29. For an all-integrated circuit the values of C_1 and C_2 may be too low to be realizable. In such cases a "double-lead compensation" as discussed in Chapter 7, Section 7.2.5 is recommended (first RC combination typically 350 Ω, 150 pF; second, 2 kΩ, 480 pF).

25 to 30 at frequencies below 100 kHz. As we know, we can use the TT-SABB, optimized for low-Q operation, to provide high pole Q's by incorporating it in the multiple-amplifier FEN configuration. Experience has shown, however, that in many communications systems second-order sections with pole Q's higher than 20 are only rarely required. In such cases it is clearly not worth going to the expense of fabricating a hybrid integrated building block incorporating the multiamplifier FEN or any other multiamplifier configuration. Instead, the TT-SABB, operating as an FRN (with $\omega_p = \omega_z$) can be used as the active twin-T in the feedback loop of the FEN,[30] as shown in Fig. 6-28a. For the active input network to the FEN, an additional hybrid integrated TT-SABB is used, as shown in Fig. 6-28b. The additional forward amplifier μ and the two resistors R_F and R_Q may be mounted either separately or on a common substrate serving as a carrier for the two TT-SABBs. It is important to remember that the active twin-T in the feedback loop of the FEN deter-

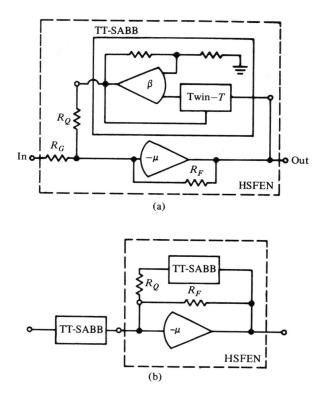

FIG. 6-28. High-selectivity FEN (HSFEN) using TT-SABB in feedback loop (a), and as additional input network (b).

30. Naturally this applies to either the MSFEN or the HSFEN.

mines the frequency characteristics of the latter and in particular its frequency stability. Since the active twin-T in the configuration of Fig. 6-28a is integrated, i.e., its stability is determined by (thick- or thin-film) passive components, no sacrifice in frequency stability results. Furthermore, this approach to manufacture is particularly economical, since it permits the use of both high- and low-Q filter networks in quantity, while only requiring the production (for example in hybrid integrated form) of TT-SABBs.

In what follows we shall examine some of the practical aspects of the FEN approach to high-Q filter design. Of these, one very important consideration is the attainable Q stability since it is primarily to improve this that we utilize the FEN, rather than simply a TT-SABB, for high-Q applications.

That we consider only the Q stability and not also the attainable frequency stability should be clear, since for all practical purposes the frequency stability depends only on the quality of the passive components. In other words, the frequency stability is determined by the temperature coefficients, tracking, and aging characteristics of the passive components, regardless of the actual circuit configuration used. This topic has already been discussed (in Section 4.1.4 of Chapter 4). This leaves the question of the attainable Q stability, which, although also discussed in general terms in Chapter 4, requires some additional treatment both for the TT-SABB as used alone in low-Q applications and when incorporated (as an FRN) in an FEN.

6.4.1 The Q Stability of the TT–SABB

We found earlier than when the TT-SABB, realized in hybrid-integrated form, is used to provide all-pole functions, it is sufficient to consider only the effects of gain drift on the pole Q. When we use the TT-SABB as an FRN we must go a little further; we must then include the effects of variations in the passive components insofar as they involve the null depth of the twin-T. As we shall see later, this becomes particularly important when the FRN is used within the multiple-amplifier FEN.

From Table 6-5 we have[31]

$$q_p' = \frac{\alpha\sqrt{(1 + r)(1 + c)(1 - \lambda)(1 - \eta)}}{[\alpha^2(1 - \lambda) + (1 - \eta)][1 - \beta(1 \pm v)] + [r + c\alpha^2](1 - \lambda)(1 - \eta)}$$

$$= \frac{q_p}{1 \pm v\dfrac{\beta}{\sqrt{(1 + r)(1 + c)}\,\hat{q}_T}q_p} \tag{6-155}$$

31. Throughout the following discussion, the upper sign on v pertains to LHP twin-T zeros, the lower sign to RHP zeros. Similarly the upper sign on Δv indicates drift of the twin-T zero into the LHP, the lower sign drift into the RHP.

where $q_p = q_p'(v = 0)$ and the other coefficients are given in Table 6-5. With variations in β and the null depth v of $\Delta\beta$ and Δv respectively, we obtain

$$\frac{\Delta q_p'}{q_p'} = S_\beta^{q_{p'}} \frac{\Delta\beta}{\beta} + S_v^{q_{p'}} \frac{\Delta v}{v} \tag{6-156}$$

With (6-155) this becomes

$$\frac{\Delta q_p'}{q_p'} = S_\beta^{q_{p'}}\left[\frac{\Delta\beta}{\beta} \mp \frac{\Delta v}{1 \pm v}\right]$$

$$= \left(\frac{q_p'}{\hat{q}} - 1\right)\left(\frac{\Delta\beta}{\beta} \mp \frac{\Delta v}{1 \pm v}\right) \tag{6-157}$$

Letting $\Delta\beta/\beta = \beta \cdot \Delta A/A^2$ and solving for β from the expression 4 given in Table 6-5 we obtain

$$\frac{\Delta q_p'}{q_p'} = \frac{\left(\dfrac{q_p'}{\hat{q}} + \dfrac{\hat{q}}{q_p'} - 2\right)}{(1 \mp v)\left[1 + \hat{q}_T\left(\dfrac{r}{\alpha} + c\alpha\right)\right]} \frac{\Delta A}{A^2} \mp \left(\frac{q_p'}{\hat{q}} - 1\right)\frac{\Delta v}{1 \pm v} \tag{6-158}$$

Notice that depending on whether the zeros are located in the LHP or the RHP, they will have an opposite effect on variations of q_p' due to gain or null-depth variations. Thus the question in which plane initially to place the zeros depends on which term in [6-158] it is more important to minimize. Unless the network is expressly designed for large v (see following discussion on the MSFEN), which would seem inadvisable here unless the anticipated variations of A and v can be closely controlled, v will be negligibly small even for a relatively shallow twin-T null. For example, for $\hat{q}$ between 0.25 and 0.5 and q_z between 25 and 50, v will vary between 0.02 and 0.005, i.e., the null depth between -34 dB and -46 dB. Typical values for the null depth stability of a hybrid integrated twin-T using tantalum components may be found elsewhere.[32] To calculate the effects of mistuning or any other error sources on the null depth, the formula (6-98) can be used.

6.4.2 The Q Stability of the MSFEN

Consider the medium-selectivity FEN (MSFEN) shown in Fig. 6-29. The passive twin-T and buffer amplifier β in the feedback loop is realizable by the TT-SABB simply by opening up the positive feedback path of an FRN. A potentially symmetrical twin-T is assumed here since the $\hat{q}_T$ value may thereby be selected close to 0.5 and the tuning procedure is simplified.

32. See Chapter 8, Section 8.4 of *Linear Integrated Networks: Fundamentals*.

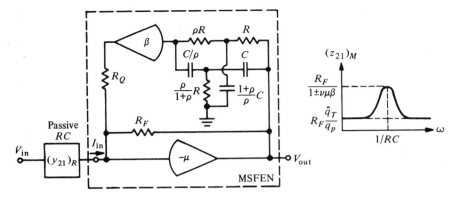

FIG. 6-29. Medium-selectivity FEN (MSFEN).

Calculating the transfer impedance of the MSFEN and assuming a twin-T in the feedback loop with finite zeros[33] we obtain

$$(z_{21})_M = -\frac{\hat{q}_T}{q_p} R_F \frac{s^2 + \dfrac{\omega_p}{\hat{q}_T} s + \omega_p^2}{s^2 + \dfrac{\omega_p}{q_p'} s + \omega_p^2} \qquad (6\text{-}159)$$

where $\hat{q}_T$ is the pole Q of the twin-T. Furthermore

$$q_p' = \frac{q_p}{1 \pm v\mu\beta} \qquad (6\text{-}160\text{a})$$

$$q_p = q_p'(v = 0) = (1 + \mu\beta)\hat{q}_T \qquad (6\text{-}160\text{b})$$

$$\mu = \frac{R_F}{R_Q} \qquad (6\text{-}160\text{c})$$

and β is the gain of the noninverting amplifier in the feedback loop. The terms μ and β always occur together in the loop-gain product

$$G_M = \mu\beta = \frac{\dfrac{q_p'}{\hat{q}_T} - 1}{1 \mp v\dfrac{q_p'}{\hat{q}_T}} \qquad (6\text{-}161)$$

33. The twin-T component values in Fig. 6-29 correspond to a perfect null. However, only slight variations from the nominal values result in a finite null. In practice, such variations always occur whether intentional or not, and a perfect null is only approximated to a greater or lesser degree.

Notice that the required loop gain decreases, the farther away from the $j\omega$ axis the twin-T zeros are in the RHP. To obtain the variation of q'_p to variations in μ, β, and v we have

$$\frac{\Delta q'_p}{q'_p} = S^{q'_p}_{G_M} \frac{\Delta G_M}{G_M} + S^{q'_p}_v \frac{\Delta v}{v} \tag{6-162}$$

where

$$\frac{\Delta G_M}{G_M} = S^{G_M}_\mu \frac{\Delta \mu}{\mu} + S^{G_M}_\beta \frac{\Delta \beta}{\beta} = \frac{\Delta \mu}{\mu} + \frac{\Delta \beta}{\beta} \tag{6-163}$$

With

$$S^{q'_p}_{G_M} = \frac{G_M}{1 + G_M} \frac{1 \mp v}{1 \pm v G_M} \tag{6-164}$$

and

$$S^{q'_p}_v = \mp \frac{v G_M}{1 \pm v G_M} \tag{6-165}$$

we obtain

$$\frac{\Delta q'_p}{q'_p} = \frac{1}{1 \mp v} \left[\left(1 - \frac{\hat{q}_T}{q'_p} \right) \left(1 \mp v \frac{q'_p}{\hat{q}_T} \right) \frac{\Delta G_M}{G_M} \mp \left(\frac{q'_p}{\hat{q}_T} - 1 \right) \Delta v \right] \tag{6-166}$$

To appreciate the effect of a finite twin-T null, we must include its influence on $\Delta G_M/G_M$. Assuming that we use two operational amplifiers with the same open-loop gain A to generate $G_M = \mu\beta$, we can rewrite (6-163) as follows:

$$\frac{\Delta G_M}{G_M} = (\mu + \beta) \frac{\Delta A}{A^2} = \left(\frac{G_M}{\beta} + \beta \right) \frac{\Delta A}{A^2} \tag{6-167}$$

and since, typically, $1 \le \beta \le 2$,

$$\frac{\Delta G_M}{G_M} \approx G_M \frac{\Delta A}{A^2} = \frac{\dfrac{q'_p}{\hat{q}_T} - 1}{1 \mp v \dfrac{q'_p}{\hat{q}_T}} \frac{\Delta A}{A^2} \tag{6-168}$$

Substituting this in (6-166) we obtain

$$\frac{\Delta q'_p}{q'_p} = \frac{1}{1 \mp v} \left[\left(\frac{q'_p}{\hat{q}_T} + \frac{\hat{q}_T}{q'_p} - 2 \right) \frac{\Delta A}{A^2} \mp \left(\frac{q'_p}{\hat{q}_T} - 1 \right) \Delta v \right] \tag{6-169}$$

This expression requires some explanation. The term involving Δv, the drift in the twin-T null depth, will be present no matter where the twin-T zeros are initially located. The quantity Δv corresponds to a real displacement of the zeros in the s-plane and may be calculated from (6-93); its value depends on the stability of the twin-T components but not on the initial null depth or phase. Naturally, a given quantity Δv will cause a much more drastic change in null depth when the twin-T zeros are in the vicinity of the $j\omega$ axis than when they are far away.[34] However the corresponding change in q_p' as given by (6-169) will be the same. The first term in the square bracket of (6-169) corresponds to the change in q_p' caused by changes in loop gain. Also this term is independent of the location of the twin-T zeros, i.e., it is valid whether the zeros are on the $j\omega$ axis or not. Only the coefficient outside the square bracket of (6-169), namely the term $(1 \mp v)^{-1}$ is directly affected by the initial twin-T zero location. Clearly RHP zeros will tend to improve the q_p stability, LHP zeros will have the opposite effect. Naturally, this is true only when $v = \hat{q}_T/q_z$ becomes large enough to have any significance. Since $\hat{q}_T$ has a maximum value of 0.5 and q_z a minimum (in the RHP) of unity, the upper limit of v in the RHP is 0.5. In attempting to approach this value, care must be taken to maintain a second-order twin-T transfer function (the pole–zero cancellation indicated in Table 6-5 must take place, i.e., $\omega_1 = \omega_1'$) without suffering a reduction in $\hat{q}_T$. A general method of achieving this optimization is available.[35] An additional advantage of placing the twin-T zeros as far right in the RHP as possible is that the actual loop gain $\mu\beta$ is thereby appreciably reduced (see (6-161)). Beside the improvement in stability (which is reflected by the $(1 \mp v)^{-1}$ term in (6-169)) this permits us to select $\beta = 1$ and also to select a relatively low value for μ. The resulting advantages in terms of usable bandwidth and ease of tuning (due to the reduction in parasitic phase shift) have been discussed earlier (e.g. Section 6.3).

It is now of interest to compare the Q stability attainable with the TT-SABB and that of the MSFEN. The gain–sensitivity product of each is given by the multiplicand of $\Delta A/A^2$ in (6-158) and (6-169), respectively. Except for the term involving $\hat{q}_T$, r, and α in the TT-SABB, and for the fact that $\hat{q}_T$ is used in (6-158) and $\hat{q}$ in (6-169), the two are identical. If β is chosen larger than unity (say two) then $\Delta G/G$ in (6-167) is essentially reduced by the same amount, and so will be the corresponding gain–sensitivity product in (6-169). Furthermore, as we saw earlier, certain limitations exist when minimizing Γ for the TT-SABB, so that in effect the lower limit given by (6-23) applies. Similar restrictions do not exist in (6-169), inasmuch as $\hat{q}_T$ can be made to approach its maximum of 0.5 arbitrarily closely, at least when $j\omega$-axis zeros are used for the twin-T. Thus Γ may be 2 to 4 times smaller for the MSFEN than for the TT-SABB. A somewhat more significant improvement may be attainable with the MSFEN by using RHP zeros which decrease both variations of A and v in the MSFEN, while increasing the variations of v in the TT-SABB. Nevertheless, even this improvement is modest and the price incurred in going from

34. For example, starting out with $j\omega$-axis zeros, the drift $\Delta v = 0.001$ will cause the attenuation at the null frequency to change from " minus infinity " to -60 dB. On the other hand, starting out with a null depth of -60 dB, the same shift Δv will cause a change in null depth of only 6 dB.
35. E. Lueder, The general second-order twin-T and its application to frequency emphasizing networks, *Bell Syst. Tech. J.* **51**, 301–316 (1972); also Chapter 8, Section 8.1.2 of *Linear Integrated Networks: Fundamentals*.

the TT-SABB to the MSFEN is an additional operational amplifier as well as a larger number of resistors and capacitors. It is therefore not surprising that even for those q_p values that lie marginally between the "low" and "high" boundaries (e.g., between 20 and 50) the alternative to the TT-SABB will generally be the high-selectivity FEN (HSFEN) rather than the MSFEN. The HSFEN is readily obtainable by connecting two TT-SABBs and a summing amplifier together and its advantages in terms of q_p stability make it useful well into the "high-q_p" range.

6.4.3 The Q–Stability of the HSFEN

Consider the high-selectivity FEN (HSFEN) shown in Fig. 6-30. The FRN in the feedback loop is realizable by the TT-SABB; it has identical pole and zero frequencies ($\omega_p = \omega_z$) and is assumed to be optimized according to the discussion in Chapter 3 (Section 3.3.2) for use in the FEN. Assuming that it also has a finite null ($v \neq 0$) we obtain the following transfer function for the HSFEN:

$$T_H(s) = -\frac{R_F}{R_G} \cdot \frac{q_L}{q_p} \cdot \frac{s^2 + \dfrac{\omega_p}{q_L'} s + \omega_p^2}{s^2 + \dfrac{\omega_p}{q_p'} s + \omega_p^2} \tag{6-170}$$

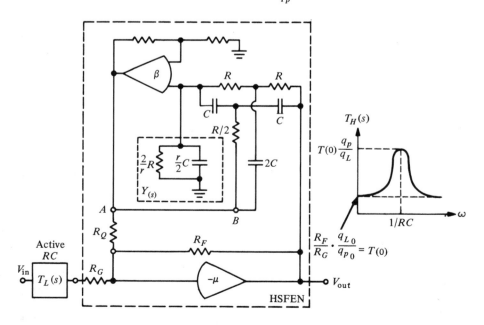

FIG. 6-30. High-selectivity FEN (HSFEN).

where

$$q'_L = \frac{q_L}{1 \pm v\left(\dfrac{\beta}{1+r}\right)\left(\dfrac{q_L}{\hat{q}_T}\right)} \tag{6-171a}$$

$$q_L = q'_L(v=0) = \frac{\hat{q}}{1 - \left(\dfrac{\beta}{1+r}\right)\left(\dfrac{\hat{q}}{\hat{q}_T}\right)} \tag{6-171b}$$

$$q'_p = \frac{q_p}{1 \pm \dfrac{\beta}{1+r}(1+\mu)v\dfrac{q_L}{\hat{q}_T}} \tag{6-171c}$$

$$q_p = q'_p(v=0) = \left(1 + \frac{\mu\beta}{1+r}\right)q_L \tag{6-171d}$$

and

$$\mu = \frac{R_F}{R_Q} \tag{6-171e}$$

Notice that q'_L and q_L correspond to the q'_p and q_p expresssions, respectively, given in Table 6-5, when $\omega_z = \omega_p$ and $r = c$. Defining the loop gain:

$$G_H = \frac{\mu\beta}{1+r} = \frac{\dfrac{q'_p}{q'_L} - 1}{1 \mp v\dfrac{q'_p}{\hat{q}_T}} \tag{6-172}$$

we have, with (6-171c),

$$q'_p = \frac{(1+G_H)q'_L}{1 \pm G_H v\dfrac{q'_L}{\hat{q}_T}} \tag{6-173}$$

and

$$\frac{\Delta q'_p}{q'_p} = S_{G_H}^{q_p'} \frac{\Delta G_H}{G_H} + S_{q_L'}^{q_p'} \frac{\Delta q'_L}{q'_L} + S_v^{q_p'} \frac{\Delta v}{v} \tag{6-174}$$

where

$$\frac{\Delta G_H}{G_H} = S_\mu^{G_H} \frac{\Delta\mu}{\mu} + S_\beta^{G_H} \frac{\Delta\beta}{\beta} + S_r^{G_H} \frac{\Delta r}{r}$$

$$= \frac{\Delta\mu}{\mu} + \frac{\Delta\beta}{\beta} - \frac{\Delta r}{1+r} \tag{6-175}$$

The last term in (6-175) is due to the variation of the resistive voltage divider comprising the series resistance of the twin-T and the resistive load to ground. It can be neglected here since we assume closely tracking film resistors. Thus

$$\frac{\Delta G_H}{G_H} \approx \frac{\Delta \mu}{\mu} + \frac{\Delta \beta}{\beta} \tag{6-176}$$

With (6-173) we obtain

$$S_{G_H}^{q_{p'}} = S_{q_L'}^{q_{p'}} = \frac{G_H}{1 + G_H} \left(1 \mp \frac{q_p'}{\hat{q}_T} v \right) \tag{6-177}$$

and

$$S_v^{q_{p'}} = \mp v \frac{G_H}{1 + G_H} \cdot \frac{q_p'}{\hat{q}_T} \tag{6-178}$$

Then, with (6-172) we obtain

$$\frac{\Delta q_p'}{q_p'} = \left(1 - \frac{q_L'}{q_p'} \right) \frac{\left(1 \mp v \dfrac{q_p'}{\hat{q}_T} \right)}{\left(1 \mp v \dfrac{q_L'}{\hat{q}_T} \right)} \left(\frac{\Delta G_H}{G_H} + \frac{\Delta q_L'}{q_L'} \right) \mp \frac{q_L'}{\hat{q}_T} \frac{\left(\dfrac{q_p'}{q_L'} - 1 \right)}{\left(1 \mp v \dfrac{q_L'}{\hat{q}_T} \right)} \Delta v \tag{6-179}$$

To derive the gain–sensitivity product of the HSFEN, we make the same assumption as in (6-167); hence,

$$\frac{\Delta G_H}{G_H} \approx G_H \frac{\Delta A}{A^2} = \frac{\dfrac{q_p'}{q_L'} - 1}{1 \mp v \dfrac{q_p'}{\hat{q}_T}} \frac{\Delta A}{A^2} \tag{6-180}$$

Substituting into (6-179) we obtain

$$\frac{\Delta q_p'}{q_p'} = \frac{1}{\left(1 \mp v \dfrac{q_L'}{\hat{q}_T} \right)} \left[\left(\frac{q_p'}{q_L'} + \frac{q_L'}{q_p'} - 2 \right) \frac{\Delta A}{A^2} \mp \left(\frac{q_p' - q_L'}{\hat{q}_T} \right) \Delta v \right] + S_{q_L'}^{q_{p'}} \frac{\Delta q_L'}{q_L'} \tag{6-181}$$

Compare the first term in the square brackets with the corresponding term for the MSFEN expression given by (6-169). Note that the two terms have the same form, namely $y = (x + x^{-1} - 2)$ where $x = q_p'/q_L'$ in the HSFEN case and $x = q_p'/\hat{q}_T$ for the MSFEN. The term y has a minimum when $x = 1$; hence, for a given q_p' value, y_{HSFEN} will be all the smaller, compared to y_{MSFEN}, the larger q_L' is selected compared to $\hat{q}_T$. However, there is a limit to how large q_L' may be made because the last term in (6-181) grows proportionally with q_L'. Thus, having optimized the null network in the feedback path of the HSFEN for a minimum gain–sensitivity product, we must still find $(q_L')_{\text{opt}}$, the optimum pole Q of the null network, such that Γ_{HSFEN}, the gain–sensitivity product of the HSFEN (and with it the Q variation given by (6-181)), is minimal. Recall that the optimized null network was found to be a standard FRN with $r = r_{opt}$ (see Chapter 3, equation (3-234)), $\hat{q}_T$ as close to 0.5 as possible, and a unity-gain amplifier (i.e., $\beta = 1$). Hence the only parameter of the null network not considered in the optimization was in fact its pole Q. In the following discussion, we shall find this optimum pole Q from the point of view of the overall HSFEN Q stability.

To find the optimum q_L we assume an infinite twin-T null, i.e., $v = 0$ and $\Delta v = 0$. This simplifies our calculation considerably without severely affecting the accuracy of the result. Furthermore we may then omit the primes from q_p' and q_L'. From (6-181) we then have

$$\frac{\Delta q_p}{q_p}\bigg|_{v=\Delta v=0} = \left(\frac{q_p}{q_L} + \frac{q_L}{q_p} - 2\right)\frac{\Delta A}{A^2} + S_{q_L}^{q_p} \cdot \frac{\Delta q_L}{q_L} \tag{6-182}$$

From (6-171d) it follows that $S_{q_L}^{q_p} = 1$. Furthermore, since we have assumed identical amplifiers for the μ and β amplifiers, i.e., $\Delta A_\mu/A_\mu^2 = \Delta A_\beta/A_\beta^2 = \Delta A/A^2$ (see (6-167)), we have (see (6-157))

$$\frac{\Delta q_L}{q_L} = \beta\left(\frac{q_L}{\hat{q}} - 1\right)\frac{\Delta A}{A^2} = \Gamma_s\frac{\Delta A}{A^2} \tag{6-183}$$

The optimum FRN in the feedback loop of an HSFEN has unity gain (i.e., $\beta = 1$) and comprises a potentially symmetrical twin-T such that

$$\Gamma_s \approx 2q_L - 1 \tag{6-184}$$

(see, e.g., Chapter 3, equation (3-236)). Thus, with (6-182) we obtain

$$\Gamma_{HSFEN} = 2q_L + \frac{q_p}{q_L} + \frac{q_L}{q_p} - 3 \tag{6-185}$$

Taking the derivative of this term with respect to q_L, equating to zero and solving for q_L, we obtain:

$$(q_L)_{opt} = \sqrt{\frac{q_p^2}{1 + 2q_p}} \approx \sqrt{\frac{q_p}{2}} \tag{6-186}$$

The last term is always valid since the HSFEN is not used unless $q_p \gg 1$. The remaining expressions for the optimum FRN for use in the HSFEN are then

$$\beta = 1$$

$$r_{opt} = \frac{1}{\sqrt{2q_p - 1}}$$

$$\hat{q}_T = \frac{1}{2}\frac{\rho}{1 + \rho}, \qquad \rho \geq 10 \tag{6-187}$$

$$\mu = \left(\frac{q_p}{q_L} - 1\right)\frac{1 + r}{\beta} \approx \frac{q_p}{q_L} = \sqrt{2q_p} = \frac{R_F}{R_Q}$$

and

$$\frac{\Delta q_p}{q_p} \approx \left[2\sqrt{2q_p} + \frac{1}{\sqrt{2q_p}} - 3 \right] \frac{\Delta A}{A^2} \qquad [6\text{-}188]$$

Some (rounded-off) values obtained for $(q_L)_{opt}$ for a typical range of q_p values are, from (6-186),

q_p	$(q_L)_{opt}$
25	3
50	5
100	7
300	12

$(q_L)_{average} = 7$

For a somewhat less optimal case than was assumed for (6-184), say, $\beta = 2$ and $\hat{q} = 0.25$, we obtain $\Gamma_S \approx 8q_L - 1$. The corresponding value for $(q_L)_{opt}$ is then $(\sqrt{q_p/2})/2$ instead of $\sqrt{q_p/2}$. Thus where the average value for $(q_L)_{opt}$ for the ideal case is approximately 7, it drops to 3.5 for the nonideal case considered. We have pointed out previously that it is desirable, in practice, to use a standardized value for q_L; this leaves the gain $\mu \approx q_p/q_L$ as a variable parameter to be set individually for the specified value of q_p. From our discussion above, it should be clear that the choice of $q_L = 5$ is therefore a useful compromise.

Returning to (6-181) and comparing it with (6-169), the improvement in Q stability attainable with the HSFEN compared to that of the MSFEN immediately becomes apparent. Because q'_L (when selected equal to 5) is over an order of magnitude larger than $\hat{q}_T$, the q_p variations due to variations in A and v are drastically reduced. Notice further that due to the multiplication of v with $q'_L/\hat{q}_T$ in the denominator of (6-181), RHP zeros will be even more effective in stabilizing q_p in the HSFEN than they were in the MSFEN.

6.4.4 Some Design Examples

At this point some design examples may be useful to illustrate the application of the TT-SABB by itself, and its incorporation in an MSFEN and an HSFEN. The principles of these three combinations and the corresponding interconnection schemes are shown in Table 6-15.

Consider for example, that we wish to design a sixth-order, synchronously tuned bandpass filter. This requires three building blocks, each of whose transfer function has the form

$$T(s) = K \frac{s}{s^2 + \dfrac{\omega_p}{q_p} s + \omega_p^2} \qquad (6\text{-}189)$$

Depending on the value of q_p we shall use a cascade of TT-SABBs, MSFENs, or HSFENs in the manner shown in Fig. 6-31.

TABLE 6-15. INTERCONNECTION OF TT-SABB IN FEN BUILDING BLOCKS

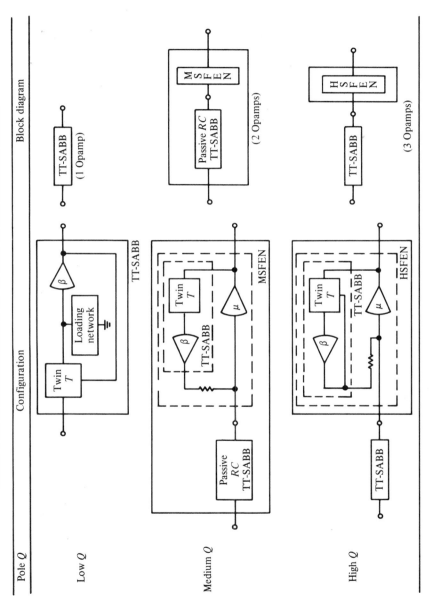

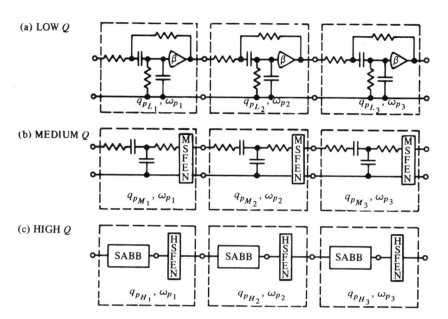

(a) LOW Q

$q_{P_{L_1}}, \omega_{P_1}$ $q_{P_{L_2}}, \omega_{P_2}$ $q_{P_{L_3}}, \omega_{P_3}$

(b) MEDIUM Q

$q_{P_{M_1}}, \omega_{P_1}$ $q_{P_{M_2}}, \omega_{P_2}$ $q_{P_{M_3}}, \omega_{P_3}$

(c) HIGH Q

SABB SABB SABB

$q_{P_{H_1}}, \omega_{P_1}$ $q_{P_{H_2}}, \omega_{P_2}$ $q_{P_{H_3}}, \omega_{P_3}$

FIG. 6-31. Sixth-order bandpass filter using: (a) TT-SABBs; (b) MSFENs; (c) HSFENs.

Using the TT-SABB $(q_p \leq 25)^{36}$ We specify, for example, that

$$\omega_p = 2\pi \cdot 1 \text{ kHz} \tag{6-190a}$$

$$q_p = 10 \tag{6-190b}$$

$$T(j\omega_p) = K \frac{q_p}{\omega_p} = 1 \tag{6-190c}$$

Having decided on a circuit configuration, we must find three network equations to satisfy these three requirements. Starting out with the general TT-SABB, we disconnect those circuit paths or components not used in the bandpass configuration. Selecting the type $\bar{B}$ bandpass circuit (it has the advantage of a capacitor shunted to ground at the high input-impedance terminal of the operational amplifier) we obtain the circuit shown in Fig. 6-32a. The disconnected circuit elements of the TT-SABB are indicated by broken lines.

In order to minimize $\beta\kappa$, we use (6-35) and obtain (in terms of the com-

---

36. The limits on q_p given in parentheses here and in the two succeeding examples are recommended but need not be adhered to rigidly. The actual circuit used will depend on the q_p stability attainable but also on numerous other practical considerations. Thus it may be advisable to use an HSFEN for a function whose q_p is 20, simply because all the other functions required are realized by HSFENs. Conversely, it may be worth stretching the capability of the TT-SABB to a q_p of, say, 30 simply because all the other functions to be supplied in a given system are acceptable with the TT-SABB.

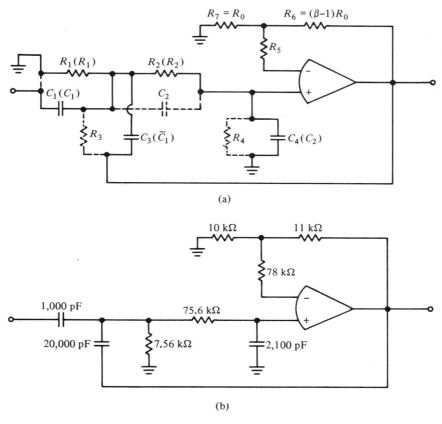

(a)

(b)

FIG. 6-32. Type $\bar{\text{B}}$ bandpass network: (a) derived from TT-SABB; (b) final circuit (for our example.)

ponent designations of Table 6-4, which are also given in parentheses in Fig. 6-32a)

$$(\beta\kappa)_{\text{BP }\bar{\text{B}}} \approx \left(1 + \frac{C_1}{\bar{C}_1}\right) \frac{C_2}{C_p} \frac{\left(1 + \dfrac{R_2}{R_1} + \dfrac{C_p}{C_2}\right)^2}{\sqrt{\dfrac{R_2\,C_p}{R_1\,C_2}}} \tag{6-191}$$

where $C_p = C_1 + \bar{C}_1$. Minimizing $\beta\kappa$ by impedance-scaling the second L section of the corresponding ladder, we select

$$\begin{aligned}
R_1 &= R \\
C_p &= C_1 + \bar{C}_1 = C \\
R_2 &= \rho R \\
C_2 &= C/\rho \\
\rho &= 10
\end{aligned} \tag{6-192}$$

In analogy to (6-38) we then obtain

$$(\beta\kappa)_{\text{BP }\bar{\text{B}}} \approx \left(1 + \frac{C_1}{\bar{C}_1}\right)\left(2 + \frac{1}{\rho}\right)^2 = 4.4\left(1 + \frac{C_1}{\bar{C}_1}\right) \tag{6-193}$$

Naturally we wish $C_1/\bar{C}_1$ to be small in order to minimize $\beta\kappa$. On the other hand we need to leave the actual value of this ratio free for the time being, in order to be able to satisfy our three network equations. From Table 6-1 and (6-192) these are

$$\omega_p = \frac{1}{\sqrt{R_1 R_2 (C_1 + \bar{C}_1) C_2}} = \frac{1}{RC} \tag{6-194a}$$

$$\beta = \frac{R_2}{\bar{C}_1}\left[\frac{C_1 + \bar{C}_1 + C_2}{R_2} + \frac{C_2}{R_1} - \sqrt{\frac{C_2(C_1 + \bar{C}_1)}{R_1 R_2}} \cdot \frac{1}{q_p}\right]$$

$$= \left(1 + \frac{C_1}{\bar{C}_1}\right)\left(2 + \frac{1}{\rho} - \frac{1}{q_p}\right) = 2\left(1 + \frac{C_1}{\bar{C}_1}\right) \tag{6-194b}$$

and

$$T(j\omega_p) = 1 = \beta\frac{C_1}{R_2 C_2(C_1 + \bar{C}_1)}\frac{q_p}{\omega_p} = \frac{\beta q_p}{1 + \frac{C_1}{\bar{C}_1}} \tag{6-194c}$$

Solving (6-194c) we find $C_1/\bar{C}_1 = [q_p(2 + 1/\rho - 1/q_p)]^{-1} = 1/20$, and from (6-194b), $\beta = 2.1$. Selecting $C_1 = 1,000$ pF and $\bar{C}_1 = 20,000$ pF, we have $C_2 = 2,100$ pF, and, from (6-194a) $R_1 = 7.56$ kΩ and $R_2 = 75.6$ kΩ. The resulting circuit is shown in Fig. 6-32b.

Note that R_5 (see Fig. 6-32a) is inserted only to compensate for the DC offset conditions of the operational amplifier. To ensure that the DC voltage drop caused by the DC currents into the inverting and noninverting input terminals is the same, R_5 must be selected such that

$$R_5 = R_1 + R_2 - R_6\|R_7 \tag{6-195}$$

As shown in Fig. 6-32b, this value is approximately 78 kΩ for our example.

As a result of the numerical values specified for this circuit, $C_1/\bar{C}_1$ turns out to be very small, which is exactly what we require to keep $\beta\kappa$ small. In our particular case we find from (6-193) that $\beta\kappa \approx 4.6$ which is very satisfactory. Should we have obtained an unacceptably large value from (6-194c) then the proper course to take would have been to select $C_1/\bar{C}_1$ so as to reduce $\beta\kappa$ to an acceptable value and to obtain the desired $T(j\omega_p)$ value by some other means. This would be the case, for example, if we wished *gain* at the peak frequency ω_p. Thus, for example, for $T(j\omega_p) = 10$ we find $C_1/\bar{C}_1 \approx 1$ and, from (6-193), $\beta\kappa \approx 8.8$. Rather than increase $\beta\kappa$ in this way and thereby deteriorate the network stability, it may be preferable to obtain the necessary gain from another stage and to select $C_1/\bar{C}_1$ equal to, say, 0.1. With our value of $\beta\kappa \approx 4.6$, an amplifier with open-loop gain A of, say, 1000 at $f_p = 1$ kHz and a variation $\Delta A/A$ of $\pm 50\%$, we may expect a stability $\Delta q_p/q_p \approx \pm 2.3\%$, i.e., on the order of two to three percent. In general, $\Delta A/A$ may be expected to be considerably less than $\pm 50\%$ with a corresponding decrease in $\Delta q_p/q_p$.

Finally, a word on the tuning of the circuit. We have here essentially the circuit shown in Fig. 6-8. The corresponding general sensitivity matrix is given in Table 6-9. As mentioned earlier, either R_1 or R_2 can be used as R_{ω_p} as each has the same effect on ω_p (since $S_{R_1}^{\omega_p} = S_{R_2}^{\omega_p} = -0.5$); however, it was also pointed out why R_1 is preferable. The closed-loop gain β is used to tune q_p; we obtain $S_\beta^{q_p} = 21$. Naturally ω_p and q_p should be tuned by phase, as already discussed in detail earlier.

Using the MSFEN ($10 \le q_p \le 50$)   The MSFEN version of the bandpass network used to provide the function (6-189) is shown in Fig. 6-33. Here we specify that

$$\omega_p = 2\pi \cdot 1 \text{ kHz} \tag{6-196a}$$

$$q_p = 30 \tag{6-196b}$$

$$T(j\omega_p) = K\frac{q_p}{\omega_p} = 1 \tag{6-196c}$$

The only difference between this and the TT-SABB case is that q_p is 30 instead of 10.

From Chapter 3, Table 3-12 the transfer admittance of the input network is given by

$$(y_{21})_R = K_R \frac{s}{s^2 + \dfrac{\omega_p}{\hat{q}} s + \omega_p^2} \tag{6-197}$$

where

$$K_R = \frac{1}{R_1 R_2 C_2} \tag{6-198a}$$

$$\omega_p = \frac{1}{\sqrt{R_1 R_2 C_1 C_2}} \tag{6-198b}$$

$$\hat{q} = \frac{\sqrt{R_1 R_2 C_1 C_2}}{(R_1 + R_2)C_1 + R_2 C_2} \tag{6-198c}$$

Referring to Fig. 6-33, the transfer impedance of the MSFEN is given by

$$(z_{21})_M = K_M \frac{s^2 + \dfrac{\omega_p}{\hat{q}_T} s + \omega_p^2}{s^2 + \dfrac{\omega_p}{q_p} s + \omega_p^2} \tag{6-199}$$

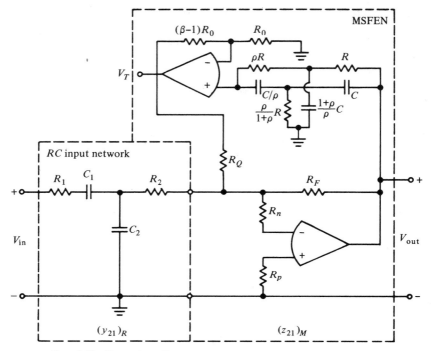

FIG. 6-33. General medium-Q bandpass network using the MSFEN.

where

$$K_M = \frac{\hat{q}_T}{q_p} R_F \tag{6-200a}$$

$$\omega_p = \frac{1}{RC} \tag{6-200b}$$

$$\hat{q}_T = \frac{\rho}{1+\rho} \cdot \frac{1}{2} \tag{6-200c}$$

$$q_p = \frac{\left(1 + \beta \dfrac{R_F}{R_Q}\right) \hat{q}_T}{1 \pm v \dfrac{R_F}{R_Q} \beta} \tag{6-200d}$$

In order for $(y_{21})_R \cdot (z_{21})_M$ to equal the specified function $T(s)$ given by (6-189), we obtain the following design conditions:

1. $K = K_R \cdot K_M$; therefore from (6-198a), and (6-200a)

$$\frac{\hat{q}_T}{q_p} \frac{R_F}{R_1 R_2 C_2} = K \tag{6-201a}$$

2. For pole–zero cancellation of $(y_{21})_R$ and $(z_{21})_M$ we obtain, from (6-198b) and (6-200b),

$$\frac{1}{(RC)^2} = \frac{1}{R_1 R_2 C_1 C_2} \tag{6-201b}$$

and from (6-198c) and (6-200c)

$$\hat{q} = \hat{q}_T \tag{6-201c}$$

i.e.,

$$\frac{\sqrt{R_1 R_2 C_1 C_2}}{(R_1 + R_2)C_1 + R_2 C_2} = \frac{\rho}{1 + \rho} \frac{1}{2} \tag{6-201d}$$

In addition, of course, we must satisfy the specifications given by (6-196). Thus, the design proceeds as follows:

1. Twin-T design: We select ρ as large as possible within the constraints of thin-film design; e.g., $\rho = 10$: then $\hat{q}_T = 0.455$. Now we select convenient C values; e.g., $C = 20,000$ pF, $C/\rho = 2000$ pF, and $C(1 + \rho)/\rho = 2,200$ pF. Consequently, from (6-196a) it follows that $R = 7.95$ kΩ, $\rho R = 79.5$ kΩ, and $\rho R/(1 + \rho) = 7.23$ kΩ.

2. RC input-network design: From (6-196c) and (6-201a) we have

$$\hat{q}_T \frac{R_F}{R_1 R_2 C_2} = \omega_p \tag{6-202}$$

It is convenient to select equal capacitors for C_1 and C_2; e.g., $C_1 = C_2 = C_0 = 10,000$ pF. Then from (6-201d) we have

$$\frac{R}{R_1 + 2R_2} \frac{C}{C_0} = \hat{q} \tag{6-203}$$

Combining this equation with (6-202) we obtain the following quadratic equation in R_2:

$$R_2^2 - \frac{1}{2\hat{q}} \frac{C}{C_0} RR_2 + \frac{\hat{q}}{2} \frac{C}{C_0} RR_F = 0 \tag{6-204}$$

Letting $R_F = 20$ kΩ and solving for R_2 (where we select the larger of the two resulting values) we obtain $R_2 = 10.62$ kΩ. From (6-203) we then have $R_1 = 13.6$ kΩ.

3. MSFEN design: From (6-200d) we have

$$R_Q = \frac{1 \mp v \dfrac{q_p}{\hat{q}_T}}{\dfrac{q_p}{\hat{q}_T} - 1} \beta R_F \tag{6-205}$$

The larger R_Q, the more loop gain is available from the μ amplifier. Two methods of increasing R_Q are evident from (6-205). We can tune the twin T for RHP zeros (i.e., positive sign in the numerator) and we can design β to be larger than unity. The latter possibility is limited, as we know, by stability considerations. Nevertheless, for the pole frequency of 1 kHz specified here, we can readily select $\beta = 2$. The method of tuning the twin-T for RHP zeros was discussed in Section 6.2.2; here, for simplicity, we shall assume a perfect null, i.e., $v = 0$. Then, from (6-205), $R_Q = 615$ Ω. Should this relatively low value of R_Q load down the β amplifier too much, either R_F can be selected larger, or a resistor T can be chosen for R_Q. Resistors R_n and R_p are used to equalize the input offset voltages resulting from the input currents. In the configuration of Fig. 6-33

$$R_p = R_n + R_F \| R_Q \tag{6-206}$$

With the numerical values computed so far, typical values would be $R_n = 10$ kΩ and $R_p = 10.6$ kΩ. The completed circuit is shown in Fig. 6-34.

A word now on the tuning of the circuit. The input RC network, having a pole Q of only 0.455, generally need not be accurately tuned. It will suffice to measure the actual capacitors C_1 and C_2 and to calculate R_1 and R_2 from (6-201d) and (6-202). After this the twin-T can be adjusted in the manner discussed earlier. If the twin-T is tuned separately for the specified null frequency (which is the pole frequency ω_p of (6-189)) the frequency dependence

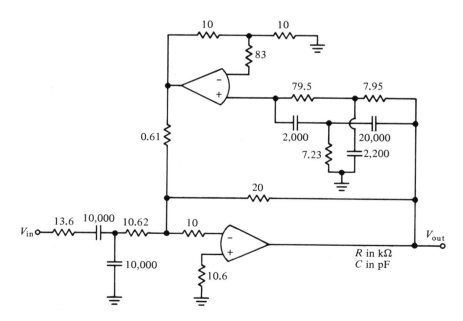

FIG. 6-34. Medium-Q bandpass network (for our example).

of the μ, or forward, amplifier may cause a slight decreasing frequency shift. To counter this, the twin-T may initially be tuned at a slightly higher frequency that ω_p and then, after the loop is closed, a functional correction can be made to reduce the frequency until it reaches the desired value. With the potentially symmetrical twin-T this amounts to no more than a slight correction (i.e., increase) of the series resistor $R_2 = \rho R$. Alternatively, the whole circuit can be tuned functionally, i.e., in the closed-loop or operating state, by monitoring the output at the terminal V_T (Fig. 6-33). From the input terminal to the output terminal V_T the MSFEN has the response of an FRN whose pole Q is q_p and for which $\omega_z = \omega_p$. Consequently the circuit can be frequency tuned as an active FRN. To tune for q_p, it is most convenient to return to the regular MSFEN terminals and, by adjusting R_Q, adjust the phase as described in Chapter 4, Section 4.4.2 (All-Pole Networks). Naturally, if a resistive-T network is used for R_Q, the possibility exists of adjusting q_p in either direction. Likewise, if β is only slightly larger than unity, then q_p can be adjusted in combination with R_Q and can be increased or decreased irrespective of the nature of R_Q.

Finally let us look at the q_p stability attainable with the MSFEN in our example. With the same amplifier as was assumed for the TT-SABB above (i.e., open-loop gain A at 1 kHz of 1000, $\Delta A/A = \pm 50\%$) and with $\nu = 0$, we obtain from (6-169), $\Delta q_p/q_p \approx \pm 3\%$. This is just about the same stability as was achieved with the TT-SABB for a pole Q three times smaller. Note that the effect of $\Delta \nu$ (i.e., twin-T null-depth variation) has been neglected in this computation. This is quite reasonable because in a hybrid integrated realization $\Delta \nu$ will remain negligibly small, since it depends on the tracking of the resistors and capacitors. Thus, for example, the typical change in twin-T null depth of a tantalum film realization will be less than 0.01%.[37]

Using the HSFEN ($20 \le q_p \le 300$) Specifying a bandpass function again here, but with a q_p higher than in either of the two previous cases we have, for example,

$$\omega_p = 2\pi \cdot 1 \text{ kHz} \tag{6-207a}$$

$$q_p = 100 \tag{6-207b}$$

$$T(j\omega_p) = K \frac{q_p}{\omega_p} = 1 \tag{6-207c}$$

Here, of course, we must use the HSFEN configuration shown in Fig. 6-35. Notice that the input stage is identical to the TT-SABB bandpass network described in the first of our examples. The only difference may be in the pole Q which can be selected somewhat closer to the optimum (in this case 7) than the value of 10 specified there. As we know, it is useful to use a standardized value (namely 5) for the pole Q of the input network.

---

37. See Chapter 8, Section 8.4 of *Linear Integrated Networks: Fundamentals*.

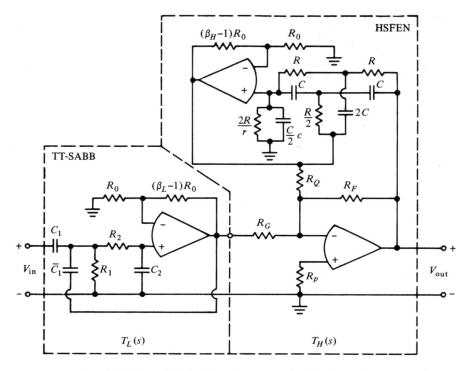

FIG. 6-35. General high-Q bandpass network using the HSFEN.

As we have previously done, we shall designate the transfer function of the input TT-SABB with the subscript L to distinguish it from the transfer function of the HSFEN connected to it (subscript H). Then the input network has the transfer function

$$T_L(s) = K_L \frac{s}{s^2 + \dfrac{\omega_p}{q_L} s + \omega_p^2} \qquad (6\text{-}208)$$

Referring to the type $\bar{B}$ bandpass network shown in Fig. 6-35 to provide this function, we obtain from Table 6-1

$$K_L = \beta_L \frac{C_1}{R_2 C_2 (C_1 + \bar{C}_1)} \qquad (6\text{-}209a)$$

$$\omega_p = \frac{1}{\sqrt{R_1 R_2 C_2 (C_1 + \bar{C}_1)}} \qquad (6\text{-}209b)$$

$$q_L = \frac{\sqrt{R_1 R_2 (C_1 + \bar{C}_1) C_2}}{R_1 (C_1 + \bar{C}_1 + C_2) + R_2 C_2 - \beta_L R_1 \bar{C}_1} \qquad (6\text{-}209c)$$

From the specifications imposed on the network we have $\omega_p = 2\pi \cdot 1$ kHz, furthermore we select $q_L = 5$. The transfer function of the HSFEN is given by

$$T_H(s) = K_H \frac{s^2 + \dfrac{\omega_p}{q_L} s + \omega_p^2}{s^2 + \dfrac{\omega_p}{q_p} s + \omega_p^2} \tag{6-210}$$

where

$$K_H = \frac{R_F}{R_G} \cdot \frac{q_L}{q_p} \tag{6-211a}$$

$$\omega_p = \frac{1}{RC} \tag{6-211b}$$

$$q_L = \hat{q}_T \frac{\sqrt{(1+r)(1+c)}}{1 + (r+c)\hat{q}_T - \beta_H} \tag{6-211c}$$

$$\hat{q}_T = \frac{1}{2} \frac{\rho}{1+\rho} \tag{6-211d}$$

$$r = c = \frac{1}{2q_L - 1} \tag{6-211e}$$

$$q_p = \left(1 + \frac{R_F}{R_Q} \cdot \frac{\beta}{1+r}\right) q_L \tag{6-211f}$$

Note that (6-211e) corresponds to an FRN optimized for minimum $\beta\kappa$ according to our discussion in Chapter 3 (Section 3.3.2). From (6-211c) and (6-211f) it is clear that a twin-T with an infinite null has been assumed here (i.e., $v = 0$). If a finite null is to be used the corresponding expressions given in (6-171) must be used. As we shall see, though, the specified q_p of 100 does not strain the capabilities of the HSFEN at all; therefore the twin-T can simply be tuned for as deep a null as possible.

In order for $T_L(s) \cdot T_H(s)$ to equal the function $T(s)$ specified by (6-189) the following design conditions must be fulfilled:

1. $K = K_L K_H$; therefore, from (6-209a) and (6-211a),

$$\beta_L \frac{C_1}{R_2 C_2 (C_1 + \bar{C}_1)} \cdot \frac{R_F}{R_G} \cdot \frac{q_L}{q_p} = K \tag{6-212}$$

2. For pole–zero cancellation of $T_L(s)$ and $T_H(s)$ we obtain, from (6-209b) and (6-211b),

$$\frac{1}{(RC)^2} = \frac{1}{R_1 R_2 C_2 (C_1 + \bar{C}_1)} \tag{6-213}$$

and, from (6-209c) and (6-211c),

$$\frac{\sqrt{R_1 R_2 (C_1 + \bar{C}_1) C_2}}{R_1 (C_1 + \bar{C}_1 + C_2) + R_2 C_2 - \beta_L R_1 \bar{C}_1} = \frac{\hat{q}_T \sqrt{(1 + r)(1 + c)}}{1 + (r + c)\hat{q}_T - \beta_H} \quad (6\text{-}214)$$

If, in addition, we now satisfy the design conditions given by (6-207), then the design procedure is as follows:

1. Designing the input network $T_L(s)$: Here we can proceed exactly as we did when designing the TT-SABB above. In fact, since ω_p is the same in both cases, we can adopt basically the same RC network, since it has already been optimized for minimum $\beta\kappa$. We therefore have $R_1 = 7.56$ kΩ, $R_2 = 75.6$ kΩ, $C_1 = 1000$ pF, $\bar{C}_1 = 20{,}000$ pF, and $C_2 = 2{,}100$ pF. With $\rho = 10$ and $q_L = 5$ we obtain, from (6-194b), $\beta_L = 1.05 \cdot 1.9 = 1.995 \approx 2$. From (6-209a) we find $K_L = 6 \cdot 10^2$. R_5, the offset resistor, is obtained from (6-195) and results in the same value as in the TT-SABB above, namely 78 kΩ. The complete input network is shown in Fig. 6-36.

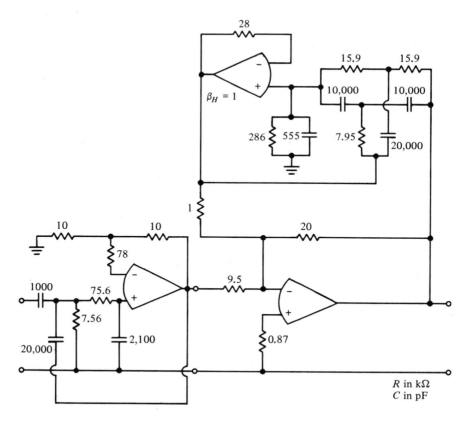

FIG. 6-36. High-Q bandpass network (for our example).

2. Designing the FRN for the HSFEN: It is convenient to prescribe the capacitors for the twin-T, which we shall assume to be symmetrical; we select $C = 10,000$ pF and obtain $R = 15.9$ kΩ. With $q_L = 5$, we have, from (6-211e), $r = 1/9 = 0.11$; hence the twin-T loading resistor $R_4 = 2R/r = 286$ kΩ and capacitor $C_4 = rC/2 = 555$ pF. From (6-211c) and (6-211e) we then obtain the closed-loop gain $\beta_H = 1$. The corresponding FRN is shown in the feedback loop of the HSFEN in Fig. 6-36. Notice that a resistor is inserted in the negative feedback loop of the β_H amplifier in order to balance the input offset voltage. Its value corresponds to the parallel value of the two resistors to ground at the positive input terminal, hence 286 k$\Omega \| 31.8$ k$\Omega \approx 28$ kΩ.

3. Designing the HSFEN: From (6-211f) we have

$$R_Q = \frac{R_F}{\dfrac{q_p}{q_L} - 1} \frac{\beta}{1 + r} \tag{6-215}$$

With $R_F = 20$ kΩ we obtain $R_Q = 1$ kΩ. Finally, from (6-207c) and (6-212) we have

$$R_G = \frac{K_L q_L}{\omega_p} R_F \tag{6-216}$$

Inserting the appropriate values (i.e., $K_L = 6 \cdot 10^2$, $q_L = 5$, $\omega_p = 2\pi \cdot 1$kHz, $R_F = 20$ kΩ) we obtain $R_G = 9.5$ kΩ. Finally, to prevent imbalance of the input offset voltage, we let

$$R_p = R_G \| R_Q \| R_F \tag{6-217}$$

which is 870 Ω.

Several comments are in order here. The design of the input network is straightforward and introduces no new insight other than demonstrating the ease with which any low-Q TT-SABB can be expanded into a high-Q circuit by combining it with an HSFEN. The trade-off inherent in this concept will also have become apparent here. The additional complexity of the HSFEN is well justified if a cascade of mainly low-Q circuits, readily designed with the TT-SABB, can be interspersed with HSFENs, which provide any high-Q poles required.

As a result of our using the optimized[38] FRN in the feedback loop of the HSFEN, the loading resistor R_4 of the twin-T is rather large, namely 286 kΩ. Whereas this is still within the limits acceptable both for thick- and thin-film hybrid integrated technology, a resistor of this value may take up more substrate area than is desirable. The obvious and simple way of reducing R_4

38. Optimization also requires a potentially symmetrical twin-T such that $\hat{q}r \rightarrow 0.5$. However, then R_4 becomes even larger, which may not be worth the resulting relatively modest decrease in the gain–sensitivity product.

is to deviate somewhat from the optimum FEN design. By letting β be somewhat larger than unity (say, 1.5 to 2) R_4 is immediately reduced in value. The resulting deterioration in q_L stability will be minimal since q_L itself has only a value of 5.

For a q_p of 100 we found that $R_Q = 1 \text{ k}\Omega$; consequently for a q_p of 200, R_Q decreases to 500 Ω and so on. This decrease has two adverse effects: first, it increases the gain required of the μ, or forward, amplifier; second, the loading of the β_H amplifier may become unacceptably severe. Raising the impedance level of R_F and R_Q may remedy the second problem but not the first. As long as we are operating at low frequencies a simple solution is to increase β_H, say to two. As we saw above, this also has the advantage of reducing R_4. However at higher frequencies this may not be feasible because of a deterioration in loop stability. It is then that the RHP twin-T zeros become particularly effective, and this virtually at no cost other than, perhaps, the slightly more complicated twin-T tuning procedure. Using RHP twin-T zeros,[39] q_p can easily be increased to 300 and, if necessary, even higher. However, as we recall, with the frequency stability attainable with the passive components, higher Q's are questionable at the very least.

This brings us to the question of tuning the circuit in Fig. 6-36. The active input network is, of course, tuned in accordance with the procedure outlined above for the TT-SABB. The same applies to the FRN in the feedback path of the HSFEN. The pole Q is then tuned by R_Q, the foreward gain by R_G. As in the case of the MSFEN, the finite gain bandwidth of the forward (μ) amplifier may cause a decrease in the pole frequency ω_p, as discussed in Chapter 4 (Section 4.2.3). To correct this error, the FRN can be pretuned to a frequency higher than ω_p by the amount ω_p/ω_g (see (4-136)) so that, when incorporated in the loop, the correct pole frequency ω_p is obtained. If a potentially symmetrical twin-T is used in the FRN, the initial FRN pole frequency can be selected sufficiently high to allow for a simple correction (by increasing R_2 of the twin-T) once the FRN is in the HSFEN loop.

6.4.5 Computer–Aided Filter Design

The high degree of standardization and the universality achievable with some of the filter building blocks described here and in Chapter 3 strongly suggest the potential for almost total filter design by computer. As an example, the general flow chart of a universal design program adapted to filter design using the TT-SABB and FEN filter combinations[40] is shown in Fig. 6-37.

39. To do so, we merely replace equations (6-211c) and (6-211f) by (6-171a) and (6-171c), respectively. Assuming, for example, RHP zeros corresponding to a null depth of -40 dB, then $v = 0.01$. For $q_p = 200$, $q_L = 5$, $\beta = 1$, $r = \frac{1}{5}$, we obtain from (6-172) $\mu = R_F/R_Q \approx 8$, as compared to $\mu = 40$ when $v = 0$. Similarly, for $q_p = 300$ we obtain $\mu = 8.5$ ($v = 0.01$) instead of $\mu = 60$ ($v = 0$).
40. R. W. Zeigler, unpublished Bell Telephone Laboratories memorandum.

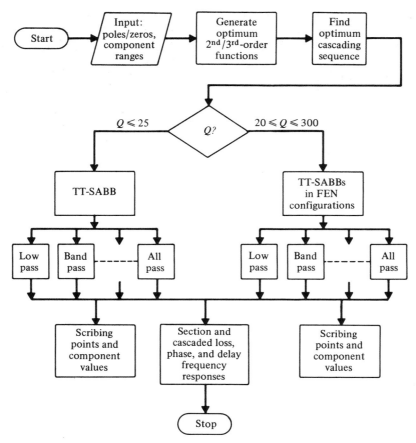

FIG. 6-37. Flow chart of universal computer program using the TT-SABB and FEN configurations for active-filter design.

The program accepts the desired transfer function in terms of its polynomials or roots, and pairs the resulting poles and zeros to minimize distortion according to the Lueder–Halfin algorithm (see Chapter 1, Section 1.3). The pole–zero pairing determines the individual second-order functions required. Depending on the specified pole Q's, use of the TT-SABB or the multiple-amplifier FEN is indicated by the program. As output data, the program provides the values of all the components required for each particular building block as well as the gain and phase response corresponding to the function in question. The program also prescribes the optimum sequence in which to cascade the individual building blocks and provides the gain and phase response consecutively at the output of each section in this cascade; the output of the last stage is the desired filter response. Note that beside the component values, also the scribing points (referred to the univesal building block layout) are given as output data. It is not difficult to envisage the potential for complete automation in the manufacture of the building blocks extended almost to the end product. For example, the actual scribing and tuning, say by laser, and the subsequent testing of the tuned circuits can be controlled by an extension of the computer program. Note also that the program chooses only between the TT-SABB as used alone ($q_p \leq 25$) or as combined in pairs in the HSFEN configuration. As mentioned earlier, this is justified when using a hybrid integrated TT-SABB to provide FEN-type filter building blocks. Since both the MSFEN and the HSFEN require two TT-SABBs, the cost of using an HSFEN—and obtaining higher q_p stability—is the additional amplifier of the input network. Naturally, if power dissipation is to be minimized, then the MSFEN should be considered for medium-valued pole Q's all the same.

7

EXTENDING THE CAPABILITIES OF ACTIVE FILTERS

INTRODUCTION

The active-filter building blocks discussed in the previous chapter were designed for use in the "medium-frequency" band (i.e., covering approximately from 100 Hz to 30 kHz). For the "high" as well as for the "low" frequencies above and below this band, certain modifications must be made and altogether other circuit techniques must frequently be considered. By so doing, it becomes possible to extend the frequency capability of active RC filters to below 1 Hz and up beyond 1 MHz. This topic will be discussed in this chapter.

We shall also address ourselves to the problem of improving the frequency stability of active RC filters. Up to this point, we have indicated that the achievable frequency stability of hybrid integrated active filters is given by the characteristics of the resistors and capacitors determining the frequency. The problem was therefore relegated to the technological domain, and a solution to the problem of improved frequency stability was tacitly anticipated from there. However, a network-oriented remedy to the problem of frequency stability also exists, namely that of introducing a certain amount of negative-feedback coupling between the second-order networks of a cascade. As a result, the frequency characteristics of the overall network may be more stable than those of the individual sections. A description of this "coupling effect" and its influence on network stability will also be given in this chapter.

7.1 HIGH-FREQUENCY FILTER BUILDING BLOCKS

In terms of active filters based on the use of commonly known, all-purpose operational amplifiers (e.g., the Fairchild μA 741, μA 748), any frequency above the voice band (i.e., beyond 3–4 kHz) may be considered "high". This is because of the inevitable phase shift of these amplifiers, which is caused by their rather limited unity-gain bandwidth. Of course, with wideband operational amplifiers (e.g., Fairchild μA 715, RCA 3015) many of the active filter types discussed so far can be used up to one, and perhaps even several, hundred kilohertz. However, in more conventional terms this still leaves us well below the desired upper frequency limit, which should extend at least up to 1 MHz. Whether the need exists for active filters in this range has already been discussed and will not concern us here. Instead, we shall examine some modifications to the building-block category discussed in Chapter 6, and consider an alternative network that seems particularly well suited to high-frequency operation.

7.1.1. Extending the Frequency Range of the FEN Building Blocks

Referring to the basic MSFEN and HSFEN topologies that were shown in Chapter 6 (Figs. 6-29 and 6-30) we recall that the unity-gain bandwidth ω_g of the β amplifier must exceed that of the μ amplifier in order to guarantee loop stability. It is assumed that both amplifiers are individually frequency compensated for a rate of closure of 6 dB-oct. The reason for this constraint was discussed in Chapter 3, (Section 3.2.2). As a result, the unity-gain bandwidth of the β amplifier determines the maximum frequency range of either of the FEN types, since the μ amplifier must be compensated in such a way that its ω_g is less than that of the β amplifier. Alternatively, an amplifier with a wider bandwidth must be used for the gain β than for the gain μ.

Modifying the FRN for Unity Gain[1]   The bandwidth of the β amplifier will be maximally large when β is equal to unity. In the MSFEN, $\beta = 1$ presents no problem other than a somewhat restricted range of q_p values. With $\beta = 2$ this range is doubled. In the HSFEN we can readily incorporate an FRN with unity gain, in particular since ω_p is required to equal ω_z (see Chapter 3, Fig. 3-30). We do suffer a slight attenuation in loop gain caused by the twin-T series components in conjunction with the loading network. However, this attenuation decreases the HSFEN loop gain by $1/(1 + r)$, an amount that will generally be insignificant. Nevertheless, in this regard it may be of interest to mention a modification of the FRN which permits unity gain to be used without incurring any attenuation in loop gain. Based

1. The FRN (frequency-rejection network) was discussed in Chapter 3, Section 3.1.2.

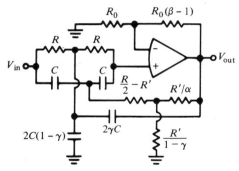

FIG. 7-1. Modified frequency-rejection network (FRN).

on the circuit transformation shown in Figs. 3-11 to 3-13 of Chapter 3, it
has the same relationship to the standard FRN as the modified all-pass
network of Fig. 3-38 has to the basic circuit of Figs. 3-34 and 3-35. This
modified FRN is shown for the case of a symmetrical twin-T in Fig. 7-1[2],
and we obtain for it the FRN-type transfer function (see (6-124)) where

$$K = \beta \tag{7-1}$$

$$\omega_z = \omega_p = \frac{1}{RC} \tag{7-2}$$

$$q_p = \frac{1}{4\left(1 - \beta \dfrac{\alpha + \gamma}{2}\right)} \tag{7-3}$$

Notice that the forward gain comprises only β. Furthermore, ω_z is equal to
ω_p in all cases, which makes the circuit particularly attractive for use in
the HSFEN but virtually useless as a general FRN. (To obtain $\omega_z \neq \omega_p$,
a twin-T loading network would again be required.) For use in the FEN,
tuning this network is actually simpler than tuning the standard FRN.
Apart from the fact that ω_p is inherently equal to ω_z, the value of the shunt
resistor of the twin-T can be increased *and* decreased because it has the
T form. This simplifies the tuning of the null. Other than that, q_p can be
tuned for in the usual manner by adjusting the closed-loop gain β.

Using an Emitter Follower in the Feedback Loop   Reducing the gain β
from two to unity will typically increase the bandwidth of the FEN by an
octave. A more effective method of increasing the available frequency range
is to replace the β amplifier by a high-frequency, high-gain transistor connected as an emitter follower. The maximum bandwidth is now no longer

2. This circuit was not included in Table 3-5 because it cannot be derived directly from the general
   circuit shown in Fig. 3-27.

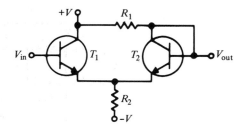

FIG. 7-2. Emitter follower, compensated for zero V_{BE} offset voltage.

determined by the emitter follower, (whose f_T may be over 100 MHz) but by the μ amplifier, which may now be compensated as loosely as if it were in a feedback loop on its own.

In order to avoid the V_{BE} offset inherent in an emitter follower, the configuration shown in Fig. 7-2 can be used. T_1 and T_2 are a dual matched pair with T_1 connected as a conventional emitter follower. Because of its V_{BE}, the emitter voltage of T_1 is approximately 0.7 V lower than its input (or base) voltage. T_2, connected as a conducting diode, adds the 0.7 V back to the emitter of T_1, thereby cancelling the level shift between input and output of the overall circuit. Resistors R_1 and R_2 are chosen in such a way that the current is divided equally between the two transistors. Thus, for symmetrical voltage supplies, R_1 is twice as large as R_2. Using this scheme, an MFSEN is shown in Fig. 7-3, an HSFEN in Fig. 7-4. Note that R_p equals $R_Q \| R_F$ in order to cancel any offset voltage. Furthermore, R_p is shunted by a capacitor C_p so that any high-frequency parasitic signals fed back to the positive terminal of the μ amplifier are shunted to ground.

The fact that the gain of the emitter follower is less than unity makes the twin-T loading network unnecessary in the HSFEN (see Fig. 7-4). As a result, we are guaranteed that $\omega_p = \omega_z$. For this advantage, however, there

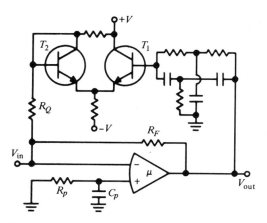

FIG. 7-3. MSFEN with zero-offset emitter follower in the feedback loop.

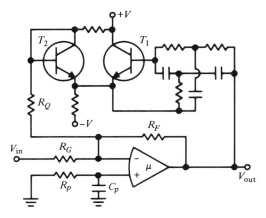

FIG. 7-4. HSFEN with zero-offset emitter follower in the feedback loop.

is the disadvantage that the emitter-follower gain can neither be adjusted nor predicted accurately. As we know, the current gain of a transistor can vary considerably from unit to unit. Of course, we can still adjust R_Q for an accurate q_p, but it is also necessary to adjust q_L if the pole–zero cancellation between the input network and the HSFEN is to be precise. This being the case, we must include a well controlled voltage amplifier in the HSFEN, as shown in Fig. 7-5. We recognize the latter to be the β amplifier of the conventional HSFEN with the important difference that it is no longer situated in the primary HSFEN feedback loop. Thus we can now select an arbitrary value for β in the corresponding TT-SABB, while still maintaining the widest possible bandwidth available in the μ-amplifier, with no regard

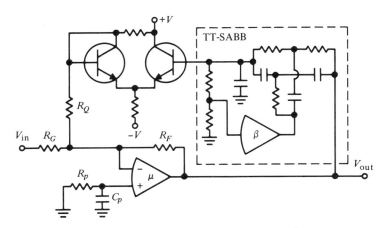

FIG. 7-5. HSFEN with zero-offset emitter follower and β amplifier.

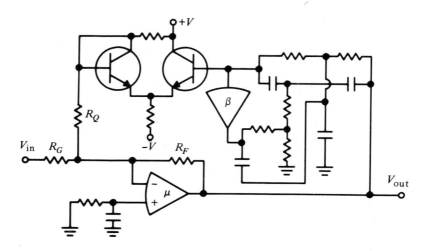

FIG. 7-6. HSFEN with zero-offset emitter follower and modified FRN.

for the ratio of the resulting unity-gain bandwidths of the two operational amplifiers. Note that the active-feedback loop now includes a complete TT-SABB in addition to the wide-band emitter follower. In order to minimize the attenuation in the HSFEN loop gain due to the twin-T loading network (which is now again necessary) the loading network is broken into two resistors in the manner shown in Fig. 7-5 (see also Chapter 3, Fig. 3-32). The loading network can be eliminated altogether if the FRN of Fig. 7-1 is used (see Fig. 7-6). However, this FRN cannot be readily derived from the TT-SABB.

Denoting the gain of the emitter follower by g we obtain the block diagram shown in Fig. 7-7. Assuming that the presence of a twin-T loading network

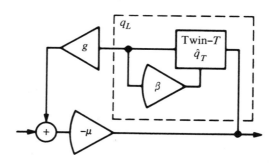

FIG. 7-7. Simplified representation of HSFEN in Fig. 7-6.

is taken into account by the values of g and β, this diagram is valid for either of the circuits in Figs. 7-5 and 7-6. The corresponding transfer function has a pole Q given by

$$q_p = (1 + \mu g)q_L \qquad (7\text{-}4)$$

where q_L is a function of β (see e.g., (6-211c)). To obtain the resulting q_p stability we note that g has now taken the place of $\beta/(1 + r)$ in the conventional HSFEN expression for q_p (see (6-171d)). Hence, for an ideally balanced twin-T we readily obtain:

$$\frac{\Delta q_p}{q_p} = \left(1 - \frac{q_L}{q_p}\right)\left(\frac{\Delta \mu}{\mu} + \frac{\Delta g}{g}\right) + \left(\frac{q_L}{\hat{q}_T} - 1\right)\frac{\Delta \beta}{\beta} \qquad (7\text{-}5)$$

For finite twin-T zeros we obtain an expression similar to that given by (6-179).

Applying the Frequency-Bypass Technique   Introducing the emitter follower in the HSFEN loop as discussed above enables us to open up the frequency compensation of the β and μ amplifiers as far as would be permitted if each occurred without the other. However, by the nature of the wide bandwidth available from the emitter follower, we could actually use considerably wider unity-gain bandwidths for β and μ than are generally available from any typical high-gain operational amplifiers. Let us therefore consider the conceptually simple method, shown in Fig. 7-8, of increasing the unity-gain

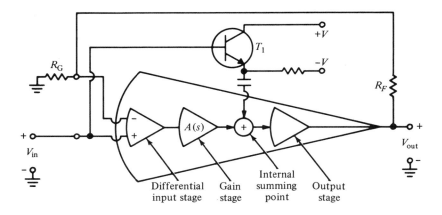

FIG. 7-8. The frequency-bypass technique.

bandwidth of a given operational amplifier. The amplifier is shown broken up into differential-input, gain, and output stages. For simplicity (but with little loss of generality) the amplifier gain is assumed to be concentrated in the middle stage. The purpose of the external transistor T_1 is to feed the high frequencies *around* the gain stage—or stages—of the amplifier, all of which are, of necessity, frequency limited. The bypassed high-frequency signals are added to the low-frequency signals at a convenient summing point within the amplifier, resulting in the modified frequency response shown in Fig. 7-9. The amplifier now has approximately unity gain up to the f_T of the bypass transistor used. The high-frequency bypass transistor is particularly effective in a noninverting operational amplifier providing low closed-loop gain. Assuming an ideal summing point (i.e., no interaction between the input signals), the open-loop gain of the operational amplifier shown in Fig. 7-8 is

$$A'(s) = A(s) + 1 \tag{7-6}$$

Assuming that $A(s)$ has only one break frequency Ω, then

$$A'(s) = \frac{A\Omega}{s + \Omega} + 1 = \frac{s + (A + 1)\Omega}{s + \Omega} \tag{7-7}$$

and the noninverting closed-loop gain $\beta(s)$ results as

$$\beta(s) = \frac{A'(s)}{1 + \frac{A'(s)}{\beta}} = \frac{\beta}{1 + \beta} \cdot \frac{s + \omega_g}{s + \frac{\omega_g}{1 + \beta}} \tag{7-8}$$

where $\beta = 1 + (R_F/R_G)$ and ω_g, the unity-gain bandwidth of the amplifier, equals $A\Omega$. In Fig. 7-9, the broken line corresponds to $\beta(s)$ when $\beta = 2$.

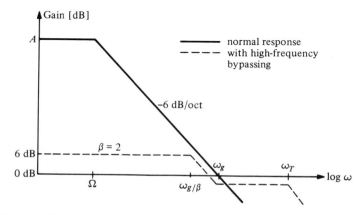

FIG. 7-9. Opamp frequency response using the frequency-bypass technique.

Since we are concerned here with the response of the operational amplifier near its cut-off frequency, it may be necessary to include the second break frequency occurring in the frequency-compensated characteristic of the open-loop gain of the operational amplifier. The first is Ω as above and the second will be very much larger, namely of the same order of magnitude as ω_g. The computations may then be carried out as above where (as would be expected) the closed-loop transfer function is found to be second-order—in fact, it possesses a zero in the vicinity of ω_g. This zero may be undesirable and can be eliminated by ensuring that the second break frequency of the operational amplifier is at least twice as large as ω_g.

In selecting a practical summing point within an operational amplifier, we will naturally be concerned to find one that is readily accessible and that requires as few additional components as possible. With some amplifiers the addition of a transistor, a bias resistor, and a coupling capacitor connected between the noninverting input terminal and a suitable internal summing point[3] as indicated in Fig. 7-8 will cause the usable frequency range to be increased considerably. Naturally the coupling capacitor must be carefully dimensioned so that its effect does not interfere with the desired frequency response.

7.1.2 The All–Pass Loop as a High–Frequency, High–Q Network

Provided that the production quantities are sufficiently high, the use of a hybrid integrated filter type particularly suitable for high-frequency applications, but unrelated to the FEN building blocks discussed above, may well be justifiable. A useful circuit type for this purpose is the all-pass loop, one version of which is shown in Fig. 7-10. Discussed as a special case of the analog-computer circuit in Chapter 3 (Section 3.3.1), its designation refers to the fact that any one of the possible circuits of this type consists essentially of two first-order all-pass circuits in a positive feedback loop. To examine

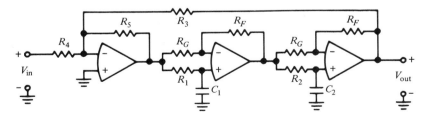

FIG. 7-10. One version of the all-pass loop.

3. See, for example, Chapter 7, Fig. 7-10 of *Linear Integrated Networks: Fundamentals*. Here a suitable summing point is the base of the output transistor T_7. Note that this point is accessible for the frequency-compensation network.

the characteristics of the all-pass loop, particularly for use at high frequencies, we therefore do well to first examine some properties of the first-order all-pass networks making up the all-pass loop.

First-Order All-pass Networks Consider the general all-pass network shown in Fig. 7-11. Including the effect of a finite source resistance R_S, the voltage transfer function from the signal source to the output is

$$\frac{V_{\text{out}}}{V_S} = \frac{Z - \dfrac{R_3 R_2}{R_1}}{Z + \dfrac{R_1 R_2 + R_S(R_1 + R_2)}{R_1}} \qquad (7\text{-}9)$$

Thus in order to obtain an all-pass characteristic we must compensate for a finite source resistance R_S by letting

$$R_3 = R_1 + R_S\left(\frac{R_1 + R_2}{R_2}\right) \qquad (7\text{-}10)$$

However, since the signal source for each all-pass network in the configuration of Fig. 7-10 is an operational amplifier, R_S will equal zero and R_3 will therefore equal R_1. To obtain a first-order all-pass network a capacitor will be used for Z.

Let us now consider the situation when Z and R_2 are interchanged in the circuit of Fig. 7-11. We then have the transfer function

$$\frac{V_{\text{out}}}{V_S} = -\frac{R_3}{R_1 + R_S} \cdot \frac{Z - \dfrac{R_1 R_2}{R_3}}{Z + R_1 \dfrac{R_2 + R_S}{R_1 + R_S}} \qquad (7\text{-}11\text{a})$$

and the input impedance is

$$Z_{\text{in}} = R_1 \frac{R_2 + Z}{R_1 + Z} \qquad (7\text{-}11\text{b})$$

Note that the frequency response of this circuit can be made insensitive to the source impedance by letting $R_1 = R_2$. Then only the multiplying constant of the transfer function is affected by R_S and, furthermore, the input

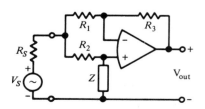

FIG. 7-11. A general all-pass network.

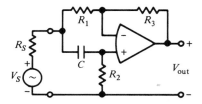

FIG. 7-12. First-order all-pass network.

impedance is constant[4]. This is very useful when the corresponding first-order all-pass network (see Fig. 7-12) is used in the RC dual version of Fig. 7-10. Whereas the source impedance of each operational amplifier is still negligibly small, the constant load seen by each operational amplifier driving a first-order all-pass section permits the resulting all-pass loop to have a larger dynamic range than the alternative, capacitive input impedances of the all-pass sections would. Since we already require that $R_2 = R_3$ to obtain an all-pass response, the first-order all-pass network with a constant input resistance (represented by Fig. 7-12) requires that $R_1 = R_2 = R_3$.

Before we examine the effect of the finite gain–bandwidth product of an operational amplifier on the all-pass loop we examine its effect on the characteristics of a first-order all-pass section. Assuming a single pole for the open-loop gain of the operational amplifier, i.e., $A(s) = A\Omega/(s + \Omega)$, and referring to the all-pass sections in Fig. 7-10, we obtain the transfer function

$$t_a(s) = -A(s) \frac{\dfrac{R_F}{R_G} s - \omega_0}{(s + \omega_0)\left(1 + \dfrac{R_F}{R_G} + A(s)\right)} \tag{7-12}$$

where $\omega_0 = 1/R_1 C_1$ (or $1/R_2 C_2$). To obtain a first-order all-pass circuit we must let $R_F = R_G$. Then with the single pole Ω of the amplifier we obtain

$$t_a(s) = -\left(\frac{s - \omega_0}{s + \omega_0}\right)\left(\frac{\omega_g/2}{s + \Omega + (\omega_g/2)}\right) \tag{7-13}$$

where $\omega_g = A\Omega$ is the gain–bandwidth product. The second bracket represents the error in the all-pass transfer function caused by the frequency dependence of the open-loop gain. Note that the additional parasitic pole has the value $\omega_g/2 + \Omega$, which is larger by Ω than half the unity-gain bandwidth. Since only unity (or close to unity) gain is required for each amplifier, devices with relatively low open-loop gain but wide bandwidths can readily be used. Hence an open-loop gain of 40 to 50 dB should be perfectly adequate, provided that the unity-gain bandwidth corresponding to a given gain–bandwidth product is thereby increased accordingly. This is shown qualitatively

4. H. J. Orchard, Active all-pass networks with constant resistance, *IEEE Trans. Circuit Theory*, **CT 20**, 177–179 (1973).

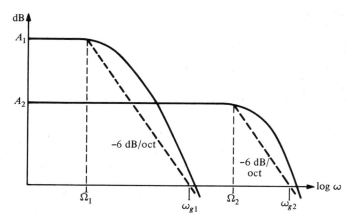

FIG. 7-13. High-gain low-frequency (A_1) and medium-gain high-frequency (A_2) response.

in Fig. 7-13. For a gain–bandwidth product of 10^7 kHz, for example, a maximum 40 dB open-loop gain would result in a unity-gain crossover frequency of 100 MHz and a parasitic pole in (7-13) on the order of 50 MHz.

Obviously, the higher the gain–bandwidth product of the amplifiers involved, the higher will be the useful frequency range of the resulting all-pass loop. Since the two or three amplifiers required in the all-pass loop must provide a loop gain of less than unity, it is clear that, for the first-order all-pass sections at least, single transistor stages may be used. The benefit to the all-pass loop will be evident when transistors with high cut-off frequencies f_T are used.

A transistor circuit that lends itself very well to first-order all-pass operation is shown in Fig. 7-14a.[5] In practice, the single transistor T may be replaced by the compound p-n-p–n-p-n transistor combination shown in

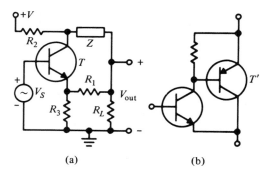

(a)   (b)

FIG. 7-14. Transistorized all-pass network (a) and compound p–n–p–n–p–n transistor (b).

5. Ibid.

Fig. 3-14b. As a result the transistor imperfections are barely detectable; for most purposes, the equivalent transistor T' may be assumed to be ideal. Making this assumption, the voltage transfer function is given by

$$\frac{V_{\text{out}}}{V_S} = \frac{R_L}{R_1 + R_L} \left(\frac{Z - \dfrac{R_1 R_2}{R_3}}{Z + R_1 \dfrac{R_2 + R_L}{R_1 + R_L}} \right) \tag{7-14}$$

where R_L is the load resistance of the network. Using a capacitor C for the reactance Z and letting

$$R_3 = \frac{R_2(R_1 + R_L)}{R_2 + R_L} \tag{7-15}$$

we obtain the characteristic of a first-order all-pass network. Letting $R_1 = R_2$, the network becomes independent of the load resistance except for its effect on the multiplying constant in (7-14). Similar reasoning as that applied to the circuit of Fig. 7-12 then implies that the circuit has a constant-resistance *source* impedance. Thus, for an all-pass network to have a constant-resistance source impedance it follows from (7-15) that we must make $R_1 = R_2 = R_3$.

The advantage of designing the transistor all-pass networks with constant source impedances is that, when the individual stages are cascaded in the all-pass loop they will operate independently of each other. This will simplify the tuning of the circuit as well as desensitize the all-pass response to ambient and aging effects. Note that the all-pass loop associated with the transistor all-pass section in Fig. 7-14 is of the "differential-output" kind whereas the loops associated with the operational amplifier all-pass sections of Figs. 7-11 and 7-12 are of the "single-ended" output kind (for differential-output and single-ended output all-pass sections, see Chapter 3, Section 3.3.1).

The Effect of Finite Gain-Bandwidth Product   Having examined the effect of the finite gain–bandwidth product on the *all-pass networks* contained in an all-pass loop, it is not difficult to evaluate this effect on the all-pass loop itself. Substituting (7-13) into the transfer function of the all-pass loop in Fig. 7-10, we obtain the transfer function

$$T(s) = \frac{g_4(s) \cdot T_a(s)}{1 + g_3(s) \cdot T_a(s)} \tag{7-16}$$

where

$$T_a(s) = \left[t_a(s) \right]^2 \Bigg|_{\Omega \ll \omega_g/2} = \left[\frac{\omega_g/2}{s + \omega_g/2} \right]^2 \cdot \left[\frac{s - \omega_0}{s + \omega_0} \right]^2 \tag{7-17}$$

and $\omega_0 = (R_1 C_1)^{-1} = (R_2 C_2)^{-1} = (RC)^{-1}$. Furthermore, $g_4(s)$ and $g_3(s)$, the gains of the inverter in the forward direction and within the loop, respectively, are given by (see Table 4-3, Chapter 4)

$$g_i(s) = -\gamma_i \cdot \frac{\dfrac{\omega_g}{1 + \gamma_i}}{s + \dfrac{\omega_g}{1 + \gamma_i}} ; \qquad i = 3, 4 \qquad (7\text{-}18)$$

where $\gamma_3 = R_5/R_3$ and $\gamma_4 = R_5/R_4$.

Notice that $T_a(s)$ is essentially the transfer function of a second-order all-pass network. Depending on its realization, it will have real or complex poles and zeros. In the case of the all-pass loop in Fig. 7-10, where it is realized by the cadcade of two first-order all-pass sections, they are real. However, this cascade could be replaced by any one of the class 4 second-order all-pass networks discussed in Chapters 2 and 3. We would thereby require one amplifier less—presumably at the cost of detrimental effects on the frequency range and tuning ease.

As discussed in detail in Chapter 3, Section 3.3.1, the resistor ratio γ_3 directly determines the pole Q of the all-pass loop. For high q_p values we have (see Chapter 3, equation (3.212))

$$\gamma_3 \approx 1 - \frac{1}{q_p}\bigg|_{q_p \gg 1} \approx 1 \qquad (7\text{-}19)$$

thus $\omega_g/(1 + \gamma_3) \approx \omega_g/2$. With (7-17) and (7-18), (7-16) then becomes

$$T(s) = -\gamma_4 \frac{1 + \gamma_3}{1 + \gamma_4} \frac{s + \dfrac{\omega_g}{2}}{s + \dfrac{\omega_g}{1 + \gamma_4}} \left[\frac{(s - \omega_0)^2}{\left(1 + \dfrac{2s}{\omega_g}\right)^3 (s + \omega_0)^2 + \gamma_3 (s - \omega_0)^2} \right] \qquad (7\text{-}20)$$

We may assume here that $2s/\omega_g \ll 1$, since the operating frequency must be well below the gain–bandwidth product. Therefore we may use the approximation

$$\left(1 + \frac{2s}{\omega_g}\right)^3 \bigg|_{s \ll \omega_g/2} \approx 1 + \frac{6s}{\omega_g} \qquad (7\text{-}21)$$

Substituting (7-21) into (7-20) and multiplying out the denominator, we obtain

$$T(s) = -\frac{\gamma_4}{1 + \gamma_4} \frac{s + \dfrac{\omega_g}{2}}{s + \dfrac{\omega_g}{1 + \gamma_4}} \frac{\omega_0 (s - \omega_0)^2}{a_3 s^3 + a_2 s^2 + a_1 s + a_0} \qquad (7\text{-}22)$$

where

$$a_3 = \frac{6}{1 + \gamma_3} \left.\frac{\omega_0}{\omega_g}\right|_{\gamma_3 \approx 1} \approx 3 \frac{\omega_0}{\omega_g} \qquad (7\text{-}23a)$$

$$a_2 = \omega_0 \left.\left(1 + \frac{12}{1 + \gamma_3} \frac{\omega_0}{\omega_g}\right)\right|_{\gamma_3 \approx 1} \approx \omega_0 \left(1 + 6 \frac{\omega_0}{\omega_g}\right) \qquad (7\text{-}23b)$$

$$a_1 = \omega_0^2 \left.\left(\frac{6}{1 + \gamma_3} \frac{\omega_0}{\omega_g} + 2 \frac{1 - \gamma_3}{1 + \gamma_3}\right)\right|_{\gamma_3 \approx 1} \approx \omega_0^2 \left(3 \frac{\omega_0}{\omega_g} + \frac{1}{q_p}\right) \qquad (7\text{-}23c)$$

$$a_0 = \omega_0^3 \qquad (7\text{-}23d)$$

and

$$q_p = \frac{1}{2} \frac{1 + \gamma_3}{1 - \gamma_3} \qquad (7\text{-}24)$$

The $[s + (\omega_g/2)]/\{s + [\omega_g/(1 + \gamma_4)]\}$ term in (7-22) will be unity, assuming that the forward gain $\gamma_4 = R_5/R_4$ is unity. The third-order polynomial in the denominator of (7-22) has the form

$$(s - p_1)(s - p_2)(s - p_3) = \left(s^2 + \frac{\omega_p'}{q_p'} s + \omega_p'^2\right)(s - p_3) \qquad (7\text{-}25)$$

where the primes on ω_p' and q_p' indicate perturbations of the nominal ω_p and q_p due to the finite gain–bandwidth product ω_g:

$$\omega_p' = \omega_p + \Delta\omega_p \qquad (7\text{-}26a)$$

$$q_p' = q_p + \Delta q_p \qquad (7\text{-}26b)$$

The finite ω_g is also responsible for the third, parasitic pole p_3. Multiplying out (7-25) and comparing the coefficients with those of the denominator of (7-22) we obtain

$$\frac{\omega_p'}{q_p'} - p_3 = \frac{a_2}{a_3} \qquad (7\text{-}27)$$

$$\omega_p'^2 - p_3 \frac{\omega_p'}{q_p'} = \frac{a_1}{a_3} \qquad (7\text{-}28)$$

$$-p_3 \omega_p'^2 = \frac{a_0}{a_3} \qquad (7\text{-}29)$$

Since the frequency of operation must be well below ω_g, and since p_3 is closely related to, and approximately equal to ω_g, we can assume that $p_3 \gg p_1, p_2$. Thus, from (7-27),

$$p_3 \approx -\frac{a_2}{a_3} = -\omega_0 \frac{1 + (6\omega_0/\omega_g)}{3\omega_0/\omega_g} \qquad (7\text{-}30)$$

This result could also have been obtained following the procedure, outlined in Chapter 4 (see equation (4-148)). Solving (7-29) for ω_p' and taking into account the fact that

$$\omega_p = \omega_0 \tag{7-31}$$

we then obtain, with (7-30),

$$\omega_p' = \frac{\omega_p}{\sqrt{1 + (6\omega_p/\omega_g)}} \tag{7-32}$$

and, assuming that $\omega_p/\omega_g \ll 1$,

$$\frac{\Delta\omega_p}{\omega_p} \approx -3\frac{\omega_p}{\omega_g} \tag{7-33}$$

Similarly, solving (7-28) for q_p' we obtain

$$q_p' = p_3 \frac{\omega_p'}{\omega_p'^2 - (a_1/a_3)} = q_p \frac{\sqrt{1 + (6\omega_p/\omega_g)}}{1 + \dfrac{18q_p}{1 + (6\omega_p/\omega_g)}\left(\dfrac{\omega_p}{\omega_g}\right)^2} \tag{7-34}$$

and, assuming that $\omega_p/\omega_g \ll 1$,

$$q_p' \approx q_p \frac{1 + (3\omega_p/\omega_g)}{1 + 18q_p\left(\dfrac{\omega_p}{\omega_g}\right)^2} \approx \frac{q_p}{1 + 18q_p\left(\dfrac{\omega_p}{\omega_g}\right)^2} \tag{7-35}$$

Thus

$$\frac{\Delta q_p}{q_p} \approx 3\frac{\omega_p}{\omega_g} - 18q_p\left(\frac{\omega_p}{\omega_g}\right)^2 \tag{7-36}$$

Note that the frequency and q_p error grows with the frequency ratio ω_p/ω_g. We can derive an upper bound on this ratio in terms of the permissible ω_p and q_p errors by examining the sensitivity of ω_p and q_p to variations in the network components. Letting

$$\varepsilon = \frac{\omega_p}{\omega_g} \tag{7-37}$$

we have

$$\frac{\Delta\omega_p'}{\omega_p'} = S_{\omega_p}^{\omega_p'}\frac{\Delta\omega_p}{\omega_p} + S_\varepsilon^{\omega_p'}\frac{\Delta\varepsilon}{\varepsilon} \tag{7-38}$$

From (7-32) we have

$$S_{\omega_p}^{\omega_p'} = 1 \tag{7-39}$$

$$S_\varepsilon^{\omega_p'} = -\frac{3\varepsilon}{1 + 6\varepsilon} \tag{7-40}$$

Assuming hybrid integrated RC components we have

$$\frac{\Delta\omega_p}{\omega_p} = -\left(\frac{\Delta R}{R} + \frac{\Delta C}{C}\right) \tag{7-41a}$$

and, since $\varepsilon = \omega_p/A_0\Omega$,

$$\frac{\Delta\varepsilon}{\varepsilon} = \frac{\Delta\omega_p}{\omega_p} - \frac{\Delta A_0}{A_0} - \frac{\Delta\Omega}{\Omega} \tag{7-41b}$$

Thus

$$\frac{\Delta\omega_p'}{\omega_p'} = -\frac{1 + 3\varepsilon}{1 + 6\varepsilon}\left(\frac{\Delta R}{R} + \frac{\Delta C}{C}\right) + \frac{3\varepsilon}{1 + 6\varepsilon}\left(\frac{\Delta A_0}{A_0} + \frac{\Delta\Omega}{\Omega}\right) \tag{7-42}$$

Similarly, from (7-35) we can write

$$\frac{\Delta q_p'}{q_p'} = S_{q_p}^{q_p'}\frac{\Delta q_p}{q_p} + S_\varepsilon^{q_p'}\frac{\Delta\varepsilon}{\varepsilon} \tag{7-43}$$

From (7-35) we have

$$S_{q_p}^{q_p'} = \frac{1}{1 + 18q_p\varepsilon^2} = \frac{q_p'}{q_p} \tag{7-44a}$$

and

$$S_\varepsilon^{q_p'} = \frac{3\varepsilon}{1 + 3\varepsilon} - \frac{36q_p\varepsilon^2}{1 + 18q_p\varepsilon^2} \approx -36q_p'\varepsilon^2 \tag{7-44b}$$

and, from (7-24),

$$\frac{\Delta q_p}{q_p} = q_p \cdot \frac{4\gamma_3}{(1 + \gamma_3)^2} \cdot \left.\frac{\Delta\gamma_3}{\gamma_3}\right|_{\gamma_3 \to 1} \approx q_p\frac{\Delta\gamma_3}{\gamma_3} \tag{7-45}$$

Thus, with (7-41) we find

$$\frac{\Delta q_p'}{q_p'} = q_p'\left[36\varepsilon^2\left(\frac{\Delta A_0}{A_0} + \frac{\Delta\Omega}{\Omega} + \frac{\Delta R}{R} + \frac{\Delta C}{C}\right) + \frac{\Delta\gamma_3}{\gamma_3}\right] \tag{7-46}$$

Both the frequency and q_p errors, as given by (7-42) and (7-46), respectively, can be reduced by selecting resistors and capacitors with compensating temperature and aging characteristics. Furthermore the variations in A_0 and Ω can also be decreased by using frequency-compensating RC components whose temperature and aging characteristics compensate those of the open-loop gain.[6]

In practice ω_g, the gain–bandwidth product of the operational amplifier used, will be given, and ε will increase with ω_p, the pole frequency of the network. As ε increases the frequency error (7-42) will remain within certain

6. See Chapter 7, Section 7.2.6 of *Linear Integrated Networks: Fundamentals.*

limits. At the same time, however, the q_p error (7-46) will increase quadratically with ε, by the amount $36q_p'\varepsilon^2$. Thus we can set an upper limit on the acceptable q_p error and derive from this the upper permissible pole frequency ω_p in relation to a given gain–bandwidth product ω_g. Selecting, for example, the criterion that

$$36q_p'\varepsilon^2 \leq 1 \tag{7-47}$$

we find that

$$\omega_p \leq \frac{\omega_g}{6\sqrt{q_p'}} \tag{7-48}$$

To put this upper bound on the pole frequency in perspective, let us compare it with the upper bounds obtained with the other analog-computer methods that were discussed in Chapter 3, Section 3-3.1.

The Upper Frequency Limit of other Analog-Computer Networks
Let us consider here the all-pass loop using differential-output operational amplifiers. Using the same terminology as above, the following two expressions for ω_p' and q_p' may be derived:[7]

$$\omega_p' = \frac{\omega_p}{\sqrt{1 + 3\varepsilon}} \tag{7-49}$$

and

$$q_p' = q_p \frac{\sqrt{1 + 3\varepsilon}}{1 + \dfrac{9q_p\varepsilon^2}{2(1 + 3\varepsilon)}} \tag{7-50}$$

Assuming that $\varepsilon \ll 1$, (7-50) can be simplified to

$$q_p' \approx \frac{q_p}{1 + 4.5q_p\varepsilon^2} \tag{7-51}$$

Calculating the frequency and q_p errors in precisely the same manner as we did for the all-pass loop using single-ended amplifiers, we obtain

$$\frac{\Delta\omega_p'}{\omega_p'} = -\frac{2 + 3\varepsilon}{2 + 6\varepsilon}\left(\frac{\Delta R}{R} + \frac{\Delta C}{C}\right) + \frac{3\varepsilon}{2 + 6\varepsilon}\left(\frac{\Delta A_0}{A_0} + \frac{\Delta\Omega}{\Omega}\right) \tag{7-52}$$

and

$$\frac{\Delta q_p'}{q_p'} = q_p'\left[\left(\alpha_1\frac{\Delta\alpha_1}{\alpha_1} - 2\alpha_2\frac{\Delta\alpha_2}{\alpha_2}\right) + 9\varepsilon^2\left(\frac{\Delta A_0}{A_0} + \frac{\Delta\Omega}{\Omega} + \frac{\Delta R}{R} + \frac{\Delta C}{C}\right)\right] \tag{7-53}$$

---

7. R. Tarmy and M. S. Ghausi, Very high-Q insensitive active RC networks, *IEEE Trans. Circuit Theory*, **CT-17**, 358–366 (1970); see equations (23a) and (23b).

where α_1 and α_2 are the resistor ratios associated with a differential-input–differential-output amplifier. Using hybrid integrated components, $\Delta\alpha_1/\alpha_1$ can be expected to equal $\Delta\alpha_2/\alpha_2$. Since $2\alpha_2 - \alpha_1 = q_p^{-1}$, we then obtain

$$\frac{\Delta q_p'}{q_p'} = q_p'\left[9\varepsilon^2\left(\frac{\Delta A_0}{A_0} + \frac{\Delta\Omega}{\Omega} + \frac{\Delta R}{R} + \frac{\Delta C}{C}\right) - \frac{1}{q_p}\frac{\Delta\alpha}{\alpha}\right] \qquad [7\text{-}54]$$

If we derive the upper limit of ω_p from the term proportional to ε^2, we specify, as in the previous case, that

$$9q_p'\varepsilon^2 \leq 1 \qquad (7\text{-}55)$$

and find that

$$\omega_p \leq \frac{\omega_g}{3\sqrt{q_p'}} \qquad [7\text{-}56]$$

Comparing (7-56) with (7-48) we find that, at least theoretically, the all-pass loop with differential-output amplifiers can be used up to a frequency twice as high as that with single-ended amplifiers. Presumably this is a result of the fact that the resistor ratio α tends to zero in the former circuit; in the latter, the resistor ratio γ_3 tends to unity. Whether this superiority in frequency range can actually be utilized in practice depends on whether the high sensitivity of the network to unsymmetrical loading of the differential-output amplifiers can be overcome (see Chapter 3, Section 3.3.1).

Finally, let us derive the upper frequency limit of one of the biquad-type analog-computer networks. With the terminology used above we obtain[8]

$$\omega_p' = \frac{\omega_p}{\sqrt{1 + 2\varepsilon}} \qquad [7\text{-}57]$$

and

$$q_p' = q_p\frac{\sqrt{1 + 2\varepsilon}}{1 - 4q_p\dfrac{\varepsilon}{1 + 2\varepsilon}} \qquad (7\text{-}58)$$

With $\varepsilon \ll 1$, this simplifies to

$$q_p' \approx \frac{q_p}{1 - 4q_p\varepsilon} \qquad [7\text{-}59]$$

Following the same procedure as above, the frequency and q_p errors result as

$$\frac{\Delta\omega_p'}{\omega_p'} = -\frac{1 + \varepsilon}{1 + 2\varepsilon}\left(\frac{\Delta R}{R} + \frac{\Delta C}{C}\right) + \frac{\varepsilon}{1 + 2\varepsilon}\left(\frac{\Delta A_0}{A_0} + \frac{\Delta\Omega}{\Omega}\right) \qquad [7\text{-}60]$$

and

$$\frac{\Delta q_p'}{q_p'} = \frac{q_p'}{q_p}\left(1 - \frac{1}{2q_p}\right)\frac{\Delta\rho}{\rho} - 4q_p'\varepsilon\left(\frac{\Delta A_0}{A_0} + \frac{\Delta\Omega}{\Omega} + \frac{\Delta R}{R} + \frac{\Delta C}{C}\right) \qquad [7\text{-}61]$$

8. Ibid, equations (32a) and (32b)

where ρ is a resistor ratio and $q_p = (1 + \rho)/2$. We derive the upper limit for ω_p from the term proportional to ε and find that, specifying

$$|4q'_p \varepsilon| \leq 1 \tag{7-62}$$

the upper frequency limit is obtained as

$$\omega_p \leq \frac{\omega_g}{4q'_p} \tag{7-63}$$

Note that the upper frequency limit is approximately $\sqrt{q'_p}$ times larger for either of the all-pass loops discussed previously than for the biquad. The reason for this superior high-frequency performance is that the two or three amplifiers in the all-pass loop each have a closed loop gain of unity or thereabout, with the attendant advantage of a correspondingly wide bandwidth for each amplifier.

7.2 LOW-FREQUENCY FILTER BUILDING-BLOCKS

By "low" frequencies we mean here frequencies anywhere between a fraction of a Hertz and, say, 100 Hz. To appreciate the problem of hybrid integrated filter realization in this frequency range we conduct the following simple computation. Assuming r equal resistors, c equal capacitors, a maximum total resistance per substrate of R_{max} and a maximum total capacitance per substrate of C_{max}, we obtain the minimum attainable critical frequency of a hybrid integrated network as

$$f_{min} = 0.159 \frac{rc}{R_{max} C_{max}} \tag{7-64}$$

Assuming a canonic second-order network, i.e., $r = c = 2$, $R_{max} = 0.5$ MΩ and $C_{max} = 0.05$ μF, we obtain $f_{min} = 25$ Hz. However, even this minimum frequency is highly optimistic, since, first, there will generally be more than two resistors and two capacitors on a substrate even if a canonic network is used, and, second, the maximum resistance and capacitance assumed is high, even if achievable, say, with tantalum thin film technology.[9] Assuming, then, that we wish to cover frequencies down to 1 Hz and below, we can fairly say that it becomes necessary to abandon our all-hybrid integrated circuit approach and to combine either film resistors with discrete capacitors or to use both component types in discrete form. Of course, we thereby lose some attainable size reduction as well as the other advantages associated with hybrid-integrated design; however, the alternative of an LCR network at these low frequencies makes even an all-discrete active equivalent highly desirable.

---

9. Using other thin-film (e.g., nichrome) or thick-film materials, higher resistor values are obtainable. However, chip capacitors of several tenths of a microfarad are very expensive, if available at all, in the high-precision (e.g., NPO) versions.

From (7-64) we readily see that, at low frequencies, the number of resistors and capacitors required for a second-order network is critical. This is true whether we intend to use integrated or discrete components. In the latter case, high-precision resistors and capacitors must be used, both of which become more costly and require more space as they become larger in value. Thus, no matter which network is selected, it will generally be advisable to make sure that it is canonic. As a consequence, a trade-off between additional active devices and fewer passive components may well be recommended at low frequencies, provided, of course, that the additional power can be supplied. Fortunately, the lower the frequency range, the more readily low- and micropower devices are available.

Using discrete components, the need no longer exists for the design of all-purpose building blocks. Any approach can be used, provided that it is canonic. As we pointed out in Chapter 5, low-frequency filters are not generally required to provide the functional diveristy that is characteristic of their voice-band counterparts. In general, bandpass- or resonator-type circuits, and possibly low-pass and band-rejection filters, will be required. If low-Q bandpass circuits are required the TT-SABB (type B) will be very suitable; for high-Q functions the biquad may be used.

One additional approach that should be considered for use at low frequencies is that of inductor simulation using gyrators, whereby the latter are realized by operational amplifier pairs. This approach, discussed only briefly in Chapter 1 (Section 1.2), will be examined in the following discussion, with a view to low-frequency filter design.

7.2.1 The Operational–Amplifier Gyrator as a Low–Frequency Filter Element

As already mentioned in Chapter 1 (Section 1.2), a useful gyrator, built from operational amplifiers, was advanced by Riordan[10] and later analyzed in detail by Orchard and Sheahan.[11] A common objection to this, or any other, operational-amplifier simulation of a gyrator is that it brings to the gyrator the limitations inherent in the operational amplifier itself. Most of these limitations result from the fact that the latter is a high-gain feedback device with, for example, the need for frequency compensation and the consequent bandwidth limitations and stability problems. However, when the opamp is used in the low-frequency range this objection hardly applies; low-cost, all-purpose operational amplifiers with internal frequency compensation are perfectly adequate here and may well provide a more economical gyrator than any other alternative.

10. R. H. S. Riordan, Simulated inductors using differential amplifiers, *Electronic Lett.* **3**, 50–51 (1967).
11. H. J. Orchard and D. F. Sheahan, Inductorless bandpass filters, *IEEE J. Solid-State Circuits*, **SC-5**, 108–118 (1970).

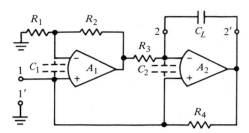

FIG. 7-15. Riordan gyrator circuit.

The general Riordan circuit is shown in Fig. 7-15 in the manner in which it is most commonly used.[12] Port 2, which is across the negative input and the output terminals of amplifier A_2, is terminated by a capacitor C; this is transformed into an inductance L at port 1. As shown, the circuit will be recognized as comprising a noninverting amplifier and an integrator connected together in a positive-feedback loop through resistor R_4. The capacitors C_1 and C_2 shown in dashed lines represent the input capacitances of the operational amplifiers, both of which are operating in the common mode.[13] Letting Ω_1 and Ω_2 be the single poles of amplifiers A_1 and A_2, respectively, both of which have the open-loop gain $A_i(s) = A_i \Omega_i / (s + \Omega_i)$ ($i = 1, 2$), we can calculate the value of the simulated inductance seen at the input of port 1. To do so we follow the procedure generally used in the analysis of nonideal gyrator–capacitor combinations.[14] Retaining only first-order terms of $1/A_i$, one obtains[15]

$$L = \frac{R_1 R_3 R_4}{R_2} C_L \left[1 + \frac{1}{A_1} \frac{(R_1 + R_2)^2}{R_1 R_2} - \frac{1}{A_2} \frac{R_1 - R_2}{R_2} \right.$$

$$\left. + \frac{1}{A_2} \frac{\omega}{\Omega_2} \left(\frac{1}{\omega C_L R_3} + \frac{\omega C_L R_3 R_1}{R_2} \right) \right]$$

$$(7\text{-}65)$$

Similarly, for the dissipation factor D, defined as $1/Q$, we obtain

$$D = \frac{1}{Q} = \omega C_1 R_1 - \frac{\omega C_2 R_3 R_1}{R_2} - \frac{1}{A_1} \frac{\omega}{\Omega_1} \frac{(R_1 + R_2)^2}{R_1 R_2}$$

$$+ \frac{1}{A_2} \frac{\omega}{\Omega_2} \frac{(R_1 - R_2)}{R_2} + \frac{1}{A_2} \left(\frac{1}{\omega C_L R_3} + \frac{\omega C_L R_3 R_1}{R_2} \right) \qquad (7\text{-}66)$$

12. See also Chapter 5, Section 5.4.3 of *Linear Integrated Networks: Fundamentals*.
13. In this mode the input resistance is sufficiently large to be neglected.
14. See, for example, Chapter 5, Section 5.4.2 of *Linear Integrated Networks: Fundamentals*.
15. Much of the analysis in this section is based on the paper mentioned above by H. J. Orchard and D. F. Sheahan.

The dissipation factor is a direct measure of the power loss in the inductor. Under ideal conditions, when A_1 and A_2 are infinite and $C_1 = C_2 = 0$, we obtain

$$L_{\text{ideal}} = \frac{C_L}{g_1 g_2} = \frac{R_1 R_3 R_4}{R_2} C_L; \qquad D = 0 \qquad \text{[7-67]}$$

i.e., the product of the gyration conductances $g_1 g_2$ is given by

$$g_1 g_2 = \frac{R_2}{R_1 R_3 R_4} \qquad \text{[7-68]}$$

and the inductor is lossless.

Since we intend to use the gyrator of Fig. 7-15 at low frequencies we can assume that Ω_1 and Ω_2 are far enough beyond our frequency of operation to be ignored. Furthermore, at low frequencies the admittances due to the parasitic capacitances C_1 and C_2 will also be negligible (typically several picofarads or less). Thus we have, from (7-65) and (7-66),

$$L = \frac{R_1 R_3 R_4}{R_2} C_L \left[1 + \frac{1}{A_1} \frac{(R_1 + R_2)^2}{R_1 R_2} - \frac{1}{A_2} \frac{R_1 - R_2}{R_2} \right] \qquad (7\text{-}69)$$

and

$$D = \frac{1}{A_2} \left(\frac{1}{\omega C R_3} + \frac{\omega C_L R_3 R_1}{R_2} \right) \qquad (7\text{-}70)$$

or, in terms of $Q = 1/D$ and (7-68),

$$Q = \frac{\omega C_L g_1 g_2}{\dfrac{1}{A_2 R_3} g_1 g_2 + \dfrac{\omega^2 C_L^2}{A_2 R_4}} \qquad (7\text{-}71)$$

We recall, now, that the general expression for the Q of a nonideal gyrator–capacitor simulated inductor is:[16]

$$Q = \frac{\omega g_1 g_2 C_L}{\Delta y y_{22} + (\omega C_L)^2 y_{11}} \qquad (7\text{-}72)$$

where the short-circuit admittance matrix of the gyrator is given as

$$[y] = \begin{bmatrix} y_{11} & g_1 \\ -g_2 & y_{22} \end{bmatrix} \qquad (7\text{-}73)$$

16. See Chapter 5, equation (5-86) of *Linear Integrated Networks: Fundamentals*.

Equating (7-72) with (7-71) we find

$$y_{11} = \frac{1}{A_2 R_4} \tag{7-74}$$

$$y_{22} = \frac{A_2 R_2}{2R_1 R_3} \left(\sqrt{1 + \frac{4R_1}{A_2^2 R_2}} - 1 \right) \Bigg|_{A_2 \gg 1} \approx \frac{1}{A_2 R_3} \tag{7-75}$$

and $g_1 g_2$ is, of course, given by (7-68). We see from (7-72) that Q is frequency dependent. The frequency of maximum Q is

$$\omega_{Q_{max}} = \frac{1}{C_L} \sqrt{\frac{y_{22} \Delta y}{y_{11}}} \approx \frac{1}{R_3 C_L} \sqrt{\frac{R_2}{R_1}} \tag{7-76}$$

With (7-76) the maximum Q then results from (7-72) as

$$Q_{max} = \frac{g_1 g_2}{2\sqrt{y_{11} y_{22}} \, \Delta y} \Bigg|_{g_1 g_2 \gg y_{11} y_{22}} \approx \frac{1}{2} \sqrt{\frac{g_1 g_2}{y_{11} y_{22}}} = \frac{A_2}{2} \sqrt{\frac{R_2}{R_1}} \tag{7-77}$$

Note that

$$g_1 g_2 = A_2^2 \frac{R_2}{R_1} y_{11} y_{22} \tag{7-78}$$

so that the approximation used in (7-77) will be valid as long as A_2 is large. In general, the Q of the simulated inductance is higher the larger the ratio $g_1 g_2 / y_{11} y_{22}$ is. Since both $g_1 g_2$ and $y_{11} y_{22}$ are inversely proportional to $R_3 R_4$, Q_{max} is independent of this resistor product; we can therefore select the value of R_3 and R_4 by some other criterion. However, to obtain y_{11} and y_{22} as small as possible, A_2, the open-loop gain of the second amplifier, should be as large as possible. Appropriate frequency compensation (such as the T compensation discussed in Chapter 6, Section 6.3.3), or high-frequency operational amplifiers should therefore be used when the frequency of operation extends beyond the low-frequency range.

In the low-frequency range a large value for A_2 can readily be taken for granted. This being the case, we can select

$$R_1 = R_2 = R \tag{7-79}$$

and still be sure of a maximum Q (which is then equal to $A_2/2$ at $\omega_{Q_{max}}$) of one, if not several, thousand. Since $R_1 = R_2 = R$ also causes one term in each of the general expressions (7-65) and (7-66) to disappear, we shall use this as a design condition. In effect, this results in A_1 operating as a noninverting amplifier with a gain of two. The absolute value of R should be compatible with this mode of operation (e.g., with respect to the input offset voltage) and also with the hybrid IC technology used.

We now come to the selection of the capacitor C_L and the resistors R_3 and R_4. In general we can add either one of two basic design criteria to our newly established condition resulting from (7-67) and (7-79)

$$L = R_3 R_4 C_L \qquad (7\text{-}80)$$

These design criteria are:

1. Maximize the dynamic range of the gyrator at the frequency of operation ω_0 (e.g., at resonance)

or:

2. Select the frequency of maximum Q, namely $\omega_{Q_{max}}$, equal to the resonance frequency ω_0.

The first criterion is important at any frequency, and in particular when the inductor, simulated by a gyrator–capacitor combination, is to be used in a series resonant circuit. The second is less important at low frequencies because a large A_2 can be counted on; however at higher frequencies it becomes increasingly important.

Designing for Maximum Dynamic Range   To establish the dynamic range of our circuit we must examine the gain of each of its amplifier stages. Referring to the generalized configuration in Fig. 7-16 and assuming high amplifier gains we obtain

$$V_2 = \left(1 + \frac{Z_2}{Z_1}\right) V_1 \qquad (7\text{-}81)$$

and

$$V_3 = V_1\left(1 + \frac{Z_5}{Z_3}\right) - V_2\left(\frac{Z_5}{Z_3}\right)$$

$$= V_1\left(1 - \frac{Z_2}{Z_1}\frac{Z_5}{Z_3}\right) \qquad (7\text{-}82)$$

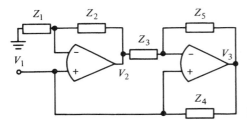

FIG. 7-16. Generalized Riordan gyrator.

As we know, the gain of the first stage is 2 and $Z_1 = Z_2 = R$. In order to optimize the dynamic range, it is desirable that the second stage also have a gain of 2. The gain of the second stage is frequency dependent, and we therefore specify a gain of 2 at the frequency ω_0. Then, with $Z_5 = 1/sC_L$, we find

$$\left| \frac{j\omega_0 C_L R_3 - 1}{j\omega_0 C_L R_3} \right| = 2 \tag{7-83}$$

or

$$\omega_0 R_3 C_L = \frac{1}{\sqrt{3}} \tag{7-84}$$

For $\omega_0 R_3 C_L$ larger than $1/\sqrt{3}$, the gain of the second stage increases and its voltage swing capabilities limit the dynamic range of the gyrator; conversely, when $\omega_0 R_3 C_L < 1/\sqrt{3}$ the limiting gyrator voltage is that appearing at the output of the first stage. Using (7-84) to obtain a value for C_L, we still require an additional condition to determine R_3. A useful one is to let R_3 and R_4 be equal. Thus, letting

$$R_3 = R_4 = R_L \tag{7-85}$$

we have, from (7-80),

$$C_L = \frac{1}{3\omega_0^2 L} \tag{7-86}$$

and

$$R_L = \sqrt{3}\,\omega_0 L \tag{7-87}$$

Note that C_L and R_L are given in terms of the two specified quantities ω_0 and L.

Designing for Maximum Q   Let us now return to criterion (2) mentioned earlier. It specifies that the maximum gyrator Q is to occur at ω_0, the frequency of operation. Thus we have, from (7-76),

$$\omega_0 R_3 C_L = \left. \sqrt{\frac{R_2}{R_1}} \right|_{R_1 = R_2} = 1 \tag{7-88}$$

and, from (7-80),

$$R_4 = \omega_0 L \tag{7-89}$$

We can use (7-88) to determine C_L and (7-85) as an additional convenient constraint to determine the resistors R_3 and R_4. We therefore obtain

$$C_L = \frac{1}{\omega_0^2 L} \tag{7-90}$$

and

$$R_L = \omega_0 L \qquad [7\text{-}91]$$

Thus, to obtain the maximum Q at ω_0 we see that R_3 should be set equal to the impedance of C_L, and R_4 equal to the impedance of L at the frequency ω_0. In addition, because we are free to choose $R_3 = R_4$, the two impedances are equal.

A comparison of (7-86) and (7-87) with (7-90) and (7-91) shows that, in order to satisfy either of the criteria for maximum dynamic range or maximum Q, the design equations for C_L and R_L differ only inasmuch as the frequency ω_0 is multiplied by a parameter λ. Thus we can write

$$C_L = \frac{1}{(\lambda\omega_0)^2 L} \qquad [7\text{-}92]$$

and

$$R_L = \lambda\omega_0 L \qquad [7\text{-}93]$$

Depending on whether the dynamic range or the inductor Q is to be maximized, λ may vary between $\sqrt{3}$ and unity, respectively. Naturally, a certain additional freedom in the choice of λ may be desirable in order to satisfy other constraints. Thus, for example, λ may be modified somewhat to provide a more convenient value for C_L (e.g., to conform with available values of thin-film or chip capacitors) or a different value for R_L (e.g., a higher value if a low-power gyrator requiring minimum current loading is used).

So far, R_1 and R_2 have been selected equal, but their absolute value R may still be chosen freely. Provided that R_L, as determined above, is a convenient value from a technology and circuit-design viewpoint, we are free to let all four resistors be equal, i.e., $R = R_L$. This is not a requirement, however, and if R_L is undesirably large it will impose an unnecessary constraint on the circuit when in hybrid integrated form, since large resistors take up a correspondingly large amount of substrate space.

Compensating for a Minimum Dissipation Factor   Let us now return briefly to the dissipation factor $D = 1/Q$ of a simulated inductor and consider additional ways of decreasing it. We shall show that this can be accomplished separately for low and high frequencies.

At low frequencies (where Ω_1 and Ω_2 are assumed much larger than ω_0) we can add an external capacitor to the parasitic capacitance C_2 (see Fig 7-15) in the following manner. Considering only the effect of C_2 and taking account of both (7-79) and (7-85) we obtain, from (7-66),

$$D = \frac{1}{A_2}\left[\frac{1}{\omega C_L R_L} + \omega C_L R_L\left(1 - A_2 \frac{C_2}{C_L}\right)\right] \qquad (7\text{-}94)$$

This expression has a minimum when C_2 is chosen such that

$$\left(1 - A_2 \frac{C_2}{C_L}\right) = \frac{1}{(\omega C_L R_L)^2} \tag{7-95a}$$

or

$$C_2 = \left(1 - \frac{1}{(\omega C_L R_L)^2}\right) \frac{C_L}{A_2} \tag{7-95b}$$

Then

$$D_{\min} = \frac{2}{A_2} \sqrt{1 - \frac{A_2 C_2}{C_L}} \tag{7-96}$$

This value is smaller than the minimum dissipation corresponding to (7-77), which is $2/A_2$. If, instead of (7-95b), $C_2 = C_L/A_2$, then

$$Q\big|_{A_2 C_2 = C_L} = A_2 \omega C_L R_L \tag{7-97}$$

that is the "coil Q" is proportional to frequency, corresponding to a resistance $R_s = R_L/A_2$ in series with an inductor $L = R_L^2 C_L$.

Should the frequency of operation be sufficiently high to warrant the inclusion of the high-frequency parasitic terms in (7-65) and (7-66), then, with (7-79) and (7-85), we obtain

$$L = R_L^2 C_L \left[1 + \frac{4}{A_1} + \frac{1}{A_2} \frac{\omega}{\Omega_2} \left(\frac{1}{\omega R_L C_L} + \omega R_L C_L\right)\right] \tag{7-98}$$

and

$$D = \omega(C_1 R - C_2 R_L) - \frac{4}{A_1} \frac{\omega}{\Omega_1} + \frac{1}{A_2} \left(\frac{1}{\omega R_L C_L} + \omega R_L C_L\right) \tag{7-99}$$

Having added an external capacitor across the parasitic capacitance C_2 to minimize the inductor losses at *low* frequencies, we can now do the same thing with C_1 in order to minimize the losses at *higher* frequencies. Depending on whether we have selected $\omega_0 R_L C_L$ for maximum dynamic range (see (7-84)) or for minimum dissipation (see (7-88)), the third term on the right side of (7-99) will equal $2.3/A_2$ or $2/A_2$, respectively. This shows, incidentally, that the increase in inductor loss incurred by optimizing the gyrator for maximum dynamic range is minimal and should generally be acceptable. At the operating frequency $\omega_0 = \Omega_1$ and above, the middle term in (7-99) is at least twice as large as $2/A_2$ and negative. By selecting C_1 such that the term $\omega_0(C_1 R - C_2 R_L)$ is positive and equal in magnitude to the negative quantity made up of the second and third terms, the dissipation factor can be cancelled at the operating frequency ω_0.

7.2.2 Design Example Using the Operational–Amplifier Gyrator

To give a numerical example using the opamp gyrator discussed above, we shall consider the design of a second-order bandpass filter. Thus we consider the function

$$T(s) = K \frac{s}{s^2 + \dfrac{\omega_p}{q_p} s + \omega_p^2} \tag{7-100}$$

where

$$\omega_p = 2\pi \cdot 10 \text{ Hz} \tag{7-101a}$$

$$q_p = 100 \tag{7-101b}$$

$$T(j\omega_p) = K \frac{q_p}{\omega_p} = 10 \tag{7-101c}$$

The simplest conventional way of realizing such a function is by designing a transistorized selective amplifier, as shown in Fig. 7-17, where the inductance L is given by a gyrator–capacitor combination and equals $R_L^2 C_L$. In order for the inductor, and with it the gyrator, to be grounded, a negative voltage supply is used. The corresponding transfer function results in the values

$$K = \frac{R_0}{R_E} \frac{1}{R_0 C_0} \tag{7-102}$$

$$\omega_p = \frac{1}{\sqrt{LC_0}} \tag{7-103}$$

$$q_p = R_0 \sqrt{\frac{C_0}{L}} \tag{7-104}$$

Furthermore

$$L = R_L^2 C_L \tag{7-105}$$

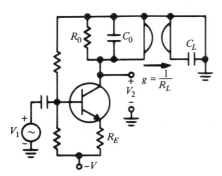

FIG. 7-17. Transistorized selective amplifier using gyrator.

With (7-103) we therefore have

$$\omega_p = \frac{1}{R_L \sqrt{C_0 C_L}} \qquad (7\text{-}106)$$

and

$$q_p = \frac{R_0}{R_L} \sqrt{\frac{C_0}{C_L}} \qquad (7\text{-}107)$$

Designing our gyrator for maximum dynamic range, we can substitute (7-87) into (7-106) and obtain an additional design equation:

$$C_0 = 3C_L \qquad (7\text{-}108)$$

Note that we have two resistors of value R_L (R_3 and R_4 in the gyrator of Fig. 7-15) and the two capacitors C_0 and C_L in our circuit. In terms of component count we are therefore back to a canonic circuit of second order and the limitation on maximum resistor and capacitor values implied by (7-64) is valid also here. As we saw from (7-64) the lowest frequency we can economically hope to approach in the foreseeable future[17] with all-hybrid-integrated circuits is on the order of 25 Hz. Thus, using our gyrator approach, or any other approach in which the lowest frequency is given by (7-64), we must at present abandon any intentions of using, say, film resistors *and* capacitors. At best, we may be able to use film resistors combined with high-quality (i.e., low-loss) discrete capacitors. Naturally, the size advantage of this approach compared to that using LC networks for the realization of the specifications given by (7-101) will still be considerable.

Returning to our design example, let us briefly examine what are the limitations of the transistorized circuit of Fig. 7-17 for the design of a high-Q bandpass circuit at low frequencies. From (7-103) and (7-104) we obtain the 3 dB bandwidth of the bandpass circuit as

$$\mathrm{BW}_{3\mathrm{dB}} = \frac{\omega_p}{q_p} = \frac{1}{R_0 C_0} \qquad (7\text{-}109)$$

With (7-108) we obtain the maximum q_p as

$$q_{p\,max} = \sqrt{3}\, \frac{R_{0\,max}}{R_{L\,max}} \qquad (7\text{-}110)$$

Furthermore, with (7-102) we have

$$T(j\omega_p) = K \frac{q_p}{\omega_p} = \frac{R_0}{R_E} \qquad (7\text{-}111)$$

With $R_{0\,max} = 1$ MΩ and $R_{L\,min}$, say, equal to $\sqrt{3} \cdot 1$ kΩ we can well realize a q_p of even 1000. However, in order to satisfy (7-101c) we then require that

17. These statements are based on industry forecasts made in the early to mid-seventies.

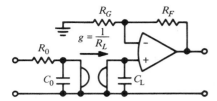

FIG. 7-18. Opamp selective amplifier using gyrator.

$R_E = 100$ kΩ. Herein lies one of the problems of using the transistorized circuit of Fig. 7-17. Because R_0 is required to be large for high q_p values, R_E will also be large unless very high gains are required at ω_p. As a result the collector current will not only depend on the specified gain and q_p but will also be very low in value at low frequencies.

A circuit that does not have these problems is the operational-amplifier version of the transistorized circuit shown in Fig. 7-18. Beside eliminating the transistor-originated disadvantages, the circuit comprises fewer components and may therefore be packaged more compactly. As a result, it may actually cost less than the transistor circuit. All the design equations for the transistorized circuit are valid here except for (7-102). Instead we now find that

$$K = \frac{R_F + R_G}{R_G} \cdot \frac{1}{R_0 C_0} \qquad (7\text{-}112)$$

With the specifications of (7-101) and assuming a maximum resistor $R_0 = 1$ MΩ, we obtain $C_0 = 1.59$ μF and, with (7-108), $C_L = 0.53$ μF. Furthermore, from (7-107), $R_L = 17.3$ kΩ. R_F and R_G are selected according to (7-101c). The resulting circuit is shown in Fig. 7-19. Note that the two noninverting amplifiers have offset-balancing resistors connected to the negative terminals.

Tuning the bandpass function is basically the same for the transistorized as for the operational-amplifier circuits. At the pole frequency, the phase is tuned for zero degrees by adjusting R_L; this essentially changes the value of

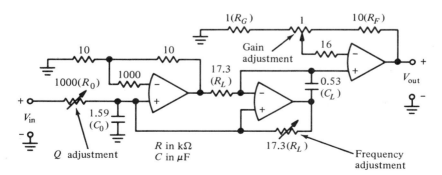

FIG. 7-19. Opamp selective amplifier using Riordan gyrator.

the simulated inductance. Either R_3 or R_4 can be used for this purpose, but since R_3 also affects the dynamic range, R_4 is used. The pole Q, q_p, is tuned by adjusting R_0 until the phase is $45°$ at the corresponding frequency. Finally, the peak gain is tuned for by adjusting the gain of the noninverting output amplifier (R_F and R_G). In the transistor version of the circuit (Fig. 7-17), R_4 and R_0 are adjusted in exactly the same way and finally the emitter resistor R_E adjusted for the gain. Note how virtually independent tuning of the network parameters is possible, particularly in the operational-amplifier version.

7.2.3 Other Operational-Amplifier Gyrator Configurations

Various other gyrator configurations using operational amplifiers have been suggested beside that of Riordan. One class of double-amplifier configurations was systematically derived by Antoniou[18] who then investigated the stability characteristics of the resulting circuits. It was found that, just after activation (by switching on the power to them), some of these configurations pass through unstable modes which may cause a latch-up condition. This happens because the gain of each amplifier, increasing from zero to its final value, causes the corresponding root locus of the network poles to cross the $j\omega$ axis. The resulting, theoretically provisional, oscillation may saturate the amplifiers, at which time the entire circuit is pulled into a permanent latch-up or unstable mode. The Riordan gyrator was found to belong to this conditionally stable group of circuits although it, and the other conditionally stable circuit variations, can be stabilized by adding back-to-back diode pairs across both ports. However, whereas this measure prevents an operational-amplifier gyrator from locking into an unstable mode, it simultaneously may severely limit its voltage handling capabilities. For this reason, Antoniou recommended two other gyrator configurations which have characteristics similar to the Riordan gyrator but are unconditionally stable (see Fig. 7-20). Note that the component count is the same and the topology

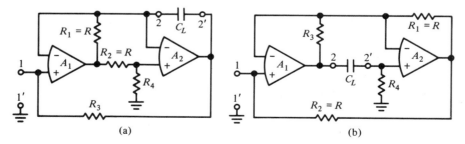

FIG. 7-20. Antoniou gyrator circuits.

18. A. Antoniou, Realization of gyrators using operational amplifiers, and their use in *RC* active-network synthesis, *Proc. IEEE*, **116**, 1838–1850 (1969).

similar to that of the Riordan circuit shown in Fig. 7-15. Also the network parameters are similar in value to those discussed in the previous section. In particular the gyrator conductances are the same. Thus, with the configurations shown, the gyrator constant is $1/g_1 g_2 = R_3 R_4$ and the inductance seen at the input port of either circuit is $L = R_3 R_4 C_L$.

7.2.4 Single–Amplifier Simulation of Grounded Inductors for Resonant Circuits

The gyrator–capacitor simulation of inductors for use in resonant (or tank) circuits invariably requires two, if not three (see Fig. 7-19) operational amplifiers. In spite of the fact that we are considering low-frequency applications, so that low-power operational amplifiers can most likely be used, the power dissipation of three amplifiers may nevertheless be unacceptably high. Furthermore, since the simulated inductor is frequently to be used in a grounded "LC" resonant circuit of moderate Q, it is natural that we should consider here the single-amplifier simulation of grounded inductors discussed in Chapter 1, Section 1.2.2. The circuits discussed there are summarized in Table 7-1. The equivalent RL configuration, its quality factor Q_L, and the equivalent inductance L_{eq} as a function of Q_L and ω are also given.

To be able to select the "best" of the four circuits listed in Table 7-1 it would be desirable to have some method of representing the "goodness" of the resulting simulated inductors with respect to, say, low-frequency operation. Indeed, such a representation would be useful for all simulated inductors including, of course, the gyrator–capacitor kind. Fortunately, methods do exist of representing graphically the frequency and Q properties of conventional inductors, and we shall show in what follows that they may readily be applied to active simulated inductors as well. We shall use the single-amplifier simulated inductors of Chapter 1 as a vehicle by which to illustrate active inductor characterization *in general*, bearing in mind that the same methods may be applied to all other active simulated inductors as well. As we shall see, one of the representations, the Q-ω_0 plot, permits simulated-inductor characterization directly in terms of the maximum and minimum resistor and capacitor values at our disposal.

The Q Contour   One method of representing the characteristics of inductors—both conventional and active—is by means of the Q contour. For conventional inductors the Q contour is used to evaluate the Q and useful frequency range of a given inductor structure, e.g., a permalloy powder toroid. The Q contour is a plot in the inductance–frequency plane; it is a locus of constant Q_L for all inductors of a given structure.

To obtain the Q contour for a *conventional inductor* we recall that any practical inductor is generally approximated by the series connection of an

TABLE 7-1. SINGLE-AMPLIFIER SIMULATION OF GROUNDED INDUCTORS

	1.	2.	3.	4.
Circuit				
Equivalent circuit	$L_{eq} = R_1 R_2 C$ $R_s = 2R_1$	$L_{eq} = R_1 R_2 C$ $R_s = R_1 + R_2$	$L_{eq} = R_1 R_2 C$ $R_p = R_1 R_2/(R_1 + R_2)$	$R_s = R_2 \dfrac{1 + \omega^2 C^2 R_1 R_2}{1 + (\omega C R_2)^2}$ $L_{eq} = \dfrac{CR_2(R_1 - R_2)}{1 + (\omega C R_2)^2}$
Q_L	$\dfrac{\omega L_{eq}}{R_s} = \dfrac{\omega R_2 C}{2}$	$\dfrac{\omega L_{eq}}{R_s} = \omega\left(\dfrac{R_1 R_2}{R_1 + R_2}\right)C$	$\dfrac{R_p}{\omega L_{eq}} = \dfrac{1}{\omega(R_1 + R_2)C}$	$\dfrac{\omega L_{eq}}{R_s} = \dfrac{\omega C(R_1 - R_2)}{1 + \omega^2 C^2 R_1 R_2}$ For large Q_L, let $R_1 \gg R_2 : Q_L \approx \dfrac{1}{2}\sqrt{\dfrac{R_1}{R_2}}$ $\tfrac{1}{2}\left(\sqrt{\dfrac{R_1}{R_2}} - \sqrt{\dfrac{R_2}{R_1}}\right)$ $\dfrac{1}{C\sqrt{R_1 R_2}}$
L_{eq}	$\dfrac{2Q_L R_1}{\omega}$	$\dfrac{Q_L(R_1 + R_2)}{\omega}$	$\dfrac{R_1 R_2/(R_1 + R_2)}{\omega Q_L}$	$\dfrac{Q_L}{\omega}\dfrac{(1 + \omega^2 C^2 R_1 R_2)}{R_2\,(1 + (\omega C R_2)^2)}$

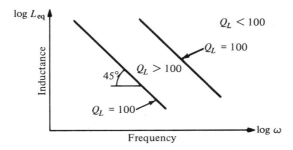

FIG. 7-21. Typical Q contour for idealized conventional inductor.

ideal inductor L and an ideal resistor R_s. The coil Q is then given by $Q_L = \omega L/R_s$ and the Q contour follows from L as a function of Q_L and ω, represented logarithmically, as shown in Fig. 7-21. Thus, ideally, the Q contour of a practical coil comprises bands in the logarithmic L-ω plane, slanted at 45°. These bands correspond to the expression $\log L = \log R_s + \log Q_L - \log \omega$, and, for a given Q_L, will differ according to the series resistance of the coil. Actually, a practical coil is frequency *dependent* with respect to both the inductance L and the series resistance R_s. This necessitates our discarding the above assumption of frequency independence and substituting the functions $L(\omega)$ and $R_s(\omega)$ for L and R, respectively. The resulting Q contour will more likely have the typical shape shown in Fig. 7-22 than that shown in Fig. 7-21.

Q_L is greater than 100 for any point inside the contour and less than 100 for any point outside. Additional contours could be drawn for larger or smaller values of Q_L, where the former would lie inside, the latter outside the contour shown. Naturally, different inductor structures will have different Q contours. The objective of improved inductor structures is to cover an ever larger portion of the L-ω plane with as high-value Q contours as possible.

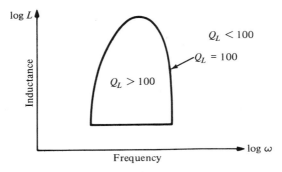

FIG. 7-22. Typical Q contour for a practical conventional inductor.

The Q contours for the *simulated inductors* shown in Table 7-1 follow directly from the expressions given for L_{eq}. For the first three circuits we obtain expressions of the form $\log L_{eq} = K \pm \log Q_L - \log \omega$, which again correspond to bands in the logarithmic L-ω plane as was shown in Fig. 7-21. The factor K follows directly from the expressions for L_{eq} given in Table 7-1. In practice the Q contour of a simulated inductor will also differ from this idealized representation; instead of being open-ended at both sides, the maximum and minimum achievable inductances will be limited by the range of component values that may be used. Furthermore, the upper-frequency end of the contour will be limited by the frequency characteristics of the active devices used. The Q contour of the fourth circuit in Table 7-1 will be slightly more complicated in that it will correspond to a function of the form $\log L_{eq} = K + \log Q_L + \log (1 + a\omega^2) - \log [\omega(1 + b\omega^2)]$, where K and the coefficients a and b again follow from the expression for L_{eq} in Table 7-1. The asymptotes of this expression can readily be plotted as in a Bode plot.

Naturally the usefulness of a given simulated inductance will depend on how large a portion of the L-ω plane is covered, compared, say to an equivalent conventional inductor. This question can be answered only in reference to a given circuit, realized by a given type and range of components.

As is to be expected, an inductor simulated by a nonideal gyrator also has a Q contour of the type shown in Fig. 7-22. This follows readily if we consider the corresponding equivalent circuit, shown in Fig. 7-23. The input admittance of this network is

$$Y_{in} = y_{11} + \frac{g^2}{y_{22} + sC} \tag{7-113}$$

Note that this circuit is quite similar to the equivalent circuit of a conventional inductor, where y_{22}/g^2 represents the copper loss and y_{11} the core loss. If we assume that y_{11}, y_{22}, and g are frequency-independent parameters, then the Q contour shown in Fig. 7-22 results.

The Q contour is a useful tool for the characterization of inductors. In the present context, however, we are more interested in the resonance properties of a simulated active inductor, connected in parallel to a capacitor, than in the characteristics of the inductor alone. For this purpose the Q-ω_0 plot is more suitable.

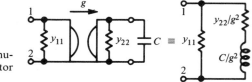

FIG. 7-23. Equivalent circuit of simulated inductor using gyrator–capacitor combination.

The Q-ω_0 Plot By deriving the Q-ω_0 plot for each, we shall in the following discussion compare the performance of the resonant tank circuits comprising the simulated inductors of Table 7-1 and a parallel capacitor C_0. The corresponding equivalent circuits are shown in Table 7-2. Also given are the expressions for Q and the resonance frequency ω_0, as well as Q_{max} and the corresponding frequency ω_M. Notice that the maximum Q depends on the permitted component spread R_{max}/R_{min} and/or C_{max}/C_{min}. To examine the frequency and Q capabilities of these tank circuits, we must therefore assume that the maximum and minimum resistor and capacitor values are given.

Let us first consider circuit 1 in Table 7-2. We can write

$$\omega_0 = \frac{1}{\sqrt{\tau_1 \tau_2}} \tag{7-114}$$

and

$$Q = \frac{1}{2}\sqrt{\frac{\tau_2}{\tau_1}} \tag{7-115}$$

where

$$\tau_1 = R_1 C_0 = \tau_{min}(1 + \Delta_1) \tag{7-116a}$$

and

$$\tau_2 = R_2 C = \tau_{min}(1 + \Delta_2) \tag{7-116b}$$

Furthermore, let

$$\tau_{min} = R_{min} C_{min} \tag{7-117a}$$

and

$$\tau_{max} = R_{max} C_{max} = \tau_{min}(1 + \Delta_{max}) \tag{7-117b}$$

so that

$$\omega_{max} = \frac{1}{\tau_{min}} \tag{7-118a}$$

and

$$\omega_{min} = \frac{1}{\tau_{max}} = \frac{1}{\tau_{min}(1 + \Delta_{max})} = \frac{\omega_{max}}{1 + \Delta_{max}} \tag{7-118b}$$

From the expressions given above we can now determine the plot of Q v. resonant frequency ω_0 as the four components R_1, R_2, C, and C_0 are varied between their limit values:

$$0 \leq \Delta_1 \leq \Delta_{max} \tag{7-119}$$
$$0 \leq \Delta_2 \leq \Delta$$

TABLE 7-2. GROUNDED RESONANT TANKS USING THE INDUCTORS OF TABLE 7-1

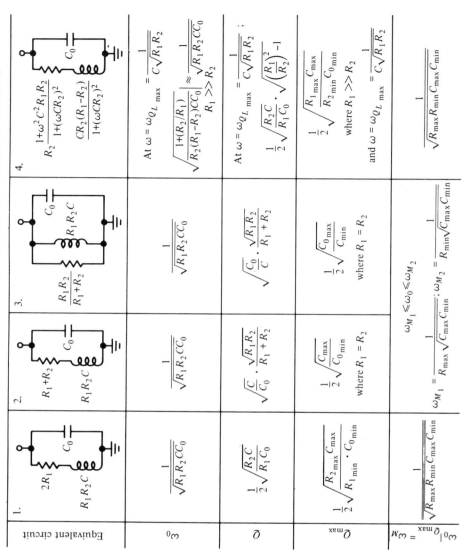

Equivalent circuit	1.	2.	3.	4.
ω_0	$\dfrac{1}{\sqrt{R_1 R_2 C C_0}}$	$\dfrac{1}{\sqrt{R_1 R_2 C C_0}}$	$\dfrac{1}{\sqrt{R_1 R_2 C C_0}}$	At $\omega = \omega_{L\,max}$: $\dfrac{1}{C\sqrt{R_1 R_2}} = \sqrt{\dfrac{1+(R_2/R_1)}{R_2(R_1-R_2)CC_0}} \approx \dfrac{1}{\sqrt{R_1 R_2 C C_0}}$ $R_1 \gg R_2$
Q	$\dfrac{1}{2}\sqrt{\dfrac{R_2 C}{R_1 C_0}}$	$\sqrt{\dfrac{C}{C_0}}\cdot\dfrac{\sqrt{R_1 R_2}}{R_1 + R_2}$	$\sqrt{\dfrac{C_0}{C}}\cdot\dfrac{\sqrt{R_1 R_2}}{R_1 + R_2}$	At $\omega = \omega_{L\,max}$: $\dfrac{1}{C\sqrt{R_1 R_2}}$; $\dfrac{1}{2}\sqrt{\dfrac{R_2 C}{R_1 C_0}}\cdot\sqrt{\left(\dfrac{R_1}{R_2}\right)^2 - 1}$
Q_{max}	$\dfrac{1}{2}\sqrt{\dfrac{R_{2max} C_{max}}{R_{1min}\cdot C_{0min}}}$	$\dfrac{1}{2}\sqrt{\dfrac{C_{max}}{C_{0min}}}$ where $R_1 = R_2$	$\dfrac{1}{2}\sqrt{\dfrac{C_{0max}}{C_{min}}}$ where $R_1 = R_2$	$\dfrac{1}{2}\sqrt{\dfrac{R_{1max} C_{max}}{R_{2min} C_{0min}}}$ where $R_1 \gg R_2$ and $\omega = \omega_{L\,max} = \dfrac{1}{C\sqrt{R_1 R_2}}$
$\omega_M = \omega_0\big|_{Q_{0\,max}}$	$\dfrac{1}{\sqrt{R_{max} R_{min} C_{max} C_{min}}}$	$\omega_{M_1} = \dfrac{1}{R_{max}\sqrt{C_{max} C_{min}}}$; $\omega_{M_1} \le \omega_0 \le \omega_{M_2}$	$\omega_{M_2} = \dfrac{1}{R_{min}\sqrt{C_{max} C_{min}}}$	$\dfrac{1}{C\sqrt{R_{max} R_{min} C_{max} C_{min}}}$

We readily find the limit values for Q, namely

$$\frac{1}{2}\sqrt{\frac{\omega_{min}}{\omega_{max}}} \le Q \le \frac{1}{2}\sqrt{\frac{\omega_{max}}{\omega_{min}}} \tag{7-120}$$

whereby Q_{min} and Q_{max} occur at the frequency

$$\omega_M = \sqrt{\omega_{max}\,\omega_{min}} \tag{7-121}$$

Furthermore

$$Q\big|_{\omega=\omega_{min}} = Q\big|_{\omega=\omega_{max}} = \tfrac{1}{2} \tag{7-122}$$

To find the value of Q for arbitrary frequencies ω_0 we must examine the expression for Q in between the frequencies ω_{min}, ω_M, and ω_{max}. This has been carried out for circuit 1 in Table 7-3. With the expressions given there, we can plot Q as a function of frequency for a given range and spread of resistor and capacitor values. Using a logarithmic scale for Q and frequency, this plot will have the form shown by the solid lines in Fig. 7-24. Naturally the Q values less than 0.5 have more theoretical than practical value, since pole Q's less than 0.5 can be obtained with purely passive RC networks.

TABLE 7-3. FINDING THE Q-ω_0 PLOT FOR THE RESONANT TANKS
OF TABLE 7-2

$\omega_{min} \leq \omega \leq \omega_M$	$\omega_M \leq \omega \leq \omega_{max}$

Circuit 1:

A: $0 \leq \Delta_1 \leq \Delta_{max}$; $\Delta_2 = \Delta_{max}$ C: $\Delta_1 = 0$; $0 \leq \Delta_2 \leq \Delta_{max}$

$$Q = \frac{1}{2}\sqrt{\frac{1 + \Delta_{max}}{1 + \Delta_1}} = \frac{1}{2}\frac{\omega_0}{\omega_{min}} \qquad\qquad Q = \frac{1}{2}\sqrt{1 + \Delta_2} = \frac{1}{2}\frac{\omega_{max}}{\omega_0}$$

B: $\Delta_1 = \Delta_{max}$; $0 \leq \Delta_2 \leq \Delta_{max}$ D: $0 \leq \Delta_1 \leq \Delta_{max}$; $\Delta_2 = 0$

$$Q = \frac{1}{2}\sqrt{\frac{1 + \Delta_2}{1 + \Delta_{max}}} = \frac{1}{2}\frac{\omega_{min}}{\omega_0} \qquad\qquad Q = \frac{1}{2}\frac{1}{\sqrt{1 + \Delta_1}} = \frac{1}{2}\frac{\omega_0}{\omega_{max}}$$

Circuits 2,3:

Sections A, B, C, D: as for circuit 1, but with Δ substituted by γ.

Section E: $Q = \dfrac{1}{2}\sqrt{\dfrac{C_{max}}{C_{min}}}$; $R_{min} \leq R \leq R_{max}$

Section F: $Q = \dfrac{1}{2}\sqrt{\dfrac{C_{min}}{C_{max}}}$; $R_{min} \leq R \leq R_{max}$

To derive the Q-ω_0 plot for circuit 2 of Table 7-2 we proceed as above, bearing in mind, however, that R_1 and R_2 are required to be equal if the maximum possible Q is to be obtained. With this constraint we can write

$$\omega_0 = \frac{1}{RC_{min}\sqrt{(1 + \gamma_1)(1 + \gamma_2)}} \qquad (7\text{-}123)$$

and, consequently

$$Q = \frac{1}{2}\sqrt{\frac{1 + \gamma_2}{1 + \gamma_1}} \qquad (7\text{-}124)$$

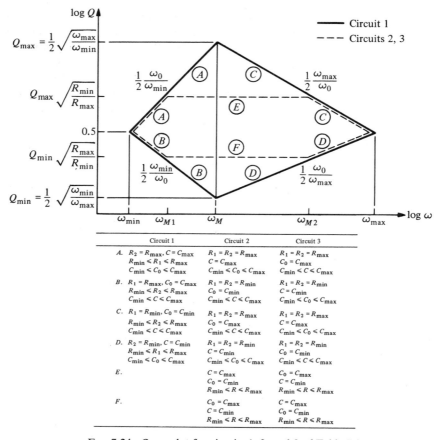

FIG. 7-24. Q-ω_0 plot for circuits 1, 2, and 3 of Table 7-2.

where

$$C_0 = C_{\min}(1 + \gamma_1)$$

$$C = C_{\min}(1 + \gamma_2)$$

$$0 \le \gamma_1 \le \gamma_{\max} \qquad\qquad (7\text{-}125)$$

$$0 \le \gamma_2 \le \gamma_{\max}$$

$$R_{\min} \le R \le R_{\max}$$

$$C_{\max} = C_{\min}(1 + \gamma_{\max})$$

Here again, we readily find the limits of Q to be

$$\frac{1}{2}\sqrt{\frac{C_{\min}}{C_{\max}}} \le Q \le \frac{1}{2}\sqrt{\frac{C_{\max}}{C_{\min}}} \qquad\qquad (7\text{-}126)$$

or, in terms of ω_{min} and ω_{max}:

$$\frac{1}{2}\sqrt{\frac{\omega_{min}}{\omega_{max}}}\sqrt{\frac{R_{max}}{R_{min}}} \le Q \le \frac{1}{2}\sqrt{\frac{\omega_{max}}{\omega_{min}}}\sqrt{\frac{R_{min}}{R_{max}}} \tag{7-127}$$

Clearly the spread of maximum to minimum Q is smaller than for circuit 1 (see (7-120)). Both maximum and minimum Q remain constant between the frequencies

$$\omega_{M1} = \frac{1}{R_{max}C_M} \tag{7-128a}$$

and

$$\omega_{M2} = \frac{1}{R_{min}C_M} \tag{7-128b}$$

where

$$C_M = \sqrt{C_{max}C_{min}} \tag{7-128c}$$

The Q-ω_0 plot resulting from the Q-ω_0 functions listed in Table 7-3 is shown by the dashed lines in Fig. 7-24. Interchanging C and C_0 in (7-125) the Q-ω_0 plot for circuit 3 is obtained. It is, of course, the same as that for circuit 2.

Finally, let us consider circuit 4 in Table 7-2. Although the expressions for Q and Q_{max} are similar to those for circuit 1, this circuit cannot cover as wide a frequency range because of the requirement that $R_2 \ll R$. Since this constraint affects the low-frequency range in particular, this circuit is of less interest to us than the others. Indeed, with the derivation of the Q-ω_0 plots for the circuits of Table 7-2 it is immediately clear that circuit 1 is by far the most useful, since it covers the largest Q range over a given frequency band. Just how dependent this Q—and frequency—range is on the range of component values at our disposal will be demonstrated by the following example.

A NUMERICAL EXAMPLE   To appreciate the foregoing discussion, we shall assume a given resistor and capacitor range and derive the corresponding Q-ω_0 plot for circuits 1, 2, and 3 of Table 7-2. Let us first consider typical component limits imposed by hybrid integrated technology. For thin-film resistors and thin-film or chip capacitors, yields and prices remain reasonable (i.e. processing and procurement are unproblematic) if we specify the following limits:

$$\begin{aligned} 100\ \Omega \le C \le 100\ k\Omega \\ 100\ pF \le C \le 20{,}000\ pF \end{aligned} \tag{7-129}$$

Consequently we obtain

$$80\ Hz \le f \le 16\ MHz \tag{7-130}$$

Then, for circuit 1,

$$f_M = 36 \text{ kHz}$$
$$Q_{max} = 225 \tag{7-131}$$
$$Q_{min} = 1.12.10^{-3}$$

and, for circuits 2 and 3,

$$f_{M1} = 1.125 \text{ kHz}$$
$$f_{M2} = 1.125 \text{ MHz}$$
$$Q_{max} = 7.11 \tag{7-132}$$
$$Q_{min} = 3.53 \cdot 10^{-2}$$

The corresponding Q-ω_0 plot is shown in Fig. 7-25. Note that even for circuit 1, the Q values at low frequencies are very modest. Using discrete components the limit values can be increased considerably. If, for example, we specify,

$$100 \ \Omega \le R \le 1 \text{ M}\Omega$$
$$100 \text{ pF} \le C \le 0.1 \ \mu\text{F} \tag{7-133}$$

then the Q-ω_0 plot is expanded as shown in Fig. 7-25. The improvement in the Q-ω_0 plot is significant, nevertheless, at frequencies below 10 Hz we are still limited to Q values below 3. One method of increasing Q without requiring a still larger spread of components is to use the technique of Q multiplication.

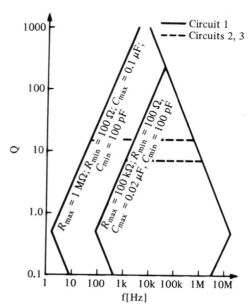

FIG. 7-25. Q-ω_0 plots for circuits 1, 2, and 3 of Table 7-2 for various component ranges.

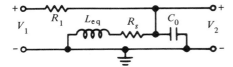

FIG. 7-26. An *RLC* bandpass network.

Q **Multiplication**   As the name implies, *Q* multiplication involves the imbedding of a low-*Q* circuit within an active network such that the pole *Q* is increased. In essence this is the basis for most active-filter techniques: the low *Q* of an *RC* network is boosted to some value larger than 0.5 by appropriate incorporation in an active configuration. Naturally, instead of the passive *RC* network we can incorporate an active network (see FEN technique), an *LC* network, or a simulated *LC* network in the active configuration. Referring to our four single-amplifier network classes (see Chapter 2), the low-*Q* network, whether *RC* or otherwise, must realize a low-pass, high-pass, FRN, or bandpass characteristic, depending on whether it is to be used in a class 1, 2, 3, or 4 configuration, respectively.

To demonstrate the *Q* multiplication technique, consider the *RLC* bandpass network shown in Fig. 7-26. The resonant tank consisting of L_{eq}, R_s, and C_0 may be realized by any of the four circuits shown in Table 7-2. Based on the corresponding Q-ω_0 plots, though, circuit 1 is the most useful. Incorporating the *RLC* bandpass in a class 4 configuration we obtain the network shown in Fig. 7-27a. Analysis of this circuit is straightforward; the expressions obtained will be similar to those of the active *RC* class 4 networks described in chapters 2 and 3. Instead of the passive quantity $\hat{q}$ we will now have the pole *Q* of the *LC* tank. Note, however, that this pole *Q* will be decreased by the presence of the parallel resistors R_a and R_b in the computation of t_{32}

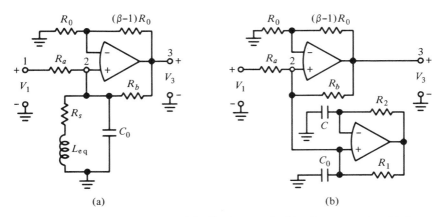

(a)   (b)

FIG. 7-27. Class 4 network with *RLC* feedback network: (a) using conventional inductor; (b) using simulated inductor.

and t_{12}, respectively. Using circuit 1 (Table 7-1) to simulate the inductor we obtain the circuit shown in Fig. 7-27b. Using a series RLC resonant circuit to provide a band-rejection network in the negative-feedback loop of an operational amplifier, class 3 networks can similarly be obtained.

7.2.5 Hybrid Integrated Low–Frequency Filters

We have demonstrated in the foregoing sections that the maximum resistor and capacitor values campatible with hybrid integrated circuit technology severely limit the lower frequency boundary down to which active networks can be realized. We pointed out that for frequencies lower than 100 Hz or so discrete components must be used. This will not be objectionable in many cases, particularly if the resulting size is compared to the LC equivalent; nevertheless, applications will no doubt arise when microcircuit size will be imperative. To comply with this requirement, we must drop one constraint that we have so far adhered to, namely that *the frequency depends only on passive components*. Of course, we thereby introduce a possible additional source of frequency drift caused by variations in the active device, but this is a consequence that cannot readily be avoided. Fortunately, it is also one that is least harmful at low frequencies.

If we permit the frequency to depend on the gain of an active device we must review the properties of class 1 and 2 networks, both of which were rejected on that premise. Based on its root locus (see Fig. 7-28) we recognize immediately that the class 2 network is well suited for low-frequency operation. As the gain β is increased beyond a certain value, the frequency decreases while the pole Q increases. Indeed if we review the general expressions for ω_p and q_p that we derived in Chapter 2 (Section 2.1.1) we have for the root locus shown in Fig. 7-28

$$\omega_p = \frac{\omega_0}{\sqrt{1 + \beta}} \tag{7-134}$$

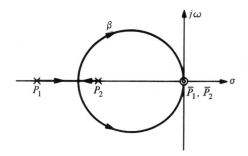

FIG. 7-28. Root locus for a class 2 network.

and

$$q_p = \hat{q}\sqrt{1 + \beta} \qquad (7\text{-}135)$$

Thus

$$\omega_p q_p = \omega_0 \hat{q} \qquad [7\text{-}136\text{a}]$$

or

$$\omega_p = \frac{\hat{q}}{q_p}\omega_0 \qquad [7\text{-}136\text{b}]$$

and

$$\beta = \left(\frac{\omega_p}{\omega_0}\right)^{-2} - 1 \qquad [7\text{-}137]$$

where β is the amplifier gain, $\hat{q}$ and ω_0 are the pole Q and frequency of the RC feedback network, and ω_p and q_p the corresponding frequency and Q of the active network. Thus for a given RC feedback network, ω_0 and $\hat{q}$ are given and the values of q_p and β necessary to decrease ω_p below ω_0 result from (7-136) and (7-137). The relationships among q_p, β, and ω_p is plotted in Fig. 7-29. The interdependence between q_p and ω_p may be undesirable in some cases; in others, attaining a low enough frequency may be more important than the value of q_p associated with it.

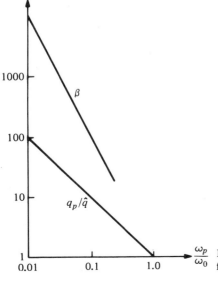

FIG. 7-29. Relationships among q_p, β and ω_p for class 2 network.

Consider, for example, the upper limits on hybrid integrated resistor and capacitor values assumed in (7-129). In order to provide the lowest frequency possible with the passive components we let $R_1 = R_2 = R_{max}$ and $C_1 = C_2 = C_{max}$ and then obtain

$$f_{0min} = \frac{1}{2\pi \cdot 100 \text{ k}\Omega \cdot 0.02 \text{ } \mu\text{F}} \approx 80 \text{ Hz} \qquad (7\text{-}138)$$

The corresponding class 2 bandpass network is shown in Fig. 7-30a. For this circuit we have:

$$\hat{q} = \frac{\sqrt{R_1 R_2 C_1 C_2}}{R_1 C_1 + (R_1 + R_2)C_2}\Bigg|_{\substack{R_1 = R_2 = R_{max} \\ C_1 = C_2 = C_{max}}} = \frac{1}{3} \qquad (7\text{-}139)$$

Thus

$$f_p q_p = f_{0min} \cdot \hat{q} = 27 \text{ Hz} \qquad (7\text{-}140)$$

In order to decrease the pole frequency by an order of magnitude with respect to f_{0min}, we require a gain

$$\beta\Bigg|_{q_p/\hat{q}=10} = \left(\frac{q_p}{\hat{q}}\right)^2 - 1 \approx 99 \qquad (7\text{-}141)$$

Whereas gains of this magnitude may be problematic at higher frequencies, in the vicinity of 8 Hz the open-loop gain of an operational amplifier may be expected to be at least 50 to 60 dB higher. One problem does arise, however; R_2 is equal to R_{max}; the gain-resistor $\beta \cdot R_2$ is therefore equal to βR_{max}, i.e., β times larger than the maximum permissible resistor value. Once again the resistive T network comes to our aid (Fig. 7-30b). We now have

$$\beta = \frac{1}{R_2}\left(R_a + R_b + \frac{R_a R_b}{R_c}\right) \qquad (7\text{-}142)$$

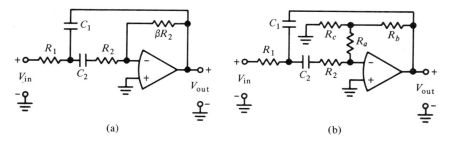

(a)   (b)

FIG. 7-30. Class 2 bandpass network: (a) with gain resistor βR_2 ; (b) with T network.

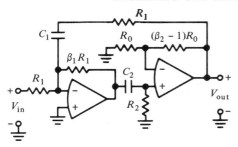

FIG. 7-31. Dual-amplifier expanded version of class 2 bandpass network.

and, assuming that we let $R_a = R_b = R_2 = R_{max}$, we find

$$R_c = \frac{R_{max}}{\beta - 2} \qquad (7\text{-}143)$$

which, for the case in question, is approximately 1 kΩ.

Naturally, as even lower frequencies are desired, the required gain increases rapidly, namely in inverse proportion to the *square* of the decrease in frequency (see Fig. 7-29). Thus, at approximately 3 Hz, β has increased to 1000 and R_c has reached the lowest permissible limit of 100 Ω. Fortunately, it is at these low frequencies that the dual-amplifier expansion described in Chapter 3 becomes feasible. The dual-amplifier expansion of a class 2 bandpass network was in fact discussed in section 3.2.1. The circuit is shown again in Fig. 7-31. At low frequencies the unity-gain bandwidths of the two amplifiers can readily be kept apart by external frequency compensation so that none of the stability problems discussed in Section 3.2.2 need occur. Since $\beta_1 R_1$ will again equal $\beta_1 R_{max}$ it will be necessary to use a resistive-T network. However, β_1 will be sufficiently small because of the distribution of β between β_1 and β_2 to permit an acceptable resistor value for R_c. The actual gain distribution will depend on the maintainable inequality between the unity-gain bandwidths of the two amplifiers. Naturally, β will be distributed as equally between the two amplifiers as the unity-gain bandwidth disparity will allow. In the ideal case $\beta_1 = \beta_2 = \sqrt{\beta}$. Provided both amplifiers are frequency compensated for a 6 dB/oct roll-off, this requires that the unity gain bandwidth of the one amplifier be at least $\sqrt{\beta}$ times larger than that of the other.

7.3 COUPLED FILTERS

We have seen previously how certain active-network topologies and techniques evolved from considerations pertaining to LC networks. Thus the impetus for inductor simulation was largely due to the observations made by Orchard.[19] As we know, he pointed out that double-terminated lossless LC structures have a particularly low sensitivity (to changes in component

19. H. J. Orchard, Inductorless filters, *Electronic Lett.*, **2**, 224–225 (June 1966).

values) inside the passband, where the loss is near zero. As a result, the replacement of inductors in conventional *LC* networks by simulated inductors (using gyrators, single-amplifier configurations, general impedance converters, etc.) was recommended and actively pursued. The multi-amplifier biquad had similar origins, although it was also influenced by the analog-computer simulation of *LC* networks. In the following discussion we shall see once again how observations made on passive *LC* networks resulted in a new approach to active filter design.

In spite of the various convincing arguments for the simulation of *LC* networks by active *RC* circuits, the cascade approach to active-filter design using individually isolated second-order building blocks has dominated the scene, at least insofar as practical, marketable hybrid integrated circuits are concerned. The reasons for this have been expounded on in previous chapters. Briefly, they are based on the efficient production and tuning methods that second-order building blocks afford, and on the fact that it is generally difficult to discredit powerful economic advantages with such niceties as improved insensitivity to component variations. Yet the fact remains that the effect of nonuniform pole frequency drift on the pass band of a high-selectivity cascaded active network may be quite unacceptable, whereas the equivalent noncascaded passive design may still be within specified limits. Thus where a given change in component values may cause the poles of two second-order isolated sections to drift in the *s*-plane as shown in Fig. 7-32a, the *same* change in component values of the equivalent passive network may cause the much smaller drift shown in Fig. 7-32b. It was this observation that led Adams[20] to reexamine the nature of *LC* ladder networks in order to find the mechanism by which their lower sensitivity is achieved. He did this with a view to applying this same mechanism, once found, to cascaded active-filter building blocks, while hoping to maintain the inherent advantages of the latter with

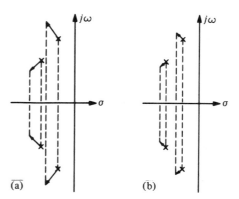

FIG. 7-32. Pole drift for given component variation: (a) active cascaded network; (b) passive *LCR* network.

20. R. L. Adams, On reduced-sensitivity active filters, *Proc. 14th Midwest Symp. Circuit Theory*, May 1971, pp. 14.3-1 to 14.3-8.

respect to production and tuning. As we shall see in what follows, his endeavours were successful. They opened the way to a new concept in active filter design—that of coupled second-order filter sections. As intended, this new concept appears to have the potential for combining the advantages of second-order filter building blocks with the characteristically low sensitivity of equivalent passive networks.

7.3.1 LC Ladder Networks Represented as Coupled Resonators [21]

Let us consider the fourth-order LCR bandpass network shown in Fig. 7-33. Summing the currents into C_1 and rewriting the resulting expression we obtain

$$V_{C_1} = \frac{1}{s}\left(-\frac{V_{C_1}}{R_S C_1} - \frac{I_{L_1}}{C_1} - \frac{I_{L_2}}{C_1} + \frac{I_S}{C_1}\right) \qquad (7\text{-}144a)$$

Similarly

$$I_{L_1} = \frac{1}{s}\frac{V_{C_1}}{L_1} \qquad (7\text{-}144b)$$

$$I_{L_2} = \frac{1}{s}\left(\frac{V_{C_1}}{L_2} - \frac{R_L}{L_2}I_{L_2} - \frac{V_{C_2}}{L_2}\right) \qquad (7\text{-}144c)$$

$$V_{C_2} = \frac{1}{s}\frac{I_{L_2}}{C_2} \qquad (7\text{-}144d)$$

and

$$V_0 = R_L I_{L_2} \qquad (7\text{-}145)$$

The equations in (7-144) with the output relation (7-145) form a state-variable[22] representation of Fig. 7-33 for the following state variables: inductor currents I_{L_1}, I_{L_2} and capacitor voltages V_{C_1} and V_{C_2}. The equations can be represented by the signal-flow graph shown in Fig. 7-34.

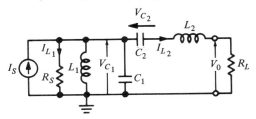

FIG. 7-33. Fourth-order LCR bandpass network.

21. The material covered here is based on published (op. cit.) and unpublished work at Bell Telephone Laboratories by R. L. Adams.
22. For an introduction to state-variable theory see P. M. Derusso, R. J. Roy and C. M. Close, *State Variables for Engineers* (New York: John Wiley & Sons, 1965).

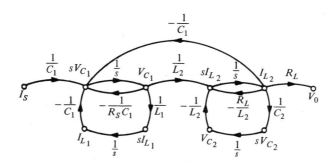

FIG. 7-34. Signal-flow graph representing LCR bandpass network of Fig. 7-33.

The important feature of Fig. 7-34 is the negative feedback path $-1/C_1$ from the output to the input. Combining the partial transfer functions sV_{C_1} to V_{C_1} and sI_{L_2} to I_{L_2}, the significance of this feedback path becomes still more apparent. The partial transfer functions have the form

$$F_i(s) = \frac{s}{s^2 + \frac{\Omega_i}{Q_i}s + \Omega_i^2} \qquad i = 1, 2 \qquad (7\text{-}146)$$

where

$$\Omega_i^2 = \frac{1}{L_i C_i} \qquad (7\text{-}147\text{a})$$

$$Q_1 = R_s\sqrt{\frac{C_1}{L_1}} \qquad (7\text{-}147\text{b})$$

and

$$Q_2 = \frac{1}{R_L}\sqrt{\frac{L_2}{C_2}} \qquad (7\text{-}147\text{c})$$

Hence the signal-flow graph of Fig. 7-34 can be qualitatively represented by the block diagram shown in Fig. 7-35. Note that we have represented our fourth-order LC bandpass network by *a cascade of two second-order bandpass networks in a negative-feedback loop*. This representation is by no means limited to the configuration shown in Fig. 7-33. Following the same procedure as above, the sixth-order bandpass filter of Fig. 7-36 can be represented by

FIG. 7-35. Block diagram corresponding to the signal-flow graph of Fig. 7-34.

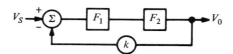

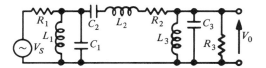

FIG. 7-36. Sixth-order LCR bandpass network.

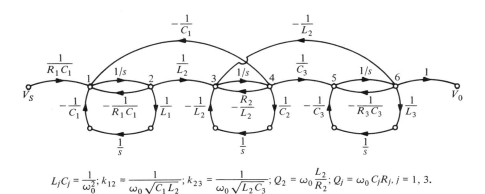

$$L_j C_j = \frac{1}{\omega_0^2}; \ k_{12} = \frac{1}{\omega_0 \sqrt{C_1 L_2}}; \ k_{23} = \frac{1}{\omega_0 \sqrt{L_2 C_3}}; \ Q_2 = \omega_0 \frac{L_2}{R_2}; \ Q_j = \omega_0 C_j R_j, \ j = 1, 3.$$

FIG. 7-37. Signal-flow graph representing the LCR bandpass network of Fig. 7-36.

the signal-flow graph of Fig. 7-37, and this in turn by the multi-loop feedback structure shown in Fig. 7-38, where

$$T(s) = \frac{V_0}{V_S} = \frac{F_1(s)F_2(s)F_3(s)}{1 + k_{12}F_1(s)F_2(s) + k_{23}F_2(s)F_3(s)} \qquad (7\text{-}148)$$

and the $F_i(s)$ functions have bandpass form. Similar representations of conventional low-pass ladder networks have been pointed out[23] and a method for synthesizing nth-order transfer functions with $j\omega$-axis zeros (conventionally realized by passive, double-terminated reactive ladder networks) using

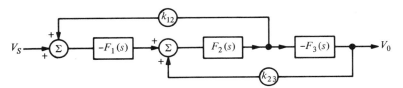

FIG. 7-38. Multi-loop feedback structure corresponding to the signal-flow graph of Fig. 7-37.

23. F. E. Girling and E. F. Good, The leapfrog or active ladder synthesis, Part 12, and Applications of the active ladder synthesis, Part 13, in Active filters, *Wireless World*, Vol. 76, July 1970, pp. 341–345 and Sept, 1970, pp. 445–450.

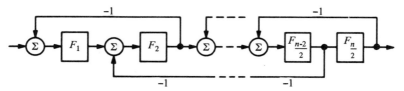

FIG. 7-39. General multi-loop feedback structure.

multiple-feedback structures that are generalizations of Fig. 7-38 has been presented.[24]

In the multiple-feedback structure of Fig. 7-38, the sensitivity to component variations is optimized if the Q of the middle section is infinite. Similarly, in the generalized structure shown in Fig. 7-39 all but the first and last section have infinite pole Q's. Thus the generalized reactive ladder network is simulated by coupled resonator sections whereby the pole Q of all intermediate sections is infinite. The reason for this is that the Q of the intermediate resonators corresponds to the unloaded Q of the ideal LC resonators of the original network. Remember that ideal reactive (i.e., LC) resonators theoretically have $j\omega$-axis poles and are, strictly speaking, unstable. In practice their natural dissipation invariably pushes the poles into the open left-half s-plane. With active resonators this inherent stabilizing effect is absent. Nevertheless, stability of the overall structure is assured because of the feedback loops coupling the resonators. In practice, as long as the Q of the intermediate sections exceeds a minimum value, the exact value may not be critical. The ensuing error can be taken into account by familiar predistortion techniques.

The essential conclusion that Adams drew from his representation of LC ladder networks as coupled resonators was that it must be the "negative-feedback loops" within the passive ladder structures that account for their superior sensitivity properties. Thus, as long as the relationships between voltages and currents of the signal-flow graph of a passive filter can be preserved, then the (sensitivity) properties of the passive-filter transfer function itself will be preserved, irrespective of whether, say, the individual resonators of the flow graph are realized in passive or active form. This has been impressively demonstrated in practice. With that, Adams indicated one possible approach to solving the problem of frequency stability in active filters—a problem that has been repeatedly pointed out in previous chapters. Where we have so far optimized frequency stability by stipulating that the pole (and zero) frequencies depend only on highly stable and temperature-coefficient-matched resistors and capacitors (and often still found the resulting frequency stability inadequate) we now find that by adding negative feedback loops between individual building blocks of a cascade, the sensitivity of the

24. G. Szentirmai, Synthesis of multiple-feedback active filters, *Bell Syst. Tech. J.*, **52**(4), 527–555 (1973).

pass-band characteristics to frequency drift can be considerably reduced. The advantage of this approach is that it can be incorporated into building-block-cascade technology. The individual building blocks are first produced as heretofore, and the feedback loops subsequently added to provide the desired response.

7.3.2 Coupled Building–Block Pairs

In designing active filters using the coupled building-block approach we must ask ourselves what the optimum transfer function of the individual second-order building blocks must be, and what value the coupling coefficients should have, to minimize the transmission sensitivity—at least in the pass-band—of the resulting network. Unfortunately a general answer to this problem does not yet exist, although Szentirmai[25] has provided one for a limited class of transfer functions. However, in the context of our discussion on the properties of hybrid-integrated networks, a question of at least as much importance is by how much we can expect to improve the pole-frequency stability of a coupled network compared to its equivalent cascade. Since the coupling takes place between building-block *pairs* we can answer this question meaningfully by examining the properties of coupled pairs of active-filter building blocks. For this purpose we must first know how arbitrary fourth-order transfer functions are to be realized by the configuration shown in Fig. 7-35. Thus, to realize a given arbitrary fourth-order transfer function $T(s)$, we must know the required functions $F_i(s)$ and the optimum coupling coefficient k. This topic will be dealt with next.

Consider the fourth-order cascaded network shown in Fig. 7-40a. The functions $T_i(s)$ are second-order and have the general form

$$T_i(s) = \left(\frac{n_2 s^2 + n_1 s + n_0}{s^2 + d_1 s + d_0}\right)_i$$

$$= K_i \frac{s^2 + \dfrac{\omega_{z_i}}{q_{z_i}} s + \omega_{z_i}^2}{s^2 + \dfrac{\omega_{p_i}}{q_{p_i}} s + \omega_{p_i}^2} \tag{7-149}$$

The overall transfer function of the cascade (subscript " ca ") is given by

$$T_{ca}(s) = \frac{V_{out}}{V_{in}} = T_1(s)T_2(s) = \frac{N(s)}{D(s)} = \frac{\sum\limits_{i=0}^{4} A_i s^i}{s^4 + \sum\limits_{i=0}^{3} B_i s^i} \tag{7-150}$$

25. Ibid.

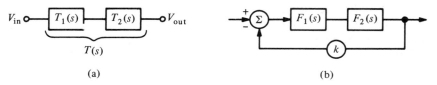

(a)   (b)

FIG. 7-40. Block Diagram of fourth-order network: (a) cascaded; (b) coupled.

or, in terms of its poles and zeros,

$$T_{ca}(s) = K_1 K_2 \frac{\prod\limits_{i=1}^{4} (s - z_i)}{\prod\limits_{i=1}^{4} (s - p_i)} \tag{7-151}$$

We now consider the coupled fourth-order configuration shown in Fig. 7-40b. The corresponding transfer function (subscript "co") is

$$T_{co}(s) = \frac{V_{out}}{V_{in}} = \frac{F(s)}{1 + kF(s)} \tag{7-152}$$

where

$$F(s) = F_1(s) \cdot F_2(s) \tag{7-153}$$

and the functions $F_i(s)$ have the general form

$$F_i(s) = g_i \frac{s^2 + \dfrac{\Omega_{z_i}}{Q_{z_i}} s + \Omega_{z_i}^2}{s^2 + \dfrac{\Omega_{p_i}}{Q_{p_i}} s + \Omega_{p_i}^2} \tag{7-154}$$

$F(s)$ is the transfer function of the open loop and has the form

$$F(s) = F_1(s) F_2(s) = \frac{n(s)}{d(s)} = \frac{\sum\limits_{i=0}^{4} a_i s^i}{s^4 + \sum\limits_{i=0}^{3} b_i s^i} \tag{7-155}$$

or, in terms of the poles and zeros,

$$F(s) = g_1 g_2 \frac{\prod\limits_{i=1}^{4} (s - Z_i)}{\prod\limits_{i=1}^{4} (s - P_i)} \tag{7-156}$$

For a given fourth-order transfer function $T(s)$ we now specify that

$$T_{ca}(s) = T_{co}(s) = T(s) \tag{7-157}$$

or, from (7-150), (7-152), and (7-155),

$$T(s) = \frac{N(s)}{D(s)} = \frac{n(s)}{d(s) + kn(s)} \qquad (7\text{-}158)$$

Thus:

$$\frac{\sum\limits_{i=0}^{4} A_i s^i}{s^4 + \sum\limits_{i=0}^{3} B_i s^i} = \frac{\sum\limits_{i=0}^{4} a_i s^i}{(1 + ka_4)s^4 + \sum\limits_{i=0}^{3} (b_i + ka_i)s^i} \qquad (7\text{-}159)$$

The coefficients A_i and B_i are given and we now wish to find the corresponding coefficients a_i, b_i and the coupling coefficient k. Comparing coefficients in (7-159) will not, in general, provide a unique closed-form solution to this problem; it involves solving a polynomial of fourth degree. Note, however, that *the numerator of $F(s)$ is the same as the given numerator $N(s)$*, so that $F(s)$ will be a function of the same type as the given function $T(s)$. For example, if $T(s)$ is realizable by two FRNs in cascade, then $F(s)$ will also consist of two FRNs in cascade whose zeros are the same as those of $T(s)$. Thus the coupling coefficient k, the poles, and the constants g_1 and g_2 of the functions $F_1(s)$ and $F_2(s)$ remain to be found. The product $g = g_1 g_2$ results directly from the expressions above:

$$g = g_1 g_2 = \frac{K_1 K_2}{1 - k K_1 K_2} = \frac{A_4}{1 - k A_4} \qquad (7\text{-}160)$$

The poles of $F_1(s)$ and $F_2(s)$ depend on k as well as on the poles and zeros of $T(s)$. From (7-152) we have

$$F(s) = \frac{n(s)}{d(s)} = \frac{T(s)}{1 - kT(s)} = \frac{N(s)}{D(s) - kN(s)} \qquad (7\text{-}161)$$

As already pointed out, the zeros of $F(s)$ are the same as those of $T(s)$, i.e., $n(s) = N(s)$. Since $N(s)$ and $D(s)$ are given by $T(s)$, we can now find all possible pole pairs of $F(s) = F_1(s)F_2(s)$ as a function of k that will satisfy $T(s)$. It will then be up to us to determine that particular set of pole pairs, and the coefficient k, that optimizes the resulting coupled structure in some specified way. The following example will serve to illustrate this procedure.

Example: A Coupled Bandpass Pair   Consider the bandpass function

$$T(s) = \frac{\omega_{p_1} \omega_{p_2} s^2}{\left(s^2 + \dfrac{\omega_{p_1}}{q_{p_1}} s + \omega_{p_1}^2\right)\left(s^2 + \dfrac{\omega_{p_2}}{q_{p_2}} s + \omega_{p_2}^2\right)} \qquad (7\text{-}162)$$

Realizing this fourth-order bandpass function by the coupled structure of Fig. 7-40b, we have, from (7-161),

$$F(s) = \frac{\omega_{p_1}\omega_{p_2} s^2}{\left(s^2 + \dfrac{\omega_{p_1}}{q_{p_1}} s + \omega_{p_1}^2\right)\left(s^2 + \dfrac{\omega_{p_2}}{q_{p_2}} s + \omega_{p_2}^2\right) - k's^2} \tag{7-163}$$

where

$$k' = k\omega_{p_1}\omega_{p_2} \tag{7-164}$$

$F(s)$ must have the form

$$F(s) = \frac{\omega_{p_1}\omega_{p_2} s^2}{\left(s^2 + \dfrac{\Omega_{p_1}}{Q_{p_1}} s + \Omega_{p_1}^2\right)\left(s^2 + \dfrac{\Omega_{p_2}}{Q_{p_2}} s + \Omega_{p_2}^2\right)} \tag{7-165}$$

Thus, comparing with (7-163) we find, after multiplying out both denominators, that the constant term in the fourth-order polynomials is unaffected by feedback. Only the s^2 coefficient is changed by the amount $-k's^2$. Thus the effect of feedback is to change the sum of the squared frequencies, while leaving the product unchanged. Consider, for example, the case that $T(s)$ is a prototype low-pass-derived bandpass filter. Then we have

$$q_{p_1} = q_{p_2} = q_{p_0} \tag{7-166a}$$

and

$$\omega_{p_1} = \frac{1}{\alpha}\omega_{p_0}, \qquad \omega_{p_2} = \alpha\omega_{p_0} \tag{7-166b}$$

where

$$\alpha > 1 \tag{7-166c}$$

We can now plot the root locus of the poles of $F(s)$ as a function of k'. This is shown in Fig. 7-41a. Note that the open-loop poles and zeros of $F(s)$, as given by (7-161), are the poles and zeros, respectively, of $T(s)$.

The reason we plot the poles of $F(s)$ corresponding to the root locus of $1 - k'T(s) = 0$ is that the poles and zeros of $T(s)$—which are the open-loop poles and zeros, respectively, of the locus—are given. Actually we are interested in the inverse root locus corresponding to $1 + k'F(s) = 0$ as given by the denominator of $T(s)$ in (7-152); it represents the actual feedback process of the coupled network more accurately. Starting out from the poles of $F(s)$, the arrows of the root locus, corresponding to increasing k', should be reversed as shown in Fig. 7-41b. In this representation we recognize the effect of feedback in the loop: as the feedback coefficient k' is decreased (i.e., negative feedback) or increased (i.e., positive feedback) from zero, the closed-loop poles leave the poles of $F(s)$ and either merge together or shift apart, respectively. The reason that we cannot plot the root locus in the form of Fig. 7-41b is that we do not know the optimum location, in terms of network stability, for the poles of $F(s)$. A method of analyzing the effect of the pole location of $F(s)$ on the stability of the overall network with a view to finding this optimum location will be discussed next.

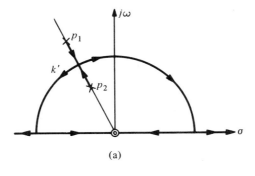

(a)

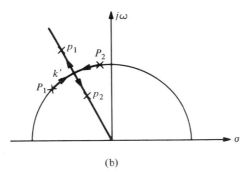

(b)

FIG. 7-41. Root loci: (a) poles of $F(s)$ as a function of k'; (b) poles of $T(s)$ as a function of k'.

7.3.3 The Frequency Stability of Coupled Pairs

We recall that the main incentive for the introduction of coupling between second-order building blocks is to reduce the frequency variation of the coupled structure as compared to that of the pure cascade. Thus if

$$\left(\frac{\Delta\omega_{pi}}{\omega_{pi}}\right)_{c} = f_{\omega_i}(k) \frac{\Delta\omega_{pi}}{\omega_{pi}}, \qquad i = 1, 2 \qquad (7\text{-}167)$$

where the subscript c is to designate the frequency variation of the coupled structure as compared to that of the cascade, then, for an appropriate coupling factor k, we should find that the coefficient

$$f_{\omega_i}(k) < 1 \qquad (7\text{-}168)$$

To find the relationship between $(\Delta\omega_{pi}/\omega_{pi})_c$ and $\Delta\omega_{pi}/\omega_{pi}$ and, incidentally, between the corresponding pole Q variations $(\Delta q_{pi}/q_{pi})_c$ and $\Delta q_{pi}/q_{pi}$, we can calculate the pole variations

$$\frac{\Delta p_i}{p_i} = \frac{\Delta\omega_{pi}}{\omega_{pi}} - j \frac{dq_{pi}/q_{pi}}{\sqrt{4q_{pi}^2 - 1}} \qquad [7\text{-}169]$$

and

$$\left(\frac{\Delta p_i}{p_i}\right)_c = \left(\frac{\Delta \omega_{p_i}}{\omega_{p_i}}\right)_c - j\frac{\left(\frac{dq_{p_i}}{q_{p_i}}\right)_c}{\sqrt{4q_{p_i}^2 - 1}}$$ [7-170]

Having obtained these expressions, the frequency and pole Q stabilities of a cascaded and coupled structure realizing the same transfer function $T(s)$ can be compared directly. This procedure will now be shown for some special cases of the coupled pair shown in Fig. 7-40b. As we shall see, more general cases become too cumbersome to be computed in closed form; they must be calculated numerically by computer. Incidentally, it is sufficient for us to confine our discussion to the poles of $T(s)$, since, as we know, coupling affects only its denominator.

To derive the pole variations of a coupled pair we compute its sensitivity matrix following the procedure outlined in Chapter 4, Section 4.3.1. With the denominator of $T(s)$ given in the form

$$D(s) = \prod_{i=1}^{4}(s - p_i) = s^4 + \sum_{j=0}^{3} B_j s^j$$ (7-171)

we obtain (see Chapter 4, equation (4-172))

$$[\Delta p] = [\tilde{Q}][\Delta B]$$ (7-172a)

or

$$\begin{bmatrix} \Delta p_1 \\ \Delta p_2 \\ \Delta p_3 \\ \Delta p_4 \end{bmatrix} = [\tilde{Q}] \begin{bmatrix} \Delta B_0 \\ \Delta B_1 \\ \Delta B_2 \\ \Delta B_3 \end{bmatrix}$$ (7-172b)

where the tilde over the $\tilde{Q}$ is to avoid confusion with the quality factor Q. The elements of the $\tilde{Q}$ matrix are given by

$$\hat{q}_{ij} = \frac{\Delta p_i}{\Delta B_j} = -\frac{(p_i)^j}{\tilde{Q}_i(p_i)}$$ (7-173)

and

$$\tilde{Q}_i(s) = \frac{D(s)}{s - p_i}$$ (7-174)

Let us now go one step further and express the variations of the coefficients B_j by the relative variations of frequency and Q. For the cascaded network we obtain

$$[\Delta p] = [\tilde{Q}][B] \cdot \left[\frac{\Delta B}{B}\right]$$

$$= [\tilde{Q}][B][S_{\omega,q}^{B_j}]\left[\frac{\Delta \omega}{\omega}, \frac{\Delta q}{q}\right]$$ (7-175)

and, for the coupled network,

$$[\Delta p]_c = [\tilde{Q}][B]\left[\frac{\Delta B}{B}\right]_c$$

$$= [\tilde{Q}][B][S^{B_j}_{\Omega,Q}]\left[\frac{\Delta\Omega}{\Omega}, \frac{\Delta Q}{Q}\right] \tag{7-176}$$

where $S^{B_j}_{\omega,q}$ and $S^{B_j}_{\Omega,Q}$ are the sensitivities to variations in frequency and Q of the coefficients of the cascaded and coupled networks, respectively. The elements of the $\tilde{Q}$ matrix in (7-175) and (7-176) are given by (7-173) and (7-174). To obtain the relative pole variations we obtain, for the cascaded network,

$$\left[\frac{\Delta p}{p}\right] = [P]^{-1}[\tilde{Q}][B][S^{B_j}_{\omega,q}]\left[\frac{\Delta\omega}{\omega}, \frac{\Delta q}{q}\right] \tag{7-177}$$

and, for the coupled network,

$$\left[\frac{\Delta p}{p}\right]_c = [P]^{-1}[\tilde{Q}][B][S^{B_j}_{\Omega,Q}]\left[\frac{\Delta\Omega}{\Omega}, \frac{\Delta Q}{Q}\right] \tag{7-178}$$

To understand these equations, we shall partially expand (7-177). We obtain

$$\frac{\Delta p_1}{p_1} = \frac{\tilde{q}_{10}}{p_1}B_0\frac{\Delta B_0}{B_0} + \frac{\tilde{q}_{11}}{p_1}B_1\frac{\Delta B_1}{B_1} + \frac{\tilde{q}_{12}}{p_1}B_2\frac{\Delta B_2}{B_2} + \frac{\tilde{q}_{13}}{p_1}B_3\frac{\Delta B_3}{B_3}$$

$$\frac{\Delta p_2}{p_2} = \frac{\tilde{q}_{20}}{p_2}B_0\frac{\Delta B_0}{B_0} + \frac{\tilde{q}_{21}}{p_2}B_1\frac{\Delta B_1}{B_1} + \frac{\tilde{q}_{22}}{p_2}B_2\frac{\Delta B_2}{B_2} + \frac{\tilde{q}_{23}}{p_2}B_3\frac{\Delta B_3}{B_3} \tag{7-179}$$

$$\vdots$$

where

$$\frac{\Delta B_0}{B_0} = S^{B_0}_{\omega_{p1}}\frac{\Delta\omega_{p1}}{\omega_{p1}} + S^{B_0}_{\omega_{p2}}\frac{\Delta\omega_{p2}}{\omega_{p2}} + S^{B_0}_{q_{p1}}\frac{\Delta q_{p1}}{q_{p1}} + S^{B_0}_{q_{p2}}\frac{\Delta q_{p2}}{q_{p2}}$$

$$\frac{\Delta B_1}{B_1} = S^{B_1}_{\omega_{p1}}\frac{\Delta\omega_{p1}}{\omega_{p1}} + S^{B_1}_{\omega_{p2}}\frac{\Delta\omega_{p2}}{\omega_{p2}} + S^{B_1}_{q_{p1}}\frac{\Delta q_{p1}}{q_{p1}} + S^{B_1}_{q_{p2}}\frac{\Delta q_{p2}}{q_{p2}} \tag{7-180}$$

$$\vdots$$

The composition of (7-178) is exactly the same, except that the $(\Delta B_j/B_j)_c$ corresponding to (7-180) have eight terms instead of four. The matrices of the type $[P]$, $[P]^{-1}$, and $[B]$ are diagonal matrices with four elements. For example:

$$[P] = \begin{bmatrix} p_1 & & & \\ & p_2 & & \\ & & p_3 & \\ & & & p_4 \end{bmatrix}; \quad [P]^{-1} = \begin{bmatrix} 1/p_1 & & & \\ & 1/p_2 & & \\ & & 1/p_3 & \\ & & & 1/p_4 \end{bmatrix} \tag{7-181}$$

The matrices of the type $[\Delta p]$, $[\Delta p/p]$, $[\Delta p]_c$, $[\Delta p/p]_c$, $[\Delta B/B]$ and $[\Delta B/B]_c$ are column matrices with four elements; for example,

$$
\left[\frac{\Delta B}{B}\right] = \begin{bmatrix} \Delta B_0/B_0 \\ \Delta B_1/B_1 \\ \Delta B_2/B_2 \\ \Delta B_3/B_3 \end{bmatrix} ; \qquad \left[\frac{\Delta p}{p}\right] = \begin{bmatrix} \Delta p_1/p_1 \\ \Delta p_2/p_2 \\ \Delta p_3/p_3 \\ \Delta p_4/p_4 \end{bmatrix} \tag{7-182a}
$$

Furthermore

$$
\left[\frac{\Delta\omega}{\omega}, \frac{\Delta q}{q}\right] = \begin{bmatrix} \dfrac{\Delta\omega_{p1}}{\omega_{p1}} \\[2ex] \dfrac{\Delta\omega_{p2}}{\omega_{p2}} \\[2ex] \dfrac{\Delta q_{p1}}{q_{p1}} \\[2ex] \dfrac{\Delta q_{p2}}{q_{p2}} \end{bmatrix} \tag{7-182b}
$$

and

$$
\left[\frac{\Delta\Omega}{\Omega}, \frac{\Delta Q}{Q}\right] = \begin{bmatrix} \dfrac{\Delta\Omega_{p1}}{\Omega_{p1}} \\[2ex] \dfrac{\Delta\Omega_{p2}}{\Omega_{p2}} \\[2ex] \dfrac{\Delta Q_{p1}}{Q_{p1}} \\[2ex] \dfrac{\Delta Q_{p2}}{Q_{p2}} \\[2ex] \dfrac{\Delta\Omega_{z1}}{\Omega_{z1}} \\[2ex] \dfrac{\Delta\Omega_{z2}}{\Omega_{z2}} \\[2ex] \dfrac{\Delta Q_{z1}}{Q_{z1}} \\[2ex] \dfrac{\Delta Q_{z2}}{Q_{z2}} \end{bmatrix} \tag{7-182c}
$$

Note that the same matrix product $[P]^{-1}[\tilde{Q}][B]$ occurs both in (7-177) and (7-178). This simplifies the computations required to compare $[\Delta p/p]$ with $[\Delta p/p]_c$. In fact, combining matrices as follows:

$$[\tilde{Q}][B][S^{B_j}_{\omega,q}] = [\mathscr{S}^{p_i}_{\omega,q}] \tag{7-183}$$

$$[\tilde{Q}][B]\,S^{B_j}_{\Omega,\varrho} = [\mathscr{S}^{p_i}_{\Omega,\varrho}] \tag{7-184}$$

we recognize that the elements of this matrix product are pole sensitivities of the familiar form $\mathscr{S}^{p_i}_{\omega_{p_i}}$, $\mathscr{S}^{p_i}_{\Omega_i}$, and so on.

In order to obtain a particular pair of pole variations $\Delta p_i/p_i$ and $(\Delta p_i/p_i)_c$ in the form of (7-169) and (7-170), we can generalize the equations given in (7-179) and obtain

$$\frac{\Delta p_i}{p_i} = \frac{1}{p_i}\sum_{j=0}^{3}\tilde{q}_{ij}B_j\frac{\Delta B_j}{B_j} \tag{7-185}$$

and

$$\left(\frac{\Delta p_i}{p_i}\right)_c = \frac{1}{p_i}\sum_{j=0}^{3}\tilde{q}_{ij}B_j\left(\frac{\Delta B_j}{B_j}\right)_c \tag{7-186}$$

where the coefficients B_j and the poles p_i are given by the denominator polynomial $D(s)$ (see (7-171)). With (7-173) and (7-174) we have

$$\begin{aligned}\tilde{q}_{ij}B_j &= -\frac{B_j(p_i)^j}{\tilde{Q}_i(p_i)}\\[2mm] &= -\left.\frac{B_j(p_i)^j(s-p_i)}{D(s)}\right|_{s=p_i} \qquad \begin{array}{l} j = 0,1,2,3\\ i = 1,2,3,4\end{array}\end{aligned} \tag{7-187}$$

so that

$$\frac{\Delta p_i}{p_i} = -\frac{1}{p_i}\left(\frac{s-p_i}{D(s)}\right)_{s=p_i}\sum_{j=0}^{3}B_j(p_i)^j\frac{\Delta B_j}{B_j}, \qquad i = 1,2,3,4 \quad [7\text{-}188]$$

and

$$\left(\frac{\Delta p_i}{p_i}\right)_c = -\frac{1}{p_i}\left(\frac{s-p_i}{D(s)}\right)_{s=p_i}\sum_{j=0}^{3}B_j(p_i)^j\left(\frac{\Delta B_j}{B_j}\right)_c, \qquad i = 1,2,3,4 \quad [7\text{-}189]$$

To obtain the coefficient variations $\Delta B_j/B_j$ and $(\Delta B_j/B_j)_c$ we require the coefficient-sensitivity matrices $[S^{B_j}_{\omega,q}]$ and $[S^{B_j}_{\Omega,\varrho}]$. These are listed in Tables 7-4 and 7-5 respectively.

Let us now rewrite (7-188) and (7-189) in the form

$$\frac{\Delta p_i}{p_i} = -\bar{C}(p_i)\sum_{j=0}^{3}\mu_j(p_i)^j \tag{7-190}$$

TABLE 7-4. COEFFICIENT-SENSITIVITY MATRIX FOR A CASCADED PAIR

$S_{\omega,q}^{B_J}$	ω_{p1}	ω_{p2}	q_{p1}	q_{p2}
B_0	$2\dfrac{\omega_{p1}^2\,\omega_{p2}^2}{B_0}$	$2\dfrac{\omega_{p1}^2\,\omega_{p2}^2}{B_0}$	0	0
B_1	$1+\dfrac{\omega_{p1}^2\omega_{p2}}{B_1 q_{p2}}$	$1+\dfrac{\omega_{p1}\omega_{p2}^2}{B_1 q_{p1}}$	$-\dfrac{\omega_{p1}\omega_{p2}^2}{B_1 q_{p1}}$	$-\dfrac{\omega_{p1}^2\omega_{p2}}{B_1 q_{p2}}$
B_2	$1+\dfrac{\omega_{p1}^2-\omega_{p2}^2}{B_2}$	$1+\dfrac{\omega_{p2}^2-\omega_{p1}^2}{B_2}$	$\dfrac{\omega_{p1}^2+\omega_{p2}^2}{B_2}-1$	$\dfrac{\omega_{p1}^2+\omega_{p2}^2}{B_2}-1$
B_3	$\dfrac{\omega_{p1}}{B_3 q_{p1}}$	$\dfrac{\omega_{p2}}{B_3 q_{p2}}$	$-\dfrac{\omega_{p1}}{B_3 q_{p1}}$	$-\dfrac{\omega_{p2}}{B_3 q_{p2}}$

where

$$\mu_j = B_j \frac{\Delta B_j}{B_j} \tag{7-191}$$

and

$$\left(\frac{\Delta p_i}{p_i}\right)_c = -\bar{C}(p_i)\sum_{j=0}^{3} v_j(p_i)^j \tag{7-192}$$

where

$$v_j = B_j\left(\frac{\Delta B_j}{B_j}\right)_c \tag{7-193}$$

$\bar{C}(p_i)$ is a complex quantity:

$$\bar{C}(p_i) = \frac{1}{p_i}\cdot\frac{s-p_i}{D(s)}\bigg|_{s=p_i} \tag{7-194}$$

and the coefficients μ_j and v_j are real. Thus we can write

$$\left(\frac{\Delta p_i}{p_i}\right)_c = \left(\frac{v_0 + v_1 p_i + v_2\,p_i^2 + v_3\,p_i^3}{\mu_0 + \mu_1 p_i + \mu_2\,p_i^2 + \mu_3\,p_i^3}\right)\frac{\Delta p_i}{p_i}$$

$$= f_{p_i}(k)\frac{\Delta p_i}{p_i} \tag{7-195}$$

The function f_{p_i}, which depends on the coupling factor k, is complex. Consider, for example a pole

$$p_i = -\frac{\omega_{p_i}}{2q_{p_i}}\left(1 - j\sqrt{4q_{p_i}^2 - 1}\right) \tag{7-196}$$

Letting

$$f_{p_i}(k) = \frac{E_1 + jE_2}{H_1 + jH_2} \qquad (7\text{-}197)$$

we then obtain

$$E_1 = v_0 - \left(\frac{\omega_{p_i}}{2q_{p_i}}\right)v_1 - 2\left(\frac{\omega_{p_i}}{2q_{p_i}}\right)^2 (2q_{p_i}^2 - 1)v_2 + 4\left(\frac{\omega_{p_i}}{2q_{p_i}}\right)^3 (3q_{p_i}^2 - 1)v_3 \qquad (7\text{-}198a)$$

and

$$E_2 = \sqrt{4q_{p_i}^2 - 1}\left[\frac{\omega_{p_i}}{2q_{p_i}} v_1 - 2\left(\frac{\omega_{p_i}}{2q_{p_i}}\right)^2 v_2 - 4\left(\frac{\omega_{p_i}}{2q_{p_i}}\right)^3 (q_{p_i}^2 - 1)v_3\right] \qquad (7\text{-}198b)$$

Naturally for $q_{p_i} \gg 1$ these expressions can be simplified accordingly. Substituting the v_j coefficients by μ_j, the expressions for H_1 and H_2 result. Now, one way of optimizing a coupled structure is to minimize the absolute value of $f_{p_i}(k)$, i.e., to find the coupling factor k and the poles of $F(s)$ such that $|f_{p_i}(k)|$ is a minimum. As we have seen, however, it is generally more important to minimize the frequency variation, i.e., the real part of $(\Delta p_i/p_i)_c$, since it is to variations of the pole frequency that the response of a network is most susceptible.

With (7-188) and (7-189) we are in a position to compare the pole variation of a cascaded structure with that of a coupled structure, both realizing a given fourth-order function $T(s)$. This will be illustrated in the following example.

Example: A Coupled Bandpass Pair Let us find the pole variations $(\Delta p_i/p_i)$ and $(\Delta p_i/p_i)_c$ for the fourth-order bandpass function $T(s) = N(s)/D(s)$ given by (7-162) and (7-166). For this case we have

$$N(s) = n(s) = \omega_{p_1}\omega_{p_2}s^2 \qquad (7\text{-}199)$$

Thus

$$\Omega_{z_1} = \Omega_{z_2} = 0 \qquad (7\text{-}200a)$$

$$A_0 = A_1 = A_3 = A_4 = 0 \qquad (7\text{-}200b)$$

and

$$A_2 = \omega_{p_1}\omega_{p_2} \qquad (7\text{-}200c)$$

Furthermore, since

$$D(s) = d(s) + kn(s) \qquad (7\text{-}201)$$

TABLE 7-5. COEFFICIENT-SENSITIVITY MATRIX FOR A COUPLED PAIR

$S_{\Omega,Q}^{B_i}$	Ω_{p_1}	Ω_{p_2}	Q_{p_1}	Q_{p_2}
B_0	$2\dfrac{1-kA_4}{B_0}\Omega_{p_1}^2\Omega_{p_2}^2$	$2\dfrac{1-kA_4}{B_0}$	0	0
B_1	$\dfrac{1-kA_4}{B_1}$ $\times\left(\dfrac{\Omega_{p_1}\Omega_{p_2}^2}{Q_{p_1}}+2\,\dfrac{\Omega_{p_1}^2\Omega_{p_2}}{Q_{p_2}}\right)$	$\dfrac{1-kA_4}{B_1}$ $\times\left(2\,\dfrac{\Omega_{p_1}\Omega_{p_2}^2}{Q_{p_1}}+\dfrac{\Omega_{p_1}^2\Omega_{p_2}}{Q_{p_2}}\right)$	$-\left(\dfrac{1-kA_4}{B_1}\right)\dfrac{\Omega_{p_1}\Omega_{p_2}^2}{Q_{p_1}}$	$-\left(\dfrac{1-kA_4}{B_1}\right)\dfrac{\Omega_{p_1}^2\Omega_{p_2}}{Q_{p_2}}$
B_2	$\dfrac{1-kA_4}{B_2}$ $\times\left(2\,\Omega_{p_1}^2+\dfrac{\Omega_{p_1}\Omega_{p_2}}{Q_{p_1}Q_{p_2}}\right)$	$\dfrac{1-kA_4}{B_2}$ $\times\left(2\Omega_{p_2}^2+\dfrac{\Omega_{p_1}\Omega_{p_2}}{Q_{p_1}Q_{p_2}}\right)$	$-\left(\dfrac{1-kA_4}{B_2}\right)\dfrac{\Omega_{p_1}\Omega_{p_2}}{Q_{p_1}Q_{p_2}}$	$-\left(\dfrac{1-kA_4}{B_2}\right)\dfrac{\Omega_{p_1}\Omega_{p_2}}{Q_{p_1}Q_{p_2}}$
B_3	$\left(\dfrac{1-kA_4}{B_3}\right)\dfrac{\Omega_{p_1}}{Q_{p_1}}$	$\left(\dfrac{1-kA_4}{B_3}\right)\dfrac{\Omega_{p_2}}{Q_{p_2}}$	$-\left(\dfrac{1-kA_4}{B_3}\right)\dfrac{\Omega_{p_1}}{Q_{p_1}}$	$-\left(\dfrac{1-kA_4}{B_3}\right)\dfrac{\Omega_{p_2}}{Q_{p_2}}$

TABLE 7-5.—(Continued)

$S_{\Omega,Q}^{B_j}$	Ω_{z_1}	Ω_{z_2}	Q_{z_1}	Q_{z_2}
B_0	$2\dfrac{kA_4}{B_0}\Omega_{z_1}^2\Omega_{z_2}^2$	$2\dfrac{kA_4}{B_0}\Omega_{z_1}^2\Omega_{z_2}^2$	0	0
B_1	$\dfrac{kA_4}{B_1}$ $\times\left(\dfrac{\Omega_{z_1}\Omega_{z_2}^2}{Q_{z_1}} + 2\,\dfrac{\Omega_{z_1}^2\Omega_{z_2}}{Q_{z_2}}\right)$	$\dfrac{kA_4}{B_1}$ $\times\left(2\,\dfrac{\Omega_{z_1}\Omega_{z_2}^2}{Q_{z_1}} + \dfrac{\Omega_{z_1}^2\Omega_{z_2}}{Q_{z_2}}\right)$	$-\dfrac{kA_4}{B_1}\dfrac{\Omega_{z_1}\Omega_{z_2}^2}{Q_{z_1}}$	$-\dfrac{kA_4}{B_1}\dfrac{\Omega_{z_1}^2\Omega_{z_2}}{Q_{z_2}}$
B_2	$\dfrac{kA_4}{B_2}\left(2\Omega_{z_1}^2 + \dfrac{\Omega_{z_1}\Omega_{z_2}}{Q_{z_1}Q_{z_2}}\right)$	$\dfrac{kA_4}{B_2}\left(2\Omega_{z_2}^2 + \dfrac{\Omega_{z_1}\Omega_{z_2}}{Q_{z_1}Q_{z_2}}\right)$	$-\left(\dfrac{kA_4}{B_2}\right)\dfrac{\Omega_{z_1}\Omega_{z_2}}{Q_{z_1}Q_{z_2}}$	$-\left(\dfrac{kA_4}{B_2}\right)\dfrac{\Omega_{z_1}\Omega_{z_2}}{Q_{z_1}Q_{z_2}}$
B_3	$\dfrac{kA_4}{B_3}\dfrac{\Omega_{z_1}}{Q_{z_1}}$	$\dfrac{kA_4}{B_3}\dfrac{\Omega_{z_2}}{Q_{z_2}}$	$-\left(\dfrac{kA_4}{B_3}\right)\dfrac{\Omega_{z_1}}{Q_{z_1}}$	$-\left(\dfrac{kA_4}{B_3}\right)\dfrac{\Omega_{z_2}}{Q_{z_2}}$

we have:

$$B_0 = \omega_{p_1}^2 \omega_{p_2}^2 = \Omega_{p_1}^2 \Omega_{p_2}^2 = \omega_{po}^4 \qquad (7\text{-}202a)$$

$$\begin{aligned} B_1 &= \frac{\omega_{p_1}^2 \omega_{p_2}}{q_{p_2}} + \frac{\omega_{p_2}^2 \omega_{p_1}}{q_{p_1}} \\[2mm] &= \frac{\Omega_{p_1}^2 \Omega_{p_2}}{Q_{p_2}} + \frac{\Omega_{p_1} \Omega_{p_2}^2}{Q_{p_1}} \\[2mm] &= \frac{\omega_{po}^3}{q_{po}} \left(\frac{1}{\alpha} + \alpha \right) \end{aligned} \qquad (7\text{-}202b)$$

$$\begin{aligned} B_2 &= \omega_{p_1}^2 + \omega_{p_2}^2 + \frac{\omega_{p_1} \omega_{p_2}}{q_{p_1} q_{p_2}} \\[2mm] &= \Omega_{p_1}^2 + \Omega_{p_2}^2 + \frac{\Omega_{p_1} \Omega_{p_2}}{Q_{p_1} Q_{p_2}} + \omega_{p_1} \omega_{p_2} k \\[2mm] &= \omega_{po}^2 \left(\frac{1}{\alpha^2} + \alpha^2 + \frac{1}{q_{po}^2} \right) \end{aligned} \qquad (7\text{-}202c)$$

$$\begin{aligned} B_3 &= \frac{\omega_{p_1}}{q_{p_1}} + \frac{\omega_{p_2}}{q_{p_2}} = \frac{\Omega_{p_1}}{Q_{p_1}} + \frac{\Omega_{p_2}}{Q_{p_2}} \\[2mm] &= \frac{\omega_{po}}{q_{po}} \left(\frac{1}{\alpha} + \alpha \right) \end{aligned} \qquad (7\text{-}202d)$$

From (7-202b) and (7-202d) it follows that

$$\Omega_{p_1} = \Omega_{p_2} = \omega_{po} \qquad (7\text{-}203)$$

Furthermore,

$$\frac{Q_{p_1} Q_{p_2}}{Q_{p_1} + Q_{p_2}} = Q_{po} = \frac{\omega_{po} q_{po}}{\omega_{p_1} + \omega_{p_2}} \qquad (7\text{-}204)$$

and, with (7-202c),

$$k = \left(\frac{q_{po}}{Q_{po}} \right)^2 + \frac{1}{q_{po}^2} - 4 - \frac{1}{Q_{p_1} Q_{p_2}} \qquad (7\text{-}205)$$

Equations (7-204) and (7-205) relate Q_1 to Q_2 as a function of the parameter k. Solving for Q_1 we obtain a quadratic equation

$$Q_{p_1}^2 + a Q_{p_1} + b = 0 \qquad (7\text{-}206)$$

where

$$a = \frac{Q_{p_2}(2q_{po}^2 - 1)}{q_{po}^2 - Q_{p_2}^2 \left(4 + k - \frac{1}{q_{po}^2} \right)} \qquad (7\text{-}207a)$$

and

$$b = \frac{Q_{p_2}^2 q_{p_0}^2}{q_{p_0}^2 - Q_{p_2}^2 \left(4 + k - \dfrac{1}{q_{p_0}^2}\right)} \tag{7-207b}$$

Solving (7-206) with (7-204) we obtain the quantities $Q_{p_1}(k)$ and $Q_{p_2}(k)$ which, of course, take on optimum values if we can find an optimum coupling factor k. The quantities ω_{p_i} and q_{p_0} are given by the desired function $T(s)$ (see (7-166)). Assuming that the solutions $Q_{p_1}(k)$ and $Q_{p_2}(k)$ are known, we obtain the coefficient sensitivities for our cascaded and coupled bandpass pair as listed in Tables 7-6 and 7-7, respectively. With (7-200a) and these sensitivities, the corresponding coefficient variations are then

$$\frac{\Delta B_j}{B_j} = \sum_{i=1}^{2} S_{\omega_{p_i}}^{B_j} \frac{\Delta \omega_{p_i}}{\omega_{p_i}} + \sum_{i=1}^{2} S_{q_{p_i}}^{B_j} \frac{\Delta q_{p_i}}{q_{p_i}} \tag{7-208}$$

and

$$\left(\frac{\Delta B_j}{B_j}\right)_c = \sum_{i=1}^{2} S_{\Omega_{p_i}}^{B_j} \frac{\Delta \Omega_{p_i}}{\Omega_{p_i}} + \sum_{i=1}^{2} S_{Q_{p_i}}^{B_j} \frac{\Delta Q_{p_i}}{Q_{p_i}} \tag{7-209}$$

Now, to simplify the following computations, let us find the pole variation for the case that only the pole *frequency* of *one* of the sections in the cascade, or of *one* of the cascaded sections in the coupled loop, varies. If, for example

$$\frac{\Delta \omega_{p_1}}{\omega_{p_1}} = \frac{\Delta q_{p_1}}{q_{p_1}} = \frac{\Delta q_{p_2}}{q_{p_2}} = 0 \tag{7-210a}$$

and

$$\frac{\Delta \Omega_{p_1}}{\Omega_{p_1}} = \frac{\Delta Q_{p_1}}{Q_{p_1}} = \frac{\Delta Q_{p_2}}{Q_{p_2}} = 0 \tag{7-210b}$$

TABLE 7-6. COEFFICIENT-SENSITIVITY MATRIX FOR CASCADED BANDPASS PAIR

$S_{\omega,q}^{B_j}$	ω_{p_1}	ω_{p_2}	q_{p_1}	q_{p_2}
B_0	2	2	0	0
B_1	$\dfrac{2 + \alpha^2}{1 + \alpha^2}$	$\dfrac{1 + 2\alpha^2}{1 + \alpha^2}$	$-\dfrac{\alpha^2}{1 + \alpha^2}$	$-\dfrac{1}{1 + \alpha^2}$
B_2	$\dfrac{\dfrac{2}{\alpha^2} + \dfrac{1}{q_{p_0}^2}}{\dfrac{1}{\alpha^2} + \alpha^2 + \dfrac{1}{q_{p_0}^2}}$	$\dfrac{2\alpha^2 + \dfrac{1}{q_{p_0}^2}}{\dfrac{1}{\alpha^2} + \alpha^2 + \dfrac{1}{q_{p_0}^2}}$	$-\dfrac{\dfrac{1}{q_{p_0}^2}}{\dfrac{1}{\alpha^2} + \alpha^2 + \dfrac{1}{q_{p_0}^2}}$	$-\dfrac{\dfrac{1}{q_{p_0}^2}}{\dfrac{1}{\alpha^2} + \alpha^2 + \dfrac{1}{q_{p_0}^2}}$
B_3	$\dfrac{1}{1 + \alpha^2}$	$\dfrac{\alpha^2}{1 + \alpha^2}$	$-\dfrac{1}{1 + \alpha^2}$	$-\dfrac{\alpha^2}{1 + \alpha^2}$

TABLE 7-7. COEFFICIENT-SENSITIVITY MATRIX FOR COUPLED BANDPASS PAIR

$S_{\Omega,Q}^{B_j}$	Ω_{p1}	Ω_{p2}
B_0	2	2
B_1	$\dfrac{2\Omega_{p1}Q_{p1}+\Omega_{p2}Q_{p2}}{\Omega_{p1}Q_{p1}+\Omega_{p2}Q_{p2}}=\dfrac{2Q_{p1}+Q_{p2}}{Q_{p1}+Q_{p2}}$	$\dfrac{\Omega_{p1}Q_{p1}+2\Omega_{p2}Q_{p2}}{\Omega_{p1}Q_{p1}+\Omega_{p2}Q_{p2}}=\dfrac{Q_{p1}+2Q_{p2}}{Q_{p1}+Q_{p2}}$
B_2	$\dfrac{2\Omega_{p1}^2\,Q_{p1}Q_{p2}+\Omega_{p1}\Omega_{p2}}{(\Omega_{p1}^2+\Omega_{p2}^2)Q_{p1}Q_{p2}+\Omega_{p1}\Omega_{p2}+k'Q_{p1}Q_{p2}}$ $=\dfrac{2Q_{p1}Q_{p2}+1}{Q_{p1}Q_{p2}(2+k)+1}$	$\dfrac{2\Omega_{p2}^2\,Q_{p1}Q_{p2}+\Omega_{p1}\Omega_{p2}}{(\Omega_{p1}^2+\Omega_{p2}^2)Q_{p1}Q_{p2}+\Omega_{p1}\Omega_{p2}+k'Q_{p1}Q_{p2}}$ $=\dfrac{2Q_{p1}Q_{p2}+1}{Q_{p1}Q_{p2}(2+k)+1}$
B_3	$\dfrac{\Omega_{p1}Q_{p2}}{\Omega_{p1}Q_{p2}+\Omega_{p2}Q_{p1}}=\dfrac{Q_{p2}}{Q_{p1}+Q_{p2}}$	$\dfrac{\Omega_{p2}Q_{p1}}{\Omega_{p1}Q_{p2}+\Omega_{p2}Q_{p1}}=\dfrac{Q_{p1}}{Q_{p1}+Q_{p2}}$

TABLE 7-7.—(*Continued*)

$S_{\Omega,Q}^{B_j}$	Q_{p1}	Q_{p2}
B_0	0	0
B_1	$-\dfrac{\Omega_{p1}\Omega_{p2}^2 Q_{p2}}{\Omega_{p1}\Omega_{p2}^2 Q_{p2} + \Omega_{p1}^2\Omega_{p2}Q_{p1}} = \dfrac{-Q_{p2}}{Q_{p1}+Q_{p2}}$	$-\dfrac{\Omega_{p1}^2\Omega_{p2}Q_{p1}}{\Omega_{p1}\Omega_{p2}^2 Q_{p2} + \Omega_{p1}^2\Omega_{p2}Q_{p1}} = \dfrac{-Q_{p1}}{Q_{p1}+Q_{p2}}$
B_2	$\dfrac{-\Omega_{p1}\Omega_{p2}}{(\Omega_{p1}^2+\Omega_{p2}^2)Q_{p1}Q_{p2}+\Omega_{p1}\Omega_{p2}Q_{p2}+k'Q_{p1}Q_{p2}}$ $= \dfrac{-1}{Q_{p1}Q_{p2}(2+k)+1}$	$\dfrac{-\Omega_{p1}\Omega_{p2}}{(\Omega_{p1}^2+\Omega_{p2}^2)Q_{p1}Q_{p2}+\Omega_{p1}\Omega_{p2}Q_{p2}+k'Q_{p1}Q_{p2}}$ $= \dfrac{-1}{Q_{p1}Q_{p2}(2+k)+1}$
B_3	$\dfrac{-\Omega_{p1}Q_{p2}}{\Omega_{p1}Q_{p2}+\Omega_{p2}Q_{p1}} = \dfrac{-Q_{p2}}{Q_{p1}+Q_{p2}}$	$\dfrac{-\Omega_{p2}Q_{p1}}{\Omega_{p1}Q_{p2}+\Omega_{p2}Q_{p1}} = \dfrac{-Q_{p1}}{Q_{p1}+Q_{p2}}$

then (7-208) yields

$$\frac{\Delta B_j}{B_j} = S_{\omega_{p2}}^{B_j} \frac{\Delta \omega_{p2}}{\omega_{p2}} \tag{7-211}$$

and (7-209) yields

$$\left(\frac{\Delta B_j}{B_j}\right)_c = S_{\Omega_{p2}}^{B_j} \frac{\Delta \Omega_{p2}}{\Omega_{p2}} \tag{7-212}$$

In the present context it is to be assumed that both the cascade and the coupled structure comprise two second-order building blocks in cascade, the former without coupling and the latter with. The pole frequencies of these building blocks will be determined only by resistors and capacitors and, if the building blocks are in hybrid integrated form, the frequency variation $\Delta \omega_{p2}/\omega_{p2}$ will be the same as $\Delta \Omega_{p2}/\Omega_{p2}$. For uniform (i.e., tracking) resistor variations $\Delta R/R$ and capacitor variations $\Delta C/C$, we can therefore write

$$\frac{\Delta \omega_{p2}}{\omega_{p2}} = \frac{\Delta \Omega_{p2}}{\Omega_{p2}} = \frac{\Delta \Omega_0}{\Omega_0} = -\left(\frac{\Delta R}{R} + \frac{\Delta C}{C}\right) \tag{7-213}$$

With (7-211), the pole variation of, say, p_1 then becomes, with (7-188),

$$\frac{\Delta p_1}{p_1} = \frac{-\Delta \Omega_0/\Omega_0}{p_1(p_1 - p_2)(p_1 - p_3)(p_1 - p_4)} \sum_{j=0}^{3} \lambda_j(\omega_{p2})(p_1)^j \tag{7-214a}$$

where

$$\lambda_j(\omega_{p2}) = B_j S_{\omega_{p2}}^{B_j} \tag{7-214b}$$

Similarly (7-189) becomes

$$\left(\frac{\Delta p_1}{p_1}\right)_c = \frac{-\Delta \Omega_0/\Omega_0}{p_1(p_1 - p_2)(p_1 - p_3)(p_1 - p_4)} \sum_{j=0}^{3} \gamma_j(\Omega_{p2})(p_1)^j \tag{7-215a}$$

where

$$\gamma_j(\Omega_{p2}) = B_j S_{\Omega_{p2}}^{B_j} \tag{7-215b}$$

With Table 7-4 and equations (7-166) and (7-202) we obtain

$$\lambda_0(\omega_{p2}) = 2\omega_{po}^4 \tag{7-216a}$$

$$\lambda_1(\omega_{p2}) = \frac{\omega_{po}^3}{q_{po}}\left(\frac{1}{\alpha} + 2\alpha\right) \tag{7-216b}$$

$$\lambda_2(\omega_{p2}) = \omega_{po}^2\left(\frac{1}{q_{po}^2} + 2\alpha^2\right) \tag{7-216c}$$

$$\lambda_3(\omega_{p2}) = \alpha \frac{\omega_{po}}{q_{po}} \tag{7-216d}$$

and with Table 7-5 and equations (7-200), (7-202), and (7-203) we obtain

$$\gamma_0(\Omega_{p2}) = 2\omega_{po}^4 \tag{7-217a}$$

$$\gamma_1(\Omega_{p_2}) = \omega_{po}^3 \left(\frac{2}{Q_{p_1}} + \frac{1}{Q_{p_2}} \right) \tag{7-217b}$$

$$\gamma_2(\Omega_{p_2}) = \omega_{po}^2 \left(2 + \frac{1}{Q_{p_1} Q_{p_2}} \right) \tag{7-217c}$$

$$\gamma_3(\Omega_{p_2}) = \frac{\omega_{po}}{Q_{p_2}} \tag{7-217d}$$

Remember that α, ω_{po}, and q_{po} are determined by the given transfer function $T(s)$. Q_{p_1} and Q_{p_2} are not determined; they depend on the coupling factor k. As discussed earlier, this dependence is contained in equations (7-204) and (7-205). It will now be the task of the designer to find the optimum coupling factor k_{opt} such that, say $(\Delta\omega_{p_1}/\omega_{p_1})_c$ is minimum, or at least sufficiently less than $\Delta\omega_{p_1}/\omega_{p_1}$. This requires comparing the real part of (7-214a) with that of (7-215a) for different values of k,—a computation well suited for the computer.

Instead of looking for the optimum coupling factor let us now consider a special case, namely that of two cascaded *identical* high-Q bandpass networks with the feedback k_0 around them. Thus beside Ω_{p_1} being equal to Ω_{p_2} (see (7-203)) we now have, with (7-204),

$$Q_{p_1} = Q_{p_2} = 2Q_{po} \tag{7-218}$$

and

$$Q_{po} = \frac{\omega_{po}}{\omega_{p_1} + \omega_{p_2}} q_{po} = \frac{\alpha}{1 + \alpha^2} q_{po} \tag{7-219}$$

Furthermore, with (7-205) we obtain:

$$k_0 = \frac{k_0'}{\omega_{po}^2} = \frac{1}{\omega_{po}^2} \left(1 - \frac{1}{4q_{po}^2} \right) \left(\omega_{p_1}^2 + \omega_{p_2}^2 - 2\omega_{po}^2 \right)$$

$$= \left(1 - \frac{1}{4q_{po}^2} \right) \left(\frac{1}{\alpha^2} + \alpha^2 - 2 \right) \tag{7-220}$$

The coefficients in (7-217) then become

$$\gamma_0(\Omega_{p_2}) = 2\omega_{po}^4 \tag{7-221a}$$

$$\gamma_1(\Omega_{p_2}) = \frac{3}{2Q_{po}} \omega_{po}^3 = 3 \left(\frac{1 + \alpha^2}{2\alpha} \right) \frac{\omega_{po}^3}{q_{po}} \tag{7-221b}$$

$$\gamma_2(\Omega_{p_2}) = \left(2 + \frac{1}{4Q_{po}^2} \right) \omega_{po}^2 = \left[2 + \left(\frac{1 + \alpha^2}{2\alpha} \right)^2 \frac{1}{q_{po}^2} \right] \omega_{po}^2 \tag{7-221c}$$

$$\gamma_3(\Omega_{p_2}) = \frac{\omega_{po}}{2Q_{po}} = \left(\frac{1 + \alpha^2}{2\alpha} \right) \frac{\omega_{po}}{q_{po}} \tag{7-221d}$$

while, of course, the coefficients λ_j in (7-216) remain unchanged.

Let us now rewrite (7-214a) and (7-215a) in the following form:

$$\frac{\Delta p_1}{p_1} = - \bar{C}(p_1) \cdot P(p_1) \tag{7-222a}$$

$$\left(\frac{\Delta p_1}{p_1}\right)_c = - \bar{C}(p_1) \cdot P_c(p_1) \tag{7-222b}$$

Then, in analogy to (7-195), we have

$$\left(\frac{\Delta p_1}{p_1}\right)_c = \frac{P_c(p_1)}{P(p_1)}\frac{\Delta p_1}{p_1} = f_{p_1}(k)\frac{\Delta p_1}{p_1} \tag{7-223}$$

The pole p_1 is given by

$$p_1 = -\frac{\omega_{p_1}}{2q_{p_1}}\left[1 - j\sqrt{4q_{p_1}^2 - 1}\right] \tag{7-224}$$

and, for $q_{p_1} \gg 1$,

$$p_1 \approx -\frac{\omega_{p_1}}{2q_{p_1}} + j\omega_{p_1} \tag{7-225}$$

In order to obtain $f_{p_1}(k)$ in the form of (7-197) we have

$$\mathrm{Re}\,(P) = \mathrm{Re}\left[\sum_{j=0}^{3}\lambda_j(\omega_{p_2})(p_1)^j\right] = H_1(p_1)$$

$$= \lambda_0 - \lambda_1\left(\frac{\omega_{p_1}}{2q_{p_1}}\right) - 2\lambda_2\left(\frac{\omega_{p_1}}{2q_{p_1}}\right)^2(2q_{p_1}^2 - 1) + 4\lambda_3\left(\frac{\omega_{p_1}}{2q_{p_1}}\right)^3(3q_{p_1}^2 - 1) \tag{7-226}$$

and, for $q_p \gg 1$,

$$\mathrm{Re}\,(P) \approx \lambda_0 - \lambda_1\left(\frac{\omega_{p_1}}{2q_{p_1}}\right) - \lambda_2\,\omega_{p_1}^2 + \tfrac{3}{2}\lambda_3\,\frac{\omega_{p_1}^3}{q_{p_1}} \tag{7-227}$$

Similarly,

$$\mathrm{Im}\,(P) = \mathrm{Im}\left[\sum_{j=0}^{3}\lambda_j(\omega_{p_2})(p_1)^j\right] = H_2(p_1)$$

$$= \sqrt{4q_{p_1}^2 - 1}\left[\lambda_1\left(\frac{\omega_{p_1}}{2q_{p_1}}\right) - 2\lambda_2\left(\frac{\omega_{p_1}}{2q_{p_1}}\right)^2 - 4\lambda_3\left(\frac{\omega_{p_1}}{2q_{p_1}}\right)^3(q_{p_1}^2 - 1)\right] \tag{7-228}$$

and, for $q_p \gg 1$,

$$\mathrm{Im}\,(P) \approx \lambda_1\omega_{p_1} - \lambda_2\frac{\omega_{p_1}^2}{q_{p_1}} - \lambda_3\,\omega_{p_1}^3 \tag{7-229}$$

Substituting the λ_j coefficients by γ_j yields the corresponding expressions for P_c and with them $E_1(p_1)$ and $E_2(p_1)$ as in (7-197). On the other hand, combining these expressions with the complex quantity $\bar{C}(p_1)$ as indicated in (7-222) and comparing the real parts, we obtain the coefficient $f\omega_1(k_0)$ as indicated in (7-167). This coefficient informs us as to the improvement, with respect to frequency stability, obtained in designing our fourth-order bandpass network by using a feedback-coupled pair of identical, high-Q, second-order bandpass networks as opposed to an equivalent cascaded, and mutually isolated, pair.

7.3.4 The Transmission Variation of Coupled Pairs

Having calculated—and compared—the relative variation of the poles of a cascaded and a coupled pair of second-order networks, it is a simple matter, at least in principle, to obtain the corresponding transmission variation. With $T(s)$ given by its poles and zeros (as in (7-151)), and recognizing that only the poles are affected by the coupling between pairs, we need consider the transmission variation only with respect to *pole* variations and obtain for the cascaded pair (subscript "*ca*")

$$\left(\frac{\Delta T(s)}{T(s)}\right)_{ca} = \sum_{i=1}^{n} S_{p_i}^{T(s)} \cdot \left(\frac{\Delta p_i}{p_i}\right)_{ca} \qquad [7\text{-}230]$$

and for the coupled pair (subscript "*co*")

$$\left(\frac{\Delta T(s)}{T(s)}\right)_{co} = \sum_{i=1}^{n} S_{p_i}^{T(s)} \left(\frac{\Delta p_i}{p_i}\right)_{co} \qquad [7\text{-}231]$$

where $n = 4$ and (see (1-150), Chapter 1)

$$S_{p_i}^{T(s)} = \frac{p_i}{s - p_i} \qquad [7\text{-}232]$$

For $s = j\omega$ the real and imaginary parts of the resulting frequency-dependent expressions correspond to the gain and phase variations respectively of the two filter structures.

Another method of obtaining the transmission variation of the cascaded and coupled pair is to start out from the basic relations pertaining to Figures 7-40a and b as given by (7-150) and (7-152), respectively. Thus:

$$\left(\frac{\Delta T(s)}{T(s)}\right)_{ca} = \frac{\Delta T_1(s)}{T_1(s)} + \frac{\Delta T_2(s)}{T_2(s)} \qquad [7\text{-}233]$$

and

$$\left(\frac{\Delta T(s)}{T(s)}\right)_{co} = (1 - kT)\left(\frac{\Delta F_1(s)}{F_1(s)} + \frac{\Delta F_2(s)}{F_2(s)}\right)$$

$$= \frac{T(s)}{F(s)}\left(\frac{\Delta F_1(s)}{F_1(s)} + \frac{\Delta F_2(s)}{F_2(s)}\right) \qquad [7\text{-}234]$$

These expressions must be the same as those given by (7-230) and (7-231), respectively. The variations $\Delta T_i/T_i$ and $\Delta F_i/F_i$ are caused by the same effects, namely by the variations (e.g. temperature drift, aging) of second-order building blocks providing the same type of function (the numerators of $T_1(s) \cdot T_2(s)$ and $F_1(s) \cdot F_2(s)$ are the same) and realized with the same technology. Alone the *poles* of $F_1(s)$ and $F_2(s)$ may be selected such that $\sum_{i=1}^{2} \Delta F_i/F_i$ is smaller than $\sum_{i=1}^{2} \Delta T_i/T_i$, at least over a prescribed frequency range. The choice of the poles of $F_i(s)$ will determine the coupling coefficient[26] k. The optimum value will minimize one or both of the terms $(1 - kT)$ and $\sum \Delta F_i/F_i$. For example one possible optimum value of k is $|1/T(j\omega)|$ within a prescribed frequency range, say over the passband.

In order to obtain either the gain or phase variation of the cascaded and coupled filter structure we can proceed as in Chapter 1, Section 1.5.1. For example, with

$$T_i(s) = \left(\frac{n_2 s^2 + n_1 s + n_0}{d_2 s^2 + d_1 s + d_0}\right)_i , \quad i = 1, 2 \tag{7-235}$$

and

$$F_i(s) = \left(\frac{n_2 s^2 + n_1 s + n_0}{D_2 s^2 + D_1 s + D_0}\right)_i , \quad i = 1, 2 \tag{7-236}$$

we obtain the gain variation by setting $s = j\omega$ and considering only the real part, namely (see (1-114), Chapter 1),

$$\Delta\alpha_{T_i} = \frac{\Delta|T_i(j\omega)|}{|T_i(j\omega)|} = \left(g_{n_2}\frac{\Delta n_2}{n_2} + g_{n_1}\frac{\Delta n_1}{n_1} + g_{n_0}\frac{\Delta n_0}{n_0}\right.$$
$$\left. - g_{d_2}\frac{\Delta d_2}{d_2} - g_{d_1}\frac{\Delta d_1}{d_1} - g_{d_0}\frac{\Delta d_0}{d_0}\right)_i \tag{7-237}$$

and

$$\Delta\alpha_{F_i} = \frac{\Delta|F_i(j\omega)|}{|F_i(j\omega)|} = \left(g_{n_2}\frac{\Delta n_2}{n_2} + g_{n_1}\frac{\Delta n_1}{n_1} + g_{n_0}\frac{\Delta n_0}{n_0}\right.$$
$$\left. - g_{D_2}\frac{\Delta D_2}{D_2} - g_{D_1}\frac{\Delta D_1}{D_1} - g_{D_0}\frac{\Delta D_0}{D_0}\right)_i \tag{7-238}$$

The positive terms in (7-237) and (7-238) are identical; the negative ones will differ depending on the g_{d_i} and g_{D_i} functions as well as on the coefficient variations $\Delta d_i/d_i$ and $\Delta D_i/D_i$. Again, the g_{D_i} functions can be optimized for minimum $\Delta\alpha_{F_i}$ by an appropriate selection of the poles of $F_i(s)$. The coefficient variations will be similar, and possibly identical, if building blocks based on the same technology are used to realize $T_i(s)$ and $F_i(s)$.

7.3.5. The Transmission Variation of General Coupled Filters

Let us now briefly consider the general nth-order coupled filter and the equivalent cascaded filter and examine how the characteristics of the two can be compared. In so doing we are not concerned with the actual coupling

---

26. Clearly, when equating the equations of the cascaded and coupled filter configurations, only one degree of freedom for optimization purposes (e.g., minimum sensitivity, minimum variation) remains. The coupling coefficient k need not necessarily be reserved for the optimization; in some cases using the parameters Ω_{Pi} or Q_{Pi} of the poles of $F_i(s)$ may lead to simpler computations.

structure; various schemes beside that shown in Fig. 7-39 have been suggested in the literature.[27] Which of them is to be preferred over another will depend on studies along the lines described in what follows.[28]

Inasmuch as the feedback coupling of second-order networks affects only the poles of the resulting transfer function $T(s)$, we need consider only the denominator sensitivity[29] and with it the variation $\Delta T(s)/T(s)$ with respect to variations of the poles p_i or the denominator coefficients B_j, where

$$T(s) = \frac{N(s)}{D(s)} = \frac{N(s)}{\displaystyle\prod_{i=1}^{n}(s-p_i)} = \frac{N(s)}{s^n + \displaystyle\sum_{j=0}^{n-1} B_j s^j} \qquad (7\text{-}239)$$

With

$$S_{B_j}^{T(s)} = -S_{B_j}^{D(s)} = -\frac{B_j s^j}{D(s)} \qquad (7\text{-}240)$$

and (7-230) and (7-232), we obtain

$$\frac{\Delta T(s)}{T(s)} = \sum_{i=1}^{n} \frac{p_i}{s-p_i} \frac{\Delta p_i}{p_i} = -\sum_{j=0}^{n-1} \frac{B_j s^j}{D(s)} \cdot \frac{\Delta B_j}{B_j} \qquad (7\text{-}241)$$

Multiplying both sides of the second equation by $D(s)$ we obtain

$$\sum_{i=1}^{n} \frac{D(s)}{s-p_i} \cdot p_i \frac{\Delta p_i}{p_i} = -\sum_{j=1}^{n-1} B_j s^j \frac{\Delta B_j}{B_j} \qquad [7\text{-}242]$$

Note that only the terms $\Delta p_i/p_i$ and $\Delta B_j/B_j$ will depend on whether a coupled structure or a simple cascade is being considered. Furthermore the terms $D(s)/(s-p_i)$ will be recognized as those designated $\tilde{Q}_i(s)$ in (7-174).

For $s = p_\nu$, all $\tilde{Q}_i(s)$ terms except the νth term $\tilde{Q}_\nu(s)$ will equal zero. Thus after solving for the relative variation of p_ν, we obtain from (7-242)

$$\frac{\Delta p_\nu}{p_\nu} = -\frac{1}{p_\nu} \cdot \frac{1}{\tilde{Q}_\nu(p_\nu)} \cdot \sum_{j=0}^{n-1} B_j \cdot (p_\nu)^j \cdot \frac{\Delta B_j}{B_j} \qquad (7\text{-}243)$$

This will be recognized as a generalization of either (7-188) or (7-189) depending on whether the $\Delta B_j/B_j$ pertain to a cascaded or a coupled filter structure. Solving for the relative pole variation of a coupled nth-order network, expressed in terms of the variation of the equivalent cascaded network, we therefore obtain

$$\left(\frac{\Delta p_\nu}{p_\nu}\right)_c = \frac{\displaystyle\sum_{j=0}^{n-1} B_j(p_\nu)^j \left(\frac{\Delta B_j}{B_j}\right)_c}{\displaystyle\sum_{j=0}^{n-1} B_j(p_\nu)^j \left(\frac{\Delta B_j}{B_j}\right)} \cdot \frac{\Delta p_\nu}{p_\nu} \qquad [7\text{-}244]$$

27. See, for example, at the end of the book, the *Additional References and Suggested Reading* list under *Coupled Filters*.
28. Based largely on unpublished material by P. Horn, Swiss Federal Institute of Technology, Zurich.
29. See Chapter 4, Section 4.3.2 equation (4-163) in *Linear Integrated Networks: Fundamentals*.

where the subscript c characterizes the parameters of the coupled filter. Clearly (7-244) can be written in the abbreviated form

$$\left(\frac{\Delta p_v}{p_v}\right)_c = f_{p_v}(k) \cdot \frac{\Delta p_v}{p_v} \tag{7-245}$$

and is seen to represent the generalized version of (7-223). Having found the $\Delta p_v/p_v$ and $(\Delta p_v/p_v)_c$ terms, where $v = 1, 2, \ldots n$, the transmission variations $(\Delta T(s)/T(s))_{ca}$ and $(\Delta T(s)/T(s))_{co}$ can be obtained according to (7-230) and (7-231) respectively.

Finally, the following useful observation can be made. *Irrespective of the filter structure*, inspection of (7-239) shows that we can write

$$B_0 = \prod_{i=1}^{n/2} \omega_{p_i}^2 \tag{7-246}$$

where we assume $T(s)$ given as a product of second-order terms (e.g., (7-162)). In the cascaded filter structure each ω_{p_i} is realized by its corresponding second-order building block T_i. Consequently we have

$$\left(\frac{\Delta B_0}{B_0}\right)_{ca} = 2 \sum_{i=1}^{n/2} \left(\frac{\Delta \omega_{p_i}}{\omega_{p_i}}\right)_{ca} \tag{7-247}$$

In the coupled filter structure, the coupling coefficient k will affect only some of the coefficients B_j in the denominator of $T(s)$. For example, in allpole bandpass filters, where the $F_i(s)$ constitute second-order bandpass functions, only the term $B_{n/2}$ will contain k. All other coefficients will depend only on the poles of the $F_i(s)$ functions, that is, on the Ω_{p_i} and Q_{p_i} values. Thus, for such cases we can write (see, for example, (7-202))

$$B_0 = \prod_{i=1}^{n/2} \Omega_{p_i}^2 \tag{7-248}$$

and

$$\left(\frac{\Delta B_0}{B_0}\right)_{co} = \sum_{i=1}^{n/2} \left(\frac{\Delta \omega_{p_i}}{\omega_{p_i}}\right)_{co} = \sum_{i=1}^{n/2} \frac{\Delta \Omega_{p_i}}{\Omega_{p_i}} \tag{7-249}$$

Now, each Ω_{p_i} will be realized by a second-order building block F_i, which, we assume, is implemented with the same technology as its cascaded equivalent T_i. Thus we can realistically assume that

$$\sum_{i=1}^{n/2} \frac{\Delta \Omega_{p_i}}{\Omega_{p_i}} = \sum_{i=1}^{n/2} \left(\frac{\Delta \omega_{p_i}}{\omega_{p_i}}\right)_{ca} \tag{7-250}$$

Consequently, with (7-249) we obtain:

$$\sum_{i=1}^{n/2} \left(\frac{\Delta \omega_{p_i}}{\omega_{p_i}}\right)_{co} = \sum_{i=1}^{n/2} \left(\frac{\Delta \omega_{p_i}}{\omega_{p_i}}\right)_{ca} \tag{7-251}$$

Hence, assuming that equivalent second-order filter building blocks are used in an nth-order cascaded and coupled structure, and that the coefficient B_0 depends on the same component types in both, then *the sum of frequency variations in the two structures will be the same.*

Consider, for example, a fourth-order allpole bandpass filter characterized by two pole frequencies ω_{p_1} and ω_{p_2} and realized once by the cascaded structure of Fig. 7-40a and then by the coupled structure of Fig. 7-40b. Assume that the pole frequency of T_1 and F_1 varies by an amount δ whereas the pole frequency of T_2 and F_2 remains unchanged. Then we have for the cascaded structure

$$\frac{\Delta\omega_{p_1}}{\omega_{p_1}} = \delta; \qquad \frac{\Delta\omega_{p_2}}{\omega_{p_2}} = 0 \qquad (7\text{-}252)$$

whereas for the coupled structure, where $\Delta\omega_{p_2}/\omega_{p_2}$ will not be zero,

$$\frac{\Delta\omega_{p_1}}{\omega_{p_1}} + \frac{\Delta\omega_{p_2}}{\omega_{p_2}} = \delta \qquad (7\text{-}253)$$

Thus (7-251) and the simple example above demonstrates one aspect of the stabilizing effect of a coupled structure. A change in the pole frequency of one section of a filter cascade will have the expected large effect on the filter response in the vicinity of that particular pole frequency; in the equivalent coupled structure the effect of the same frequency change will be distributed over *all* pole frequencies of the filter, thereby affecting the filter response over a wider frequency range, but affecting it less severely at any one particular frequency.

APPENDIX A
Calculating the Q-factor for Various Second-Order Networks

Some confusion often exists with respect to the definition of the quality factor Q of a network. This is evident from the fact that the definitions vary quite widely in textbooks on network analysis and synthesis.[1] To prevent even more confusion as a result of our use of the factor q, in addition to Q, throughout this text, we go through the following detailed discussion on the definitions of Q and q, and on the values of Q as a function of q, for various second-order networks.

Consider the second-order bandpass network whose transmission function is given by

$$T_{BP}(s) = K_{BP} \cdot \frac{s}{(s - p_1)(s - p_2)} = K_{BP} \cdot \frac{s}{s^2 + 2\sigma s + \omega_n^2} \tag{A.1}$$

The corresponding pole-zero diagram is shown in Figure A.1, the frequency response in Figure A.2. Referring to these two figures, Q has been defined by $\omega_c/2\sigma$, $\omega_n/2\sigma$ which is identical with $1/2\zeta$, and $\omega_0/(\omega_h - \omega_l)$. Only when σ is very small, i.e., when the poles are very close to the $j\omega$ axis, are these terms approximately equal. However in RC active circuit design, where passive RC networks are often considered by themselves, the expressions above can vary significantly.

It seems sufficient to distinguish between two different terms for the characterization of active RC networks, namely the *inverse damping factor q* and the quality factor or *frequency selectivity*[2] Q. They are related to one another but are only identical in special cases.

The inverse damping factor is a theoretical term that determines the location of a particular pole or zero in the s-plane. It is given by:

$$q = \frac{\omega_n}{2\sigma} = \frac{1}{2\zeta} \tag{A.2}$$

where ω_n is the undamped natural frequency, σ is the distance of the pole from the $j\omega$-axis and ζ is the cosine of the angle between the pole and the negative real axis. The frequency selectivity Q is a practical, easily measurable term defined as:

$$Q = \frac{\omega_0}{\omega_h - \omega_l} \tag{A.3}$$

---

1. In conventional LCR networks an additional cause of ambiguity is the necessary differentiation between the quality factor Q of reactive components and the frequency selectivity Q of, say, a tuned LCR tank.
2. Q, of course, comes from the term "quality factor," however this designation is generally reserved for reactive components. Thus one speaks of the quality factor Q_L of an inductor or Q_C of a capacitor. These may, in turn, influence the frequency selectivity Q of a tank circuit in which they are used.

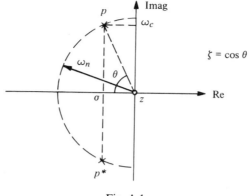

Fig. A.1

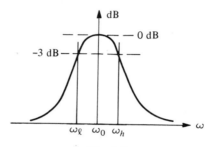

Fig. A.2

ω_0 is the frequency at which the steady state transmission is maximum and $(\omega_h - \omega_l)$ is the bandwidth between the two frequencies at which the maximum transmitted signal power is halved or the signal level reduced by 3 dB. The high and low half-power frequencies are the geometric mean of the peak frequency, namely:

$$\omega_0^2 = \omega_h \cdot \omega_l \tag{A.4}$$

The theoretical term q is quite general and can be applied to any second-order network for which it specifies the location of the corresponding complex conjugate pole or zero pair. On the other hand, by nature of its practical definition *the frequency selectivity Q is only applicable sensibly to bandpass and bandstop* networks.[3]

A.I SECOND-ORDER BANDPASS NETWORKS

The inverse damping factor q results directly from the location of the poles of the specified transmission function and requires no further elaboration. As we know, it can also be used to characterize the zeros of a network. In both cases q increases as the pole or zero approaches the $j\omega$-axis.

---

3. As we have seen in the text, it can also be used to characterize the frequency response of a higher-order network in the vicinity of a dominant high-Q pole; here the response resembles that of a second-order bandpass network closely.

The frequency selectivity Q has so far only been defined in physical terms and it remains to be shown that it is directly related to the inverse damping factor. To do so, we must first find an expression for the frequency of peak transmission in terms of the transmission poles.

1. Peak Frequency for Complex Conjugate Pole Pair

Let us consider the pole-zero diagram of the transmission function given by (A.1); it is redrawn in Figure A.3. The corresponding frequency response can be expressed in terms of the vectors shown in Figure A.3 as follows:

$$|T(j\omega)| = K_{BP} \cdot \frac{\nu}{\mu_1 \mu_2} \tag{A.5}$$

The area A of the triangle formed by the three points p_1, p_2 and $s = j\omega$ can be expressed by:

$$A = \tfrac{1}{2}\mu_1 \mu_2 \sin \varphi \tag{A.6}$$

and also by:

$$A = \tfrac{1}{2}(2\omega_c \sigma) \tag{A.7}$$

Equating (A.6) and (A.7) we find:

$$\frac{1}{\mu_1 \mu_2} = \frac{\sin \varphi}{2\omega_c \sigma} \tag{A.8}^4$$

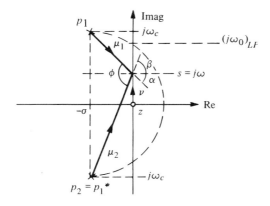

Fig. A.3

4. But for a constant scaling factor, this is the frequency response of a second-order lowpass filter expressed as a function of the angle φ. The peak value $K_{LP}/2\omega_c \sigma$ occurs when $\varphi = 90°$. The corresponding peak frequency $[\omega_0]_{LP}$ occurs at the intersection between the $j\omega$-axis and the circle whose diameter is the line segment between the complex conjugate pole pair, (see Figure A.3). Therefore $[\omega_0]_{LP}$ equals $\sqrt{\omega_c^2 - \sigma^2}$.

Referring again to Figure A.3:

$$\varphi = \alpha + \beta \tag{A.9}$$

$$\text{tg } \alpha = \frac{\omega_c - \nu}{\sigma} \tag{A.10}$$

$$\text{tg } \beta = \frac{\omega_c + \nu}{\sigma} \tag{A.11}$$

therefore

$$\text{tg } \varphi = \frac{2\dfrac{\omega_c}{\sigma}}{1 - \left(\dfrac{\omega_c^2 - \nu^2}{\sigma^2}\right)} \tag{A.12}$$

Solving for ν, we obtain

$$\nu = [\omega_c^2 - \sigma^2 + 2\omega_c \sigma \cdot \cot \varphi]^{1/2} \tag{A.13}$$

and with (A.5), (A.8), and (A.13) we find:

$$|T(j\omega)| = a[b + \cot \varphi]^{1/2} \cdot \sin \varphi \tag{A.14}$$

where

$$a = \frac{K_{BP}}{\sqrt{2\omega_c \sigma}} \tag{A.15}$$

and

$$b = \frac{\omega_c^2 - \sigma^2}{2\omega_c \sigma} \tag{A.16}$$

To obtain the angle φ_0 at which the transmission is maximum the derivative of (A.14) with respect to φ is taken and we obtain:

$$\frac{d|T(j\omega)|}{d\varphi} = \frac{a}{2}\left[\frac{2(b + \cot \varphi)\cos \varphi - (\sin \varphi)^{-1}}{(b + \cot \varphi)^{1/2}}\right] \tag{A.17}$$

Setting (A.17) equal to zero we obtain:

$$b \cdot 2 \sin \varphi \cos \varphi + 2 \cos^2 \varphi - 1 = 0$$

or

$$b \cdot \sin 2\varphi + \cos 2\varphi = 0$$

With (A.16) this becomes:

$$\text{tg } 2\varphi = \frac{2\dfrac{\omega_c}{\sigma}}{1 - \left(\dfrac{\omega_c}{\sigma}\right)^2} \tag{A.18}$$

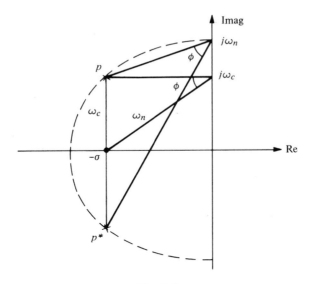

Fig. A.4

or

$$\operatorname{tg} \varphi_0 = \frac{\omega_c}{\sigma} \qquad (A.19)$$

Substituting this value in (A.13) we find:

$$\nu_0 = \omega_0 = \omega_n \qquad (A.20)$$

The peak frequency of a second-order bandpass filter comprising a complex conjugate pole pair is thus equal to the undamped natural frequency ω_n. This is shown graphically in Figure A.4. The peak transmission value follows from (A.1) as:

$$\overline{T}_{\text{BP}} = \frac{K_{\text{BP}}}{2\sigma} = \frac{K_{\text{BP}}}{\omega_n} \cdot q \qquad (A.21)$$

2. Peak Frequency for Negative–Real Pole Pair

If the poles of the transmission function given by (A.1) are negative real, the corresponding pole-zero diagram is of the type shown in Figure A.5. The frequency response, expressed in terms of the vectors shown in the figure, are again given by (A.5). Similarly the area A of the triangle formed by the points p_1, p_2, and $s = j\omega$ can be expressed by (A.6) and by:

$$A = \tfrac{1}{2}(p_2 - p_1) \cdot \nu \qquad (A.22)$$

Letting

$$(p_2 - p_1) = \rho \qquad (A.23)$$

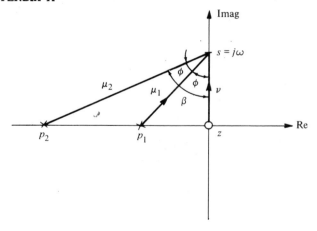

Fig. A.5

and equating (A.6) and (A.22) we find:

$$\frac{1}{\mu_1 \mu_2} = \frac{1}{\rho \cdot \nu} \cdot \sin \varphi \tag{A.24}$$

Substituting this expression in (A.5) we get:

$$|T(jw)| = \frac{K_{BP}}{\rho} \cdot \sin \varphi \tag{A.25}$$

$T(s)$ can now be expressed in terms of ν. Referring to Figure A.5.

$$\text{tg } \alpha = \frac{p_1}{\nu} \tag{A.26}$$

$$\text{tg } \beta = \frac{p_2}{\nu} \tag{A.27}$$

and

$$\varphi = \beta - \alpha \tag{A.28}$$

Therefore:

$$\text{tg } \varphi = \frac{\dfrac{p_2 - p_1}{\nu}}{1 + \dfrac{p_1 p_2}{\nu^2}} \tag{A.29}$$

From (A.1) we have:

$$p_1 p_2 = \omega_n^2 \tag{A.30}$$

and with (A.23), (A.29) becomes:

$$\text{tg } \varphi = \frac{\rho \nu}{\nu^2 + \omega_n^2} \tag{A.31}$$

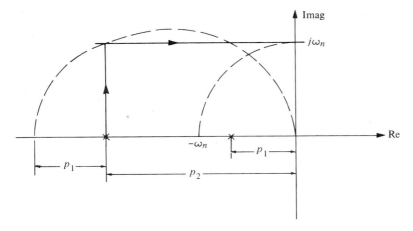

Fig. A.6

Since

$$\sin \varphi = \frac{\text{tg } \varphi}{\sqrt{1 + \text{tg}^2 \varphi}} \tag{A.32}$$

(A.25) becomes

$$|T(jw)| = K_{\text{BP}} \cdot \frac{\nu}{[(\nu^2 + \omega_n^2)^2 + \rho^2 \nu^2]^{1/2}} \tag{A.33}$$

Taking the derivative of (A.33) and setting it equal to zero we can again solve for the value of $\nu = \nu_0$ for which (A.33) has a maximum. We find

$$\nu_0 = \omega_0 = \omega_n \tag{A.20}$$

This is the same result as was found for the case of complex conjugate poles. The peak transmission value is therefore also given by (A.21).

It may be of interest to obtain the peak frequency $\omega_0 = \omega_n$ graphically. For the complex conjugate pole case, it is, of course, simply the distance from either pole to the origin. In the negative real pole case it is the graphical solution to (A.30). This is given by the familiar geometrical construction shown in Figure A.6.

3. The 3 dB—Bandwidth

Having obtained the peak frequency ω_0, the half-power bandwidth

$$BW = \omega_h - \omega_l \tag{A.34}$$

remains to be calculated. From (A.1) the amplitude response is given by

$$|T(j\omega)| = K_{\text{BP}} \cdot \frac{\omega}{[(\omega_n^2 - \omega^2)^2 + 4\sigma^2 \omega^2]^{1/2}} \tag{A.35}$$

The half-power frequencies for this expression can be obtained by setting it equal to $\bar{T}_{BP}/\sqrt{2}$. From (A.35) and (A.21) these frequencies are given by the two solutions to the quadratic equation:

$$\omega^2 + 2\sigma\omega - \omega_n^2 = 0 \tag{A.36}$$

which are

$$\omega_h = \sqrt{\omega_n^2 + \sigma^2} + \sigma \tag{A.37a}$$

and

$$\omega_l = \sqrt{\omega_n^2 + \sigma^2} - \sigma \tag{A.37b}$$

Forming the product of ω_h and ω_l we get

$$\omega_h \cdot \omega_l = \omega_n^2 \tag{A.38}$$

which, with (A.20) proves the statement of geometric symmetry expressed by (A.4). The 3 dB bandwidth follows from (A.37) namely

$$BW = \omega_h - \omega_l = 2\sigma \tag{A.39}$$

This expression is valid whether the two transmission poles are complex conjugate or negative real. In the latter case, however, it is more convenient to express the bandwidth in terms of the real poles p_1 and p_2 than in terms of σ. From (A.1) we find

$$BW = 2\sigma = p_1 + p_2 \tag{A.40}$$

Comparing (A.20), (A.30) and (A.40) with (A.2) and (A.3) we find that for second-order bandpass networks of the general form (A.1):

$$Q_{BP} = q = \frac{\omega_n}{2\sigma} = \frac{\sqrt{p_1 p_2}}{p_1 + p_2} \tag{A.41}$$

This relation applies accurately whether the two transmission poles are complex conjugate or negative real.

A.II FREQUENCY EMPHASIZING NETWORKS

The transmission function of an FEN is given by:

$$T_F(s) = K_F \cdot \frac{s^2 + 2\sigma_z s + \omega_z^2}{s^2 + 2\sigma_p s + \omega_p^2} \tag{A.42}$$

where $\sigma_p < \sigma_z$.

To derive the steady-state frequency characteristic of $T_F(s)$, it is useful to introduce the normalized angular frequencies

$$\Omega_z = \frac{\omega}{\omega_z}, \qquad \Omega_p = \frac{\omega}{\omega_p}$$

and the inverse damping factor given by (A.2). The amplitude response then follows as:

$$|T_F(j\Omega)| = K_F \cdot \frac{\omega_z}{\omega_p} \cdot \left[\frac{\left(\dfrac{1}{\Omega_z} - \Omega_z\right)^2 + \dfrac{1}{q_z^2}}{\left(\dfrac{1}{\Omega_p} - \Omega_p\right)^2 + \dfrac{1}{q_p^2}} \right]^{1/2} \tag{A.43}$$

where $q_p > q_z$.

For the symmetrical FEN, $\omega_z = \omega_p$, that is $\Omega_z = \Omega_p = \Omega$ and (A.43) becomes:

$$|T_F(j\Omega)| = K_F \cdot \left[\frac{\left(\dfrac{1}{\Omega} - \Omega\right)^2 + \dfrac{1}{q_z^2}}{\left(\dfrac{1}{\Omega} - \Omega\right)^2 + \dfrac{1}{q_p^2}} \right]^{1/2} \tag{A.44}$$

As under Section A.1, it can be shown that the peak value of (A.44) occurs at $\Omega = 1$ and is given by

$$\hat{T}_F = \frac{q_p}{q_z} K_F \tag{A.45}$$

The half-power frequencies of (A.44) can be obtained by setting it equal to $\bar{T}_F/\sqrt{2}$ thus:

$$\frac{\left(\dfrac{1}{\Omega} - \Omega\right)^2 + \dfrac{1}{q_z^2}}{\left(\dfrac{1}{\Omega} - \Omega\right)^2 + \dfrac{1}{q_p^2}} = \frac{1}{2}\left(\frac{q_p}{q_z}\right)^2 \tag{A.46}$$

Introducing:

$$q_{\text{FEN}}^2 = q_p^2 - 2q_z^2 \tag{A.47}$$

we obtain the following quadratic equation in Ω:

$$\Omega^2 + \frac{\Omega}{q_{\text{FEN}}} - 1 = 0 \tag{A.48}$$

This is the normalized version of (A.36). Thus, the two -3 dB frequencies in normalized form are given by

$$\Omega_h = \sqrt{\frac{1}{4q_{\text{FEN}}^2} + 1} + \frac{1}{2q_{\text{FEN}}} \tag{A.49}$$

and

$$\Omega_l = \sqrt{\frac{1}{4q_{\text{FEN}}^2} + 1} - \frac{1}{2q_{\text{FEN}}} \tag{A.50}$$

These two frequencies satisfy the condition for geometrical symmetry, namely

$$\Omega_h \cdot \Omega_l = 1 \tag{A.51}$$

The frequency selectivity of the FEN results from (A.49) and (A.50) as

$$Q_{FEN} = \frac{1}{\Omega_h - \Omega_l} = q_{FEN} = \sqrt{q_p^2 - 2q_z^2} \qquad (A.52)$$

Since $q_p > q_z$ for the FEN, the expression under the square root cannot, become negative.

A.III FREQUENCY REJECTION NETWORKS

We start out again here with (A.42) and (A.43) except that now $\sigma_p > \sigma_z$ or, in terms of the inverse damping factor, $q_p < q_z$. Considering only the symmetrical FRN, we then obtain the amplitude response given by (A.44). This is maximum at zero and very high frequencies where it approaches the constant value K_F. Thus, the half-power frequencies are obtained by setting (A.44) equal to $K_F/\sqrt{2}$, namely:

$$\frac{\left(\dfrac{1}{\Omega} - \Omega\right)^2 + \dfrac{1}{q_z^2}}{\left(\dfrac{1}{\Omega} - \Omega\right)^2 + \dfrac{1}{q_p^2}} = \frac{1}{2} \qquad (A.53)$$

Multiplying out and using the substitution

$$q_{FRN}^2 = \frac{q_p^2 q_z^2}{q_z^2 - 2q_p^2} \qquad (A.54)$$

we obtain an equation in Ω of the form (A.48). Thus, all subsequent results are formally the same as under A.II except that q_{FEN} must be replaced by q_{FRN}. Substituting (A.54) into (A.52), the frequency selectivity of the FRN results as:

$$Q_{FRN} = \frac{1}{\Omega_h - \Omega_l} = \frac{q_p q_z}{\sqrt{q_z^2 - 2q_p^2}} \qquad (A.55)$$

Because $q_p < q_z$ for the FRN, this expression always remains real.

An FRN with a perfect null is obtained when $\sigma_z = 0$, that is, when $q_z = \infty$. Letting q_z approach infinity in (A.55), we obtain:

$$Q_{FRN} = q_p \qquad (A.56)$$

A.IV PASSIVE BANDPASS NETWORK IN CASCADE WITH HSFEN

One way of *approximating* a high-Q second-order bandpass network is to cascade a *passive RC* ladder network with an HSFEN (High Selectivity Frequency Emphasizing Network), see Chapters 3 and 6. One amplifier is thereby saved compared to the exact realization. The corresponding transmission function is of the form:

$$T(s) = K \cdot \frac{s}{s^2 + \dfrac{\omega_n}{q_1} s + \omega_n^2} \cdot \frac{s^2 + \dfrac{\omega_n}{q_2} s + \omega_n^2}{s^2 + \dfrac{\omega_n}{q_3} s + \omega_n^2} \qquad (A.57)$$

q_1 pertains to the poles of the *RC* ladder network and q_3 is determined by the poles of the specified selective network. Being a high-*Q* network, q_3 is by definition much larger than q_1. Furthermore the HSFEN is characterized by the fact that its zeros are conjugate complex, in other words q_2 is also larger than q_1. Thus:

$$q_1 < q_2 < q_3 \tag{A.58}$$

Equation (A.57) can be written as follows:

$$T(s) = T_{BP}(s) \cdot T_E(s) \tag{A.59}$$

where

$$T_{BP}(s) = K_{BP} \cdot \frac{s}{s^2 + \dfrac{\omega_n}{q_3} s + \omega_n^2} \tag{A.60}$$

$$T_E(s) = K_E \cdot \frac{s^2 + \dfrac{\omega_n}{q_2} s + \omega_n^2}{s^2 + \dfrac{\omega_n}{q_1} s + \omega_n^2} \tag{A.61}$$

and

$$K = K_{BP} \cdot K_E \tag{A.62}$$

$T_{BP}(s)$ is the exact transfer function desired, $T_E(s)$ constitutes an error in the resulting frequency response. Obviously if q_2 were equal to q_1, as in the case of the MSFEN (Medium Selectivity Frequency Emphasizing Network), then the *Q* of the overall network would be equal to q_3. Since this is not the case here, the *Q* of the overall network represented by the transmission function (A.57) must be calculated separately.

The normalized amplitude response corresponding to (A.57) is given by

$$|T(j\Omega)| = \frac{\dfrac{K}{\omega_n}}{\left[\left(\dfrac{1}{\Omega} - \Omega\right)^2 + \dfrac{1}{q_3^2}\right]^{1/2}} \cdot \left[\frac{\left(\dfrac{1}{\Omega} - \Omega\right)^2 + \dfrac{1}{q_2^2}}{\left(\dfrac{1}{\Omega} - \Omega\right)^2 + \dfrac{1}{q_1^2}}\right]^{1/2} \tag{A.63}$$

The peak value occurring for $\Omega = 1$ is given by

$$\overline{T} = \frac{K}{\omega_n} \cdot \frac{q_1}{q_2} \cdot q_3 \tag{A.64}$$

The -3 dB (i.e., half-power) frequencies can now be obtained as before. For convenience, we first introduce the substitution:

$$w = \left(\frac{1}{\Omega} - \Omega\right) \tag{A.65}$$

Setting (A.63) equal to $\overline{T}/\sqrt{2}$ and making the above substitution we find:

$$\frac{2}{w^2 q_3^2 + 1} \cdot \frac{w^2 q_2^2 + 1}{w^2 q_1^2 + 1} = 1 \tag{A.66}$$

This can be multiplied out to give

$$w^4 + w^2 \left[\frac{q_1^2 + q_3^2 - 2q_2^2}{q_1^2 q_3^2} \right] - \frac{1}{q_1^2 q_3^2} = 0 \tag{A.67}$$

With the new substitutions

$$z = w^2 \tag{A.68}$$

$$Q_1^2 = \frac{q_1^2 q_3^2}{q_1^2 + q_3^2 - 2q_2^2} \tag{A.69}$$

and

$$Q_2^2 = q_1 q_3 \tag{A.70}$$

(A.67) becomes

$$z^2 + \frac{z}{Q_1^2} - \frac{1}{Q_2^4} = 0 \tag{A.71}$$

The roots of this equation can be found by inspection. The negative root has no physical meaning. Thus we are left with the solution:

$$z = \frac{1}{2} \left[\sqrt{\frac{1}{Q_1^4} + \frac{4}{Q_2^4}} - \frac{1}{Q_1^2} \right] \tag{A.72}$$

With (A.65) and (A.68) this becomes:

$$\left(\frac{1}{\Omega} - \Omega \right)^2 = \frac{1}{2} \left[\sqrt{\frac{1}{Q_1^4} + \frac{4}{Q_2^4}} - \frac{1}{Q_1^2} \right] \tag{A.73}$$

With the substitution

$$q_{\text{HF}}^2 = \frac{2}{\left[\sqrt{\dfrac{1}{Q_1^4} + \dfrac{4}{Q_2^4}} - \dfrac{1}{Q_1^2} \right]} \tag{A.74}$$

(A.73) becomes

$$\Omega^2 + \frac{\Omega}{q_{\text{HF}}} - 1 = 0 \tag{A.75}$$

This equation is identical with the FEN equation (A.48). From (A.52), (A.69), (A.70) and (A.74) we therefore find for the Q of the bandpass function approximated by (A.57):

$$Q_{\text{HSFEN}} = \frac{\sqrt{2}\, q_1 q_3}{[\sqrt{(q_1^2 + q_3^2 - 2q_2^2)^2 + 4q_1^2 q_3^2} - (q_1^2 + q_3^2 - 2q_2^2)]^{1/2}} \tag{A.76}$$

or written in another form:

$$Q_{\text{HSFEN}}^2 = \frac{2q_1^2 q_3^2}{q_1^2 + q_3^2 - 2q_2^2} \cdot \frac{1}{\sqrt{1 + \dfrac{4q_1^2 q_3^2}{(q_1^2 + q_3^2 - 2q_2^2)^2}} - 1} \tag{A.77}$$

As to be expected, for the limiting case that $q_1 = q_2$ the above equations give

$$Q_{\text{HSFEN}}\bigg|_{q_1 = q_2} = q_3 \tag{A.78}$$

In general q_1 and q_2 will be given and the value of q_3 will be of interest that corresponds to a specified frequency selectivity Q_{HSFEN}. Solving (A.77) for q_3 we find:

$$q_3 = Q_{\text{HSFEN}} \left[\frac{\dfrac{Q_{\text{HSFEN}}^2}{q_1^2} + 2\left(\dfrac{q_2}{q_1}\right)^2 - 1}{\dfrac{Q_{\text{HSFEN}}^2}{q_1^2} + 1} \right]^{1/2} \tag{A.79}$$

Using the terminology of the main text we have:

$$q_1 \equiv \hat{q} \tag{A.80a}$$

$$q_2 = q_z \tag{A.80b}$$

$$q_3 = q_p \tag{A.80c}$$

Assuming, for example, a symmetrical twin-T in the HSFEN and a unity-gain noninverting amplifier ($\beta = 1$) we have:

$$\hat{q} = 0.25 \tag{A.81}$$

and

$$q_z = \frac{1}{2}\frac{1+r}{r} \tag{A.82}$$

where

$$r = \frac{2R}{R_L} \tag{A.83}$$

Substituting (A.80) to (A.83) in (A.79) we obtain:

$$q_p = Q_{\text{HSFEN}} \left[\frac{16Q_{\text{HSFEN}}^2 + 8\left(\dfrac{1+r}{r}\right)^2 - 1}{16\,Q_{\text{HSFEN}}^2 + 1} \right]^{1/2} \tag{A.84}$$

A typical value for r is 0.1 in which case $q_z = 5.5$. Substituting these values into (A.79) we have:

$$q_p = Q_{\text{HSFEN}} \left[\frac{16Q_{\text{HSFEN}}^2 + 967}{16Q_{\text{HSFEN}}^2 + 1} \right]^{1/2} \tag{A.85}$$

Expression (A.84) has been plotted in Figure A.7a and A.7b with the resistor ratio r as parameter. For r values larger than 0.05, the required inverse damping factor q_p is approximately equal to the specified frequency selectivity Q_{HSFEN}. For r values smaller than 0.05, q_p is required to be appreciably larger than the desired frequency selectivity. This is to make up for the error function defined by (A.61); note that the latter has FRN characteristics and therefore affects the desired bandpass characteristic adversely. In Figure A.8 the ratio of q_p to Q_{HSFEN} has been plotted on semilog paper for different values of the parameter r.

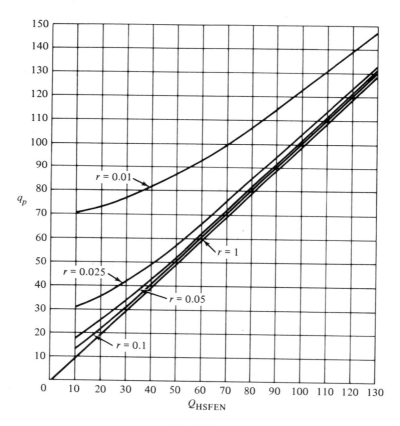

Fig. A.7a

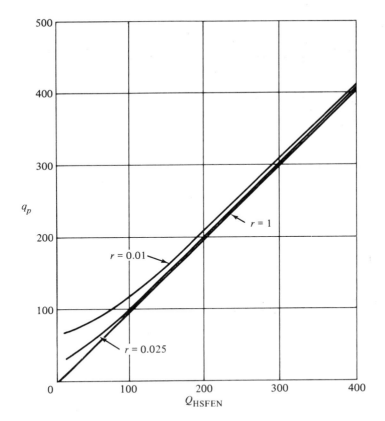

Fig. A.7b

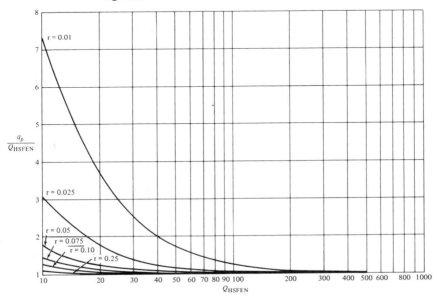

Fig. A.8

A.V IDENTICAL BANDPASS NETWORKS IN CASCADE

Finally let us consider the Q of n identical cascaded second-order bandpass networks.[5] The transfer function of such a cascade is

$$T_n(s) = \prod_{i=1}^{n} T_i(s) \tag{A.86}$$

where each $T_i(s)$ has the form

$$T_i(s) = K \frac{s}{s^2 + \dfrac{\omega_n}{Q_0} s + \omega_n^2} \tag{A.87}$$

Normalizing the peak value $|T_i(j\omega_n)|$ to unity, we find

$$K = \frac{\omega_n}{Q_0} \tag{A.88}$$

Normalizing the center frequency ω_n to unity, we must now find the frequencies ω_l and ω_h for which $|T_n(j\omega)|$ is 3 dB below the peak value (i.e., unity). Thus we must solve the equation:

$$\left[\frac{\dfrac{\omega}{Q_0}}{\sqrt{(1 - \omega^2)^2 + \left(\dfrac{\omega}{Q_0}\right)^2}} \right]^n = \frac{1}{\sqrt{2}} \tag{A.89}$$

and obtain:

$$\omega_l = 1 + \alpha - [\alpha(\alpha + 2)]^{1/2} \tag{A.90}$$

$$\omega_h = 1 + \alpha + [\alpha(\alpha + 2)]^{1/2} \tag{A.91}$$

where

$$\alpha = \frac{2^{1/n} - 1}{2Q_0^2} \tag{A.92}$$

With the center frequency ω_n normalized to unity the Q of $T_n(s)$, designated Q_n, is

$$Q_n = \frac{1}{\omega_h - \omega_l} \tag{A.93}$$

With (A.90) and (A.91) we obtain

$$Q_n = \frac{Q_0}{\sqrt{2^{1/n} - 1}} \tag{A.94}$$

Thus Q_2 of a cascaded bandpass pair is approximately 1.58 times higher, Q_3 of a cascaded triple 1.95 times higher, than that of the corresponding single stage, Q_0.

5. See, for example: S. C. Dutta Roy: "On Q of Cascaded Identical Resonators," *Proc. IEEE*, **61**, No. 6, June 1973, p. 790.

APPENDIX B
Maximum Q of Second-Order RC Networks

Consider the cascade connection of the two networks n_a and n_b shown in Figure B.1. Multiplying the transmission parameters of the two networks and converting these to open-circuit impedance parameters, the overall z-matrix results as:

$$[z]_{ab} = \frac{1}{z_{22a} + z_{11b}} \cdot \begin{bmatrix} z_{11a}z_{11b} + \Delta z_a & z_{12a}z_{12b} \\ z_{21a}z_{21b} & z_{22a}z_{22b} + \Delta z_b \end{bmatrix} \tag{B.1}$$

The overall voltage transfer ratio $T(s)$ of n_a in cascade with n_b is then given by:

$$T(s) = \frac{V_{\text{OUT}}}{V_{\text{IN}}} = \frac{z_{21}}{z_{11}} = \left(\frac{z_{21a}z_{21b}}{z_{11a}z_{11b}}\right) \cdot \frac{1}{1 + \dfrac{\Delta z_a}{z_{11a}z_{11b}}} = \frac{T'(s)}{1 + \varepsilon(s)} \tag{B.2}$$

where

$$T'(s) = \left(\frac{z_{21a}z_{21b}}{z_{11a}z_{11b}}\right), \quad \varepsilon(s) = \frac{\Delta z_a}{z_{11a}z_{11b}}$$

and $\Delta z_a = z_{11a}z_{22a} - z_{12a}z_{21a}$. If we consider only passive RC networks then $z_{12a} = z_{21a}$ and $z_{12b} = z_{21b}$. Limiting ourselves still further by considering only second order networks, (B.2) must have the form:

$$T(s) = \frac{N(s)}{(s + p_1)(s + p_2)} = \frac{N(s)}{s^2 + \dfrac{p}{\hat{q}}s + p^2} \tag{B.3}$$

where $p_1 p_2 = p^2$ and $\hat{q} = p/(p_1 + p_2)$. The pole Q, $\hat{q}$ approaches its maximum value 0.5 as p_1 and p_2 merge into a double pole at p. Let us assume that n_a and n_b are first order networks providing p_1 and p_2, respectively. In order to approach the maximum $\hat{q}$, (B.2) must therefore approximate as closely as possible the form:

$$T'(s) = \frac{N(s)}{(s + p)^2} = \frac{N_a N_b}{(s + p)^2} = \left(\frac{z_{21a}}{z_{11a}}\right)\left(\frac{z_{21b}}{z_{11b}}\right) \tag{B.4}$$

One way of satisfying this condition is to let $\Delta z_a = 0$. It can be shown that this results in a nonrealizable 2-port for n_a. There is a way of approximating (B.4) however, if we let:

$$z_{22a} \ll z_{21a} \tag{B.5}$$

and

$$z_{11b} \gg z_{21a} \tag{B.6}$$

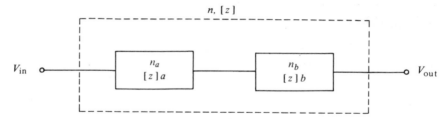

Fig. B.1

Then (B.2) becomes:

$$T(s) \approx \left(\frac{z_{21a}z_{21b}}{z_{11a}z_{11b}}\right) \cdot \frac{1}{1 - \dfrac{z_{21a}^2}{z_{11a}z_{11b}}} \tag{B.7}$$

Due to the residue condition which relates the transfer impedance of a network with its driving point impedances,[1] z_{21a} decreases with z_{22a}. This ensures a small error quantity $z_{21a}^2/z_{11a}z_{11b}$ in (B.7) when z_{11b} is large.

MAXIMIZING Q BY IMPEDANCE MISMATCHING

It should be clear that conditions (B.5) and (B.6) in effect constitute the isolation of network n_a from n_b by *mismatching* the impedance levels of the two networks such as to minimize the loading of n_a by n_b. One common method of achieving such an isolation is to connect a buffer amplifier between the two networks as shown in Figure B.2. In this way the transfer function (B.4) is automatically obtained.

Another practical way of isolating the two networks from one another is to multiply the impedance level of network n_b by a factor $\rho \gg 1$. Then from (B.2):

$$T(s) = \rho \frac{z_{21a}z_{21b}}{\rho z_{11a}z_{11b} + z_{11a}z_{22a} - z_{21a}^2} \approx \frac{z_{21a}z_{21b}}{z_{11a}z_{11b}} \tag{B.8}$$

Clearly, this method is more economical since no buffer amplifier is required. The potentially symmetrical twin-T is based on this method[2] whereby the right side of the symmetrical twin-T is impedance scaled by a factor ρ. As ρ increases, the pole Q of the twin-T approaches 0.5. Similarly the pole Q of any other symmetrical configuration (e.g., bridged-T) can be made to approach 0.5 by making it potentially symmetrical by a large impedance scaling factor ρ.

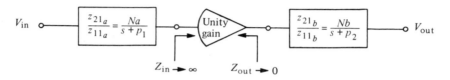

Fig. B.2

1. See Chapter 1, Section 1.3.4, *Linear Integrated Networks: Fundamentals.*
2. See, for example, Chapters 3 and 8, Sections 3.5 and 8.1.2, respectively, of *Linear Integrated Networks: Fundamentals.*

1. Symmetrical and Potentially Symmetrical Networks

In general, if n_a and n_b are passive and symmetrical, then $z_{21a} = z_{21b}$, $z_{11a} = z_{22b}$ and $z_{22a} = z_{11b}$. The resulting impedance matrix results from (B.1) as:

$$[z_{ab}]_S = \begin{bmatrix} z_{11a} - \dfrac{z_{21a}^2}{2z_{22a}} & \dfrac{z_{21a}^2}{2z_{22a}} \\[3mm] \dfrac{z_{21a}^2}{2z_{22a}} & z_{11a} - \dfrac{z_{21a}^2}{2z_{22a}} \end{bmatrix} \tag{B.9}$$

The potentially symmetrical network, in which n_b is scaled by ρ results as:

$$[z_{ab}]_{PS} = \begin{pmatrix} z_{11a} - \dfrac{1}{1+\rho} \dfrac{z_{21a}^2}{z_{22a}} & \dfrac{\rho}{1+\rho} \cdot \dfrac{z_{21a}^2}{z_{22a}} \\[3mm] \dfrac{\rho}{1+\rho} \dfrac{z_{21a}^2}{z_{22a}} & \rho z_{11a} - \dfrac{\rho^2}{1+\rho} \cdot \dfrac{z_{21a}^2}{z_{22a}} \end{pmatrix} \tag{B.10}$$

The voltage transfer ratio of the potentially symmetrical network is then:

$$T(s) = \frac{\rho}{1+\rho} \left(\frac{z_{21a}^2}{z_{11a} z_{22a}} \right) \cdot \frac{1}{1 - \dfrac{1}{1+\rho} \dfrac{z_{21a}^2}{z_{11a} z_{22a}}} \tag{B.11}$$

This is the same form as (B.2), therefore:

$$\varepsilon(s) = - \frac{1}{1+\rho} \cdot \frac{z_{21a}^2}{z_{11a} z_{22a}} \tag{B.12}$$

As ρ becomes much larger than unity, $\varepsilon(s)$ goes to zero and $\hat{q}$ approaches 0.5.

2. Ladder Networks

A commonly used second-order RC configuration is the ladder network. Referring to Figure B.3, the overall impedance matrix is given by:

$$[z]_{ab} = \frac{1}{Z_{a_2} + Z_{b_1} + Z_{b_2}} \begin{bmatrix} (Z_{a_1} + Z_{a_2})(Z_{b_1} + Z_{b_2}) + Z_{a_1} Z_{a_2} & Z_{a_2} Z_{b_2} \\ Z_{a_2} Z_{b_2} & Z_{a_2} Z_{b_2} + Z_{b_1} Z_{b_2} \end{bmatrix} \tag{B.13}$$

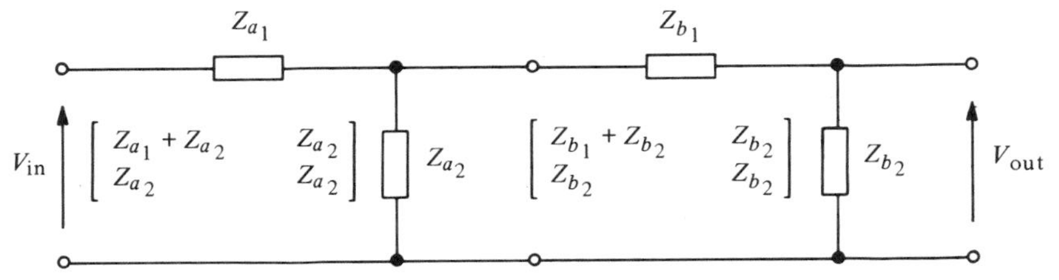

Fig. B.3

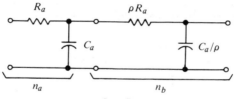

a. Low Pass

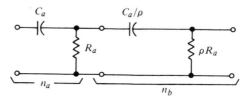

b. High Pass

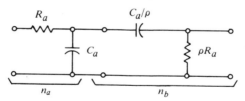

c. Band Pass

Fig. B.4

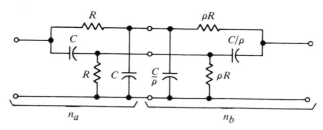

Fig. B.5

The voltage transfer ratio follows as:

$$T(s) = \frac{Z_{a_2}Z_{b_2}}{(Z_{a_1}+Z_{a_2})(Z_{b_1}+Z_{b_2})} \cdot \frac{1}{1+\dfrac{Z_{a_1}Z_{a_2}}{(Z_{a_1}+Z_{a_2})(Z_{b_1}+Z_{b_2})}} \tag{B.14}$$

Clearly (B.5) cannot be satisfied here, since $z_{22a} = z_{21a}$. In order to approximate double poles we see from (B.14) that we must let:

$$(Z_{b_1} + Z_{b_2}) \gg Z_{a_1} \| Z_{a_2} = Z_{ap} \tag{B.15}$$

In other words the elements of network n_b in series must be much larger than the elements of n_a in parallel. This condition automatically satisfies the general condition (B.6) and minimizes the loading of network n_a by n_b. For the special case that n_b is identical to n_a except that it is impedance scaled by the factor ρ, we have from (B.14):

$$T(s) = \left(\frac{Z_{a_2}}{Z_{a_1} + Z_{a_2}}\right)^2 \cdot \frac{1}{1 + \dfrac{1}{\rho}\dfrac{Z_{ap}}{Z_{as}}} \tag{B.16}$$

where $Z_{ap} = Z_{a_1} Z_{a_2}/(Z_{a_1} + Z_{a_2})$ and $Z_{as} = Z_{a_1} + Z_{a_2}$.

The amount of mismatch between networks n_a and n_b required to minimize the loading of n_a by n_b, and thereby to maximize $\hat{q}$, can be obtained as follows. Comparing the denominators in (B.3) and (B.4) we can write:

$$(s + p_1)(s + p_2) = (s + p)^2 + \mu ps \tag{B.17}$$

where $\mu \geq 0$. Since $p_1 p_2 = p^2$, we obtain with (B.3):

$$\mu = \frac{p_1 + p_2}{p} - 2 = \frac{1}{\hat{q}} - 2 = \kappa - 2 \tag{B.18}$$

where

$$\kappa = 1/\hat{q}. \tag{B.19}$$

Comparing (B.2) and (B.17) we have, with (B.18):

$$\varepsilon(s) = (\kappa - 2)\frac{ps}{(s + p)^2} \tag{B.20}$$

Consider for example the three ladder networks shown in Figure B.4. In each network, n_b is impedance scaled by ρ and we have from (B.16) and (B.20):

$$\rho = \frac{1}{\kappa - 2} \tag{B.21}$$

Thus to obtain a κ value of 2.2 we must use an impedance scaling factor $\rho = 5$, since $\kappa = 2 + 1/\rho$.

The relation between ρ and κ for other network types follows in a similar way. Perhaps the most important is the potentially symmetrical twin-T (see Figure B.5) for which it is easy to show that

$$\rho = \frac{2}{\kappa - 2} \tag{B.22}$$

To obtain a κ value of 2.2, ρ must here equal 10. Conversely, for a ρ value of 5, $\kappa = 2(1 + \rho)/\rho = 2.4$.

ADDITIONAL REFERENCES AND SUGGESTED READING

Approximation Theory

R. W. Daniels: *Approximation Methods for Electronic Filter Design*, McGraw-Hill Book Co., New York, 1974.

G. C. Temes and S. K. Mitra, Eds.,: *Modern Filter Theory and Design*, Chapter 2, John Wiley & Sons, New York, 1973.

Computer-Aided Network Optimization and Design

D. A. Calahan: *Computer-Aided Network Design*, McGraw-Hill Book Co., New York, 1972.

F. F. Kuo and J. F. Kaiser, Eds.,: *System Analysis by Digital Computer*, Chapters 5 and 6, John Wiley & Sons, New York, 1966.

F. F. Kuo and W. G. Magnuson, Jr., Eds., *Computer-Oriented Circuit Design*, Chapter 5, Prentice Hall, Englewood Cliffs, N.J., 1969.

G. Szentirmai, Ed., *Computer-Aided Filter Design*, IEEE Press, Inc., New York, 1973.

G. C. Temes and S. K. Mitra, Eds., *Modern Filter Theory and Design*, Chapter 6, John Wiley & Sons, New York, 1973.

J. Vlach, *Computerized Approximation and Synthesis of Linear Networks*, John Wiley & Sons, New York, 1969.

Filter Tables and Filter Design

E. Christian and E. Eisenmann, *Filter Design Tables and Graphs*, John Wiley & Sons, New York 1966.

G. E. Hansell, *Filter Design and Evaluation*, Van Nostrand Reinhold Co., New York, 1969.

D. S. Humpherys, *The Analysis, Design, and Synthesis of Electrical Filters*, Prentice Hall, Englewood Cliffs, N.J., 1970.

J. K. Skwirzynski, *Design Theory and Data for Electrical Filters*, Van Nostrand Reinhold Co., New York, 1965.

A. I. Zverev, *Handbook of Filter Synthesis*, John Wiley & Sons, New York, 1967.

LC-Filter Simulation Using General Impedance Converters (GIC'S) and Frequency-Dependent Negative Resistances (FDNR's)

A. Antoniou, Bandpass Transformation and Realization Using Frequency-Dependent Negative-Resistance Elements, *IEEE Trans. on Circuit Theory*, Vol. CT-18, pp. 297–299, 1971.

B. B. Bhattacharyya, W. B. Mikhael and A. Antoniou, Design of RC-Active Networks by Using Generalized Immittance Converters, *Proc. Int. Symp. on Circuit Theory*, pp. 290–294, 1973.

677

L. T. Bruton, Nonideal Performance of Two-Amplifier Positive-Impedance Converters, *IEEE Trans. on Circuit Theory*, Vol. CT-17, pp. 541-549, 1970.

L. T. Bruton and A. B. Haase, Sensitivity of Generalized Immitance Converter Embedded Ladder Structures, *IEEE Trans. on Circuits and Systems*, Vol. CAS-21, pp. 245–250, 1974.

———, High Frequency Limitations of RC-Active Filters Containing Simulated-L and FDNR Elements, *Int. J. Circuit Theory and Applications*, **2**, pp. 187–194, 1974.

J. Gorski-Popiel, RC-Active Synthesis Using Positive-Immitance Converters, *Electron. Letters*, **3**, pp. 381–382 (August, 1967).

W. E. Heinlein and W. H. Holmes, *Active Filters for Integrated Circuits*, Springer-Verlag, New York, 1974.

S. K. Mitra, Ed., *Active Inductorless Filters*, IEEE Press, Inc., New York, 1971.

K. Panzer, Active Bandpass Filter with a Minimum Number of Capacitances Using Impedance Converters" (in german), *Nachrichtentechnische Zeitschrift (NTZ)*, **27**, pp. 379–382, 1974.

J. M. Rollett and C. Nightingale, Exact Synthesis of Active Lowpass FDNR Filters, *Electron. Letters*, **10**, pp. 34–35, (February, 1974).

H. R. Trimmel, Realization of Canonical Bandpass Filters with Frequency-Dependent and Frequency-Independent Negative Resistances, *Proc. IEEE Int. Symp. on Circuit Theory*, pp. 134–137, 1973.

Noise in Active Filters

A. Fettweis, On Noise Performance of Capacitor-Gyrator Filters, *Int. J. of Circuit Theory and Applications*, **2**, pp. 181–186, 1974.

W. E. Heinlein and W. H. Holmes, *Active Filters for Integrated Circuits*, Springer-Verlag, New York, 1974.

F. N. Trofimenkoff, D. H. Treleaven and L. T. Bruton, Noise Performance of RC-Active Quadratic Filter Sections, *IEEE Trans. on Circuit Theory*, Vol. CT-20, pp. 524–532, 1973.

J. O. Voorman and D. Blom, Noise in Gyrator-Capacitor Filters, *Philips Research Reports*, No. 26, pp. 114–133, 1971.

J. Zurada and M. Bialko, Noise and Dynamic Range of Active Filters with Operational Amplifiers, *Proc. IEEE Int. Symp. on Circuit Theory*, pp. 677–681, 1974.

Active Filter Design Using Integrated Circuits

A. Antoniou and K. S. Naidu, Modeling of a Gyrator Circuit, *IEEE Trans. on Circuit Theory*, Vol. CT-20, pp. 533–540, 1973.

A. Forsén and L. Kristiansson, Analysis of Nonlinear Model for Operational Amplifiers in Active RC Networks, *Int. J. Circuit Theory and Applications*, **2**, pp. 13–22, 1974.

P. R. Gray and R. G. Meyer, Recent Advances in Monolithic Operational Amplifier Design, *IEEE Trans. on Circuits and Systems*, Vol. CAS-21, pp. 317–327, 1974.

W. E. Heinlein and W. H. Holmes, *Active Filters for Integrated Circuits*, Springer-Verlag, New York, 1974.

S. K. Mitra, Ed., *Active Inductorless Filters*, IEEE Press, Inc., New York, 1971.

A. S. Sedra, Generation and Classification of Single Amplifier Filters, *Int. J. of Circuit Theory and Applications*, **2**, pp. 51–67, 1974.

H. O. Voorman and A. Biesheuvel, An Electronic Gyrator, *IEEE J. of Solid State Circuits*, Vol. SC-7, pp. 469–474, 1972.

Active Filter Dependence on the Gain-Bandwidth Product

D. Åckerberg and K. Mossberg, A Versatile Active RC Building Block with Inherent Compensation for the Finite Bandwidth of the Amplifier, *IEEE Trans. on Circuits and Systems*, Vol. CAS-21, pp. 75–78, 1974.

A. Budak and D. M. Petrela, Frequency Limitations of Active Filters Using Operational Amplifiers, *IEEE Trans. on Circuit Theory*, Vol. CT-19, pp. 322–329, 1972.

W. B. Mikhael and B. B. Bhattacharyya, A Practical Design for Insensitive RC-Active Filters, *Proc. IEEE Int. Symp. on Circuits and Systems*, pp. 205–209, 1974.

R. Srinivasagopalan and G. O. Martens, A Comparison of a Class of Active Filters with Respect to the Operational-Amplifier Gain-Bandwidth Product, *IEEE Trans. on Circuits and Systems*, Vol. CAS-21, pp. 377–381, 1974.

P. Vogel, Active RC Network Realization of Stable High-Q Transfer Functions (in german), AGEN Mitteilungen, Zurich, No. 16, pp. 23–38, December, 1973.

G. Wilson, Y. Bedri and P. Bowron, RC-Active Networks with Reduced Sensitivity to Amplifier Gain-Bandwidth Product, *IEEE Trans. on Circuits and Systems*, Vol. CAS-21, pp. 618–626, 1974.

Coupled Filters

G. Hurtig, The Primary Resonator Block Technique of Filter Synthesis, *Proc. Int. Filter Symposium*, p. 84, 1972.

————, Voltage Tunable Multipole Bandpass Filters, *Proc. IEEE Int. Symp. on Circuits and Systems*, pp. 569–572, 1974.

K. R. Laker and M. S. Ghausi, Synthesis of a Low-Sensitivity Multiloop Feedback Active RC Filter, *IEEE Trans. on Circuits and Systems*, Vol. CAS-21, pp. 252–259, 1974.

K. R. Laker, M. S. Ghausi and J. J. Kelly, Minimum Sensitivity Leapfrog Active Filters, *Proc. IEEE Int. Symp. on Circuits and Systems*, pp. 201–204, 1974.

J. Tow and Y. L. Kuo, Coupled-Biquad Active Filters, *Proc. IEEE Int. Symp. on Circuit Theory*, pp. 164–167, 1972.

J. Tow, Design and Evaluation of Shifted Companion Form (Follow the Leader) Active Filters, *Proc. IEEE Int. Symp. on Circuits and Systems*, pp. 656–660, 1974.

Large Change Sensitivity and Statistical Circuit Design

Bell System Technical Journal, Special Issue on *Statistical Circuit Design*, **50**, No. 4 (April, 1971).

D. Hilberman, An Approach to the Sensitivity and Statistical Variability of Biquadratic Filters, *IEEE Trans. on Circuit Theory*, Vol. CT-20, pp. 382–390, 1973.

B. A. Shenoi, Optimum Variability Design and Comparative Evaluation of Thin-Film RC Active Filters, *IEEE Trans. on Circuits and Systems*, Vol. CAS-21, pp. 263–267, 1970.

R. Spence, *Linear Active Networks*, Chapter 10, Wiley-Interscience, John Wiley & Sons Ltd., London, 1970.

INDEX